Classical Complex Analysis

A Geometric Approach — Vol. 2

Classical Complex Analysis

A Geometric Approach — Vol. 2

I-Hsiung Lin

National Taiwan Normal University, Taiwan

World Scientific

NEW JERSEY · LONDON · SINGAPORE · BEIJING · SHANGHAI · HONG KONG · TAIPEI · CHENNAI

Published by

World Scientific Publishing Co. Pte. Ltd.

5 Toh Tuck Link, Singapore 596224

USA office: 27 Warren Street, Suite 401-402, Hackensack, NJ 07601

UK office: 57 Shelton Street, Covent Garden, London WC2H 9HE

British Library Cataloguing-in-Publication Data
A catalogue record for this book is available from the British Library.

ISBN-13 978-981-4271-28-8
ISBN-13 978-981-4271-29-5 (pbk)

Typeset by Stallion Press
Email: enquiries@stallionpress.com

Printed in Singapore.

To my wife Hsiou-o and my children
Zing, Ting, Ying and Fei

Preface

Complex analysis, or roughly equivalently the theory of analytic functions of one complex variable, budded in the early ages of Gauss, d'Alembert, and Euler as a main branch of mathematical analysis. In the 19th century, Cauchy, Riemann, and Weierstrass laid a rigorous mathematical foundation for it (see Ref. [47]). Nourished by the joint effort of generations of brilliant mathematicians, it grows up into one of the remarkable branches of exact science, and serves as a prototype or model of other theories concerned with generalizations of analytic functions such as Riemann surfaces, analytic functions of several complex variables, quasiconformal and quasiregular mappings, complex dynamics, etc. Its methods and theory are widely used in many branches of mathematics, ranging from analytic number theory to fluid mechanics, elasticity theory, electrodynamics, string theory, etc.

Elementary complex analysis stands as a discipline to the whole mathematical training. This book is designed for beginners in this direction, especially for upper level undergraduate and graduate mathematics majors, and to those physics (or engineering) students who are interested in more theoretically oriented introduction to the subject rather than only in computational skills. The content is thus selective and its level of difficulty should be then adequately arranged.

Beside its strong intuitive flavor, it is the geometric (mapping) properties, derived from or characterized the analytic properties, that makes the theory of analytic functions differ so vehemently from that of real analysis and so special yet restrictive in applications. This is the reason why I favor a geometric approach to the basics. The degree of difficulty, as a whole, is not higher than that of L.V. Ahlfors's classic *Complex Analysis* [1]. But I try my best to give detailed and clear explanations to the theory as much as possible. I hope that the presentation will be less arduous in order to be more available to not-so-well-prepared or not-so-gifted students and be easier for self-study. Neither the greatest possible generality nor the most up-to-date terminologies is our purpose. Please refer to Berenstein and Gay

[9] for those purposes. I would consider my purpose fulfilled if the readers are able to acquire elementary yet solid fundamental classical results and techniques concerned.

Knowledge of elementary analysis, such as a standard calculus course including some linear algebra, is assumed. In many situations, mathematical maturity seems more urgent than purely mathematical prerequisites. Apart from these the work is self-contained except some difficult theorems to which references have been indicated. Yet for clearer and thorough understanding where one stands for the present in the whole mathematical realm and for the ability to compare with real analysis, I suggest readers get familiar with the theory of functions of two real variables.

Sketch of the Contents

If one takes a quick look at the Contents or read over Sketch of the Content at the beginning of each chapter and then s/he will have an overall idea about the book.

A complex number is not just a plane vector but also carries by itself the composite motion of a one-way stretch and rotation, and hence, is a two-dimensional "number". They constitute a field but cannot be ordered. Mathematics based on them is the one about similarity in global geometric sense and is the one about conformality in local infinitesimal sense. *Chapter 1* lays the algebraic, geometric and point-set foundations barely needed in later chapters.

Just as one experienced in calculus, we need to know some standard *elementary* complex-valued *functions* of a complex variable before complex differentiation and integration are formally introduced. It is the *isolated-zero principle* (see (3) and (4) in (3.4.2.9)) that makes many of their *algebraic properties* or algebraic identities similar to their real counterparts. Owing to the complex plane $\mathbf{C}$ having the same topological structure as the Euclidean plane $\mathbf{R}^2$, their *point-set properties* (such as continuity and convergence) are the same as the real ones, too. It is the *geometric mapping properties* owned by these elementary functions that distinguish them from the real ones and make one feel that complex analysis is not just a copy of the latter. In particular, the local and global single-valued continuous branches of arg z are deliberately studied, and then, prototypes of Riemann surfaces are introduced. *Chapter 2* tries to figure out, though loosely and vaguely organized, the common analytic and geometric properties owned by these individual elementary functions and then, to foresee what properties a general analytic function might have.

A complex valued function $f(z)$, defined in a domain (or an open set) Ω, is called *analytic* if any one of the following equivalent conditions is satisfied:

1. $f(z)$ is differentiable everywhere in Ω (Chap. 3).
2. $f(z)$ is continuous in Ω and $\int_{\partial \Delta} f(z)dz = 0$ for any triangle Δ contained in Ω (Chaps. 3 and 4).
3. For each fixed point $z_0 \in \Omega$, $f(z)$ can be expressed as a convergent power series $\sum_{n=0}^{\infty} a_n(z - z_0)^n$ in a neighborhood of z_0 (Chap. 5).

An analytic function $f(z)$ infinitesimally, via the Cauchy–Riemann equations, appears as a conformal mapping in case $f'(z) \neq 0$ and $\overline{f(z)}$ can be interpreted as the velocity field of a solenoidal, irrotational flow (see (3.2.4.3)). *Chapter 3* develops the most fundamental and important analytic and geometric properties, both locally and globally, which an analytic function might possess. The most subtle one, among all, is that a function $f(z)$ analytic at z_0 can always be written as $f(z) = f(z_0) + (z - z_0)\varphi(z)$ where $\varphi(z)$ is another function analytic at z_0. From this, many properties, such as the isolated-zero principle, maximum–minimum modulus principle and the open mapping property, inverse and implicit function theorems, can be either directly or indirectly deduced. This chapter also studies some global theorems such as Schwarz's Lemma, symmetry principle, argument principle and Rouché's theorem and their illustrative examples.

After proving homotopic and homologous forms of Cauchy integral theorem, most part of Chapter 4 is devoted to the residue theorem and its various applications in evaluating integrals, summation of series, and the Fourier and Laplace transforms.

Chapter 5 starts with various local power series representations of an analytic function and analytic continuation of power series which eventually lead to the monodromy theorem. Besides power series representation, an entire function can also be factorized as an infinite product of its zeros such as polynomials do, and meromorphic function can be expanded into partial fractions via its poles as rational functions. The most remarkable example is Euler's gamma function $\Gamma(z)$ and its colorful properties. Riemann zeta function is only sketched. The essential limit process in the whole chapter is the method of *local uniform convergence*. Weierstrass's theorem and Montel's normality criterion are two of the most fundamental results in this direction. Both are used to prove Picard's theorems via the elliptic modular function. These classical theorems can also be obtained by Schwarz–Ahlfor's Lemma, a geometric theorem.

Riemann mapping theorem initiated the study of global geometric mapping properties of univalent analytic functions, simply called univalent mappings. Schwarz–Christoffel formulas provide fruitful examples for the theorem. As a consequence, *Chapter 6* solves Dirichlet's problem for an open disk, a Jordan domain and hence, a class of general domains via Perron's method. This, in turn, is adopted to determine the canonical mappings and canonical domains for finitely connected domains.

Based on our intuitive and descriptive knowledge about Riemann surfaces of particularly chosen multiply-valued functions, scattered from Chaps. 2–6, *Chapter* 7 tries to give a formal, rigorous yet concise introduction to *abstract* Riemann surfaces. We will cover the fundamental group, covering surfaces and covering transformations, and finally highlight the proof of the uniformization theorem of Riemann surfaces via available classical methods, even though most recently it admits a purely differential geometric one-page proof [17].

Almost all sections end up with *Exercises A* for getting familiar with the basic techniques and applications; most of them also have *Exercises B* for practice of combining techniques and deeper thinking or applications; few of them have *Exercises C* for extra readings of a paper or *Appendixes A–C*.

As far as starred sections are concerned, see "How to use the book" below.

Features of the Book

(1) *Style of Writing.* As a textbook for beginners, I try to introduce the concepts clearly and the whole theory gradually, by giving definite explanations and accompanying their geometric interpretations whenever possible. Geometric points of view are emphasized. There are about 546 figures and many of them are particularly valid or meaningful only for complex variable but not for reals. Most definitions come out naturally in the middle of discussions, while main results obtained after a discussion are summarized and are numbered along with important formulas.

(2) *Balance between Theory and Examples.* As an introductory text or reference book to beginners, how to grasp and consolidate the basic theory and techniques seems more important and practical than to go immediately to deeper theories concerned. Therefore there are sufficient

amount of examples to practice main ideas or results. Exercises are usually divided into parts A and B; the former is designed to familiarize the readers with the established results, while the latter contains challenging exercises for mature and minded readers. Both examples and exercises are classic and are benefited very much from Refs. [31, 52, 58, 60, 80]. What should be mentioned is that many exercise problems in Ref. [1] have been adopted as illustrative examples in this text.

(3) *Careful Treatment of Multiple-Valued Functions.* Owing to historical and pedagogical reasons, complex analysis is conventionally carried out in the (one-layer) classical complex plane. Later development shows that the most natural place to do so is multiple-layer planes, the so-called Riemann surfaces or one-dimensional complex manifolds in its modern terminology. Multiple-valuedness is a difficult subject to most beginners and most introductory books just avoid or sketch it by focusing on $\sqrt[n]{z}$ and $\log z$ only. To provide intuitive feeling toward abstract Riemann surfaces in Chap. 7, Chaps. 2–6 take no hesitation to treat multiple-valued analytic functions whenever possible in the theory and in the illustrative examples. Once the trouble-maker $\arg z$, the origin of multiple-valuedness, is tamed (see Sec. 2.7.1), what is left is much easier to handle with. Also we construct many (purely descriptive and nonrigorous) Riemann surfaces or their line complexes of specified multiple-valued functions, merely for purposes of clearer illustration, wherever we feel worthy to do so.

(4) *Emphasis on the Difference between Real and Complex Analyses.* The origin of all these differences comes from the very character of what a complex number is (see the second paragraph inside the title Sketch of the Contents). This fact reflects, upon differentiating process, in the aspects of algebra, analysis as well as in geometry (see (3.2.2) for short; Secs. 3.2.1–3.2.3 for details).

(5) *Paving the Way to Advanced study.* The contents chosen are so arranged that they will provide solid background knowledge to further study in fields mentioned in the first paragraph of this Preface. Besides, the book contains more materials than what is required in a Ph.D. qualifying examination for complex analysis.

How to Use the Book (a Suggestion to the Readers)?

The book is rich in contents, examples, and exercises when comparing to other books on complex analysis of the same level. It is designed for a variety of usages and motivations for advanced studies concerned.

The whole content is divided into two volumes: Volume 1 contains Chaps. 1–4 and Appendix A, while Volume 2 contains Chaps. 5–7 and Appendixes B and C.

May I have the following suggestions for different proposes:

Chapters 1 and 2 are preparatory. Except those basic concepts such as limits and functions needed, topics in these two chapters could be selective, up to one's taste.

(1) As an introductory text for undergraduates
    Sections 2.5.2, 2.5.4, 2.6, 2.7.2, 2.7.3 (sketch only), 2.9 (sketch only); 3.2.2, 3.3.1 (only basic examples and $\sqrt[n]{z}$, $\log z$), 3.3.2, 3.4.1–3.4.4, 3.5.1–3.5.5, 3.5.7 (sketch only), 4.8, 4.9, 4.10 (sketch only), 4.11 (sketch only), 4.12.1–4.12.3, 5.3.1, 5.4.1, 5.5.2 (optional and sketch).

    As a whole, examples and Exercises A should be selective. Minded readers should try more, both examples and exercises, and pay attentions to more elementary multiply-values functions and their Riemann surfaces, if possible.

    In a class, the role played by a lecture to select topics is crucial.
(2) As a beginning graduate text
    With a solid understanding of materials in (1), the following topics are added: Secs. 2.8, 3.4.5, 3.4.6, 4.1–4.7, 4.12.3A–4.15 (selective), 5.1.3, 5.2, 5.3.2, 5.5–5.6 (selective), 5.8.1–5.8.3; Chap. 6 except 6.6.4.

    Examples and Exercises A (even Exercise B) should be emphasized. Of course, the adding or deleting some topics are still possible.
(3) To readers whose are interested in Riemann surfaces
    Pay more attention to multiply-valued functions and their descriptive Riemann surfaces such as Secs. 2.7, 3.3.3, 3.4.7, 3.5.6, 5.1, 5.2, and end up with the whole Chap. 7.
(4) Several complex variables
    Sections 3.4.7, 3.5.6; Chap. 7.
(5) Quasiconformal mappings and complex dynamical systems
    Section 3.2.3, Example 2 in Sec. 3.5.5; Secs. 5.3.4, 5.8; 6.6.4; Chap. 7, and Appendix C.
(6) As a general reference book supplement to other books on complex analysis.

Acknowledgments

The following students in Mathematics Department helped to type my hand-written manuscript:

Jing-ya Shui; Ya-ling Zhan; Yu-hua Weng; Ming-yang Kao; Wei-ming Su; Wen-jie Li; Shuen-hua Liang; Shi-wei Lin; D.C. Peter Hong; Hsuan-ya Yu; Yi-hsuan Lin; Ming-you Chin; Che-wei Wu; Cheng-han Yang; Kuo-han Tseng; Yi-ting Tsai; Yi-chai Li; Po-tsu Lin; Hsin-han Huang.

Yan-yu Chen graphed all the figures that appeared in this book. Yan-yu Chen, Aileen Lin, Wen-jie Li, and Ming-yang Kao helped to edit the final manuscript for printing. Here may I pay my sincerest thanks to all of them. Without their unselfish dedication, this book definitely could have not been published so soon.

Also, teaching assistant Jia-ming Ying helped to improve and correct partially my English writing. My colleagues Prof. Tian-yu Tsai proof-read the entire manuscript, and Prof. Yu-lin Chang proof-read Chaps. 5–7 and adopted parts of the content in his graduate course on complex analysis. Both of them pointed out many mistakes and gave me valuable suggestions. It would be my pleasure to express my gratitude toward them for their kindest help.

As usual, teaching assistant Ching-yu Yang did all the computer work for the several editions of the manuscript. Ms. Zhang Ji and Ms. Tan Rok Ting, editors in World Scientific, copy-edited the whole book with carefulness and expertise. Thank them so much.

I-hsiung Lin
2009/1/21
Taipei, Taiwan, China

Contents

CHAPTER 5

Fundamental Theory: Sequences, Series, and Infinite Products

Introduction

A (single-valued) function $f(z) : \Omega$ (an open set) $\to \mathbf{C}$ is said to be *analytic* in Ω if, for each point $z_0 \in \Omega$, $f(z)$ can be expressed locally by a power series $\sum_0^\infty a_n(z - z_0)^n$ in a neighborhood of z_0. The various power series representations of the same $f(z)$ at various points in Ω are connected by either direct or indirect *analytic continuation* (see (5.1.2.2), (5.1.3.5), (5.1.3.6), and Sec. 5.2). The starting points in this approach to the theory are the contents of Sec. 3.3.2; in particular, that of (3.3.2.4). Its main advantage over the differentiation (Chap. 3) and the integration (Chap. 4) methods lies on the fact that it can be easily applied to the study of the *local behaviors* of analytic functions (see, for instance, Secs. 5.1.2 and 3.5.1) and, hence, the introduction of the concept of *abstract Riemann surfaces* (see Section 3.3.3, Chap. 7, and, of course, the monumental book [81] by H. Weyl).

Sketch of the Content

Section 5.1 (and Sec. 3.3.2) investigates mainly the properties of an analytic function $f(z)$ defined by a *single* convergent power series $\sum_0^\infty a_n(z - z_0)^n$ in its open disk of convergence and on the boundary circle. In this restricted yet special case, it is easy to obtain important results such as the max–min modulus principle, and the Cauchy integral theorem and formula. The existence of at least one *singular point* on the circle of convergence presents unexpected complicated situations about the boundary behavior of a power series. And *Abel's limit theorem* in this direction is fundamental.

A convergent power series can be *analytically continued* through its *regular points* (if exist) on the circle of convergence to as far as possible in the complex plane. The resulted function is usually multiple-valued in its

domain, and, indeed, will be *single-valued* if *the domain is simply-connected.*
Section 5.2 highlights the proof of the *monodromy theorem.*

The important role played by the *local uniform convergence* of a
sequence or a series of analytic functions is formally introduced and studied
in Sec. 5.3. In addition to the basic *Weierstrass's theorem*, some other criteria, originated mainly from the maximum principles, for local uniform
convergence are also derived; they include *Vitali's* and *Montel's theorems*
(5.3.3.10). This kind of convergence *preserves* analyticity and the number
of zeros (*Hurwitz's theorem*), and creates wonderful phenomena about the
fixed points of an analytic function and its iterate functions.

Mittag–Leffler's partial fractions theorem extends the partial fraction
expansion of a rational function, as shown in (2.5.3.5), to meromorphic
functions defined on the entire plane. Section 5.4 adopts three methods to
achieve this, including Cauchy's residue method and the $\overline{\partial}$-operator method
(see Exercises B of Sec. 5.4.1).

Section 5.5 extends the factor product expression of a polynomial, as
shown in (2.5.1.2), to entire functions, usually known as *Weierstrass's factorization theorem*, including the canonical form, genus and *Hadamard's
order theorem*. The introduction of infinite products is sketched and
preparatory.

The *gamma function* $\Gamma(z)$(see Sec. 5.6) and the *Riemann zeta function*
$\zeta(z)$ (see Sec. 5.7) are two important illustrative examples of the materials
in both Secs. 5.4 and 5.5. Various representations and characteristic properties, including *Stirling's formula*, of $\Gamma(z)$ are studied; while, that of $\zeta(z)$
is only sketched.

The concept of *normal* sequence (family) of analytic functions is the
version for functions of Bolzano–Weierstrass's theorem for bounded infinite
point set or sequence. Section 5.8 devotes to the study of the criteria for
normality and its applications. The *elliptic modular function* is adopted to
prove the important *Montel's normality criterion* which, in turn, is used
to prove the famous *Picard's first and second theorems*. In Sec. 5.8.4, various types of *Schottky's theorem* are introduced and provide another proofs
of Montel's and Picard's theorems. Also, L. V. Ahlfors (1938) extended
Schwarz–Pick's lemma (see (3.4.5.2)) to *Schwarz–Ahlfor's lemma* (5.8.4.14),
in the content of Poincaré's metric and curvature, which provides differential geometric proofs to some fundamental results such as Livioulle's theorem, Schottky's theorem, Montel's and Picard's theorems. To the end (see
Sec. 5.8.5), we present some results in complex dynamical system as another
meaningful application of the concept of normality.

5.1   Power Series

Before we start, it is supposed that the readers are familiar with the content
of Sec. 3.3.2; in particular, (3.3.2.3) and (3.3.2.4). We will use these results
directly.

5.1.1   *Algorithm of power series*

What we are going to do here is to provide a detailed account about (5)
in (4.8.1) and the readers are urged to review examples presented there.
Also, the readers might refer to Chap. 1 of Ref. [16] for a discussion of the
algebra of *formal* power series.

Given two power series

$$f(z) = \sum_0^\infty a_n(z - z_0)^n, \quad |z - z_0| < r_1 \text{ (the radius of convergence)};$$

$$g(z) = \sum_0^\infty b_n(z - z_0)^n, \quad |z - z_0| < r_2 \text{ (the radius of convergence)}.$$

Three easier operations are as follows:

The identity operation:

$$\sum_0^\infty a_n(z - z_0)^n = \sum_0^\infty b_n(z - z_0)^n, \quad |z - z_0| \le \min(r_1, r_2)$$

$$\Leftrightarrow a_n = b_n, \quad \text{for } n \ge 0. \tag{5.1.1.1}$$

The addition and subtraction operations:

$$f(z) \pm g(z) = \sum_0^\infty (a_n \pm b_n)(z - z_0)^n, \quad |z - z_0| \le \min(r_1, r_2). \tag{5.1.1.2}$$

The Cauchy product operation: By absolute convergence of both series
in $|z - z_0| \le \min(r_1, r_2)$, then

$$f(z)g(z) = \sum_{n=0}^\infty (a_n b_0 + a_{n-1} b_1 + \cdots + a_0 b_n)(z - z_0)^n,$$

$$|z - z_0| \le \min(r_1, r_2). \tag{5.1.1.3}$$

Proofs are left to the readers.

The following is divided into four subsections.

Section (1) The division operation

Suppose $b_0 = g(z_0) \neq 0$. Since the zeros of $g(z)$ in $|z - z_0| < r_2$ are isolated (see (3) in (3.4.2.9) or (5.1.2.4)), let z_1 be a zero of $g(z)$ which is nearest to z_0. Let $r = \min(r_1, r_2, |z_1 - z_0|)$. On $|z - z_0| < r$, $f(z)$ converges absolutely and so does $g(z) = \sum_0^\infty b_n (z - z_0)^n$ which does not have any zero there. Hence,

$$\frac{f(z)}{g(z)} = \frac{\sum_0^\infty a_n (z - z_0)^n}{\sum_0^\infty b_n (z - z_0)^n}, \quad |z - z_0| < r$$

is analytic and thus, can be represented by a power series $\sum_0^\infty c_n (z - z_0)^n$ (see (3.4.2.6) and Sec. 4.8). *The algorithm of division* of two power series $f(z)$ and $g(z)$ means how to determine the coefficients c_n from the known coefficients a_n and b_n.

Via the absolute convergence of the power series concerned in $|z - z_0| < r$, (5.1.1.3) suggests that

$$\sum_0^\infty a_n (z - z_0)^n = \left(\sum_0^\infty b_n (z - z_0)^n \right) \left(\sum_0^\infty c_n (z - z_0)^n \right)$$

$$= \sum_0^\infty (b_n c_0 + b_{n-1} c_1 + \cdots + b_0 c_n)(z - z_0)^n, \quad |z - z_0| < r$$

$$\Rightarrow \text{(by (5.1.1.1))} \; c_0 b_0 = a_0$$

$$\vdots$$

$$c_0 b_n + c_1 b_{n-1} + \cdots + c_{n-1} b_1 + c_n b_0 = a_n, \quad n \geq 1.$$

Since $b_0 \neq 0$, one can solve out $c_0, c_1, \ldots, c_n, \ldots$ successively or use Cramer's rule to do this.

Summarize the above as

The division of two power series. Suppose $f(z) = \sum_0^\infty a_n (z - z_0)^n$ and $g(z) = \sum_0^\infty b_n (z - z_0)^n$ have positive radii of convergence. In case $b_0 = g(z_0) \neq 0$, there is an $r > 0$ so that $\frac{f(z)}{g(z)}$ is analytic in $|z - z_0| < r$ and

$$\frac{f(z)}{g(z)} = \sum_0^\infty c_n (z - z_0)^n, \quad |z - z_0| < r,$$

where the coefficients are given by

$$c_0 = \frac{a_0}{b_0};$$

$$c_n = \frac{1}{b_0^{n+1}} \begin{vmatrix} b_0 & 0 & 0 & \cdots & a_0 \\ b_1 & b_0 & 0 & \cdots & a_1 \\ b_2 & b_1 & b_0 & \cdots & a_2 \\ \vdots & \vdots & \vdots & & \vdots \\ b_n & b_{n-1} & b_{n-2} & \cdots & a_n \end{vmatrix}, \quad n \geq 1. \qquad (5.1.1.4)$$

Refer to Examples 5 and 6 in Sec. 4.8.

Section (2) The composite operation

Given two power series

$$f(z) = \sum_0^\infty a_n(z - z_0)^n, \quad |z - z_0| < r \, (r > 0),$$

and

$$g(w) = \sum_0^\infty b_n(w - w_0)^n, \quad |w - w_0| < \rho \, (\rho > 0)$$

where $w_0 = f(z_0) = a_0$. Substitute $w = f(z)$ in $g(w)$, the resulting

$$g(f(z)) = \sum_0^\infty b_n(f(z) - w_0)^n$$

is called *the power series of the composite* $g \circ f$ of $f(z)$ and $g(z)$. By invoking formally and purely algebraic operations:

1. expand $(f(z) - a_0)^n = (\sum_{m=1}^\infty a_m(z - z_0)^m)^n$, $n \geq 1$, and
2. collect all the coefficients of $(z - z_0)^n$, $n \geq 1$, in $\sum_{n=0}^\infty b_n(\sum_{m=1}^\infty a_m(z - z_0)^m)^n$, $n \geq 1$, together and denote it by c_n,

then we obtain a new power series for $g(f(z))$

$$\sum_{n=0}^\infty c_n(z - z_0)^n,$$

called the *formal* substitution of $\sum_{n=0}^\infty a_n(z - z_0)^n$ into $\sum_{n=0}^\infty b_n(w - w_0)^n$. "Formal" means purely algebraic process, without recourse to the limit processes, such as convergence and the validity of rearrangement of terms, etc.

Our main result is

The composite operation. Let $f(z)$ and $g(w)$ be as above. Then the composite function $g(f(z))$ has a power series representation $\sum_{n=0}^{\infty} c_n(z - z_0)^n$ at z_0, obtained by the formal substitution of

$$\sum_{n=0}^{\infty} a_n(z - z_0)^n \text{ into } \sum_{n=0}^{\infty} b_n(w - w_0)^n, \quad \text{where } w_0 = f(z_0) = a_0.$$

$$(5.1.1.5)$$

Sketch of proof. Since $\lim_{z \to z_0} f(z) = f(z_0) = w_0 = a_0$, for $0 < \varepsilon < \rho$, there is a $0 < \delta \le r$ such that $|f(z) - w_0| \le \rho - \varepsilon$ whenever $|z - z_0| \le \delta$. Then, in case $|z - z_0| \le \delta$,

$$g(f(z)) = \sum_{n=0}^{\infty} b_n(f(z) - w_0)^n = \sum_{n=0}^{\infty} h_n(z), \quad h_n(z) = b_n(f(z) - w_0)^n, \quad n \ge 0$$

holds, and

1. each $h_n(z)$ is analytic in $|z - z_0| \le \delta$, $n \ge 0$, and
2. the series $\sum_{n=0}^{\infty} h_n(z)$, as a series in w and then in z, converges uniformly in $|z - z_0| \le \delta$.

According to (5.1.1.7), $g(f(z))$ is analytic at z_0 and the coefficient c_k of its power (Taylor) series expansion at z_0 is given by $\sum_{n=1}^{\infty} \frac{1}{k!} h_n^{(k)}(z_0)$ for $k \ge 1$. As a matter of fact,

1. $f(z) - a_0 = \sum_{k=1}^{\infty} a_k(z - z_0)^k$ converges absolutely in $|z - z_0| \le \delta$, and
2. by successive application of (5.1.1.3),

$$b_n(f(z) - a_0)^n = b_n \left(\sum_{k=1}^{\infty} a_k(z - z_0)^k \right)^n = \sum_{k=1}^{\infty} b_n a_{kn}(z - z_0)^k, \quad n \ge 1.$$

Consequently,

$$g(f(z)) = b_0 + \sum_{k=1}^{\infty} \left\{ \sum_{n=1}^{\infty} b_n a_{kn} \right\} (z - z_0)^k.$$

$\square$

Take, for instance,

$$\sqrt{\cos z} = 1 - \frac{1}{4}z^2 - \frac{1}{96}z^4 - \frac{19}{5760}z^6 - \cdots, \quad |z| < \frac{\pi}{2}, \quad \text{where } \sqrt{\cos 0} = 1;$$

$$e^{z \sin z} = 1 + z^2 + \frac{1}{3}z^4 + \cdots, \quad |z| < \infty.$$

Section (3) The inverse function operation

Suppose that

$$f(z) = \sum_0^\infty a_n(z - z_0)^n, \quad |z - z_0| < r \ (r > 0)$$

satisfies $f'(z_0) = a_1 \neq 0$. According to the inverse function theorem (see (3.4.7.1) or (3.5.1.4)), there are $\delta > 0 (0 < \delta \leq r)$ and $\rho > 0$ such that $w = f(z) : \{|z - z_0| < \delta\} \to \{|w - w_0| < \rho\}$, $w_0 = f(z_0)$, is a univalent analytic function and the inverse function $z = f^{-1}(w) : \{|w - w_0| < \rho\} \to \{|z - z_0| < \delta\}$ is an analytic function, too.

How can we find the power (Taylor) series expansion of f^{-1} at w_0? All one needs to do is to let $f^{-1}(w) = \sum_{n=0}^\infty b_n(w - w_0)^n$, $|w - w_0| < \rho$, and then, try to apply (5.1.1.5) to $f^{-1} \circ f(z) = z$. We summarize as

The inverse function operation. Suppose $f(z) = \sum_0^\infty a_n(z - z_0)^n$ has a positive radius r of convergence and $a_1 = f'(z_0) \neq 0$. Let $w_0 = f(z_0) = a_0$. Then there exists a power series $g(w) = \sum_0^\infty b_n(w - w_0)^n$, $|w - w_0| < \rho \ (\rho > 0)$, such that

$$(g \circ f)(z) = z, \quad |z - z_0| < r,$$

where the coefficients b_n, $n \geq 0$, satisfy the following recursive relations: $g(w_0) = b_0 = z_0$

$$a_1 b_1 = 1,$$

$$a_1^2 b_2 + a_2 b_1 = 0,$$

$$a_1^3 b_3 + 2a_1 a_2 b_2 + a_3 b_1 = 0,$$

$$\vdots$$

$$a_1^n b_n + P_n(a_1, \ldots, a_{n-1}, b_1, \ldots, b_{n-1}) = 0,$$

$$\vdots$$

in which $P_n(a_1, \ldots, a_{n-1}, b_1, \ldots, b_{n-1})$ is a polynomial in $a_1, \ldots, a_{n-1}$ and $b_1, \ldots, b_{n-1}$.   (5.1.1.6)

For instance,

$$w = \operatorname{Arc tan} z = z - \frac{1}{3}z^3 + \frac{1}{5}z^5 - \frac{1}{7}z^7 + \cdots, \quad |z| < 1, \text{ and}$$

$$z = \tan w = w + \frac{1}{3}w^3 + \frac{2}{15}w^5 + \frac{1}{7}w^7 + \cdots, \quad |w| < \frac{\pi}{2}.$$

$$w = e^z = \sum_{n=0}^{\infty} \frac{z^n}{n!}, \text{ and } z = \text{Log}(1+w) = \sum_{n=1}^{\infty} (-1)^{n-1} \frac{z^n}{n}, \quad |z| < 1,$$

where Log $1 = 0$.

$$w = \sin z = \sum_{n=0}^{\infty} (-1)^{2n+1} \frac{1}{(2n+1)!} z^{2n+1}, \quad |z| < \infty, \text{ and}$$

$$z = \text{Arc}\sin z = z + \frac{1}{2} \cdot \frac{1}{3} z^3 + \frac{1 \cdot 3}{2 \cdot 4} \cdot \frac{1}{5} z^5 + \frac{1 \cdot 3 \cdot 5}{2 \cdot 4 \cdot 6} \cdot \frac{1}{7} z^7 + \cdots, \quad |z| < 1.$$

What is Arccos z?

Section (4) The double series operation

This is the

Weierstrass double series theorem. Suppose

$$f_n(z) = \sum_{k=0}^{\infty} a_{nk}(z - z_0)^k, \quad a_{nk} = \frac{f_n^{(k)}(z_0)}{k!}, \quad k \geq 0,$$

all converge in $|z - z_0| < r(r > 0)$ for $n \geq 1$. Also, suppose

$$\sum_{n=1}^{\infty} f_n(z)$$

converges uniformly on each compact subset of $|z - z_0| < r$ to a function $f(z)$. Then,

1. $f(z)$ is analytic in $|z - z_0| < r$, and
2. $f(z) = \sum_{k=0}^{\infty} a_k(z - z_0)^k$, $|z - z_0| < r$, where $a_k = \sum_{n=1}^{\infty} a_{nk} = \sum_{n=1}^{\infty} \frac{f_n^{(k)}(z_0)}{k!}$, $k \geq 0$.

In short,

$$f(z) = \sum_{n=1}^{\infty} \left(\sum_{k=0}^{\infty} a_{nk}(z - z_0)^k \right) = \sum_{k=0}^{\infty} \left(\sum_{n=1}^{\infty} a_{nk} \right)(z - z_0)^k. \quad (5.1.1.7)$$

This is a rather special case of a general theorem by Weierstrass, to be formally stated and proved in (5.3.1.1) and (5.3.1.2). See also Exercise A(3) of Sec. 5.1.2.

Example 1 (*J. H. Lambert, 1913*). Show that

$$\sum_{n=1}^{\infty} \frac{z^n}{1-z^n} = \sum_{n=1}^{\infty} d(n)z^n, \quad |z| < 1,$$

where $d(n)$ denotes the number of positive divisors of n. The series in the right is called the *Lambert series*.

Solution. For any $0 < \rho < 1$, in $|z| < \rho$, we have

$$\left| \frac{z^n}{1-z^n} \right| \le \frac{|z|^n}{1-|z|^n} \le \frac{\rho^n}{1-\rho^n} \le \frac{\rho^n}{1-\rho}$$

$$\Rightarrow \left(\text{since } \sum \rho^n \text{ converges} \right) \sum_{n=1}^{\infty} \frac{z^n}{1-z^n} \text{ converges locally uniformly in}$$

$|z| < 1$.

On the other hand,

$$f_n(z) = \frac{z^n}{1-z^n} = \sum_{k=1}^{\infty} z^{nk}, \quad |z| < 1, \text{ for } n \ge 1.$$

The result follows by (5.1.1.7).

Example 2. The *Euler φ-function* $\varphi(n)$ is defined to be the number of positive integers, both less than n (a positive integer) and relatively prime to n. Designate $\varphi(1) = 1$. Find the Taylor series expansion of

$$f(z) = \sum_{n=1}^{\infty} \varphi(n) \frac{z^n}{1-z^n}, \quad |z| < 1$$

at $z_0 = 0$ and its radius of convergence. Also, write $f(z)$ out explicitly.

Solution. Since $\lim_{n\to\infty} \varphi(n) = \infty$, $\sum_{n=1}^{\infty} \varphi(n)$ diverges. Since $\varphi(n) \le n$, $\overline{\lim}_{n\to\infty} \varphi(n)^{\frac{1}{n}} \le 1$ holds and, thus, the series $\sum_{n=1}^{\infty} \varphi(n)z^n$ has its radius of convergence not less than 1.

In this case, $|\varphi(n)z^n/(1-z^n)| \le |\varphi(n)z^n|/1 - |z|$ for $|z| < 1$ and thus, $\sum_{n=1}^{\infty} \varphi(n)z^n/(1-z^n)$ does converge in $|z| < 1$. If $\sum_{n=1}^{\infty} \varphi(n)z^n/(1-z^n)$ would converge in $|z| > 1$, so does $\sum_{n=1}^{\infty} \varphi(n)/(1-z^n)$ there.

Consequently, for $|z| > 1$,

$$\sum_{n=1}^{\infty} \frac{\varphi(n)}{1 - z^n} - \sum_{n=1}^{\infty} \frac{\varphi(n)z^n}{1 - z^n} = \sum_{n=1}^{\infty} \varphi(n)$$

is convergent, a contradiction. Therefore,

$$\sum_{n=1}^{\infty} \frac{\varphi(n)z^n}{1 - z^n}$$

does converge only in $|z| < 1$ and locally uniformly there.

On the other hand,

$$\frac{z^n}{1 - z^n} = \sum_{k=1}^{\infty} z^{nk} \quad \text{converges in } |z| < 1, \text{ for } n \geq 1.$$

$$\Rightarrow \text{(by (5.1.1.7))} \sum_{n=1}^{\infty} \frac{\varphi(n)z^n}{1 - z^n} = \sum_{n=1}^{\infty} \varphi(n) \left(\sum_{k=1}^{\infty} z^{nk} \right) = \sum_{k=1}^{\infty} \left(\sum_{n|k} \varphi(n) \right) z^k$$

$$= \sum_{k=1}^{\infty} kz^k = \frac{z}{(1 - z)^2}, \quad |z| < 1.$$

Example 3. The *Riemann zeta function* is given by

$$\zeta(z) = \sum_{n=1}^{\infty} \frac{1}{n^z} (n^z = e^{z \operatorname{Log} n} \text{ and } \operatorname{Log} n = \log n, \text{ the real logarithm of } n).$$

Show that the series converges locally uniformly in $\operatorname{Re} z > 1$ to the function $\zeta(z)$ which is thus analytic there. Try to find the Taylor series expansion of $\zeta(z)$ at $z_0 = 2$ and its radius of convergence. Also, refer to Sec. 5.7.

Solution. Choose any fixed $\sigma > 0$. On the half-plane $\operatorname{Re} z \geq 1 + \sigma$, $|n^z| = e^{(\operatorname{Re} z) \log n} \geq e^{(1+\sigma) \log n} = n^{(1+\sigma)}$.

Therefore,

$$\sum_{n=1}^{\infty} \frac{1}{|n^z|} \leq \sum_{n=1}^{\infty} \frac{1}{n^{(1+\sigma)}} \quad \text{on } \operatorname{Re} z \geq 1 + \sigma$$

$$\Rightarrow \text{(by Weierstrass test)} \sum_{n=1}^{\infty} \frac{1}{n^z} \text{ converges absolutely and}$$

uniformly on $\operatorname{Re} z \geq 1 + \sigma$.

By (5.1.1.7), $\zeta(z)$ is thus analytic in $\operatorname{Re} z > 1$.

On the other hand,

$$\frac{1}{n^z} = e^{-z\log n} = \sum_{k=0}^{\infty} \frac{1}{k!}\frac{d^k}{dz^k}\{e^{-z\log n}\}_{z_0=2}(z-2)^k$$

$$= \frac{1}{n^2}\sum_{k=0}^{\infty}\frac{(-1)^k}{k!}(\log n)^k(z-2)^k, \quad z \in \mathbf{C} \text{ for } n \geq 1$$

$$\Rightarrow \text{ (by(5.1.1.7)) } \zeta(z) = \sum_{n=1}^{\infty}\frac{1}{n^2}\sum_{k=0}^{\infty}\frac{(-1)^k}{k!}(\log n)^k(z-2)^k$$

$$= \frac{\pi^2}{6} + \sum_{k=1}^{\infty}\left\{\frac{(-1)^k}{k!}\sum_{n=1}^{\infty}\frac{(\log n)^k}{n^2}\right\}(z-2)^k, \quad |z-2| < 1$$

where $\sum_{n=1}^{\infty}\frac{1}{n^2} = \frac{\pi^2}{6}$ has been used (see Example 1 of Sec. 4.15.1).

*Example 4. Let

$$w = f(z) = 2\frac{z-t}{z^2-1} \quad (t \in \mathbf{C} \text{ is a parameter and } t \neq \pm 1).$$

Use this to justify (3.5.6.4) in case $z_0 = t$ and $w_0 = f(z_0) = 0$, and derive that

$$\frac{1}{\sqrt{1-2tw+w^2}} = \sum_{n=0}^{\infty} p_n(t)w^n, \quad \sqrt{1} = 1$$

where $p_n(t)$ is a polynomial in t of degree n, satisfying the *Rodrigue formula*

$$p_n(t) = \frac{1}{2^n n!}\left\{\frac{d^n}{dz^n}(z^2-1)^n\right\}_{z=t} = \frac{1}{2^n n!}\frac{d^n}{dz^n}(t^2-1)^n, \quad n \geq 0.$$

They are called *the Legendre polynomials* and $\frac{1}{\sqrt{1-2tw+w^2}}$ *the generating function* for them.

Try to determine its radius of convergence of the series if $t = t_0 \neq \pm 1$ is given.

The readers should consult Exercises B for more information concerned with the Legendre polynomials.

Solution. Note that $f(z_0) = w_0 = 0$ if $z_0 = t$. Now

$$f'(z) = \frac{2(-z^2 + 2zt - 1)}{(z^2 - 1)^2} \Rightarrow f'(z_0) = \frac{2}{t^2 - 1} \neq 0.$$

Observe that

$$\frac{z - z_0}{f(z) - w_0} = \frac{z - t}{f(z)} = \frac{z^2 - 1}{2}$$

$\Rightarrow$ (by (3.5.6.4)) The inverse function $z = f^{-1}(w)$ of $w = f(z)$ is given by

$$z = f^{-1}(w) = t + \sum_{n=1}^{\infty} \frac{1}{n!} \left\{ \frac{d^{n-1}}{dz^{n-1}} \left(\frac{z^2 - 1}{2} \right)^n \right\}_{z=t} w^n. \qquad (*_1)$$

The series in the right is a power series in w with t viewed as a parameter. For any fixed $K > 0$, the series will converge uniformly in $|t| \leq K$ if $|w|$ is sufficiently small, and, in this case, the series also converges uniformly for such w. To see this, recall that the coefficient a_n of the series can be represented by (compare to $(*_3)$ in the proof of (3.5.6.4))

$$a_n = \frac{1}{n!} \left\{ \frac{d^{n-1}}{dz^{n-1}} \left(\frac{z^2 - 1}{2} \right)^n \right\}_{z=t} = \frac{1}{2n\pi i} \int_{|z-t|=1} \left(\frac{z^2 - 1}{2} \right)^n \frac{1}{(z - t)^n} dz$$

$$(*_2)$$

$$\Rightarrow |a_n| \leq \frac{1}{2n\pi} \int_{|z-t|=1} \frac{|z^2 - 1|^n}{2^n} \cdot \frac{1}{|z - t|^n} |dz|$$

$$\leq \frac{1}{2n\pi} \cdot \frac{(K + 2)^{2n}}{2^n} \cdot 2\pi \leq \left\{ \frac{(K + 2)^2}{2} \right\}^n, \qquad n \geq 1.$$

Choose any θ, $0 < \theta < 1$. Consider

$$|w| < \delta = \frac{2\theta}{(K + 2)^2}.$$

$$\Rightarrow |a_n w^n| < \theta^n, \qquad n \geq 1.$$

This suggests that the series $(*_1)$ converges uniformly in $|t| \leq K$ if $|w| < \delta$, and, at the same time, converges uniformly in $|w| < \delta$ if $|t| \leq K$.

For fixed w, $|w| < \delta$, differentiate $(*_1)$ termwise with respect to t, $|t| \le K$, and we obtain

$$\frac{\partial z}{\partial t} = 1 + \sum_{n=1}^{\infty} \frac{1}{n!} \left\{ \frac{d^n}{dz^n} \left(\frac{z^2 - 1}{2} \right)^n \right\}_{z=t} w^n.$$

This is a new series converging uniformly either in t or in w as long as $|t| \le K$ and $|w| < \delta$. On the other hand, solve $w = \frac{2(z-t)}{z^2-1} = f(z)$ and, then,

$$z = f^{-1}(w) = \frac{1 - \sqrt{1 - 2tw + w^2}}{w} \quad (t \ne \pm 1), \quad \sqrt{1} = 1.$$

Note that $\sqrt{1} = 1$ guarantees that $f^{-1}(0) = t$ holds. It follows

$$\frac{\partial z}{\partial t} = \frac{1}{\sqrt{1 - 2tw + w^2}} = \sum_{n=0}^{\infty} p_n(t)w^n, \quad p_0(t) = 1. \qquad (*_3)$$

Fix $t_0 \notin [-1, 1]$. We try to determine the radius of convergence of the series

$$\frac{1}{\sqrt{1 - 2t_0 w + w^2}} = \sum_{n=0}^{\infty} p_n(t_0)w^n. \qquad (*_4)$$

We know, from the previous discussion, that this series converges uniformly in $|w| < \delta = \frac{2\theta}{(|t_0|+2)^2}$ $(0 < \theta < 1)$ and thus, represents the analytic function $\frac{1}{\sqrt{1-2t_0w+w^2}}(\sqrt{1} = 1)$ there. But $\frac{1}{\sqrt{1-2t_0w+w^2}}$ has a Taylor series expansion at $w_0 = 0$ whose radius of convergence is equal to the shortest distance R from 0 to the singular points of the function (see the comment followed in (4.8.1) or (1) in (3.4.2.6)). To find this R, solve $1 - 2t_0w + w^2 = 0$ and get $t_0 = \frac{1}{2}(w_0 + \frac{1}{w_0})$ where $w_0 = t_0 \pm \sqrt{t_0^2 - 1}$. It is well-known (see (5) in (2.5.5.4)) that $t = \frac{1}{2}(w + \frac{1}{w})$ maps $|w| < \rho\,(0 < \rho < 1)$ univalently and conformally onto the exterior of the ellipse

$$\frac{(\operatorname{Re} t)^2}{\left[\frac{1}{2}\left(\rho + \frac{1}{\rho}\right)\right]^2} + \frac{(\operatorname{Im} t)^2}{\left[\frac{1}{2}\left(\rho - \frac{1}{\rho}\right)\right]^2} = 1$$

and the circle $|w| = \rho$ onto the ellipse itself. Since each point t in the t-plane, other than these lying on $[-1, 1]$, lies on exactly only one such an ellipse, we just *choose* $\rho_0 = |w_0|\,(0 < \rho_0 < 1)$ so that the corresponding ellipse shown above passes through t_0. By the uniqueness of the Taylor series expansion at a point, the required radius of convergence R for $(*_4)$ is ρ_0.

What happens to R if $t_0 \in (-1, 1)$? It is $R = 1$. Justify this.

Exercises A

(1) Show that the coefficients c_n of the series

$$\frac{1}{1 - z - z^2} = \sum_{n=0}^{\infty} c_n z^n$$

satisfy the recursive relations: $c_0 = c_1 = 1$, $c_n = c_{n-1} + c_{n-2}$ $(n \geq 2)$. Find c_n explicitly and determine the radius of convergence. c_n are called the *Fibonacci numbers*.

(2) If

$$\frac{A + Bz + Cz^2}{a_0 + a_1 z + a_2 z^2 + a_3 z^3} = \sum_{n=0}^{\infty} c_n z^n, \quad a_0 \neq 0,$$

determine c_0, c_1, and c_2, and set up the recursive relations among c_n, c_{n-1}, c_{n-2}, c_{n-3} $(n \geq 3)$.

(3) Find an entire function $f(z)$ satisfying

$$z^2 f''(z) + z f'(z) + z^2 f(z) = 0, \quad z \in \mathbf{C}.$$

Note. One of the solutions is *the Bessel function of order* 0, defined by

$$J_0(z) = 1 + \sum_{n=1}^{\infty} \frac{(-1)^n}{(2 \cdot 4 \cdot 6 \cdots 2n)^2} z^{2n}, \quad |z| < \infty.$$

While, *the Bessel function of order m* is defined by

$$J_m(z) = \sum_{n=0}^{\infty} \frac{(-1)^n}{n!(n + m)!} \left(\frac{z}{2}\right)^{2n+m}, \quad |z| < \infty.$$

Try to show that they all are entire functions.

(4) Show that, for $|z| < 1$,

$$e^{(1-z)^{-1}} = e \left\{ 1 + z + \left[\frac{1}{1!} + \frac{1}{2!}\right] z^2 + \left[\frac{1}{1!} + \frac{2}{2!} + \frac{1}{3!}\right] z^3 \right.$$
$$\left. + \cdots + \left[\sum_{k=1}^{n} C_{k-1}^{n-1} \cdot \frac{1}{k!}\right] z^n + \cdots \right\}$$

whose radius of convergence is equal to 1.

(5) Suppose $\log(1 + e^z) = \log 2 + \sum_{n=0}^{\infty} \frac{c_n}{n+1} z^{n+1}$ in a neighborhood of 0. Try to show that $c_0 = \frac{1}{2}$, $c_0 + n c_1 + n(n-1) c_2 + \cdots + n! c_{n-1} + 2 n! c_n = 1$ and $c_{2n+1} = 0$ for $n \geq 1$.

(6) Develop each of the following functions in powers of z up to the term z^5.

    (a) $(1+z)^z = e^{z \log(1+z)}$, $\log 1 = 0$.   (b) e^{e^z},   (c) $e^z \log(1+z)$, $\log 1 = 0$.

(7) Let $\zeta(z)$ be the Riemann ζ-function (see Example 3). Show that

$$(1 - 2^{1-z})\zeta(z) = 1^{-z} - 2^{-z} + 3^{-z} + \cdots$$

and the right series represents an analytic function in $\operatorname{Re} z > 0$.

(8) Show that

$$\sum_{n=1}^{\infty} \frac{nz^n}{1-z^n} = \sum_{n=1}^{\infty} \frac{z^n}{(1-z^n)^2}, \quad |z| < 1.$$

(9) Prove that

$$\sum_{k=1}^{\infty} \frac{2z}{z^2 - k^2\pi^2} = -2 \sum_{n=1}^{\infty} \left(\sum_{k=1}^{\infty} \frac{1}{(k\pi)^{2n}} \right) z^{2n-1}, \quad |z| < \pi$$

and that the series in the right diverges at $z = \pi$ (so, the radius of convergence is π).

(10) If $\sum_{n=1}^{\infty} a_n$ converges, show that $\sum_{n=1}^{\infty} a_n z^n / (1 - z^n)$ converges in $|z| < 1$ and $|z| > 1$; if $\sum_{n=1}^{\infty} a_n$ diverges, show that $\sum_{n=1}^{\infty} a_n z^n / (1-z^n)$ converges in the open disk of convergence of $\sum_{n=1}^{\infty} a_n z^n$ but diverges outside it.

(11) Find the Taylor series expansion of

$$\sum_{k=1}^{\infty} a_k \frac{z^k}{1 - z^k}$$

at $z_0 = 0$ and determine its radius of convergence.

(12) *The Hermite polynomials $H_n(t)$, $n \geq 1$, are defined by*

$$e^{2tw - w^2} = \sum_{n=0}^{\infty} \frac{H_n(t)}{n!} w^n.$$

Show that

(a) $H_n'(t) = 2n H_{n-1}(t)$, $n \geq 1$.
(b) $H_{n+1}(t) - 2t H_n(t) + 2n H_{n-1}(t) = 0$, $n \geq 1$.
(c) $H_n''(t) - 2t H_n'(t) + 2n H_n(t) = 0$, $n \geq 0$.
(d) $H_n(t) = (-1)^n e^{t^2} \frac{d^n}{dt^n} (e^{-t^2})$.

For fixed $t_0 \in \mathbf{C}$, determine the radius of convergence of $\sum_{n=0}^{\infty} \frac{H_n(t_0)}{n!} w^n$.

Exercises B

The Legendre polynomials

$$P_n(t) = \frac{1}{2^n n!} \frac{d^n}{dt^n} (t^2 - 1)^n$$

$$= \frac{(-1)^n}{2^n} \sum_{k=0}^{n} (-1)^k C_k^n (1+t)^k (1-t)^k, \quad t \in \mathbf{C}, \ n \geq 1$$

have the following basic properties.

(1) $P_n(t)$ is a polynomials of degree n in t whose coefficient of t^n is $\frac{(2n)!}{2^n (n!)^2}$. $P_n(t)$ is an even or odd function according to n is even or odd, respectively. Moreover, $P_n(t)$, $n \geq 1$, satisfy the recursive relations:

$$P_1(t) = t;$$

$$2P_2(t) - 3tP_1(t) + 1 = 0;$$

$$(n+1)P_{n+1}(t) - (2n+1)tP_n(t) + nP_{n-1}(t) = 0, \quad n \geq 2.$$

Hence, it follows that

$p_0(t) = 1$, $p_1(t) = t$, $p_2(t) = \frac{1}{2}(3t^2 - 1)$, $p_3(t) = \frac{1}{2}(5t^3 - 3t)$, $p_4(t) = \frac{1}{2}(35t^4 - 30t^2 + 3), \ldots$.

(2) $p_n(1) = 1$ and $p_n(-1) = (-1)^n$. $p_n(t) = 0$ has real roots only and has n distinct real roots in $(-1, 1)$.

(3) $p_n(t)$ satisfies the following ordinary differential equation of second order:

$$(t^2 - 1)p_n''(t) + 2tp_n'(t) - n(n+1)p_n(t) = 0.$$

In addition,

$$p_n(t) = p_{n+1}'(t) - 2tp_n'(t) + p_{n-1}'(t);$$

$$(2n+1)p_n(t) = p_{n+1}'(t) + p_{n-1}'(t), \quad t \in \mathbf{C}.$$

(4) $p_n(t)$, $n \geq 0$, are *orthogonal* in $[-1, 1]$, namely,

$$\int_{-1}^{1} p_m(t)p_n(t)dt = 0, \quad m \neq n.$$

5.1.2 *Basic properties of an analytic function defined by a power series in an open disk*

Suppose the analytic function $f(z)$ is given by

$$f(z) = \sum_{n=0}^{\infty} a_n(z - z_0)^n, \quad |z - z_0| < r \text{ (the radius of convergence)}.$$

$$(5.1.2.1)$$

Here, we will try our best to develop some basic properties of such an $f(z)$ as listed in Secs. 3.4.2 and 3.4.4 without recourse to the theory of integration introduced in Sec. 2.9. Of course, readers might skip materials here and go directly to Sec. 5.3.

The following is divided into four sections.

Section (1) The Taylor series expansion

Fix a point $z_1 \neq z_0$, where $0 < |z_1 - z_0| < r$. By the binomial expansion,

$$(z - z_0)^n = (z - z_1 + z_1 - z_0)^n = \sum_{k=0}^{n} C_k^n (z_1 - z_0)^{n-k} (z - z_1)^k, \quad n \geq 1$$

$$\Rightarrow f(z) = \sum_{n=0}^{\infty} a_n (z - z_0)^n$$

$$= \sum_{n=0}^{\infty} a_n \left(\sum_{k=0}^{n} C_k^n (z_1 - z_0)^{n-k} (z - z_1)^k \right), \quad |z - z_0| < r. \quad (*_1)$$

Let

$$z_{nk} = \begin{cases} a_n C_k^n (z_1 - z_0)^{n-k} (z - z_1)^k, & k \leq n \\ 0, & k > n \end{cases}, \quad \begin{array}{l} 0 \leq k \leq n \text{ and} \\ n = 0, 1, 2, \ldots. \end{array}$$

Then $(*_1)$ is the double series $\sum_{n=0}^{\infty} (\sum_{k=0}^{\infty} z_{nk})$. To change the order of summation, we need to show that $\sum z_{nk}$ converges absolutely in $|z - z_1| < r - |z_1 - z_0|$ (refer to Exercise B(4) of Sec. 1.7). Indeed,

$$\sum |z_{nk}| \leq \sum_{n=0}^{\infty} |a_n| \left(\sum_{k=0}^{n} C_k^n |z_1 - z_0|^{n-k} |z - z_1|^k \right)$$

$$= \sum_{n=0}^{\infty} |a_n| (|z_1 - z_0| + |z - z_1|)^n < \infty,$$

where $|z - z_1| < r - |z_1 - z_0|$ (see Fig. 5.1 in which $r_{z_0} = r$). Hence, by $(*_1)$,

$$f(z) = \sum_{k=0}^{\infty} \frac{(z - z_1)^k}{k!} \left[\sum_{n=k}^{\infty} n(n-1) \cdots (n - k + 1)(z_1 - z_0)^{n-k} a_n \right]$$

$$= \sum_{k=0}^{\infty} \frac{f^{(k)}(z_1)}{k!} (z - z_1)^k, \quad |z - z_1| < r - |z_1 - z_0| \quad (*_2)$$

and this is the Taylor series expansion of $f(z)$ at z_1.

We summarize the above as

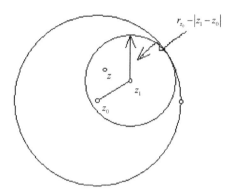

Fig. 5.1

The Taylor series expansion. Let $f(z)$ be as in (5.1.2.1). Fix any point z_1, $|z_1 - z_0| < r$. Then the Taylor series expansion of $f(z)$ at z_1 is given by

$$f(z) = \sum_{k=0}^{\infty} \frac{f^{(k)}(z_1)}{k!}(z - z_1)^k, \quad |z - z_1| < r_{z_1},$$

whose radius of convergence r_{z_1} satisfies $r - |z_1 - z_0| \le r_{z_1} \le r + |z_1 - z_0|$. It is understood that $r_{z_1} = \infty$ if $r = \infty$ holds.   (5.1.2.2)

This result is going to be the starting point of direct analytic continuation of power series, to be formally introduced in Sec. 5.2.

Section (2) The interior uniqueness theorem and the max–min modulus principle, etc.

Fix a point a in $|z - z_0| \le r$. Suppose $f'(a) = \cdots = f^{(k-1)}(a) = 0$ yet $f^{(k)}(a) \ne 0$ for $k \ge 1$. Then, according to (5.1.2.2),

$$f(z) = f(a) + \frac{f^{(k)}(a)}{k!}(z - a)^k(1 + \varphi(z)),$$

where

$$\varphi(z) = \sum_{n=1}^{\infty} b_n(z - a)^n \text{ with } b_n = \frac{k!}{f^{(k)}(a)} \cdot \frac{f^{(n+k)}(a)}{(n + k)!},$$

$$n \ge 1, \ |z - a| < r - |a - z_0|. \quad (5.1.2.3)$$

This factorization of $f(z)$ in a neighborhood of a is crucial in obtaining the following remarkable properties of analytic functions, *especially those defined in open disks through power series.*

First, as in (3.4.2.9),

The zeros. Let $f(z)$ be as in (5.1.2.1). Then

(1) *The isolated-zero principle*: In case $f(z) \not\equiv 0$, then $f(z)$ has only isolated zeros in $|z - z_0| < r$, namely, if $f(a) = 0$, then there is a $\delta > 0$ so that $f(z) \neq 0$ in $0 < |z - a| < \delta$.

$\Leftrightarrow$ (2) *The interior uniqueness principle*: If $f(z)$ has a sequence z_k of zeros in $|z - z_0| < r$ such that z_k, $k \geq 1$, has a *limit point in* $|z - z_0| < r$, then $f(z) \equiv 0$ throughout $|z - z_0| < r$. In particular, $f(z) \equiv 0$ if $f(z)$ is identically equal to zero in a nonempty open subset of $|z - z_0| < r$.

$$(5.1.2.4)$$

This leads immediately to the following

Principle of (direct) analytic continuation of power series. Let

$$f(z) = \sum_0^{\infty} a_n(z - z_0), \quad |z - z_0| < r; \quad g(z) = \sum_0^{\infty} b_n(z - z_0), \quad |z - z_0| < r.$$

Then

(1) $f(z) \equiv g(z)$, $|z - z_0| < r$.

$\Leftrightarrow$ (2) $a_n = b_n$, $n \geq 0$.

$\Leftrightarrow$ (3) There is a point a in $|z - z_0| < r$ such that $f^{(n)}(a) \equiv g^{(n)}(a)$, $n \geq 0$.

$\Leftrightarrow$ (4) There is a sequence z_k of points in $|z - z_0| < r$, converging to a point in $|z - z_0| < r$, such that $f(z_k) \equiv g(z_k)$, $k \geq 1$.

$\Leftrightarrow$ (5) There is a subdomain in $|z - z_0| < r$ on which $f(z) = g(z)$ holds.

$$(5.1.2.5)$$

Proofs are left to the readers.

Second, exactly like the process how $(*_1)$–$(*_4)$ in Sec. 3.4.4 lead to (3.4.4.1), we have

The max–min modulus principle. Let $f(z)$ be as in (5.1.2.1).

(1) *The maximum principle.* If $f(z)$ is *not* a constant function in $|z - z_0| < r$, then $|f(z)|$ cannot attain its local maximum at any point in $|z - z_0| < r$. Equivalently, if $f(z)$ is *also* continuous on $|z - z_0| \leq r < \infty$, then $|f(z)|$ attains its global maximum on $|z - z_0| \leq r$ at a boundary point of the disk.

(2) *The minimum principle.* If $f(z)$ is *not* a constant function and $f(z) \neq 0$ throughout $|z - z_0| < r$, then $|f(z)|$ cannot attain its local minimum at any point in $|z - z_0| < r$. Equivalently, if $f(z)$ is *also* continuous on

$|z - z_0| \leq r < \infty$, then $|f(z)|$ attains its global minimum on $|z - z_0| \leq r$ at a boundary point of the disk.   (5.1.2.6)

Let

$$M(\rho) = \max_{|z-z_0|\leq\rho} |f(z)| \text{ and } m(\rho) = \min_{|z-z_0|\leq\rho} |f(z)|, \ 0 \leq \rho < r. \quad (5.1.2.7)$$

Then $M(\rho)$ is a nondecreasing continuous function of ρ in $(0, r)$ and it is strictly increasing if $f(z)$ is nonconstant; while $m(\rho)$ is a nonincreasing continuous function of ρ in $(0, r)$ if $f(z) \neq 0$ on $|z - z_0| < r$ and it is strictly decreasing if $f(z)$ is also nonconstant. Refer to Exercise A(6) of Sec. 3.4.4. As a bonus, these results help us to imagine what the graph $|f(z)|$ might look like in the Euclidean three-dimensional space without graphing it practically. For instance, suppose $f(z)$ is not a constant in $|z - z_0| < r$ and fix a point a within. Then $|f(a)|$ is neither a local maximum nor a local minimum, if $f(a) \neq 0$. For any ρ, $0 < \rho < r - |a - z_0|$, there are points z_1 and z_2 on $|z - a| = \rho$ so that $|f(z_1)| < |f(a)| < |f(z_2)|$. By the intermediate value property of a continuous function on a connected set, there is at least one point z_ρ on $|z - a| = \rho$ so that $|f(z_\rho)| = |f(a)|$ holds. Thus, the *level curve* $|f(z)| = |f(a)|$, namely,

$$\{z | |z - z_0| < r \text{ and } |f(z)| = |f(a)|\}$$

is nonempty and leads to the boundary $|z - z_0| = r$. Refer to Exercise A(7) of Sec. 3.4.4.

Thirdly, (5.1.2.3) can also be used to study the local behavior of $f(z)$ at a point. No detail will be given here and the readers may refer to (3.5.1.8) for information.

Section (3) Fourier series and power series

Suppose $F(\theta) : \mathbf{R} \to \mathbf{C}$ is a continuous function with period 2π, namely, $F(\theta + 2\pi) = F(\theta)$, for all $\theta \in \mathbf{R}$. For any integer n, the *n-th Fourier coefficient* of $F(\theta)$ is defined by

$$c_n(F) = \frac{1}{2\pi} \int_0^{2\pi} F(\theta)e^{-in\theta} d\theta.$$

Note that

$$c_n(1) = \frac{1}{2\pi} \int_0^{2\pi} e^{-in\theta} d\theta = \begin{cases} 1, & n = 0 \\ 0, & n \neq 0 \end{cases}.$$

The complex series

$$\sum_{n=-\infty}^{\infty} c_n(F)e^{in\theta}, \quad \theta \in \mathbf{R}$$

is called the *Fourier series* of $F(\theta)$ with $s_n(\theta) = \sum_{k=-n}^{n} c_n(F)e^{ik\theta}$, $n = 0, 1, 2, \ldots$, its *partial sums*. In case $\lim_{n\to\infty} s_n(\theta)$ *converges* to a finite complex number, then we say that $\sum_{n=-\infty}^{\infty} c_n(F)e^{in\theta}$ converges to this limit.

Let $f(z)$ be as in (5.1.2.1). Choose ρ, $0 \le \rho < r$. Then $F(\theta) = f(z_0 + \rho e^{i\theta}) = \sum_0^{\infty} a_n \rho^n e^{in\theta}$, $0 \le \theta \le 2\pi$, has its Fourier coefficients

$$c_n(F) = \frac{1}{2\pi} \int_0^{2\pi} \left(\sum_{k=0}^{\infty} a_k \rho^k e^{ik\theta} \right) e^{-in\theta} d\theta$$

$$= \frac{1}{2\pi} \sum_{k=0}^{\infty} a_k \rho^k \int_0^{2\pi} e^{i(k-n)\theta} d\theta$$

(by the uniform convergence on $|z - z_0| \le \rho$)

$$= \begin{cases} a_n \rho^n, & n \ge 0 \\ 0, & n < 0 \end{cases}. \tag{5.1.2.8}$$

In this case, the partial sum $s_n(\theta) = \sum_{k=-n}^{n} c_n(F)e^{ik\theta} = \sum_{k=0}^{n} a_k \rho^k e^{ik\theta}$ is just the partial sum of the power series $\sum_0^{\infty} a_n(z - z_0)^n$ at $z = z_0 + \rho e^{i\theta}$, which converges uniformly to $f(z_0 + \rho e^{i\theta})$ on $|z - z_0| = \rho$. Consequently,

$$\frac{1}{2\pi} \int_0^{2\pi} |f(z_0 + \rho e^{i\theta})|^2 d\theta = \lim_{n\to\infty} \frac{1}{2\pi} \int_0^{2\pi} |s_n(\theta)|^2 d\theta.$$

Since

$$|s_n(\theta)|^2 = s_n(\theta)\overline{s_n(\theta)} = \sum_{k,l=0}^{n} a_k \overline{a_l} \rho^{k+l} e^{i(k-l)\theta}$$

$$\Rightarrow \int_0^{2\pi} |s_n(\theta)|^2 = \sum_{k,l=0}^{n} a_k \overline{a_l} \rho^{k+l} \int_0^{2\pi} e^{i(k-l)\theta} d\theta = 2\pi \sum_{k=0}^{n} |a_k|^2 \rho^{2k}.$$

Summarize the above as (see Exercise B(2) of Sec. 3.4.2)

Some equality and inequalities. Let $f(z)$ be as in (5.1.2.1).

(1) *Parseval's identity:* For $0 \le \rho < r$,

$$\frac{1}{2\pi} \int_0^{2\pi} |f(z_0 + \rho e^{i\theta})|^2 d\theta = \sum_{n=0}^{\infty} |a_n|^2 \rho^{2n}.$$

(2) *Gutzmer's inequality:* Let $M(\rho) = \max_{|z-z_0|\le\rho} |f(z)|$, $0 \le \rho < r$. Then

$$\sum_{n=0}^{\infty} |a_n|^2 \rho^{2n} \le M(\rho)^2, \quad 0 \le \rho < r.$$

(3) *Cauchy inequality:*

$$|f^{(n)}(z_0)| \le \frac{n! M(\rho)}{\rho^n}, \quad 0 < \rho < r, \ n = 0, 1, 2, \ldots. \tag{5.1.2.9}$$

It follows easily *the maximum modulus principle* (moreover, $M(\rho) = |a_k|\rho^k$ for some k and $0 < \rho < r \Leftrightarrow f(z) = a_k(z - z_0)^k$), *Liouville's theorem* (see (3.4.2.10)), *the fundamental theorem of algebra* (see (3.4.2.12)) and even *Schwarz's lemma* (see Sec. 3.4.5).

Section (4) Cauchy's integral theorem and formula

As we have already mentioned after (3.4.2.1), for any rectifiable closed curve γ lying entirely in $|z - z_0| < r$, the locally uniform convergence of $\sum_0^\infty a_n(z - z_0)^n$ to $f(z)$ gives the fact that

$$\int_\gamma f(z)dz = \sum_{n=0}^{\infty} a_n \int_\gamma (z - z_0)^n dz = 0 \text{ (see Example 2 in Sec. 2.9).}$$

$$\tag{5.1.2.10}$$

This is *Cauchy's integral theorem* in an open disk.

Suppose $f(z)$ is also continuous on the closed disk $|z - z_0| \le r$. Then,

$$\frac{1}{2\pi} \int_0^{2\pi} f(z_0 + re^{i\theta})e^{-in\theta} d\theta$$

$$= \lim_{\rho \to r^-} \frac{1}{2\pi} \int_0^{2\pi} f(z_0 + \rho e^{i\theta})e^{-in\theta} d\theta \text{ (by uniform continuity)}$$

$$= \begin{cases} a_n r^n, & n \ge 0 \\ 0, & n < 0 \end{cases} \text{ (by (5.1.2.8)),}$$

$$\Rightarrow f(z) = \sum_{n=0}^{\infty} a_n(z - z_0)^n = \sum_{n=0}^{\infty} (z - z_0)^n \frac{1}{2\pi r^n} \int_0^{2\pi} f(z_0 + re^{i\theta})e^{-in\theta} d\theta$$

$$= \frac{1}{2\pi} \int_0^{2\pi} f(z_0 + re^{i\theta}) \sum_{n=0}^{\infty} \left(\frac{z - z_0}{r} e^{-i\theta}\right)^n d\theta$$

(z is fixed and $|z - z_0| < r$; uniform convergence)

$$= \frac{1}{2\pi} \int_0^{2\pi} f(z_0 + re^{i\theta}) \frac{1}{1 - \dfrac{z - z_0}{r} e^{-i\theta}} d\theta.$$

Summarize the above as

Cauchy's fundamental theorems. Let $f(z)$ be as in (5.1.2.1). Also, suppose $f(z)$ is *continuous* on $|z - z_0| \leq r < \infty$.

(1) *Integral theorem.* For any rectifiable closed curve γ in $|z - z_0| \leq r$,

$$\int_\gamma f(z)dz = 0.$$

(2) *Integral formula.*

$$f(z) = \frac{1}{2\pi} \int_0^{2\pi} \frac{f(z_0 + re^{i\theta})re^{i\theta}}{z_0 + re^{i\theta} - z} d\theta \underset{\text{(def.)}}{=} \frac{1}{2\pi i} \int_{|\zeta - z_0| = r} \frac{f(\zeta)}{\zeta - z} d\zeta, \quad |z - z_0| < r.$$

Moreover,

$$f^{(n)}(z) = \frac{n!}{2\pi i} \int_{|\zeta - z_0| = r} \frac{f(\zeta)}{(\zeta - z)^{n+1}} d\zeta, \quad |z - z_0| < r, \ n = 0, 1, 2, \ldots.$$

(3) Conversely, if $\varphi(z)$ is continuous on $|z - z_0| = r$. Then

$$f(z) \underset{\text{(def.)}}{=} \frac{1}{2\pi i} \int_{|\zeta - z_0| = r} \frac{\varphi(\zeta)}{\zeta - z} d\zeta, \quad |z - z_0| < r$$

$$= \sum_{n=0}^{\infty} a_n (z - z_0)^n, \text{ where } a_n = \frac{1}{2\pi i} \int_{|\zeta - z_0| = r} \frac{\varphi(\zeta)}{(\zeta - z_0)^{n+1}} d\zeta, \ n \geq 0,$$

is analytic in $|z - z_0| < r$. \hfill (5.1.2.11)

The readers should compare these results to (3.4.2.1), (3.4.2.4), (3.4.2.1)', (3.4.2.4)', (3.4.2.16), and (4.7.1).

Exercises A

(1) (a) Prove the inequality for radii of convergence in (5.1.2.2).

   (b) Show that the radius r_z of convergence is a continuous function of z in $|z - z_0| < r$, namely, $|r_{z_1} - r_{z_2}| \leq |z_1 - z_2|$ if $|z_1 - z_2| < \max(r_{z_1}, r_{z_2})$.

(2) Prove (5.1.2.4), (5.1.2.5), (5.1.2.6), (5.1.2.9), and (5.1.2.11) in detail.

(3) Try to use (5.1.2.11) to prove (5.1.1.7). Since the partial sum $s_n(z) = \sum_{k=1}^{n} f_k(z)$ converges locally uniformly to $f(z)$ in $|z - z_0| < r$, we may prove (5.1.1.7) just in the case of a *sequence* $f_n(z)$ converging locally uniformly to $f(z)$ in $|z - z_0| < r$. Notice the following steps.

(a) Choose $\rho > 0$ so that $|z - z_0| \leq \rho < r$. Try to show that

$$f(z) = \frac{1}{2\pi i} \int_{|\zeta - z_0| = \rho} \frac{f(\zeta)}{\zeta - z} d\zeta, \quad |z - z_0| < \rho.$$

(b) Starting from $f'_n(z) = \frac{1}{2\pi i} \int_{|\zeta-z_0|=\rho} \frac{f_n(\zeta)}{(\zeta-z)^2} d\zeta$, try to show that

$$\lim_{n\to\infty} f'_n(z) = \frac{1}{2\pi i} \int_{|\zeta-z_0|=\rho} \frac{f(\zeta)}{(\zeta-z)^2} d\zeta = f'(z), \quad |z-z_0| < \rho$$

locally uniformly in $|z - z_0| < \rho$.

Refer to (5.3.1.1) if necessary.

5.1.3   *Boundary behavior of a power series on its circle of convergence*

Suppose the analytic function $f(z)$ is given by (5.1.2.1). On the circle of convergence $|z - z_0| = r$, the power series $\sum_0^\infty a_n(z - z_0)^n$ might diverge everywhere, or converge everywhere, or converge at some points and diverge elsewhere. For instance,

1. $\sum_{n=0}^\infty z^n (|z| < 1)$ diverges everywhere along $|z| = 1$;
2. $\sum_{n=1}^\infty \frac{z^n}{n} (|z| < 1)$ diverges at $z = 1$ only and converges conditionally at all other points of $|z| = 1$. For, if $0 < \theta < 2\pi$ and $z = e^{i\theta}$,

$$\sum_{n=1}^\infty \frac{z^n}{n} = \sum_{n=1}^\infty \frac{\cos n\theta}{n} + i \sum_{n=1}^\infty \frac{\sin n\theta}{n}$$

     and its real and imaginary part series converge by Dirichlet test (see (2.3.5) and Example 1 below).
3. $\sum_{n=1}^\infty \frac{z^n}{n^2} (|z| < 1)$ converges absolutely and uniformly on $|z| = 1$.

Note that the sum function $f(z)$ might be well-defined and analytic on a subset outside $|z - z_0| < r$. For example, $\sum_{n=0}^\infty z^n$ represents $\frac{1}{1-z}$ only on $|z| < 1$, even though $\frac{1}{1-z}$ is analytic in $\mathbf{C} - \{1\}$. It is the analytic continuation of $\sum_{n=0}^\infty z^n$ throughout $\mathbf{C} - \{1\}$ that produces $\frac{1}{1-z}$ (see Example 2 below).

     The following is divided into five subsections.

Section (1) Abel's limit theorem

Suppose $\sum_0^\infty a_n(z - z_0)^n$ converges at a boundary point ζ, $|\zeta - z_0| = r$. Is there any possible relation between the sum $\sum_0^\infty a_n(\zeta - z_0)^n$ and the sum function $f(z)$ in $|z - z_0| < r$?

     Recall that, in the real case, if $f(x) = \sum_0^\infty a_n(x - x_0)^n$ converges locally uniformly in an open interval (a, b) and if the series $\sum_0^\infty a_n(b - x_0)^n$ converges, then $f(x)$ is continuous at b from the left and $f(b) = \lim_{x\to b-} f(x) =$

$\sum_{n=0}^{\infty} a_n(b-x_0)^n$ holds. The complex case will imitate this result. Using w to replace $z - z_0$ in $\sum_0^{\infty} a_n(z-z_0)^n$, we have indeed

Abel's limit theorem. *Suppose* $f(z) = \sum_0^{\infty} a_n z^n$ *has a positive radius* r *of convergence. If there is a boundary point* z_0, $|z_0| = r$ *such that* $\sum_0^{\infty} a_n z_0^n$ *converges, then*

$$\lim_{z \in D_\delta \to z_0} f(z) = f(z_0) = \sum_0^{\infty} a_n z_0^n$$

holds, where D_δ *is a Stolz domain (see Example 2 of Sec. 1.4.2 and Fig. 1.12). The convergence is even locally uniformly in* $D_\delta \cup \{z_0\}$. *In particular, if* $t \in [0, 1)$ *is real, then*

$$\lim_{t \to 1^-} f(tz) = f(z_0). \tag{5.1.3.1}$$

Historically, this is the first theorem studying the relation of values of a certain kind of function in a domain and on its boundary. This result can be extended to formulate a method for the summability of series. The converse of (5.1.3.1) is not more true. For instance, $\sum_0^{\infty} z^n = \frac{1}{1-z}(|z| < 1)$ is the case at the point $z_0 = -1 : \sum_0^{\infty} (-1)^n$ diverges, while $\lim_{z \to -1} \frac{1}{1-z} = \frac{1}{2}$ exists.

Proof. Let $\theta_0 = \operatorname{Arg} z_0$. Substituting $z = re^{i\theta_0} w = z_0 w$ into the original series, then $\sum_{n=0}^{\infty} a_n z^n = \sum_0^{\infty} b_n w^n$, where $b_n = a_n r^n e^{in\theta_0} = a_n z_0^n$ for $n \geq 0$. The resulted series $\sum_0^{\infty} b_n w^n$ has the radius 1 of convergence and converges at $w = 1$, namely, $\sum_0^{\infty} b_n = \sum_0^{\infty} a_n z_0^n$ converges. If this theorem is valid for $\sum_0^{\infty} b_n w^n$ at $w = 1$, then it is also valid for $\sum_{n=0}^{\infty} a_n z^n$ at z_0. Hence, we may assume from the very beginning that $\sum_{n=0}^{\infty} a_n z^n$ has 1 as its radius of convergence, $z_0 = 1$ and even $\sum_0^{\infty} a_n = 0$, after a readjustment of the term a_0.

Let $A_n = a_0 + a_1 + \cdots + a_n$ for $n \geq 0$. In case $|z| < 1$, applying Abel's law of summation (see (2.3.6)), we have

$$\sum_{n=0}^{k} a_n z^n = (1 - z) \sum_{n=0}^{k-1} A_n z^n + a_k z^k$$

$$\Rightarrow (\text{since } \lim_{k \to \infty} a_k z^k = 0 \text{ if } |z| < 1) \ f(z) = (1 - z) \sum_0^{\infty} A_n z^n, |z| < 1.$$

What remains is to estimate the right side so that $f(z) \to 0$ wherever $z \in D_\delta$ and $z \to 1$.

Since $\sum_0^\infty a_n = 0$, $\lim_{n\to\infty} A_n = 0$. For any given $\varepsilon > 0$, there exists an n_0 so that

$$|A_n| < \varepsilon, \quad n \geq n_0$$

$$\Rightarrow \left| \sum_{n=n_0}^\infty A_n z^n \right| \leq \sum_{n=n_0}^\infty |A_n||z|^n \leq \varepsilon \sum_{n=n_0}^\infty |z|^n < \frac{\varepsilon}{1 - |z|}, \quad |z| < 1.$$

Now, suppose $z \in D_\delta$ $(0 \leq \delta < \frac{\pi}{2})$, then by Example 2 of Sec. 1.4.2,

$$|1 - z| \leq \frac{2}{\cos \delta}(1 - |z|)$$

$$\Rightarrow \left| (1 - z) \sum_{n=n_0}^\infty A_n z^n \right| \leq \frac{\varepsilon|1 - z|}{1 - |z|} \leq \frac{2\varepsilon}{\cos \delta}, \quad z \in D_\delta. \tag{$*_1$}$$

On the other hand, letting $\varepsilon > 0$ and n_0 be as above, there is an $\eta > 0$ so that

$$\left| (1 - z) \sum_{n=0}^{n_0-1} A_n z^n \right| < \varepsilon, \quad \text{if } |1 - z| < \eta \text{ and } z \in D_\delta. \tag{$*_2$}$$

Combining $(*_1)$ and $(*_2)$, we have

$$|f(z)| \leq \left| (1 - z) \sum_{n=0}^{n_0-1} A_n z^n \right| + \left| (1 - z) \sum_{n=n_0}^\infty A_n z^n \right|$$

$$< \varepsilon \left(1 + \frac{2}{\cos \delta} \right), \quad \text{if } z \in D_\delta \text{ and } |1 - z| < \eta$$

$$\Rightarrow \lim_{z \in D_\delta \to 1} f(z) = 0.$$

This finishes the proof. $\qquad\qquad\qquad\qquad\qquad\qquad\qquad\qquad\qquad\qquad$ □

As a routine application, we illustrate the following

Example 1. Show that

$$\sum_{n=1}^\infty \frac{\cos n\theta}{n} = -\log \left| 2 \sin \frac{\theta}{2} \right|, \quad 0 < |\theta| \leq \pi \text{ or } 0 < \theta < 2\pi;$$

$$\sum_{n=1}^\infty \frac{\sin n\theta}{n} = \frac{\pi - \theta}{2}, \quad 0 < \theta < 2\pi.$$

Solution. It is well-known that the series

$$\log \frac{1}{1-z} = \sum_0^\infty \frac{z^n}{n}, \quad |z| < 1$$

converges conditionally at points on $|z| = 1$ except at $z = 1$ (try to work out the details by using Dirichlet test). On the other hand, if $|z| = 1$ and $z \neq 1$, set $z = e^{i\theta}$, $0 < \theta < 2\pi$ and we have

$$\log \frac{1}{1-z} = -\log(1 - e^{i\theta}) = -\log|1 - e^{i\theta}| - i\operatorname{Arg}(1 - e^{i\theta})$$

$$= -\log\left|2\sin\frac{\theta}{2}\right| + i\left(\frac{\pi}{2} - \frac{\theta}{2}\right).$$

The results follow immediately by using (5.1.3.1).

Section (2) Regular points of a power series

Let

$$f(z) = \sum_{n=0}^\infty a_n(z - z_0)^n, \quad |z - z_0| < r_{z_0} \text{ (the radius of convergence).}$$
$$(5.1.3.2)$$

For any fixed point z_1 in the disk of convergence, according to (5.1.2.2), the Taylor series expansion $\sum_{n=0}^\infty \frac{f^{(n)}(z_1)}{n!}(z - z_1)^n$ of $f(z)$ at z_1 has its radius r_{z_1} of convergence satisfying the inequality

$$r_{z_0} - |z_1 - z_0| \leq r_{z_1} \leq r_{z_0} + |z_1 - z_0|. \qquad (5.1.3.3)$$

In case $r_{z_1} > r_{z_0} - |z_1 - z_0|$ (see Fig. 5.2), then the power series $\sum_{n=0}^\infty \frac{f^{(n)}(z_1)}{n!}(z - z_1)^n$ defines an analytic function $g(z)$ in $|z - z_1| < r_{z_1}$. Therefore, by (5.1.2.2),

$$g(z) = \sum_{n=0}^\infty \frac{f^{(n)}(z_1)}{n!}(z - z_1)^n = f(z) \text{ in } |z - z_1| < r_{z_0} - |z_1 - z_0|.$$

Consequently, by (5.1.2.5),

$$g(z) = f(z) \text{ in } \{|z - z_0| < r_{z_0}\} \cap \{|z - z_1| < r_{z_1}\}.$$

Define a new function

$$F(z) = \begin{cases} f(z), & |z - z_0| < r_{z_0} \\ g(z), & |z - z_1| < r_{z_1} \end{cases}. \qquad (5.1.3.4)$$

Then $F(z)$ is well-defined in the union of these two open disks and analytic there. It is in this sense that we say $g(z)$ extends the analyticity of $f(z)$ beyond the circle $|z - z_0| = r_{z_0}$ of convergence and into the outside of

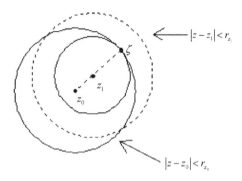

$$\longleftarrow |z - z_1| < r_{z_1}$$

$$|z - z_0| < r_{z_0}$$

Fig. 5.2

it along the radius $\frac{z_1 - z_0}{|z_1 - z_0|}$ $(z_1 \neq z_0)$. In short, $g(z)$ is said to be a *direct analytic continuation* of $f(z) = \sum_{n=0}^{\infty} a_n(z - z_0)^n$ *to the point* z_1.

We summarize the above as

The regular point of a power series. Let the power series be as in (5.1.3.2). Suppose z_1 is a point in $|z - z_0| < r_{z_0}$ so that the power series $g(z) = \sum_{n=0}^{\infty} \frac{f^{(n)}(z_1)}{n!}(z - z_1)^n$ has its radius of convergence $r_{z_1} > r_{z_0} - |z_1 - z_0|$. Then $g(z)$ is a direct analytic continuation of $f(z)$ and the tangent point ζ of the circles $|z - z_1| = r_{z_0} - |z_1 - z_0|$ and $|z - z_0| = r_{z_0}$ (see Fig. 5.2) is called a *regular point* of the power series $f(z) = \sum_0^{\infty} a_n(z - z_0)^n$. $\hspace{2cm}$ (5.1.3.5)

A point, lying on the circle of convergence, which is not a regular point, is called a *singular point* of the power series. For details, see Section (3).

We illustrate two examples.

Example 2. Show that $\sum_0^{\infty} z^n$, $|z| < 1$, can be extended to the global analytic function $\frac{1}{1-z}$ on $\mathbf{C} - \{1\}$ by a finite number of successive direct analytic continuation if any point $z \in \mathbf{C} - \{1\}$ is given. 1 is the only singular point of the series.

Proof. Pick up any point a in $\mathbf{C} - \{1\}$ and connect it to the origin o via a Jordan curve γ in $\mathbf{C} - \{1\}$. Fix a point z_1 on γ that lies in $|z| < 1$. Exactly like the process from $(*_1)$ to $(*_2)$ in the proof of (5.1.2.2), $\sum_0^{\infty} z^n$ can be analytically continued to z_1 directly, namely,

$$\frac{1}{1 - z} = \sum_{n=0}^{\infty} \frac{1}{(1 - z_1)^{n+1}}(z - z_1)^n, \quad |z - z_1| < |1 - z_1|.$$

See Fig. 5.3.

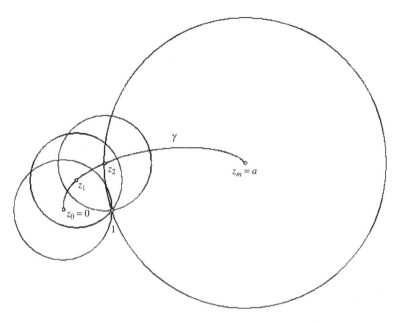

Fig. 5.3

Again, choose a point z_2 on γ that lies in $|z - z_1| < |1 - z_1|$ but not in $|z| < 1$ and continue analytically to z_2 as

$$\frac{1}{1-z} = \sum_{n=0}^{\infty} \frac{1}{(1-z_2)^{n+1}} (z - z_2)^n, \quad |z - z_2| < |1 - z_2|.$$

After a finite number of such a process (since, as a point set, γ is compact), we will reach the given point a and

$$\frac{1}{1-z} = \sum_{n=0}^{\infty} \frac{1}{(1-a)^{n+1}} (z - a)^n, \quad |z - a| < |1 - a|. \qquad (*_3)$$

In this case, we say that $(*_3)$ *is obtained by analytic continuation from* $\sum_0^{\infty} z^n$, $|z| < 1$, *along the curve* γ.

As a matter of fact, choose another Jordan curve σ in $\mathbf{C} - \{1\}$ connecting o to a (or choose different intermediate points other than $z_1, z_2, \ldots$, along γ) and continue $\sum_0^{\infty} z^n$, $|z| < 1$, analytically along σ to the point a. By the uniqueness property (see (5) in (5.1.2.5)), we will get the same series expansion $(*_3)$. Hence, the resulted $\frac{1}{1-z}$ is globally well-defined in $\mathbf{C} - \{1\}$.

Example 3. Given the power series $(\log 1 = 0)$

$$\log(1+z) = \sum_{n=1}^{\infty} \frac{(-1)^{n-1}}{n} z^n, \quad |z| < 1.$$

(a) Show that $\log 2 - \sum_{n=1}^{\infty} \frac{1}{n} \left(\frac{1-z}{2}\right)^n$ is a direct analytic continuation of this series.

(b) Find its direct analytic continuation at the point 2. Find the resulting power series after continuing analytically this new series along the circle $|z+1| = 3$ (one circulation and returning to the point 2).

(c) Is it possible to extend the original series into a *single-valued* analytic function in $\mathbf{C} - \{1\}$? In $\mathbf{C} - (-\infty, -1]$?

Solution. (a) The series has radius 2 of convergence. Sum the series as

$$\log 2 - \sum_{n=1}^{\infty} \frac{1}{n} \left(\frac{1-z}{2}\right)^n = \log 2 + \sum_{n=1}^{\infty} \frac{(-1)^{n-1}}{n} \left(\frac{z-1}{2}\right)^n, \quad \left|\frac{z-1}{2}\right| < 1$$

$$= \log 2 + \log\left(1 + \frac{z-1}{2}\right)$$

$$= \log(1+z), \quad |z-1| < 2.$$

Since $|z| < 1$ is contained in $|z - 1| < 2$, this series is a direct analytic continuation of the original series. See Fig. 5.4.

(b) The series in (a) is the Taylor series of $\log(1 + z)$ at $z = 1$, valid in $|z - 1| < 2$. Since 2 is contained in $|z - 1| < 2$, according to (5.1.2.2), the

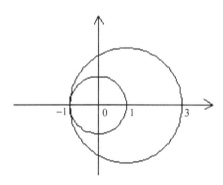

Fig. 5.4

Taylor series expansion of $\log(1 + z)$ at $z = 2$ is given by

$$\log(1 + z) = \sum_{n-0}^{\infty} \frac{1}{n!} \frac{d^n}{dz^n} \log(1 + z)|_{z=2}(z - 2)^2$$

$$= \log 3 + \sum_{n=1}^{\infty} \frac{(-1)^{n-1}(n - 1)!}{n!} \frac{1}{3^n}(z - 2)^n$$

$$= \log 3 + \sum_{n=1}^{\infty} \frac{(-1)^{n-1}}{n} \left(\frac{z - 2}{3}\right)^n, \quad |z - 2| < 3. \qquad (*_4)$$

This is also a direct analytic continuation of the series $\sum_0^{\infty} \frac{(-1)^n}{n} z^n$, $|z| < 1$, since $|z| < 1$ is contained in $|z - 2| < 3$.

Continue analytically the power series $(*_4)$ along the circle $|z + 1| = 3$ up to the point $z = -1 + 3e^{i\theta}$, $0 \le \theta \le 2\pi$, and we get the corresponding Taylor series expansion (θ is fixed), see Fig. 5.5.

$$f_\theta(z) = \log(3e^{i\theta}) + \sum_{n=1}^{\infty} \frac{1}{n!} \frac{d^n}{dz^n} \log(1 + z)|_{z=-1+3e^{i\theta}} \cdot (z + 1 - 3e^{i\theta})^n$$

$$= \log 3 + i\theta + \sum_{n=1}^{\infty} \frac{(-1)^{n-1}}{n} \left(\frac{z + 1 - 3e^{i\theta}}{3e^{i\theta}}\right)^n, \quad |z + 1 - 3e^{i\theta}| < 3.$$

When z starts from 2 and winds once along the circle $|z + 1| = 3$ in the counterclockwise direction and then returns to 2, f_θ continues analytically from f_0 to

$$f_{2\pi}(z) = \log 3 + 2\pi i + \sum_{n=1}^{\infty} \frac{(-1)^{n-1}}{n} \left(\frac{z - 2}{3}\right)^n = f_0(z) + 2\pi i, \quad |z - 2| < 3.$$

This result holds for general θ, namely, $f_{2\pi+\theta}(z) = f_\theta(z) + 2\pi i$, and even more,

$$f_{2n\pi+\theta}(z) = f_\theta(z) + 2n\pi i, \quad n = 0, \pm 1, \pm 2, \dots, \quad 0 \le \theta \le 2\pi.$$

(c) Fix any point $a \in \mathbf{C} - \{-1\}$. Choose a branch of $\log(1 + a)$, say the principal branch $\text{Log}(1 + a)$.

Then, as in (b),

$$f(z) = \text{Log}(1 + a) + \sum_{n=1}^{\infty} \frac{(-1)^{n-1}}{n} \left(\frac{z - a}{1 + a}\right)^n, \quad |z - a| < |1 + a| \qquad (*_5)$$

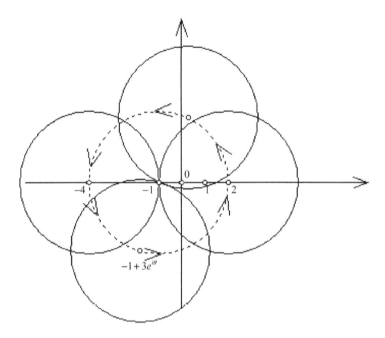

Fig. 5.5

is a direct analytic continuation of the original power series. Continue $(*_5)$ along a chosen Jordan curve γ in $\mathbf{C}-\{-1\}$, with a as the initial and terminal point. In case $-1 \in \text{Ext } \gamma$, the resulting series after one circulation will be coincident with $(*_5)$; while, if $-1 \in \text{Int } \gamma$, the resulting series becomes $f(z) + 2\pi i$ and it will be $f(z) + 2n\pi i$ after n circulations.

Hence, $\sum_1^\infty \frac{(-1)^{n-1}}{n} z^n$, $|z| < 1$, can only be extended into the multiple-valued function $\log(1+z)$ in $\mathbf{C}-\{-1\}$. Yet, when restricted to the simply connected domain $\mathbf{C} - (-\infty, -1]$, the resulted global analytic function is the principal branch $\text{Log}(1+z)$, $\log 1 = 0$. See Sec. 5.2.2.

Section (3) Singular points of a power series

Based on (5.1.3.5), we formally define

A singular point of a power series. Let the power series be as in (5.1.3.2) (see also (5.1.3.3)). Suppose ζ is a point on the circle of convergence, namely, $|\zeta - z_0| = r_{z_0}$, such that there is a point z_1 along the radius $\overline{z_0\zeta}$ ($0 < |z_1 - z_0| < r_{z_0}$) causing the Taylor series expansion of $f(z)$ at z_1 having its radius of convergence equal to $r_{z_1} = r_{z_0} - |z_1 - z_0|$. In this case, the Taylor

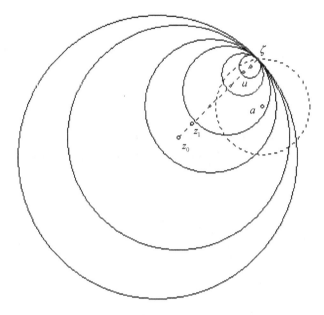

Fig. 5.6

series of $f(z)$ at any point $a, |a - \zeta| < r_{z_0}$, has its radius r_a of convergence satisfying

$$r_a \leq |a - \zeta|.$$

In particular, if a lies on $\overline{z_0 \zeta}$, then $r_a \leq |a - \zeta|$ holds. Such a point ζ is called a *singular point* of the series (5.1.3.2). See Fig. 5.6.   (5.1.3.6)

If a is in $|z - z_0| < r_{z_0}$ and moves toward ζ, then $r_a \to 0^+$ holds. According to (5.1.3.5), it is not possible to continue the power series analytically beyond the circle of its convergence through the point ζ. In short, it is impossible to redefine $f(z)$ at ζ so that it turns out to be analytic there.

Does each power series have at least one singular point?

The one in Example 2 has $\zeta = 1$ and the one in Example 3 has $\zeta = -1$ as their singular points, respectively. The power series $e^z = \sum_0^\infty \frac{1}{n!} z^n$ has $\zeta = \infty$, the only boundary point in $\mathbf{C}^*$, as its unique singular point. In general

The existence of singular point. Each power series, having a positive radius of convergence, will definitely have at least one singular point on its circle of convergence.   (5.1.3.7)

It is possible that every point on the circle of convergence is a singular point (see Example 5 below). In this case, the circle of convergence is called the *natural boundary* of the power series.

Proof. Let $f(z) = \sum_0^\infty a_n(z - z_0)^n$ be as in (5.1.3.2). Denote $|z - z_0| < r_{z_0}$ by D.

Suppose on the contrary that each point ζ on $|z - z_0| = r_{z_0}$ is a regular point. According to (5.1.3.5), to each point ζ on $|z - z_0| = r_{z_0}$, there is an open disk D_ζ with center at ζ on which there exists an analytic function $g_\zeta(z)$ satisfying

$$g_\zeta(z) = f(z), \quad z \in D \cap D_\zeta, \quad \zeta \in \partial D.$$

Now, $\{D_\zeta\}_{\zeta \in \partial D}$ forms an open covering of the compact set ∂D. Thus, there are finitely many disks $D_{\zeta_1}, \ldots, D_{\zeta_K}$ so that $\partial D \subseteq \Omega = \cup_{j=1}^k D_{\zeta_j}$. Let $\delta = \text{dist}(\partial D, \partial \Omega)$. $\delta > 0$ and the open disk $|z - z_0| < r_{z_0} + \delta$ lies entirely in $D \cup \Omega$.

Define a function $F : \Omega \to \mathbf{C}$ by

$$F(z) = g_{\zeta_j}(z), \quad z \in D_{\zeta_j}, \ 1 \leq j \leq k.$$

$F(z)$ is a single-valued analytic function in Ω. To see this, suppose $D_{\zeta_{j_1}} \cap D_{\zeta_{j_2}} \neq \emptyset$ holds. Since $D \cap D_{\zeta_{j_1}} \cap D_{\zeta_{j_2}} \neq \emptyset$, on which we have

$$g_{\zeta_j}(z) = f(z) = g_{\zeta_2}(z)$$

$\Rightarrow$ (by the interior uniqueness principle, see (2) in (5.1.2.4))

$$g_{\zeta_j}(z) = g_{\zeta_2}(z), \quad z \in D_{\zeta_{j_1}} \cap D_{\zeta_{j_2}}.$$

Hence $F(z)$ is well-defined and single-valued.

Since $F(z) = f(z)$ on $\Omega \cap D$, this means that $f(z)$ can be continued analytically from D into $\Omega \cup D$ which contains $|z - z_0| < r_{z_0} + \delta$. According to (3.4.2.6), $f(z)$ can be expanded into its Taylor series at z_0, having its radius of convergence at least equal to $r_{z_0} + \delta$. Yet, by (5.1.2.4) or (5.1.2.5), this series should be the original series $\sum_{n=0}^\infty a_n(z - z_0)^n$. This is impossible because the radius of convergence of $\sum_{n=0}^\infty a_n(z - z_0)^n$ is r_{z_0} but not $r_{z_0} + \delta$.

Hence, $\sum_{n=0}^\infty a_n(z - z_0)^n$ has at least one singular point on its circle of convergence.   □

Here comes two subsequent *questions*: How a singular point destroys the analyticity of a function at a point? And how to pinpoint singular points on the circle of convergence?

As for *question* 1: Recall (see (5.1.3.5) and (3.3.2.4)) that, if ζ is a regular point of $f(z) = \sum_{n=0}^{\infty} a_n(z - z_0)^n$, $|z - z_0| < r_{z_0}$, then there exist a $r_\zeta > 0$ and a power series $g(z) = \sum_{n=0}^{\infty} b_n(z - \zeta)^n$, $|z-\zeta| < r_\zeta$, so that $g(z) = f(z)$ in $|z - z_0| < r_{z_0}$ and $|z-\zeta| < r_\zeta$. In this case, $g(z)$ is infinitely differential at ζ and even satisfies Cauchy's inequality (see (3) in (5.1.2.9)). Consequently, any one of the following conditions will guarantee that *a boundary point ζ is a singular point*: z always represents a point in $|z - z_0| < r_{z_0}$.

1. If $z \to \zeta$, $f(z)$ does not approach a finite limit or $f(z) \to \infty$ (this destroys the well-definedness of g at ζ or the continuity of g at ζ, respectively).
2. Even though $\lim_{z \to \zeta} f(z)$ exists as a finite limit, $\lim_{z \to \zeta} f'(z)$ does not exist (this destroys the differentiability of g at ζ).
3. Even though $\lim_{z \to \zeta} f'(z)$ exists, there is a positive integer $k \geq 2$ so that $\lim_{z \to \zeta} f^{(k)}(z)$ does not exist (this violates the infinite differentiability of g at ζ). (5.1.3.8)

These are some sufficient conditions for singular points. Much more worse condition could happen at a singular point. See Ref. [44], Vol. I, p. 133.

As for *question* 2: We give a simple

Criterion for singular point (Vivanti–Pringsheim theorem). Suppose $a_n \geq 0$ for $n \geq 0$. If the power series $\sum_0^{\infty} a_n z^n$ has its radius r of convergence satisfying $0 < r < \infty$, then $z = r$ is a singular point. (5.1.3.9)

See Exercises A(9)–(11) for some generalizations of this theorem and Exercise A(8) for another proof. See also Ref. [76], pp. 214–217.

Proof. Using $\frac{z}{r}$ to replace z if necessary, we may suppose $r = 1$.

Suppose on the contrary that 1 is a regular point. According to (5.1.3.5), there is a point ρ, $0 < \rho < 1$, so that the Taylor series of $f(z) = \sum_0^{\infty} a_n z^n$, $|z| < 1$, at ρ

$$\sum_{k=0}^{\infty} (z - \rho)^k \left\{ \sum_{n=k}^{\infty} a_k C_k^n \rho^{n-k} \right\} = \sum_{k=0}^{\infty} \frac{f^{(k)}(\rho)}{k!} (z - \rho)^k$$

has its radius of convergence $r_\rho > 1 - \rho$. Then, there is a $\delta > 0$ so that the right series converges absolutely at $z = 1 + \delta$. Substituting $z = 1 + \delta$ into the left side of the above series, which is a double series with nonnegative terms, and changing the order of summation, we have a convergent series

$$\sum_{n=0}^{\infty} a_n \left\{ \sum_{k=0}^{n} C_k^n (1 + \delta - \rho)^k \rho^{n-k} \right\} = \sum_{n=0}^{\infty} a_n (1 + \delta)^n, \quad 1 + \delta > 1$$

contradicting to the assumption that 1 is the radius of convergence of $\sum_0^\infty a_n z^n$.   □

Some illustrative examples are provided in the following.

Take, for instance,

$$e^z = \sum_{n=0}^{\infty} \frac{1}{n!} z^n, \quad |z| < \infty.$$

In case $z = x$ is real, then $\lim_{x \to \infty} e^x = \infty$ and $\lim_{x \to -\infty} e^x = 0$ show that ∞ is a (and the only) singular point of the *power series*. Note that, ∞ is also the only singular point of the *entire function* e^z (see (2.2.5)). Similarly, in the case of $\sin z = \sum_0^\infty (-1)^n \frac{1}{(2n+1)!} z^{2n+1}$, $|z| < \infty$, and $\cos z = \sum_0^\infty (-1)^n \frac{1}{(2n)!} z^{2n}$, $|z| < \infty$, ∞ is the only singular point of both series and, at the same time, is the only singular point of both entire functions.

On the other hand, the power series

$$-\log(1 - z) = \sum_{n=1}^{\infty} \frac{1}{n} z^n \ (\log 1 = 0), \quad |z| < 1 \qquad (*_6)$$

has 1 as its only singular point. Observe that $\lim_{z \to 1} (-\log(1 - z)) = \infty$ holds, where $|z| < 1$. As an analytic function, $-\log(1-z) = -\text{Log}(1-z)$ has each point along the cut $(-\infty, -1]$ as its singular point (see (2.2.5)); but, according to Example 3, any *branch power series* of $-\log(1 - z)$ *obtained by analytic continuation* of the power series $(*_6)$ will always have only 1 as its unique singular point.

Two less obvious examples are listed in Examples 4 and 5.

Example 4. Let p be a real number. Try to use

$$f_p(z) = \sum_{n=1}^{\infty} \frac{1}{n^p} z^n, \quad |z| < 1$$

to explain (5.1.3.8).

Explanation. The radius of convergence is 1. According to (5.1.3.9), 1 is a singular point of the power series; as a matter of fact, it is the unique one (see Exercise A(17)). In case $p = 1$, $f_1(z) = -\log(1 - z)$ is the one given by $(*_6)$.

If $|z| < 1$, then

$$f_p^{(k)}(z) = \sum_{n=k}^{\infty} n^{-p} \cdot n(n-1) \cdots (n-k+1) z^{n-k}. \qquad (*_7)$$

To see if the above right series converges or diverges at $z = 1$, observe that

$$\frac{n^{-p} \cdot n(n-1) \cdots (n-k+1)}{n^{-p+k}} = \left(1 - \frac{1}{n}\right) \cdots \left(1 - \frac{k-1}{n}\right) \to 1 \text{ as } n \to \infty$$

and that $\sum_1^{\infty} n^{-p+k}$ converges if $p > k + 1$; diverges if $p \leq k + 1$. So does the series in $(*_7)$. Hence, by Abel's limit theorem (5.1.3.1),

$$\lim_{z=x \to 1-} f_p^{(k)}(z) = \begin{cases} a \text{ finite positive real number,} & p > k + 1 \\ \infty, & p \leq k + 1 \end{cases}.$$

This fact proves again that 1 is a singular point of the power series defining $f_p(z)$.

Note. More information can be obtained from $f_p(z) = \sum_1^{\infty} n^{-p} z^n$, $|z| < 1$.

Case 1. $p \leq 0$: $\sum_1^{\infty} n^{-p} z^n$ diverges at every point on the circle $|z| = 1$, yet each point on $|z| = 1$ other than $z = 1$ is a regular point of the power series.

Case 2. $0 < p \leq 1$: $\sum_1^{\infty} n^{-p} z^n$ conditionally converges at every point z on $|z| = 1$, $z \neq 1$, and diverges at $z = 1$; uniformly converges on every compact subset of $|z| = 1$, $z \neq 1$ (by Dirichlet test (see (2.3.5))). As a consequence, $\sum_1^{\infty} n^{-p} z^n$ represents a continuous function $h_p(z)$ on $|z| \leq 1$, $z \neq 1$, which is identical with $f_p(z)$ in $|z| < 1$. Indeed,

$$h_p(z) = \begin{cases} f_p(z), & |z| < 1 \\ \lim_{\zeta \to z} f_p(\zeta), & |z| = 1, z \neq 1 \text{ and } |\zeta| \leq 1, \zeta \neq 1. \\ \lim_{\zeta \to 1} f_p(\zeta) = \infty, & z = 1 \text{ and } 0 \leq \zeta < 1 \end{cases}$$

When restricted to $|z| = 1$, $h_p(z)$ is called the *boundary value function* of $f_p(z)$ on $|z| = 1$.

Case 3. $p > 1$: $\sum_1^{\infty} n^{-p} z^n$ converges absolutely and uniformly on $|z| = 1$, even though $z = 1$ is a singular point of the power series. In this case,

$\sum_1^\infty n^{-p} z^n$ represents, on $|z| \leq 1$, the continuous function

$$h_p(z) = \begin{cases} f_p(z), & |z| < 1 \\ \lim\limits_{\zeta \to z} f_p(z), & |z| = 1 \text{ and } |\zeta| \leq 1 \end{cases}$$

which is called the *boundary value function* of $f_p(z)$ on $|z| = 1$.

Example 5. Show that

$$f(z) = \sum_{n=0}^\infty z^{2^n}, \quad |z| < 1$$

has $|z| = 1$ as its natural boundary, namely, each point on $|z| = 1$ is a singular point.

Based on Vivante–Pringsheim theorem (5.1.3.9) and its generalization (see Exercise A(9)), we give a

Proof. By (5.1.3.9), 1 is a singular point of the series. While, any 2^k-th root $\zeta = e^{2l\pi i/2^k} (0 \leq l \leq 2^k - 1)$ of the unity 1 is also a singular point. To see this, for $n \geq k$, then $\zeta^{2^n} = (\zeta^{2^k})^{2^{n-k}} = 1^{2^{n-k}} = 1 > 0$ shows that ζ is indeed a singular point, by Exercise A(9). It is well-known that the set $\{e^{2l\pi i/2^k} | 0 \leq l \leq 2^k - 1 \text{ and } k \geq 0\}$ is dense in $|z| = 1$. Because the set of singular points of a power series deems to be a closed set (see Exercise A(12)), it follows that each point on $|z| = 1$ is a singular point. $\square$

Another proof (a direct one). Fix any point z_0, $|z_0| = \frac{1}{2}$. If we can show that the series

$$g(z) = \sum_{n=0}^\infty \frac{f^{(n)}(z_0)}{n!} (z - z_0)^n \qquad (*_8)$$

has its radius r_{z_0} of convergence equal to $\frac{1}{2}$, then, by (5.1.3.6), the end point of the radius passing z_0 of the circle $|z| = 1$ is a singular point of the original series.

Suppose on the contrary that $r_{z_0} > \frac{1}{2}$. Note that there is a point of the form $\zeta = e^{2l\pi i/2^k}$ $(0 \leq l \leq 2^k - 1, k > 0)$, lying on $|z| = 1$, that lies in the disk $|z - z_0| < r_{z_0}$. For the series $(*_8)$,

$$\lim_{r \to 1} g(r\zeta) = g(\zeta), \quad 0 < r < 1$$

$$\Rightarrow \text{(since } g(z) = f(z) \text{ on } |z| < 1 \text{ and } |z - z_0| < r_{z_0})$$

$$\lim_{r \to 1} f(r\zeta) = g(\zeta), \quad 0 < r < 1.$$

Well, for such a k, separate $f(z)$ into two parts as

$$f(r\zeta) = \sum_{n=0}^{k} r^{2^n} \zeta^{2^n} + \sum_{n=k+1}^{\infty} r^{2^n} \zeta^{2^n}.$$

Since $\zeta^{2^n} = 1$ for any $n \geq k$, choose any integer $N \geq k + 1$,

$$\sum_{n=k+1}^{\infty} r^{2^n} \zeta^{2^n} = \sum_{n=k+1}^{\infty} r^{2^n} > \sum_{n=k+1}^{N} r^{2^n} > (N - k - 1) r^{2^N}$$

$$\Rightarrow \lim_{r \to 1} \sum_{n=k+1}^{\infty} r^{2^n} \zeta^{2^n} = \infty$$

$$\Rightarrow \lim_{r \to 1} f(r\zeta) = \infty$$

contradicting to the fact that $g(\zeta)$ is a finite complex number. $\square$

*Section (4) Comments on convergence, divergence and regular, singular points

There is no obvious connection, in general, as might be seen in Example 4, between the convergence or divergence of a power series at a boundary point ζ, $|\zeta - z_0| = r_{z_0}$, and the regularity or singularity of that power series at ζ. It is even possible that a power series converges uniformly and is infinitely differentiable on its natural boundary (see Exercises A(3) and (20)).

A power series could diverge at its regular points except under some additional restriction on its coefficients. For instance,

Fatou's theorem. *Suppose $f(z) = \sum_0^{\infty} a_n z^n$ has radius 1 of convergence and $\lim_{n \to \infty} a_n = 0$ holds. Then, the series converges at its regular point (if exists); moreover, it converges uniformly on its set of regular points.*   (5.1.3.10)

See Refs. [58], Vol. II, pp. 404–408 or [76], pp. 217–220 for the proof. Lusin constructed a power series satisfying the assumption of Fatou's theorem yet diverging everywhere on $|z| = 1$.

Generally speaking, the coefficients a_n for all sufficiently large n determine completely the distribution of singular points on the circle $|z| = r$ of convergence. For instance, when these a_n which are not equal to zero are evenly distributed, then $z = r$ turns out, in many cases, to be the only singular point (see (5.1.3.9) and Example 4). Furthermore, suppose

$$f(z) = \sum_{0}^{\infty} a_{nk} z^{nk} \quad (k, \text{a fixed positive integer}), \quad a^{nk} \geq 0$$

has the radius 1 of convergence. In this case, the number of the missing terms between two consecutive terms is equal to a constant, namely, $k(n+1) - kn = k$, and the k-th roots of 1 are singular points of the series (see "Another proof" for Example 5). Even more, if the distribution of nonzero a_n is badly irregular, some additional singular points might be added, as Example 5 shows. As a matter of fact, Example 5 is a special case of the following

Hadamard's gap theorem. *Suppose a sequence n_k, $k \geq 0$, of positive integers satisfying:*

1. $n_0 < n_1 < \cdots < n_k < n_{k+1} < \cdots$, *and* $\lim_{k \to \infty} n_k - \infty$;
2. *there is a positive number α such that $n_{k+1} - n_k \geq \alpha n_k$, $k \geq 0$.*

If the power series $f(z) = \sum_{k=0}^{\infty} a_{n_k} z^{n_k}$ has a positive radius r of convergence, then $|z| = r$ is the natural boundary of the series. In particular, in case

$$\lim_{k \to \infty} \frac{n_{k+1}}{n_k} = 1 \quad \text{or} \quad n_{k+1} = \lambda_k n_k \ (\lambda_k, a positive integer), \quad k \geq 0$$

then $|z| = r$ is the natural boundary.   (5.1.3.11)

Sketch of proof (Faber, Mordall). Replacing $f(z)$ by $f(re^{i\theta}z)$ for some constant θ, we may suppose that $r = 1$ and $z = 1$ is a regular point of the series, and we want to get a contradiction.

According to the definition of a regular point (5.1.3.5), there is an analytic function $F(z)$ in $|z| < 1$ and in a neighborhood of 1 so that $F(z) = f(z)$ on $|z| < 1$. Choose a positive integer $p > \frac{1}{\alpha}$. Let

$$z = \frac{1}{2}(w^p + w^{p+1}).$$

Note that $|w| \leq 1$ is mapped into $|z| \leq 1$, and into $|z| < 1$ if $|w| \leq 1$, except that $z = 1$ if $w = 1$. Then

$$\varphi(w) = F\left(\frac{1}{2}(w^p + w^{p+1})\right)$$

is analytic on $|w| \leq 1$ and has the power series expansion in w as

$$\varphi(w) = \sum_{k=0}^{\infty} a_{n_k} \left(\frac{w^p + w^{p+1}}{2} \right)^{n_k} = \sum_{k=0}^{\infty} a_{n_k} \sum_{l=0}^{n_k} C_l^{n_k} \frac{w^{pn_k+l}}{2^{n_k}} = \sum_{k=0}^{\infty} b_k w^k.$$
$$(*_9)$$

Since $p > \frac{1}{\alpha}$ implies that $1 < \frac{1}{\alpha} \left(\frac{n_{k+1}}{n_k} - 1 \right) < p \left(\frac{n_{k+1}}{n_k} - 1 \right)$, namely, $pn_k + n_k < pn_{k+1}$, we see that *every* power of w in $(*_9)$ occurs at most once and is then obtained by simply rearranging the terms in the middle expression. This series $(*_9)$ has its radius of convergence greater than 1. It will converge at a real point $r_0 > 1$, and the series for $f(z)$ would be convergent at the point $\frac{1}{2}(r_0^p + r_0^{p+1}) > 1$, a contradiction. This proves (5.1.3.11). $\square$

A slight modification of this method can be adopted to prove *Ostrowski's overconvergence theorem* (see Exercise B(5)). Hadamard's gap theorem can be used to prove *Fatou–Pólya's theorem*, saying that "most" of the power series with positive radii of convergence cannot be continued analytically beyond their circles of convergence (see Exercise B(8)). For related information, see Refs. [71], pp. 246–252; [25], pp. 126–133, 153–157; [76], Chap. VII, in particular, pp. 214–224.

*Section (5) The boundary value function of a power series

Case 3 $(p > 1)$ in Example 4 gives a concrete illustration.

The most commonly adopted definition for the boundary value at a point on the circle of convergence is provided by the one stated in Abel's limit theorem (5.1.3.1): Let $f(z) = \sum_{n=0}^{\infty} a_n (z - z_0)^n$, $|z - z_0| < r$ (the radius of convergence) and ζ be a point on $|z - z_0| = r$. Denote by D_δ ($0 \leq \delta < \frac{\pi}{2}$) a Stolz domain with vertex at ζ. If the limit

$$h(\zeta) = \lim_{z \in D_\delta \to \zeta} f(z) \qquad (5.1.3.12)$$

exists as a finite complex number or ∞, then $h(\zeta)$ is called *the nontangial boundary value* of $f(z)$ at ζ as z approaches ζ from inside the circle $|z - z_0| = r$. In case $h(\zeta)$ is defined everywhere on $|z - z_0| = r$, it is called *the nontangential boundary value function* of $f(z)$. In particular,

$$\lim_{t \to 1^-} f(t\zeta) = h(\zeta), \qquad 0 \leq t < 1 \qquad (5.1.3.13)$$

holds. $h(\zeta)$, defined in this manner, is called *the radial boundary value function* of $f(z)$.

The fundamental yet important result concerning the radial boundary value is the following

Fatou's theorem. *Suppose* $f(z) = \sum_0^\infty a_n z^n$, $|z| < 1$ *(the radius of convergence), is a bounded analytic function. Then the radial boundary value function*

$$h(e^{i\theta}) = \lim_{r \to 1^-} f(re^{i\theta}), \quad 0 \le r < 1$$

exists almost everywhere in $0 \le \theta \le 2\pi$.   (5.1.3.14)

Here, "almost everywhere (a.e.)" means that, except a set of Lebesgue measure zero in $0 \le \theta \le 2\pi$, $h(e^{i\theta})$ exists.

Sketch of proof. Suppose $|f(z)| \le M$ (a constant), $|z| < 1$. According to Gutzmer's inequality (see (2) in (5.1.2.9)), we obtain

$$\sum_0^\infty |a_n|^2 \rho^{2n} \le M^2, \quad 0 \le \rho < 1$$

$$\Rightarrow \lim_{\rho \to 1^-} \sum_0^\infty |a_n|^2 \rho^{2n} = \sum_0^\infty |a_n|^2 \le M$$

$\Rightarrow$ (by Riesz–Fischer's theorem (see, for instance, Ref. [76], p. 423))

$\sum_0^\infty a_n e^{in\theta}$ is the Fourier series of a Lebesgue square integrable

function $h(e^{i\theta})$, $0 \le \theta \le 2\pi$.

Moreover, then $\sum_0^\infty a_n e^{in\theta}$ is *Cesaro summable* to $h(e^{i\theta})$ almost everywhere in $0 \le \theta \le 2\pi$, namely, denoting $s_n = \sum_{k=0}^n a_k e^{ik\theta}$, $n \ge 0$,

$$\lim_{n \to \infty} \frac{1}{n+1}(s_0 + s_1 + \cdots + s_n) = h(e^{i\theta}), \quad \text{a.e.}$$

This fact, in turn, implies that $\sum_0^\infty a_n e^{in\theta}$ is *Abel summable* to $h(e^{i\theta})$ almost everywhere in $0 \le \theta \le 2\pi$, namely,

$$h(e^{i\theta}) = \lim_{\rho \to 1^-} f(\rho e^{i\theta}) = \lim_{\rho \to 1^-} \sum_0^\infty a_n \rho^n e^{in\theta}, \quad \text{a.e.}$$

This last result is, in general, due to Frobenius and is a special kind of Tauberian theorem.   □

Based on (5.1.3.14), we have a remarkable

F. Riesz and M. Riesz's theorem. *Suppose* $f(z) = \sum_0^\infty a_n z^n$ *is a bounded analytic function in* $|z| < 1$. *If its radial boundary value function* $h(e^{i\theta})$ *assumes zero value on a set of **positive** Lebesgue measure in* $0 \le \theta \le 2\pi$, *then*

$$f(z) \equiv 0, \quad |z| < 1.$$ (5.1.3.15)

This result can be viewed as a kind of uniqueness theorem for bounded analytic functions defined in an open disk. Also, its content generalizes that of (2) in (5.1.2.4).

For detailed account about (5.1.3.14) and (5.1.3.15), refer to Refs. [14], Vol. II, pp. 43–51; [64], pp. 198–201, 204–205; [19], Chap. 2.

Exercises A

(1) Suppose a_n, $n \geq 0$, is a real sequence and the series $\sum_0^\infty a_n z^n$ has radius 1 of convergence. Show that the limit

$$\lim_{x \to 1^-} f(x) = \lim_{x \to 1^-} \sum_0^\infty a_n x^n$$

exists if and only if $\sum_0^\infty a_n$ converges, and, in this case, $\sum_0^\infty a_n = \lim_{x \to 1^-} f(x)$.

(2) Use Abel's limit theorem to show that

(a) $\displaystyle\sum_0^\infty \frac{\cos(2n+1)\theta}{2n+1} = \frac{1}{2}\log\left|\cot\frac{\theta}{2}\right|, \quad 0 < |\theta| < \pi;$

$\displaystyle\sum_0^\infty \frac{\sin(2n+1)\theta}{2n+1} = \frac{\pi}{4}, \quad 0 < \theta < \pi.$

(b) $\displaystyle\sum_0^\infty (-1)^{n+1}\frac{\cos n\theta}{n} = \log(2\cos\frac{\theta}{2}), \quad -\pi < \theta < \pi;$

$\displaystyle\sum_0^\infty (-1)^{n+1}\frac{\sin n\theta}{n} = \frac{\theta}{2}, \quad -\pi < \theta < \pi.$

(3) Investigate the boundary behavior of each of the following series on its circle of convergence: (conditional, absolute, uniform) convergence or divergence; regular or singular point.

(a) $\displaystyle\sum_{n=1}^\infty \frac{z^{pn}}{n}$ (p, a fixed positive integer).

(b) $\displaystyle\sum_{n=2}^\infty \frac{(-1)^n}{\log n} z^{3n-1}.$

(c) $\displaystyle\sum_{n=1}^\infty \frac{z^{n!}}{n^2}.$

(4) Given the power series

$$1 + \sum_{n=1}^\infty (-1)^{n-1}\frac{(2n)!}{2^{2n}(n!)^2} \cdot \frac{(z-1)^n}{2n-1}, \quad |z-1| < 1.$$

This is the Taylor series expansion of $\sqrt{z}\,(\sqrt{1}=1)$ at $z=1$. Denote by γ the unit circle $z(t)=e^{2\pi it}$, $0\le t\le 1$. Refer to Example 1 in Sec. 3.3.3.

(a) Continue the power series analytically along γ up to the point $z(t)$ and find the resulting power series expansion with center at $z(t)$, $0<t<1$.

(b) In (a), find the power series at $z(1)$ after one circulation along γ.

(c) Find the power series at $z(1)$ after two consecutive circulations along γ in the same direction.

(d) Try to use (b) and (c) to construct the Riemann surface for $\sqrt{z}$.

(5) Show that the functions defined, respectively, by

$$\sum_{n=0}^{\infty} a^n z^n; \quad \sum_{n=0}^{\infty}(-1)^n \frac{(1-a)^n z^n}{(1-z)^{n+1}} \quad (a \neq 0, 1)$$

are direct analytic continuation of each other.

(6) Suppose $f(z)=\sum_0^\infty a_n z^n$ has $r=1$ as its radius of convergence. Let $z=\frac{w}{w+1}$ so that

$$f(z)=f\left(\frac{w}{w+1}\right)=F(w)=\sum_0^\infty b_n w^n$$

has ρ as its radius of convergence.

(a) Show that $\rho \geq \frac{1}{2}$, and $\rho=\frac{1}{2}$ if $z=-1$ is a singular point of $f(z)$.

(b) In case $\frac{1}{2}<\rho<1$, then $f(z)$ can be analytically continued into the part common to $|z|>1$ and the inside of the Apollonius circle $\left|\frac{z}{z-1}\right|=\rho$.

(c) If $\rho=1$, then $f(z)$ can be analytically continued into the half-plane $\operatorname{Re} z<\frac{1}{2}$.

(d) If $\rho>1$, then $f(z)$ can be analytically continued into the outside of $\left|\frac{z}{z-1}\right|=\rho$.

(e) If $\rho=\infty$, then $f(z)$ can be analytically continued into $\mathbf{C}-\{1\}$.

(7) Prove (5.1.3.6) in detail.

(8) *Another proof* of *Vivanti–Pringsheim theorem* (5.1.3.9). According to (5.1.3.7), let $e^{i\theta_0}$ $(0\le\theta_0\le 2\pi)$ be a singular point of $f(z)$ on $|z|=1$ (here, it is assumed that 1 is the radius of convergence). $f(z)$ has the

Taylor series expansion at $re^{i\theta_0}$ $(0 < r < 1)$ as

$$f(z) = \sum_0^\infty \frac{f^{(n)}(re^{i\theta_0})}{n!}(z - re^{i\theta_0})^n, \quad |z - re^{i\theta_0}| < 1 - r$$

(radius of convergence).

By assumption that $a_n \geq 0$, $n \geq 0$, show that

$$|f^{(n)}(re^{i\theta_0})| \leq |f^{(n)}(r)|, \quad n \geq 0$$

and, then, try to deduce that the Taylor series expansion of $f(z)$ at r has its radius of convergence equal to $1 - r$.

(9) An *extension* of *Vivanti–Pringsheim theorem*. Suppose $f(z) = \sum_0^\infty a_n z^n$ has its radius of convergence equal to 1. Let z_0 be a point on the circle of convergence, namely, $|z_0| = 1$, so that $a_n z_0^n \geq 0$ for $n \geq n_0$ for some n_0. Show that z_0 is a singular point.

(10) An *extension* of *Vivanti–Pringsheim theorem*. Let $r = 1$ be the radius of convergence of both $f(z) = \sum_0^\infty a_n z^n$ and $g(z) = \sum_0^\infty (\mathrm{Re}\, a_n) z^n$. In case $\mathrm{Re}\, a_n \geq 0$ for all sufficiently large n, show that $z = 1$ is a singular point of $f(z)$.

(11) An *extension* of *Vivanti–Pringsheim theorem*.

  (a) Let $f(z) = \sum_0^\infty \alpha_n z^n$ and $g(z) = \sum_0^\infty \beta_n z^n$ be series with real coefficients, both having their radii of convergence not less than 1. If 1 is a singular point of $\sum_0^\infty (\alpha_n + i\beta_n) z^n$, show that 1 is a singular point of both $f(z)$ and $g(z)$.

  (b) Use (a) to reprove Exercise (10).

  (c) Suppose $f(z) = \sum_0^\infty a_n z^n$ has 1 as its radius of convergence and $|\mathrm{Im}\, a_n| \leq M \mathrm{Re}\, a_n$, $n \geq 0$, for some positive constant M. Use (a) to show that 1 is a singular point of $f(z)$.

(12) Show that the set of singular points of a power series is a closed subset of its circle of convergence.

(13) *Dienes theorem*. Suppose $f(z) = \sum_0^\infty a_n z^n$ has the radius of convergence equal to 1 and $|\mathrm{Arg}\, a_n| \leq \alpha < \frac{\pi}{2}$, $n \geq 1$, for some constant $\alpha \geq 0$. Show that 1 is a singular point of the series.

(14) Let $f(z) = \sum_0^\infty a_n z^n$ be a power series with real coefficients and having 1 as its radius of convergence. Suppose $s_n = \sum_{k=0}^n a_k \to \infty$ or $-\infty$ as $n \to \infty$. Show that $z = 1$ is a singular point of the series. Yet the result is not necessarily true if the assumption is changed to $|s_n| \to \infty$ as $n \to \infty$. For instance, $\frac{1}{(1+z)^3} = 1 - 3z + \cdots + (-1)^n$

$\frac{1}{2}(n+1)(n+2)z^n+\cdots$, $|z|<1$, is a case in which $|s_n|\sim\frac{n^2}{4}$ (approximate value).

(15) Let r_{z_0} $(0<r_{z_0}<\infty)$ be the radius of convergence of the power series $f(z)=\sum_0^\infty a_n(z-z_0)^n$. Then, $|z-z_0|=r_{z_0}$ is its natural boundary if and only if

$$r_{z_0+\frac{1}{2}r_{z_0}e^{i\theta}}=\frac{1}{2}r_{z_0},\quad 0\le\theta\le 2\pi,$$

where $r_{z_0+\frac{1}{2}r_{z_0}e^{i\theta}}$ is the radius of convergence of the Taylor series expansion of $f(z)$ at $z_0+\frac{1}{2}r_{z_0}e^{i\theta}$, $0\le\theta\le 2\pi$, namely,

$$\overline{\lim_{k\to\infty}}\left|\sum_{n=k}^\infty C_k^n a_n\left(\frac{1}{2}r_{z_0}e^{i\theta}\right)^n\right|^{\frac{1}{k}}=1,\quad 0\le\theta\le 2\pi.$$

(16) Suppose $r=1$ is the radius of convergence of $f(z)=\sum_0^\infty a_nz^n$. Then $z=1$ is a singular point of the series if and only if

$$\overline{\lim_{n\to\infty}}\sqrt[n]{|b_n|}=2,$$

where $b_n=\sum_{k=0}^n C_k^n a_k$; in case $z=e^{i\theta}$ is a singular point, then a_k, $0\le k\le n$, in the expression of b_n should be replaced by $e^{ik\theta}a_k$.

(17) Try to use Exercise (15) or (16) to prove that the power series in Example 4 has only one singular point, namely, $z=1$.

(18) Let p be a positive integer and $h_p(t)=t(t-1)(t-2)\cdots(t-p+1)$. Try to show that $t^m=h_m(t)+a_{1,m}h_{m-1}(t)+\cdots+a_{m-1,m}h_1(t)$, m is a positive integer, by induction, where each coefficient $a_{j,m}$, $1\le j\le m-1$, is a positive integer. Then use this fact or some other method to show that

$$\sum_{n=1}^\infty n^p z^n=\frac{Q_p(z)}{(1-z)^{p+1}},\quad |z|<1,$$

where $Q_p(z)$ is a polynomial of degree p, with coefficients all positive integers. Finally, show that $z=1$ is the only singular point of the series.

(19) Try to use Exercise (16) to redo Example 5.

(20) Try to use Hadamard's gap theorem (5.1.3.11) to show that $|z|=1$ is the natural boundary for each of the following power series. And discuss the convergence (including absolute, conditional and uniform)

or divergence on $|z| = 1$ for each series.

(a) $\sum_0^\infty \frac{1}{2^n} z^{2^n}$.

(b) $\sum_0^\infty z^{n!}$.

(c) $\sum_0^\infty \frac{1}{2^n} z^{n!}$.

(d) $\sum_0^\infty \frac{1}{n^2} z^{n!}$.

(e) $\sum_0^\infty \frac{1}{n!} z^{2^n}$.

Exercises B

(1) Show that the power series

$$\sum_{n=1}^\infty \frac{(-1)^{[\sqrt{n}]}}{n} z^n, \quad |z| < 1$$

conditionally converges at each point on $|z| = 1$.

(2) Let $a_n \in \mathbf{C}$, $n \geq 1$, and na_n form a bounded sequence. In this case, $f(z) = \sum_0^\infty a_n z^n$ has its radius of convergence at least equal to 1.

(a) Show that f is bounded in the real interval $[0, 1]$ if and only if the sequence $\sum_{k=0}^n a_k$, $n \geq 0$, is bounded.

(b) Show that f is bounded in $|z| < 1$ if and only if the sequence $\sum_{k=0}^n a_k z^k$, $n \geq 0$, is uniformly bounded on $|z| = 1$, namely,

$$\sup_{n \geq 0} \sup_{|z|=1} \left| \sum_{k=0}^n a_k z^k \right| < \infty.$$

(3) Consider the power series (see Ref. [44], Vol. I, p. 133)

$$f(z) = \sum_0^\infty \frac{1}{n!} z^{2^n}, \quad |z| < 1.$$

(a) Try to imitate Example 5 to show that $|z| = 1$ is its natural boundary (compare to Exercise A(20)(e)).

(b) Show that the power series

$$f^{(k)}(z) = \sum_{n=0}^\infty \frac{1}{n!} 2^n (2^n - 1) \cdots (2^n - k + 1) z^{2^n - k}, \quad |z| < 1$$

converges at $z = 1$ (here, it is understood that the coefficient is deemed as zero if $2^n < k$). Try to use Abel's limit theorem to show that the radial limit

$$\lim_{\zeta \to 1^-} f^{(k)}(\zeta) = \sum_{n=0}^\infty \frac{1}{n!} 2^n (2^n - 1) \cdots (2^n - k + 1) \underset{\text{(def.)}}{=} b_k$$

exists.

(c) Show that the power series $\sum_0^\infty \frac{b_n}{n!}(z-1)^n$ has 0 as its radius of convergence. Use this to show again that $z = 1$ is a singular point of $f(z)$.

(d) $\sum_0^\infty \frac{1}{n!}z^{2^n}$ converges absolutely and uniformly on $|z| \le 1$ to a continuous function $h(z)$, while the series in (b) does so on $|z| \le 1$ to a continuous function $g(z)$. Show that,

$$
h^{(k)}(z) = \begin{cases} f^{(k)}(z), & |z| < 1 \\[2mm] \displaystyle\lim_{\substack{\zeta \to z \\ |\zeta| < 1}} \frac{h^{(k-1)}(\zeta) - h^{(k-1)}(z)}{\zeta - z} = \displaystyle\lim_{\substack{\zeta \to z \\ |\zeta| \le 1}} f^{(k)}(\zeta), & |z| = 1 \end{cases}
$$

and $h^{(k)}(z) = g(z)$, $|z| \le 1$, $k \ge 0$.

(4) Show that the Lambert series (see Example 1 in Sec. 5.1.1) has $|z| = 1$ as its natural boundary.

(5) *Ostrowski's overconvergence theorem.* Suppose $r = 1$ is the radius of convergence of $f(z) = \sum_0^\infty a_n z^n$. Suppose there exist two sequences of positive integers $n_1 < n_2 < \cdots < n_k < n_{k+1} < \cdots$, $m_1 < m_2 < \cdots < m_k < m_{k+1} < \cdots$, and a positive number α satisfying:

1. $a_n = 0$ for $n_k < n < m_k$, $k \ge 1$;
2. $m_k - n_k \ge \alpha n_k$, $k \ge 1$.

Then, the subsequence

$$
s_{n_k}(z) = \sum_{l=0}^{n_k} a_l z^l, \quad k \ge 1
$$

of the partial sum sequence $s_n(z) = \sum_{l=0}^n a_l z^l$, $n \ge 1$, of the power series converges uniformly in a neighborhood of each regular point (if exists!) of $f(z)$.

Note. The original partial sum sequence $s_n(z)$, $n \ge 1$, converges (uniformly) at most in $|z| \le 1$. This theorem says that a certain subsequence of it does converge at some points outside $|z| \le 1$ if $f(z)$ has a regular point. See Exercise (6) below.

(6) Given the series

$$
F(z) = \sum_{n=1}^\infty \frac{(z(1-z))^{4^n}}{2^{4^n}}.
$$

(a) Show that the series converges uniformly in each compact subset of $|z| < 1$ and $|z-1| < 1$. Hence, $F(z)$ is analytic there (see (5.1.1.7)).

(b) Expand the series into the power series $f(z) = \sum_0^\infty a_n z^n$ in z.

(c) Determine the radius of convergence and regular points for the power series in (b).

(7) Try to use Exercise (5) to reprove Hadamard's gap theorem (5.1.3.11).
(8) *Fatou Pólya's theorem.* Suppose the power series $f(z) = \sum_0^\infty a_n z^n$ has a positive radius r of convergence. Then, there exist real numbers λ_n, either 1 or -1, so that the power series

$$\sum_{n=0}^\infty \lambda_n a_n z^n$$

has $|z| = r$ as its natural boundary.

5.2   Analytic Continuation

Based on (5.1.2.2) and as illustrated by Examples 2 and 3 in Sec. 5.1.3, an analytic function $f(z)$ defined by a power series $\sum_{n=0}^\infty a_n (z - z_0)^n$ can be analytically continued beyond its circle of convergence $|z - z_0| = r_{z_0} > 0$, if this circle is not the natural boundary, to as far as possible in the complex plane.

For simplicity, denote, for the moment, the Taylor series expansion of $f(z)$ at a by $f(z; a) = \sum_0^\infty \frac{f^{(n)}(a)}{n!} (z - a)^n$ and the corresponding open disk of convergence by $D_a : |z - a| < r_a (r_a > 0)$. Note that $f(z; z_0) = f(z)$ and $D_{z_0} : |z - z_0| < r_{z_0}$.

Choose $z_1, z_2 \in D_{z_0}$ so that $D_{z_1} \cap D_{z_2} \neq \emptyset$. Hence $(D_{z_1} \cap D_{z_2}) \cap D_{z_0} \neq \emptyset$ (see the shaded part of Fig. 5.7) and

$$f(z; z_1) = f(z) = f(z; z_2), \quad z \in (D_{z_1} \cap D_{z_2}) \cap D_{z_0}$$

$$\Rightarrow \text{(by (4) in (3.4.2.9) or (5.1.2.5))} \; f(z; z_1) = f(z; z_2), \quad z \in D_{z_1} \cap D_{z_2}.$$

This suggests that, on the domain

$$\Omega_{z_0} = \bigcup_{z \in D_{z_0}} D_z \supsetneq D_{z_0},$$

it is possible to define a unique analytic function $F(z)$ satisfying

$$F(z)|_{D_{z_0}} = f(z; z_0) = f(z).$$

In other words, the analyticity of $f(z; z_0) = f(z)$ on D_{z_0} can be extended to the analyticity of another function $F(z)$ on a larger domain Ω_{z_0}. Since $f(z)$ has at least one singular point on ∂D_{z_0}, then $\partial \Omega_{z_0}$ contains at least one point on ∂D_{z_0} (see Fig. 5.6).

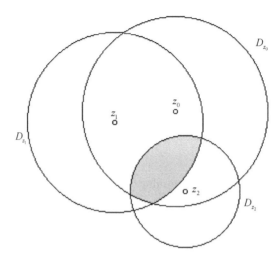

Fig. 5.7

For instance, in the case $(\text{Log}(1-z)$ is single-valued in $\mathbf{C}-[1,\infty))$

$$(1-z)\text{Log}(1-z)+z=\sum_{n=2}^{\infty}\frac{1}{n(n-1)}z^{n},\quad |z|<1,$$

D_0 : $|z|<1$ and $f(z)=f(z;0)=(1-z)\text{Log}(1-z)+z$. And the corresponding Ω_0 is the interior of a cardioid shown in Fig. 5.8. Note that 1 is a singular point of $f(z)$ and is a boundary point of Ω_0, and $F(z)=(1-z)\text{Log}(1-z)+z$, $z\in\Omega_0$.

The same process can be repeated to continue $F(z)$ on Ω_{z_0} analytically to still larger domains.

This section tries to explain the above process in a concise and precise manner. Section 5.2.1 introduces the concept of the analytic continuation of an *analytic germ* along a curve, formalized and unified the ideas in Secs. 3.3.3, 5.1.2, and 5.1.3. While, Sec. 5.2.2 proves the *monodromy theorem*; in particular, analytic continuation in a simply connected domain always leads to *global single-valued* analytic functions.

For the sake of general reference, we formally give the following definitions.

Suppose f is analytic in a domain Ω, usually denoted as (f,Ω), and is called an analytic *function element*. Let D be a larger domain containing Ω. If it is possible to find an analytic function F on D such that $F(z)=f(z)$ for all $z\in\Omega$, then $f(z)$ is said to be able to *analytically continue* to $D-\Omega$ and $F(z)$ is a *direct analytic continuation* of $f(z)$ from Ω into D.

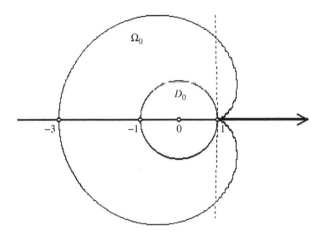

Fig. 5.8

Suppose f_k is analytic in a domain Ω_k for $k = 1, 2$. If $\Omega_1 \cap \Omega_2 \neq \emptyset$ and $f_1(z) = f_2(z)$ holds throughout $\Omega_1 \cap \Omega_2$, then the function $F : \Omega_1 \cup \Omega_2 \to \mathbf{C}$ defined by

$$F(z) = \begin{cases} f_1(z), & z \in \Omega_1 \\ f_2(z), & z \in \Omega_2 \end{cases} \tag{5.2.1}$$

is well-defined and analytic, and is a direct analytic continuation of both $f_1(z)$ and $f_2(z)$. In this case, we say that the function elements (f_1, Ω_1) and (f_2, Ω_2) are *direct analytic continuation* of each other.

Two function elements (f_0, Ω_0) and (f, Ω) are said to be *analytic continuation* of each other if there exists a finite number of function elements (f_j, Ω_j), $1 \leq j \leq n$, so that, in the following ordering

$$(f_0, \Omega_0), (f_1, \Omega_1), (f_2, \Omega_2), \ldots, (f_{j-1}, \Omega_{j-1}), (f_j, \Omega_j), \ldots, (f_n, \Omega_n) = (f, \Omega) \tag{5.2.2}$$

each precedent one (f_{j-1}, Ω_{j-1}) is a *direct* analytic continuation of the successive one (f_j, Ω_j) for $1 \leq j \leq n$. See Fig. 5.9, where Ω_j is a disk D_j for $0 \leq j \leq n$.

In case $f(z) = \sum_0^\infty a_n(z - z_0)^n$ with $D : |z - z_0| < r_{z_0}$ as its disk of convergence, the function element (f, D) or f is specifically called an (analytic) *function germ* or just a *germ* for simplicity. The function elements in Fig. 5.9 are germs. *Note* that the term "germ" we defined here is much less formal than usually seen in the literature. See, for instance, Ref. [1], p. 285, for a formal definition.

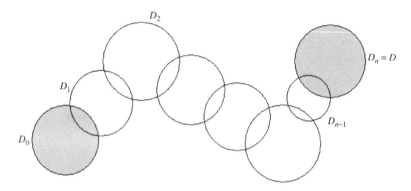

Fig. 5.9

5.2.1   Analytic continuation along a curve

Let (f_j, D_j), $0 \leq j \leq n$, be germs instead of function elements in the definition (5.2.2). It is possible that the center of some D_{j-1} does not lie in D_j for some $j, 1 \leq j \leq n$. If this does happen, one may insert some additional germs between (f_{j-1}, D_{j-1}) and (f_j, D_j) so that the centers of both D_{j-1} and D_j lie in $D_{j-1} \cap D_j$ for each $j, 1 \leq j \leq n$. See Fig. 5.10. In this case, call $D_0, D_1, \ldots, D_{n-1}, D_n$ (in this ordering) a *chain* of *disks* and f_n is the analytic function obtained from f_0 by *the analytic continuation along the chain of disks*.

In Fig. 5.10, let z_j be the center of the disk D_j, $0 \leq j \leq n$. Then the polygonal curve $\overline{z_0 z_1} + \overline{z_1 z_2} + \cdots + \overline{z_{n-1} z_n}$ is a continuous curve $z = z(t)$: $[0, 1] \to \mathbf{C}$ having the following properties: Suppose $0 = t_0 < t_1 < \cdots < t_{n-1} < t_n = 1$ so that $z(t_j) = z_j$ for $0 \leq j \leq n$. Then,

1. for each $t \in [0, 1]$, there corresponds a germ $(f_{z(t)}, D_{z(t)})$: in case $t \in [t_{j-1}, t_j]$, $1 \leq j \leq n$, $(f_{z(t)}, D_{z(t)})$ is a germ obtained by direct analytic continuation of $(f_{z_{j-1}}, D_{z_{j-1}}) = (f_{j-1}, D_{j-1})$, and

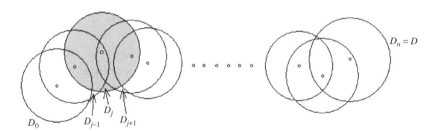

Fig. 5.10

2. there is a $\delta > 0$ so that, if $s, t \in [0, 1]$ and $|s - t| < \delta$, $(f_{z(s)}, D_{z(s)})$
   and $(f_{z(t)}, D_{z(t)})$ are direct analytic continuation of each other: indeed,
   $\delta = \min_{1 \le j \le n} |t_j - t_{j-1}|$ will work.

The detail is left to the readers.

We summarize the above as the

Equivalent statements of analytic continuation. Let (f_0, D_0) and (f, D) be
two analytic function germs, and z_0 and z be the centers of the open disks
D_0 and D, respectively. Then,

    (1) (f_0, D_0) and (f, D) are analytic continuation of each other (see
         Fig. 5.9 and (5.2.2)).

$\Leftrightarrow$ (2) (f_0, D_0) and (f, D) are analytic continuation of each other via a
         chain of disks (see Fig. 5.10).

$\Leftrightarrow$ (3) There is a continuous curve $z = z(t) : [0, 1] \to \mathbf{C}$ connecting $z_0 = z(0)$ to $z = z(1)$ and satisfying:

         1. to each $t \in [0, 1]$, there is a germ $(f_{z(t)}, D_{z(t)})$ at $z(t)$ so that
            $(f_{z(0)}, D_{z(0)}) = (f_0, D_0)$ and $(f_{z(1)}, D_{z(1)}) = (f, D)$, and
         2. (consistence) there is a $\delta > 0$ so that, for $s, t \in [0, 1]$ with
            $|s - t| < \delta$, the germs $(f_{z(s)}, D_{z(s)})$ and $(f_{z(t)}, D_{z(t)})$ are direct
            analytic continuation of each other.

         Note that $z(t)$ is the center of the disk $D_{z(t)}$ and, in 2, the subcurve
         $z|_{[s,t]}$ lies in the set $D_{z(s)} \cap D_{z(t)}$.

In the case of (3), $\{f_{z(t)}\}_{t \in [0,1]}$ is called *an analytic continuation* of f_0 *along
the curve* $z = z(t)$, and (f_0, D_0) and (f, D) are *analytic continuation of each
other along the curve* or *along the chain of disks on the curve* (in view of
(2)). See (5.2.1.2) below.   (5.2.1.1)

Remind that, as a point set, $z([0, 1])$ is a compact subset of $\mathbf{C}$, and thus,
(3) $\Rightarrow$ (1) follows obviously.

In what follows, we develop *basic properties* of analytic continuation.

First, the

Uniqueness. Suppose $\{f_{z(t)}\}_{t \in [0,1]}$ and $\{g_{z(t)}\}_{t \in [0,1]}$ be two analytic continuations of f_0 along the same curve $z = z(t) : [0, 1] \to \mathbf{C}$. Then, on their
common domain,

$$f_{z(t)} = g_{z(t)} \qquad\qquad (5.2.1.2)$$

for each $t \in [0, 1]$.

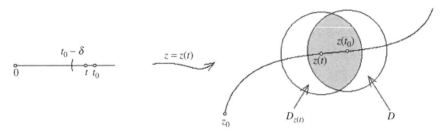

Fig. 5.11

It is in this sense that we are able to say that $\{f_{z(t)}\}_{t\in[0,1]}$ is *the* analytic continuation of the function f_0 or the germ (f_0, D_0) along the curve.

Proof. Let $A = \{t \in [0,1] | f_{z(s)} = g_{z(s)}, 0 \le s \le t\}$. $0 \in A$ because $f_{z(0)} = g_{z(0)} = f_0$ on the common domain of $f_{z(0)}$ and $g_{z(0)}$. Let $t_0 = \sup A$. Then $0 \le t_0 \le 1$. It is routine to show that $t_0 \in A$ and $t_0 = 1$ hold.

Let $\delta' > 0$ be the δ in (3)2 in (5.2.1.1), corresponding to $\{g_{z(t)}\}_{t\in[0,1]}$. Denote $\delta_0 = \min\{\delta, \delta'\}$. Choose any $t \in A$ such that $|t - t_0| < \delta_0$. Then $f_{z(t)} = g_{z(t)}$ on $D_{z(t)} \cap \Delta_{z(t)}$, where $\Delta_{z(t)}$ is the open disk associated to the germ $(g_{z(t)}, \Delta_{z(t)})$. See Fig. 5.11. Let D denote the open disk with center at $z(t_0)$ and radius the minimum of the radii of $D_{z(t_0)}$ and $\Delta_{z(t_0)}$. Since both $f_{z(t_0)}$ and $g_{z(t_0)}$ are direct analytic continuations of $f_{z(t)} = g_{z(t)}$, it follows that

$$f_{z(t_0)} = g_{z(t_0)} = f_{z(t)} \text{ on } D_{z(t)} \cap \Delta_{z(t)} \cap D$$

$$\Rightarrow \text{(by Eq. (5.1.2.5)) } f_{z(t_0)} = g_{z(t_0)} \quad \text{on } D \subseteq D_{z(t_0)}, \Delta_{z(t_0)}$$

$$\Rightarrow f_{z(t_0)} = g_{z(t_0)} \quad \text{on } D_{z(t_0)} \cap \Delta_{z(t_0)}.$$

This shows that $t_0 \in A$.

In case $t_0 < 1$, choose t so that $t_0 < t < 1$ and $|t_0 - t| < \delta_0$ (as above). By (3)2 in (5.2.1.1), both $f_{z(s)}$ and $g_{z(s)}$ are direct analytic continuations of $f_{z(t_0)} = g_{z(t_0)}$ for each $s, t_0 \le s \le t$. Adopting argument similar to the last paragraph, then $f_{z(s)} = g_{z(s)}$ for $t_0 \le s \le t$; in particular, $f_{z(t)} = g_{z(t)}$ holds. Hence, $t \in A$ yet $t > t_0$, contradicting to the definition of t_0. Consequently, $t_0 = 1$ should hold.   □

Remark 1 (The resulted multiple-valued function). In (5.2.2), it is quite possible that there exist j and k with $k - j \ge 2$ so that $D_j \cap D_k \ne \phi$ on which $f_j(z) \ne f_k(z)$ throughout. Take, for instance, in Fig. 5.5, consider

(f_o, D_0) where $D_0 : |z - 2| < 3$ and (f_θ, D_θ) where $\frac{3\pi}{2} \le \theta \le 2\pi$ and $D_\theta : |z + 1 - 3e^{i\theta}| < 3$. Then $f_\theta(z) \ne f_0(z)$ for all $z \in D_\theta \cap D_0$.

Such as in this case, the function

$$F(z) : \Omega = D_0 \cup D_1 \cup \cdots \cup D_n \to \mathbf{C}$$

defined by $F(z) = f_j(z)$ for $z \in D_j$, $0 \le j \le n$, is *multiple-valued* in general. Usually, one has to consider *single-valued analytic branches* of $F(z)$ on Ω (see Sec. 2.7) or to construct the so-called *Riemann surface* on which $F(z)$ turns out to be single-valued (see Secs. 2.7 and 3.3.3). Meanwhile, the monodromy theorem in Sec. 5.2.2 says that this kind of multiple-valuedness would not happen in a simply connected domain.

For some related examples, see Exercises A(1)–(5). Minded readers should work out Exercises B. □

Second,

The simultaneous analytic continuation of an analytic function and its derivative. Let $z = z(t) : [0, 1] \to C$ be a curve and (f_{z_0}, D_{z_0}), where $z_0 = z(0)$, be a germ. Then

(f_{z_0}, D_{z_0}) is analytically continuable along the curve.

$\Leftrightarrow (f'_{z_0}, D_{z_0})$ is analytically continuable along the curve.   (5.2.1.3)

Proof. Only the sufficiency is needed to be proved.

Adopt (2) in (5.2.1.1). Suppose (g_j, D_j), $0 \le j \le n$, are germs where $g_0 = f'_{z_0}$ and $\{D_j\}_{j=0}^n$ is a chain of disks. Also, suppose (f_0, D_0) can be analytically continued to (f_k, D_k) via $D_0 = D_{z_0}, D_1, \ldots, D_k$ for some k, $0 \le k < n$. We try to show that (f_0, D_0) can be analytically continued to (f_{k+1}, D_{k+1}) via $D_0, D_1, \ldots, D_k, D_{k+1}$.

It is known that $f'_{z_0} \equiv g_0$ on D_0, and $f'_j \equiv g_j$ on D_j for $0 \le j \le k$. Choose any analytic function G_{k+1} on D_{k+1} such that $G'_{k+1} = g_{k+1}$ (see Sec. 4.1). It follows that $G'_{k+1} - f'_k = g_{k+1} - g_k \equiv 0$ on $D_k \cap D_{k+1}$; and thus, $G_{k+1} - f_k = c$ (a constant) on $D_k \cap D_{k+1}$. Choose $f_{k+1} = G_{k+1} - c$ on D_{k+1}.

Then (f_{k+1}, D_{k+1}) is the required one. This is because $f'_{k+1} = G'_{k+1} = g_{k+1} = g_k = f'_k$ on $D_k \cap D_{k+1}$ and, then, $f_{k+1} = f_k + c_0$ (a constant) and, eventually, $f_{k+1} = f_k$ on $D_k \cap D_{k+1}$ by readjusting the constant c if necessary. □

As a trivial example, let us go back to Examples 2 and 3 in Sec. 5.1.3. It is known that

$$\left(\sum_{n=0}^{\infty} z^n, |z| < 1\right) = \left(\frac{1}{1-z}, |z| < 1\right)$$

can be analytically continued to obtain a global single-valued analytic function $\frac{1}{1-z}$ on $\mathbf{C} - \{1\}$. And at the same time, so does

$$\left(\sum_{n=0}^{\infty} \frac{z^n}{n}, |z| < 1\right) = (-\mathrm{Log}(1-z), |z| < 1)$$

to obtain a global multiple-valued analytic function $-\log(1-z)$ on $\mathbf{C} - \{1\}$.

Remark 2 (The advantage of the proof of (5.2.1.3)). Suppose (f_0, D_0) and (f, D) are analytic continuation of each other.

Let $w = f_0(z)$ be a (local) solution of an ordinary differential function

$$a_n(z)\frac{d^n w}{dz^n} + a_{n-1}(z)\frac{d^{n-1}w}{dz^{n-1}} + \cdots + a_1(z)\frac{dw}{dz} + a_0(z) = 0,$$

where

$$a_n(z), a_{n-1}(z), \ldots, a_1(z), a_0(z)$$

are assumed to be entire functions, on D_0. Then the same proof as that for (5.2.1.3) will guarantee that $w = f(z)$ is a solution of the equation on D.

Similarly, let $w = f_0(z)$ be a root of an algebraic function

$$P(z, w) = a_0(z) + a_1(z)w + \cdots + a_{n-1}(z)w^{n-1} + a_n(z)w^n,$$

where

$$a_0(z), a_1(z), \ldots, a_{n-1}(z), a_n(z)$$

are entire functions, on D_0, namely, $P(z, f_0(z)) = 0$ for all $z \in D_0$ (and then, $f_0(z)$ is an algebraic function on D_0). Then, so is $w = f(z)$ on D, namely, $P(z, f(z)) = 0$ on D.

For formal and detailed account, refer to Ref. [1], pp. 300–306 and 308–321.   □

As an extension of the inequality that appeared in (5.1.2.2), we have

The continuity of the radii of convergence. Let $\{(f_{z(t)}, D_{z(t)})\}$ be the analytic continuation of the germ (f_0, D_0) along the curve $z = z(t) : [0, 1] \to \mathbf{C}$. Denote by $r_{z(t)}$ the radius of the disk $D_{z(t)}$ for $t \in [0, 1]$.

(1) $|r_{z(s)} - r_{z(t)}| \leq |z(s) - z(t)|$ whenever $s, t \in [0, 1]$ and $|s - t| < \delta$ (see (3) in 5.2.1.1); thus, $r_{z(t)}$ is uniformly continuous on $[0, 1]$ except $r_{z(t)} = \infty$.

(2) It follows that

$$r = \min_{0 \leq t \leq 1} r_{z(t)} > 0.$$

(3) There is a $\delta_0 > 0$ so that $|z(s) - z(t)| < r$ if $s, t \in [0, 1]$ and $|s - t| < \delta_0$; in particular,

$$|z(s) - z(t)| < \min(r_{z(s)}, r_{z(t)})$$

whenever $|s - t| < \delta_0$.

(3) shows that $f_{z(s)}$ and $f_{z(t)}$ are not only direct analytic continuation of each other, but also $f_{z(s)}$ is the Taylor series expansion of $f_{z(t)}$ at $z(s) \in D_{z(t)}$ and vice versa.   (5.2.1.4)

The proof is left as Exercise A(6).
Moreover, there are some

Operational relations between curves and analytic continuation. Let (f, D) be the germ obtained by analytic continuation from the germ (f_0, D_0) along the curve $z = z(t) : [0, 1] \to \mathbf{C}$ (recall that $f_0 = f_{z(0)}$ and $f = f_{z(1)}$).

(1) "Inverse curve" operation: f_0 is obtainable from f by analytic continuation along the curve $z = z(1 - t) : [0, 1] \to \mathbf{C}$.

(2) "Change of variables" operation: Let $t = t(s) : [a, b] \to [0, 1]$ be a change of variable. Then $\{f_{z(t(s))}\}$ is an analytic continuation of f_0 along the curve $z = z(t(s)) : [a, b] \to \mathbf{C}$.

(3) "Composite of curves" operation: Suppose $\{g_{w(t)}\}$, $g_{w(0)} = f$, is an analytic continuation of $f = f_{z(1)}$ along the curve $w = w(t)$. Then the original f_0 has an analytic continuation $\{h_{\eta(t)}\}$ along the curve $\eta = \eta(t) : [0, 2] \to \mathbf{C}$, where

$$h_{\eta(t)} = \begin{cases} f_{z(t)}, & 0 \leq t \leq 1 \\ g_{w(t-1)}, & 1 \leq t \leq 2 \end{cases}.$$ (5.2.1.5)

Proof is left as Exercise A(7).
Finally,

The analytic continuability along two sufficiently-closed curves. Let $\{f_{z(t)}\}$ be the analytic continuation of f_0 along the curve $z = z(t) : [0, 1] \to \mathbf{C}$. Let $r > 0$ be as in (2) of (5.2.1.4). Suppose $w = w(t) : [0, 1] \to \mathbf{C}$ is another curve satisfying:

1. (common initial point) $w(0) = z(0)$;
2. (sufficiently-closed) $|w(t) - z(t)| < \frac{r}{4}$, $0 \le t \le 1$.

Then

(1) f_0 has an analytic continuation $\{g_{w(t)}\}$ along $w = w(t)$, $0 \le t \le 1$, with $g_{w(0)} = f_0$, and
(2) in case $w(1) = z(1)$ holds, then the germs at $t = 1$ obtained by analytic continuation of f_0 along these two curves are identical, namely,

$$g_{w(1)} = f_{z(1)}. \tag{5.2.1.6}$$

Proof. Try to use (2) in (5.2.1.1) to justify these statements.

Recall that $r = \min_{0 \le t \le 1} r_{z(t)}$ and $r > 0$. There is a partition $0 = t_0 < t_1 < \cdots < t_{n-1} < t_n = 1$ of $[0,1]$ such that

$$|z(t) - z(t_{j-1})| < \frac{r}{4} \text{ whenever } t_{j-1} \le t \le t_j \text{ for } 1 \le j \le n.$$

Hence $D_{z(t_j)}$: $|z - z(t_j)| < r_{z(t_j)}$, $0 \le j \le n$, form a chain of disks, and the germ $(f_0, D_0) = (f_{z(0)}, D_{z(0)})$ is analytically continued to $(f_{z(1)}, D_{z(1)}) = (f_{z(t_n)}, D_{z(t_n)})$ via the successive direct analytic continuations $(f_{z(t_1)}, D_{z(t_1)}), \ldots, (f_{z(t_{n-1})}, D_{z(t_{n-1})})$.

For any t, $t_{j-1} \le t \le t_j$ and $1 \le j \le n$, by assumption

$$|w(t) - z(t_{j-1})| \le |w(t) - z(t)| + |z(t) - z(t_{j-1})| < \frac{r}{4} + \frac{r}{4} = \frac{r}{2}.$$

This indicates that $w(t) \in D_{z(t_{j-1})}$ holds for $t_{j-1} \le t \le t_j$ and $1 \le j \le n$. Use γ_j to denote the line segment joining $z(t_j)$ and $w(t_j)$ for $1 \le j \le n$. Note that γ_j lies completely in $D_{z(t_j)}$ for $1 \le j \le n$.

The key point in the argument relies mainly on the fact that, the direct analytic continuation to the same point along any curve in the open disk of a germ always lead to the same germ (see (5.1.2.2) and (5.1.2.5), or the implication of (3) in (5.2.1.4)).

Now, consider the disk $D_{z(0)} = D_0$. (f_0, D_0) can be analytically continued to $(f_{z(t_1)}, D_{z(t_1)})$ along $z = z(t)$ from $z(0)$ as well as to $(g_{w(t_1)}, D_{w(t_1)})$ along $w = w(t)$ from $w(0)$. If we continue $(g_{w(t_1)}, D_{w(t_1)})$ from $w(t_1)$ to $z(t_1)$ along γ_1^{-1}, then we get $(f_{z(t_1)}, D_{z(t_1)})$ because all the processes proceed in the same open disk D_0 of the germ (f_0, D_0). See Fig. 5.12.

Then, proceed to the disk $D_{z(1)}$. $(f_{z(t_1)}, D_{z(t_1)})$ can be analytically continued to $(f_{z(t_2)}, D_{z(t_2)})$ along $z = z(t)$ and $(g_{w(t_1)}, D_{w(t_1)})$ to $(g_{w(t_2)}, D_{w(t_2)})$ along $w = w(t)$. Based on what we obtained in the last paragraph and (5.2.1.5), if we continue $(f_{z(t_1)}, D_{z(t_1)})$ from $z(t_1)$ to $w(t_1)$

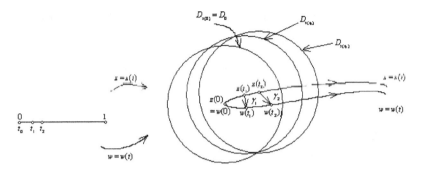

Fig. 5.12

along γ_1 and then to $w(t_2)$ along $w = w(t)$ and then to $z(t_2)$ along γ_2^{-1}, we will obtain the same $(f_{z(t_2)}, D_{z(t_2)})$.

Repeat this process after a finite number of steps and we will prove the statements.   □

Exercises A

(1) Let
$$f_1(z) = \sqrt{|z|}e^{\frac{i\theta}{2}}, \quad z \in D_1 = \mathbf{C} - [0, \infty) \text{ with } 0 < \operatorname{Arg} z = \theta < 2\pi,$$
and
$$f_2(z) = \sqrt{|z|}e^{\frac{i\theta}{2}}, \quad z \in D_2 = \mathbf{C} - (-\infty, 0] \text{ with } -\pi < \operatorname{Arg} z = \theta < \pi.$$
Then, $f_1(z) \equiv f_2(z)$ on $\operatorname{Im} z > 0$, while $f_1(z) \neq f_2(z)$ for all z with $\operatorname{Im} z < 0$. So (f_1, D_1) and (f_2, D_2) are not direct analytic continuation of each other. Try to show that (f_1, D_1) and (f_2, D_2) are analytic continuation of each other by showing some intermediate analytic germs between them.

(2) Given two power series
$$f_1(z) = \sum_{n=1}^{\infty} \frac{z^n}{n} \quad \text{and} \quad f_2(z) = \pi i + \sum_{n=1}^{\infty} \frac{(-1)^n}{n}(z - 2)^n.$$

(a) Determine their radii of convergence and show that their disks of convergence do not intersect each other.

(b) Show that $f_1(z)$ can be analytically continued to $f_2(z)$ by presenting a third analytic function $f_3(z)$ so that both $f_1(z)$ and $f_2(z)$ are direct analytic continuations of $f_3(z)$.

(3) Shows that the series
$$\sum_{n=1}^{\infty}\left(\frac{1}{1 - z^{n+1}} - \frac{1}{1 - z^n}\right)$$

represents an analytic function $f_1(z)$ on $|z| < 1$ and another one $f_2(z)$ on $|z| > 1$, yet $f_1(z)$ and $f_2(z)$ cannot be analytically continued to each other.

(4) Do the same problem as Exercise (3) for the series

$$\frac{1}{1-z} + \sum_{n=0}^{\infty} \frac{z^{2^n}}{z^{2^{n+1}} - 1}.$$

(5) Let $\sum_{n=1}^{\infty} a_n$ be a series with $a_n > 0$ for all $n \geq 1$. Let r_n, $n \geq 1$, be the sequence formed by all the rational numbers. Show that the series

$$f(z) = \sum_{n=1}^{\infty} \frac{a_n}{z - r_n}$$

is analytic in $\mathrm{Im}\, z > 0$ and in $\mathrm{Im}\, z < 0$, respectively, but the analytical functions then represented cannot be analytically continued to each other.

(6) Prove (5.2.1.4) in detail.

(7) Prove (5.2.1.5) in detail.

(8) Try to use (3) in (5.2.1.1) to reprove (5.2.1.6).

Exercises B

Try to use the concept of analytic continuation to formally define the Riemann surface of the multiple-valued function $F(z)$ mentioned in Remark 1 and justify that it is coincident with the one given by (2.7.2.8). It is beneficial to refer to Ref. [81], or, in less arduous, to Ref. [75], Chap. 3.

5.2.2   *Homotopy and monodromy theorem*

Roughly speaking, that a curve is continuously deformed into another curve without leaving a given open set during the process is termed as a *homotopy* between the two curves. See Sec. 4.2.2 or Exercises B of Sec. 2.7.1 for formal definition.

Section 4.2.3 (see also Exercise B(2)(d) of Sec. 2.7.1) discussed the homotopic invariance of winding numbers.

Here, in this section, we will see that the analytic continuability is another homotopic invariance which is specifically called the monodromy theorem in the realm of complex analysis.

Our main result is

The homotopic invariance of the analytic continuation or Monodromy theorem. Let γ and $\tilde{\gamma}$ be homotopic with fixed endpoints z_0 and z_1 in a domain

Ω, and

$$H : [0, 1] \times [0, 1] \rightarrow \Omega$$

be the homotopy between γ and $\tilde{\gamma}$. Suppose a germ (f_0, D_0) at z_0 can be analytically continued along each curve $H(\cdot, \tau) : [0, 1] \rightarrow \Omega$ for $0 \leq \tau \leq 1$. Then the continuation of (f_0, D_0) along γ and $\tilde{\gamma}$ lead to the same germ at z_1.   (5.2.2.1)

This result can be restated in terms of covering surface (see (2.7.2.9) and Sec. 7.5.2, in particular, (7.5.2.3)).

Proof. The proof is based on (5.2.1.6).

Let (f, D) be the germ obtained by analytic continuation of (f_0, D_0) along γ. Let

$$A = \{\tau \in [0, 1] |$$

the germ obtained by continuation of (f_0, D_0) along the curve

$$H(t, \tau), 0 \leq t \leq 1, \text{ is } (f, D)\}.$$

Since $0 \in A$, A is nonempty and is bounded above by 1. Consider $\tau_0 = \sup A$. Then $0 \leq \tau_0 \leq 1$.

Try to show that $\tau_0 \in A$ and $\tau_0 = 1$ (refer to the proof of (5.2.1.2)).

To see $\tau_0 \in A$: According to (5.2.1.6), there is an $\varepsilon > 0$ such that, for any curve $\alpha = \alpha(t) : [0, 1] \rightarrow \Omega$ satisfying:

1. $\alpha(t)$ and $H(t, \tau_0)$ having the same end points, namely, $\alpha(0) = H(0, \tau_0) = z_0$ and $\alpha(1) = H(1, \tau_0) = z_1$;
2. $|\alpha(t) - H(t, \tau_0)| < \varepsilon$, $0 \leq t \leq 1$,

then the continuations of (f_0, D_0) along $\alpha(t)$ and $H(t, \tau_0)$ lead to the same germ at z_1. For this ε, by the uniform continuity of $H(t, \tau)$ on $[0, 1] \times [0, 1]$, there is a $\delta > 0$ so that

$$|H(t, \tau) - H(t, \tau_0)| < \varepsilon \text{ whenever } t \in [0, 1] \text{ and } \tau, \tau_0 \in [0, 1] \text{ with } |\tau - \tau_0| < \delta.$$

Choose $\tau_1 \in A$ so that $\tau_0 - \delta < \tau_1 < \tau_0$. Set the aforementioned $\alpha(t) = H(t, \tau_1)$. Then, by this choice of $\alpha(t)$ and the related result, it follows that $\tau_0 \in A$.

To see $\tau_0 = 1$: In case $\tau_0 < 1$, choose τ_2 such that $\tau_0 < \tau_2 < \min\{1, \tau_0 + \delta\}$ and then set $\alpha(t) = H(t, \tau_2)$. Hence, the continuation of (f_0, D_0) along $H(t, \tau_0)$ and $H(t, \tau_2)$ will lead to the same germ (f, D), contradicting to the definition of τ_0. $\square$

How to realize the monodromy theorem (5.2.2.1) in a simply connected domain Ω?

In Example 3 in Sec. 4.2.2, a simply connected domain is characterized as a domain in which each closed curve is homotopic to a constant curve. Equivalently, a domain is simply connected *if* and only if any two curves with the same end points in them is always homotopic.

This observation will lead to the important

Monodromy theorem in a simply connected domain. Let Ω be a simply connected domain and $D \subseteq \Omega$ be an open disk with center z_0. Suppose a germ (f, D) can be analytically continued along any curve with initial point z_0 in Ω. Then there is a unique analytic element (F, Ω) so that

$$F(z) = f(z), \quad z \in D. \tag{5.2.2.2}$$

Of course, $F(z)$ is single-valued and analytic in Ω.

Proof. Fix any point $z \in \Omega$. Construct any curve $z = z(t) : [0, 1] \to \Omega$ so that $z(0) = z_0$ and $z(1) = z$. Continue (f, D) analytically along $z = z(t)$ to z and get the germ $(f_{z(1)}, D_{z(1)})$. Define $F : \Omega \to \mathbf{C}$ by

$$F(z) = f_{z(1)}(z), \quad z \in D_{z(1)}.$$

F is well defined: If $\tilde{\gamma}$ is another curve joining z_0 to z in Ω and $(\tilde{f}_{z(1)}, \tilde{D}_{z(1)})$ is the germ at z obtained by continuation of (f, D) along $\tilde{\gamma}$, then it is necessary that $(\tilde{f}_{z(1)}, \tilde{D}_{z(1)}) = (f_{z(1)}, D_{z(1)})$. Otherwise, even though the composite curve $\tilde{\gamma} \circ (-\gamma)$ of $\tilde{\gamma}$ and the inverse curve $-\gamma$ of γ is homotopic to the point $z(1)$, yet the continuation of $(f_{z(1)}, D_{z(1)})$ along $\tilde{\gamma} \circ (-\gamma)$ will lead to $(\tilde{f}_{z(1)}, \tilde{D}_{z(1)})$ at $z(1)$, not itself, and this contradicts (5.2.2.1).

F is analytic by its very definition and is unique by (3.4.2.9) or (5.1.2.4). $\square$

Exercises A

(1) Let

$$f(z) = \sum_{n=1}^{\infty} \frac{z^n}{n^2}, \quad |z| < 1.$$

Suppose Ω is a simply connected domain and $1 \notin \Omega$ so that $\Omega \cap \{|z| < 1\} \neq \phi$. In case $\Omega \cap \{|z| < 1\}$ and $\Omega \cup \{|z| < 1\}$ are simply connected, show that $(f, |z| < 1)$ can be analytically continued along any curve in $\Omega \cup \{|z| < 1\}$.

(2) Let $\Omega \subseteq \mathbf{C} - \{0\}$ be a simply connected domain.

   (a) Show that there is an analytic function $f : \Omega \to \mathbf{C}$ satisfying $e^{f(z)} = z, z \in \Omega$.

   (b) Show that any function in (a) should be of the form $f + 2n\pi i$, $n = 0, \pm 1, \pm 2, \ldots$.

(3) Let $\Omega \subseteq \mathbf{C} - \{0\}$ be a simply connected domain. If $g : \Omega \to \mathbf{C}$ is analytic and $g(z) \neq 0$ throughout Ω, do the same problems as in Exercise (2) by using $g(z)$ to replace z.

5.3   Local Uniform Convergence of a Sequence or a Series of Analytic Functions

Readers are supposed to be familiar with the content of Sec. 2.3, where the definition, basic criteria, and operational properties about uniform convergence of a sequence or a series of complex-valued functions have been sketched.

Uniform convergence, as learned in real analysis, preserves nice properties of the functions that appeared in the sequence, such as boundedness, continuity, differentiability, and integrability. So does the uniform convergence of a sequence of *analytic functions* but much richer in content, owing to the Cauchy integral theorem.

As we proceed, we will see that many meaningful results, such as continuity and analyticity, need only the concept of local but not global uniform convergence. Hence, we formally state the

Local uniform convergence. Suppose $f_n(z) : \Omega$ (domain or open set)$\to \mathbf{C}$, $n \geq 1$. Then

   (1) For each open disk $D \subseteq \Omega$ with its closure $\bar{D}$ still lying in Ω, $f_n(z)$ converges uniformly on D to a function $f(z) : \Omega \to \mathbf{C}$.

$\Leftrightarrow$ (2) $f_n(z)$ converges uniformly on *each* compact subset of Ω to a function $f(z) : \Omega \to \mathbf{C}$.

In this case, we say that the sequence f_n or $f_n(z)$ converges *locally uniformly* in Ω or converges on *compacta* in Ω to $f(z)$ and $f(z)$ is called the *limit function* of the sequence.   (5.3.1)

Another advantage of local uniform convergence is that each domain of $f_n(z)$ is subject to the following flexibility: Suppose a domain Ω_n is the domain of definition of f_n for $n \geq 1$, satisfying

$$\Omega_1 \subseteq \Omega_2 \subseteq \cdots \subseteq \Omega_n \subseteq \cdots .\tag{5.3.2}$$

Then, for each $z \in \Omega = \bigcup_{n=1}^{\infty} \Omega_n$, there is an n_0 so that $z \in \Omega_n$ if $n \geq n_0$. Hence, it is meaningful to talk about the pointwise, even local uniform, convergence of the sequence $f_n(z)$ in Ω after disregarding a finite number of terms.

Let $f_n(z)$, $n \geq 1$, be defined on the same domain (or open set) Ω. The series

$$\sum_{n=1}^{\infty} f_n(z) \quad \text{or} \quad \sum f_n \tag{5.3.3}$$

is said to *converge locally uniformly* in Ω to a function $f(z)$ if the partial sum sequence $S_n(z) = \sum_{n=1}^{k} f_k(z)$ does in Ω to $f(z)$. In this case, $f(z)$ is called the *sum* (function) of the series.

The following is divided into four subsections.

Local uniform convergence of a sequence or a series of analytic functions preserves the *analyticity* of its members. Section 5.3.1 proves this basic important theorem due to Weierstrass.

Local uniform convergence also "preserves" the number of zeros under some restricted conditions. This is Hurwitz's theorem in Sec. 5.3.2.

Section 5.3.3 presents some criteria for local uniform convergence that do not have suitable counterparts in real analysis. The last Sec. 5.3.4 studies some basic properties about the fixed points of an analytic function and its iterate functions. A continuation of Sec. 5.3.4 will be Sec. 5.8.5.

5.3.1   *Analyticity of the limit function: Weierstrass's theorem*

Suppose $f_n(z)$, $n \geq 1$, are analytic on a domain Ω and converges locally uniformly on Ω to a function $f(z)$.

To see the *analyticity* of $f(z)$ on Ω: Let γ be any rectifiable Jordan closed curve in Ω so that $\overline{\text{Int } \gamma} \subseteq \Omega$ holds. Since, as a point set, γ is compact and then, $f_n(z)$ is uniformly convergent on γ to $f(z)$, it follows that

$$\int_{\gamma} f_n(z)dz = 0, \quad n \geq 1$$

$$\Rightarrow \int_{\gamma} f(z)dz = \lim_{n \to \infty} \int_{\gamma} f_n(z)dz = 0.$$

By Morera's theorem (3.4.2.13), $f(z)$ is thus analytic in Ω.

However, Cauchy's integral formula provides another proof for this fact
and even more: Choose any closed disk $|z - z_0| \leq r$ containing in Ω. Then

$$f_n(z) = \frac{1}{2\pi i} \int_{|\zeta - z_0| = r} \frac{f_n(\zeta)}{\zeta - z} d\zeta, \quad |z - z_0| < r \quad \text{and} \quad n \geq 1$$

$\Rightarrow$ (Since $f_n(z)$ converges uniformly to $f(z)$ on $|z - z_0| \leq r$.) Letting
$n \to \infty$, thus

$$f(z) = \frac{1}{2\pi i} \int_{|\zeta - z_0| = r} \frac{f(\zeta)}{\zeta - z} d\zeta, \quad |z - z_0| < r.$$

This shows that $f(z)$ is analytic in $|z - z_0| < r$ (see (4.7.1)), and hence,
$f(z)$ is analytic in Ω.

Similarly,

$$f_n'(z) = \frac{1}{2\pi i} \int_{|\zeta - z_0| = r} \frac{f_n(\zeta)}{(\zeta - z)^2} d\zeta, \quad |z - z_0| < r \quad \text{and} \quad n \geq 1.$$

$\Rightarrow$ Letting $n \to \infty$, $\quad f'(z) = \frac{1}{2\pi i} \int_{|\zeta - z_0| = r} \frac{f(\zeta)}{(\zeta - z)^2} d\zeta, \quad |z - z_0| < r.$

This shows that $f_n'(z)$ converges pointwise to $f'(z)$ on Ω. As a matter of
fact, the convergence is locally uniformly. To see this, choose $0 < \rho < r$.
Then

$$f_n'(z) - f'(z) = \frac{1}{2\pi i} \int_{|\zeta - z_0| = r} \frac{f_n(\zeta) - f(\zeta)}{(\zeta - z)^2} d\zeta \quad \text{on } |z - z_0| \leq \rho, \quad n \geq 1$$

$$\Rightarrow |f_n'(z) - f'(z)| \leq \frac{1}{2\pi} \max_{|\zeta - z_0| = r} |f_n(\zeta) - f(\zeta)| \cdot \frac{2\pi r}{(r - \rho)^2} \to 0$$

$$\text{as } n \to \infty \quad \text{on } |z - z_0| \leq \rho.$$

Thus, $f_n'(z)$ converges uniformly to $f'(z)$ on $|z - z_0| \leq \rho$ and hence, locally
uniformly to $f'(z)$ on Ω.

Repeat the above process and we are able to show that $f_n^{(k)}(z)$ converges
locally uniformly to $f^{(k)}(z)$ on Ω for each $k \geq 1$.

We conclude as the following basic

Weierstrass's theorem. Let $f_n(z)$, $n \geq 1$, be analytic functions on a
domain Ω.

(1) Suppose the sequence $f_n(z)$ converges locally uniformly in Ω to the
function $f(z)$. Then

    1. $f(z)$ is analytic in Ω, and
    2. $f_n^{(k)}(z)$ converges locally uniformly to $f^{(k)}(z)$ on Ω, for each $k \geq 1$.

(2) Suppose the series $\sum_{n=1}^{\infty} f_n(z)$ converges locally uniformly on Ω to the function $f(z)$. Then

    1. $f(z)$ is analytic in Ω, and

    2. $\sum_{n=1}^{\infty} f_n^{(k)}(z) = f^{(k)}(z)$ locally uniformly on Ω, for each $k \geq 1$.

$$(5.3.1.1)$$

Note that (1) is still valid for a sequence $f_n(z) : \Omega_n \to \mathbf{C}$ satisfying the restriction (5.3.2). For an extension of this theorem, refer to Exercises B(1) and (2); for other types of convergence, refer to (5.3.1.6) and (5.3.1.7) in Exercises B.

A restatement of (5.3.1.1) is the following

Weierstrass's double series theorem. Let $f_n(z) = \sum_{k=0}^{\infty} a_{nk}(z - z_0)^k$, $|z - z_0| < r$ for $n \geq 1$.

(1) Suppose $f_n(z)$ converges locally uniformly in $|z - z_0| < r$ to the function $f(z)$. Then

    1. For each $k \geq 0$, the sequence a_{nk}, $n \geq 1$, converges to a_k as $n \to \infty$, and

    2. $f(z) = \sum_{k=0}^{\infty} a_k(z - z_0)^k$, $|z - z_0| < r$.

Namely,

$$f(z) = \lim_{n \to \infty} \left\{ \sum_{k=0}^{\infty} a_{nk}(z - z_0)^k \right\}$$

$$= \sum_{k=0}^{\infty} \left\{ \lim_{n \to \infty} a_{nk} \right\} (z - z_0)^k, \quad |z - z_0| < r.$$

(2) Suppose $\sum_{n=1}^{\infty} f_n(z) = f(z)$ locally uniformly in $|z - z_0| < r$. Then

    1. For each $k \geq 0$, the series $\sum_{n=1}^{\infty} a_{nk} = a_k$ converges, and

    2. $f(z) = \sum_{k=0}^{\infty} a_k(z - z_0)^k$, $|z - z_0| < r$.

Namely,

$$f(z) = \sum_{n=1}^{\infty} \left\{ \sum_{k=0}^{\infty} a_{nk}(z - z_0)^k \right\}$$

$$= \sum_{k=0}^{\infty} \left\{ \sum_{n=1}^{\infty} a_{nk} \right\} (z - z_0)^k, \quad |z - z_0| < r. \qquad (5.3.1.2)$$

We have used (5.3.1.1) or (5.3.1.2) before in (5.1.1.7).

Observe that, by using (5.3.1.1),

$$\lim_{n\to\infty} f_n^{(k)}(z_0) = f^{(k)}(z_0), \quad k \geq 0$$

$$\Leftrightarrow \lim_{n\to\infty} k! a_{nk} = f^{(k)}(z_0), \quad k \geq 0$$

$$\Leftrightarrow \lim_{n\to\infty} a_{nk} = \frac{f^{(k)}(z_0)}{k!} = a_k, \quad k \geq 0.$$

This is (1). (2) holds similarly.

Proof. A direct proof is as follows: Fix ρ, $0 < \rho < r$. Then

$$a_{mk} - a_{nk} = \frac{1}{k!}[f_m^{(k)}(z_0) - f_n^{(k)}(z_0)], \quad k \geq 0$$

$\Rightarrow$ (by using Cauchy's integral formula or, directly, by Cauchy's inequality (5.1.2.9))

$$|a_{mk} - a_{nk}| \leq \frac{\|f_m - f_n\|_\infty}{\rho^k}, \quad k \geq 0,$$

$$\text{where } \|f_m - f_n\|_\infty = \sup_{|z-z_0|=\rho} |f_m(z) - f_n(z)|. \tag{$*_1$}$$

Hence, a_{nk} is Cauchy for each $k \geq 0$ and thus, $\lim_{n\to\infty} a_{nk} = a_k$ exists, $k \geq 0$. On the other hand, the uniform convergence of $f_m(z)$ on $|z - z_0| \leq \rho$ to $f(z)$ means that $\|f_m - f\|_\infty \to 0$ as $m \to \infty$. Let $m \to \infty$ in $(*_1)$ and we have

$$|a_{nk} - a_k| \leq \frac{\|f_n - f\|_\infty}{\rho^k}, \quad k \geq 0, \quad n \geq 1.$$

Fix any δ, $0 < \delta < \rho$. Then, on $|z - z_0| \leq \delta$,

$$|a_{nk}(z - z_0)^k - a_k(z - z_0)^k| \leq \frac{\delta^k}{\rho^k}\|f_n - f\|_\infty, \quad k \geq 0, \quad n \geq 1$$

$$\Rightarrow \sum_{k=0}^{\infty} |a_k(z - z_0)^k| \leq \|f_n - f\|_\infty \sum_{k=0}^{\infty} \left(\frac{\delta}{\rho}\right)^k + \sum_{k=0}^{\infty} |a_{nk}(z - z_0)^k|, \quad \text{and}$$

$$\left|\sum_{k=0}^{\infty} a_{nk}(z - z_0)^k - \sum_{k=0}^{\infty} a_k(z - z_0)^k\right| \leq \|f_n - f\|_\infty \sum_{k=0}^{\infty} \left(\frac{\delta}{\rho}\right)^k,$$

$$\to 0 \quad \text{as } n \to \infty, \quad |z - z_0| \leq \delta < \rho < r.$$

This shows that $\sum_{k=0}^{\infty} a_k(z - z_0)^k$ converges on $|z - z_0| \leq \delta$ and

$$\lim_{n \to \infty} f_n(z) = \sum_{k=0}^{\infty} \{\lim_{n \to \infty} a_{nk}\}(z - z_0)^k$$

$$= \sum_{k=0}^{\infty} a_k(z - z_0)^k \quad \text{holds on } |z - z_0| \leq \rho,$$

which, in turn, implies that this result is valid in $|z - z_0| < r$. $\qquad\square$

Four examples are provided.

Example 1. Let

$$f_n(z) = z \exp\left(-\frac{1}{2}n^2 z^2\right), \quad z \in \mathbf{C}.$$

Show that $f_n(z)$ converges uniformly to the function $f(z) \equiv 0$ on the real axis but does not converge locally uniformly to 0 in $\mathbf{C}$.

This example indicates that (5.3.1.1) is not necessarily valid in its full content if the set Ω on which uniform convergence occurs is *not* an open set in the plane. See Exercises A(2)–(4) for more illustrations.

Solution. When $z = x$ is restricted to reals, then

$$\sup_{x \in R} |f_n(x)| = f_n\left(\frac{1}{n}\right) = \frac{1}{n} \exp\left(-\frac{1}{2}\right) \to 0 \quad \text{as } n \to \infty.$$

Thus, $f_n(x)$ converges uniformly to $f(x) \equiv 0$ in $\mathbf{R}$. On the other hand, for any fixed $r > 0$,

$$\max_{|z| \leq r} |f_n(z)| = \max_{|z| = r} |f_n(z)| \geq |f_n(ir)| = r \exp\left\{\frac{1}{2}n^2 r^2\right\} \to \infty \quad \text{as } n \to \infty.$$

Hence, $f_n(z)$ does not converge uniformly to 0 on $|z| \leq r$. Note that, $f_n(z)$ converges pointwise to 0 in $\mathbf{C}$.

On the contrary, suppose that $f_n(z)$ would converge uniformly to $f(z) = 0$ on $|z| < r$. According to (1)2 in (5.3.1.1), $f_n'(0) \to f'(0) = 0$ as $n \to \infty$. But $f_n'(z) = (1 - n^2 z^2) \exp(-\frac{1}{2}n^2 z^2)$ and $f_n'(0) = 1 \to 1$ as $n \to \infty$, a contradiction. Hence, it is not possible that the uniform convergence of $f_n(z)$ on $|z| < r$ could happen.

Example 2. Show that

$$\lim_{n \to \infty} \left(1 + \frac{z}{n}\right)^n = e^z, \quad z \in \mathbf{C}$$

locally uniformly in $\mathbf{C}$.

Proof. Example 1 in Sec. 1.7 had showed that $(1+\frac{z}{n})^n$ converges pointwise to e^z in $\mathbf{C}$. And we adopted this as one of the definition for e^z (see (1.7.5)). An alterative way to realize this fact is as follows. Choose the principal branch $\text{Log}(1+\frac{z}{n})$ of $\log(1+\frac{z}{n})$ in $\mathbf{C}^* - [\infty, -n]$ for $n \geq 1$. Then

$$\text{Log}\left(1+\frac{z}{n}\right) = \sum_{k=1}^{\infty} \frac{(-1)^{k-1}}{k}\left(\frac{z}{n}\right)^k \quad \text{locally uniformly in } |z| < n$$

$\Rightarrow$ For fixed $z \in \mathbf{C}$ and n large enough, by (2.3.3),

$$\lim_{n\to\infty} \text{Log}\left(1+\frac{z}{n}\right)^n = \lim_{n\to\infty} n\,\text{Log}\left(1+\frac{z}{n}\right)$$

$$= \lim_{n\to\infty}\left\{z + \sum_{k=2}^{\infty} \frac{(-1)^{k-1}}{k}\frac{z^k}{n^{k-1}}\right\}$$

$$= z + \sum_{k=2}^{\infty} \frac{(-1)^{k-1}}{k} \lim_{n\to\infty} \frac{z^k}{n^{k-1}} = z \quad \text{pointwise.}$$

$\Rightarrow \lim_{n\to\infty}\left(1+\frac{z}{n}\right)^n = e^z$ pointwise in $\mathbf{C}$.

To see local uniform convergence: Consider $|z| \leq R$ and let $n > R$. Then, on $|z| \leq R$,

$$\left|\text{Log}\left(1+\frac{z}{n}\right)^n - z\right| \leq \sum_{k=2}^{\infty} \frac{1}{k}\frac{R^k}{n^{k-1}} = n\sum_{k=2}^{\infty} \frac{1}{k}\left(\frac{R}{n}\right)^k$$

$$= n\left\{-\log\left(1-\frac{R}{n}\right) - \frac{R}{n}\right\} \to 0 \quad \text{as } n \to \infty. \qquad (*_2)$$

Thus, $\text{Log}(1+\frac{z}{n})^n$ converges uniformly to z on $|z| \leq R$, and then, locally uniformly in $\mathbf{C}$ to $f(z) = z$ as claimed above.

The other method: Let $m, n > R \geq |z|$. By $(*_2)$,

$$\left|\text{Log}\left(1+\frac{z}{m}\right)^m - \text{Log}\left(1+\frac{z}{n}\right)^n\right| \leq m\left\{-\log\left(1-\frac{R}{m}\right) - \frac{R}{m}\right\}$$

$$+ n\left\{-\log\left(1-\frac{R}{n}\right) - \frac{R}{n}\right\} \to 0 \quad \text{as } m, n \to \infty.$$

Cauchy condition (2.3.2) says that $f_n(z) = \text{Log}(1+\frac{z}{n})^n$ converges uniformly in $|z| \leq R$, and hence, locally uniformly to an analytic function $f(z)$ in $\mathbf{C}$. Now, $f_n'(z) = \frac{1}{1+z/n}$ converges locally uniformly to $f'(z)$ in $\mathbf{C}$. Then $f'(z) \equiv 1$ on $\mathbf{C}$ and thus, $f(z) = z + a$ for some constant a. Since $\lim_{n\to\infty} f_n(0) = f(0) = 0$, $a = 0$ holds and $f(z) = z$, $z \in \mathbf{C}$. $\qquad \square$

Example 3. Suppose $\sum_{n=1}^{\infty} |f_n(z)|$ converges locally uniformly in a set Ω.

(1) Shows that $\sum_{n=1}^{\infty} f_n(z)$ also converges locally uniformly in Ω but not conversely.
(2) In case Ω is a *domain* and each $f_n(z)$ is *analytic* in Ω for $n \geq 1$. Then, for each positive integer k,

$$\sum_{n=1}^{\infty} |f_n^{(k)}(z)|$$

converges locally uniformly in Ω.

Proof. (1) This follows easily by Cauchy condition. While, $\sum_{n=1}^{\infty} \frac{z^n}{n}$ on the set $\Omega = \{z | |z| = 1 \text{ but } z \neq 1\}$ is such a case (refer to Example 1 in Sec. 5.1.3 and try to use Dirichlet test in (2.3.5)).

(2) By (1) and (5.3.1.1), $\sum_1^{\infty} f_n(z)$ converges locally uniformly in the domain Ω and so does $\sum_1^{\infty} f_n^{(k)}(z)$ for each $k \geq 1$. Choose a closed disk $|z - z_0| \leq r$ containing in Ω. Then

$$f_n^{(k)}(z) = \frac{k!}{2\pi i} \int_{|\zeta - z_0| = r} \frac{f_n(\zeta)}{(\zeta - z)^{k+1}} d\zeta, \quad |z - z_0| < r \quad \text{for } n \geq 1$$

$$\Rightarrow |f_n^{(k)}(z)| \leq \frac{k!}{2\pi} \frac{1}{(\frac{r}{2})^{k+1}} \int_{|\zeta - z_0| = r} |f_n(\zeta)||d\zeta| \quad \text{on } |z - z_0| \leq \frac{r}{2}$$

$$\Rightarrow \text{For} \quad n \geq m \geq 1,$$

$$\sum_{l=m}^{n} |f_l^{(k)}(z)| \leq \frac{k!}{2\pi} \left(\frac{2}{r}\right)^{k+1} \int_{|\zeta - z_0| = r} \sum_{l=m}^{n} |f_l(\zeta)||d\zeta| \quad \text{on } |z - z_0| \leq \frac{r}{2}$$

$$\leq rk! \left(\frac{2}{r}\right)^{k+1} \sup_{|\zeta - z_0| = r} \sum_{l=m}^{n} |f_l(z)| \quad \text{on } |z - z_0| \leq \frac{r}{2}.$$

The result follows by using Cauchy condition.   □

Example 4. Let $f_n(z)$, $n \geq 1$, be analytic in a simply connected domain Ω and let $f_n(z)$ converge locally uniformly to $f(z)$ in Ω.

(1) Fix any point z_0. Then the primitive sequence $\int_{z_0}^{z} f_n(\zeta)d\zeta$ (see Sec. 4.1) converges locally uniformly to a primitive $\int_{z_0}^{z} f(\zeta)d\zeta$ of $f(z)$ in Ω.
(2) Let $F_n(z)$ be a primitive of $f_n(z)$ in Ω for $n \geq 1$. Suppose $F_n(z_0)$ converges for some point $z_0 \in \Omega$. Then $F_n(z)$ converges locally uniformly in Ω to a function $F(z)$ which is a primitive of $f(z)$.

Note that the result in (1) is still valid in a multiply-connected domain Ω once everything is restricted to a simply connected subdomain D. More precisely, choosing a single-valued branch $F(z)$ on D of the may-be multiple-valued function $\int_{z_0}^{z} f(\zeta)d\zeta$, then there is a branch $F_n(z)$ on D of $\int_{z_0}^{z} f_n(\zeta)d\zeta$, for each $n \geq 1$, so that $F_n(z)$ converges to $F(z)$ locally uniformly in D.

Proof. (1) is obvious.

(2) Fix any point $z \in \Omega$. Let γ be any rectifiable curve joining z_0 to z in Ω. Then the function

$$F(z) = F(z_0) + \int_{\gamma} f(\zeta)d\zeta$$

is well-defined and is a primitive of $f(z)$ (see (4.1.2)), where $F(z_0) = \lim_{n\to\infty} F_n(z_0)$. Observe that $F_n(z) = F_n(z_0) + \int_{\gamma} f_n(\zeta)d\zeta$ converges uniformly to $F(z)$ in any closed disk containing in Ω. $\qquad\square$

Remark (The local uniform convergence in terms of function space). Let Ω be a domain in $\mathbf{C}$. There is a sequence of compact sets K_n, $n \geq 1$, satisfying:

1. $K_n \subseteq \Omega$ and any compact subset K of Ω is contained in one of these K_n;
2. $K_n \subseteq K_{n+1}$, $n \geq 1$, and
3. $\Omega = \bigcup_{n=1}^{\infty} K_n$.

Such a sequence K_n is called an *exhaustion* of Ω. For instance, the sequence

$$K_n = \left\{ z \in \Omega \,\|\, |z| \leq n \quad \text{and disk}\,(z, \partial\Omega) \geq \frac{1}{n} \right\}, \quad n \geq 1$$

is such an exhaustion, and, even more, satisfying:

4. $K_n \subseteq \operatorname{Int} K_{n+1}$, $n \geq 1$, and
5. each component of $\mathbf{C} - K_n$ is contained in a component of $\mathbf{C} - \Omega$; in particular, each $\mathbf{C} - K_n$ is a connected set for $n \geq 1$ if $\mathbf{C} - \Omega$ does not have bounded component.

Then, we figure out the following

Local uniform convergence in terms of function space. Let Ω be a domain (or an open set) in $\mathbf{C}$ and (X, d) a metric space (see Exercises B of

Secs. 1.8 and 1.9). Consider the function space

$$F(\Omega; X) = \{\text{function } f : \Omega \to X\}.$$

(1) For $f, g \in F(\Omega; X)$, define

$$\rho(f, g) = \sum_{n=1}^{\infty} \frac{1}{2^n} \sup_{z \in k_n} \frac{d(f(z), g(z))}{1 + d(f(z), g(z))}.$$

Then $(F(\Omega; X), \rho(,))$ is a metric space and is complete if (X, d) is.

(2) For a sequence f_n in $F(\Omega; X)$ and $f \in F(\Omega; X)$, then

$$\rho(f_n, f) \to 0 \text{ (i.e., } converges \ in \ F(\Omega; X) \text{ according to } \rho(,))$$

$\Leftrightarrow$ f_n converges to f *locally uniformly* in Ω (in the sense of (5.3.1)).

In case $X = \mathbf{C}$ and d the usual Euclidean metric $\| \|$ (or any normed space $(X, \| \|)$), the metric $\rho(,)$ also has the

(3) *parallel invariance*: For $f, g, h \in F'(\Omega; X)$

$$\rho(f - h, g - h) = \rho(f, g). \tag{5.3.1.3}$$

Proof is left as Exercise B(3). In case $F(\Omega; \mathbf{C})$ is the vector space consisting of the *analytic functions* defined on Ω, then for *any* metric $\rho(,)$ possessing the property (3), the *differential operator* $D : F(\Omega; \mathbf{C}) \to F(\Omega; \mathbf{C})$ defined by

$$D(f) = f'$$

is always continuous. This claim is an easy consequence of (1)2 in (5.3.1.1).   □

Exercises A

(1) Suppose $f_n(z)$ is analytic in $|z - z_0| < r$ and continuous on $|z - z_0| \le r$ for each $n \ge 1$. Let $\varphi(z)$ be continuous on $|z - z_0| = r$ such that $\lim_{n \to \infty} \int_{|\zeta - z_0| = r} |f_n(\zeta) - \varphi(\zeta)||d\zeta| = 0$ holds (this is the case if $f_n(\zeta)$ converges to $\varphi(\zeta)$ uniformly on $|\zeta - z_0| = r$). Show that $f_n(z)$ converges to the analytic function

$$f(z) = \frac{1}{2\pi i} \int_{|\zeta - z_0| = r} \frac{\varphi(\zeta)}{\zeta - z} d\zeta$$

locally uniformly in $|z - z_0| < r$.

(2) Given the series

$$\sin z + \sum_{n=2}^{\infty} \left(\frac{\sin nz}{n} - \frac{\sin(n-1)z}{(n-1)} \right).$$

(a) Show that this series converges to 0 pointwise in $\mathbf{C}$, and uniformly on the real axis.

(b) Differentiating the series termwise and replacing z by the real variable x, show that the resulted series $\cos x + \sum_{n=2}^{\infty} (\cos nx - \cos(n-1)x)$ diverges at $x \neq 2k\pi$ ($k = 0, \pm 1, \pm 2, \ldots$) and converges to 1 at $x = 2k\pi$.

(c) Show that the original series does no converge uniformly in any open set containing points lying on the real axis.

(3) Show that the series $\sum_{n=1}^{\infty} \frac{\cos nz}{n^2}$ converges only on the real axis and uniformly there. Hence, there is no domain in $\mathbf{C}$ on which its sum function will be analytic.

(4) Show that the series $\sum_{n=1}^{\infty} \frac{\sin n^2 z}{n^2}$ converges uniformly on the real axis and diverges elsewhere. In case $z = x$ is real, is it possible to differentiate the series termwise to obtain a convergent series $\sum_{1}^{\infty} \cos n^2 x$?

(5) Show that

$$\sum_{n=1}^{\infty} \frac{z^n}{(1-z^n)(1-z^{n+1})} = \begin{cases} \dfrac{z}{(1-z)^2}, & |z| < 1 \\[2mm] \dfrac{1}{(1-z)^2}, & |z| > 1 \end{cases} \qquad \text{locally uniformly.}$$

(6) Show that $\sum_{n=0}^{\infty} \frac{\sin nz}{3^n}$ converges to an analytic function $f(z)$ locally uniformly in the strip $|\operatorname{Im} z| < \log 3$. Calculate $f'(0)$.

(7) Show that the series

$$\sum_{n=1}^{\infty} \frac{1}{q^n z + q^{-n} z^{-1} - 2}, \qquad 0 < q < 1$$

converges locally uniformly in the ring domain $q < |z| < q^{-1}$ and diverges elsewhere.

(8) Show that $\sec \pi z$ can be expanded, in the half-plane $\operatorname{Im} z > 0$, as

$$\sec \pi z = 2 \sum_{n=0}^{\infty} (-1)^n e^{(2n+1)\pi i z}$$

and that the series in the right converges to $\sec \pi z$ absolutely and locally uniformly in $\operatorname{Im} z > 0$.

(9) Show that the series

$$\sum_{n=0}^{\infty} \left(\frac{1-z}{1+z} \right)^n$$

converges absolutely and locally uniformly in $\operatorname{Re} z > 0$ and diverges
everywhere in $\operatorname{Re} z \le 0$. Sum the series.

(10) Let a be a nonzero constant. Show that $\sum_{n=-\infty}^{\infty} a^n e^{-|n|z}$ con-
verges absolutely and locally uniformly in $\operatorname{Re} z > |\log|a||$. Sum the
series.

(11) Show that the series $\sum_{n=-\infty}^{\infty} (n^2 + 1)^n e^{n^2 z}$ converges absolutely and
locally uniformly in $\operatorname{Re} z < 0$ but diverges elsewhere.

(12) Show that $\sum_{n=0}^{\infty}(1 - \cos \frac{z}{n})$ defines an entire function.

(13) Show that $\sum_{n=0}^{\infty} \frac{1}{n!} \cos nz$ defines an entire function $f(z)$. Find this
$f(z)$ explicitly.

(14) Show that $\sum_{n=1}^{\infty} (z^{2^n} - z^{-2^n})^{-1}$ converges locally uniformly in $\mathbf{C} - \{|z| = 1\}$. Sum the series.

(15) Show that

$$\frac{1}{1-z} + \sum_{n=1}^{\infty} \frac{z^{2^{n-1}}}{z^{2^n} - 1} = \begin{cases} 1, & |z| < 1 \\ 0, & |z| > 1 \end{cases}$$

locally uniformly in $\mathbf{C} - \{|z| = 1\}$.

(16) Show that the series $\sum_{n=1}^{\infty} \operatorname{Arc} \sin(n^{-2}z)$ converges locally uniformly
and defines an analytic function in $\mathbf{C} - [1, \infty) - (-\infty, -1]$. Then deduce
that the series

$$\sum_{n=1}^{\infty} \frac{1}{\sqrt{n^4 - z^2}}$$

defines an analytic function in $\mathbf{C} - [1, \infty) - (-\infty, -1]$.

(17) Example 2 in Sec. 2.3 showed that the series $\sum_{n=1}^{\infty} \frac{(-1)^{n-1}}{z+n}$ converges
locally uniformly to an analytic function $f(z)$ in $\mathbf{C} - \{-1, -2, -3, \ldots\}$.
Find the Taylor series expansion of $f(z)$ at $z = 0$.

(18) Show that the series

$$\sum_{n=1}^{\infty} \frac{1}{z^2 + n^2}$$

converges absolutely and locally uniformly to an analytic function $f(z)$
in $\mathbf{C} - \{ni | n = \pm 1, \pm 2, \ldots\}$.

(a) For each nonintegral real number $r > 0$, show that

$$\int_{|z|=r} f(z)dz = 0;$$

$$\int_{|z-ki|=r} f(z)dz = \pi \sum_{|k-r|<n<k+r} \frac{1}{n}, \quad k = 1, 2, 3, \ldots.$$

(b) Show that a primitive of $f(z)$ in $\mathbf{C} - \{z | \mathrm{Re}\, z = 0 \text{ and } |z| \geq 1\}$ is given by

$$F(z) = \sum_{n=1}^{\infty} \frac{1}{n} \mathrm{Arc} \tan \frac{z}{n}.$$

(19) Show that the series

$$\frac{1}{2z^2} + \sum_{n=1}^{\infty} \frac{1}{z^2 - n^2}$$

converges locally uniformly to an analytic function $g(z)$ in $\Omega = \mathbf{C} - \{0, \pm 1, \pm 2, \ldots\}$ and then, the function $F(z) = -2zg(z)$ is a primitive of the function

$$f(z) = \sum_{n=-\infty}^{\infty} \frac{1}{(z-n)^2}$$

in Ω.

(20) Suppose $f(z)$ is analytic in $|z| < 1$ and $f'(0) = 0$ holds. Try to use Cauchy's integral formula and Schwarz's Lemma, respectively, to show that the series

$$\sum_{n=1}^{\infty} f(z^n)$$

converges locally uniformly to an analytic function $g(z)$ in $|z| < 1$. If it is also assumed that $|f(z)| \leq 1$ in $|z| < 1$, show that $|g'(0)| \leq 1$ and try to figure out when $|g'(0)| = 1$ will be true.

(21) Given a sequence of polynomials

$$p_n(z) = a_{n,d}z^d + a_{n,d-1}z^{d-1} + \cdots + a_{n,1}z + a_{n,0}, \quad a_{n,j} \in \mathbf{C}$$

$$\text{for } 0 \leq j \leq d \quad \text{and} \quad n \geq 1,$$

whose degrees are not larger than d, a nonnegative integer. Show that

(a) $p_n(z)$ converges uniformly in $\mathbf{C}$;

⇔ (b) there exist $(d + 1)$ *distinct* points $z_0, z_1, z_2, \ldots, z_d$ so that the sequences

$$p_n(z_k), \quad n \geq 1$$

converge for $0 \leq k \leq d$;

⇔ (c) for each k, $0 \leq k \leq d$, the coefficient sequence a_{nk} converges as $n \to \infty$.

In this case, let $a_k = \lim_{n \to \infty} a_{nk}$ for $0 \leq k \leq d$. Then $a_d z^d + a_{d-1} z^{d-1} + \cdots + a_1 z + a_0$ is the limit of $p_n(z)$, $n \geq 1$.

(22) Let $p_n(z)$, $n \geq 1$, be a sequence of polynomials. Show that

(a) $p_n(z)$ converges uniformly in $\mathbf{C}$;

⇔ (b) there is a positive integer N and a sequence α_n so that

$$p_n(z) = p_N(z) + \alpha_n, \quad n > N.$$

In this case, the limit function is $p_N(z) + \lim_{n \to \infty} \alpha_n$. Note that this result is not more true if "uniform convergence" is replaced by "locally uniform convergence". Example 2 provides such an example.

Exercises B

(1) *An extension of Weierstrass's theorem (5.3.1.1).* Let $-\infty < a \leq t < b < +\infty$ and $\{f_t(z)\}_{t \in [a,b)}$ be a family of analytic functions defined on a domain Ω. Suppose $f_t(z)$, $t \in [a, b)$, converges to $f(z)$ *locally uniformly* in Ω as $t \to b^-$. Then

1. $f(z)$ is analytic in Ω, and
2. for each integer $k \geq 1$, $f_t^{(k)}(z)$, $t \in [a, b)$, converges to $f^{(k)}(z)$ locally uniformly in Ω as $t \to b^-$.   (5.3.1.4)

Note. For each compact set K in Ω and $\varepsilon > 0$, if there is a number $\alpha(K, \varepsilon)$, $a \leq \alpha(K, \varepsilon) < b$, so that $|f_t(z) - f'_{t'}(z)| < \varepsilon$ for all t, $t' \geq \alpha(K, \varepsilon)$, and $z \in K$, then $\{f_t(z)\}$ is said to converge *locally uniformly* in Ω as $t \to b^-$. In this case, for each sequence $t_n \in [a, b) \to b$, $f_{t_n}(z)$ converges to a function $f(z)$ locally uniformly in Ω. This $f(z)$ is well-defined in the sense that its definition is independent of the choice of t_n and is called the *limit* of $f_t(z)$ on Ω, denoted as

$$f(z) = \lim_{t \to b^-} f_t(z), \quad z \in \Omega.$$

(2) *Improper integral of Cauchy type* (see Sec. 4.7), an application of Exercise (1). Let γ be a locally rectifiable curve in $\mathbf{C}$, unbounded at one or both of its end points (see (2.4.3)). Suppose $\varphi(z)$ is continuous (or Lebesgue integrable) along γ so that $\int_\gamma |\varphi(z)||dz| < \infty$ (namely, the improper integral is absolutely convergent). Then

1. the function defined by

$$f(z) = \frac{1}{2\pi i} \int_\gamma \frac{\varphi(\zeta)}{\zeta - z} d\zeta, \quad \zeta \in \mathbf{C} - \gamma$$

is analytic on each component of $\mathbf{C} - \gamma$, and

2. for each integer $k \geq 1$,

$$f^{(k)}(z) = \frac{k!}{2\pi i} \int_\gamma \frac{\varphi(\zeta)}{(\zeta - z)^{k+1}} d\zeta, \quad z \in \mathbf{C} - \gamma. \tag{5.3.1.5}$$

Also, refer to Exercises B of Sec. 5.3.3 for an extension of (4.7.4) in Chap. 4.

(3) Prove (5.3.1.3) in detail.

In what follows, we try to introduce various types of convergence for sequence of complex-valued functions defined on a general metric space, with (5.3.1) as a special case. Let (X, d) be a metric space. Let $f_n : X \to \mathbf{C}$, $n \geq 1$, be a sequence of functions and $f : X \to \mathbf{C}$ be a function. Five types of convergence are defined as follows.

1. If $\|f_n - f\|_\infty = \sup_{x \in X} |f_n(x) - f(x)| \to 0$ as $n \to \infty$, f_n is said to *converge uniformly* to f in X.

2. For each $x \in X$, if there is an open neighborhood O_x on which f_n converges to f uniformly, then f_n is said to *converge* to f *locally uniformly* in X.

3. If f_n converges to f uniformly on each compact subset of X, then f_n is said to *converge* to f *compactly* in X.

4. For each $x \in X$, if there is an open neighborhood O_x such that

$$\sum_{n=1}^\infty \|f_n\|_{O_x} < \infty, \quad \text{where } \|f_n\|_{O_x} = \sup_{y \in O_x} |f_n(y)| \text{ for } n \geq 1,$$

then the series $\sum f_n$ is said to *converge* to f *normally* in X.

Note. In case $X = \mathbf{C}$, think about the Weierstrass M-test (see (2.3.5)).

5. If for each convergent sequence x_n in X, the sequences $f_n(x_n)$ always converges in $\mathbf{C}$, then f_n is said to *converge continuously* in X. $\tag{5.3.1.6}$

Try to do the following problems.

(4) In a general metric space (X, d), show that

$$\text{unif. convergence} \Rightarrow \text{local uniform} \Rightarrow \text{compact}$$

	convergence	convergence
	$\Uparrow$ Exercise (7)	$\Uparrow\Downarrow$ Exercise (5)(d) (if the limit function is known to be continuous)
	(series) normal	continuous convergence
	convergence	

In case (X, d) is a *locally compact* metric space (namely, each point has an open neighborhood whose closure is compact), show that

1. compact convergence $\Rightarrow$ local uniform convergence, and
2. if $\sum_{n=1}^{\infty} \|f_n\|_K < \infty$ for each compact set K in X, then $\sum_1^{\infty} f_n$ converges normally.

(5) Suppose f_n is continuously convergent in X.

   (a) Show that f_n converges *pointwise* to a *limit function* f such that $\lim_{n \to \infty} f_n(x_n) = f(x)$ for each sequence $x_n \in X \to x \in X$, namely, $f_n \to f$ *continuously* in X.
   (b) Suppose f_n converges to f continuously in X and f_{n_k}, $k \geq 1$, is any subsequence of f_n. Then, for each $x_k \to x$ as $k \to \infty$, $\lim_{k \to \infty} f_{n_k}(x_k) = f(x)$ holds, namely, $f_{n_k} \to f$ continuously in X.
   (c) If $f_n \to f$ continuously in X, show that f is continuous in X (even though f_n may not be continuous).
   (d) Show that compact convergence $\Leftrightarrow$ continuous convergence. In the case of necessity, it is supposed that the limit function is known to be continuous.
   (e) Suppose that a sequence of continuous functions f_n converges to a continuous function f compactly in a domain Ω in $\mathbf{C}$ and another sequence of continuous functions g_n converges to a continuous function g compactly in a domain D in $\mathbf{C}$. In case $f_n(\Omega) \subseteq D$, $n \geq 1$, and $f(\Omega) \subseteq D$ holds, show that $g_n \circ f_n \to g \circ f$ compactly in Ω.

(6) Let

$$f_n(z) = e^{i\theta_n} \frac{z - z_n}{1 - \bar{z}_n z}, \quad |z_n| < 1, \quad 0 \leq \theta_n \leq 2\pi \quad \text{for } n \geq 1.$$

Show that

(a) f_n converges compactly in $|z| < 1$;

⇔ (b) there are three *distinct* points α, β, and γ in $|z| < 1$ so that the sequences $\{f_n(\alpha)\}$, $\{f_n(\beta)\}$, and $\{f_n(\gamma)\}$ all converge;

⇔ (c) the sequences θ_n, $n > 1$, and z_n, $n \geq 1$, all converge.

In this case, the limit function is either a constant or a bilinear fractional transformation.

(7) Given a series $\sum_{n=1}^{\infty} f_n(x)$. It is assumed that the series is normally convergent to $f(x)$ in X in (a)–(d) below.

(a) For any subsequence n_k, $k \geq 1$, of positive integers, show that $\sum_{k=1}^{\infty} f_{n_k}(x)$ converges normally in X.

(b) (*Rearrangement theorem*) For any permutation $\tau : \mathbf{N}$ (set of positive integers) $\to \mathbf{N}$ (namely, τ is a one-to-one and onto mapping), $\sum_{n=1}^{\infty} f_{\tau(n)}(x)$ converges normally to $f(x)$ in X.

(c) (*Regrouping theorem*) Let $\mathbf{N} = \bigcup_{k \geq 0} \mathbf{N}_k$ be a union of nonempty, disjoint subsets $\mathbf{N}_k$ of $\mathbf{N}$. Then the series $\sum_{n \in \mathbf{N}_k} f_n(x)$ converges normally to a function $g_k(x) : X \to \mathbf{C}$ for each $k \geq 0$, and the series $\sum_{k=0}^{\infty} g_k(x)$ converges normally to $f(x)$ in X.

(d) (*Product theorem*) Suppose $\sum_{m=1}^{\infty} g_m(x)$ converges normally to $g(x)$ in X. Try to rearrange the product functions $f_n(x)g_m(x)$, $1 \leq n$, $m < \infty$, into a new sequence $h_1(x), h_2(x), \ldots, h_l(x), \ldots$. Then $\sum_{l=1}^{\infty} h_l(x)$ converges to $f(x)g(x)$ normally in X.

(e) (*Bracket theorem*) Suppose $\sum_{n=1}^{\infty} f_n(x)$ converges to $f(x)$ locally uniformly in X. Given any fixed point $x \in X$. Show that there is an open neighborhood O_x of x, satisfying: there is a strictly increasing sequence $1 = n_1 < n_2 < \cdots < n_k < \cdots$ of positive integers such that the series $\sum_{k=1}^{\infty} F_k(x)$, where $F_k(x) = f_{n_k}(x) + f_{n_k+1}(x) + \cdots + f_{n_{k+1}-1}(x)$ for $k \geq 1$, converges to $f(x)$ and $\sum_{k=1}^{\infty} \|F_k\|_{O_x} < \infty$.

(f) Show that $\sum_{n=1}^{\infty} (-1)^{n+1} \frac{1}{z+n}$ converges locally uniformly but not normally in $\mathbf{C} - \{-1, -2, -3, \ldots\}$. Refer to Example 2 in Sec. 2.3.

(g) Show that $\sum_{n=1}^{\infty} \frac{z^{2n}}{1-z^n}$ converges locally uniformly and normally in $|z| < 1$.

(8) Suppose $f_n(z)$ and $g_n(z)$, $n \geq 1$, are analytic functions in a domain Ω in $\mathbf{C}$. Also, suppose $\sum_{1}^{\infty} f_n(z) = f(z)$ and $\sum_{1}^{\infty} g_n(z) = g(z)$ normally

in Ω. Note that $f(z)$ and $g(z)$ are analytic in Ω.

(a) (*Weierstrass differentiation theorem*) For each integer $k \geq 1$, $\sum_1^\infty f_n^{(k)}(z) = f^{(k)}(z)$ normally in Ω. Refer to Example 3.

(b) (*Product theorem*) Let $h_l(z)$, $l \geq 1$, be as in Exercise (7)(d). Then $\sum_{l=1}^\infty h_l(z) = f(z)g(z)$ normally in Ω. In particular, if $h_l(z) = \sum_{m+n=l} f_m(z)g_n(z)$ is the *Cauchy product* for $l \geq 1$, then $\sum_{l=1}^\infty h_l(z) = f(z)g(z)$ normally in Ω.

  Note. This result is not necessarily true for local uniform convergence. For instance, $f_n(z) = g_n(z) = \frac{(-1)^n}{\sqrt{n+1}}$ is such an example.

(c) Suppose $f_n(z)$ has a primitive $F_n(z)$ in Ω for $n \geq 1$. In case there is a point $z_0 \in \Omega$ such that $\sum_{n=1}^\infty |F_n(z_0)| < \infty$ holds, then $\sum_{n=1}^\infty F_n(z)$ converges normally in Ω.

(d) If $\sum_0^\infty a_n(z - z_0)^n$ has a positive radius r of convergence, then the series converges normally in $|z - z_0| < r$ (refer (3.3.2.3)).

(9) Let $f(z)$ be analytic at 0 so that the series $\sum_{n=1}^\infty f^{(n)}(z)$ converges absolutely at $z = 0$. Show that $f(z)$ is an entire function, and the series converges normally in $\mathbb{C}$.

The concepts of local uniform and normal convergence introduced in (5.3.1.6) can be extended to sequence or series whose terms are meromorphic functions. Let $f_n(z)$, $n \geq 1$, be meromorphic in a domain Ω of the complex plane.

6. Suppose, for each compact subset K of Ω, there is a positive integer $N(K)$ such that

   a. the poles of $f_n(z)$ lie in $\Omega - K$ for all $n \geq N(K)$, and
   b. the series $\sum_{n=N(K)}^\infty f_n(z)$ converges uniformly in K.

   Then, $\sum_1^\infty f_n(z)$ is said to converge *locally uniformly* in Ω.

7. If condition b in 6 is replaced by

   b'. The series $\sum_{n=N(K)}^\infty \|f_n\|_{\infty,K} < \infty$, where $\|f_n\|_{\infty,K} = \sup_{z \in K} |f_n(z)|$, $n \geq 1$, while reserving condition a, then $\sum_1^\infty f_n(z)$ is said to converge *normally* in Ω.   (5.3.1.7)

Observe that, in both cases, the set of poles of $f_n(z)$, $n \geq 1$, forms a discrete closed subset of Ω. This follows easily by condition a and the exhaustion of Ω by compact subsets (see the Remark before (5.3.1.3)). As before, normal convergence implies local uniform convergence but not conversely. See Exercise (7)(f). A series of the type

$$\sum_{n=-\infty}^\infty f_n(z) = \sum_{n=-\infty}^{-1} f_n(z) + \sum_0^\infty f_n(z), \qquad (5.3.1.8)$$

is said to converge *locally uniformly* or *normally* if both $\sum_{-\infty}^{-1} f_n(z)$
and $\sum_{0}^{\infty} f_n(z)$ do, respectively.

Try to do the following problems.

(10) Show that the following series

$$\sum_{n=1}^{\infty} \left(\frac{1}{z+n} - \frac{1}{n} \right); \quad \sum_{n=1}^{\infty} \left(\frac{1}{z-n} + \frac{1}{n} \right); \quad \sum_{n=0}^{\infty} \frac{1}{(z+n)^k} \ (k \geq 2);$$

$$\sum_{n=0}^{\infty} \frac{1}{(z-n)^k} (k \geq 2)$$

all converge normally in **C**. Hence, the combined series

$$\sum_{\substack{n=-\infty \\ n \neq 0}}^{\infty} \left(\frac{1}{z+n} - \frac{1}{n} \right) = \sum_{n=1}^{\infty} \frac{2z}{z^2 - n^2} \quad \text{and} \quad \sum_{n=-\infty}^{\infty} \frac{1}{(z+n)^k} \ (k \geq 2)$$

converge normally in **C**, too.

(11) Show that the series

$$\sum_{n=0}^{\infty} \left(\frac{2^n}{z - 2^n} + 1 \right) \quad \text{and} \quad \sum_{n=1}^{\infty} \left[\frac{n^2}{(z-n)^2} - 1 - \frac{2z}{n} \right]$$

converge normally in **C**.

(12) Suppose $f_n(z)$ and $g_n(z)$, $n \geq 1$, are meromorphic functions in a domain Ω.

(a) (*Convergence theorem*) Suppose $\sum f_n(z)$ converges locally uniformly (or normally) in Ω. Then there exists a unique meromorphic function $f(z)$ in Ω satisfying: For any open set O in Ω, if N is a positive integer such that $f_n(z)$ does not have poles in O for $n \geq N$, then the series $\sum_{n=N}^{\infty} f_n|_O$ of analytic terms converges locally uniformly (or normally) to an analytic function $F(z)$ in O so that

$$f|_O = \sum_{n=1}^{N-1} f_n|_O + F.$$

In particular, $f(z)$ is analytic in $\Omega - \{\text{poles of } f_n(z) \text{ for } n \geq 1\}$. In this sense we denote $f(z) = \sum f_n(z)$.

(b) Suppose $f(z) = \sum f_n(z)$ and $g(z) = \sum g_n(z)$ locally uniformly (or normally) in Ω. Then

$$\alpha f(z) + \beta g(z) = \sum (\alpha f_n(z) + \beta g_n(z)), \quad \alpha, \beta \in \mathbf{C}$$

locally uniformly (or normally) in Ω.

(c) *(Rearrangement theorem)* Suppose $f(z) = \sum f_n(z)$ normally in Ω and $\tau : N \to N$ is a permutation. Then, so does $\sum f_{\tau(n)}(z) = f(z)$ normally in Ω.

(d) *(Termwise differentiation theorem)* Suppose $f(z) = \sum f_n(z)$ locally uniformly (or normally) in Ω. For each integer $k \geq 1$, $f^{(k)}(z) = \sum f_n^{(k)}(z)$ locally uniformly (or normally) in Ω.

(e) But, the product theorem in Exercise (7)(d) is not necessarily true in this case. This is because the condition a in (5.3.1.7) may be false for product series $\sum h_l(z)$. Yet $\sum h_l(z)$ still converges normally in the set $\Omega - \{\text{poles of } f_n(z) \text{ and } g_n(z), n \geq 1\}$.

5.3.2　Zeros of the limit function: Hurwitz's theorem

Local uniform convergence preserves the number of zeros of analytic functions in the following sense.

Hurwitz's Theorem (1889). *Suppose a sequence $f_n(z)$ of analytic functions in a domain Ω converges locally uniformly to $f(z)$ in Ω. Then*

(1) *Let O be any bounded open set in Ω so that $\bar{O} \subseteq \Omega$ holds and $f(z) \neq 0$ on the boundary ∂O. Then there is a positive integer n_0 so that $f_n(z)$ and $f(z)$ have the same number of zeros in O for each $n \geq n_0$.*

$\Leftrightarrow$ (2) *In case $f(z)$ is not a constant, then for each point $z_0 \in \Omega$, there are a positive integer n_0 and a sequence $z_n \in \Omega$, $n \geq 1$, such that*

1. $\lim_{n \to \infty} z_n = z_0$, and
2. $f_n(z_n) = f(z_0)$ for $n \geq n_0$. 　　　　　　　　(5.3.2.1)

Three proofs will be given below. The first two are basic and direct; while in the third one, we will show that (2) does hold by itself and then try to show that (2) $\Leftrightarrow$ (1) holds. See Exercise B(1) for an extension of (5.3.2.1).

Proof 1 (Use the Rouché's theorem, see Sec. 3.5.4). First, suppose O is an open disk. Now $\min_{z \in \partial O} |f(z)| = \varepsilon > 0$ since $f(z) \neq 0$ on ∂O. There is a positive integer n_0 so that

$$|f_n(z) - f(z)| < \varepsilon, \quad z \in \partial O \text{ for } n \geq n_0$$

$$\Rightarrow |f_n(z) - f(z)| < |f(z)| \text{ on } \partial O \text{ for } n \geq n_0.$$

Rouché's theorem says that $f_n(z)$, $n \geq n_0$, and $f(z)$ have the same number of zeros in O.

For a general bounded open set O in Ω with $\overline{O} \subseteq \Omega$, $f(z)$ has only finitely many zeros in O because $f(z) \neq 0$ on ∂O. So there are finitely many disjoint open disks $O_1, \ldots, O_n$, all contained in O, such that $f(z) \neq 0$ on the compact set $A = \overline{O} - \bigcup_{k=1}^{n} O_k$. Since $f_n(z)$ converges uniformly to $f(z)$ in A, there is a positive integer n_1 so that

$$\|f_n(z) - f(z)\| < \min_{z \in A} |f(z)|(> 0), \quad z \in A \text{ for } n \geq n_1.$$

This shows that $f_n(z) \neq 0$ on A for $n \geq n_1$. Apply the known result in the last paragraph to each O_k, $1 \leq k \leq n$, and the required result follows.   □

Proof 2 (Based on the counting formula for zeros, see (3.5.3.3), and Weierstrass's theorem, see (5.3.1.1)). Let O be an open disk for the moment. Since $\min_{z \in \partial O} |f(z)| > 0$, so $\frac{1}{f_n(z)}$ converges to $\frac{1}{f(z)}$ uniformly on ∂O. Also, $f_n'(z)$ converges to $f'(z)$ uniformly on ∂O by Weierstrass's theorem.

Hence,

$$\lim_{n \to \infty} \frac{1}{2\pi i} \int_{\partial O} \frac{f_n'(z)}{f_n(z)} dz = \frac{1}{2\pi i} \int_{\partial O} \frac{f'(z)}{f(z)} dz$$

holds. Recall that

$$\frac{1}{2\pi i} \int_{\partial O} \frac{f_n'(z)}{f_n(z)} dz = \text{the number } m_n \text{ of zeros of } f_n(z) \text{ in } O;$$

$$\frac{1}{2\pi i} \int_{\partial O} \frac{f'(z)}{f(z)} dz = \text{ the number } m \text{ of zeros of } f(z) \text{ in } O.$$

Since m_n, $n \geq 1$, and m are nonnegative integers and $\lim_{n \to \infty} m_n = m$, it follows that $m_n = m$ for $n \geq n_0$, where n_0 is a suitable integer.   □

For general bounded open set O, see the later part of Proof 1.

Proof 3 (To prove the validity of (2) in (5.3.2.1): Need to use the modulus minimum principle, (see (2) in (5.1.2.6) or (3.4.4.2))). Using $f_n(z) - f(z_0)$ to replace $f_n(z)$, we may suppose that $f(z_0) = 0$ in advance. Since $f(z) \not\equiv 0$, the isolatedness of zeros of $f(z)$ shows that there is a $\rho > 0$ so that $|z - z_0| \leq \rho$ is contained in O and $f(z) \neq 0$ in $0 < |z - z_0| \leq \rho$. Since $f_n(z)$ converges to $f(z)$ uniformly in the set $\{|z - z_0| = \rho\} \cup \{z_0\}$, there is an $n_0 \geq 1$ so that

$$|f_n(z) - f(z)| < \frac{\varepsilon}{2} \text{ on } |z - z_0| = \rho \quad \text{and} \quad z = z_0 \text{ for } n \geq n_0,$$

where $\varepsilon = \min_{|z-z_0|=\rho} |f(z)| > 0$. In particular,

$$|f_n(z_0)| < \frac{\varepsilon}{2} = \varepsilon - \frac{\varepsilon}{2} = \min_{|z-z_0|=\rho} |f(z)| - \frac{\varepsilon}{2}$$

$$\leq |f(z)| - \frac{\varepsilon}{2}, \quad \text{if } |z - z_0| = \rho$$

$$< |f_n(z)|, \quad \text{if } |z - z_0| = \rho \quad \text{for } n \geq n_0$$

$$\Rightarrow |f_n(z_0)| < \min_{|z-z_0|=\rho} |f_n(z)|, \quad n \geq n_0.$$

By the minimum principle, $f_n(z)$ has at least one zero z_n in the disk $|z - z_0| < \rho$ for $n \geq n_0$, namely $f_n(z_n) = f(z_0) = 0$ for $n \geq n_0$. It comes that $\lim_{n\to\infty} z_n = z_0$ should hold. Otherwise, a subsequence z_{n_k} of z_n will converge to a point z_0^* satisfying $0 < |z^* - z_0| \leq \rho$. In this case, $\lim_{k\to\infty} f_{n_k}(z_{n_k}) = f(z_0^*) = 0$, contradicting to the definition of ρ. This finishes the proof of (2) in (5.3.2.1). $\qquad\square$

To prove (2) $\Rightarrow$ (1): Based on induction and the basic factorization formula (5.1.2.3) or (3.4.2.9) of analytic functions. Let O be an open set in Ω. Then, the *number m* (counting multiplicities) of zeros of $f(z)$ on O is finite because, otherwise, $f(z) \equiv 0$ on $\overline{\Omega}$ and this will contradict to the assumption that $f(z) \neq 0$ along ∂O.

In case $m = 0$: Let $\varepsilon = \min_{z \in \overline{O}} |f(z)|$. Then $\varepsilon > 0$ and there is an $n_0 \geq 1$ such that

$$|f_n(z) - f(z)| < \varepsilon \quad \text{for all } z \in \overline{O} \text{ and } n \geq n_0.$$

This implies that $f_n(z) \neq 0$ on $\overline{O}$ for all $n \geq n_0$.

Assume the result holds for $m - 1$, where $m \geq 1$.

Let $z_0 \in O$ be a zero of $f(z)$. According to (2) in (5.3.2.1), there is an integer $n_1 \geq 1$ so that $f_n(z)$ has a zero z_n in O for each $n \geq n_1$ and $\lim_{n\to\infty} z_n = z_0$ holds. For each of such $n \geq n_1$, there are analytic functions $g_n(z)$ and $g(z)$ on Ω such that

$$f_n(z) = (z - z_n)g_n(z), \quad n \geq n_1, \quad \text{and}$$

$$f(z) = (z - z_0)g(z) \qquad\qquad\qquad (*)$$

hold throughout Ω. Since $z - z_n$ converges to $z - z_0$ uniformly in Ω, by assumption, $g_n(z)$ converges to $g(z)$ locally uniformly in Ω. $(*)$ says that $g(z) \neq 0$ along ∂O and $g(z)$ has $(m-1)$ zeros in O. By inductive assumption, there is an integer $n_0 \geq n_1$ so that $g_n(z)$ has exactly $(m - 1)$ zeros in O for $n \geq n_0$. It follows that $f_n(z)$ has exactly m zeros in O for $n \geq n_0$.

To prove (1) ⇒ (2). We may suppose $f(z_0) = 0$ when $f(z)$ is replaced by $f(z) - f(z_0)$. All one needs to do is to show that, *if $f(z)$ is not identically equal to zero, then each zero of $f(z)$ is the limit point of a sequence z_n,* where $f_n(z_n) = 0$ for each $n \geq 1$.

Let z_0 be a zero of $f(z)$. Choose $\rho > 0$ so that $|z - z_0| \leq \rho$ is contained in Ω and $f(z) \neq 0$ in $0 < |z - z_0| \leq \rho$. Let $\rho_k = 2^{-k}\rho$ for $k \geq 0$. According to (1) in (5.3.2.1), there is an integer n_k so that $f_n(z)$ has a zero $z_{n,k}$ in $|z - z_0| < \rho_k$ for each $n \geq n_k$ and $k \geq 0$. May choose $n_{k+1} > n_k$ for $k \geq 0$. Define a sequence z_n in Ω by

$$z_n = z_{n,k}, \quad n_k < n \leq n_{k+1}, \quad k \geq 0.$$

Then $z_n \to z_0$ and $f_n(z_n) = 0 \to 0 = f(z_0)$. This finishes the proof. □

Even though $f_n(z)$ is univalent in Ω for each $n \geq 1$, the limit function $f(z)$ in (5.3.2.1) could be identically equal to zero. For instance,

$$f_n(z) = \frac{z}{n}, \quad z \in \Omega = \mathbf{C} - \{0\}$$

is univalent in Ω and $f_n(z) \neq 0$ everywhere in Ω, yet $f_n(z)$ converges to the zero function locally uniformly in Ω. While $g_n(z) = \frac{e^z}{n} \neq 0$ in $\mathbf{C}$ for $n \geq 1$, and $g_n(z)$ converges to 0 locally uniformly in $\mathbf{C}$. Anyway, we have the following

Two special cases of Hurwitz's theorem. Suppose $f_n(z) : \Omega \to \mathbf{C}$ is analytic for $n \geq 1$ and the sequence $f_n(z)$ converges to $f(z)$ locally uniformly in Ω.

(1) If $f_n(z)$ does not have zeros in Ω for each $n \geq 1$, then the limit function $f(z)$ is either identically equal to zero in Ω or does not have any zero in Ω.

(2) In case each $f_n(z)$ is univalent in Ω, then $f(z)$ is either a constant or is univalent in Ω.   (5.3.2.2)

(1) is obvious. As for (2), fix any point $z_0 \in \Omega$. Since $f_n(z) - f_n(z_0)$ converges to $f(z) - f(z_0)$ locally uniformly in $\Omega - \{z_0\}$, by (1), $f(z) - f(z_0)$ does not have zero in $\Omega - \{z_0\}$ in case $f(z)$ is not a constant. Hence, $f(z) \neq f(z_0)$ for all $z \in \Omega$ and $z \neq z_0$. This *proves* the claim.

We illustrate three examples.

Example 1. Example 2 in Sec. 5.3.1 showed that $e^z = \lim_{n \to \infty}(1 + \frac{z}{n})^n$ locally uniformly in $\mathbf{C}$. Try to use this fact to show that $e^z \neq 0$ in $\mathbf{C}$.

Solution. Suppose there exists a point $z_0 \in \mathbf{C}$ so that $e^{z_0} = 0$. Since $f(z) = e^z$ is not a constant, there is an $R > 0$ so that $e^z \neq 0$ on $0 < |z - z_0| \leq R$.

By (5.3.2.1), there is an integer $n_0 \geq 1$ so that $f_n(z) = (1 + \frac{z}{n})^n$ has the same number of zero as $f(z)$ does in $|z - z_0| < R$ for each $n \geq n_0$. Yet, for all sufficiently large n such that $|-n - z_0| = |n + z_0| > R$, $f_n(z) \neq 0$ in $|z - z_0| < R$, a contradiction. Hence, $e^z \neq 0$ for all $z \in \mathbf{C}$.

Example 2. Let $a_n > a_{n+1} > 0$ for $n \geq 0$. Show that

$$f(z) = \sum_0^\infty a_n z^n$$

is analytic in $|z| < 1$ and $f(z) \neq 0$ there.

Solution. $\lim_{n \to \infty} a_n = \rho \geq 0$ exists. In case $\rho > 0$, then $\lim_{n \to \infty} \frac{a_n}{a_{n+1}} = 1$ shows that the series has its radius of convergence equal to 1; in case $\rho = 0$, then $\overline{\lim}_{n \to \infty} a_n^{\frac{1}{n}} \leq 1$ and the radius of convergence is not less than 1. Hence, $\sum_0^\infty a_n z^n$ is analytic in $|z| < 1$.

According to Eneström–Kakeya theorem (2.5.1.8), $S_n(z) = \sum_{k=0}^n u_k z^k \neq 0$ on $|z| < 1$ for each $n \geq 1$. Since $S_n(z)$ converges to $f(z)$ locally uniformly in $|z| < 1$ and $f(0) = a_0 > 0$, by (1) in (5.3.2.2), $f(z) \neq 0$ throughout $|z| < 1$.

Example 3. Let $\varphi(t) : [0, 1] \to \mathbf{R}$ be a strictly increasing continuous function. Show that

$$f(z) = \int_0^1 \varphi(t) \cos zt \, dt$$

is an entire function and can only have real zeros.

Solution. Let

$$f_n(z) = \sum_{k=1}^n \frac{1}{n} \varphi\left(\frac{k}{n}\right) \cos \frac{kz}{n}, \quad z \in \mathbf{C} \quad \text{and} \quad n \geq 1.$$

Try to show that $f_n(z)$ converges to $f(z)$ locally uniformly in $\mathbf{C}$. Hence $f(z)$ is entire.

Choose any compact set K in $\mathbf{C}$. The function $F(z, t) = \varphi(t) \cos zt$ is uniformly continuous on the compact set $K \times [0, 1]$. Hence, for a given $\varepsilon > 0$, there is a $\delta > 0$ so that

$$|F(z, t) - F(z, t')| < \varepsilon, \quad z \in K \quad \text{and} \quad t, t' \in [0, 1] \text{ with } |t - t'| < \delta.$$

Choose n large enough so that $\frac{1}{n} < \delta$ (say $n \geq n_0 \geq [\frac{1}{\delta}] + 1$). Then, for $z \in K$,

$$f(z) - f_n(z) = \sum_{k=1}^{n} \int_{(k-1)/n}^{k/n} \varphi(t) \cos ztdt - \sum_{k=1}^{n} \int_{(k-1)/n}^{k/n} \varphi\left(\frac{k}{n}\right) \cos \frac{kz}{n} dt$$

$$= \sum_{k=1}^{n} \int_{(k-1)/n}^{k/n} \left[F(z,t) - F\left(z, \frac{k}{n}\right) \right] dt,$$

$$\Rightarrow |f(z) - f_n(z)| \leq \sum_{k=1}^{n} \int_{(k-1)/n}^{k/n} \left| F(z,t) - F\left(z, \frac{k}{n}\right) \right| dt < \varepsilon \cdot \frac{1}{n} \cdot n = \varepsilon,$$

$$z \in K \quad \text{and} \quad n \geq n_0.$$

This shows that $f_n(z)$ converges to $f(z)$ uniformly in K, and hence, locally uniformly in $\mathbf{C}$.

Since $f(0) = \int_0^1 \varphi(t) dt > 0$, $f(z)$ is not identically equal to zero in $\mathbf{C}$.

In case $f(z_0) = 0$ for same $z_0 \in \mathbf{C}$, by (2) in (5.3.2.1), there is an integer n_0 so that $f_n(z)$ has a zero z_n for $n \geq n_0$ and $\lim_{n\to\infty} z_n = z_0$ holds. Returning to Example 3 in Sec. 3.5.3, wherein let

$$\theta = \frac{z}{n},$$

$$a_k = \frac{1}{n} \varphi\left(\frac{k}{n}\right), \quad k = 1, 2, \ldots, n.$$

Then $0 < a_1 < a_2 < \cdots < a_n$ (and choose any a_0, $0 < a_0 < a_1$ and let $a_0 \to 0$). It follows that $f_n(z) = 0$ has only real roots and hence, the limit equation $f(z) = 0$ can only have real roots (see Exercise A(3)).

Exercises A

(1) Show that all the zeros of $f_n(z) = 1 + \frac{1}{z} + \frac{1}{2!z^2} + \cdots + \frac{1}{n!z^n}$ lie in $|z| < \rho$ for any fixed $\rho > 0$ if n is large enough.

(2) Fix $0 < \rho < 1$. Show that $f_n(z) = 1 + 2z^2 + 3z^2 + \cdots + nz^{n-1}$ does not have zeros in $|z| < \rho$ if n is large enough.

(3) Suppose entire functions $f_n(z)$, $n \geq 1$, have only real zeros and converges locally uniformly to $f(z)$ in $\mathbf{C}$. Show that $f(z)$ can have only real zeros.

Note. Such an example is provided by the Bessel function $J_0(z)$ defined by

$$J_0(z) = \lim_{n \to \infty} P_n\left(1 - \frac{z^2}{n^2}\right), \quad \text{locally uniformly in } \mathbf{C},$$

where

$$P_n(z) = 1 + \sum_{k=1}^{n} (-1)^k \frac{(n+k)(n+k-1)\cdots(n-k+1)}{(k!)^2} \left(\frac{1-z}{2}\right)^k, \quad n \geq 1,$$

whose zeros are all real.

Exercises B

(1) (*Extension of Hurwitz's theorem*) Let Ω be an open set in $\mathbf{C}$ and Λ be a topological space (see Exercise B(2) in Sec. 1.8). Suppose a function

$$f : \Omega \times \Lambda \to \mathbf{C}$$

is continuous such that, for each $\lambda \in \Lambda$, the function $f_\lambda : \Omega \to \mathbf{C}$ defined by $f_\lambda(z) = f(z, \lambda)$, $z \in \Omega$, is analytic.

   (a) Suppose z_0 is a zero of multiplicity $k(1 \leq k < \infty)$ of $f_{\lambda_0}(z)$. Then there is an $\varepsilon_0 > 0$ so that, for each ε, $0 < \varepsilon \leq \varepsilon_0$, there exists an open neighborhood V_ε of λ_0 such that $f_\lambda(z)$ has exactly k zeros in $|z - z_0| < \varepsilon$ for each $\lambda \in V_\varepsilon$.

   (b) Let $\lambda_0 \in \Lambda$ and γ be a rectifiable Jordan closed curve in Ω such that the closure $\overline{\text{Int}\,\gamma} \subseteq \Omega$. Then there is an open neighborhood V of λ_0 so that, for any $\lambda \in V$, $f_\lambda(z)$ and $f_{\lambda_0}(z)$ have the same number of zeros in $\text{Int}\,\gamma$.

(2) (*Inverse function theorem of analytic sequence*, due to Carathéodory (1932)) Suppose $f_n(z)$, $n \geq 1$, are analytic in a neighborhood Ω of z_0 and converges uniformly to an analytic function $f(z)$ on Ω, satisfying $f'(z_0) \neq 0$. Then, there exist an open neighborhood U of z_0 in Ω, an open neighborhood V of $w_0 = f(z_0)$ and a positive integer N satisfying:

   1. $f_n(z)$, $n \geq N$, and $f(z)$ are univalent in U;
   2. $f_n(U)$, $n \geq N$, and $f(U)$ contain V, and
   3. the inverse functions $f_n^{-1}(w)$, $n \geq N$, converge to $f^{-1}(w)$ uniformly in V.

*5.3.3   *Some sufficient criteria for local uniform convergence*

The maximum modulus principle and some inequalities derived from it, such as the three-circles theorem (3.4.4.4). Schwarz's Lemma (3.4.5.1) and

Hadamard–Borel–Carathéodory inequality (see Example 3 in Sec. 3.4.5), etc., can be used to establish sufficient conditions for local uniform convergence of analytic sequence. Note that uniform convergence in the content of real analysis lacks these counterparts.

The simplest one among all is the following easy consequence of the maximum modulus principle.

A Criterion. *Let $f_n(z)$, $n \geq 1$, be analytic in a bounded domain Ω and continuous on $\overline{\Omega}$.*

Then,

(1) $f_n(z)$ (or $\sum_1^\infty f_n(z)$) converges uniformly in $\overline{\Omega}$;

⇔ *(2) $f_n(z)$ (or $\sum_1^\infty f_n(z)$) converges uniformly along the boundary $\partial\Omega$.*

$$(5.3.3.1)$$

In what follows, we review some remarkable criteria for local uniform convergence. Only sketch, but no proofs in most cases, is given. Refer to Refs. [12] and [35] for detailed account.

The following is divided into four subsections.

Section (1) Local uniform approximation by Lagrange's interpolation polynomials

Let $f(z)$ be analytic in $|z - z_0| < r$ and continuous on $|z - z_0| \leq r$. Given n distinct points $z_1, \ldots, z_n$ in $|z - z_0| < r$. Then the polynomial $L_{n-1}(z)$ of degree $n - 1 (n \geq 2)$ having the property $L_{n-1}(z_k) = f(z_k)$, $1 \leq k \leq n$, is uniquely given by

$$L_{n-1}(z) = \sum_{k=1}^n f(z_k) \prod_{\substack{l=1 \\ l \neq k}}^n \frac{(z - z_l)}{(z_k - z_l)}, \quad \text{if } z \in \mathbf{C}$$

$$= \sum_{k=1}^n \frac{f(z_k)}{w_n'(z_k)} \cdot \frac{w_n(z)}{z - z_k}, \quad \text{where } w_n(z) = \prod_{k=1}^n (z - z_k), \quad \text{if } z \in \mathbf{C}$$

$$= \frac{1}{2\pi i} \int_{|\zeta - z_0| = r} \frac{f(\zeta)}{w_n(\zeta)} \cdot \frac{w_n(\zeta) - w_n(z)}{\zeta - z} d\zeta \quad \text{(by residue theorem)},$$

$$\text{if } |z - z_0| < r.$$

$$(*_1)$$

$$\Rightarrow f(z) - L_{n-1}(z) = \frac{w_n(z)}{2\pi i} \int_{|\zeta - z_0| = r} \frac{f(\zeta)}{w_n(\zeta)(\zeta - z)} d\zeta, \quad |z - z_0| < r.$$

$$\Rightarrow |f(z) - L_{n-1}(z)| \leq \max_{|\zeta - z_0| = r} \left| \frac{w_n(z)}{w_n(\zeta)} \right| \cdot \frac{Mr}{r - |z - z_0|}, \quad |z - z_0| < r \quad \text{and}$$

$$M = \max_{|\zeta - z_0| = r} |f(\zeta)|. \tag{*2}$$

Observe that

$$\left| \frac{z - z_k}{\zeta - z_k} \right| \leq \frac{|z - z_0| + |z_k - z_0|}{|\zeta - z_0| - |z_k - z_0|} = \frac{|z - z_0| + |z_k - z_0|}{r - |z_k - z_0|}, \quad 1 \leq k \leq n$$

$$\Rightarrow \overline{\lim_{k \to \infty}} \left| \frac{z - z_k}{\zeta - z_k} \right| \leq \frac{|z - z_0|}{r} < 1, \quad |z - z_0| < r,$$

under the assumption that $\lim_{k \to \infty} z_k = z_0$. *In this case, it follows that the sequence* $L_{n-1}(z)$ *will converge to* $f(z)$ *locally uniformly in* $|z - z_0| < r$ *(see* (*3) *below).*

Now, suppose $f_m(z)$, $m \geq 1$, are analytic in $|z - z_0| < r$ and continuous on $|z - z_0| \leq r$. Also, suppose $f_m(z)$, $m \geq 1$, are *uniformly bounded* in $|z - z_0| \leq r$, namely,

$$\sup_{m \geq 1} \sup_{|z - z_0| \leq r} |f_m(z)| = M < \infty. \tag{5.3.3.2}$$

Let z_k, $|z_k - z_0| < r$, be distinct and $\lim_{k \to \infty} z_k = z_0$ hold. For each fixed m, from (*2), we have on $|z - z_0| < r$

$$|f_m(z) - L_{n-1}^{(m)}(z)| \leq \frac{Mr}{r - |z - z_0|} \prod_{k=1}^{k_0 - 1} \left[\frac{|z - z_0| + |z_k - z_0|}{r - |z_k - z_0|} \right]$$

$$\times \left(\frac{|z - z_0|}{r} \right)^{n - k_0} \quad \text{if } k \geq k_0 > 0 \quad \text{and} \quad n > k_0.$$

$$\tag{*3}$$

$\Rightarrow$ For fixed $m \geq 1$, $0 < \rho < r$, and $\varepsilon > 0$, there is an integer $n_1 = n_1(m, \rho, \varepsilon) > 0$ so that for $n \geq n_1$,

$$|f_m(z) - L_{n-1}^{(m)}(z)| < \frac{\varepsilon}{2}, \quad |z - z_0| < \rho. \tag{*4}$$

On the other hand,

$$L_{n-1}^{(m)}(z) = \sum_{k=1}^{n} \frac{f_m(z_k)}{w_n'(z_k)} \frac{w_n(z)}{z - z_k}, \quad n \geq 2.$$

Under the *additional* assumption that $\lim_{m \to \infty} f_m(z_k) = 0$ for each $k \geq 1$, for fixed $1 \leq k \leq n$ and $\varepsilon > 0$, there is an integer $m_{\varepsilon,k} > 0$ so that for $m \geq m_{\varepsilon,k}$

$$|f_m(z_k)| < \frac{\varepsilon}{L}, \quad \text{where } L = 2 \sum_{k=1}^{n} \frac{r^{n-1}}{|w_n'(z_k)|}, \quad 1 \leq k \leq n$$

$$\Rightarrow |L_{n-1}^{(m)}(z)| < \frac{\varepsilon}{2} \quad \text{if } |z - z_0| \leq \rho \quad \text{and} \quad m \geq \max_{1 \leq k \leq n} m_{\varepsilon,k}. \qquad (*_5)$$

Combining $(*_4)$ and $(*_5)$, for $0 < \rho < r$ and $\varepsilon > 0$, there is an integer $N = N(\rho, \varepsilon) > 0$ so that for $m \geq N$,

$$|f_m(z)| \leq |f_m(z) - L_{n-1}^{(m)}(z)| + |L_{n-1}^{(m)}(z)| < \varepsilon$$

$$\text{if } n \text{ is large enough and } |z - z_0| \leq \rho.$$

This means that $f_m(z) \to 0$ locally uniformly in $|z - z_0| < r$.

In conclusion, we have a

Convergence theorem (I). *Suppose $f_n(z)$ is analytic in $|z - z_0| < r$ and continuous on $|z - z_0| \leq r$ for each $n \geq 1$. Also suppose that,*

1. *$f_n(z)$, $n \geq 1$, are uniformly bounded (see (5.3.3.2)) in $|z - z_0| \leq r$;*
2. *z_k, $k \geq 1$, are distinct points in $|z - z_0| < r$ satisfying $\lim_{k \to \infty} z_k = z_0$, and*
3. *$\lim_{n \to \infty} f_n(z_k) = 0$ for each $k \geq 1$.*

Then $f_n(z)$ converges to 0 locally uniformly in $|z - z_0| < r$. $\qquad$ (5.3.3.3)

Section (2) An application of Hadamard's three-circles theorem (3.4.4.4)

Suppose $f(z)$ is analytic in $|z - z_0| \leq 4R$, where $R > 0$. Then

$$\max_{|z-z_0| \leq 2R} |f(z)| = \max_{|z-z_0|=2R} |f(z)|$$

$$\leq \left[\max_{|z-z_0|=R} |f(z)| \right]^{1-\alpha} \left[\max_{|z-z_0|=4R} |f(z)| \right]^{\alpha},$$

$$\text{where } \alpha = \frac{\log 2R - \log R}{\log 4R - \log R} = \frac{\log 2}{\log 4}. \qquad (*_6)$$

Observe that the constant $\alpha = \frac{\log 2}{\log 4}$ is good for *any* function analytic in $|z - z_0| \leq 4R$ so that $(*_6)$ remains valid.

Let $f(z)$ be analytic in a domain Ω. Fix any closed disk $|z - z_0| \leq 4R$ containing in Ω and choose any z_1 in an open subset O of Ω. Construct a sequence of open disks $D(z_j, R_j)$: $|z - z_j| < R_j$, $1 \leq j \leq n$, with $z_n = z_0$ and $R_n = R$, satisfying

1. $D(z_1, R_1) \subseteq O$;
2. the closed disks $\overline{D}(z_j, 4R_j) \subseteq \Omega$ for $1 \leq j \leq n$, and
3. $D(z_{j+1}, R_{j+1}) \subseteq D(z_j, 2R_j)$ for $1 \leq j \leq n - 1$.

Then, applying $(*_6)$ to $R_1 \leq |z - z_1| \leq 4R_1$, and under the assumption that $\sup_{z \in \Omega} |f(z)| = 1$, we have

$$
\sup_{D(z_2, R_2)} |f(z)| \leq \max_{|z - z_1| = 2R_1} |f(z)| \leq \left[\max_{|z - z_1| = R_1} |f(z)| \right]^{1-\alpha} \left[\max_{|z - z_1| = 4R_1} |f(z)| \right]^{\alpha}
$$

$$
\leq \left[\sup_{D(z_1, R_1)} |f(z)| \right]^{1-\alpha} \leq \sup_{z \in O} |f(z)|^{1-\alpha}
$$

$\Rightarrow$ (repeating the same process to $R_2 \leq |z - z_2| \leq 4R_2$)

$$
\sup_{D(z_3, R_3)} |f(z)| \leq \left[\sup_{|z - z_2| = R_2} |f(z)| \right]^{1-\alpha}
$$

$$
\leq \left[\sup_{D(z_2, R_2)} |f(z)| \right]^{1-\alpha} \leq \left[\sup_{z \in O} |f(z)| \right]^{(1-\alpha)^2}.
$$

$\Rightarrow \cdots$

$$
\sup_{D(z_n, R_n)} |f(z)| = \sup_{D(z_0, R)} |f(z)| \leq \left[\sup_{z \in O} |f(z)| \right]^{(1-\alpha)^{n-1}}, \qquad \alpha = \frac{\log 2}{\log 4}. \quad (*_7)
$$

A standard compactness argument will then guarantee

An inequality. Let Ω be a domain, O a nonempty open subset, and K a compact subset. Then, there is constant $c = c(\Omega, O, K)$, $0 < c < 1$, such that, for *any* function $f(z)$ analytic in Ω,

$$
\sup_{z \in K} |f(z)| \leq \left[\sup_{z \in O} |f(z)| \right]^{1-c} \left[\sup_{z \in \Omega} |f(z)| \right]^{c}. \qquad (5.3.3.4)
$$

This kind of inequalities appears frequently and plays an essential role in the regularity problems in partial differential equations.

As an application of (5.3.3.4), we have a

Convergence theorem (II) (Stieltjes, 1894). *Let $f_n(z)$, $n \geq 1$, be analytic in a domain Ω. Suppose*

1. *$f_n(z)$ converges uniformly in a nonempty open subset O of Ω, and*
2. *$f_n(z)$, $n \geq 1$, are uniformly bounded in each compact subset of Ω.*

Then $f_n(z)$ converges locally uniformly in Ω. (5.3.3.5)

Proof. To prove this, choose any compact set K in Ω and a closed disk $\overline{D}$ in O. Construct a bounded domain U so that $\overline{D} \cup K \subseteq \overline{U}$ (compact) $\subseteq \Omega$. Apply (5.3.3.3) to U with $c = c(U, D, K)$.
Then

$$\sup_{z \in K} |f_m(z) - f_n(z)| \leq \left[\sup_{z \in D} |f_m(z) - f_n(z)| \right]^{1-c} \left[\sup_{z \in U} |f_m(z) - f_n(z)| \right]^c$$

says everything needed.

By the way, via the concept of subharmonic function (refer to Sec. 6.4, if needed), one can prove

A generalized maximum principle (A. Ostrowski (1922/23)). Let Γ_1 and Γ_2 be two nondegenerated complimentary arcs on $|z| = 1$ (namely, $\Gamma_1 \neq \phi$, $\Gamma_2 \neq \phi$, $\Gamma_1 \cap \Gamma_2 = \phi$, and $\Gamma_1 \cup \Gamma_2$ is the circle $|z| = 1$). Let $f(z)$ be an analytic function in $|z| < 1$ satisfying

$$\overline{\lim_{z \to \zeta}} |f(z)| \leq \begin{cases} M_1, & \zeta \in \overline{\Gamma}_1 \\ M_2, & \zeta \in \overline{\Gamma}_2 \end{cases},$$

where $0 \leq M_1 \leq M_2 < \infty$. Then, for each r, $0 < r < 1$, there is a number $\lambda(r)$, $0 < \lambda(r) < 1$, so that

$$\sup_{|z| \leq r} |f(z)| \leq M_1^{\lambda(r)} M_2^{1-\lambda(r)}.$$ (5.3.3.6)

This $\lambda(r)$ is valid for any such analytic function $f(z)$. Thus, it follows easily the following

Convergence theorem (III) (P. Montel, 1927). *Let Γ_1 and Γ_2 be nondegenerated complimentary arcs on $|z| = 1$. Suppose $f_n(z)$, $n \geq 1$, are analytic in $|z| < 1$ and continuous on the set $\{|z| < 1\} \cup \Gamma_1$. Also, suppose*

1. *$f_n(z)$, $n \geq 1$, are uniformly bounded in $|z| < 1$;*
2. *$f_n(z)$ converges locally uniformly on Γ_2.*

Then $f_n(z)$ converges locally uniformly in $|z| < 1$. (5.3.3.7)

Section (3) How can Schwarz's lemma help?

Fix a point z_0, $|z_0| < 1$. Then

$$\left| \frac{z - z_0}{1 - \overline{z_0}z} \right| \leq \frac{|z| + |z_0|}{1 + |z_0||z|} \text{ (see (3.4.5.3))}, \quad |z| < 1$$

$$= 1 - (1 - |z|) \cdot \frac{1 - |z_0|}{1 + |z_0||z|} < 1 - (1 - |z|)\frac{1 - |z_0|}{1 + |z|}$$

$$< \exp\left\{ -(1 - |z_0|)\frac{1 - |z|}{1 + |z|} \right\}$$

$$\text{(using } 1 - t < e^{-t}, \; t > 0), \quad |z_0| < 1 \quad \text{and} \quad |z| < 1. \quad (*_8)$$

Suppose $f(z)$ is a bounded analytic function in $|z| < 1$, say $|f(z)| \leq M$ there. Let $z_1, \ldots, z_n$ be n distinct points in $|z| < 1$ so that $|f(z_k)| \leq \alpha$ for $1 \leq k \leq n$. Let $L_{n-1}(z)$ be Lagrange's polynomial given by $(*_1)$. Observe that $L_{n-1}(z) - f(z)$ has zeros at $z_1, \ldots, z_n$ and

$$\frac{L_{n-1}(z) - f(z)}{B_n(z)}, \quad \text{where } B_n(z) = \prod_{k=1}^{n} \frac{z - z_k}{1 - \overline{z_k}z}$$

is analytic in $|z| < 1$ (see (3.4.2.17)). Apply the maximum principle to this function on $|z| \leq r < 1$ and we have

$$\left| \frac{L_{n-1}(z) - f(z)}{B_n(z)} \right| \leq \sup_{|z| \leq r} \left| \frac{L_{n-1}(z) - f(z)}{B_n(z)} \right|$$

$$= \sup_{|z| = r} \left| \frac{L_{n-1}(z) - f(z)}{B_n(z)} \right|, \quad |z| < r.$$

$\Rightarrow$ Letting $r \to 1$ and observing that $|B_n(z)| \to 1$ as $|z| = r \to 1$,

$$\left| \frac{L_{n-1}(z) - f(z)}{B_n(z)} \right| \leq \sup_{|z| < 1} |L_{n-1}(z) - f(z)|, \quad |z| < 1$$

$$\leq \alpha c_n + M, \quad \text{where } c_n = \sum_{k=1}^{n} \left[\prod_{\substack{l=1 \\ l \neq k}}^{n} \frac{2}{|z_l - z_k|} \right].$$

Owing to $(*_8)$, we have

A generalization of the interior uniqueness theorem (see (4) in (3.4.2.9) or (2) in (5.1.2.4)). Let $f(z)$ be analytic in $|z| < 1$ and $|f(z)| \leq M$ there. Let

$z_1, \ldots, z_n$ be n distinct points in $|z| < 1$. Suppose $|f(z_k)| \leq \alpha$ for $1 \leq k \leq n$. Then there is a positive constant $c_n = c_n(z_1, \ldots, z_n)$ so that

$$|f(z)| \leq 2\alpha c_n + M \exp \left\{ \left[\sum_{k=1}^{n} (1 - |z_k|) \right] \frac{|z| - 1}{|z| + 1} \right\}, \quad |z| < 1.$$

In particular, if $z_1, z_2, \ldots, z_n, \ldots$ are zeros of $f(z)$ in $|z| < 1$ such that $\sum_{k=1}^{\infty} (1 - |z_k|) = \infty$, then $f(z) \equiv 0$ in $|z| < 1$.   (5.3.3.8)

As an application, we have a

Convergence theorem (IV) (Löwner and Radó, 1915). *Let $f_n(z)$ be analytic in $|z| < 1$ and $z_1, z_2, \ldots, z_k, \ldots$ be a sequence of distinct points in $|z| < 1$. Suppose*

1. *$\sum_{k=1}^{\infty} (1 - |z_k|) = \infty$;*
2. *$f_n(z)$, $n \geq 1$, are uniformly bounded in $|z| < 1$ (see (5.3.3.1)), and*
3. *for each $k \geq 1$, $\lim_{n \to \infty} f_n(z_k)$ exists.*

Then $f_n(z)$ converges locally uniformly in $|z| < 1$.   (5.3.3.9)

In case z_k converges to a point in $|z| < 1$, then condition 1 is automatically fulfilled and thus this result can be derived from (5.3.3.3) by suitable adjustment (see Exercise A(1)). Privalov (1924) relaxed condition 2 to $\sup_{n \geq 1} \sup_{0 < r < 1} \int_0^{2\pi} \log^+ |f_n(re^{i\theta})| d\theta < \infty$, while preserving 1 and 3, to obtain the same result. Refer to Ref. [35], Theorem 4, p. 397 for a further generalization by Ostrowski.

Proof. To prove (5.3.3.9), for $\varepsilon > 0$ and $0 < r < 1$, choose $N \geq 1$ so that

$$M \exp \left\{ \left[\sum_{k=1}^{N} (1 - |z_k|) \right] \frac{r - 1}{r + 1} \right\} < \varepsilon \quad \text{with } |f_n(z)| \leq M \quad \text{for } n \geq 1,$$

and then choose the corresponding constant $c_N > 0$ as in (5.3.3.8) so that

$$\max_{1 \leq k \leq N} |f_m(z_k) - f_n(z_k)| < \frac{\varepsilon}{C_N}, \quad m, n \geq N.$$

Consequently, by the inequality in (5.3.3.8),

$$|f_m(z) - f_n(z)| \leq 2C_N \frac{\varepsilon}{C_N} + \varepsilon = 3\varepsilon, \quad |z| \leq r \quad \text{and} \quad m, n \geq N.$$

Cauchy criterion says what remains.

Our main result is the following

Convergence theorem (V). *Let $f_n(z)$, $n \geq 1$, be analytic in a domain Ω. Then*

(1) (*Vitali, 1903,1904; Porter, 1904, 1905*) *Suppose*

    1. $f_n(z)$, $n \geq 1$, *are locally uniformly bounded in Ω (namely, uniformly bounded on each compact subset of Ω), and*

    2. $f_n(z)$ *converges pointwise on a subset E of Ω which has a limit point in Ω.*

    Then $f_n(z)$ converges locally uniformly in Ω.

$\Leftrightarrow$ (2) (*P. Montel, 1907*) *Suppose $f_n(z)$, $n \geq 1$, are locally uniformly bounded in Ω. Then $f_n(z)$ has a subsequence $f_{n_k}(z)$, $k \geq 1$, converging locally uniformly in Ω.*   (5.3.3.10)

(2) is still valid even if Ω is only a nonempty open set. In what follows, we try to prove the validity of (1) via (5.3.3.9) (see Exercise A(2) for a second proof), and then the equivalence of (1) and (2). For another proof of (2) via the concepts of equicontinuity and normal family, see Sec. 5.8.1. Exercises A(3) and (4) will give two applications of (5.3.3.10).

To prove (1). Let $A = \{z \in \Omega | f_n(z), n \geq 1, \text{converges}\}$. Then $E \subseteq A$. Let Ω_0 be the set of the limit points of A in Ω. Then, $\Omega_0 \neq \phi$ holds.

Suppose $z_0 \in \Omega_0$ and the closed disk $|z - z_0| \leq r$ is contained in Ω. Choose a sequence z_k of distinct points in A such that $|z_k - z_0| < r$ for $k \geq 1$ and $\lim_{k \to \infty} z_k = z_0$. In this case, $\sum_{k=1}^{\infty} (1 - |z_k|) = \infty$ holds and it follows by (5.3.3.9) that $f_n(z)$ converges locally uniformly in $|z - z_0| < r$. Consequently, $\{|z - z_0| < r\} \subseteq \Omega_0 \cap A$ holds and thus Ω_0 is a relatively open set in Ω. Since Ω_0 is relatively closed in Ω which is a connected set, hence $\Omega_0 = \Omega$ and then $\{|z - z_0| < r\} \subseteq A$. Therefore $\Omega_0 = \Omega \subseteq A$ and, indeed, $A = \Omega$ holds. By using (5.3.3.9) again, $f_n(z)$ converges locally uniformly in Ω.

To prove (1) $\Rightarrow$ (2). We may suppose that $E = \{z_1, z_2, \ldots, z_k, \ldots\}$ is a countable set.

Since $\{f_n(z_1)\}_{n=1}^{\infty}$ is bounded, it contains a convergent subsequence $f_{n_k^{(1)}}(z_1)$, $k \geq 1$. Similarly, the boundedness of $f_{n_k^{(1)}}(z_2)$, $k \geq 1$, implies that it has a convergent subsequence $f_{n_k^{(2)}}(z_2)$, $k \geq 1$. Repeat this process. Then, each sequence $f_{n_k^{(l)}}(z_{l+1})$, $k \geq 1$, has a convergent subsequence

$f_{n_k^{(l+1)}}(z_{l+1})$, $k \geq 1$, so that

$$f_{n_k^{(l+1)}}(z), \quad k \geq 1, \text{ converges on } \{z_1, \ldots, z_{l+1}\} \quad \text{for } l \geq 0.$$

By *Cantor's diagonal process*, the sequence $f_{m_k}(z)$, $k \geq 1$, defined by

$$f_{m_k}(z) = f_{n_k^{(k)}}(z), \quad k \geq 1$$

is a subsequence of the original sequence and converges pointwise on E. By (1), it follows that $f_{m_k}(z)$, $k \geq 1$, converges locally uniformly in Ω.

Note that each *open* set O in $\mathbf{C}$ can be expressed as a countable union of connected open sets. A second usage of the Cantor diagonal process will extend the known result on each component to the whole set O.

To prove (2) $\Rightarrow$ (1). Suppose there is a point $z_0 \in \Omega$ so that $f_n(z_0)$, $n \geq 1$, diverges.

Since $\{f_n(z_0)\}_{n=1}^{\infty}$ is bounded, there are $\alpha, \beta \in \mathbf{C}$, $\alpha \neq \beta$, and two subsequences $f_{n_k^{(1)}}(z_0)$, $k \geq 1$, and $f_{n_k^{(2)}}(z_0)$, $k \geq 1$, of it such that

$$\lim_{k \to \infty} f_{n_k^{(1)}}(z_0) = \alpha \quad \text{and} \quad \lim_{k \to \infty} f_{n_k^{(2)}}(z_0) = \beta.$$

Since the two subsequences $f_{n_k^{(1)}}(z)$, $k \geq 1$, and $f_{n_k^{(2)}}(z)$, $k \geq 1$, are locally uniformly bounded in Ω, by (2), it follows that each has a subsequence $f_{m_k^{(1)}}(z) \to f^{(1)}(z)$ and $f_{m_k^{(2)}}(z) \to f^{(2)}(z)$ locally uniformly in Ω, respectively. Moreover, $f^{(1)}(z_0) = \alpha \neq \beta = f^{(2)}(z_0)$. On the other hand, choose a sequence z_l, $l \geq 1$, of distinct points from E, converging to a point in Ω. Then, by assumption 2 in (1),

$$\lim_{k \to \infty} \{f_{m_k^{(1)}}(z_l) - f_{m_k^{(2)}}(z_l)\} = 0 = f^{(1)}(z_l) - f^{(2)}(z_l) \quad \text{for } l \geq 1.$$

By the interior uniqueness theorem, $f^{(1)}(z) \equiv f^{(2)}(z)$ in Ω, a contradiction.

Therefore, the sequence $f_n(z)$ converges everywhere in Ω. And, finally, (5.3.3.9) guarantees the desired result. $\square$

Section (4) Finally, applications of Hadamard–Borel–Carathéodory inequality

Recall the *inequality* (see Example 3 in Sec. 3.4.5): Suppose $f(z)$ is analytic in $|z| < R$ and continuous on $|z| \leq R$ (or, $\operatorname{Re} f(z)$ is just assumed to be bounded above in $|z| \leq R$). Then

$$\sup_{|z|=r} \operatorname{Re} f(z) \leq \frac{R-r}{R+r} \operatorname{Re} f(0) + \frac{2r}{R+r} \sup_{|z|=R} \operatorname{Re} f(z);$$

$$\sup_{|z|=r} |f(z)| \le \frac{R+r}{R-r}|f(0)| + \frac{2r}{R-r}\sup_{|z|=R} \operatorname{Re} f(z), \quad 0 \le r < R.$$

$$(5.3.3.11)$$

A direct application of (5.3.3.11) is the following

Convergence theorem (VI). *Suppose* $f_n(z)$, $n \ge 1$, *are analytic in a domain* Ω. *Let* $z_0 \in \Omega$ *be a point so that the sequence* $f_n(z_0)$ *converges. Then, any one of following conditions will imply that the sequence* $f_n(z)$ *converges locally uniformly in* Ω:

1. $\operatorname{Re} f_n(z)$, $n \ge 1$, *converges locally uniformly in* Ω.
2. $\operatorname{Re} f_n(z) \le \operatorname{Re} f_{n+1}(z)$ *on* Ω *for each* $n \ge 1$.
3. $|f_n(z)| \le |f_{n+1}(z)|$ *on* Ω *for each* $n \ge 1$ *and* $\lim_{n\to\infty} f_n(z_0) \ne 0$ *holds.*

$$(5.3.3.12)$$

Proof. Let

$$A = \{\zeta \in \Omega | \ f_n(\zeta) \text{ converges}\}, \quad \text{and}$$

$$O = \{\zeta \in \Omega | \ f_n(z) \text{ converges uniformly in an open neighborhood of } \zeta\}.$$

$$(*_9)$$

Since $z_0 \in A$, $A \ne \phi$. Also, $O \subseteq A$ holds.

Fix any point $\zeta \in A$. Choose $R > 0$ so that $|z - \zeta| \le R$ is contained in Ω. Apply (5.3.3.11) to $(f_n - f_m)(z)$, $n \ge m \ge 1$, in $|z - \zeta| \le R$. We have, for $n \ge m \ge 1$,

$$\sup_{|z-\zeta|=\frac{R}{2}} |f_n(z) - f_m(z)| \le \frac{R + R/2}{R - R/2}|f_n(\zeta) - f_m(\zeta)|$$

$$+ \frac{2 \cdot R/2}{R - R/2}\sup_{|z-\zeta|=R} \operatorname{Re}(f_m(z) - f_n(z))$$

$$\Rightarrow \sup_{|z-\zeta|\le\frac{R}{2}} |f_n(z) - f_m(z)| \le 3|f_n(\zeta) - f_m(\zeta)|$$

$$+ 2 \sup_{|z-\zeta|\le R} \operatorname{Re}(f_m(z) - f_n(z)), \quad (*_{10})$$

by the maximum modulus principle.

Condition 1 will show, by $(*_{10})$, that $f_n(z)$ converges uniformly in $|z - \zeta| \le \frac{R}{2}$. Hence

$$\left\{|z - \zeta| < \frac{R}{2}\right\} \subseteq A \subseteq O,$$

$$(*_{11})$$

whenever $\zeta \in A$ and $|z - \zeta| \leq R$ is contained in Ω. Similarly, Condition 2 implies that $\sup_{|z-\zeta| \leq \frac{R}{2}} |f_n(z) - f_m(z)| \leq 3|f_n(\zeta) - f_m(\zeta)|$ holds because $\operatorname{Re} f_m(z) \leq \operatorname{Re} f_n(z)$ if $n \geq m \geq 1$. In this case, $(*_{11})$ is still true.

In conclusion, both cases show that $A = O \neq \phi$ holds as an open subset of Ω. On the other hand, fix any point $\zeta_0 \in \Omega$ and choose a curve $z = z(t)$: $[0, 1] \to \Omega$, connecting $z_0 = z(0)$ to $\zeta_0 = z(1)$. Let $B = \{t \in [0, 1]|\ z(t) \in O\}$. $B \neq \phi$ since $0 \in B$. Let $t_0 = \sup B$. Then $0 < t_0 \leq 1$. A standard argument will show that $t_0 \in B$ and $t_0 = 1$ (see the proof of (5.2.1.2)): Choose $R > 0$ so that $|\zeta - z(t_0)| < R$ is contained in Ω. Take a point $t_1 \in B$ so that $|z(t_1) - z(t_0)| < \frac{R}{4}$. Then apply $(*_9)$ to $|\zeta - z(t_1)| < \frac{R}{2}$ to show that $(*_{10})$ holds in this case, namely, $\{|\zeta - z(t_1)| < \frac{R}{4}\} \subseteq O$. Hence, $t_0 \in B$ holds. Consequently, $\Omega = O$ and the claimed result follows.

As for condition 3: We may suppose $f_1(z_0) \neq 0$ without loss of generality. Then $f_1(z)$ has isolated zeros and the set $\Omega_0 = \Omega - f_1^{-1}(0)$ is still a domain. Just like $(*_9)$, consider the sets A_0 and O_0 with Ω_0 replacing Ω there. Since $f_1(z_0) \neq 0$, $z_0 \in A_0$, and $A_0 \neq \phi$, $O_0 \subseteq A_0$ holds.

To set up result such as $(*_{11})$ for Ω_0, choose $\zeta \in A_0$ and $R > 0$ so that $|z - \zeta| \leq R$ is contained in Ω_0. Condition 3 indicates that $f_n(z) \neq 0$ on $|z - \zeta| < R$ for each $n \geq 1$. Let $g_n(z)$ be a branch of $\operatorname{Log} f_n(z)$ on $|z - \zeta| < R$ (see (3.3.1.8)), namely, $e^{g_n(z)} = f_n(z)$ for $n \geq 1$ and $|z - \zeta| < R$. Observe now that $\operatorname{Re} g_n(z) \leq \operatorname{Re} g_{n+1}(z)$ on $|z - \zeta| < R$ for $n \geq 1$. Applying the known result (case 2) to $g_n(z)$, $n \geq 1$, then $g_n(z)$ converges locally uniformly in $|z - \zeta| < \frac{R}{2}$ and so does $f_n(z)$ owing to the continuity of the exponential function. It follows that

$$\left\{ |z - \zeta| < \frac{R}{2} \right\} \subseteq A_0 \subseteq O_0.$$

By connectivity of Ω_0, then $\Omega_0 = O_0 = A_0$ holds and thus, $f_n(z)$ converges locally uniformly in Ω_0.

In case $\zeta_0 \in f_1^{-1}(0)$: There is an $R > 0$ so that $f_1(z) \neq 0$ in $0 < |z - \zeta_0| \leq R$ which is assumed to be contained in Ω. Then $\{|z - \zeta_0| = R\} \subseteq \Omega_0$. By the result from the last paragraph, $f_n(z)$ converges uniformly along the circle $|z - \zeta_0| = R$ and then, uniformly in the disk $|z - \zeta_0| \leq R$ by using (5.3.3.1).

This ends up the proof. $\square$

For another application of (5.3.3.11), we set up a preliminary *inequality* as follows: Suppose $f(z)$ is analytic in $|z| \leq R$ and $f(z) \neq 0$ there. Then

$$\sup_{|z|=r} |f(z)| \leq |f(0)|^{\frac{R-r}{R+r}} \left[\sup_{|z|=R} |f(z)| \right]^{\frac{2r}{R+r}}, \quad 0 \leq r \leq R. \quad (5.3.2.13)$$

To see this, choose $\rho > R$ so that $f(z)$ is analytic in $|z| < \rho$ on which $f(z) \neq 0$. Choose a branch $g(z)$ of $\log f(z)$ in $|z| < R$, namely, $e^{g(z)} = f(z)$. Applying (5.3.3.11) to $g(z)$, we have

$$\sup_{|z|=r} \operatorname{Re} g(z) \leq \frac{2r}{R+r} \sup_{|z|=R} \operatorname{Re} g(z) + \frac{R-r}{R+r} \operatorname{Re} g(0), \quad \text{where } |z| = r \leq R.$$

Observe that e^t is an increasing function of t and that $|f(z)| = e^{\operatorname{Re} g(z)}$. The inequality follows easily.

(5.3.3.13) can be used to obtain the following

Convergence theorem (VII) (Montel, 1912). *Suppose $f_n(z)$, $n \geq 1$, are analytic in a domain Ω and satisfy:*

1. *$f_n(z) \neq 0$ in Ω for each $n \geq 1$;*
2. *$f_n(z)$, $n \geq 1$, are locally uniformly bounded; and*
3. *There is a point $z_0 \in \Omega$ so that $\lim_{n \to \infty} f_n(z_0) = 0$.*

Then $f_n(z)$ converges to 0 locally uniformly in Ω. $\qquad\qquad$ (5.3.3.14)

This result can provide (2) in (5.3.2.2) another proof. See Exercise A(5). The proof of (5.3.3.14) is exactly the same as that of (5.3.3.12).

Sketch of proof. Suppose $\zeta \in \Omega$ so that $f_n(\zeta)$ converges to zero. Choose $R > 0$ so that $|z - \zeta| \leq R$ is contained in Ω. Let $M = \sup_{n \geq 1} \sup_{|z-\zeta| \leq R} |f_n(z)|$. Note that $M < \infty$. Choose $r = \frac{R}{2}$ in (5.3.3.13) and we have

$$\sup_{|z-\zeta| \leq \frac{R}{2}} |f_n(z)| \leq |f_n(\zeta)|^{\frac{R-R/2}{R+R/2}} \left\{ \sup_{|z-\zeta|=R} |f_n(z)| \right\}^{\frac{2 \cdot R/2}{R+R/2}}$$

$$\leq M^{\frac{2}{3}} |f_n(\zeta)|^{\frac{1}{3}}, \quad n \geq 1.$$

This means that $f_n(z)$ converges to 0 uniformly in $|z - \zeta| < \frac{R}{2}$. Then, try to finish the proof. $\qquad\qquad\square$

Exercises A

Try to fill up the gaps in the proofs of these theorems in this section. Again, refer to Refs. [12] and [35], if necessarily, for detailed account and more information concerned. Then, try to do the following problems.

(1) In (5.3.3.9), suppose that z_k converges to a point in $|z| < 1$. Then $\sum_{k=1}^{\infty} (1 - |z_k|) = \infty$ holds true obviously. Try to use (5.3.3.3) or its proof to show (5.3.3.9) under this assumption.

(2) Try to use (5.3.3.3) (refer to Exercise (1)) and (5.3.3.5) to give (1) in (5.3.3.10) another proof.

(3) Use (5.3.3.10) to prove the following theorem (Montel, 1910): Let $f_n(z)$, $n \geq 1$, be analytic in an open set O in **C**. Suppose

  1. $f_n(z)$, $n \geq 1$, are locally uniformly bounded in O, and
  2. each convergent subsequence of $f_n(z)$ converges to the same function $f(z)$ in O.

  Then $f_n(z)$ converges to $f(z)$ locally uniformly in O.

(4) Use (2) in (5.3.3.10) and Exercise (3) to prove the following statement: Suppose $f_n(z) = \sum_{k=0}^{\infty} a_{nk} z^k$, $n \geq 1$, are analytic in $|z| < 1$ and $f_n(z)$, $n \geq 1$, are locally uniformly bounded in $|z| < 1$. Then,

  (a) $f_n(z)$, $n \geq 1$, converges to $f(z)$ locally uniformly.
  $\Leftrightarrow$ (b) $\lim_{n \to \infty} a_{nk} = a_k$, for each $k \geq 0$. In this case, $f(z) = \sum_{k=0}^{\infty} a_k z^k$.

(5) Try to use (5.3.3.14) to prove (2) in (5.3.2.2).

(6) (Osgood, 1901–1902) Let $f_n(z)$, $n \geq 1$, be analytic in an open set Ω and converge pointwise to a function $f(z)$ in Ω. Then, there is an open dense subset Ω_0 of Ω on which $f_n(z)$ converges locally uniformly to $f(z)$. Hence, $f(z)$ is analytic in Ω_0.

(7) Let $f(z)$ be analytic and bounded in the angular domain $-\theta_0 < \operatorname{Arg} z < \theta_0$ $(\theta_0 > 0)$. Suppose $\lim_{\substack{x \to 0 \\ x > 0}} f(x) = a$ holds. Show that, for each $0 < \alpha < \theta_0$, $\lim_{z \to 0} f(z) = a$ uniformly in $-\alpha \leq \operatorname{Arg} z \leq \alpha$.

Exercises B

The following Exercises (1) and (2) are extensions of (4.7.4) and (5.3.1.5). Ω always denotes a domain in the z-plane and γ a continuous curve in the w-plane. $\Phi(z,t) : \Omega \times \gamma \to$ **C** is considered as a complex-valued function of two complex variables z and t.

(1) Suppose that γ is rectifiable such that

  1. for each $t \in \gamma$, $\Phi(z,t)$ is analytic in $z \in \Omega$;
  2. for each $z \in \Omega$, $\Phi(z,t)$ is continuous in $t \in \gamma$, and
  3. the family of analytic functions $\{\Phi(z,t)\}_{t \in \gamma}$ is locally uniformly bounded in Ω, namely, $\sup_{\substack{z \in K \\ t \in \gamma}} |\Phi(z,t)| < \infty$ for each compact set K in Ω.

  Then

$$F(z) = \int_{\gamma} \Phi(z,t) dt$$

is analytic in Ω. Moreover, for each $z \in \Omega$, $\frac{\partial^n \Phi(z,t)}{\partial z^n}$ is continuous in $t \in \gamma$ for each $n \geq 1$ and

$$F^n(z) = \int_\gamma \frac{\partial^n \Phi(z,t)}{\partial z^n} dt, \quad z \in \Omega \quad \text{for } n \geq 1.$$

Note. In case $\Phi(z,t)$ is *continuous* as a function of $(z,t) \in \Omega \times \gamma$, then conditions 2 and 3 are automatically fulfilled. Under this circumstance, if γ is a real interval $[a, b]$, then this is nothing new but (4.7.4).

(2) Suppose γ is *locally* rectifiable with parametric equation $t = t(\tau)$: $[a, b) \to \mathbf{C}$ and passes through the point ∞, namely, $\lim_{\tau \to b^-} t(\tau) = \infty$. Denote by γ_c the subcurve $t = t(\tau)$, $a \leq \tau \leq c < b$. Also, suppose that

1. as in Exercise (1)1;
2. as in Exercise (1)2;
3. for each $c \in [a, b)$, the family $\{\Phi(z,t)\}_{t \in \gamma_c}$ is locally uniformly bounded in Ω, and
4. for each $z \in \Omega$, the limit

$$\lim_{\tau \to b^-} \int_{\gamma_c} |\Phi(z,t)||dt| = \int_\gamma |\Phi(z,t)||dt|$$

exists as a finite number and is locally bounded as a function in $z \in \Omega$.

Then

$$F(z) = \int_\gamma \Phi(z,t)dt$$

is analytic in $z \in \Omega$. Moreover, for each fixed $z \in \Omega$, $\frac{\partial^n \Phi(z,w)}{\partial z^n}$ is continuous in $t \in \gamma$ for each $n \geq 1$ and

$$F^n(z) = \int_\gamma \frac{\partial^n \Phi(z,t)}{\partial z^n} dt, \quad z \in \Omega \text{ for } n \geq 1.$$

*5.3.4 *An application: The fixed points of an analytic function and its iterate functions*

Let Ω be a nonempty set in $\mathbf{C}$ and $f : \Omega \to \Omega$ be a function mapping Ω into itself. A point $z_0 \in \Omega$ is called a *fixed point* of f if $f(z_0) = z_0$. The *iterates* of f are defined and denoted by, for simplicity,

$$f^0 = \text{id (the identity map of } \Omega),$$

$$f^1 = f,$$

$$f^n = f \circ f^{n-1}, \quad n \geq 1.$$

In what follows, Ω is *always assumed* to be an *open set* while f is *analytic* in Ω, mapping Ω into itself. All functions concerned are supposed to be analytic unless otherwise stated.

Let z_0 be a fixed point of f. The fixed point is classified according to $f'(z_0)$, called the *multiplier* of f at z_0, as follows:

1. *attractive* if $|f'(z_0)| < 1$; *superattracting* if $f'(z_0) = 0$;
2. *repellent* or *repulsive* if $|f'(z_0)| > 1$;
3. *indifferent* if $|f'(z_0)| = 1$; more precisely, *rational neutral* if $|f'(z_0)| = 1$ and $f'(z_0)^n = 1$ for some integer n, and *irrational neutral* if $|f'(z_0)| = 1$ and $f'(z_0)^n \neq 1$ for any integer n. (5.3.4.1)

Two functions $f : \Omega \to \Omega$ and $g : O(open) \to O$ are said to be (conformally) *conjugate* to each other if there is a univalent mapping $\Phi : \Omega \to O$ so that $g = \Phi \circ f \circ \Phi^{-1}$, i.e.,

$$\Phi(f(z)) = g(\Phi(z)), \quad z \in \Omega \qquad (5.3.4.2)$$

referred to as *Schröder's equation* (1871). In this case, $g^n = \Phi \circ f^n \circ \Phi^{-1}$ and thus, f^n and g^n are also conjugate for each $n \geq 1$, and so are f^{-1} and g^{-1}, when both are defined. Moreover, Φ maps fixed points of f to fixed points of g and vice versa, and the multipliers at the corresponding fixed points are equal. This Φ plays as a *coordinatizing map*, specifically called the *conjugation*, and f and g can be regarded as the same function viewed in different coordinate systems related by this Φ.

In what follows, we will *sketch* some results about the existence of fixed points and the univalence of an analytic function, mostly in terms of the iterates of the function via the concept of local uniform convergence. For details, see Ref. [12]. The study of the behavior of analytic functions under iteration constitutes the content of *complex dynamics*, a field originated with the Newton–Raphson iteration method for approximating roots (see, for example, Ref. [7], Chap. 11, and Sec. 5.8.5 for some fundamental results) and of current research interest. One may refer to Ref. [15] and the references therein for its analytic aspects.

The following is divided into four subsections.

Section (1) Linear fractional transformations as preliminary examples

Here, readers are supposed to be familiar with the content of Sec. 2.5.4 and the exercises there, in particular, Exercises A(9) and B(1). We will have no hesitation to adopt materials presented there. Also, the families of Steiner's and Appollonius's circles introduced in Sec. 1.4.4 are helpful, too, in the understanding of this subsection.

Given

$$w = \frac{az+b}{cz+d}, \quad \text{with normalized } ad - bc = 1.$$

$w = w(z)$ has either two distinct fixed points z_1 and z_2 or one coincident fixed point z_0 except the identity mapping $w = z$.

The case that $c = 0$ and $z_1 = \frac{b}{d-a}$, $z_2 = \infty$: Then

$$w = \frac{az+b}{d}, \quad ad = 1 \tag{$*_1$}$$

$$\Leftrightarrow w - z_1 = k(z - z_1), \text{ with } k = \frac{a}{d} = w'(z_1) \text{ and } k^{-1} = w'(\infty),$$

the multipliers.

$$\Leftrightarrow \text{(canonical form) } \eta = k\zeta, \text{ where } \zeta = z - z_1 \text{ and } \eta = w - z_1. \tag{$*_2$}$$

Note that $\zeta = \Phi(z) = z - z_1$ acts as the conjugation of $w = f(z) = \frac{az+b}{d}$ to $\eta = g(\zeta) = k\zeta$ in the sense (5.3.1.2): $g(\zeta) = \Phi \circ f \circ \Phi^{-1}(\zeta)$. The advantage of $(*_2)$, namely, $\eta = g(\zeta)$, over $(*_1)$, namely, $w = f(z)$, is that it is easy to write down the iterates g^n:

$$\eta = k^n\zeta, \quad n \geq 1.$$

In case $w = f(z)$ is *hyperbolic* ($k > 0$ but $k \neq 1$) or *loxodromic* ($k = |k|e^{i\theta}$, $|k| > 0$, $|k| \neq 1$, and $\theta \neq 2n\pi$ for integers n), $g^n(\zeta)$ converges to 0 or ∞ uniformly in any bounded neighborhood of 0 or ∞, respectively. Consequently, so does $f^n(z)$ to z_1 or z_2 as $n \to \infty$. In case $g^n(\zeta) = g(\zeta)$ for some integer $n \geq 2$, then it is necessary that $k^n = 1$, and thus, $w = f(z)$ is *elliptic* ($|k| = 1$ and $k \neq 1$).

The case that $c \neq 0$ and $z_1 \neq z_2$: Then

$$w = \frac{az+b}{cz+d}, \quad ad - bc = 1 \quad \text{and} \quad (a-d)^2 + 4bc \neq 0 \tag{$*_3$}$$

$$\Leftrightarrow \frac{w - z_1}{w - z_2} = k\frac{z - z_1}{z - z_2}, \text{ with } w'(z_1) = k = \frac{a - cz_1}{a - cz_2} \text{ and } w'(z_2) = k^{-1},$$

the multipliers.

$$\Leftrightarrow \text{(canonical form) } \eta = k\zeta, \text{ where } \zeta = \frac{z - z_1}{z - z_2} \text{ and } \eta = \frac{w - z_1}{w - z_2}. \tag{$*_4$}$$

Hence $\zeta = \Phi(z) = \frac{z-z_1}{z-z_2}$ is the conjugation of $w = f(z) = \frac{az+b}{cz+d}$ to $\eta = g(\zeta) = k\zeta$. Similar explanations are still valid in this case as in the last paragraph.

The case of one coincident fixed point z_0: If $c = 0$, then $z_0 = \infty$ and $w = w(z)$ reduces to $w = z \pm b$, $b \neq 0$. If $c \neq 0$, then $z_0 = \frac{a-d}{2c}$. In this case,

$$w = \frac{az+b}{cz+d}, \quad ad - bc = 1, \quad \text{and} \quad a+d = \pm 2 \tag{$*_5$}$$

$$\leftrightarrow \frac{1}{w - z_0} = \frac{1}{z - z_0} \pm c \text{ (it is } c \text{ if } a + d = 2 \text{ and } -c \text{ if } a + d = -2\text{)}.$$

$$\leftrightarrow \text{(canonical form) } \eta = \zeta \pm c, c \neq 0, \text{ where } \zeta = \frac{1}{z - z_0}$$

$$\text{and } \eta = \frac{1}{w - z_0}. \tag{$*_6$}$$

Now, $\zeta = \Phi(z) = \frac{1}{z - z_0}$ is the conjugation of $w = f(z) = \frac{az+b}{cz+d}$ to $\eta = g(\zeta) = \zeta \pm c$. In this case, the iterates g^n: $\eta = \zeta \pm nc$, $n \geq 1$. $g^n(\zeta)$ converges to ∞ locally uniformly (according to the spherical chord metric) in $\mathbf{C}$. As a consequence, the original $f^n(z)$ converges to $z_0 = \frac{a-d}{2c}$ locally uniformly in $\mathbf{C}$. Observe that z_0 is a *parabolic* fixed point.

For the sake of reference and completion, we list the following three types of popular geometries and leave the details or proofs to the readers or refer to *Appendix B*. Recall that the set of all linear fractional transformations forms a group under composite of mappings (see (1) in (2.5.4.13)) and is called the (*general*) *linear group* on the plane. When the symmetric mapping $z \rightarrow \bar{z}$ is added, the resulted one is called the *generalized linear group*.

Parabolic (or Euclidean) geometry. The group of rigid motions

$$G_P = \{e^{i\theta} z + b | \theta \in \mathbf{R}, b \in \mathbf{C}\}$$

is a subgroup of the linear group. It is composed of the following two fundamental motions:

1. rotation (elliptic type): $w = e^{i\theta} z$;
2. translation (parabolic type): $w = z + b$.

The distance $|z_1 - z_2|$ between two points z_1 and z_2 in $\mathbf{C}$ is *invariant* under G_p.

Note. Sometimes, G_p is replaced by the group $\widetilde{G_p}$ of *Euclidean motions*, generated by adding $z \rightarrow \bar{z}$ to G_p. $\widetilde{G_p}$ is a subgroup of the generalized linear group. $\tag{5.3.4.3}$

Elliptic (or spherical) geometry. The unitary group

$$G_e = \left\{ \frac{az - b}{\bar{b}z + \bar{a}} \middle| a, b \in \mathbf{C} \quad \text{and} \quad |a|^2 + |b|^2 = 1 \right\}$$

is a subgroup of the linear group. Elements in G_e are always of the elliptic type:

$$\frac{w - z_0}{1 + \overline{z_0}w} = e^{i\theta}\frac{z - z_0}{1 + \overline{z_0}z}, \quad \text{where } z_0 = \frac{\operatorname{Im} a + \sqrt{(\operatorname{Im} a)^2 + |b|^2}}{\overline{b}}i \quad \text{and} \quad \theta \in \mathbf{R}$$

(see Exercise B(2) of Sec. 1.6). Consequently,

$$\left|\frac{z_1 - z_2}{1 + \overline{z_2}z_1}\right|, \quad z_1, z_2 \in \mathbf{C}^* \quad \left(\text{it is } \frac{1}{|z_1|} \quad \text{if } z_2 = \infty\right)$$

is an *invariant* under G_e and so is

$$d_e(z_1, z_2) = \begin{cases} 2\tan^{-1}\dfrac{|z_1 - z_2|}{|1 + \overline{z_2}z_1|}, & z_1, z - 2 \in \mathbf{C} \\[3mm] 2\tan^{-1}\dfrac{1}{|z_1|}, & z_1 \in \mathbf{C} \quad \text{and} \quad z_2 = \infty \end{cases}.$$

Under $d_e(,)$, the Riemann sphere or $\mathbf{C}^*$ is a complete metric space with the great circles (or parts of it) as geodesics. $\qquad(5.3.4.4)$

Hyperbolic (or non-Euclidean or Lobachevski) geometry. The group of non-Euclidean motions

$$G_h = \left\{e^{i\theta}\frac{z - a}{1 - \overline{a}z}\,\middle|\,\theta \in \mathbf{R}, \quad \text{and} \quad a \in \mathbf{C} \quad \text{and} \quad |a| < 1\right\}$$

is a subgroup of the linear group (sometimes, $z \to \overline{z}$ is adjoined). Elements in G_h are classified into three types:

1. *Non-Euclidean rotation* (elliptic type with one fixed point in $|z| < 1$):

$$\frac{w - a}{1 - \overline{a}w} = e^{i\theta}\frac{z - a}{1 - \overline{a}z} \quad (|a| < 1; a \text{ is a fixed point}).$$

2. *Non-Euclidean parallelism* (hyperbolic type with two distinct fixed points on $|z| = 1$):

$$\frac{w - z_1}{w - z_2} = k\frac{z - z_1}{z - z_2} \quad (z_1 \neq z_2, |z_1| = |z_2| = 1).$$

   Note that, by $(*_4)$, the iterates $w^n(z)$, $n \geq 1$, converge pointwise to one of these two fixed points.

3. *Non-Euclidean limit rotation* (parabolic type with only one fixed point on $|z| = 1$):

$$\frac{1}{w - z_0} = \frac{1}{z - z_0} + \lambda \quad (|z_0| = 1).$$

Consequently,

$$\left| \frac{z_1 - z_2}{1 - \overline{z_2} z_1} \right|, \quad |z_1| < 1, \quad |z_2| < 1$$

is an *invariant* under G_h and so is

$$d_h(z_1, z_2) = \frac{1}{2} \log \frac{|1 - \overline{z_2} z_1| + |z_1 - z_2|}{|1 - \overline{z_2} z_1| - |z_1 - z_2|}, \quad |z_1| < 1, \quad |z_2| < 1.$$

The unit disk $|z| < 1$, endowed with $d_h(,)$, is a complete metric space in which, a circular arc, orthogonal to the boundary $|z| = 1$, is a geodesic.

$$(5.3.4.5)$$

One can construct these three types of geometries via the bundles of circles or the Hermitian matrices of order 2 (see *Appendix B*). Refer to Carathéodory's classic [14] and Lin [56], Vol. 2, while the later also presents another linearly algebraic method to construct these geometries.

Before we go to general description of fixed points of an analytic function, to be sketched in Sections (2)–(4) below, we try to characterize a *necessary condition* for the existence of a fixed point.

Suppose f is analytic at z_0 and $f(z_0) = z_0$ holds. By definition of iteration,

$$(f^n)'(z_0) = f'(f^{n-1}(z_0))(f^{n-1})'(z_0)$$

$$= f'(z_0)(f'(z_0))^{n-1} \text{ (by induction)}$$

$$= (f'(z_0))^n, \quad n \geq 1. \qquad (*_7)$$

Choose $r > 0$ so that $f(z)$ is analytic in $|z - z_0| \leq r$. Then, for any integer $k \geq 1$,

$$(f^n)^{(k)}(z_0) = \frac{k!}{2\pi i} \int_{|\zeta - z_0| = r} \frac{f^n(\zeta)}{(\zeta - z_0)^{k+1}} d\zeta$$

$$= \frac{k!}{2\pi i} \int_0^{2\pi} \frac{f^n(z_0 + re^{i\theta})}{(re^{i\theta})^{k+1}} rie^{i\theta} d\theta$$

$$\Rightarrow |(f^n)^{(k)}(z_0)| \leq \frac{k!}{r^k} M, \text{ where } M = \max_{|z - z_0| \leq r} |f(z)|$$

$$\Rightarrow (\text{setting } k = 1) \ |(f^n)'(z_0)| = |f'(z_0)|^n \leq \frac{M}{r}$$

$$\Rightarrow |f'(z_0)| \leq \sqrt[n]{\frac{M}{r}}$$

$$\Rightarrow (\text{letting } n \to \infty) \; |f'(z_0)| \leq 1. \tag{5.3.4.6}$$

This is the desired condition for $f(z_0) = z_0$ to be true.

Section (2) Attractive and repulsive fixed points

Suppose z_0 is an attractive fixed point for f. Choose $r > 0$ so that f is analytic in $|z - z_0| \leq r$ on which $|f'(z)| \leq \alpha < 1$, where $\alpha > 0$ is a constant. By mean-value theorem (3.4.1.1) or just writing $f(z) = f(z_0) + \int_{z_0}^{z} f'(\zeta)d\zeta$ in $|z - z_0| < r$, we have

$$|f(z) - z_0| \leq \alpha|z - z_0|, \quad |z - z_0| < r \tag{$*_8$}$$

$$\Rightarrow |f^n(z) - z_0| \leq \alpha|f^{n-1}(z) - z_0| \leq \cdots \leq \alpha^n|z - z_0|,$$
$$|z - z_0| < r \quad \text{for } n \geq 1.$$

As a consequence, the iterate sequence $f^n(z)$ converges to z_0 uniformly in $|z - z_0| < r$.

Conversely, suppose $f^n(z)$ converges to z_0 locally uniformly in $|z - z_0| < r$. In particular, $f^n(z_0) = f \circ f^{n-1}(z_0) \to z_0$. By continuity of f at z_0, it follows that $f(z_0) = z_0$. On the other hand, Weierstrass's theorem (5.3.1.1) says that $(f^n(z))' = (f'(z))^n \to 0$ locally uniformly in $|z - z_0| < r$; in particular, $(f'(z_0))^n \to 0$ as $n \to \infty$. It follows that $|f'(z_0)| < 1$ holds and thus z_0 is an attractive fixed point of f.

As a matter of fact, $(*_8)$ says geometrically that the closure of the image of the open disk $|z - z_0| < r$, under f, is contained in a smaller closed disk $|z - z_0| \leq \alpha r$. This suggests that we are able to relax the condition $(*_8)$ a little in order to determine whether there is a fixed point for f.

To see this, now suppose $f : \Omega$ (a domain) $\to \Omega$ is analytic such that $\overline{f(\Omega)}$ is a *compact* set (of course, contained in Ω). Let $\Omega_n = f^n(\Omega)$, $n \geq 1$, with $\Omega_0 = \Omega$. Note that $\Omega_1 \subseteq \overline{\Omega_1} \subseteq \Omega$. Then, $\overline{\Omega_{n+1}} = \overline{f(\Omega_n)} \subseteq f(\overline{\Omega_n})(\text{compact}) \subseteq f(\Omega_{n-1}) = \Omega_n$ for $n \geq 1$. It follows, by Cantor's intersection theorem (see Exercise A(1) of Sec. 1.9), that

$$K = \bigcap_{n=1}^{\infty} \Omega_n = \bigcap_{n=1}^{\infty} \overline{\Omega}_n$$

is a nonempty compact set. We try to show that $K = \{z_0\}$ and z_0 is the only fixed point of such a f.

Since $\overline{f(\Omega)}$ is compact, so $f^n(z)$, $n \geq 1$, are uniformly bounded in Ω. By (2) of (5.3.3.10), $f^n(z)$, $n \geq 1$, has a subsequence $f^{n_k}(z)$ converging to an analytic function $g(z)$ locally uniformly in Ω. Then $g(\Omega) \subseteq K$ holds. On the other hand, for each $w \in K$ and, then, for each $n \geq 0$, $w \in \Omega_{n+1} = f^n(f(\Omega)) = f^n(\Omega_1)$, so there is a point $w_n \in \Omega_1$ so that $w = f^n(w_n)$. The compactness of $\overline{\Omega_1}$ shows that w_{n_k}, $k \geq 1$ has a convergent subsequence w_{m_j} converging to a point $w_0 \in \overline{\Omega_1}$. Since the subsequence $f^{m_j}(z)$, $j \geq 1$, of $f^{n_k}(z)$, $k \geq 1$, converges to g uniformly in $\overline{\Omega_1}$, (refer to Exercise B(4) of Sec. 5.3.1)

$$\lim_{j \to \infty} f^{m_j}(w_{m_j}) = g(w_0) = w$$

$$\Rightarrow w \in g(\overline{\Omega_1}) \subseteq g(\Omega),$$

$$\Rightarrow K \subseteq g(\Omega).$$

Combining together, $g(\Omega) = K$ is a nonempty compact set in Ω and thus, $g(\Omega) = \{z_0\}$ should hold; otherwise, the nonconstantness of $g(z)$ would imply that $g(\Omega)$ is an open set in Ω, a contradiction (see (1) in (3.4.4.4)). Consequently, by Exercise A(3) of Sec. 5.3.3, $f^n(z)$ converges to z_0 locally uniformly in Ω and thus, $f(z_0) = z_0$ holds.

What are the conjugation and the canonical form for $f(z)$ near such an attractive fixed point z_0? It is the local factorization formula $f(z) = z_0 + f'(z_0)(z - z_0) + O((z - z_0)^2)$ in a neighborhood of z_0 (see (5.1.2.3)) that enables us to solve this problem.

After the substitution of $z - z_0$ by w, we may suppose that $z_0 = 0$, for simplicity. For the sake of later reference, we consider a little general situation

$$f(z) = f'(0)z + a_k z^k g(z), \quad k \geq 2 \qquad (*_9)$$

where $g(z)$ is analytic in a neighborhood of 0, $g(0) = 1$, and $a_k \neq 0$. By induction,

$$f^n(z) = (f'(0))^n z + a_k z^k g_n(z), \quad \text{where}$$

$$g_1(z) = g(z), \quad \text{and}$$

$$g_{n+1}(z) = f'(0)g_n(z) + [(f'(0))^n + a_k z^{k-1} g_n(z)]^k g(f^n(z)), \quad n \geq 1.$$

$$(*_{10})$$

It is understood henceforth that, if $(*_9)$ holds in $|z| < r = r_1$, then, since $f^n(0) = 0$, it is possible to choose $0 < r_{n+1} < r_n$ so that $f^n(\{|z| < r_{n+1}\}) \subseteq \{|z| < r_n\}$ and $g_{n+1}(z)$ is analytic in $|z| < r_{n+1}$ for $n \geq 1$.

For this moment, let $k = 2$ in $(*_9)$ and suppose $o < |f'(0)| < 1$. Then, setting

$$\Phi_n(z) = \frac{f^n(z)}{(f'(0))^n} = \frac{f^n(z)}{(f^n)'(0)}, \quad n \geq 1 \tag{$*_{11}$}$$

$$\Rightarrow \Phi_n \circ f = \frac{f^{n+1}(z)}{(f'(0))^n} = f'(0)\Phi_{n+1}, \quad n \geq 1,$$

$$\Rightarrow \Phi \circ f = f'(0)\Phi \quad \text{or} \quad \Phi \circ f \circ \Phi^{-1}(\zeta) = f'(0)\zeta \quad \text{if } \Phi_n \to \Phi \quad \text{as } n \to \infty.$$

Then Φ is a conjugation of f to the canonical form $\eta = f'(0)\zeta$ (see $(*_2)$). To show the convergence of Φ_n, observe that

$$|f(z) - f'(0)z| \leq M|z|^2, \quad \text{if } |z| \leq r \ (r > 0 \text{ is small enough}) \text{ and}$$

$$M > 0 \text{ is a constant.}$$

$$\Rightarrow |f(z)| \leq (|f'(0)| + Mr)|z|, \quad \text{if } |z| \leq r$$

$$\Rightarrow (\text{by induction}) \ |f^n(z)| \leq (|f'(0)| + Mr)^n|z|, \quad \text{if } |z| \leq r$$

$$\Rightarrow |\Phi_{n+1}(z) - \Phi_n(z)| = \frac{|f^n(f(z)) - f'(0)f^n(z)|}{|f'(0)|^{n+1}}$$

$$= \frac{|f(f^n(z)) - f'(0)f^n(z)|}{|f'(0)|^{n+1}} \leq \frac{M|f^n(z)|^2}{|f'(0)|^{n+1}}$$

$$\leq \frac{(|f'(0)| + Mr)^{2n}}{|f'(0)|^{n+1}} M|z|^2$$

$$\leq \frac{Mr^2}{|f'(0)|}\delta^n \quad \text{if } r > 0 \text{ is so small that}$$

$$\delta = \frac{(|f'(0)| + Mr)^2}{|f'(0)|} < 1 \quad \text{and} \quad |z| \leq r.$$

Cauchy criterion says that the sequence $\Phi_n(z)$, $n \geq 1$, converges to an analytic function $\Phi(z)$ uniformly in $|z| \leq r$. Since $\Phi_n'(0) = 1 \to \Phi'(0) = 1$, $\Phi(z)$ is not a constant, and (2) in (5.3.2.2) says that $\Phi(z)$ indeed is univalent in $|z| < r$.

Is $\Phi(z)$ the unique one? Suppose Φ and Ψ are two conjugations of $f(z)$ to the canonical forms $g(\zeta) = f'(0)\zeta$. Then

$$\Phi \circ f \circ \Phi^{-1} = \Psi \circ f \circ \Psi^{-1} = g$$

$$\Rightarrow \Phi^{-1} \circ g \circ \Phi = \Psi^{-1} \circ g \circ \Psi$$

$$\Rightarrow (\Psi \circ \Phi^{-1}) \circ g \circ (\Psi \circ \Phi^{-1})^{-1} = g.$$

Let $p(z) = \Psi \circ \Phi^{-1}(z) = \sum_{n=1}^{\infty} a_n z^n$ with $a_1 \neq 0$. Then $p(g(z)) = g(p(z))$ means that $p(f'(0)\zeta) = f'(0)p(\zeta)$. After equating coefficients of both sides, we obtain

$$a_n f'(0)^n = f'(0)a_n, \quad n \geq 1 \qquad (*_{12})$$

$$\Rightarrow a_n = 0 \quad \text{for } n \geq 2.$$

Hence, $\Psi \circ \Phi^{-1}(z) = a_1 z$, or equivalently, $\Psi = a_1 \Phi$ where $a_1 \neq 0$ is a constant.

We summarize the above as the

Attractive fixed point. Let $f : \Omega$ (domain) $\rightarrow \Omega$ be analytic and $z_0 \in \Omega$.

(1) *Existence.*

     1. $f(z_0) = z_0$ and $|f'(z_0)| < 1$.

  $\Leftrightarrow$ 2. The iterates $f^n(z)$ converges to z_0 locally uniformly in an open neighborhood of z_0.

  $\Leftarrow$ 3. (Ritt, 1920–1921) $\overline{f(\Omega)}$ is a compact subset of Ω.

In case 2, $f^n(z)$ converges to z_0 locally uniformly in Ω if Ω is a *bounded* domain (see (5.3.3.10) and the *Note* below); in case 3, $f^n(z)$ converges to a *unique* fixed point z_0 locally uniformly in Ω.

(2) *Conjugation* and *canonical form* (G. Koenigs, 1884). Suppose $0 < |f'(z_0)| < 1$. There is an univalent map $\zeta = \Phi(z)$, unique up to a nonzero multiplicative scale factor, of a neighborhood of z_0 onto a neighborhood of 0 which conjugates $f(z)$ to the canonical form $g(\zeta) = f'(z_0)\zeta$. $\qquad\qquad\qquad\qquad\qquad\qquad\qquad\qquad (5.3.4.7)$

Note. If the complement $\mathbf{C} - \Omega$ contains at least *two distinct* points, then Montel's theorem (see (5.8.3.1)) says that $f^n(z)$, $n \geq 1$, form a *normal family*. Under this circumstance, the case 2 still guarantees that $f^n(z)$ converges to z_0 locally uniformly in Ω. Also, that $\overline{f(\Omega)}$ is compact is *not* a necessary condition for the existence of an attractive fixed point. For instance, $f(z) = az$, where $0 < |a| < 1$, is such a case in $\Omega = \mathbf{C}$.

Remark (Basin of an attractive fixed point and the analytic continuation of the canonical form to the basin). Adopt notations in (5.3.4.7).

Suppose z_0 is an attractive fixed point of f. The set

$$A(z_0) = \{z \mid f^n(z) \text{ is defined for } n \geq 1 \quad \text{and} \quad f^n(z) \rightarrow z_0\} \qquad (5.3.4.8)$$

is called the *basin of attraction* of z_0. It is open by its very definition. The component of $A(z_0)$ that contains z_0 is called the *immediate* basin of attraction of z_0.

In this case, Schröider's equation turns out to be

$$\Phi \circ f = \lambda \Phi \quad \text{with } \lambda = f'(z_0)$$

which induces the equation $\Phi(f^n(z)) = \lambda^n \Phi(z)$ in a neighborhood $|z - z_0| < r$ of z_0 for each $n \geq 1$. Fix any $z \in A(z_0)$. Let $n(z)$ be the least positive integer so that $|f^{n(z)}(z) - z_0| < r$. By $(*_{11})$, then

$$\frac{\Phi(f^{n(z)}(z))}{\lambda^{n(z)}} = \frac{\Phi(f^n(z))}{\lambda^n}, \quad n \geq n(z)$$

is *well-defined* and can be *defined* as the value $\Phi(z)$ of Φ at z. In this sense, $\Phi(z)$ can be analytically continued from $|z - z_0| < r$ to the whole basin $A(z_0)$ and still satisfies the functional equation

$$\Phi(f(z)) = f'(z_0)\Phi(z), \quad z \in A(z_0). \tag{5.3.4.9}$$

The extended Φ is not more univalent in general. Since $\Phi(z_0) = 0$ is known, hence $\Phi(z) = 0$ if and only if $f^n(z) = z_0$ for some $n \geq 1$. The inverse function Φ^{-1} is thus multiple-valued. The branch of Φ^{-1} that maps 0 to z_0 can be continued until a critical point of f is met or leave the original domain of f. The Riemann surface for Φ^{-1} is a branched covering surface over $\mathbf{C}$ (see Definition 3 (2.7.2.9) in Exercise A of Sec. 2.7.1 or 7.5), with its branch points resulted from the critical points of Φ which, by (5.3.4.9), are composed of the critical points of f and f^{-1}.

By a *critical point* of f, we mean a point where f ceases to be univalent. More precisely, in case f is a rational function, the critical points of f consist of these solutions of $f'(z) = 0$ and poles of order two or higher and they are all *algebraic* (see the Note in Exercise A(4) of Sec. 2.5.3 or (3.5.1.11)). The corresponding *critical values* (namely, the images under f of critical points) constitute the *algebraic branch points* for the Riemann surface for f^{-1}. If f is a transcendental meromorphic function (such as $\tan z$ and $\cot z$), in addition to the algebraic ones, the Riemann surface for f^{-1} also has *transcendental branch points* which are the exceptional or asymptotic values of f. Refer to Sec. 2.7 for concrete examples. □

We present an

Example 1. Find the fixed points of $f(z) = \frac{z(z+a)}{1+az}$ $(-1 < a < 0)$ and try to explain them.

Solution. Solve

$$f(z) = z, \quad \text{namely} \quad \frac{z(z+a)}{1-az} = z,$$

and we get the fixed points 0 and $\frac{1-a}{1+a}$. Since $f'(z) = \frac{-az^2 + 2a \, z + a}{(1-az)^3}$, $f'(0) = -a < 1$, and 0 is an attractive fixed point. While $f'(\frac{1-a}{1+a}) = -2\frac{a^3 + a^2 + a - 1}{(1+a^2)^2} > -(a^3 + a^2 + a - 1) > 1$ on the real interval $(-1, 0)$, $\frac{1-a}{1+a}$ is a repulsive fixed point of f.

According to Exercise B(1) of Sec. 2.5.3, $w = f(z)$ is a two-to-one map of $|z| < 1$ onto $|w| < 1$ and maps $|z| = 1$ onto $|w| = 1$. This can also be seen by using Rouché's theorem, observing that $f(z)$ has two distinct simple zeros in $|z| < 1$ and, for any fixed w with $|w| < 1$, $|w| < 1 = |f(z)|$ holds along $|z| = 1$.

Since $|f(z)| < 1$ if $|z| < 1$, it follows obviously that the basin $A(0)$ of attraction of f at 0 is the whole open disk $|z| < 1$. The conjugation $\zeta = \Phi(z)$ is univalent in an open neighborhood of 0. How about the extended $\Phi(z)$ in $|z| < 1$?

Fixed any $w \in \mathbf{C}$. Choose n large enough so that $(-a)^n w$ lies in the open unit disk. Consider the n-th iterate $f^n(z)$. Along $|z| = 1$, $|(-a)^n w| < 1 = |f^n(z)|$ holds. By Rouché's theorem, $f^n(z) - (-a)^n w = 0$ has finitely many distinct solutions in $|z| < 1$. Consequently, by $(*_{11})$, $\Phi_n(z) = \frac{f^n(z)}{(-a)^n}$ assumes w in $|z| < 1$ at finitely many distinct points. Since $\Phi_n(z)$ converges to the conjugation $\Phi(z)$ locally uniformly in an open neighborhood of 0 (see (5.3.4.7)), Hurwitz's theorem (5.3.2.1) says that $\Phi(z)$ assumes w in $|z| < 1$, even infinitely many times. Hence, the extended Φ is an infinite-to-one map of $|z| < 1$ onto the whole plane $\mathbf{C}$.

Note that, by exactly the same argument, the extended Φ will map the basin $A(z_0)$ onto the whole plane $\mathbf{C}$ if the original f is a rational function. $\square$

In case z_0 is a *repulsive fixed point* of f, since $|f'(z_0)| > 1$, $w = f(z)$ is univalent in a neighborhood of z_0 and hence, $w = f^{-1}(z)$ exists and is univalent in a neighborhood of z_0. Recall that $(f^{-1})'(z_0) = \frac{1}{f'(z_0)}$. It follows that (refer to (5.1.1.6))

$$(f^{-1})(z) = z_0 + \frac{z - z_0}{f'(z_0)} + \cdots$$

has an attractive fixed point at z_0 with the multiplier $\frac{1}{f'(z_0)}$. According to (5.3.4.7), there is a univalent mapping Φ conjugating $f^{-1}(z)$ to the canonical form $g(\zeta) = \frac{\zeta}{f'(z_0)}$, namely,

$$\Phi(f^{-1}(z)) = \frac{\Phi(z)}{f'(z_0)}, \quad \text{in a neighborhood of } z_0.$$

$$\Leftrightarrow \Phi(f(z)) = f'(z_0)\Phi(z), \quad \text{in a neighborhood of } z_0. \qquad (5.3.4.10)$$

Hence, *the same map* Φ *provides a conjugation of* $f(z)$ *to* $g(\zeta) = f'(z_0)\zeta$.

Example 2. Find the fixed points of $f(z) = (1+z)^m - 1$, where $m \geq 2$ is an integer.

Solution. 0 and ∞ are fixed points of f. $f'(0) = m$, so 0 is a repulsive fixed point, while $f'(\infty) = 0$ (see Remark in Sec. 3.3) and ∞ is a superattractive one which is intuitively clear.

To find the conjugation at 0: Consider the inverse function $f^{-1}(z) = (z+1)^{\frac{1}{m}} - 1$, single-valued in the deleted domain $\mathbf{C} - (\infty, -1]$. Observe that $f^{-n}(z) = (z+1)^{\frac{1}{m^n}} - 1$ for $n \geq 1$. According to $(*_{11})$ and Example 2 in Sec. 1.7,

$$\Phi_n(z) = \frac{f^{-n}(z)}{((f^{-1})'(0))^n} = m^n[(z+1)^{\frac{1}{m^n}} - 1] \to \Phi(z) = \mathrm{Log}(z+1),$$

$$\text{as } n \to \infty,$$

locally uniformly in $\mathbf{C} - (\infty, -1]$. $\Phi(z) = \mathrm{Log}(z+1)$, $z \in \mathbf{C} - (\infty, -1]$, is the conjugation of $f^{-1}(z)$ to the canonical form $g(z) = \frac{1}{m}z$; in fact,

$$\Phi(f^{-1}(z)) = \mathrm{Log}(f^{-1}(z) + 1) = \mathrm{Log}(z+1)^{\frac{1}{m}} = \frac{1}{m}\mathrm{Log}(z+1)$$

$$= \frac{1}{m}\Phi(z), \quad z \in \mathbf{C} - (\infty, -1].$$

$$\Leftrightarrow \Phi(f(z)) = \mathrm{Log}(f(z) + 1) = \mathrm{Log}(z+1)^m = m\mathrm{Log}(z+1)$$

$$= m\Phi(z), \quad z \in \mathbf{C} - (\infty, -1].$$

The same $\Phi(z) = \mathrm{Log}(z+1)$ is the conjugation of $f(z)$ to the canonical form $h(z) = mz$. Equivalently, the monomial z^m $(m \geq 2)$ has a repulsive fixed point at $z = 1$ and $g(z) = \mathrm{Log}\,z$, $z \in (\infty, 0]$, conjugates it to $h(z) = mz$. $\qquad \square$

In case z_0 is a *superattractive fixed point* of $f(z) = z_0 + \sum_{k=p}^{\infty} a_k(z-z_0)^k$, where $p \geq 2$ and $a_p \neq 0$, then an univalent map $\zeta = \Phi(z)$ conjugates $f(z)$ to $g(\zeta) = \zeta^p$. This Φ is unique, up to a multiplicative scale factor which is a $(p-1)$th root of unity (due to L. E. Boettcher (1904)). In the case of *neutral fixed points*, the situations turn out to be more complicated and dependent (on f). For all of these, refer to Ref. [15], pp. 33–51. Examples 1 and 2 above are adopted from this book.

Anyway, in what remains in this section, we try to sketch what happens to $f(z)$ if $|f'(z_0)| = 1$ or $f'(z_0) = 1$ holds in (5.3.4.6). Then, we try to use the existence of fixed points or the convergent behavior of some subsequence of the iterates $f^n(z)$ to characterize whether a mapping $f : \Omega \to \Omega$ is univalent and onto or not so.

Section (3) The case that $|f'(z_0)| = 1$ in (5.3.4.6) and "univalent and onto" property and more

Suppose $f : \Omega$ (domain) $\to \Omega$ is analytic and $z_0 \in \Omega$ is a fixed point of f. Also suppose that $f'(z_0) = 1$ holds. We try to show that $f(z) = z$ throughout Ω.

Let there be a smallest integer $k \geq 2$ such that $f^{(k)}(z_0) \neq 0$. For simplicity, suppose $z_0 = 0$ after a translation $z \to z + z_0$. Then $(*_9)$ becomes

$$f(z) = z + a_k z^k g(z), \ a_k \neq 0, \ g(0) = 1, \text{ and } g \text{ is analytic at } z_0 = 0.$$

$\Rightarrow$ (by $(*_{10})$) $f^{(n)}(z) = z + a_k z^k g_n(z)$, $g_n(0) = n$, and g_n is analytic at 0 with $g_1 = g$, $n \geq 1$.

Differentiate the above expression k times with respect to z and then, set $z = 0$. We get

$$(f^n)^{(k)}(0) = k! a_k g_n(0) = nk! a_k \neq 0, \quad k \geq 2, \quad n \geq 1.$$

Contradicting to the fact that $|(f^n)^k(0)| \leq \frac{k!}{r^k} M$ (see the formula two lines above (5.3.4.6)) if n is large enough. This very fact shows that $f^{(k)}(0) = 0$ for $k \geq 2$. It follows that $f(z) = z$ in a neighborhood of 0 and thus, throughout the whole Ω by the interior uniqueness theorem (see (3.4.2.9) or (5.1.2.4))).

What happens if $|f'(z_0)| = 1$ holds, instead of $f'(z_0) = 1$? We try to show that f is univalent and onto Ω itself if the complement $\mathbf{C} - \Omega$ contains at least two distinct points.

In case Ω is a bounded domain, then $f^n(z)$, $n \geq 1$, are uniformly bounded in Ω. In the general case, Montel's normality criterion (see (5.8.3.1)) says that $f^n(z)$, $n \geq 1$, constitute a normal family. In either case, $f^n(z)$ contains a subsequence $f^{n_j}(z)$, $j \geq 1$, converges to an analytic function $h(z)$ locally uniformly in Ω (refer to (2) in (5.3.3.10)). Since $(f^n)'(z_0) = (f'(z_0))^n$ for $n \geq 1$ (see $(*_7)$) and $\lim_{j \to \infty} (f^{n_j})'(z_0) = h'(z_0)$

(see (5.3.1.1)), hence

$$|(f^{n_j})'(z_0)| = 1, \quad j \geq 1$$

$$\Rightarrow |h'(z_0)| = 1.$$

In particular, $h(z)$ is not a constant. According to (1) in (5.3.4.12) below, f is necessarily univalent and onto.

Conversely, if $f : \Omega \to \Omega$ is univalent and onto, and $f(z_0) = z_0$ for some $z_0 \in \Omega$, then, by (5.3.4.6), $|f'(z_0)| \leq 1$ and $|(f^{-1})'(z_0)| \leq 1$ hold simultaneously. Since $(f^{-1})'(z_0) = f'(z_0)^{-1}$, $|f'(z_0)| = 1$ holds.

As far as the univalent and onto property is concerned, there is another sufficient condition. Suppose $f^n(z)$, $n \geq 1$, has a subsequence $f^{n_j}(z)$, $j \geq 1$, converging to the identity mapping id on Ω locally uniformly. Then $f(z)$ must be *univalent* and *onto* on Ω. To see this, suppose $z_1, z_2 \in \Omega$ so that $f(z_1) = f(z_2)$ holds. Then, $f^{n_j}(z_1) = f^{n_j}(z_2)$ automatically holds, too, for each $j \geq 1$. Letting $j \to \infty$, then id$(z_1) =$id(z_2) just means that $z_1 = z_2$. This is the univalence. In case $f(z)$ is not onto, namely, $f(\Omega) \subsetneqq \Omega$ holds. Consequently, $f^n(\Omega) \subseteq f(\Omega) \subsetneqq \Omega$ holds for each $n \geq 2$. Choose any $w_0 \in \Omega - f(\Omega)$. It follows that $w_0 \in \Omega - f^n(\Omega)$ for $n \geq 2$; in particular, $f^{n_j}(z) - w_0 \neq 0$ in Ω for $j \geq 1$. Since $f^{n_j}(w_0) - w_0 \to$ id$(w_0) - w_0 = w_0 - w_0 = 0$ as $j \to \infty$, by (5.3.3.14), $f^{n_j}(z) - w_0$ converges to 0 locally uniformly in Ω. This fact results in id$(z) = w_0$ in Ω, which is not possible. Hence $f(\Omega) = \Omega$ and f is thus onto.

We summarize as the following

"Univalent and onto" property of an analytic function mapping a domain into itself. Let $f : \Omega$ (domain) $\to \Omega$ be analytic.

(1) If $z_0 \in \Omega$ is a fixed point of f such that $f'(z_0) = 1$ holds, then $f(z) = z$, the identity mapping of Ω.
(2) In case the complement $\mathbf{C} - \Omega$ contains at least two distinct points and $z_0 \in \Omega$ is a fixed point of f, then

$$|f'(z_0)| = 1$$

$$\Leftrightarrow f(z) \text{ maps } \Omega \text{ onto itself univalently.}$$

In particular, if $w = f(z) : \Omega \to \Omega$ is univalent and onto, $f(z_0) = z_0 \in \Omega$ and $f'(z_0) > 0$, then $f(z) = z$ for all $z \in \Omega$.
(3) Suppose the iterate sequence $f^n(z)$, $n \geq 1$, contains a subsequence converging to the identity mapping on Ω, locally uniformly in Ω. Then f maps Ω onto itself univalently.

(4) In case $\mathbf{C} - \Omega$ contains at least two points and f has two distinct fixed points in Ω, then f maps Ω onto itself univalently. (5.3.4.11)

An illustrative example for (2) is provided by the non-Euclidean rotation stated in (5.3.4.5). The bilinear transformation $f(z) = \frac{az+b}{cz+a}$ ($c \neq 0$ and $a^2 + bc \neq 0$) maps $\mathbf{C} - \{\frac{a}{c}\}$ onto itself univalently with two distinct fixed points $\frac{a+\sqrt{a^2+bc}}{c}$, each having its multiplier of absolute value equal to 1. Note that, even though $f(z) = az(|a| \neq 1)$ maps $\mathbf{C}$ onto itself univalently and $f(0) = 0$, yet $|f'(0)| = |a| \neq 1$ holds. The assumption in (3) is not a necessary condition for "univalent and onto". For instance, $f(z) = az(|a| > 1)$: $\mathbf{C} \to \mathbf{C}$ is such a case. So does in (4) as the non-Euclidean parallel motion in (5.3.4.5) indicates. Observe that $f(z) = az(|a| \neq 1)$ maps $\mathbf{C} - \{0\}$ onto itself univalently but $f(z)$ does not possess any fixed point; while $g(z) = \frac{1}{z}$ maps the same domain $\mathbf{C} - \{0\}$ univalently onto itself, and $g(z)$ has two distinct fixed points ± 1. (4) follows obviously (5.3.4.12).

Section (4) The cases that $f(z)$ fails to be univalent or onto

Suppose Ω is a *bounded* domain in $\mathbf{C}$. In what follows, $f(z)$ is *assumed* to be an analytic function, not univalent on Ω or even univalent but not onto Ω.

Suppose the iterate sequence $f^n(z)$, $n \geq 1$, has a subsequence $f^{n_j}(z)$, $j \geq 1$, converging pointwise and hence locally uniformly (see (1) in (5.3.3.10)) to a nonconstant analytic function $g(z)$ in Ω. As a nonempty open subset of $\bar{\Omega}$, $g(\Omega)$ contains points in Ω. More precisely, $g(\Omega) \subseteq \Omega$ should hold. Otherwise, there is a point $z_0 \in \Omega$ so that $g(z_0) = w_0 \notin \Omega$. Applying Hurwitz's theorem (see (1) in (5.3.2.2)) to $f^{n_j}(z) - w_0$ which converges locally uniformly to $g(z) - w_0$ in Ω, we will deduce that $g(z) = w_0$ for all $z \in \Omega$, a contradiction.

Let $m_j = n_{j+1} - n_j$, $j \geq 1$. According to (2) in (5.3.3.10), $f^{m_j}(z)$, $j \geq 1$, has a subsequence $f^{m_{j_k}}(z)$, $k \geq 1$, converging locally uniformly in Ω to an analytic function $h(z)$.

Fix any point $w_0 \in g(\Omega)$. There exists a point $z_0 \in \Omega$ so that $w_0 = g(z_0)$. Since $f^{n_j}(z_0) \to g(z_0) = w_0$, so does $f^{n_{j_k}}(z_0)$ as $k \to \infty$; hence,

$$\lim_{k \to \infty} f^{m_{j_k}} \circ f^{n_{j_k}}(z_0) = \lim_{k \to \infty} f^{n_{j_k}+1}(z_0) = w_0 \qquad (*13)$$

On the other hand, choose $r > 0$ so that $|z - z_0| \leq r$ is contained in Ω. Since $f^{m_j}(z)$, $j \geq 1$, converges uniformly to $h(z)$ on the compact set $\cap_{k \geq 1}^{\infty} f^{n_k}(\{|z - z_0| \leq r\}) \neq \emptyset$, it follows that

$$\lim_{k \to \infty} f^{m_{j_k}} \circ f^{n_{j_k}}(z_0) = h(w_0)$$

holds. Combining together,

$$h(w) = w, \quad w \in g(\Omega) \subseteq \Omega. \tag{$*_{14}$}$$

By the interior uniqueness theorem, $h(w) = w$ throughout on Ω. Yet (3) in (5.3.4.11) says, under this circumstance, that $f(z)$ is both univalent and onto Ω, a contradiction to our assumption.

In conclusion, under our assumption, any convergent subsequence of $f^n(z)$, $n \geq 1$, should converge to a constant function.

In addition to the previous assumption, *suppose* that $f(z)$ has a fixed point z_0 in Ω. Since $f^n(z_0) = z_0$ holds for $n \geq 1$, by the last paragraph, any convergent subsequence of $f^n(z)$, $n \geq 1$, will always converge locally uniformly in Ω to the point z_0. So does the original iterate sequence $f^n(z)$, $n \geq 1$, as indicated in Exercise A(3) of Sec. 5.3.3.

As a matter of fact, the boundedness of Ω can be replaced by the assumption that *the complement* $\mathbf{C} - \Omega$ *contains at least two distinct points.* Under this circumstance, $f^n(z)$, $n \geq 1$, form a *normal family* (see (5.8.3.1)) which means that any subsequence of $f^n(z)$ contains another subsequence converging locally uniformly in Ω. In order to prove this claim, observe that the afore-mentioned $f^{m_j}(z)$, $j \geq 1$, has a subsequence $f^{m_{j_k}}(z)$, $k \geq 1$, converging locally uniformly in Ω to ∞ *or* to an analytic function $h(z)$. Since, for a fixed point $z_0 \in \Omega$, the set $K = \{h(z_0) \cup \{f^{n_j}(z_0)|j \geq 1\}$ is a compact set in Ω, the functional relation $f^{m_{j_k}} \circ f^{n_{j_k}} = f^{n_{j_k}+1}$ shows that $f^{m_{j_k}}(z)$, $k \geq 1$, cannot converge to ∞ in K but converges to $h(z)$ instead. Consequently, $f^{m_{j_k}}(z)$, $k \geq 1$, converges locally uniformly in Ω to $h(z)$. And finally, the same process from $(*_{13})$ to $(*_{14})$ will prove the claim.

Summarize the above as

The cases that $f(z)$ fails to be univalent or onto. Let Ω be a domain in $\mathbf{C}$ whose complement $\mathbf{C} - \Omega$ contains at least two distinct points. Suppose $f : \Omega \to \Omega$ is analytic, neither univalent nor onto. Then,

(1) any convergent subsequence of the iterates $f^n(z)$, $n \geq 1$, converges pointwise (and hence, locally uniformly) to a constant function in Ω;

(2) in case $f(z)$ has a fixed point z_0 in Ω, the iterates $f^n(z)$, $n \geq 1$, themselves converge to z_0 locally uniformly in Ω.   (5.3.4.12)

Recall that (4) in (5.3.4.11) is an obvious consequence of results stated here.

More concrete result can be obtained if the domain Ω is an open disk. We state as

The case (5.3.4.12) where Ω is the open unit disk. Let $f : \{|z| < 1\} \to \{|z| < 1\}$ be analytic, neither univalent nor onto. Then,

(1) either $f(z)$ has only one fixed point z_0 in $|z| < 1$, and the iterate sequence $f^n(z)$ converges to z_0 locally uniformly in $|z|$,

(2) or $f(z)$ does not have any fixed point in $|z| < 1$, and $f^n(z)$ converges to a constant locally uniformly in $|z| < 1$ such that $|f^n(z)| \to 1$ locally uniformly, too.   (5.3.4.13)

(1) is an easy consequence of the uniform boundedness of $f^n(z)$, $n \geq 1$, (5.3.3.10) and (5.3.4.12). A complete proof of the former part in (2) needs to use the concepts of Lebesgue measure zero and Fatou's radial limit theorem (see (5.1.3.4)) and one may refer to Ref. [12] and, *Craig and Macintyre: Inequalities for functions regular and bounded in a circle. Pac. J. Math. 20 (1967), 449–454.* For the later part of (2), see Exercise A(9). Via Riemann mapping theorem (see Chap. 6), these results are still valid in a simply connected domain $\Omega \subsetneq \mathbf{C}$. For extension to multiply-connected domains, see *M. Heins: On the iteration of functions which are analytic and single-valued in a given multiply-connected region. Am. J. Math. 63 (1941), 461–480.*

Exercises A

(1) Let $f(z) = \frac{z-a}{1-az}$, $-1 < a < 0$. Show that f has an attractive fixed point at 1 with multiplier $\frac{1+a}{1-a}$ and a repulsive fixed point at -1 with multiplier $\frac{1-a}{1+a}$. Also, show that $f^n(z) \to 1$ in $|z| < 1$ by the following two methods:

1. Rewrite $w = f(z)$ as $\frac{w-1}{w+1} = \frac{1+a}{1-a}\frac{z-1}{z+1}$ (see $(*_3)$ and $(*_4)$).
2. Show that $f_n(z) = \frac{z-a_n}{1-a_n z}$, where $a_1 = a$ and $a_n = \frac{a+a_{n-1}}{1+aa_{n-1}}$, $n \geq 2$, and then try to see if $\lim_{n \to \infty} a_n$ exists.

(2) Prove (5.3.4.3) in detail.
(3) Prove (5.3.4.4) in detail.
(4) Prove (5.3.4.5) in detail.
(5) Let $f(z) = z^2 - 2$.

   (a) Show that $f(z)$ has a superattractive fixed point at ∞. Also, $\Phi(\zeta) = \zeta + \frac{1}{\zeta}$ ($|\zeta| > 1$) conjugates $f(z)$ to $g(\zeta) = \zeta^2$ so that the basin of attraction of ∞ for $f(z)$ is the set $A(\infty) = \mathbf{C}^* - [-2, 2]$.
   (b) The set of nonnormality, namely,

$$\Im(f) = \{z \in \mathbf{C}^* \,|\, \text{the iterates } f^n(z), n \geq 1, \text{do not form a}$$

$$\text{normal family}\},$$

   is called the *Julia set* of an analytic function f; and the complement of it is called the *Fatou set* $F(f)$. $F(f)$ is open while $\Im(f)$ is closed. In this case, show that $F(f) = A(\infty)$ and $\Im(f) = [-2, 2]$.

(6) The polynomial $T_n(z) = \cos(n \operatorname{Arc} \cos z)$ is called the nth *Tchebychev polynomial*. Set the monic polynomial $f(z) = 2T_n(\frac{z}{2})$. Show that $z = h(\zeta) = \zeta + \frac{1}{\zeta}$ conjugates $f(z)$ to $g(\zeta) = \zeta^n (n \geq 2)$. Find the basin of attraction at ∞ for $g(\zeta)$, its Fatou set and Julia set. What are the corresponding information for $f(z)$?

(7) Suppose $\alpha > 0$, $\beta > 0$, and $\alpha + \beta = 1$. Set

$$f(z) = \alpha z + \frac{\beta}{z}, \quad z \in \Omega = \mathbf{C} - \{0\}.$$

Show $f^n(z)$, $n \geq 1$, converges locally uniformly to 1 in $\operatorname{Re} z > 0$ and to -1 in $\operatorname{Re} z < 0$. Determine the basins of attraction, Julia set, and Fatou set of f. In case $f(z)$ is redefined as $\alpha z + \frac{\beta}{z}$, where α and β are nonzero complex numbers (refer to Exercise B(1) of Sec. 2.5.5), what happens to results obtained in the previous case?

(8) Let $\Omega = \{z \mid |z| < 1, \operatorname{Re} z > 0 \text{ and } \operatorname{Im} z > 0\}$, the part of $|z| < 1$ that lies in the first quadrant. Show that $w = f(z) = e^{\frac{1}{2}\pi i z}$ maps Ω into Ω and has a fixed point in Ω. Use this to show that i_n converges as $n \to \infty$, where $i_1 = i$ and $i_{n+1} = i^{i_n} = e^{\frac{1}{2}\pi i i_n}$, $n \geq 1$.

(9) Prove that $|f^n(z)| \to 1$ locally uniformly in $|z| < 1$ as stated in (2) of (5.3.4.13).

Exercises B

(1) Let Ω be a domain whose complement $\mathbf{C} - \Omega$ contains at least two distinct points. Let $a \in \Omega$ be a fixed point. Set

$G_a = \{$an analytic function f mapping Ω onto itself univalently

and preserving a fixed$\}$.

(a) Show that G_a is a group under composition and is group isomorphic to a subgroup of the multiplicative group $|z| = 1$. Hence, G_a is abelian.

(b) Suppose $f_n(z) \in G_a$, $n \geq 1$, and converges to a function $f(z)$. Show that $f(z) \in G_a$.

(c) In case $f_n(z) \in G_a$, $n \geq 1$, and $f_n'(a)$ converges, then $f_n(z)$ converges to a function $f(z) \in G_a$.

(d) Define a mapping $\Phi : G_a \to \{|z| = 1\}$ by

$$\Phi(f(z)) = f'(a).$$

Then Φ is continuous (namely, once $f_n(z) \to f(z)$ locally uniformly, then so is $f_n'(a) \to f'(a)$) and so is $\Phi^{-1} : \Phi(G_a) \to G_a$. Consequently,

G_a can only be isomorphic to

1. either a finite cyclic subgroup of $|z| = 1$,
2. or the whole unit circle $|z| = 1$.

Note. By Riemann mapping theorem (see Chap. 6), in case Ω is a simply connected domain whose boundary has at least two distinct points, G_a is necessary isomorphic to $|z| = 1$; on the contrary, Aumann and Carathéodory (1934) proved that G_a is isomorphic to a finite cyclic subgroup if Ω is not so.

(2) (Shields, 1964) Let $\mathfrak{F}$ be a family of functions $f(z)$ satisfying the following properties:

1. $f(z) : \{|z| \leq 1\} \to \{|z| \leq 1\}$ is continuous;
2. $f(z)$ is analytic in $|z| < 1$, and
3. $f \circ g(z) = g \circ f(z)$, $|z| \leq 1$, for all such $f(z)$ and $g(z)$.

Show that members in $\mathfrak{F}$ have a common fixed point.

(3) Let $f : \Omega \to \Omega$ be analytic, where Ω is a domain whose complement $\mathbf{C} - \Omega$ contains at least two distinct points. Suppose there is a point $z_0 \in \Omega$ so that the sequence $f^n(z_0)$ converges to a point w_0 in Ω. Then

1. either f is univalent and onto Ω,
2. or $f(w_0) = w_0$ and the iterate sequence $f^n(z)$ converges to w_0 locally uniformly in Ω.

5.4   Meromorphic Functions: Mittag–Leffler's Partial Fractions Theorem

It might be constructive and instructive to learn Sec. 5.5.2 (in particular, (5.5.2.10)) in advance rather than the material in this section, at least, for logical reasons. But for some technical problems in handing the infinite product representations concerning entire function with simple zeros (see Examples in Sec. 5.5.2 and the content of Sec. 5.4.2), it is harmless, for pedagogical purpose, to sketch ahead some properties of meromorphic functions.

A meromorphic function $f(z)$ in the whole plane can be classified as one of the following three types according to its behavior at ∞:

1. $f(z)$ has a pole or removable singularity at ∞.

   $\Leftrightarrow$ $f(z) = \frac{p(z)}{q(z)}$, the quotient of two relatively prime polynomials (see Example in Sec. 2.1, the rational functions in Sec. 2.5.3 and Example 6 in Sec. 4.10.2).

$\Leftrightarrow$ $f(z) = g(z) + c + \sum_{j=1}^{k} g_j(\frac{1}{z-a_j})$, where $g(z)$ and $g_j(z)$, $1 \le j \le k$, are polynomials (see the partial fractions expansion in (2.5.3.5) and Example 6 in Sec. 4.10.2).

2. $f(z)$ has an essential singularity at ∞ (a transcendental one, and hence, has only finite many poles in $\mathbf{C}$).

$\Leftrightarrow$ $f(z) = g(z) + \sum_{j=1}^{k} g_j(\frac{1}{z-a_j})$, where $g(z)$ is a transcendental entire function; $g_j(z)$, $1 \le j \le k$, are polynomials.

3. $f(z)$ has ∞ as the limit of its poles (a transcendental one).   (5.4.1)

The main theme in Sec. 5.4.1 is to expand function $f(z)$ of type 3 into a convergent series whose terms are principal parts $R_n(\frac{1}{z-a_n})$ at poles a_n, $n \ge 1$, such as

$$g(z) + \sum_{n=1}^{\infty} \left(R_n \left(\frac{1}{z - a_n} \right) - P_n(z) \right) \tag{5.4.2}$$

where $g(z)$ is entire; $P_n(z)$, $n \ge 1$, are suitably chosen polynomials acting as convergence-producing terms.

Under some constrained conditions on the growth of $f(z)$, Cauchy used the residue methods to show that $g(z)$ in (5.4.2) indeed is a polynomial. This is shown in Sec. 5.4.2, in addition to some illustrative examples.

5.4.1 Mittlag–Leffler's partial fractions expansion for meromorphic functions

Given a sequence a_n, $n \ge 1$, of complex numbers with $\lim_{n\to\infty} a_n = \infty$. Let $R_n(z)$, $n \ge 1$, be polynomials without constant term. We try to construct a meromorphic function in $\mathbf{C}$ with poles at a_n, $n \ge 1$, and the corresponding singular or principal parts $R_n(\frac{1}{z-a_n})$, $n \ge 1$.

Suppose $a_n \ne 0$ for some fixed n. It is well-known already (see (3.3.2.4)) that

$$\frac{1}{z - a_n} = -\frac{1}{a_n} \sum_{k=0}^{\infty} (\frac{z}{a_n})^k \text{ converges absolutely}$$

$$\text{and locally uniformly in } |z| < |a_n|. \tag{$*_1$}$$

$\Rightarrow$ $\frac{1}{(z-a_n)^l}$ can be represented as a power series converging absolutely and locally uniformly in $|z| < |a_n|$ for $l \ge 1$.

$\Rightarrow$ $R_n(\frac{1}{z-a_n})$ can be represented as a power series converging absolutely and locally uniformly in $|z| < |a_n|$.

Let $P_n(z)$ be the first k_n-th partial sums of the power series expansion of $R_n(\frac{1}{z-a_n})$ so that

$$\left| R_n\left(\frac{1}{z-a_n}\right) - P_n(z) \right| \leq \frac{1}{2^n} \quad \text{on } |z| \leq \frac{1}{2}|a_n|. \tag{$*_2$}$$

Try to show that this $P_n(z)$ will work as claimed in (5.4.2).

To see this, let $R > 0$ be given. There is an integer $N \geq 1$ so that $|a_n| > 2R$ for $n \geq N$.

Now, consider the closed disk $|z| \leq R$ on which $|z| < \frac{1}{2}|a_n|$ for each $n \geq N$. By $(*_1)$ and $(*_2)$,

$$\sum_{n=N}^{\infty} [R_n(\frac{1}{z-a_n}) - P_n(z)] \text{ converges absolutely and uniformly in}$$

$|z| \leq R$ to an analytic function. $\tag{$*_3$}$

$$\Rightarrow R_0\left(\frac{1}{z}\right) + \sum_{n=1}^{N-1} \left[R_n\left(\frac{1}{z-a_n}\right) - P_n(z) \right]$$

$$+ \sum_{n=N}^{\infty} \left[R_n\left(\frac{1}{z-a_n}\right) - P_n(z) \right]$$

$$= R_0\left(\frac{1}{z}\right) + \sum_{n=1}^{\infty} \left[R_n\left(\frac{1}{z-a_n}\right) - P_n(z) \right]$$

converges absolutely and locally uniformly to an analytic function in

$|z| < R$ except at poles $a_1, \ldots, a_N$ and possibly at 0 with

the singular part $R_0(\frac{1}{z})$. $\tag{$*_4$}$

In conclusion, we obtain the following

Mittag–Leffler's partial fractions theorem. Let a_n, $n \geq 1$, be a sequence of distinct complex numbers with $\lim_{n\to\infty} a_n = \infty$ and let $R_n(z)$, $n \geq 1$, be polynomials without constant terms.

(1) By suitably chosen polynomials $P_n(z)$, $n \geq 1$, the series

$$\sum_{n=1}^{\infty} \left[R_n\left(\frac{1}{z-a_n}\right) - P_n(z) \right]$$

converges absolutely and locally uniformly in $\mathbf{C} - \{a_n | n \geq 1\}$ to a meromorphic function with poles at a_n, $n \geq 1$, and the associated singular parts $R_n(\frac{1}{z-a_n})$.

(2) The most general meromorphic function of this kind is of the form

$$f(z) = g(z) + \sum_{n=1}^{\infty} \left[R_n \left(\frac{1}{z - a_n} \right) - P_n(z) \right]$$

where $g(z)$ is an entire function.   (5.4.1.1)

Refer to (7.3.3.6) for a general setting.

To practice the process of proof for (5.4.1.1), we give two examples at the present. See Sec. 5.4.2 for more concrete examples.

Example 1. Let a_n, $n \geq 1$, be a sequence of distinct nonzero complex numbers converging to ∞.

(a) Find a meromorphic function in $\mathbf{C}$ with simple poles at a_n, $n \geq 1$, and the corresponding singular parts $\frac{1}{z-a_n}$, $n \geq 1$.
(b) Do (a) again with "simple" replaced by "double".

Solution. (a) In $(*_1)$, choose

$$P_n(z) = -\left(\frac{1}{a_n} + \frac{z}{a_n^2} + \cdots + \frac{z^{n-1}}{a_n^n} \right), \quad |z| < |a_n| \quad \text{for } n \geq 1.$$

In this case, on $|z| \leq \frac{1}{2}|a_n|$, $(*_2)$ becomes (recall that $R_n(z) = z$ for $n \geq 1$)

$$\left| R_n \left(\frac{1}{z - a_n} \right) - P_n(z) \right| = \left| \sum_{k=n+1}^{\infty} \frac{z^{k-1}}{a_n^k} \right| \leq \sum_{k=n+1}^{\infty} \frac{|z|^{k-1}}{|a_n|^k} \leq \frac{1}{|a_n|} \sum_{k=n+1}^{\infty} \frac{1}{2^{k-1}}$$

$$= \frac{1}{|a_n|} \cdot \frac{1}{2^{n-1}}, \quad n \geq 1.$$

Fix $R > 0$. Choose an integer $N \geq 1$ so that $|a_n| > 2R$ if $n \geq N$. Then, on the set $|z| \leq R$,

$$\sum_{n=k}^{l} \left| R_n \left(\frac{1}{z - a_n} \right) - P_n(z) \right| \leq \sum_{n=k}^{l} \frac{1}{|a_n|} \frac{1}{2^{n-1}} \leq \frac{1}{|a_k|} \sum_{n=k}^{l} \frac{1}{2^{n-1}}$$

for all $l \geq k \geq N$. Here it is tacitly assumed that $|a_n| \leq |a_{n+1}|$ for $n \geq 1$. Hence $(*_3)$ holds and so does $(*_4)$. Consequently, the most general meromorphic function required is in the form

$$f(z) = g(z) + \sum_{n=1}^{\infty} \left(\frac{1}{z - a_n} + \frac{1}{a_n} + \frac{z}{a_n^2} + \cdots + \frac{z^{n-1}}{a_n^n} \right), \qquad (5.4.1.2)$$

where $g(z)$ is entire.

Suppose if it is possible to find a *largest nonnegative integer* α for which the series

$$\sum_{n=1}^{\infty} \frac{1}{|a_n|^{\alpha}} \tag{$*_5$}$$

diverges. Reconsider the estimate in $(*_2)$ as, setting $P_n(z) = -\left(\frac{1}{a_n} + \frac{z}{a_n^2} + \cdots + \frac{z^{\alpha-1}}{a_n^{\alpha}}\right)$,

$$\left| R_n\left(\frac{1}{z-a_n}\right) - P_n(z) \right| \le \sum_{k=\alpha+1}^{\infty} \frac{|z|^{k-1}}{|a_n|^k} = \frac{|z|^{\alpha}}{|a_n|^{\alpha+1}} \sum_{k=0}^{\infty} \left(\frac{|z|}{|a_n|}\right)^k$$

$$\le \frac{|z|^{\alpha}}{|a_n|^{\alpha+1}} \sum_{k=0}^{\infty} \left(\frac{1}{2}\right)^k = 2\frac{|z|^{\alpha}}{|a_n|^{\alpha+1}}, \quad \text{if } |a_n| > 2|z|.$$

In this case, $(*_3)$ still holds in $|z| \le R$. As a result, *under the condition* $(*_5)$, the most general form can be simplified as

$$f(z) = g(z) + \sum_{n=1}^{\infty} \left(\frac{1}{z-a_n} + \frac{1}{a_n} + \frac{z}{a_n^2} + \cdots + \frac{z^{\alpha-1}}{a_n^{\alpha}}\right) \tag{5.4.1.3}$$

in which the adding polynomials are of the same degree α.

(b) Owing to the local uniform convergence of the series (5.4.1.2) and (5.4.1.3) in the set $\mathbf{C} - \{a_1, a_2, \ldots, a_n, \ldots\}$, the termwise differentiation is permissible. Hence, the required one is of the form

$$f(z) = \begin{cases} g(z) + \displaystyle\sum_{n=1}^{\infty} \left(\frac{1}{(z-a_n)^2} - \frac{1}{a_n^2} - \frac{2z}{a_n^3} - \cdots - \frac{(n-1)z^{n-2}}{a_n^n}\right), \\ \qquad \text{in case of (5.4.1.2)} \\[4pt] g(z) + \displaystyle\sum_{n=1}^{\infty} \left(\frac{1}{(z-a_n)^2} - \frac{1}{a_n^2} - \frac{2z}{a_n^3} - \cdots - \frac{(\alpha-1)z^{\alpha-2}}{a_n^{\alpha}}\right), \\ \qquad \text{in case of (5.4.1.3)} \end{cases}$$

$$\tag{5.4.1.4}$$

where $g(z)$ is entire.

In particular, if

$$\sum_{n=1}^{\infty} \frac{1}{|a_n|^3} < \infty, \tag{$*_6$}$$

then (5.4.1.4) is then simplified to

$$f(z) = g(z) + \sum_{n=1}^{\infty} \left(\frac{1}{(z-a_n)^2} - \frac{1}{a_n^2}\right), \quad \text{where } g(z) \text{ is entire,} \tag{5.4.1.5}$$

disregarding whether or not the series $\sum_{1}^{\infty} \frac{1}{|a_n|^2}$ diverges.

Example 2 (Ref. [1], p. 188). Show that

$$\frac{\pi^2}{\sin^2 \pi z} = \sum_{n=-\infty}^{\infty} \frac{1}{(z-n)^2}, \quad z \in \mathbf{C} - \{0, \pm 1, \pm 2, \ldots\}. \tag{5.4.1.6}$$

Then, deduce that (refer to Exercise B(10) of Sec. 5.3.1)

$$\pi \cot \pi z = \lim_{n \to \infty} \sum_{k=-n}^{n} \frac{1}{z-k} = \frac{1}{z} + \sum_{n \neq 0} \left(\frac{1}{z-n} + \frac{1}{n} \right) = \frac{1}{z} + \sum_{n=1}^{\infty} \frac{2z}{z^2 - n^2};$$

$$\frac{\pi}{\sin \pi z} = \lim_{n \to \infty} \sum_{k=-n}^{n} (-1)^k \frac{1}{z-k} = \frac{1}{z} + \sum_{n=1}^{\infty} (-1)^n \frac{2z}{z^2 - n^2}. \tag{5.4.1.7}$$

Solution. $\frac{\pi^2}{\sin^2 \pi z}$ has double poles at integral points $n = 0, \pm 1, \pm 2, \ldots$.

The singular part at $z = 0$ is $\frac{1}{z^2}$. To see this, perform some algorithmic operations to $\sin \pi z = \sum_{n=0}^{\infty} \frac{(-1)^n}{(2n+1)!} (\pi z)^{2n+1}$ (see Sec. 5.1.1) to find the Laurent series expansion of $\frac{\pi^2}{\sin^2 \pi z}$ at $z = 0$ and get the result. Alternatively, find the following limits:

$$\lim_{z \to 0} \frac{\pi^2 z^2}{\sin^2 \pi z} = \lim_{z \to 0} \frac{2\pi z}{\sin 2\pi z} = 1,$$

$$\operatorname{Re} s \left(\frac{\pi^2}{\sin^2 \pi z}; 0 \right) = \lim_{z \to 0} \frac{d}{dz} \frac{\pi^2 z^2}{\sin^2 \pi z} = \lim_{z \to 0} z \left(\frac{\pi^2}{\sin^2 \pi z} - \frac{1}{z^2} \right) = 0;$$

and the singular part is $\frac{0}{z} + \frac{1}{z^2} = \frac{1}{z^2}$. Since $\sin^2 \pi (z - n) = \sin^2 \pi z$, the singular part at $z = n$ is then $\frac{1}{(z-n)^2}$ for $n = 0, \pm 1, \pm 2, \ldots$.

Consider the series (refer to (5.3.1.7) and (5.3.1.8) and Exercise A(19) of Sec. 5.3.1)

$$\sum_{n=-\infty}^{\infty} \frac{1}{(z-n)^2}.$$

By comparing to the series $\sum_{1}^{\infty} \frac{1}{n^2}$, it easily seems that this series converges absolutely and locally uniformly in $\mathbf{C} - \{0, \pm 1, \pm 2, \ldots\}$. Hence, we set

$$\frac{\pi^2}{\sin^2 \pi z} = g(z) + \sum_{n=-\infty}^{\infty} \frac{1}{(z-n)^2} \tag{$*_7$}$$

where $g(z)$ is entire, and try to show that $g(z)$ is a constant via Liouville's theorem.

Observe that $g(z)$ has period 1 since both the function $\frac{\pi^2}{\sin^2 \pi z}$ and the series in $(*_7)$ do. Thus it is sufficient to show that $g(z)$ is bounded on $0 \leq \operatorname{Re} z \leq 1$, and, in particular, is bounded on the set $A = \{z \in \mathbf{C} | 0 \leq \operatorname{Re} z \leq 1$ and $|\operatorname{Im} z| \geq 1\}$.

Recall that $|\sin^2 \pi z| \geq \sinh^2 \pi y > \sinh^2 \pi$ (see (2.6.2.6)) for $z = x + iy \in A$. Hence

$$\left| \frac{\pi^2}{\sin^2 \pi z} \right| \leq \frac{\pi^2}{\sinh^2 \pi y} \leq \frac{\pi^2}{\sinh^2 \pi} < \infty \quad \text{in } A.$$

While $\sum_{n=-\infty}^{\infty} \frac{1}{(z-n)^2}$ converges absolutely and uniformly in A by observing that

$$\sum_{-\infty}^{\infty} \frac{1}{|z - n|^2} \leq 2 \sum_{0}^{\infty} \frac{1}{n^2 + y^2} \leq 2 \sum_{0}^{\infty} \frac{1}{n^2 + 1} < \infty \quad \text{for } z \in A.$$

$(*_7)$ then implies that $g(z)$ is bounded in A which, in turn, implies that $g(z)$ is bounded in $\mathbf{C}$. $g(z)$ must then reduce to a constant c. On the other hand,

$$\lim_{\substack{z \in A \to \infty}} |g(z)| \leq \lim_{|y| \to \infty} \left\{ \frac{\pi^2}{\sinh^2 \pi y} + 2 \sum_{0}^{\infty} \frac{1}{n^2 + y^2} \right\} = 0.$$

It follows that $g(z) = c = 0$ in $\mathbf{C}$. This proves (5.4.1.6).

Termwise integration of (5.4.1.6) is permissible because the series is locally uniformly convergent in $\mathbf{C} - \{0, \pm 1, \pm 2, \ldots\}$. The resulted expression is

$$\pi \cot \pi z = c + \frac{1}{z} + \sum_{n \neq 0} \left(\frac{1}{z - n} + \frac{1}{n} \right) \tag{$*_8$}$$

for some constant c. Note that $\sum_{n \neq 0} \frac{1}{z-n}$ diverges and the constant term $\frac{1}{n}$ of the Taylor series expansion of $\frac{1}{z-n}$ is added to each $\frac{1}{z-n}$ in order to guarantee that the series in $(*_8)$ converges absolutely and locally uniformly in $\mathbf{C} - \{0, \pm 1, \pm 2, \ldots\}$. The absolute convergence grants that the terms corresponding to n and $-n$ can be bracketed together. Then $(*_8)$ becomes

$$\pi \cot \pi z = c + \frac{1}{z} + \sum_{n \neq 0} \left(\frac{1}{z - n} + \frac{1}{n} \right) = c + \frac{1}{z} + \sum_{n=1}^{\infty} \frac{2z}{z^2 - n^2}. \tag{$*_9$}$$

The last expression has the advantage that both $\pi \cot \pi z$ and $\sum_{n=1}^{\infty} \frac{2z}{z^2-n^2}$ are odd functions of z and thus the constant $c = 0$.

Argument similar to the last paragraph shows that

$$\lim_{n\to\infty} \sum_{k=-n}^{n} \frac{(-1)^k}{z-k} = \frac{1}{z} + \sum_{n=1}^{\infty} (-1)^n \frac{2z}{z^2-n^2} \qquad (*10)$$

represents a meromorphic function in $\mathbf{C} - \{0, \pm 1, \pm 2, \ldots\}$. Note that the series $\sum_{-\infty}^{\infty} \frac{(-1)^k}{z-k}$ converges uniformly in $\mathbf{C} - \{0, \pm 1, \pm 2, \ldots\}$ but converges absolutely nowhere (see Example 2 in Sec. 2.3), while the series in the right converges absolutely and locally uniformly there. Since

$$\sum_{-(2n+1)}^{2n+1} \frac{(-1)^k}{z-k} = \sum_{k=-n}^{n} \frac{1}{z-2k} - \sum_{k=-n-1}^{n} \frac{1}{z-1-2k}$$

$$\to \frac{\pi}{2} \cot \frac{\pi z}{2} - \frac{\pi}{2} \cot \frac{\pi(z-1)}{2} = \frac{\pi}{\sin \pi z}$$

as $n \to \infty$, the sum of the series $(*10)$ is then $\frac{\pi}{\sin \pi z}$.

Exercises A

(1) Try to use (5.4.1.4) in Example 1 to reprove (5.4.1.6).
(2) Try to use (5.4.1.4) in Example 1 to reprove the two expressions in (5.4.1.7) and then, deduce the validity of (5.4.1.6).
(3) Use the partial fractions expansion for $\frac{\pi}{\sin \pi z}$ to find the one for $\frac{1}{\cos \pi z}$, and deduce that

$$\frac{\pi}{4} = 1 - \frac{1}{3} + \frac{1}{5} - \frac{1}{7} + \cdots .$$

(4) Sum the following series

$$\sum_{-\infty}^{\infty} \frac{1}{z^3 - n^3}, \quad \sum_{-\infty}^{\infty} \frac{1}{(z+n)^2 + a^2}.$$

(5) Use the following two methods:

1. the Taylor series expansion of $z \cot z$ at $z = 0$ and the partial fractions expansion of $\cot z$ (refer to Exercise A(5) of Sec. 4.8);
2. the Laurent series and partial fractions expansions of $(e^z - 1)^{-1}$ at $z = 0$ (refer to Example 5 of Sec. 4.8),

to show that

$$\sum_{n=1}^{\infty} \frac{1}{n^{2k}} = 2^{2k-1} \frac{B_{2k}}{(2k)!} \pi^{2k}, \quad k \geq 1,$$

where B_{2k} are Bernoulli's numbers. In particular, deduce that

$$\sum_{n=1}^{\infty} \frac{1}{n^2} = \frac{\pi^2}{6}; \quad \sum_{n=1}^{\infty} \frac{1}{n^4} = \frac{\pi^4}{90}; \quad \sum_{n=1}^{\infty} \frac{1}{n^6} = \frac{\pi^6}{945}.$$

Exercises B

Generalized Cauchy's integral formula (2.9.13) (or (4.11.1.8)) can be used to prove a weak form of the Mittag–Leffler's theorem (5.4.1.1). In (1)–(3) that follows, readers are supposed to be familiar with the content of (2.9.13) and its meaning or its proof.

(1) (*One-dimensional $\bar{\partial}$-problem*) Let $\varphi \in C^1$ (**C**) be a complex-valued function with compact support (i.e., the set supp $\varphi = \overline{\{z | \varphi(z) \neq 0\}}$ is compact). Set

$$u(z) = \frac{1}{2\pi i} \iint_{\mathbf{C}} \frac{\varphi(\zeta)}{\zeta - z} d\zeta \wedge d\bar{\zeta}.$$

Then $u \in C^1(\mathbf{C})$ and $\frac{\partial u}{\partial \bar{z}}(z) = \varphi(z)$, $z \in \mathbf{C}$. Try the following steps.

(a) Fix $z \in \mathbf{C}$. Let $\zeta - z = \eta$. Then

$$u(z) = \frac{1}{2\pi i} \iint_{\mathbf{C}} \frac{\varphi(\eta + z)}{\eta} d\eta \wedge d\bar{\eta}.$$

Observe that $u \in C^0(\mathbf{C})$ since $\frac{1}{\eta}$ is integrable on each compact in the η-plane. Choose $h \in \mathbf{R}$ and $h \neq 0$. Since $\varphi \in C^1(\mathbf{C})$ with compact support , $\frac{1}{h}[\varphi(\eta + z + h) - \varphi(\eta + z)] \to \frac{\partial \varphi}{\partial \alpha}(\eta + z)$ uniformly in η and z as $h \to 0$, where $\eta = \alpha + i\beta$. Therefore, if $z = x + iy$,

$$\frac{\partial u}{\partial x}(z) = \frac{1}{2\pi i} \iint_{\mathbf{C}} \frac{1}{\eta} \frac{\partial \varphi}{\partial \alpha}(\eta + z) d\eta \wedge d\bar{\eta};$$

$$\frac{\partial u}{\partial y}(z) = \frac{1}{2\pi i} \iint_{\mathbf{C}} \frac{1}{\eta} \frac{\partial \varphi}{\partial \beta}(\eta + z) d\eta \wedge d\bar{\eta}, \quad \text{where } \eta = \zeta - z$$

and then $u \in C^1(\mathbf{C})$.

(b) Note that

$$\frac{\partial u}{\partial \bar{z}}(z) = \frac{1}{2} \left(\frac{\partial u}{\partial x} + i \frac{\partial u}{\partial y} \right)(z) = \frac{1}{2\pi i} \iint_{\mathbf{C}} \frac{1}{\eta} \frac{\partial \varphi}{\partial \bar{\eta}} d\eta \wedge d\bar{\eta}$$

$$= \frac{1}{2\pi i} \iint_{\mathbf{C}} \frac{\partial \varphi}{\partial \bar{\zeta}}(\zeta) \frac{1}{\zeta - z} d\zeta \wedge d\bar{\zeta}$$

$$= \frac{1}{2\pi i} \iint_{|z| < R + \varepsilon} \frac{\partial \varphi}{\partial \bar{\zeta}}(\zeta) \frac{1}{\zeta - z} d\zeta \wedge d\bar{\zeta},$$

where $R > 0$ is chosen so that supp $\varphi \subseteq \{|z| < R\}$ and $\varepsilon > 0$. Then, try to use (2.9.13) to finish the proof.

Note. In case $\varphi \in C^k(\mathbf{C})$ with compact support, where k is a positive integer or ∞, then $u \in C^k(\mathbf{C})$ holds. Even if supp φ is a countable union of disjoint compact sets, this result is still valid, too.

(2) (*Mittag–Leffler's theorem*) Given a sequence a_n, $n \geq 1$, in $\mathbf{C}$ with $|a_n| \leq |a_{n+1}|$, $n \geq 1$, and $\lim_{n \to \infty} a_n = \infty$, and polynomials $R_n(z)$, $n \geq 1$, without constant terms. Then there is a meromorphic function $f(z)$ in $\mathbf{C}$ with poles at a_n and the principal parts $R_n(\frac{1}{z-a_n})$, $n \geq 1$. Try the following steps.

(a) For each $n \geq 1$, choose an open disk D_n with center a_n so that $D_n \cap D_m = \emptyset$ if $n \neq m$. Let O_n be a smaller concentric disk with D_n, $n \geq 1$. Construct a C^∞ function φ_n on $\mathbf{C}$ so that $\varphi_n(z) = 1$ on O_n and $\varphi_n(z) = 0$ on $\mathbf{C} - D_n$ (how?). Set

$$\varphi(z) = \sum_{n=1}^{\infty} \varphi_n(z) R_n\left(\frac{1}{z - a_n}\right), \quad z \in \mathbf{C} - \{a_1, a_2, \ldots\};$$

$$g(z) = \begin{cases} \dfrac{\partial \varphi}{\partial \bar{z}}(z), & z \in \mathbf{C} - \{a_1, a_2, \ldots\} \\ 0, & z = a_n \text{ for } n \geq 1 \end{cases}.$$

Show that $g \in C^\infty(\mathbf{C})$.

(b) Use Exercise (1) to solve $\frac{\partial v}{\partial \bar{z}} = g$. Let $v \in C^\infty(\mathbf{C})$ be such a solution. Set

$$f(z) = \varphi(z) - v(z), \quad z \in \mathbf{C}.$$

Try to show that $f(z)$ is a required one.

(3) (*An interpolation theorem*) Let $\{z_1, z_2, \ldots\}$ be a discrete point set in $\mathbf{C}$ and $n_1, n_2, \ldots$ be a sequence of positive integers. Given a complex sequence $a_{j,k}$ ($j \geq 1, 0 \leq k \leq n_j - 1$), then there exists an entire function $f(z)$ satisfying:

$$f^{(k)}(z_j) = k! a_{j,k}, \quad j \geq 1, \quad 0 \leq k \leq n_j - 1.$$

Namely, given a sequence z_j, $j \geq 1$, and the coefficients of the first n_j terms of a Taylor series expansion at each point z_j, then there is an entire function having such a Taylor series expansion at each of these points. Try the following steps.

(a) Let $g(z)$ be an entire function having a zero of order n_j at z_j for each $j \geq 1$ (see (5.5.2.2)). Since $\{z_1, z_2, \ldots\}$ is discrete, there exists a sequence of positive numbers $r_1, r_2, \ldots$ so that the disks $|z - z_j| < 2r_j, j \geq 1$, are pairwise disjoint. For each j, construct a C^∞ function φ_j on $\mathbf{C}$ satisfying $0 \leq \varphi_j(z) \leq 1$ on $\mathbf{C}$ and

$$\varphi_j(z) = \begin{cases} 1, & |z - z_j| \leq r_j \\ 0, & |z - z_j| \geq 2r_j \end{cases}.$$

Set

$$f(z) = \sum_{j \geq 1} P_j(z)\varphi_j(z) - g(z)\varphi(z), \quad z \in \mathbf{C}$$

where $p_j(z) = \sum_{0 \leq k \leq n_j - 1} a_{j,k}(z - z_j)^k$ for $j \geq 1$, are polynomials and φ is a certain C^∞ function on $\mathbf{C}$ to be so chosen that $\frac{\partial f}{\partial \bar{z}} = 0$, namely, f turns out to be entire. Explain why $\sum_{j \geq 1}$ has a meaning!

(b) Note that

$$\frac{\partial f}{\partial \bar{z}} = 0$$

$$\Leftrightarrow \sum_{j \geq 1} P_j(z)\frac{\partial \varphi_j}{\partial \bar{z}} = g(z)\frac{\partial \varphi}{\partial \bar{z}}.$$

Let $h(z) = \sum_{j \geq 1} P_j(z)\frac{\partial \varphi_j}{\partial \bar{z}}$ which is identically to zero on $\bigcup_{j \geq 1} \{|z - z_j| < r_j\} \cup \bigcup_{j \geq 1}\{|z - z_j| \geq 2r_j\}$ and whose compact support is a pairwise disjoint union of compact sets. Designate $\frac{h(z)}{g(z)} = 0$ at $z = z_j$ for $j \geq 1$. Then $\frac{h}{g}$ is C^∞ on $\mathbf{C}$.

(c) According to Exercise (1), let φ be a C^∞ solution of the $\bar{\partial}$-equation $\frac{\partial \varphi}{\partial \bar{z}} = \frac{h}{g}$. For such a φ, f is entire. Note that φ is analytic on each disk $|z - z_j| < r_j$ for $j \geq 1$. Then, try to show that $f^{(k)}(z_j) = p_j^{(k)}(z_j) = k!a_{j,k}$.

(4) (*A generalized version of Mittag–Leffler's theorem* (5.4.1.1).) In (5.4.1.1), use a domain Ω (even an open set) in $\mathbf{C}$ to replace the complex plane $\mathbf{C}$ there, and suppose a_n, $n \geq 1$, is a sequence of distinct points in Ω without limit points in Ω. Then the result is still valid in Ω with $g(z)$ in (2) being analytic in Ω. Note the difference between (5.4.1.1) and (b) below.

We may assume that $a_n \neq 0$ for $n \geq 1$, and $\Omega \neq \mathbf{C}$. Let r_n be the radius of the largest open disk with center at a_n that is contained in Ω. In this case, there is a point $b_n \in \partial\Omega$ such that $|b_n - a_n| = r_n$ for each $n \geq 1$. Try the following steps.

(a) In case $|a_n|r_n \geq 1$ for each $n \geq 1$: If $|a_n|$, $n \geq 1$, is bounded, then a subsequence a_{n_k} converges to a point a_0 in $\mathbf{C}$. Hence

$$|z - a_{n_k}| \geq r_{n_k} \geq \frac{1}{|a_{n_k}|}, k \geq 1, \quad \text{and} \quad z \in \mathbf{C} - \Omega$$

$$\Rightarrow |z - a_0| \geq \frac{1}{|a_0|}, \quad z \in \mathbf{C} - \Omega$$

$$\Rightarrow a_0 \neq 0 \quad \text{and} \quad a_0 \in \Omega. \tag{$*_{11}$}$$

This contradiction leads to the fact that $|a_n| \to \infty$ as $n \to \infty$ and Ω is *unbounded*. Exactly the same process from $(*_1)$ to $(*_4)$ will work for what remains.

(b) In case $|a_n|r_n < 1$ for each $n \geq 1$: If $\underline{\lim} r_n = r_0 > 0$, there is a subsequence $r_{n_k} \to r_0$. Since $|a_{n_k}| < r_{n_k}^{-1} \leq r_0^{-1}$, $k \geq 1$, it follows that a_{n_k} (say itself) converges to a point $a_0 \in \mathbf{C}$. Argument similar to $(*_{11})$ results in a contradictory fact that $a_0 \in \Omega$. Therefore $r_n \to 0$ as $n \to \infty$.

Recall that $b_n \in \partial\Omega$ and $|b_n - a_n| = r_n$ for $n \geq 1$. Note that $R_n(\frac{1}{z-a_n})$ is analytic in $D_n = \{z | r_n < |z - b_n| \leq \infty\}$. Then

$$\frac{1}{z - a_n} = \sum_{k=1}^{\infty} \frac{(a_n - b_n)^{k-1}}{(z - b_n)^k} \quad \text{(compare to } (*_1)), \quad z \in D_n \tag{$*_1'$}$$

$$\Rightarrow R_n \left(\frac{1}{z - a_n} \right) \text{ can be represented as a power series in}$$

$(z - b_n)^{-1}$, converging absolutely and locally uniformly in D_n.

$$\Rightarrow \text{There is a polynomial } P_n(z) = \sum_{k=1}^{k_n} a_k z^k \text{ (without constant}$$

term) so that

$$\left| R_n \left(\frac{1}{z - a_n} \right) - P_n \left(\frac{1}{z - b_n} \right) \right| < \frac{1}{2^n}$$

(compare to $(*_2)$),

$$\text{for } z \in D_n' = \{z | 2r_n < |z - b_n| \leq \infty\} \tag{$*_2'$}$$

what remains is the same as in $(*_3)$ and $(*_4)$.

(c) In general case, consider $\{a_n | n \geq 1\} = \{a_n | |a_n|r_n \geq 1\} \cup \{a_n | |a_n|r_n < 1\}$. Then, apply (a) and (b) to these two subsets, respectively. Note that one of these two subsets might be a finite set for which the result is obviously true.

Refer to Ref. [58], vol. III, p. 84.

(5) State and prove Exercise (4) in case Ω is a domain (or open set) in $\mathbf{C}^*$ so that $\infty \in \Omega$. Recall (see (4.10.2.2)) that the singular part of a meromorphic function at pole ∞ is of the form $a_n z^n + \cdots + a_1 z$.

(6) Given a domain Ω in $\mathbf{C}$, show that there is a function meromorphic on Ω but not on any larger domain. *Hint.* Let $a_n, n \geq 1$, be a sequence of points in Ω without limit points in Ω which is dense in the boundary $\partial\Omega$. Then use Exercise (4).

Exercises C

Pick up a book about Riemann surface, such as Refs. [6, 32, 33, 75], and study the Riemann–Roch theorem and then see how to construct analytic, meromorphic or harmonic functions on Riemann surfaces.

5.4.2 *Cauchy's residue method*

Oeuvres complètes d'Augustin Cauchy (2) 7, 324–344, Paris 1882–1974, adopted the residue method to expand a meromorphic function into partial functions.

Cauchy's method. Let $f(z)$ be meromorphic in $\mathbf{C}$. Suppose there exists a system of rectifiable closed Jordan curves $\gamma_n, n \geq 1$, satisfying the following conditions:

1. $0 \in \operatorname{Int} \gamma_1$, and each γ_n is contained in the interior $\operatorname{Int} \gamma_{n+1}$ of γ_{n+1} and does not pass through the poles of $f(z)$ for each $n \geq 1$.
2. The distance $d_n = \operatorname{dist}(0, \gamma_n) \to \infty$ as $n \to \infty$.
3. The quotient of the length $L(\gamma_n)$ of γ_n to the distance d_n remains bounded, namely,

$$\frac{L(\gamma_n)}{d_n} \leq A \text{ (constant)}, \quad n \geq 1.$$

Also, suppose that

$$|f(z)| \leq M|z|^p \quad \text{on } \gamma_n$$

for all sufficiently large n, where $M > 0$ is a constant and p is a nonnegative integer.

(1) *If $f(z)$ is analytic at 0,*

$$f(z) = \sum_{j=0}^{p} \frac{f^{(j)}(0)}{j!} z^j + \lim_{n \to \infty} \sum_{\gamma_n} \left\{ R_k(z) - \sum_{j=0}^{p} \frac{R_k^{(j)}(0)}{j!} z^j \right\}$$

locally uniformly in $\mathbf{C} - \{\text{poles of } f(z)\}$, where $\sum_{\gamma_n}$ denotes the summation over the singular parts $R_k(z)$ of these poles of $f(z)$ lying in the interior Int γ_1 and then adding up these $R_k(z)$ related to poles between γ_1 and γ_2, and so on, up to poles between γ_{n-1} and γ_n. In case the following infinite series converges absolutely in $\mathbf{C} - \{\text{poles of } f(z)\}$, $f(z)$ can also be expressed as

$$f(z) = \sum_{j=0}^{p} \frac{f^{(j)}(0)}{k!} z^k + \sum_{k=1}^{\infty} \left\{ R_k(z) - \sum_{j=0}^{p} \frac{R_k^{(j)}(0)}{j!} z^j \right\},$$

where $\sum_{k=1}^{\infty}$ denotes the summation related to all the poles a_k, $k \geq 1$, of $f(z)$.

(2) If $f(z)$ has a pole at 0 with the singular part $R_0(z)$, then apply (1) to the function $f(z) - R_0(z)$ instead of $f(z)$.   (5.4.2.1)

Refer to (5.4.2.2) and (5.4.2.3) for some simplified variations of (5.4.2.1).

Proof. Fix $n \geq 1$ and consider a point z in Int γ_n other than these poles of $f(z)$ lying in Int γ_n. Then the function $\frac{f(\zeta)}{\zeta - z}$ in ζ is analytic in Int γ_n except at the point z and the poles of $f(\zeta)$ lying in Int γ_n. By the residue theorem (4.11.3.2) or the more elementary Cauchy's integral formula, we have

$$f(z) - \sum_{\gamma_n} R_k(z) = \frac{1}{2\pi i} \int_{\gamma_n} \frac{f(\zeta)}{\zeta - z} d\zeta. \tag{$*_1$}$$

Since $\frac{1}{\zeta - z} = \sum_{j=0}^{p} \frac{z^j}{\zeta^{j+1}} + \frac{1}{\zeta - z}(\frac{z}{\zeta})^{p+1}$, it follows that

$$f(z) = \sum_{\gamma_n} R_k(z) + \sum_{j=0}^{p} \frac{z^j}{2\pi i} \int_{\gamma_n} \frac{f(\zeta)}{\zeta^{j+1}} d\zeta + \frac{z^{p+1}}{2\pi i} \int_{\gamma_n} \frac{f(\zeta)}{\zeta^{p+1}(\zeta - z)} d\zeta. \tag{$*_2$}$$

Applying the integral of Cauchy's type (see (4.7.1)) to $(*_1)$ or differentiating both sides of $(*_2)$ j times, then

$$\frac{1}{2\pi i} \int_{\gamma_n} \frac{f(\zeta)}{\zeta^{j+1}} d\zeta = \frac{1}{2\pi i} \frac{1}{j!} \frac{d^j}{dz^j} \left(\int_{\gamma_n} \frac{f(\zeta)}{\zeta - z} d\zeta \right) \Bigg|_{z=0}$$

$$= \frac{1}{j!} \left\{ f^{(j)}(z) - \sum_{\gamma_n} R_k^{(j)}(z) \right\} \Bigg|_{z=0}$$

$$= \frac{1}{j} \left\{ f^{(j)}(0) - \sum_{\gamma_n} R_k^{(j)}(0) \right\}, \quad 1 \leq j \leq p.$$

On the other hand, since $\lim_{n\to\infty} d_n = \infty$, we can choose n large enough so that $d_n > |z|$ (remind that z is a fixed point in Int γ_n). Under this circumstance,

$$\left| \frac{1}{2\pi i} \int_{\gamma_n} \frac{f(\zeta)}{\zeta^{p+1}(\zeta - z)} d\zeta \right| \leq \frac{1}{2\pi} \cdot \frac{M d_n^p L(\gamma_n)}{d_n^{p+1}(d_n - |z|)}$$

$$\leq \frac{MA}{2\pi(d_n - |z|)} \to 0 \quad \text{as } n \to \infty;$$

and then, as $n \to \infty$,

$$\frac{z^{p+1}}{2\pi i} \int_{\gamma_n} \frac{f(\zeta)}{\zeta^{p+1}(\zeta - z)} d\zeta \to 0 \text{ uniformly on compact sets in } \mathbf{C} - \{\text{poles}\}.$$

Substituting the above three estimates into $(*_2)$, the claims follows. $\qquad\square$

If the meromorphic function $f(z)$ has only simple poles as we will encounter in many practical cases, (5.4.2.1) can be modified a little to an easier form.

A special case of (5.4.2.1). Let $f(z)$ be meromorphic in $\mathbf{C}$ with nonzero simple poles at a_n, $n \geq 1$, satisfying $0 < |a_1| \leq |a_2| \leq \cdots \leq |a_n| \leq \cdots$. Suppose $f(z)$ satisfies *one* of the following conditions:

Condition 1. 1, 2, 3 in (5.4.2.1) plus $\sup_{z \in \gamma_n} |f(z)| \leq M$ (constant) $< +\infty$.

Condition 2. There exists a system of circles (or squares or rectangles) γ_n : $|z| = R_n, n \geq 1$, with increasing R_n and $\lim_{n\to\infty} R_n = \infty$ and satisfying:

1. Each γ_n does not pass the poles of $f(z)$.
2. $\sup_{z \in \gamma_n} |f(z)| \leq M < \infty$ for $n \geq 1$.

Then

(1) If $f(z)$ is analytic at 0,

$$f(z) = f(0) + \sum_{n=1}^{\infty} \operatorname{Re} s(f(z); a_n) \left(\frac{1}{z - a_n} + \frac{1}{a_n} \right)$$

locally uniformly in $\mathbf{C} - \{a_1, a_2, \ldots\}$.
(2) If $f(z)$ has a simple pole at 0 with singular part $\frac{r_0}{z}$, apply (1) to the function $f(z) - \frac{r_0}{z}$ instead of $f(z)$. $\qquad$ (5.4.2.2)

Sketch of proof. In the case of Condition 1, fix z inside γ_n other than zero and possible poles a_k of $f(z)$. Compute

$$\mathrm{Re}\,s\left(\frac{f(\zeta)}{\zeta(\zeta-z)};z\right) = \frac{f(z)}{z};$$

$$\mathrm{Re}\,s\left(\frac{f(\zeta)}{\zeta(\zeta-z)};0\right) = \lim_{\zeta\to 0}\zeta\cdot\frac{f(\zeta)}{\zeta(\zeta-z)} = -\frac{f(0)}{z};$$

$$\mathrm{Re}\,s\left(\frac{f(\zeta)}{\zeta(\zeta-z)};a_k\right) = \lim_{\zeta\to a_k}(\zeta-a_k)\cdot\frac{f(\zeta)}{\zeta(\zeta-z)}$$

$$= \frac{1}{a_k(a_k-z)}\lim_{\zeta\to a_k}(\zeta-a_k)f(\zeta)$$

$$= \mathrm{Re}\,s(f(\zeta);a_k)\frac{1}{a_k(a_k-z)}$$

$$= -\frac{1}{z}\mathrm{Re}\,s(f(\zeta);a_k)\left(\frac{1}{z-a_k}+\frac{1}{a_k}\right).$$

Now, by the residue theorem,

$$\frac{1}{2\pi i}\int_{\gamma_n}\frac{f(\zeta)}{\zeta(\zeta-z)}d\zeta$$

$$= \frac{f(z)}{z} - \left\{\frac{f(0)}{z} + \frac{1}{z}\sum_{|a_k|<R_n}\mathrm{Re}\,s(f(z);a_k)\left(\frac{1}{z-a_k}+\frac{1}{a_k}\right)\right\} \qquad (*_3)$$

On the other hand, if n is large enough such that $d_n > |z|$ holds, then

$$\left|\frac{1}{2\pi i}\int_{\gamma_n}\frac{f(\zeta)}{\zeta(\zeta-z)}d\zeta\right| \leq \frac{1}{2\pi}\cdot\frac{L(\gamma_n)}{d_n(d_n-|z|)}\cdot\sup_{\zeta\in\gamma_n}|f(\zeta)| \to 0 \quad \text{as } n\to\infty,$$

and uniformly in compact subsets of $\mathbf{C} - \{\text{poles}\}$. $\qquad (*_4)$

In the case of Condition 2, $(*_4)$ is replaced by the following estimate:

$$\left|\frac{1}{2\pi i}\int_{\gamma_n}\frac{f(\zeta)}{\zeta(\zeta-z)}d\zeta\right| \leq \frac{1}{2\pi}\cdot\frac{2\pi R_n}{R_n(R_n-|z|)}\cdot\sup_{\zeta\in\gamma_n}|f(\zeta)| \to 0 \quad \text{as } n\to\infty,$$

and uniformly in compact subsets of $\mathbf{C} - \{\text{poles}\}$. $\qquad (*_5)$

$\qquad\qquad\qquad\qquad\qquad\qquad\qquad\qquad\qquad\qquad\qquad\qquad\qquad\qquad\square$

As a second modification of (5.4.2.1), we have

A general form of (5.4.2.1). Let $f(z)$ be mermorphic in $\mathbf{C}$ with simple poles at a_n, $n\geq 1$, satisfying $0 < |a_1| \leq |a_2| \leq \cdots \leq |a_n| \leq \cdots$. Let γ_n, $n\geq 1$, be as in (5.4.2.1), with properties 1, 2, and 3 such that $\lim_{n\to\infty}\int_{\gamma_n}\frac{f(\zeta)}{\zeta-z}d\zeta = 0$

for $z \in \text{Int } \gamma_n - \{\text{poles}\}$, or γ_n be as in Condition 2 in (5.4.2.2) with the property 2 replaced by $\sup_{z \in \gamma_n} |f(z)| \to 0$ as $n \to \infty$. Then

$$f(z) = \lim_{n \to \infty} \sum_{\gamma_n} \frac{1}{z} \frac{1}{a_k} \text{Re } s(f(z); a_k) = \sum_{n=1}^{\infty} \text{Ro } o(f(z), u_n) \frac{1}{z - a_n},$$

$$z \in \mathbf{C} - \{\text{poles}\}.$$

More general, if a_n, $n \geq 1$, are just poles of $f(z)$ with the corresponding singular parts $R_n(z)$ and

$$\lim_{n \to \infty} \int_{\gamma_n} \frac{|f(z)|}{1 + |z|} |dz| = 0, \quad z \in \text{Int } \gamma_n - \{\text{poles}\},$$

then

$$\lim_{n \to \infty} \left\{ f(z) - \sum_{k=1}^{n} R_k(z) \right\} = 0, \quad z \in \mathbf{C} - \{\text{poles}\}. \tag{5.4.2.3}$$

See Ref. [31], p. 262, for a proof of the latter part.

We give two illustrative examples.

Example 1. Show that (compare to (5.4.1.7))

$$\frac{1}{\sin z} = \frac{1}{z} + \sum_{n=1}^{\infty} \frac{(-1)^n 2z}{z^2 - n^2 \pi^2}$$

absolutely and locally uniformly in $\mathbf{C} - \{n\pi | n = 0, \pm 1, \pm 2, \ldots\}$. Then, deduce that

$$\frac{\cos z}{\sin^2 z} = \sum_{n=-\infty}^{\infty} \frac{(-1)^n}{(z - n\pi)^2}, \quad \text{and}$$

$$\frac{\tan z}{z} = \frac{\prod_{n=1}^{\infty}(1 - z^2/n^2\pi^2)}{\prod_{n=1}^{\infty}(1 - 4z^2/(2n - 1)^2\pi^2)}$$

absolutely and locally uniformly in $\mathbf{C} - \{n\pi | n = 0, \pm 1, \pm 2, \ldots\}$.

Solution. We use four methods in the following.

Method 1. Applying Condition 1 in (5.4.2.2) to $f(z) = \frac{1}{\sin z} - \frac{1}{z}$, where $\frac{1}{z}$ is the singular part of $\frac{1}{\sin z}$ at the simple pole $z = 0$. Then $f(0) = 0$ and

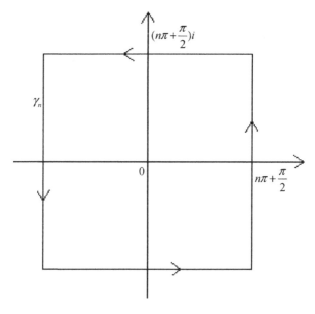

Fig. 5.13

$f(z)$ is analytic at 0. Also, $z_n = n\pi$, $n = 0, \pm 1, \pm 2, \ldots$, are simple poles of $f(z)$ and, for $n \geq 1$,

$$\operatorname{Re} s(f(z); n\pi) = \lim_{z \to z_n} (z - z_n) f(z) = \left. \frac{z - \sin z}{(z \sin z)'} \right|_{z=z_n} = (-1)^n.$$

Let γ_n be the square with vertices at $\pm(n\pi + \frac{\pi}{2})(1 \pm i)$ for $n \geq 1$ (see Fig. 5.13). γ_n, $n \geq 1$, obviously satisfy conditions 1, 2, and 3 in (5.4.2.1). To show that $\sup_{\gamma_n} |f(z)|$ are uniformly bounded (and, indeed, converge to 0 uniformly), let us recall the identity $|\sin z|^2 = \sinh^2 y + \sin^2 x = \frac{1}{2}(\cosh 2y - \cos 2x)$, where $z = x + iy$. Along the vertical sides of γ_n, $\sin^2 x = 1$ and

$$\left| \frac{1}{\sin z} \right| \leq \frac{1}{\sqrt{1 + \sinh^2 \beta}} = \frac{1}{\cosh \beta} \left(\beta = \pm \left(n\pi + \frac{\pi}{2} \right) \right) \to 0 \quad \text{as } n \to \infty;$$

while along the horizontal sides of γ_n, $|\cos 2x| \leq 1$ and

$$\left| \frac{1}{\sin z} \right| \leq \frac{2}{\cosh 2\beta - 1} \left(\beta = \pm \left(n\pi + \frac{\pi}{2} \right) \right) \to 0 \quad \text{as } n \to \infty.$$

The claims follows.

According to (5.4.2.2),

$$\frac{1}{\sin z} = \frac{1}{z} + 0 + \sum_{n=-\infty}^{\infty} {}'(-1)^n \left\{ \frac{1}{z - n\pi} + \frac{1}{n\pi} \right\} \quad \text{(omitting the term } n = 0)$$

$$= \frac{1}{z} + \sum_{n=1}^{\infty} \frac{(-1)^n 2z}{z^2 - n^2 \pi^2} \quad \text{(by grouping terms } n \text{ and } -n \text{ together owing}$$

to absolute convergence).

Method 2. Apply condition 1 in (5.4.2.2) to $f(z) = \frac{z}{\sin z}$ which is analytic at $z = 0$ with $f(0) = 1$. Adopt the same γ_n, $n \geq 1$, as in Fig. 5.13. On the vertical sides of γ_n,

$$\left| \frac{z}{\sin z} \right| \leq \frac{2|\beta|}{\cosh |\beta|} \quad \left(\beta = \pm \left(n\pi + \frac{\pi}{2} \right) \right) \to 0 \quad \text{as } n \to \infty;$$

on the horizontal sides of γ_n,

$$\left| \frac{z}{\sin z} \right| \leq \frac{4|\beta|}{\cosh 2\beta - 1} \quad \left(\beta = \pm \left(n\pi + \frac{\pi}{2} \right) \right) \to 0 \quad \text{as } n \to \infty.$$

So, $\sup_{\gamma_n} |f(z)|$, $n \geq 1$, are uniformly bounded. Since $\operatorname{Re} s(f(z); z_n) = (-1)^n z_n$, the result follows.

Method 3. Apply the former part of (5.4.2.3) to $f(z) = \frac{1}{\sin z}$ and adopt the same γ_n, $n \geq 1$, as in Fig. 5.13. Observe that, if $z \in \operatorname{Int} \gamma_n - \{\text{poles}\}$,

$$\left| \int_{\gamma_n} \frac{f(\zeta)}{\zeta - z} d\zeta \right| \leq \int_{\gamma_n} \frac{|d\zeta|}{|\zeta - z||\sin \zeta|} \leq \frac{1}{d_n} \int_{\gamma_n} \frac{|d\zeta|}{|\sin \zeta|},$$

$$\text{where } d_n = \operatorname{dist}(z, \gamma_n). \tag{$*6$}$$

Along the lower horizontal side L_n : $\zeta = t - \frac{\pi}{2}(2n + 1)i$, setting $\beta_n = \frac{\pi}{2}(2n + 1)$,

$$\int_{L_n} \frac{|d\zeta|}{|\sin \zeta|} \leq \int_{-1}^{1} \frac{\beta_n dt}{\sqrt{\sinh^2 \beta_n + \sin^2 \beta_n t}} \leq \frac{2\beta_n}{\sinh \beta_n} \to 0 \quad \text{as } n \to \infty.$$

Similarly, the integral along the upper horizontal side approaches 0 as $n \to \infty$. Along the right vertical side R_n : $\zeta = \beta_n + it$,

$$\int_{R_n} \frac{|d\zeta|}{|\sin \zeta|} \leq 4\beta_n \int_0^1 \frac{dt}{\cosh 2\beta_n t} = 4 \tan^{-1} e^{2t} \big|_0^{\beta_n}$$

$$= 4 \left(\tan^{-1} e^{2\beta_n} - \frac{\pi}{4} \right) \to \pi \quad \text{as } n \to \infty.$$

So does the integral along the left vertical side of γ_n. Putting these results into $(*_6)$, then

$$\lim_{n\to\infty} \int_{\gamma_n} \frac{f(\zeta)}{\zeta - z} d\zeta = 0.$$

Hence, by (5.4.2.3),

$$\frac{1}{\sin z} = \sum_{-\infty}^{\infty} \operatorname{Re} s\left(\frac{1}{\sin z}; n\pi\right) \cdot \frac{1}{z - n\pi} = \sum_{-\infty}^{\infty} \frac{(-1)^n}{z - n\pi}$$

(refer to Example 2 in Sec. 2.3.)

$$= \sum_{0}^{\infty} \frac{(-1)^n}{z - n\pi} + \sum_{n=1}^{\infty} \frac{(-1)^n}{z + n\pi} = \frac{1}{z} + \sum_{n=1}^{\infty} (-1)^n \left[\frac{1}{z - n\pi} + \frac{1}{z + n\pi}\right]$$

(refer to (5.3.1.8))

$$= \frac{1}{z} + \sum_{1}^{\infty} (-1)^n \frac{2z}{z^2 - n^2\pi^2}. \qquad (*_7)$$

Method 4. Apply Condition 2 in (5.4.2.2) to $f(z) = \frac{1}{\sin z}$ and adopt, instead, the circles $C_n : |z| = R_n = n\pi + \frac{\pi}{2}$ for $n \geq 1$. See Fig. 5.14.

Fix any $0 < r < \frac{\pi}{2}$. Construct a circle with center at $z_n = n\pi$ $(n = 0, \pm 1, \pm 2, \ldots)$ and radius r. Let $\Omega = \mathbf{C} - \bigcup_{n=-\infty}^{\infty} \{|z - z_n| < r\}$. We try to

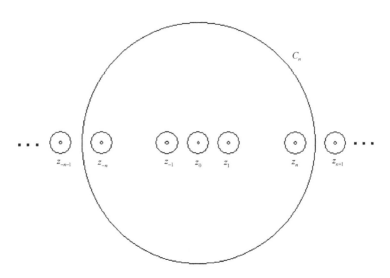

Fig. 5.14

show that $f(z)$ is bounded in Ω and hence, $\sup_{z \in C_n} |f(z)| \leq M < \infty$ for all $n = 0, \pm 1, \pm 2, \ldots$. See Fig. 5.14.

Observe that, if $z = x + iy$,

$$\left| \frac{1}{\sin z} \right| \leq \frac{2}{|c^{-y} - e^y|} = 2\operatorname{csch}|y|$$

$$\leq 2\operatorname{csch} a, \text{ in case } z \in \Omega \cap \{|\operatorname{Im} z| \geq a\} \text{ for any } a > 0.$$

Since $\left| \frac{1}{\sin z} \right|$ is bounded in the set formed by the intersection of Ω and the interior of the rectangle with vertices at $\pm \frac{3\pi}{2} \pm ai$, by the periodic property of it, we know that $\left| \frac{1}{\sin z} \right|$ is bounded in $\Omega \cap \{|\operatorname{Im} z| \leq a\}$. Hence, there is a constant $M = M(r) > 0$ so that $\left| \frac{1}{\sin z} \right| \leq M$ for all $z \in \Omega$; in particular, $\sup_{z \in C_n} \left| \frac{1}{\sin z} \right| \leq M < \infty$ for $n = 0, \pm 1, \pm 2, \ldots$. The final result $(*_7)$ follows (5.4.2.2).

The series $\sum_{-\infty}^{\infty} (-1)^{n-1} \frac{1}{(z-n\pi)^2}$ obtained by termwise differentiation of the series representation for $\frac{1}{\sin z}$ is locally uniformly convergent in $\mathbf{C} - \{n\pi | n = 0, \pm 1, \pm 2, \ldots\}$. Therefore, it is permissible to differentiate both sides to deduce that

$$\frac{\cos z}{\sin^2 z} = \sum_{-\infty}^{\infty} (-1)^n \frac{1}{(z - n\pi)^2}, \quad z \neq n\pi \quad \text{for } n = 0, \pm 1, \pm 2, \ldots.$$

Also, termwise integration of both sides of the series representation for $\frac{1}{\sin z}$ gives rise to

$$\operatorname{Log} \tan \frac{z}{2} = \operatorname{Log} C + \operatorname{Log} \frac{z}{2} + \sum_{n=1}^{\infty} (-1)^n \operatorname{Log} \left(1 - \frac{z^2}{n^2 \pi^2} \right),$$

$$z \neq n\pi \quad \text{for } n = 0, \pm 1, \pm 2, \ldots,$$

where C is a constant, and all the logarithmic functions are chosen to be the principal branches in their common domain. In turn, this expression can be written in the infinite product (refer to Sec. 5.5.1)

$$\frac{\tan z/2}{z/2} = C \prod_{n=1}^{\infty} \frac{(1 - z^2/4n^2\pi^2)}{(1 - z^2/(2n-1)^2\pi^2)}, \quad z \neq 2n\pi \quad \text{for } n = 0, \pm 1, \pm 2, \ldots.$$

Letting $z \to 0$ and observing that both $\frac{\tan z/2}{z/2}$ and the infinite product approach 1, so $C = 1$ holds. Replace z by $2z$ and we obtain

$$\frac{\tan z}{z} = \frac{\prod_{n=1}^{\infty} (1 - z^2/n^2\pi^2)}{\prod_{n=1}^{\infty} (1 - 4z^2/(2n-1)^2\pi^2)}, \quad z \neq n\pi \quad \text{for } n = 0, \pm 1, \pm 2, \ldots.$$

Example 2. Show that

$$\cot z = \frac{1}{z} + \sum_{n=1}^{\infty} \frac{2z}{z^2 - n^2\pi^2}, \quad z \in \mathbf{C} - \{n\pi | n = 0, \pm 1, \pm 2, \ldots\}, \quad \text{or}$$

$$\pi \cot \pi z = \frac{1}{z} + \sum_{-\infty}^{\infty}{}' \left(\frac{1}{z-n} + \frac{1}{n} \right) = \frac{1}{z} + \sum_{n=1}^{\infty} \frac{2z}{z^2 - n^2},$$

$$z \in \mathbf{C} - \{0, \pm 1, \pm 2, \ldots\}$$

absolutely and locally uniformly for the series $\sum_1^{\infty}$, where $\sum{}'_{-\infty}^{\infty}$ denotes the series with the term $n = 0$ omitted. Also, deduce that

$$\frac{1}{\sin^2 z} = \sum_{-\infty}^{\infty} \frac{1}{(z-n)^2}, \quad z \neq 0, \pm 1, \pm 2, \ldots;$$

$$\frac{1}{\sin z} = \frac{1}{z} + \sum_{n=1}^{\infty} \frac{(-1)^n 2z}{z^2 - n^2\pi^2} \quad \text{(see Example 1)}, z \neq 0, \pm\pi, \ldots.$$

Solution. Two methods are provided.

Method 1. $\pi \cot \pi z$ has simple poles at $z = 0, \pm 1, \pm 2, \ldots$. Observe that $\pi \cot \pi z$ has the singular part $\frac{1}{z}$ at $z = 0$. Hence, consider the function $f(z) = \pi \cot \pi z - \frac{1}{z}$.

Choose γ_n the square with the vertices $\pm(n + \frac{1}{2}) \pm i(n + \frac{1}{2})$ for $n \geq 1$. Setting $z = n + \frac{1}{2} + iy$, $-\infty < y < \infty$,

$$|\cot \pi z| = \left| \cot \pi \left(n + \frac{1}{2} + iy \right) \right| = |\tan i\pi y| = \left| \frac{e^{-\pi y} - e^{\pi y}}{e^{-\pi y} + e^{\pi y}} \right| \leq 1;$$

while, for $z = x + i(n + \frac{1}{2})$, $-\infty < x < \infty$,

$$|\cot \pi z| \leq \frac{1 + e^{-(2n+1)\pi}}{1 - e^{-(2n+1)\pi}} \leq \frac{1 + e^{-3\pi}}{1 - e^{-3\pi}}.$$

Consequently,

$$\sup_{z \in \gamma_n} |\pi \cot \pi z| \leq \pi \cdot \frac{1 + e^{3\pi}}{1 - e^{-3\pi}} = M < \infty$$

$$\Rightarrow \sup_{z \in \gamma_n} |f(z)| \leq M + \frac{1}{n + 1/2} \leq M + 1, \quad n \geq 1.$$

Thus, (5.4.2.2) with Condition 2 is applicable to $f(z)$. Now,

$$f(0) = 0;$$

$$\operatorname{Re} s(f(z); n) = \pi \cdot \frac{\cos n\pi}{\pi \cos n\pi} = 1, \quad \text{for } n \geq 1$$

$$\Rightarrow \pi \cot \pi z - \frac{1}{z} = \sum_{n=1}^{\infty} \left[\left(\frac{1}{z-n} + \frac{1}{n} \right) + \left(\frac{1}{z+n} - \frac{1}{n} \right) \right]$$

$$= \sum_{n=1}^{\infty} \frac{2z}{z^2 - n^2}, \quad z \neq 0, \pm 1, \pm 2, \dots.$$

Method 2. Consider, for fixed ζ,

$$f(z) = \frac{\pi \cot \pi z}{\zeta^2 - z^2}, \quad \zeta \neq 0, \pm 1, \pm 2, \dots,$$

which has simple poles at $z = \pm\zeta$ and $z = 0, \pm 1, \pm 2, \dots$.

Choose γ_n the rectangle with vertices at $\pm(\frac{1}{2} + n) \pm ni$ for $n \geq 1$. These γ_n satisfy 1, 2, and 3 in (5.4.2.1). To see the uniform boundedness of $f(z)$ along γ_n's, setting $z = \pm(\frac{1}{2} + n) + iy$ for the vertical sides of γ_n's, then

$$|\cot \pi z| \leq \frac{|\sinh \pi y|}{\sqrt{1 + \sinh^2 \pi y}} \leq 1;$$

while, along the horizontal sides $z = x \pm ni$,

$$|\cot \pi z| \leq \frac{\sqrt{1 + \sinh^2 n\pi}}{|\sinh n\pi|} \leq 1 + \frac{1}{\sinh n\pi} \leq 1 + \frac{1}{n\pi} \leq 2.$$

Hence,

$$\sup_{z \in \gamma_n} |f(z)| \leq \frac{2\pi}{|n^2 - |\zeta|^2|} \to 0 \quad \text{as } n \to \infty.$$

And thus, (5.4.2.2) with Condition 1 is applicable to this $f(z)$ in the following modified sense.

Observe that

$$\operatorname{Re} s(f(z); \pm\zeta) = \frac{-\pi \cot \pi\zeta}{2\zeta};$$

$$\operatorname{Re} s(f(z); n) = \frac{1}{\zeta^2 - n^2}$$

$$\text{for } n = 0, \pm 1, \pm 2, \dots.$$

Fix any point ζ in $\mathbf{C} - \{0, \pm 1, \pm 2, \dots\}$ and choose n large enough such that $\pm\zeta$ belongs to Int γ_n. By the residue theorem,

$$\frac{1}{2\pi i} \int_{\gamma_n} f(z)dz = -\frac{\pi \cot \pi\zeta}{\zeta} + \sum_{|k|<n} \frac{1}{\zeta^2 - k^2}$$

$$= -\frac{\pi \cot \pi\zeta}{\zeta} + \sum_{k=1}^{n-1} \frac{2}{\zeta^2 - k^2} + \frac{1}{\zeta^2}. \tag{$*8$}$$

The integral in the left side has the estimate

$$\left| \frac{1}{2\pi i} \int_{\gamma_n} f(z) dz \right| \le \frac{1}{2\pi} \sup_{z \in \gamma_n} |f(z)| \cdot \text{the length of } \gamma_n$$

$$\le \frac{8n+2}{|n^2 - |\zeta|^2|} \to 0 \quad \text{as } n \to \infty.$$

Let $n \to \infty$ in $(*_8)$ and the result follows.

Termwise differentiation of the partial fraction expansion for $\pi \cot \pi z$ gives in

$$-\pi^2 \csc^2 \pi z = -\frac{1}{z^2} - \sum_{|n| \ge 1} \frac{1}{(z-n)^2} = -\sum_{-\infty}^{\infty} \frac{1}{(z-n)^2}$$

$$\Rightarrow \frac{1}{\sin^2 z} = \sum_{-\infty}^{\infty} \frac{1}{(z - n\pi)^2}, \quad z \ne 0, \pm\pi, \pm 2\pi, \dots.$$

On the other hand, via the identity $\frac{\pi}{\sin \pi z} = \frac{\pi}{2} \cot \frac{\pi z}{2} - \frac{\pi}{2} \cot \frac{\pi(z-1)}{2}$ and the relations

$$\frac{\pi}{2} \cot \frac{\pi z}{2} = \lim_{n \to \infty} \sum_{k=-n}^{n} \frac{1}{z - 2k}; \quad \frac{\pi}{2} \cot \frac{\pi(z-1)}{2} = \lim_{n \to \infty} \sum_{k=-n}^{n} \frac{1}{z - (2k+1)},$$

it follows that

$$\frac{\pi}{\sin \pi z} = \lim_{n \to \infty} \sum_{k=-n}^{n} \frac{(-1)^k}{z - k} = \frac{1}{z} + \sum_{n=1}^{\infty} \frac{(-1)^n 2z}{z^2 - n^2}$$

and the claim for $\frac{1}{\sin z}$ holds.

Exercises A

(1) (Extension of Method 2 in Example 2.) Suppose $f(z)$ and $g(z)$ satisfy the following conditions:

1. $f(z)$ is meromorphic in $\mathbf{C}$ and has simple poles at $z_n = a + n$ ($n = 0, \pm 1, \pm 2, \dots$) with residue r_n and nowhere else.
2. $g(z)$ is analytic in $\mathbf{C}$, except isolated singularities, such that $\lim_{z \to \infty} z g(z) = 0$ holds. For instance, $g(z) = \frac{p(z)}{q(z)}$ where $p(z)$ and $q(z)$ are relatively prime polynomials and the degree of $q(z)$ is at least two greater than that of $p(z)$.
3. There is a constant $b \ge 0$ with $b \ne \operatorname{Re} a$ so that $f(z)$ is uniformly bounded along the rectangles γ_n with vertices at $\pm(b+n) \pm ni$ for $n \ge 1$, namely, $\sup_{n \ge 1} \sup_{z \in \gamma_n} |f(z)| < \infty$.

Then

$$\lim_{m\to\infty} S_m = -\sum_n r_n g(a+n),$$

where S_m denotes the sum of residues of $g(z)f(z)$ at its singularities in Int γ_m, and $\sum_n$ denotes the sum over these n for which $a+n$ is a simple pole of $f(z)$ yet $g(z)$ is analytic at $a+n$.

(2) Imitate Method 2 in Example 2 and Exercise (1) to redo Example 1.

(3) Set $C_n : |z| = n + \frac{1}{2}$ for $n \geq 1$ and redo Example 2.

(4) Try to use at least two distinct methods (say, model after Example 2) to prove the following expansions.

(a) $\frac{1}{\tan z} = \frac{1}{z} + \sum_{n=1}^{\infty} \frac{2z}{\left[\frac{(2n-1)\pi}{2}\right]^2 - z^2}$. What is $\frac{\pi}{2}\cot\frac{\pi z}{2}$?

(b) $\frac{1}{\cos z} = \pi\sum_{n=1}^{\infty}(-1)^n \frac{2n-1}{z^2 - \left[\frac{(2n-1)\pi}{2}\right]^2}$. What is $\frac{\pi}{2}\sec\frac{\pi z}{2}$? Sum the series $\sum_{n=1}^{\infty}\frac{(-1)^{n-1}}{2n-1}$.

(c) $\tan\frac{\pi z}{2} + \sec\frac{\pi z}{2} = -\frac{4}{\pi}\sum_{n=1}^{\infty}\frac{1}{z+(-1)^n(2n-1)}$.

(d) $\tanh z = \sum_{n=1}^{\infty}\frac{2z}{z^2 + \left[\frac{(2n-1)\pi}{2}\right]^2}$.

(e) $\pi\coth\pi z = \frac{1}{z} + 2z\sum_{n=1}^{\infty}\frac{1}{z^2+n^2}$.

(f) $\frac{1}{\sinh z} = \frac{1}{z} + \sum_{n=1}^{\infty}\frac{(-1)^n 2z}{z^2+n^2\pi^2}$.

(g) $\frac{\pi^2}{\cos^2\pi z} = \sum_{-\infty}^{\infty}\frac{1}{\left(z-\frac{1}{2}-n\right)^2}$.

(h) $\frac{z}{e^z-1} = 1 - \frac{z}{2} + \sum_{n=1}^{\infty}\frac{2z^2}{z^2+4n^2\pi^2}$.

(5) Show that

$$\frac{1}{\sin z \sinh z} = \frac{1}{z^2} + \sum_{n=1}^{\infty}\frac{(-1)^n 4n\pi z^2}{(z^4 - n^4\pi^4)\sinh n\pi}.$$

And, then, deduce that

$$\frac{1}{\cosh z - \cos z} = \frac{1}{z^2} + \pi z^2\sum_{n=0}^{\infty}\frac{(-1)^n n}{\sinh n\pi}\cdot\frac{1}{\frac{1}{4}z^4 + n^4\pi^4}.$$

(6) Let γ_n denote the square with vertices at $\pm n\pi \pm n\pi i$ for $n \geq 1$. Consider the integral

$$\int_{\gamma_n} \frac{\sin\zeta}{\zeta(\zeta - z)\cos\zeta}\,d\zeta$$

to show that (refer to Example 1)

$$\frac{\sin z}{\cos^2 z} = \sum_{n=-\infty}^{\infty} (-1)^n \frac{1}{z - \left(n + \frac{1}{2}\right)\pi}.$$

(7) Let λ_n, $n \geq 1$, be the positive roots of the equation $\tan z = z$ (refer to Exercise A(10) of Sec. 3.5.4). Show that

$$\frac{z\sin z}{\sin z - z\cos z} = \frac{3}{z} + \sum_{n=1}^{\infty} \frac{2z}{z^2 - \lambda_n^2}.$$

(8) Suppose $0 < a < 1$. Show that

$$\frac{e^{az}}{e^z - 1} = \frac{1}{z} + \sum_{n=1}^{\infty} \frac{2z\cos 2na\pi - 4n\pi\sin na\pi}{z^2 + 4n^2\pi^2}.$$

Exercises B

(1) Suppose a meromorphic function $f(z)$ in $\mathbf{C}$ has only simple poles $z_1, z_2, \ldots$ satisfying $0 < |z_1| \leq |z_2| \leq \cdots$. Let γ_n, $n \geq 1$, be a system of circles or squares, with centers at 0 and passing no poles, such that

1. $\gamma_n \subseteq \mathrm{Int}\,\gamma_{n+1}$, for $n \geq 1$, and
2. $\mathrm{Int}\,\gamma_n$ contains only poles $z_1, \ldots, z_n$ but not others.

In case there are constants $M > 0$ and $\alpha < 1$ so that $|f(z)| \leq M|z|^\alpha$ along γ_n for all sufficiently large n, then

$$f(z) = f(0) + \sum_{n=1}^{\infty} \mathrm{Re}\,s(f(z); z_n)\left(\frac{1}{z - z_n} + \frac{1}{z_n}\right), \quad z \in \mathbf{C} - \{z_1, z_2, \ldots\}.$$

Then, try to prove the following expressions, where $-1 < a < 1$.

(a) $\dfrac{\pi e^{\pi az}}{\sinh \pi z} = \sum_{n=-\infty}^{\infty} \dfrac{(-1)^n(z\cos n\pi a - n\sin n\pi a)}{z^2 + n^2}$.

(b) $\dfrac{\pi\cosh \pi az}{\sinh \pi z} = \sum_{n=-\infty}^{\infty} \dfrac{(-1)^n z\cos n\pi a}{z^2 + n^2}$.

(c) $\pi\sin \pi az \csc \pi z = \sum_{n=-\infty}^{\infty} \dfrac{(-1)^n n\sin n\pi a}{z^2 - n^2}$.

(d) $\pi\cos \pi az \csc \pi z = \sum_{n=-\infty}^{\infty} \dfrac{(-1)^n z\cos n\pi a}{z^2 - n^2}$.

(2) Conditions stated in Exercise (1) are altered to: the poles $z_1, z_2, \ldots$ are not integers and $\alpha < -1$ while the others remains unchanged. Show that

$$\sum_{n=-\infty}^{\infty} f(n) = -\pi \sum_{n=1}^{\infty} \operatorname{Re} s(f(z); z_n) \cot \pi z_n;$$

$$\sum_{n=-\infty}^{\infty} (-1)^n f(n) = -\pi \sum_{n=1}^{\infty} \operatorname{Re} s(f(z); z_n) \csc \pi z_n.$$

Then, try to use the first identity to prove that

$$1 + \frac{\pi z(\sin 2\pi z + \sinh 2\pi z)}{\cosh 2\pi z - \cos 2\pi z} = 8 \sum_{n=0}^{\infty} \frac{z^4}{4z^4 + n^4};$$

and the second one to prove that

$$\pi^2 \cot \pi z \csc \pi z = \sum_{n=-\infty}^{\infty} \frac{(-1)^n}{(z+n)^2};$$

$$\frac{\pi}{2z^3}(\csc \pi z + \cosh \pi z) = \sum_{n=-\infty}^{\infty} \frac{(-1)^n}{z^4 - n^4}.$$

(3) Try to prove (5.4.2.3), at least the former part there.

5.5   Entire Functions: Weierstrass's Factorization Theorem and Hadamard's Order Theorem

Recall that an entire function is a function analytic in the plane $\mathbf{C}$. Such a function $f(z)$ has a Taylor series expansion at 0 as

$$f(z) = \sum_{0}^{\infty} a_n z^n, \quad |z| < \infty, \qquad (*_1)$$

which converges absolutely and locally uniformly in $\mathbf{C}$. Moreover, $\lim_{n \to \infty} \sqrt[n]{|a_n|} = 0$ holds (see (3.3.2.2) and (3.3.2.4)).

An entire function $f(z)$ can behave at infinity in the following three possible ways:

1. $f(z)$ has a removable singularity at ∞.
   $\Leftrightarrow a_n = 0$ for $n \geq 1$ in $(*_1)$.
   $\Leftrightarrow f(z) = a_0$, a constant (see (3.4.2.10) and Sec. 4.10.1).
2. $f(z)$ has a pole of order $k \geq 1$ at ∞.
   $\Leftrightarrow a_n = 0$ for $n \geq k+1$ in $(*_1)$

$\Leftrightarrow f(z) = a_0 + a_1 z + \cdots + a_k z^k \ (a_k \neq 0)$ is a polynomial of degree k (see Example 6 in Sec. 3.4.2 and (4.10.2.2)).

3. $f(z)$ has an essential singularity at ∞, and then such a $f(z)$ is called a *transcendental entire function*; such as e^z, $\cos z$, and $\sin z$.

$\Leftrightarrow a_n \neq 0$ for infinitely many n in $(*_1)$ (see Sec. 4.10.3).

What interests us mostly in this section is the transcendental one.

The fundamental theorem of algebra (see (3.4.2.12)) characterizes another important expression for polynomials in term of their zeros, namely,

$$f(z) = a_k(z - z_1) \cdots (z - z_k) = a_k \prod_{n=1}^{k} (z - z_n). \qquad (*_2)$$

As the limit function of local uniform convergence of its partial sums in $(*_1)$, it is desirable and tempting to expect that a general entire function would have a similar representation by means of an infinite product.

To start with, observe that:

$f(z)$ is an entire function without zeros.
$\Leftrightarrow f(z)$ is of the form $e^{g(z)}$, where $g(z)$ is an entire function. $\qquad (5.5.1)$

To see the necessity, note that, as a primitive of $\frac{f'(z)}{f(z)}$ which is entire,

$$g(z) = \int_{z_0}^{z} \frac{f'(z)}{f(z)} dz$$

is well-defined in $\mathbf{C}$ (see Sec. 4.1) and hence, is entire, too. We infer that the derivative of $f(z)e^{-g(z)}$ is identically equal to zero in $\mathbf{C}$ and hence (see (3.4.1.3)), $f(z)$ is a constant multiple of $e^{g(z)}$ where the constant can be absorbed into $g(z)$.

In case only finitely many zeros happen, we have that:

An entire function $f(z)$ has only finitely many zeros at $0, a_1, \ldots, a_k$ ($a_n \neq 0$ for $1 \leq k \leq n$), of respective order $m, m_1, \ldots, m_k$.
$\Leftrightarrow f(z)$ is of the form

$$e^{g(z)} z^m \left(1 - \frac{z}{a_1}\right)^{m_1} \cdots \left(1 - \frac{z}{a_k}\right)^{m_k} = e^{g(z)} z^m \prod_{n=1}^{k} \left(1 - \frac{z}{a_n}\right)^{m_n},$$

where $g(z)$ is an entire function. $\qquad (5.5.2)$

Again, to see the necessity, consider the polynomial $p(z) = z^m(z - a_1)^{m_1} \cdots (z - a_k)^{m_k}, \ = Az^m \prod_{n=1}^{k}(1 - \frac{z}{a_n})^{m_n}, \ A \neq 0.$ (3.4.2.15) or

(4.10.1.1) says that the function

$$h(z) = \frac{f(z)}{p(z)}$$

has removable singularities at each of the points $0, a_1, \ldots, a_k$. Then, $h(z)$ is an entire function without zeros. The result follows from (5.5.1) with the constant A absorbed by $g(z)$.

Now, suppose a nonzero entire function $f(z)$ has infinitely many zeros (such as $\cos z$, $\sin z$ or $e^z - w_0$). These zeros are countably infinite in number and do not have finite limit points. We arrange these zeros, except $z = 0$ if 0 is a zero of $f(z)$, in the following way:

$$a_1, a_2, \ldots, a_n, \ldots \quad \text{(counting multiplicities), with}$$

$$|a_1| \le |a_2| \le \cdots \le |a_n| \le \cdots, \quad \text{and}$$

$$\lim_{n \to \infty} |a_n| = \infty. \tag{5.5.3}$$

Then, it is reasonable to expect that $f(z)$ can be represented and characterized via its zeros by the expression

$$e^{g(z)} z^m \prod_{n=1}^{\infty} \left(1 - \frac{z}{a_n}\right), \quad \text{where } m \ge 0 \quad \text{and} \quad g(z) \text{ is entire.} \tag{5.5.4}$$

Unfortunately, the infinite product within might be divergent. And a convergence factor, in general, should be attached to each factor $1 - \frac{z}{a_n}$ in order to ascertain the (local uniform) convergence.

Section 5.5.1 will sketch the convergence of infinite product $\prod_{n=1}^{\infty}(1 + a_n)$ and the (local) uniform convergence of infinite product $\prod_{n=1}^{\infty}(1 + f_n(z))$ with complex-valued function terms. These material are preparatory for the next two sections.

Section 5.5.2 formally justifies the validity of (5.5.4), after multiplying a convergent factor $e^{p_n(z)}$ ($p_n(z)$, a polynomial) to each $1 - \frac{z}{a_n}$, $n \ge 1$. This is Weierstrass's factorization theorem. The canonical product and its genus, and an entire function of finite genus are introduced.

If the polynomial $p_n(z)$ in each multiplying factor $e^{p_n(z)}$ can be chosen to be of the same degree (genus) and $g(z)$ in $e^{g(z)}$ reduces to a polynomial, the representation (5.5.4) will turn out to be considerably more interesting and useful. It was Hadamard who first pointed out how the rate of growth of the function can be used to solve these problems. For an entire function $f(z)$, the most important characteristic of it is the maximum modulus $M(r) = \max_{|z|=r} |f(z)|$. In terms of $M(r)$, Sec. 5.5.3 will introduce the order $\lambda = \varlimsup_{r \to \infty} \frac{\log \log M(r)}{\log r}$ of an entire function and prove Hadamard's order theorem, an inequality $h \le \lambda \le h + 1$ relating the order λ and the genus h.

5.5.1   *Infinite products (of complex numbers and functions)*

Given a sequence $z_n \in \mathbf{C}$, $n \ge 1$, do the formal *infinite product* of them as

$$\prod_{n=1}^{\infty} z_n = z_1 z_2 \cdots z_n \cdots \quad \text{or simply as} \quad \prod z_n \tag{5.5.1.1}$$

where z_n $(n \ge 1)$ is called the n-th *term*. The product $P_n = z_1 z_2 \cdots z_n$ of former n terms is called the n-th *partial product*.

$\prod_{n=1}^{\infty} z_n$ is said to be *convergent* if, for some $m \ge 1$, $z_n \ne 0$ for $n \ge m$, and if the partial products $z_{m+1} \cdots z_n$ $(n \ge m + 1)$ converges to a *nonzero* finite number P as $n \to \infty$. In this case, $\prod_{n=1}^{\infty} z_n$ is said to *converge* to $z_1 \cdots z_m P$, call the *product* of $z_1, z_2, \ldots, z_n, \ldots$, and denote it as

$$\prod_{n=1}^{\infty} z_n = z_1 \cdots z_m P.$$

Note that $\prod_{n=1}^{\infty} z_n = 0$ if and only if some of its terms is equal to zero. On the contrary, if $z_{m+1} \cdots z_n \to 0$ as $n \to \infty$ or does not converge to a finite limit, then $\prod_{n=1}^{\infty} z_n$ is called *divergent*.

Suppose $\prod_1^{\infty} z_n$ is convergent. Let m be as above. For $n \ge m + 1$,

$$z_n = \frac{z_{m+1} \cdots z_{n-1} z_n}{z_{m+1} \cdots z_{n-1}} \to \frac{P}{P} = 1 \quad \text{as } n \to \infty.$$

So it is preferable to rewrite (5.5.1.1) in the form

$$\prod_{n=1}^{\infty} (1 + a_n) \quad \text{or} \quad \prod (1 + a_n) \tag{5.5.1.2}$$

and $a_n \to 0$ becomes a necessary condition for convergence.

Section (1) Convergence and absolute convergence of infinite product of numbers

It is convenient, in practice, to use tests of convergence for series (see Sec. 2.3) to check if a given infinite product is convergent or not. Without loss of its generality, we assume henceforth that each $a_n \ne -1$ for $n \ge 1$ in (5.5.1.2) during the process of proofs in the following.

First, let $P_n = (1 + a_1) \cdots (1 + a_n)$ and $S_n = \sum_{k=1}^{n} \mathrm{Log}(1 + a_k)$, where, for $n \ge 1$, $\mathrm{Log}(1 + a_n)$ with $-\pi < \mathrm{Arg}(1 + a_n) \le \pi$, is the principal branch of $\log(1 + a_n)$. Then

$$P_n = e^{S_n}$$

shows that, if $S_n \to S$ as $n \to \infty$, then $P_n \to e^S$ and $\prod(1 + a_n)$ converges to the product e^S.

Is the converse true? To see this, let $\lim_{n\to\infty} P_n = P \ (\neq 0, \infty)$. Choose the principal branch $\text{Log } P$, $-\pi < \text{Arg } P \leq \pi$. It is possible (see (1.7.3)) to choose $\arg P_n$ so that $\text{Arg } P - \pi < \arg P_n < \text{Arg } P + \pi$ if n is large enough. For such n, then

$$S_n = \sum_{k=1}^{n} \text{Log } (1 + a_k) = \log P_n + 2h_n \pi i, \quad \log P_n = \log |P_n| + i \arg P_n,$$

where h_n is an integer to be determined later on. Hence

$$\text{Log } (1 + a_{n+1}) = S_{n+1} - S_n = \log P_{n+1} - \log P_n + 2(h_{n+1} - h_n)\pi i$$

$$\Rightarrow \text{Arg } (1 + a_{n+1}) = \arg P_{n+1} - \arg P_n + 2(h_{n+1} - h_n)\pi.$$

Since $\text{Log } (1 + a_{n+1}) \to \text{Log } 1 = 0$ and $\log P_{n+1}, \log P_n \to \text{Log } P$, there is an integer $n_0 \geq 1$ so that $|\text{Arg } (1 + a_{n+1})| < \frac{2\pi}{3}$, $|\arg P_{n+1} - \text{Arg } P| < \frac{2\pi}{3}$, and $|\arg P_n - \text{Arg } P| < \frac{2\pi}{3}$ if $n \geq n_0$. Consequently,

$$2|(h_{n+1} - h_n)|\pi < 3 \cdot \frac{2\pi}{3} = 2\pi, \quad n \geq n_0$$

$$\Rightarrow |(h_{n+1} - h_n)| < 1, \quad n \geq n_0.$$

Since h_n are integers, $h_{n+1} = h_n$ for $n \geq n_0$. Let $h = h_n$, $n \geq n_0$. Then $\lim_{n\to\infty} S_n = \text{Log } P + 2h\pi i$ and $\sum_{n=1}^{\infty} \text{Log } (1 + z_n)$ converges (to this limit).

Combining together, we have proved the following

Equivalent for convergence of infinite product.

$$\prod_{n=1}^{\infty} (1 + a_n) \text{ converges (to } P \neq 0).$$

$$\Leftrightarrow \text{ the series } \sum_{n=1}^{\infty} \text{Log } (1 + a_n) \text{ converges (to } \text{Log } P + 2h\pi i$$

for some integer h).

where each $\text{Log } (1 + a_n)$ is the principal branch of $\log (1 + a_n)$, $n \geq 1$.

$$(5.5.1.3)$$

Can (5.5.1.3) be further reduced to the convergence of $\sum_{n=1}^{\infty} a_n$? Simple examples say that it is negative in general. For instance,

$$\prod_{n=1}^{\infty} \frac{n}{n+1} = 1, \text{ while } \sum_{n=1}^{\infty} \left(\frac{n}{n+1} - 1 \right) \text{ diverges to } -\infty;$$

$$\prod_{n=1}^{\infty}\left(1+\frac{(-1)^{n+1}}{n^p}\right) \text{ diverges if } 0 < p \le \frac{1}{2}, \text{ while } \sum_{n=1}^{\infty}\left(\frac{(-1)^{n+1}}{n^p}\right)$$

converges (see Exercise A(3)(e));

$$\prod_{n=2}^{\infty}\left(1+\frac{(-1)^{n+1}}{n}\right) \text{ converges } \left(\text{to } \frac{1}{2}\right) \text{ and } \sum_{n=2}^{\infty}\left(\frac{(-1)^{n+1}}{n}\right)$$

converges (to what sum?).

However, in the realm of absolute convergence, the answer is in the affirmative.

To justify this claim, note that

$$\lim_{z \to 0} \frac{\text{Log}(1+z)}{z} = 1$$

$\Rightarrow$ For $\varepsilon > 0$, there is a $\delta = \delta(\varepsilon) > 0$ so that, for $0 < |z| < \delta$,

$$(1 - \varepsilon)|z| < |\text{Log}(1+z)| < (1 + \varepsilon)|z|. \tag{$*_1$}$$

Applying this inequality to each $\text{Log}(1 + z_n)$, $n \ge 1$, it follows immediately that $\sum |z_n|$ and $\sum |\text{Log}(1 + z_n)|$ converges or diverges simultaneously.

Also, the inequality

$$1 + x \le e^x \text{ for } x \ge 0$$

$\Rightarrow |a_1| + \cdots + |a_n| \le (1 + |a_1|) \cdots (1 + |a_n|) \le e^{|a_1| + \cdots + |a_n|}, \quad n \ge 1. \tag{$*_2$}$

Hence, $\sum |a_n|$ and $\prod (1 + |a_n|)$ converges or diverges simultaneously, too.

We summarize the above as

The absolute convergence of infinite product.

  1. $\prod(1 + |a_n|)$ converges.
$\Leftrightarrow$ 2. $\sum |a_n|$ converges.
$\Leftrightarrow$ 3. $\sum |\text{Log}(1 + a_n)|$ converges.

In this case, $\prod(1 + a_n)$ is said to be *absolutely convergent*. Moreover, absolute convergence has the following properties.

(1) Absolute convergent infinite product does converge.
(2) The order of terms in an absolute convergent infinite product can be altered arbitrarily without affecting the absolute convergence and the product.   (5.5.1.4)

The last two statements can be proved by known results from series and the very definitions introduced above.

Sketch of proof. (1) let $P_n = \prod_{k=1}^{n}(1 + a_k)$ and $P'_n = \prod_{k=1}^{n}(1 + |a_k|)$. Then

$$|P_n - P_{n-1}| \leq P'_n - P'_{n-1}(P_0 = 0) \tag{$*_3$}$$

$$\Rightarrow \left(\text{Since } \sum_{n=1}^{\infty}(P'_n - P'_{n-1}) = \lim_{n\to\infty} P'_n \text{ converges.} \right)$$

$$\sum_{n=1}^{\infty}(P_n - P_{n-1}) = \lim_{n\to\infty} P_n = P \text{ converges (absolutely)}.$$

To see $P \neq 0$, observe that, since $\lim_{n\to\infty}(1+a_n) = 1$ and $\sum |a_n|$ converges,

$$\sum_{n=1}^{\infty} \left| \frac{a_n}{1 + a_n} \right| < \infty$$

$\Rightarrow$ (by statements 1 and 2, and the result just proved)

$$\prod_{n-1}^{\infty} \left(1 - \frac{a_n}{1 + a_n} \right) = \lim_{n\to\infty} \prod_{k=1}^{n} \left(1 - \frac{a_k}{1 + a_k} \right) = \lim_{n\to\infty} \frac{1}{\prod_{k=1}^{n}(1 + a_k)}$$

$$= \frac{1}{P}(\neq 0, \infty) \text{ converges.}$$

Hence $P \neq 0$ holds.

(2) Let $\pi : \{1, 2, \ldots, n, \ldots\} \to \{1, 2, \ldots, n, \ldots\}$ be a permutation (namely, one-to-one and onto map). Since $\sum |a_n| < \infty \Rightarrow \sum |a_{\pi(n)}| < \infty$ holds, by (1), then $\prod(1 + a_{\pi(n)}) = P'$ converges. Fix any n, choose m large enough so that $P'_m = \prod_{k=1}^{m}(1 + a_{\pi(k)})$ contains all the factors appearing in P_n. Then $m \geq n$ and

$$P'_m = P_n(1 + a_{\alpha_1})(1 + a_{\alpha_2}) \cdots (1 + a_{\alpha_k}), \quad \alpha_1, \alpha_2, \ldots, \alpha_k \geq n + 1$$

$$\Rightarrow \left| \frac{P'_m}{P_n} - 1 \right| \leq (1 + |a_{\alpha_1}|)(1 + |a_{\alpha_2}|) \cdots (1 + |a_{\alpha_k}|) - 1$$

$$\leq e^{|a_{\alpha_1}| + \cdots + |a_{\alpha_k}|} - 1, \quad \text{by } (*_2)$$

$$\leq e^{\sum_{l=n+1}^{\infty} |a_l|} - 1 \to e^0 - 1 = 0 \quad \text{as } n \to \infty$$

(and hence $m \to \infty$)

$$\Rightarrow \frac{P'}{P} = 1 \quad \text{or} \quad P' = P.$$

This finishes the proof. $\square$

We illustrate two examples.

Example 1. If $\sum_{n=1}^{\infty} |z_n|^2$ converges, then $\prod_{n=1}^{\infty} \cos z_n$ converges absolutely.

Solution. Note that $\lim_{z \to 0} \frac{\cos z - 1}{z^2} = \lim_{z \to 0} \frac{-\sin z}{2z} = -\frac{1}{2}$. Hence $\sum |z_n|^2$ and $\sum |\cos z_n - 1|$ converge simultaneously. It follows by (5.5.1.4) that the product converges absolutely.

Example 2. The product

$$\prod_{n=1}^{\infty} \left(1 + \frac{(-1)^{n-1}}{n^z} \right), \quad n^z = e^{z \log n}$$

converges in the half-plane $\operatorname{Re} z > \frac{1}{2}$ and absolutely in $\operatorname{Re} z > 1$.

Solution. Using $\operatorname{Log}(1+z) = z - \frac{1}{2}z^2 + \cdots + (-1)^{n-1}\frac{1}{n}z^n + \cdots$, $|z| < 1$, let

$$\operatorname{Log}\left(1 + \frac{(-1)^{n-1}}{n^z} \right) = z_n + R_n(z), \quad \text{where}$$

$$z_n = \frac{(-1)^{n-1}}{n^z} \quad \text{and} \quad R_n(z) = \sum_{k=2}^{\infty} (-1)^{k-1} \frac{z_n^k}{k}$$

$$\text{if } n \text{ is chosen so that } |z_n| < 1. \tag{$*_4$}$$

In case $\operatorname{Re} z > \frac{1}{2}$, then

$$|z_n| = \frac{1}{n^{\operatorname{Re} z}} \leq \frac{1}{\sqrt{n}} < \frac{1}{2} \quad \text{if } n \geq 4$$

$$\Rightarrow |R_n(z)| \leq \sum_{k=2}^{\infty} \frac{1}{k} |z_n|^k = |z_n|^2 \sum_{k=2}^{\infty} \frac{1}{k} |z_n|^{k-2}$$

$$\leq |z_n|^2 \left(1 + \frac{1}{2} + \frac{1}{2^2} + \cdots \right) \leq 2|z_n|^2 \quad \text{if } n \geq 4.$$

Hence,

$$\sum_{n=1}^{\infty} z_n \text{ and } \sum_{n=1}^{\infty} |z_n|^2 \text{ converge if } \operatorname{Re} z > \frac{1}{2}; \sum_{n=1}^{\infty} |z_n| \text{ converges if } \operatorname{Re} z > 1.$$

$$\Rightarrow \sum_{n=4}^{\infty} |R_n(z)| \leq 2 \sum_{n=1}^{\infty} |z_n|^2 < \infty \quad \text{if } \operatorname{Re} z > \frac{1}{2}.$$

The claimed results follow from $(*_4)$, (5.5.1.3), and (5.5.1.4).

Section (2) (Local) Uniform convergence of $\prod(1 + f_n(z))$

Let us turn to infinite product of complex-valued functions.

Suppose $f_n(z)$: Ω(domain) $\to$ **C**, for $n \geq 1$. If $z_o \in \Omega$ such that $\prod_1^\infty (1 + f_n(z_o))$ converges (see Example 2 above), then the infinite product

$$\prod_{n=1}^\infty (1 + f_n(z)) \quad \text{or} \quad \prod (1 + f_n(z)) \qquad (5.5.1.5)$$

is said to *converge* at z_o. The set of points in Ω where (5.5.1.5) converges is not necessarily open or connected, in general, and is called the *domain of convergence* of the product. The *product* $f(z)$ defined on this domain of convergence (if exists) is a function and is denoted by

$$f(z) = \prod_{n=1}^\infty (1 + f_n(z)) \quad \text{or} \quad \prod (1 + f_n(z)). \qquad (5.5.1.6)$$

If the partial product $P_n(z) = \prod_{k=1}^n (1 + f_k(z))$ converges uniformly to $f(z)$ on a subset of Ω, then (5.5.1.6) is said to *converge uniformly* to $f(z)$ on the set; and *converge locally uniformly* to $f(z)$ in the domain Ω if it converges uniformly to $f(z)$ on each compact subset of Ω.

(5.5.1.3) *and* (5.5.1.4) *have obvious counterparts for (local) uniform convergence* (namely, these results are still valid if convergence in (5.5.1.3) and absolute convergence in (5.4.1.4) are replaced by uniform convergence, respectively). See Exercise A(8). Anyway, we have a concise

Necessary and sufficient condition for absolute and uniform convergence.

(1) $\prod_{n=1}^\infty (1 + f_n(z))$ converges absolutely and locally uniformly in a domain Ω

$\Leftrightarrow$ (2) $\sum_{n=1}^\infty f_n(z)$ converges absolutely and locally uniformly in Ω.

Let $f(z) = \prod(1 + f_n(k))$ in Ω, in this case. Then,

1. $z_o \in \Omega$ is a zero of $f(z) \Leftrightarrow z_o$ is a zero of at least one factor $1 + f_n(z)$ in the product, and
2. The order of zero of $f(z)$ at z_o is equal to the sum of the orders of zeros of the finitely many factors having z_o as their common zeros.

It is tacitly understood henceforth that absolute convergence and local uniform convergence are considered only on subsets of Ω where the zeros of $f(z)$ are deleted.   (5.5.1.7)

In case $|f_n(z)|$ is sufficiently small (if n is large enough), then $(*_1)$ says

$$(1 - \varepsilon)|f_n(z)| < | \text{Log } (1 + f_n(z))| < (1 + \varepsilon)|f_n(z)|$$

$$\Rightarrow (1 - \varepsilon) \sum_{k=n}^{n+p} |f_k(z)| \leq \sum_{k=n}^{n+p} |\text{Log } (1 + f_k(z))| < (1 + \varepsilon) \sum_{k=n}^{n+p} |f_k(z)|, \quad p \geq 0.$$

And (5.5.1.7) follows obviously. Alternatively, one can imitate $(*_3)$ to present another proof. The details are left to the readers.

What is important for our purpose is the following

Infinite product of analytic functions. Let $f_n(z) : \Omega$ (domain) $\to$ **C** be analytic for $n \geq 1$.

Suppose $\sum_{n=1}^{\infty} f_n(z)$ converges absolutely and locally uniformly in Ω. Then,

(1) $\prod_{n=1}^{\infty}(1 + f_n(z))$ converges absolutely and locally uniformly in Ω to an analytic function $f(z)$ (recall that the behavior of convergence is performed on the set $\Omega - \{z|f(z) = 0\}$).

(2) Also,

$$\frac{f'(z)}{f(z)} = \sum_{n=1}^{\infty} \frac{f_n'(z)}{1 + f_n(z)} \quad \text{absolutely and locally uniformly in}$$

$$\Omega - \{z|f(z) = 0\}. \tag{5.5.1.8}$$

Sketch of proof. (1) is an easy consequence of (5.5.1.7) and Weierstrass's theorem (5.3.1.1).

$$\text{Set } f(z) = \prod_{n=1}^{\infty}(1 + f_n(z)) = (1 + f_1(z)) \cdots (1 + f_n(z)) \quad F_n(z), \quad \text{where}$$

$$F_n(z) = \prod_{k=n+1}^{\infty} (1 + f_k(z)), \quad z \in \Omega.$$

Choose any compact set $K \subseteq \Omega - \{z|f(z) = 0\}$. By logarithmic differentiation (see Exercise A(8) of Sec. 3.3.1), then

$$\frac{f'(z)}{f(z)} = \sum_{k=1}^{n} \frac{f_k'(z)}{1 + f_k(z)} + \frac{F_n'(z)}{F_n(z)}, \quad z \in K.$$

Since $P_m(z) = (1 + f_{n+1}(z)) \cdots (1 + f_m(z))$ converges to $F_n(z)$ uniformly on K as $m \to \infty$, so does $P_m'(z)$ to $F_n'(z)$ on K, again by (5.3.1.1).

Then

$$\frac{F_n'(z)}{F_n(z)} = \lim_{m\to\infty} \frac{P_m'(z)}{P_m(z)} = \lim_{m\to\infty} \sum_{k=n+1}^{m} \frac{f_k'(z)}{1+f_k(z)} = \sum_{k=n+1}^{\infty} \frac{f_k'(z)}{1+f_k(z)}$$

uniformly on K.

$$\Rightarrow \frac{f'(z)}{f(z)} = \sum_{n=1}^{\infty} \frac{f_n'(z)}{1+f_n(z)} \quad \text{pointwise (and, indeed, locally uniformly)}$$

in $\Omega - \{z|f(z) = 0\}$.

What remains is to prove the absolute and local uniform convergence of the above right series in $\Omega - \{z|f(z) = 0\}$.

Let K be a closed disk $|z-z_o| \leq r$, contained in $\Omega - \{z|f(z) = 0\}$. There is an $n_0 \geq 1$ so that $|f_n(z)| \leq \frac{1}{2}$ on K for $n \geq n_0$; in turn, $|1+f_n(z)| \geq \frac{1}{2}$ on K for $n \geq n_0$. Since $f(z) \neq 0$ on K, so do $1+f_n(z) \neq 0$ for $1 \leq n \leq n_0$, and there is some $\alpha > 0$ so that $|1+f_n(z)| \geq \alpha$ on K for $1 \leq n \leq n_0$. Set $M^{-1} = \min(\frac{1}{2}, \alpha)$. Then

$$|1+f_n(z)| \geq \frac{1}{M}, \quad z \in K, \quad n \geq 1$$

$$\Rightarrow \frac{|f_n'(z)|}{|1+f_n(z)|} \leq M|f_n'(z)|, \quad z \in K, \quad n \geq 1.$$

Try to show that $\sum |f_n'(z)|$ converges uniformly on K; then, $\sum \frac{f_n'(z)}{1+f_n(z)}$ converges absolutely and locally uniformly on K. Using Cauchy's integral formula,

$$f_n'(z) = \frac{1}{2\pi i} \int_{|\zeta - z_o| = r} \frac{f_n(\zeta)}{(\zeta - z)^2} d\zeta, \quad |z - z_o| \leq \frac{r}{2}, \quad n \geq 1$$

$$\Rightarrow \sum_{k=n}^{n+p} |f_k'(z)| \leq \frac{1}{2\pi} \int_{|\zeta - z_o| = r} \frac{1}{|\zeta - z|^2} \sum_{k=n}^{n+p} |f_k(\zeta)| d\zeta$$

$$\leq \frac{4}{r} \max_{|\zeta - z_o| = r} \left(\sum_{k=n}^{n+p} |f_k(\zeta)| \right), \quad |z - z_o| \leq \frac{r}{2} \quad \text{and} \quad p \geq 0.$$

Then Cauchy criterion justifies the claim. And the final result follows by routine argument on compactness of sets. □

Since $\sum_{n=1}^{\infty} \frac{(-1)^{n-1}}{n^z}$ converges absolutely and locally uniformly in $\text{Re}\, z > 1$, so does the infinite product in Example 2. One more

Example 3. Determine the domain of absolute and local uniform convergence for each of the following infinite products.

(1) $\prod_{n=1}^{\infty} \cos \frac{z}{n}$.

(2) $\prod_{n=1}^{\infty} \frac{(\sin z/n)}{z/n}$.

(3) (Ref. [1], p. 193) $\prod_{n=1}^{\infty} (1 + \frac{z}{n}) e^{-\frac{z}{n}}$.

(4) $\prod_{n=1}^{\infty} (1 + \frac{e^{-nz}}{n^2})$.

(5) $\prod_{n=1}^{\infty} (1 + \frac{1}{n} \sin \frac{z}{n})$.

Solution. (1) Note that

$$\lim_{n\to\infty} \frac{\cos z/n - 1}{z^2/n^2} = \lim_{n\to\infty} \frac{-\sin z/n}{2 \cdot z/n} = -\frac{1}{2}$$

and that $\sum_{n=1}^{\infty} \frac{z^2}{n^2}$ converges absolutely and locally uniformly in **C**. According to (5.5.1.7), so does the infinite product $\prod_{n=1}^{\infty} \cos \frac{z}{n}$ in **C**.

(2) Note that

$$\lim_{z\to 0} \frac{(\sin z/z) - 1}{z^2} = \lim_{z\to 0} \frac{\sin z - z}{z^3} = -\frac{1}{6}.$$

As in (1), the product converges absolutely and locally uniformly in **C**.

(3) Note that

$$\lim_{z\to 0} \frac{(1+z)e^{-z} - 1}{z^2} = \lim_{z\to 0} \frac{e^{-z} - (1+z)e^{-z}}{2z} = \lim_{z\to 0} \frac{-e^{-z}}{2} = -\frac{1}{2}.$$

So the product converges absolutely and locally uniformly in **C**.

(4) Note that

$$\left| \frac{e^{-nz}}{n^2} \right| = \frac{e^{-n\operatorname{Re} z}}{n^2}, \quad n \geq 1$$

and that the series $\sum_{1}^{\infty} \frac{e^{-n\operatorname{Re} z}}{n^2}$ converges if and only if $\operatorname{Re} z \geq 0$. Hence, the product converges absolutely and locally uniformly in $\operatorname{Re} z \geq 0$.

(5) Note that

$$\lim_{n\to\infty} \frac{\frac{1}{n} \cdot \sin z/n}{z/n^2} = \lim_{n\to\infty} \frac{\sin z/n}{z/n} = 1.$$

The product converges absolutely and locally uniformly in **C**.

Exercises A

(1) Suppose $a_n \neq 0$ for $n \geq 1$ and designate $-\pi < \operatorname{Arg} a_n \leq \pi$. Show that $\prod_{n=1}^{\infty} a_n$ and $\sum_{n=1}^{\infty} \operatorname{Log} a_n$ converge or diverge simultaneously. What happens if $0 \leq \operatorname{Arg} a_n < 2\pi$ is assumed for each $n \geq 1$? If $\alpha \leq \operatorname{Arg} a_n < \alpha + 2\pi$ $(\alpha < 0)$?

(2) Suppose both $\prod a_n$ and $\prod b_n$ converge. Determine if the following products converge or diverge: (a) $\prod(a_n + b_n)$. (b) $\prod(a_n - b_n)$. (c) $\prod a_n b_n$. (d) $\prod \frac{a_n}{b_n}$.

(3) Prove the following statements.

   (a) Let $a_n \geq 0$ for $n \geq 1$. Then $\prod(1+a_n)$ converges $\Leftrightarrow \sum a_n$ converges; and $\prod(1 - a_n)$ converges $\Leftrightarrow \sum a_n$ converges.

   (b) Let $0 \leq a_n < 1$ for $n \geq 1$. Then $\sum a_n$ diverges $\Rightarrow \prod(1 - a_n)$ diverges to 0; but $\sum a_n$ and $\prod(1 + a_n)$ diverge or converge simultaneously.

   (c) If $\sum a_n, \sum a_n^2, \dots, \sum a_n^{k-1}$ $(k \geq 2)$ and $\sum |a_n|^k$ converge, so does $\prod(1 + a_n)$.

   (d) Suppose $a_n \in \mathbf{R}$ for $n \geq 1$ and $\sum a_n$ converges. Then $\prod(1 + a_n)$ converges or diverges to 0 according to $\prod a_n^2$ converges or diverges, respectively.

   (e) Suppose $a_n \in \mathbf{R}$ for $n \geq 1$ and $\sum a_n$ converges, while $\sum |a_n|$ diverges. Then $\prod(1 + a_n)$ converges $\Leftrightarrow \sum a_n^2$ converges.

   (f) Suppose $\sum |a_n|^2 < \infty$. Then $\prod(1 + a_n)$ converges $\Leftrightarrow \sum a_n$ converges.

(4) (a) Let $a_n = (-1)^{n-1} \frac{1}{\sqrt{n}}$, $n \geq 1$. Show that $\sum a_n$ converges and $\sum a_n^2$ diverges, while $\prod(1 + a_n)$ diverges to 0.

   (b) Let $a_{2n-1} = \frac{-1}{\sqrt{n+1}}$ and $a_{2n} = \frac{1}{\sqrt{n+1}} + \frac{1}{n} + 1$ for $n \geq 1$. Show that $\sum a_n$ and $\sum a_n^2$ diverge, while $\prod(1 + a_n)$ converges.

(5) Prove the identity $\sin \theta = 2^k \sin \frac{\theta}{2^k} \prod_{n=1}^{k} \cos \frac{\theta}{2^n}$ for $k \geq 1$ and then, deduce that $\prod_{n=1}^{\infty} \cos \frac{\theta}{2^n} = \frac{\sin \theta}{\theta}$. Also, show that

$$\frac{2}{\sqrt{2}} \cdot \frac{2}{\sqrt{2+\sqrt{2}}} \cdot \frac{2}{\sqrt{2+\sqrt{2+\sqrt{2}}}} \cdots = \frac{\pi}{2}.$$

(6) Show that $\prod_{n=1}^{\infty} e^{\frac{1}{n}} (1 + \frac{1}{n})^{-1} = e^{\gamma}$, where $\gamma = \lim_{n \to \infty} (\sum_{k=1}^{n} \frac{1}{k} - \log n)$ is the Euler's constant.

(7) Find the domain of absolute and local uniform convergence for each of the following products.

   (a) $\prod_{n=1}^{\infty} (1 - z^n)$.

   (b) $\prod_{n=2}^{\infty} [1 - (1 - \frac{1}{n})^{-n} z^{-n}]$.

(c) $\prod_{n=1}^{\infty}\left[1+\left(1+\frac{1}{n}\right)^{n^2}z^n\right]$.

(d) $\prod_{n=1}^{\infty}(1+a_n z)$, where $\sum_{n=1}^{\infty}|a_n|<\infty$.

(e) $\prod_{n=1}^{\infty}\left(1+\frac{1}{z+n}\right)e^{-\frac{1}{n}}$.

(f) $\prod_{n=1}^{\infty}\left(1-\frac{1}{n^z}\right)$.

(g) $\prod_{p}\left(1-\frac{1}{p^z}\right)$, where p runs through all the primes.

(8) Prove (5.5.1.3) and (5.5.1.4) with convergence and absolute convergence replaced by (local) uniform convergence.

(9) Show that

$$\prod_{n=1}^{\infty}(1+z^{2^n})=\frac{1}{1-z},\quad |z|<1$$

both absolutely and locally uniformly.

(10) Let $|h|<1$. Show that $\theta(z)=\prod_{n=1}^{\infty}(1+h^{2n-1}e^z)(1+h^{2n-1}e^{-z})$ is an entire function satisfying the functional equation $\theta(z+2\log h)=h^{-1}e^{-z}\theta(z)$.

(11) Suppose $\sum|a_n|<\infty$. Show that

$$\prod_{n=1}^{\infty}(1+a_n z)=1+z\sum_{1}^{\infty}a_n+z^2\sum_{m\neq n}a_m a_n+\cdots.$$

(12) Let $|a|>1$ and

$$f(z)=\prod_{n=0}^{\infty}\left(1+\frac{z}{a^n}\right).$$

(a) Show that the product converges to the analytic function $f(z)$ absolutely and locally uniformly in $\mathbf{C}$.

(b) Show that $f(az)=(1+az)f(z)$, $z\in\mathbf{C}$.

(c) Show that

$$f(z)=1+\sum_{n=1}^{\infty}\frac{a^n z^n}{(a-1)(a^2-1)\cdots(a^n-1)},\quad z\in\mathbf{C}.$$

(13) Suppose it is already known that $\sin\pi z=\pi z\prod_{n=1}^{\infty}(1-\frac{z^2}{n^2})$.

(a) Show that the product converges absolutely and locally uniformly in $\mathbf{C}$.

(b) Use (5.5.1.8) to prove the following identities:

1. $\sum_{n=1}^{\infty} \frac{1}{n^2} = \frac{\pi^2}{6}$.

2. $\sum_{n=1}^{\infty} \frac{1}{n^4} = \frac{\pi^4}{90}$.

3. $\prod_{n=1}^{\infty} \left(\frac{2n}{2n-1} \cdot \frac{2n}{2n+1} \right) = \frac{\pi}{2}$, the *Wallis formula*.

(14) Let $a_n \in \mathbf{C}$, $n \geq 1$, be such that $0 < |a_n| < 1$ for each n.

(a) Show that the *Blaschke product*

$$\prod_{n=1}^{\infty} \frac{|a_n|}{a_n} \frac{a_n - z}{1 - \bar{a}_n z}$$

converges absolutely and locally uniformly in $|z| < 1$ if and only if $\sum_{n=1}^{\infty}(1 - |a_n|) < \infty$. In this case, the product $f(z)$ is analytic in $|z| < 1$, with absolute value ≤ 1. Also, all the zeros of $f(z)$ in $|z| < 1$ occur only at a_n, $n \geq 1$.

(b) In case $a_n \neq a_m$ if $n \neq m$, show that $\sum_{1}^{\infty}(1 - |a_n|^2) < \infty$ is a sufficient condition for the Blaschke product to be convergent.

Exercises B

(1) Let

$$f_n(z) = \frac{z(z+1)(z+2)\cdots(z+n)}{n!n^z}, \quad z \in \mathbf{C}.$$

Show that $f_n(z)$ converges to an analytic function $f(z)$ uniformly in $\mathbf{C}$ which assumes zeros only at $0, -1, -2, \ldots, -n, \ldots$. Set $\Gamma(z) = [f(z)]^{-1}$, $z \in \mathbf{C} - \{0, -1, -2, \ldots\}$. Show that

1. $\Gamma(z+1) = z\Gamma(z)$.
2. $\Gamma(n) = (n-1)!$ for any positive integer n.

5.5.2   *Weierstrass's factorization theorem*

Let us go back and start with (5.5.3) and (5.5.4).

We will show that there exist polynomials $p_n(z)$, $n \geq 1$, such that

$$\prod_{n=1}^{\infty} \left(1 - \frac{z}{a_n}\right) e^{p_n(z)} \text{converges absolutely and locally uniformly in } \mathbf{C}.$$

$\Leftrightarrow$ (By (5.5.1.3), (5.5.1.4), and Exercise A(8).) The series $\sum_{n=1}^{\infty} r_n(z)$, where $r_n(z) = \log\left(1 - \frac{z}{a_n}\right) + p_n(z)$ with $\log\left(1 - \frac{z}{a_n}\right)$ so chosen that $-\pi < \operatorname{Im} r_n(z) \leq \pi$ for $n \geq 1$, converges absolutely and locally uniformly in $\mathbf{C}$. $(*_1)$

To this end, choose any fixed $R > 0$. There is an integer N so that $|a_n| > R$ if $n \geq N$ (and thus, $|a_n| \leq R$ if $n < N$) and we consider only these terms with $|a_n| > R$. Then

$$\left| \frac{z}{a_n} \right| < 1 \quad \text{if } |z| \leq R \text{ (and, of course, } n \geq N)$$

$$\Rightarrow \text{Log}\left(1 - \frac{z}{a_n}\right) = -p_n(z) + r_n(z), \quad |z| \leq R, \quad \text{where}$$

$$p_n(z) = \frac{z}{a_n} + \frac{1}{2}\left(\frac{z}{a_n}\right)^2 + \cdots + \frac{1}{m_n}\left(\frac{z}{a_n}\right)^{m_n}, \quad \text{and}$$

$$r_n(z) = -\sum_{k=m_n+1}^{\infty} \frac{1}{k}\left(\frac{z}{a_n}\right)^k.$$

The remainder $r_n(z)$ has the estimate, in absolute value,

$$|r_n(z)| \leq \sum_{k=m_n+1}^{\infty} \frac{1}{k}\left(\frac{R}{|a_n|}\right)^k < \frac{1}{m_n+1}\left(\frac{R}{|a_n|}\right)^{m_n+1} \sum_{k=1}^{\infty}\left(\frac{R}{|a_n|}\right)^{k-1}$$

$$= \frac{1}{m_n+1}\left(\frac{R}{|a_n|}\right)^{m_n+1}\left(1 - \frac{R}{|a_n|}\right)^{-1}$$

$$\leq \frac{2}{m_n+1}\left(\frac{R}{|a_n|}\right)^{m_n+1} \quad \text{if } |a_n| > 2R \quad \text{for } n \text{ large enough.} \quad (*_2)$$

It is possible to choose m_n so that the series

$$\sum_{n=1}^{\infty} \frac{1}{m_n+1}\left(\frac{R}{|a_n|}\right)^{m_n+1} \quad \text{converges for } \textit{any fixed } R > 0. \quad (5.5.2.1)$$

Say, let $m_n = n$ for $n \geq 1$. For any $R > 0$, since $\lim_{n \to \infty}|a_n| = \infty$, then $|a_n| > 2R$ will hold for all sufficiently large n, say $n \geq k$. Then, in this case,

$$\sum_{n=k}^{\infty} \frac{1}{n+1}\left(\frac{R}{|a_n|}\right)^{n+1} \leq \sum_{n=k}^{\infty}\left(\frac{1}{2}\right)^{n+1} < \infty.$$

Under this circumstance, $(*_2)$ implies the following two facts wanted. One is $r_n(z) \to 0$ as $n \to \infty$ if $|z| \leq R$, and hence $r_n(z)$ has its imaginary part between $-\pi$ and π as soon as n is large enough. The other is that the series $\sum r_n(z)$ converges absolutely and (locally) uniformly in $|z| \leq R$. This finishes the proof of $(*_1)$.

Paying attention to the explanation within (5.5.1.7), we summarize the above as

Weierstrass's factorization theorem. There exists an entire function with arbitrary zeros a_n: see (5.5.1) and (5.5.2) for the case of finitely many zeros and the following for infinitely many zeros. Suppose the zeros are $a_1, a_2, \ldots, a_n, \ldots$ (counting multiplicities) with $0 < |a_1| \leq |a_2| \leq \cdots \leq |a_n| \leq \cdots$ and $\lim_{n\to\infty} |a_n| = \infty$. Then each entire function with these and only these zeros and a possible zero at 0 of order $m \geq 0$ is of the form

$$f(z) = e^{g(z)} z^m \prod_{n=1}^{\infty} \left(1 - \frac{z}{a_n}\right) e^{\frac{z}{a_n} + \frac{1}{2}\left(\frac{z}{a_n}\right)^2 + \cdots + \frac{1}{m_n}\left(\frac{z}{a_n}\right)^{m_n}}, \quad z \in \mathbf{C},$$

where the m_n are certain integers (see (5.5.2.1)), and $g(z)$ is an entire function. Recall that the infinite product in the right converges to $f(z)$ absolutely and locally uniformly in $\mathbf{C}$. $\qquad$ (5.5.2.2)

The preceding proof has shown that, if there is an integer $h \geq 0$ so that

$$\sum_{n=1}^{\infty} \frac{1}{|a_n|^{h+1}} < \infty \qquad\qquad (5.5.2.3)$$

$\Rightarrow$ (compare to (5.5.2.1)) $\displaystyle\sum_{n=1}^{\infty} \frac{1}{h+1} \left(\frac{R}{|a_n|}\right)^{h+1}$ converges for all fixed $R > 0$.

$\Rightarrow \displaystyle\prod_{n=1}^{\infty} \left(1 - \frac{z}{a_n}\right) e^{\frac{z}{a_n} + \frac{1}{2}\left(\frac{z}{a_n}\right)^2 + \cdots + \frac{1}{h}\left(\frac{z}{a_n}\right)^h}, \quad$ if $h > 0 \quad$ or $\quad \displaystyle\prod_{n=1}^{\infty} \left(1 - \frac{z}{a_n}\right),$

if $h = 0$, converges absolutely and locally uniformly in $\mathbf{C}$ to an entire function. $\qquad$ (5.5.2.5)

Assume that h is the smallest nonnegative integer for which the series (5.5.2.3) converges; then the corresponding expression (5.5.2.5) is called the *canonical product* associated with the sequence a_n, and this h is the *genus* of the canonical product. Note that the canonical product associated with given a_n, $n \geq 1$ is thereby uniquely determined. See Exercise B(1) for a generalized version of this theorem.

Suppose the infinite product in the expression (5.5.2.2) for $f(z)$ is canonical with genus h and $g(z)$ reduces to a polynomial; the function $f(z)$ is then said to be of *finite genus* and the

$$\text{genus of } f(z) \underset{\text{(def.)}}{=} \max\{h, \text{ degree of } g(z)\}. \qquad (5.5.2.6)$$

For instance, an entire function of

$$\text{genus zero:} \quad Az^m \prod_{n=1}^{\infty} \left(1 - \frac{z}{a_n}\right) \quad \text{if } \sum \frac{1}{|a_n|} < \infty;$$

$$\text{genus one:} \quad Ae^{\alpha z} z^m \prod_{n=1}^{\infty} \left(1 - \frac{z}{a_n}\right) e^{\frac{z}{a_n}} \quad \text{if } \sum \frac{1}{|a_n|} = \infty$$

$$\text{and } \sum \frac{1}{|a_n|^2} < \infty, \quad \text{or}$$

$$Ae^{\alpha z} z^m \prod_{n=1}^{\infty} \left(1 - \frac{z}{a_n}\right) \quad \text{if } \sum \frac{1}{|a_n|} < \infty. \qquad (5.5.2.7)$$

Readers are urged to write out forms of entire functions of genus three. Also see Exercise A(11).

We list two simple corollary of (5.5.2.2).

As an easy consequence of (5.5.1.8), we have

The logarithmic derivative of a canonical product. Given a canonical product

$$C(z) = \prod_{n=1}^{\infty} E_n(z) \quad \text{where } E_n(z) = \left(1 - \frac{z}{a_n}\right) e^{\frac{z}{a_n} + \frac{1}{2}\left(\frac{z}{a_n}\right)^2 + \cdots + \frac{1}{h}\left(\frac{z}{a_n}\right)^h}$$

$$\text{if } h > 0 \quad \text{or } E_n(z) = 1 - \frac{z}{a_n} \quad \text{if } h = 0.$$

Then, in $\mathbf{C} - \{a_1, a_2, \ldots, a_n, \ldots\}$,

$$\frac{C'(z)}{C(z)} = \sum_{n=1}^{\infty} \frac{E_n'(z)}{E_n(z)} \quad \text{absolutely and locally uniformly.} \qquad (5.5.2.8)$$

In case $M(z)$ is meromorphic in the whole plane, by (5.5.2.2), there exists an entire function $f_1(z)$ with the poles of $M(z)$ for zeros. The product $M(z)f_1(z)$ has removable singularity at each pole of $M(z)$ and hence is an entire function $f_2(z)$. It follows that $M(z) = \frac{f_2(z)}{f_1(z)}$. Then we obtain

The expression for meromorphic function in $\mathbf{C}$. A function is meromorphic in $\mathbf{C}$ if and only if it is the quotient of two entire functions (with the denominator not identically equal to zero). (5.5.2.9)

As an easy consequence of (5.4.2.2) and (5.4.2.3), we have the following counterpart for infinite product.

A special case of Weierstrass's factorization theorem (5.5.2.2). Suppose an entire function $f(z)$ satisfying $f(0) \neq 0$ and having simple zeros $a_1, a_2, \ldots, a_n, \ldots$ so arranged as $0 < |a_1| \leq |a_2| \leq \cdots \leq |a_n| \leq \cdots$ that $\lim_{n\to\infty} |a_n| = \infty$.

(1) Let γ_n be as in (5.4.2.1) with properties 1, 2, and 3 so that $\lim_{n\to\infty}\int_{\gamma_n}\frac{f'(\zeta)}{(\zeta-z)f(\zeta)}d\zeta = 0$ for $z \in \text{Int }\gamma_n - \{\text{poles of }\frac{f'(z)}{f(z)}\}$. Then

$$f(z) = f(0)\prod_{n=1}^{\infty}\left(1 - \frac{z}{a_n}\right) \text{ locally uniformly in } \mathbf{C} - \{a_1, a_2, \ldots\}.$$

(2) Suppose $\frac{f'(z)}{f(z)}$ satisfies either condition 1 or 2 in (5.4.2.2). Then

$$f(z) = f(0)e^{\frac{f'(0)}{f(0)}z}\prod_{n=1}^{\infty}\left(1 - \frac{z}{a_n}\right)e^{\frac{z}{a_n}}$$

locally uniformly in $\mathbf{C} - \{a_1, a_2, \ldots\}$.

Note that, in these infinite product expressions, factors corresponding to zeros between γ_n and γ_{n+1}, should be considered as one "term" in the product.   (5.5.2.10)

Proof is left to the readers. See Exercise A(1) for application.
Four examples are illustrated.

Example 1. (Ref. [1], p. 197). Shows that
$\sin \pi z = \pi z \prod_{n=1}^{\infty}\left(1 - \frac{z^2}{n^2}\right) = \pi z \prod'^{\infty}_{n=-\infty}\left(1 - \frac{z}{n}\right)e^{\frac{z}{n}}$ absolutely and locally uniformly in $\mathbf{C}$, where $\prod'^{\infty}_{n=-\infty}$ means the term $n = 0$ is omitted in the product. Try to deduce:

1. *Wallis product:* $\frac{\pi}{2} = \lim_{n\to\infty}\frac{2^2 \cdot 4^2 \cdots (2n-2)^2}{3^2 \cdot 5^2 \cdots (2n-1)^2}\cdot 2n$;

2. $\prod_{n=1}^{\infty}\left(1 + \frac{z^2}{n^2} + \frac{z^4}{n^4}\right) = \frac{\sin \pi\omega z \cdot \sin \pi\bar{\omega}z}{\pi^2 z^2}$, where $\omega = e^{\frac{2\pi i}{3}}$;

3. in particular, $\prod_{n=1}^{\infty}\left(1 + \frac{1}{n^2} + \frac{1}{n^4}\right) = \frac{1+\cosh\sqrt{3}\pi}{2\pi^2}$.

More generally, in case α is not an integer, then

$$\sin \pi(z + \alpha) = \sin \pi\alpha\, e^{\pi z \cot \pi\alpha}\prod_{n=-\infty}^{\infty}\left(1 + \frac{z}{n + \alpha}\right)e^{-\frac{z}{n+\alpha}}$$

$$= \sin \pi\alpha\left(1 + \frac{z}{\alpha}\right)\prod'^{\infty}_{n=-\infty}\left(1 + \frac{z}{n + \alpha}\right)e^{-\frac{z}{n}}.$$

Solution.

Method 1 (based on (5.5.2.7)). Observe that $\sin \pi z$ has simple zeros at integers n $(n = 0, \pm 1, \ldots)$. Since $\sum \frac{1}{|n|} = \infty$ while $\sum \frac{1}{|n|^2} < \infty$, according to (5.5.2.7), there is an entire function $g(z)$ so that (refer to Example 3(3) in Sec. 5.5.1)

$$\sin \pi z = z e^{g(z)} \prod_{n \neq 0} \left(1 - \frac{z}{n}\right) e^{\frac{z}{n}} \text{ absolutely and locally uniformly in } \mathbf{C}. \quad (*_3)$$

Taking the logarithmic derivatives on both sides, we find

$$\pi \cot \pi z = \frac{1}{z} + g'(z) + \sum_{n \neq 0} \left(\frac{1}{z-n} + \frac{1}{n}\right)$$

$\Rightarrow$ (by Example 2 of Sec. 5.4.2 or adopting the method used in

Example 2 of Sec. 5.4.1) $g'(z) \equiv 0$

$\Rightarrow g(z) \equiv c$, a constant.

Letting $z \to 0$ in $(*_3)$ and observing that $\lim_{z \to 0} \frac{\sin \pi z}{\pi z} = 1$, we obtain the constant $g(z) = c = \log \pi$ and the result follows.

Method 2 (based on (5.5.2.10)). Set $f(z) = \frac{\sin \pi z}{\pi z}$, $z \in \mathbf{C}$. $f(z)$ is entire and has simple zeros at $z = \pm n$ $(n = 1, 2, 3, \ldots)$. Note that the derivative of $f(z)$

$$\frac{f'(z)}{f(z)} = \pi \cot \pi z - \frac{1}{z}$$

satisfies Condition 1 in (5.4.2.2) (refer to Method 1 in Example 2 of Sec. 5.4.2). Hence, the required expression follows directly from (2) in (5.5.2.10), namely,

$$\frac{\sin \pi z}{\pi z} = \prod_{n=-\infty}^{\infty}{}' \left(1 - \frac{z}{n}\right) e^{\frac{z}{n}}$$

$$= \prod_{n=1}^{\infty} \left\{ \left(1 - \frac{z}{n}\right) e^{\frac{z}{n}} \cdot \left(1 + \frac{z}{n}\right) e^{-\frac{z}{n}} \right\} = \prod_{n=1}^{\infty} \left(1 - \frac{z^2}{n^2}\right). \quad (*_4)$$

Caution again that $\frac{\sin \pi z}{\pi z}$ cannot be expressed as $\prod'^{\infty}_{-\infty} \left(1 - \frac{z}{n}\right)$ because of the divergence of this product.

Set $z = \frac{1}{2}$ in $(*_4)$ and the Wallis product follows.

Also, by $(*_4)$,

$$\prod_{n=1}^{\infty} \left(1 - \frac{\omega^2 z^2}{n^2}\right) = \frac{\sin \pi \omega z}{\pi \omega z}; \quad \prod_{n=1}^{\infty} \left(1 - \frac{\bar{\omega}^2 z^2}{n^2}\right) = \frac{\sin \pi \bar{\omega} z}{\pi \bar{\omega} z}.$$

The absolute convergence of both products permits us to multiply these two expressions termwise together (see (5.5.1.4)) to obtain the claimed identity. Set $z = 1$ in the identity and we get the third one.

Exactly the same methods as before still work to express $\sin \pi(z+\alpha)$ as the claimed product. Note that, in this case, γ_n is the square with center at α and vertices at $\alpha \pm (n+\frac{1}{2})(1+i)$ for each $n \geq 1$. For the second expression for $\sin \pi(z+\alpha)$, just use Example 2 of Sec. 5.4.2 as follows:

$$\frac{\sin \pi(z+\alpha)}{\sin \pi \alpha} = \left(1+\frac{z}{\alpha}\right) \exp z \left[\pi \cot \pi \alpha - \frac{1}{\alpha}\right] \cdot \prod_{-\infty}^{\infty}{}' \left(1+\frac{z}{n+\alpha}\right)$$

$$\times \exp\left[-\frac{z}{n}+\frac{z}{n}-\frac{z}{\alpha+n}\right]$$

$$= \left(1+\frac{z}{\alpha}\right) \exp z \left[\pi \cot \pi \alpha - \frac{1}{\alpha} - \sum_{-\infty}^{\infty}{}' \left(\frac{1}{n+\alpha}-\frac{1}{n}\right)\right]$$

$$\times \prod_{-\infty}^{\infty}{}' \left(1+\frac{z}{n+\alpha}\right) \exp^{-\frac{z}{n}}$$

$$= \left(1+\frac{z}{\alpha}\right) \prod_{-\infty}^{\infty}{}' \left(1+\frac{z}{n+\alpha}\right) \exp^{-\frac{z}{n}}.$$

Example 2. Show that

$$\cos \pi z = \prod_{n=0}^{\infty} \left(1 - \frac{z^2}{(n+1/2)^2}\right); \qquad \frac{\sin \pi z}{\pi z (1-z)} = \prod_{n=1}^{\infty} \left(1+\frac{z-z^2}{n+n^2}\right).$$

Solution. $f(z) = \cos \pi z$ has simple zeros at $z = \pm(n+\frac{1}{2})$ for $n = 0, 1, 2, \ldots$. Let γ_n, $n \geq 1$, be the square with vertices at $\pm n(1+i)$. Along the vertical side $z = \pm n + iy$, $-n \leq y \leq n$:

$$|\cos \pi z|^2 = \cosh^2 \pi y \geq 1; \quad |\sin \pi z|^2 = \sinh^2 \pi y \leq \sinh^2 n\pi$$

$$\Rightarrow \left|\frac{f'(z)}{f(z)}\right|^2 = \pi^2 \tanh^2 \pi y \leq \pi^2.$$

Along the horizontal side $z = x \pm in$, $-n \leq x \leq n$:

$$|\cos \pi z|^2 \geq \cosh^2 n\pi - 1; \quad |\sin \pi z|^2 \leq \sinh^2 n\pi + 1$$

$$\Rightarrow \left|\frac{f'(z)}{f(z)}\right| \leq \pi^2 \frac{\sinh^2 n\pi + 1}{\cosh^2 n\pi - 1} \to \pi^2 \quad \text{as } n \to \infty.$$

Hence $\sup_{z \in \gamma_n} |\frac{f'(z)}{f(z)}| \leq M < \infty$ for all $n \geq 1$. Since $f(0) = 1$ and $f'(0) = 0$, it follows from (2) in (5.5.2.10) that

$$\cos \pi z = \prod_{n=-\infty}^{\infty} \left(1 - \frac{z}{n+1/2}\right) e^{-z/(n+1/2)}$$

(absolutely by Example 3(3) of Sec. 5.5.1)

$$= \prod_{n=0}^{\infty} \left(1 - \frac{z^2}{(n+1/2)^2} \right)$$

(see also the explanation in (5.5.2.10)). $(*5)$

Set $\zeta = z - \frac{1}{2}$ to replace z in $(*5)$ and we obtain

$$\cos \pi \zeta = \prod_{n=0}^{\infty} \left(1 - \frac{\zeta^2}{(n+1/2)^2} \right) = (1 - 4\zeta^2) \prod_{n=1}^{\infty} \left(1 - \frac{\zeta^2}{(n+1/2)^2} \right)$$

$$\Rightarrow \quad \frac{\cos \pi \zeta}{1 - 4\zeta^2} = \prod_{n=1}^{\infty} \frac{1}{(n+1/2)^2} \left[\left(n + \frac{1}{2} \right)^2 - \left(z - \frac{1}{2} \right)^2 \right]$$

$$= \prod_{n=1}^{\infty} \frac{n^2 + n}{(n+1/2)^2} \left(1 + \frac{z - z^2}{n + n^2} \right). \qquad (*6)$$

Letting $z = \zeta + \frac{1}{2} \to 0$ in $(*6)$ and observing that

$$\lim_{\zeta \to -\frac{1}{2}} \frac{\cos \pi \zeta}{1 - 4\zeta^2} = \lim_{\zeta \to -\frac{1}{2}} \frac{-\pi \sin \pi \zeta}{-8\zeta^2} = \frac{\pi}{4} = \prod_{n=1}^{\infty} \frac{n^2 + n}{(n+1/2)^2}$$

$$\Rightarrow \quad \left(\text{set } \zeta = z - \frac{1}{2} \text{ in the left of } (*6) \right) \qquad \frac{\sin \pi z}{\pi z (1 - z)}$$

$$= \frac{4 \cos \pi \zeta}{\pi (1 - 4\zeta^2)} = \prod_{n=1}^{\infty} \left(1 + \frac{z - z^2}{n + n^2} \right).$$

Example 3. (Ref. [1], p. 197). Let a_n be a sequence of distinct nonzero complex numbers converging to ∞ and A_n, $n \geq 1$, be arbitrary complex numbers. Find an entire function $f(z)$ such that $f(a_n) = A_n$, $n \geq 1$.

Solution. Using (5.5.2.2), construct an entire function $g(z)$ with simple zeros at a_n, $n \geq 1$, namely,

$$g(z) = \prod_{n=1}^{\infty} \left(1 - \frac{z}{a_n} \right) e^{\frac{z}{a_n} + \frac{1}{2} \left(\frac{z}{a_n} \right)^2 + \cdots + \frac{1}{m_n} \left(\frac{z}{a_n} \right)^{m_n}}, \quad z \in \mathbf{C}.$$

Then $g'(a_n) \neq 0$ for each $n \geq 1$. Observe that the function

$$\frac{A_n}{g'(a_n)} \cdot \frac{g(z)}{z - a_n}$$

has a removable singularity at a_n and may thus be assigned the value A_n at $z = a_n$. It is entire for each $n \geq 1$.

Try to choose suitable constants r_n so that the series

$$f(z) = \sum_{n=1}^{\infty} \frac{A_n}{g'(a_n)} \cdot \frac{g(z)}{z - a_n} e^{r_n (z - a_n)} \qquad (*7)$$

converges locally uniformly in $\mathbf{C}$. To see this, fix any $R > 0$. There is an integer $N > 0$ so that $|a_n| > R$ if $n \geq N$. On $|z| \leq \frac{R}{2}$ and $n \geq N$, $|z - a_n| \geq R - \frac{R}{2} = \frac{R}{2}$ holds; consequently, at least one of $\mathrm{Re}\,(z - a_n)$ and $\mathrm{Im}(z - a_n)$ should be not less than $\frac{R}{2}$. Now

$$\left| \frac{A_n}{g'(a_n)} \cdot \frac{g(z)}{z - a_n} e^{r_n(z - a_n)} \right| \leq \frac{2}{R} \cdot \max_{|z| \leq \frac{R}{2}} |g(z)| \cdot \left| \frac{A_n}{g'(a_n)} \right|$$

$$\times\, e^{r_n \mathrm{Re}(z - a_n)}, \quad \text{if } r_n \text{ is real};$$

$$\text{or} \ \leq \frac{2}{R} \cdot \max_{|z| \leq \frac{R}{2}} |g(z)| \cdot \left| \frac{A_n}{g'(a_n)} \right| e^{i r_n \mathrm{Im}\,(z - a_n)}, \quad \text{if } i r_n \text{ is real}.$$

Set $\alpha_n = \left| \frac{A_n}{g'(a_n)} \right|$ for $n \geq 1$. Suppose that we are able to choose constants $r_n, n \geq 1$ so that

$$\sum_{n=N}^{\infty} \left| \frac{A_n}{g'(a_n)} e^{r_n(z-a)} \right| \leq \sum_{n=N}^{\infty} \alpha_n e^{r_n \cdot \frac{R}{2}} < \infty \quad \text{on } |z| \leq \frac{R}{2}. \tag{$*_8$}$$

This means that, with r_n chosen as in $(*_8)$, the series in $(*_7)$ does converge locally uniformly in $\mathbf{C}$ and hence represents an entire function $f(z)$. This $f(z)$ is obviously a required one. Unfortunately, such constants r_n are not avilable all the times.

An alternative way than $(*_7)$ is to find a meromorphic function with simple poles at a_n, $n \geq 1$, and the corresponding principal parts $\frac{A_n}{g'(a_n)} \cdot \frac{1}{z-a_n}$, $n \geq 1$, namely,

$$M(z) = \sum_{n=1}^{\infty} \left(\frac{A_n}{g'(a_n)} \cdot \frac{1}{z - a_n} - p_n(z) \right),$$

where $p_n(z)$, $n \geq 1$, are suitably chosen polynomials (see (5.4.1.1)). Then the entire function

$$f(z) = g(z)M(z) \tag{$*_9$}$$

satisfies the condition that $f(a_n) = \lim_{z \to a_n} g(z)M(z) = \frac{g'(a_n)A_n}{g'(a_n)} = A_n$, $n \geq 1$.

Example 4. (Ref. [1], p. 198). Suppose $f(z)$ is of genus 0 or 1 with real zeros, and $f(z)$ is real for real z. Then all the zeros of $f'(z)$ are also real, and between any two consecutive zeros of $f(z)$ there lies precisely only one zero of $f'(z)$.

This is a well-known result for polynomial with all its zeros real which can be easily shown by using the mean-value theorem for differentiation. The extension to the entire function of genus 0 or 1 was due to Laguerre. However, the function $(z+1)e^{z^2}$, which is of genus 2, shows that it is no more true for entire functions of genus ≥ 2.

Proof. Such an entire function of genus 0 or 1 can be expressed as (see Exercise A(11))

$$f(z) = e^{\alpha z + \beta} z^m \prod_{n=1}^{\infty} \left(1 - \frac{z}{a_n}\right) e^{\frac{z}{a_n}}, \qquad (*10)$$

where $m \geq 0$ and a_n, $n \geq 1$, are real. By assumption that $f(z)$ is real for real z, α within should be real, too. Differentiating $(*10)$ logarithmically or using (5.5.1.8) and (5.5.2.8) directly, we have

$$\frac{f'(z)}{f(z)} = \alpha + \frac{m}{z} + \sum_{n=1}^{\infty} \left(\frac{1}{z-a_n} + \frac{1}{a_n}\right), \quad \text{locally uniformly in}$$

$$\mathbf{C} - \{a_n, n \geq 1\} \cup \{0\}. \qquad (*11)$$

$$\Rightarrow \operatorname{Im}\frac{f'(z)}{f(z)} = -(\operatorname{Im} z)\left(\frac{m}{|z|^2} + \sum_{n=1}^{\infty} \frac{1}{|z-a_n|^2}\right).$$

Then $f'(z) = 0$ implies that $\operatorname{Im}\frac{f'(z)}{f(z)} = 0$ which, in turn, implies that $\operatorname{Im} z = 0$ should hold. Hence $f'(z) = 0$ can hold for real z only.

Differentiate $(*11)$ again to obtain

$$\frac{d}{dz}\frac{f'(z)}{f(z)} = -\frac{m}{z^2} - \sum_{n=1}^{\infty} \frac{1}{(z-a_n)^2}, \quad z \in \mathbf{C} - \{a_n, n \geq 1\} \cup \{0\},$$

which is real and negative for real z. As a strictly decreasing function from $+\infty$ to $-\infty$ as z increases through real values between two consecutive zeros of $f(z)$, $\frac{f'(z)}{f(z)}$, and thus $f'(z)$ assumes the zero value once between these two zeros. $\qquad \square$

Exercises A

(1) Try to adopt the following two methods:

    1. (5.5.2.10), and

    2. the infinite product expansion for $\sin \pi z$ (see Example 1)

to prove the following expansions. Determine the genus for each entire function.

(a) $\sinh \pi z = \pi z \prod_{n=1}^{\infty} \left(1 + \frac{z^2}{n^2} \right)$.

(h) $\cosh z \quad \cos z = z^2 \prod_{n=1}^{\infty} \left(1 + \frac{z^4}{4n^4 \pi^4} \right)$.

(c) $e^z - 1 = z e^{\frac{z}{2}} \prod_{n=1}^{\infty} \left(1 + \frac{z^2}{4n^2 \pi^2} \right)$.

(d) $e^{az} - e^{bz} = (a - b) z e^{\frac{1}{2}(a+b)z} \prod_{n=1}^{\infty} \left(1 + \frac{(a-b)^2 z^2}{4n^2 \pi^2} \right)$.

(2) Find the product function for each of the following infinite products.

(a) $\prod_{n=2}^{\infty} \left(1 + \frac{z^4}{n^4} \right)$.

(b) $\prod_{n=2}^{\infty} \left(1 - \frac{z^4}{n^4} \right)$ and show that $\prod_{n=2}^{\infty} \left(1 - \frac{1}{n^4} \right) = \frac{1}{8\pi} (e^{\pi} - e^{-\pi})$.

(c) $\prod_{n=1}^{\infty} \left(1 + \frac{z^6}{n^6} \right)$ and show that $\prod_{n=1}^{\infty} \left(1 + \frac{1}{n^6} \right) = \frac{1}{2} \cdot \frac{\sinh \pi}{\pi^3} (\cosh \pi - \cos \sqrt{3}\pi)$.

(d) $\prod_{n=1}^{\infty} \left(1 - \frac{z^6}{n^6} \right)$.

(3) Let λ_n be the positive root of the equation $\tan z = z$ in the interval $(n\pi, n + \frac{1}{2}\pi)$ for $n = 1, 2, \ldots$. Show that

$$\sin z - z \cos z = \frac{1}{3z^2} \prod_{n=1}^{\infty} \left(1 - \frac{z^2}{\lambda_n^2} \right)$$

(compare to Exercise A(7) of Sec. 5.4.2).

(4) Show that

$$\frac{\sin 3z}{\sin z} = - \prod_{n=-\infty}^{\infty} \left(1 - \frac{4z^2}{(n\pi + z)^2} \right), \quad z \neq n\pi \quad \text{for } n = 0, \pm 1, \pm 2, \ldots.$$

(5) Show that $\cos \frac{\pi z}{4} - \sin \frac{\pi z}{4} = (1 - z)(1 + \frac{z}{3})(1 - \frac{z}{5}) \ldots$.

(6) Determine the genus of $\cos \sqrt{z}$.

(7) If $f(z)$ is of genus h, how large and how small can the genus of $f(z^2)$ be?

(8) Derive the results of Example 1 in Sec. 5.4.1 directly from Weierstrass's factorization theorem.

(9) Deduce the periodicity of $\sin z$ from its infinite product expansion (refer to Example 1).

(10) Extend Example 3 to the case where the sequence a_n contains the number zero.

(11) Show that an entire function $f(z)$ of genus h can be represented in the form

$$f(z) = z^m e^{g(z)} \prod_n \left(1 - \frac{z}{a_n}\right) e^{\frac{z}{a_n} + \frac{1}{2}\left(\frac{z}{a_n}\right)^2 + \cdots + \frac{1}{h}\left(\frac{z}{a_n}\right)^h},$$

where $m \geq 0$ and $g(z)$ is a polynomial of degree $\leq h$. Note that the infinite product, in this case, is not necessarily a canonical one since its genus might be strictly less than h.

(12) Try to find an entire function $g(z)$ so that $\frac{\sin iz}{e^{2z}-1} = e^{g(z)}$.

(13) Deduce Mittag–Leffler's theorem (5.4.1.1) from Weierstrass's theorem (5.5.2.5).

Exercises B

(1) *A generalized version of Weierstrass's factorization theorem (5.5.2.2).* (5.5.2.2) is still valid in a domain (or open set) Ω in $\mathbf{C}$. More precisely, let a_n, $n \geq 1$, be a sequence of points in Ω without limit points in it, and m_n, $n \geq 1$, be positive integers. Then there exists an analytic function $f : \Omega \to \mathbf{C}$ with these and only these zeros at a_n, $n \geq 1$, the zero at a_n being of the order m_n for each $n \geq 1$. Moreover, the quotient of any two such functions is analytic and zero-free in Ω.
   Hint. Model after the proofs of (5.5.2.2) and Exercise B(4) of Sec. 5.4.1. Refer to Ref. [58], Vol. I, p. 86.

(2) Given a domain Ω in $\mathbf{C}$, show that there is a function analytic in Ω but cannot be analytically continued outside Ω.
   Hint. Compare to Exercise B(6) of Sec. 5.4.1.

5.5.3   *Hadamard's order theorem*

Let $f(z)$ be a nonconstant entire function and $M(r) = \max_{|z|=r}|f(z)|$ for $r > 0$. Then $M(r)$ is a strictly increasing continuous function in r and $\lim_{r \to \infty} M(r) = \infty$ holds (see Exercise A(6) of Sec. 3.4.4) .

In case $f(z)$ is a polynomial of degree $n \geq 1$: since $\lim_{z \to \infty} \frac{f(z)}{z^n} \neq 0$, there are constants $M_1 \geq M_2 > 0$ so that

$$M_2 r^n \leq M(r) \leq M_1 r^n \quad \text{for } r \geq r_0 \quad \text{(some positive constant)}.$$

$$\Rightarrow \lim_{r \to \infty} \frac{\log M(r)}{\log r} = n \quad \text{or} \quad M(r) \sim r^n \quad \text{as } r \to \infty. \qquad (*_1)$$

This is to say, the order of growth of $M(r)$ is the same as that of r^n as $n \to \infty$.

Well, in case $f(z)$ is a transcendental function: we do have a remarkable fact that

$$\lim_{r\to\infty} \frac{\log M(r)}{\log r} = \infty, \quad \text{or equivalently} \quad \varliminf_{r\to\infty} \frac{\log M(r)}{\log r} = \infty. \qquad (*_2)$$

To see this, suppose on the contrary that

$$\varliminf_{r\to\infty} \frac{\log M(r)}{\log r} = \alpha < \infty$$

$\Rightarrow$ For each fixed $\varepsilon > 0$, there is a sequence $r_n \to \infty$ so that

$$\frac{\log M(r_n)}{\log r_n} < \alpha + \varepsilon \quad \text{for } n \geq 1.$$

Exactly as in the proof of Example 6 of Sec. 3.4.2, it follows that $f(z)$ is a polynomial of degree $\leq [\alpha]$, contradicting to the assumption that $f(z)$ is transcendental. Consequently, no distinction can be made among transcendental entire functions if $(*_2)$ or the power function r^n is chosen as a measure of rate of their growth.

Here comes naturally to compare $M(r)$ with e^{r^n}.

If there is a number $\rho > 0$ so that

$$M(r) < e^{r^\rho} \quad \text{for all sufficiently large } r, \qquad (5.5.3.1)$$

then $f(z)$ is said to be of *finite order*; otherwise, of *infinite order*.

Suppose, for the moment, that $f(z)$ is of finite order. Set

$$\lambda = \inf \rho \quad \text{where } \rho\text{'s are these that satisfy (5.5.3.1).} \qquad (5.5.3.2)$$

$\Leftrightarrow$ For any $\lambda_1 > \lambda$, there is an $r_0 > 0$ so that whenever $r > r_0$,

$$M(r) < e^{r^{\lambda_1}} \quad \text{or} \quad \frac{\log \log M(r)}{\log r} < \lambda_1;$$

for any $\lambda_2 < \lambda$, there is a sequence $r_n \to \infty$ such that

$$M(r_n) > e^{r_n^{\lambda_2}} \quad \text{or} \quad \frac{\log \log M(r_n)}{\log r_n} > \lambda_2.$$

$\Leftrightarrow$ (Since $\lambda_2 < \lambda < \lambda_1$, and λ_1 and λ_2 are arbitrary.)

$$\lambda = \varlimsup_{r\to\infty} \frac{\log \log M(r)}{\log r}. \qquad (5.5.3.3)$$

λ, defined either by (5.5.3.2) or (5.5.3.3), is called the *order* (of growth) of the entire function $f(z)$ of finite order. Note that the definition (5.5.3.3) is still valid for infinite order. For instance,

a polynomial is of order 0 (because

$$M(r) < Ar^n < e^{r^\rho} \text{ for any} \rho > 0);$$

e^{z^k} (k, a positive integer) is of order k;

$\sin z$ and $\cos z$ are of order 1;

$\cos \sqrt{z}$ is of order $\dfrac{1}{2}$ $\left(\text{because, setting } z = re^{i\theta},\right.$

$$\left|\cos \sqrt{z}\right| \sim \cosh\left(r^{\frac{1}{2}} \sin \frac{\theta}{2}\right) \sim e^{r^{\frac{1}{2}} \sin \frac{\theta}{2}}\right). \tag{5.5.3.4}$$

For our claimed purpose, we need a preparatory result relating the modulus $|f(z)|$ on a circle to the moduli of the zeros of $f(z)$ inside the circle. It is the

Jensen's formula. Let $f(z)$ be analytic on $|z| \leq r$ with zeros $a_1, \ldots, a_n$ (multiple zeros being repeated) in $|z| < r$ and $f(0) \neq 0$. Then

$$\log |f(0)| = -\sum_{k=1}^{n} \log \frac{r}{|a_k|} + \frac{1}{2\pi} \int_0^{2\pi} \log |f(re^{i\theta})| d\theta.$$

In general, for $|z| < r$,

$$\log |f(z)| = -\sum_{k=1}^{n} \log \left| \frac{r^2 - \overline{a_k} z}{r(z - a_k)} \right| + \frac{1}{2\pi} \int_0^{2\pi} \operatorname{Re} \frac{re^{i\theta} + z}{re^{i\theta} - z} \log |f(re^{i\theta})| d\theta,$$

called the *Poisson–Jensen formula*. (5.5.3.5)

As a matter of fact, Exercise B(1) of Sec. 3.4.4 shows that $|f(0)| \leq \frac{M(r)}{r^n}|a_1 \cdots a_n|$ and $\int_0^r \frac{v(\rho)}{\rho} d\rho \leq \log \frac{M(r)}{|f(0)|}$, where $v(\rho)$ counts the number of zeros of f on $|z| \leq \rho < r$. Can you figure these out now?

Proof. First, suppose $f(z) \neq 0$ on $|z| \leq r$. Then $\log |f(z)|$ is harmonic in $|z| \leq r$ and hence (see (3.4.3.2) and (3.4.3.5)),

$$\log |f(0)| = \frac{1}{2\pi} \int_0^{2\pi} \log |f(re^{i\theta})| d\theta. \tag{$*_3$}$$

In general, we have Poisson's formula (see Exercise B(4) of Sec. 3.4.2 or (6.3.2.1))

$$\log |f(z)| = \frac{1}{2\pi} \int_0^{2\pi} \operatorname{Re} \frac{re^{i\theta} + z}{re^{i\theta} - z} \log |f(re^{i\theta})| d\theta, \quad |z| < r. \tag{$*_4$}$$

$(*_3)$ remains valid even if $f(z)$ has zeros on $|z| = r$. It is sufficient to assume that $f(z)$ has only one zero $re^{i\theta_0}$ along $|z| = r$. Then $\frac{f(z)}{z - re^{i\theta_0}}$ has a removable singularity at $re^{i\theta_0}$ and thus, is analytic on $|z| \leq r$ and free of zeros there. Applying $(*_3)$ to this a new function, we obtain

$$\log |f(0)| - \log r = \frac{1}{2\pi} \int_0^{2\pi} \log |f(re^{i\theta})| d\theta - \frac{1}{2\pi} \int_0^{2\pi} \log |re^{i\theta} - re^{i\theta_0}| d\theta.$$

All one needs to do is to show that

$$\log r = \frac{1}{2\pi} \int_0^{2\pi} \log r |e^{i\theta} - e^{i\theta_0}| d\theta$$

$$\Leftrightarrow \int_0^{2\pi} \log |e^{i\theta} - e^{i\theta_0}| d\theta = 0 \quad \text{or}$$

$$\int_{-\theta_0}^{2\pi - \theta_0} \log |1 - e^{i\psi}| d\psi = 0 \quad (\text{with } \theta - \theta_0 = \psi)$$

$$\Leftrightarrow \int_0^{2\pi} \log |1 - e^{i\theta}| d\theta = \int_0^{2\pi} \log \left| e^{\frac{i\theta}{2}} - e^{\frac{-i\theta}{2}} \right| d\theta = \int_0^{2\pi} \log \left| 2i \sin \frac{\theta}{2} \right| d\theta$$

$$= \int_0^{\pi} \log |2i \sin \theta| d\theta = 0$$

$$\Leftrightarrow \int_0^{\pi} \log \sin \theta d\theta = -\pi \log 2. \tag{$*_5$}$$

This last integral was established in Example 4 of Sec. 4.12.1.

Returning to the general case, the function

$$F(z) = f(z) \prod_{k=1}^{n} \frac{r^2 - \bar{a}_k z}{r(z - a_k)}$$

is analytic on $|z| \le r$, free from zeros in $|z| < r$, and $|F(z)| = |f(z)|$ on $|z| = r$. Applying $(*_3)$ to this $F(z)$, the required identity follows; while, by invoking $(*_4)$ to $F(z)$, the Poisson–Jensen formula follows, too. $\qquad\square$

The genus (see (5.5.2.6) and Exercise A(11) there) and the order (see (5.5.3.3)) are closely related in the following

Hadamard's order theorem. *The genus h and the order λ of an entire function (of finite order) satisfy the inequality*

$$h \le \lambda \le h + 1. \tag{5.5.3.6}$$

The following proof is based on that appeared in Ref. [1], pp. 209–211.

Proof. $\lambda \le h + 1$: It is harmless to suppose that $h < \infty$. In this case, the entire function $f(z)$ can be expressed as

$$f(z) = z^m e^{g(z)} \prod_{n=1}^{\infty} E_h \left(\frac{z}{a_n} \right), \quad \text{with } E_h(w) = (1 - w) e^{\sum_{l=1}^{h} \frac{1}{l} w^l}, \quad h \ge 1,$$

and

$$E_0(w) = 1 - w, \tag{$*_6$}$$

where $m \geq 0$ and $g(z)$ is a polynomial of degree $\leq h$ (refer to Exercise A(11) of Sec. 5.5.2).

By (5.5.3.4), z^m is of order 0 and $e^{g(z)}$ has order $\leq h$. Since the order of a product cannot exceed the orders of its factors (see Exercise A(1)), it is sufficient to show that

$$P(z) = \prod_{n=1}^{\infty} E_h\left(\frac{z}{a_n}\right) \tag{$*_7$}$$

is of order $\leq h+1$.

We need the following estimate for $E_h(w)$ in $(*_6)$:

$$\log|E_h(w)| \leq (2h+1)|w|^{h+1}, \quad w \in \mathbf{C}. \tag{$*_8$}$$

To begin with, observe that

$$\log|E_h(w)| \leq \log|E_{h-1}(w)| + |w|^h \quad \text{for any } w \in \mathbf{C} \quad \text{and} \quad h \geq 1. \tag{$*_9$}$$

While, if $|w| < 1$, we have, by using $\log(1-w) = -\sum_{l=1}^{\infty}\frac{1}{l}w^l$,

$$\log|E_h(w)| \leq \sum_{l=h+1}^{\infty}\frac{1}{l}|w|^l \leq \frac{1}{h+1}\sum_{l=h+1}^{\infty}|w|^l = \frac{1}{h+1}\cdot\frac{|w|^{h+1}}{1-|w|}$$

$$\Rightarrow (1-|w|)\log|E_h(w)| \leq |w|^{h+1}. \tag{$*_{10}$}$$

Now, $(*_8)$ is proved by induction on h:

$$h = 0 : \log|E_0(w)| = \log|1-w| \leq \log(1+|w|) \leq |w|.$$

Assume $(*_8)$ holds with $h-1$ in place of h: $\log|E_{h-1}(w)| \leq (2h-1)|w|^h$, $w \in \mathbf{C}$.

The case h: Then

$$\log|E_h(w)| \leq \log|E_{h-1}(w)| + |w|^h \quad \text{(by $(*_9)$)}$$
$$\leq 2h|w|^h \quad \text{(by inductive assumption)}$$
$$\leq (2h+1)|w|^{h+1}, \quad \text{if } |w| \geq 1;$$

if $|w| < 1$, then

$$\log|E_h(w)| \leq |w|\log|E_h(w)| + |w|^{h+1} \quad \text{(by $(*_{10})$)}$$
$$\leq |w|\log|E_{h-1}(w)| + 2|w|^{h+1} \quad \text{(by $(*_9)$)}$$
$$\leq (2h+1)|w|^{h+1} \quad \text{(by inductive assumption)}.$$

Returning to $(*_7)$, we have, by using $(*_8)$, that

$$\log|P(z)| = \sum_{n=1}^{\infty} \log\left|E_h\left(\frac{z}{a_n}\right)\right| \le (2h+1)|z|^{h+1} \sum_{n=1}^{\infty} \frac{1}{|a_n|^{h+1}}.$$

Since $\sum \frac{1}{|a_n|^{h+1}} < \infty$ (the essential hypothesis, owing to the definition of a genus), it follows from (5.5.3.4) that the order of $P(z)$ is $\le h+1$.

Proof. $h \le \lambda$: We may assume that $\lambda < \infty$. Let $h = [\lambda]$, the largest integer $\le \lambda$. Then $\lambda < h+1$. We need to show that $\sum \frac{1}{|a_n|^{h+1}} < \infty$ so that the genus of the canonical product in (5.5.2.5) is $\le h$ and that the entire function $g(z)$ in (5.5.2.2) is of degree $\le h$. In what follows, we assume $f(0) \ne 0$ and (5.5.3.5) is needed.

For the first part, let $v(\rho)$ count the number of zeros of $f(z)$ on $|z| \le \rho$. Setting $r = 2\rho$ in Jensen's formula and omitting the terms where $|a_n| > \rho$, then

$$\sum_{k=1}^{n} \log\frac{2\rho}{|a_k|} = \sum_{|a_n| \le \rho} + \sum_{|a_n| > \rho} = -\log|f(0)| + \frac{1}{2\pi}\int_0^{2\pi} \log|f(2\rho e^{i\theta})|d\theta$$

$$\Rightarrow v(\rho)\log 2 \le \frac{1}{2\pi}\int_0^{2\pi} \log|f(2\rho e^{i\theta})|d\theta - \log|f(0)|$$

$$\le \log M(2\rho) - \log|f(0)|. \tag{$*_{11}$}$$

Now, the definition (5.5.3.3) for λ is equivalent to the fact that

$$\log M(\rho) \le \rho^{\lambda + \frac{1}{2}\varepsilon} \quad \text{for any } \varepsilon > 0 \text{ if } \rho \text{ is sufficiently large.}$$

$$\Rightarrow \log M(2\rho) \le (2\rho)^{\lambda + \frac{\varepsilon}{2}} \quad \text{for any } \varepsilon > 0$$

$$\Rightarrow (\text{by } (*_{11}))(\log 2)\frac{v(\rho)}{\rho^{\lambda + \varepsilon}} \le (2\rho)^{\lambda + \frac{\varepsilon}{2}} \cdot \frac{1}{\rho^{\lambda + \varepsilon}} - \frac{\log|f(0)|}{\rho^{\lambda + \varepsilon}} \to 0 \quad \text{as } \rho \to \infty$$

$$\Rightarrow \lim_{\rho \to \infty} \frac{v(\rho)}{\rho^{\lambda + \varepsilon}} = 0 \quad \text{for every } \varepsilon > 0. \tag{$*_{12}$}$$

Arrange $|a_n|$, $n \ge 1$, so that $|a_1| \le |a_2| \le \cdots \le |a_n| \le \cdots$. Then

$$n \le v(|a_n|) < |a_n|^{\lambda + \varepsilon}, \text{ by } (*_{12}) \text{ if } n \ge N \quad (\text{some positive integer})$$

$$\Rightarrow \sum_{n \ge N} \frac{1}{|a_n|^{h+1}} \le \sum_{n \ge N} n^{-\frac{h+1}{\lambda + \varepsilon}} < \infty \quad \text{if } \varepsilon \text{ is so chosen that } \lambda + \varepsilon < h+1.$$

For the second part, it is sufficient to show that $g^{(h+1)}(z) \equiv 0$ in $\mathbf{C}$. Combining $(*_6)$ and $(*_7)$ and disregarding the factor z^m, we have

$$f(z) = e^{g(z)} P(z)$$

$$\Rightarrow g(z) = \log f(z) - \log P(z), \quad z \in \mathbf{C} - \{a_n, n \geq 1\}$$

$$\Rightarrow g^{(h+1)}(z) = \frac{d^h}{dz^h} \frac{f'(z)}{f(z)} - \frac{d^h}{dz^h} \frac{P'(z)}{P(z)}, \quad z \in \mathbf{C} - \{a_n, n \geq 1\}. \quad (*_{13})$$

What remains is to compute the h-th derivatives of $\frac{f'(z)}{f(z)}$ and $\frac{P'(z)}{P(z)}$.

Recall (see (1) in (3.4.3.4)) that the operation $\frac{\partial}{\partial x} - i\frac{\partial}{\partial y}$ is the most natural way of passing a harmonic function $u(z)$ to an analytic function $u_x(z) - iu_y(z)$. Applying this operation to both sides of the Poisson–Jensen formula (5.5.3.5), then direct computation shows that, in $|z| < \rho$,

$$\frac{f'(z)}{f(z)} = \sum_1^{v(\rho)} \frac{1}{z - a_n} + \sum_1^{v(\rho)} \frac{\bar{a}_n}{\rho^2 - \bar{a}_n z} + \frac{1}{2\pi} \int_0^{2\pi} \frac{2\rho e^{i\theta}}{(\rho e^{i\theta} - z)^2} \log|f(\rho e^{i\theta})| d\theta$$

$$\Rightarrow \frac{d^h}{dz^h} \frac{f'(z)}{f(z)} = -h! \sum_1^{v(\rho)} \frac{1}{(a_n - z)^{h+1}} + h! \sum_1^{v(\rho)} \frac{\bar{a}_n^{h+1}}{(\rho^2 - \bar{a}_n z)^{h+1}}$$

$$+ \frac{(h+1)!}{2\pi} \int_0^{2\pi} \frac{2\rho e^{i\theta}}{(\rho e^{i\theta} - z)^{h+2}} \log|f(\rho e^{i\theta})| d\theta. \quad (*_{14})$$

And we try to let $\rho \to \infty$ in $(*_{14})$. Note that (see (3.4.2.7) or (4.6.2))

$$\int_0^{2\pi} \frac{\rho e^{i\theta}}{(\rho e^{i\theta} - z)^{h+2}} d\theta = \frac{1}{i} \int_{|\zeta|=\rho} \frac{\zeta}{(\zeta - z)^{h+2}} d\zeta = 0 \quad \text{for } z \text{ in } |z| < \rho$$

$\Rightarrow$ (the last integer in $(*_{14})$)

$$\left| \frac{1}{2\pi} \int_0^{2\pi} \frac{2\rho e^{i\theta}}{(\rho e^{i\theta} - z)^{h+2}} \log|f(\rho e^{i\theta})| d\theta \right|$$

$$= \left| \frac{1}{2\pi} \int_0^{2\pi} \frac{2\rho e^{i\theta}}{(\rho e^{i\theta} - z)^{h+2}} (\log M(\rho) - \log|f(\rho e^{i\theta})|) d\theta \right|$$

$$\leq 2^{h+3} \rho^{-h-1} \cdot \frac{1}{2\pi} \int_0^{2\pi} \log \frac{M(\rho)}{|f(\rho e^{i\theta})|} d\theta, \quad \text{if } |z| < \frac{\rho}{2}$$

$$\leq 2^{h+3} \rho^{-h-1} (\log M(\rho) - \log|f(0)|)$$

$$\left(\text{since } \log|f(0)| \leq \frac{1}{2\pi} \int_0^{2\pi} \log|f(\rho e^{i\theta})| d\theta, \quad \text{by Jensen's formula} \right)$$

$$\longrightarrow 0 \text{ as } \rho \to \infty \quad \text{(since the genus } \lambda < h + 1\text{)}. \quad (*_{15})$$

On the other hand, in case $|a_n| \leq \rho$ and $|z| < \frac{\rho}{2}$,

$$\left| \frac{\bar{a}_n^{h+1}}{(\rho^2 - \bar{a}_n z)^{h+1}} \right| \leq \frac{\rho^{h+1}}{\left(\frac{\rho^2}{2}\right)^{h+1}} = \left(\frac{2}{\rho}\right)^{h+1}$$

$$\Rightarrow \left| \sum_1^{\upsilon(\rho)} \frac{\bar{a}_n^{h+1}}{(\rho^2 - \bar{a}_n z)^{h+1}} \right| \leq \sum_1^{\upsilon(\rho)} \left(\frac{2}{\rho}\right)^{h+1}$$

$$= 2^{h+1} \frac{\upsilon(\rho)}{\rho^{h+1}} \to 0 \text{ as } \rho \to \infty \quad (\text{see } (*_{12})). \qquad (*_{16})$$

Letting $\rho \to \infty$ in $(*_{14})$, and using $(*_{15})$ and $(*_{16})$, we obtain

$$\frac{d^h}{dz^h} \frac{f'(z)}{f(z)} = -h! \sum_1^\infty \frac{1}{(a_n - z)^{h+1}}, \quad z \in \mathbf{C} - \{a_n, \ n \geq 1\}. \qquad (*_{17})$$

As for $\frac{d^h}{dz^h} \frac{P'(z)}{P(z)}$ where $P(z)$ is given by $(*_7)$, note that (see $(5.5.1.8)$)

$$\frac{P'(z)}{P(z)} = \sum_{n=1}^\infty \frac{E_h'\left(\frac{z}{a_n}\right)}{E_h\left(\frac{z}{a_n}\right)}, \quad \text{locally uniformly in } \mathbf{C} - \{a_n, \ n \geq 1\}.$$

$$\Rightarrow \frac{d^h}{dz^h} \frac{P'(z)}{P(z)} = -h! \sum_1^\infty \frac{1}{(a_n - z)^{h+1}}$$

$$\text{(by Weierstrass's theorem } (5.3.1.1)), \qquad (*_{18})$$

where $z \in \mathbf{C} - \{a_n, n \geq 1\}$.

Compare $(*_{17})$ and $(*_{18})$. We find from $(*_{13})$ that $g^{(h+1)}(z) \equiv 0$ in $\mathbf{C}$. Therefore, $g(z)$ is a polynomial of degree $\leq h$ (see Example 6 in Sec. 3.4.2). The theorem is thus proved. $\qquad \square$

$(5.5.3.9)$ will give a somewhat different proof for the first inequality in $(5.5.3.6)$ without using Poisson–Jensen's formula explicitly. Exercise A(2) states another from of Hadamard's theorem.

To the end, we give two corollaries of Hadamard's theorem.

A special case of Picard's little theorem (see $(4.10.3.3)$ or $(5.8.3.11)$). *A nonconstant entire function of finite order assumes every finite complex number with at most one exceptional value.* $\qquad (5.5.3.7)$

Proof. Suppose such a function $f(z)$ does not assume two distinct complex numbers a and b. Then $f(z) - a$ is also of finite order and $f(z) - a \neq 0$ for all $z \in \mathbf{C}$. Consequently, $f(z) - a = e^{g(z)}$ for some entire function $g(z)$ (see (5.5.1)). By Hadamard's theorem, $g(z)$ is a polynomial whose degree is not greater than the order of $f(z)$. Yet $g(z)$ does not assume the value $\log(b-a)$, contradicting to the fundamental theorem of algebra (see (3.4.2.12)). □

The second one is

A value-distribution theorem. *An entire function of fractional order assumes every finite complex number infinitely many times.* (5.5.3.8)

Proof. For such a function $f(z)$, $f(z)$ and $f(z) - a$ $(a \in \mathbf{C})$ have the same order. It is thus sufficient to show that $f(z)$ has infinitely many zeros. If not, suppose $f(z)$ has only finite many zeros $a_1, \ldots, a_n$. Hadamard's theorem says that $f(z)$ is of the form $e^{g(z)}(z - a_1) \cdots (z - a_n)$, where $g(z)$ is a polynomial of degree m which is not greater than the order of $f(z)$. Well, the order of $e^{g(z)}$ is m (see (5.5.3.4)) and is equal to that of $f(z)$ (see Exercise A(3)), a fractional by assumption. This is a contradiction. □

Exercises A

(1) Suppose $f(z)$ and $g(z)$ are entire functions of finite orders λ_1 and λ_2, respectively. Show that the orders of $f(z)g(z)$ and $f(z) + g(z)$ are $\leq \max(\lambda_1, \lambda_2)$. In case $\lambda_1 \neq \lambda_2$, the order of $f(z) + g(z)$ is just equal to $\max(\lambda_1, \lambda_2)$.

(2) Suppose A, B, and α are positive constants so that an entire function $f(z)$ satisfies $|f(z)| \leq e^{A|z|^{\alpha}+B}$ for all large $|z|$. Show that the order of $f(z)$ is $\leq \alpha$.

(3) Let $f(z)$ be entire of finite order λ and $p(z)$ be a polynomial. Show that the order of $p(z)f(z)$ is still λ.

(4) Find the order of e^{e^z}.

Exercises B

(1) Let $f(z)$ be an entire function of finite order λ and $a_n \neq 0$, $n \geq 1$, be its zeros. Then
$$\sum_{n=1}^{\infty} \frac{1}{|a_n|^{\lambda+\varepsilon}} < \infty$$
for each $\varepsilon > 0$. Try the following steps: Choose $0 < \alpha < \varepsilon$. Suppose $|a_1| \leq |a_2| \leq \cdots \leq |a_n| \leq \cdots$.

1. Set $Q_n(z) = (z - a_1) \cdots (z - a_n)$. Then
$$\frac{1}{2\pi i} \int_{|\zeta|=R} \frac{f(\zeta)}{\zeta Q_n(\zeta)} d\zeta = \frac{f(0)}{a_1 \cdots a_n} \quad \text{if } |a_n| < R.$$

2. $M(R) \leq e^{R^{\lambda+\alpha}}$ if $R > 0$ is large enough. Choose $|z| = R = 3|a_n|$. Then

$$\frac{|f(0)|}{|a_1 \cdots a_n|} \leq \frac{e^{(3|a_n|)^{\lambda+\alpha}}}{2^n |a_n|^n}$$

$$\Rightarrow n \leq \frac{1}{\log 2}(3|a_n|)^{\lambda+\alpha} = A_\lambda |a_n|^{\lambda+\alpha}, \quad \text{if } f(0) = 1 \text{ is assumed.}$$

3. Note that $\frac{1}{|a_n|^{\lambda+\varepsilon}} \leq A_\lambda^\delta \frac{1}{n^\delta}$, where $\delta = \frac{\lambda+\varepsilon}{\lambda+\alpha} > 1$ and $\sum \frac{1}{n^\delta} < \infty$.

Let $\delta = \inf \alpha$ where $\alpha > 0$ are such that $\sum \frac{1}{|a_n|^\alpha} < \infty$. This δ is called the *index of convergence* of zeros of $f(z)$. Then $\delta \geq 0$. As a matter of fact, we have

$$\delta - 1 \leq h \quad \text{(the genus of the canonical product associated}$$

$$\text{with the zeros } a_n)$$

$$\leq \delta \quad \text{(the index of convergence of } a_n)$$

$$\leq \lambda \quad \text{(the order of } f(z)). \tag{5.5.3.9}$$

$\delta \leq \lambda$ follows from Exercise (1). If δ is not an integer, then $h = [\delta]$; if δ is an integer and $\sum \frac{1}{|a_n|^\delta} < \infty$, then $h = \lambda - 1$, and if $\sum \frac{1}{|a_n|^\delta} = +\infty$, $h = \lambda$ holds.

(2) Let $f(z)$ be entire with finite order λ. Then

$$f(z) = z^m e^{g(z)} P(z), \tag{5.5.3.10}$$

where $m \geq 0$, $g(z)$ is a polynomial of degree $\leq \lambda$ and

$$P(z) = \prod_{n=1}^\infty \left(1 - \frac{z}{a_n}\right) e^{\frac{z}{a_n} + \frac{1}{2}\left(\frac{z}{a_n}\right)^2 + \cdots + \frac{1}{h}\left(\frac{z}{a_n}\right)^h}$$

is the canonical product associated with the zeros a_n, $n \geq 1$, of $f(z)$. As a beginning, note that (5.5.3.9) and (5.5.2.2) guarantee that $f(z) = z^m e^{g(z)} P(z)$ with $g(z)$ an *entire* function only. Try the following steps to prove that $g(z)$ is a polynomial of degree $\leq \lambda$.

1. Fix any $R > 1$. Choose an integer $N \geq 1$ so that $|a_n| \leq R$ if $n \leq N$ and $|a_n| > R$ if $n > N$. Let

$$f(z) = f_1(z) f_2(z),$$

where

$$f_1(z) = z^m \prod_{n=1}^{N} \left(1 - \frac{z}{a_n}\right) \quad \text{and}$$

$$f_2(z) = \exp\left\{g(z) + \sum_{n=1}^{N} F_h\left(\frac{z}{a_n}\right)\right\} \prod_{n=N+1}^{\infty} \left(1 - \frac{z}{a_n}\right) e^{F_h\left(\frac{z}{a_n}\right)} \quad \text{with}$$

$$F_h(w) = \sum_{l=1}^{h} \frac{1}{l} w^l.$$

It is tacitly understood that $F_h(w) = 0$ if $h = 0$.

2. On $|z| = 2R$, $M(2R) \geq |f(z)| \geq |f_2(z)|$ holds. Consequently, $|f_2(z)| \leq M(2R)$ if $|z| < R$.

3. Rewrite $f_2(z) = e^{\varphi(z)}$ if $|z| < R$, where

$$\varphi(z) = g(z) + \sum_{n=1}^{N} F_h\left(\frac{z}{a_n}\right) + \sum_{n=N+1}^{\infty} \left\{\text{Log}\left(1 - \frac{z}{a_n}\right) + F_h\left(\frac{z}{a_n}\right)\right\}.$$

Owing to (5.5.2.1) with $m_n + 1$ replaced by h, the series $\sum_{n=N+1}^{\infty}$ converges uniformly in $|z| < R$.

4. Fix an integer $m > \lambda \geq h$. Differentiate $\varphi(z)m$ times to get

$$\varphi^{(m)}(z) = g^{(m)}(z) + \sum_{n=N+1}^{\infty} \frac{(-1)^m (m-1)!}{(z - a_n)^m}.$$

Try to show that $g^{(m)}(z) \equiv 0$. According to

$$\left|\frac{\varphi^{(m)}(0)}{m!}\right| \leq \frac{4 \max_{|z|=r} \text{Re}\,\varphi(z) - 2\text{Re}\,\varphi(0)}{r^m}, \quad 0 < r < R$$

$$\Rightarrow \text{(using step 2 and letting } r \to R) \left|\frac{\varphi^{(m)}(0)}{m!}\right|$$

$$\leq \frac{4 \log M(2R) - 2\text{Re}\,\varphi(0)}{R^m}$$

$$\Rightarrow \left(\text{noting } \frac{\varphi^{(m)}(0)}{m!} = \frac{g^{(m)}(0)}{m!} - \prod_{n=N+1}^{\infty} \frac{1}{m a_n^m}\right)$$

$$\left|\frac{g^{(m)}(0)}{m!}\right| \leq \left|\frac{\varphi^{(m)}(0)}{m!}\right| + \prod_{n=N+1}^{\infty} \frac{1}{m|a_n|^m} \to 0 \quad \text{as } R \to \infty.$$

In the last process, we need to use $M(2R) < e^{(2R)^{\lambda+\varepsilon}}$ for $0 < \varepsilon < m - \lambda$ and R large.

5.6   The Gamma Function $\Gamma(z)$

Euler's gamma function $\Gamma(x)$, where x is a real variable, extends the factorial $n! = n \cdot (n-1)! = n \cdot (n-1) \cdots 3 \cdot 2 \cdot 1$ to nonintegral x in the sense that $\Gamma(x+1) = x\Gamma(x)$ for $x \neq 0, -1, -2, \ldots$ In advanced calculus courses, one can learn the following basic properties of $\Gamma(x)$:

(1) *Euler's second integral:*

$$\Gamma(x) = \int_0^\infty e^{-t} t^{x-1} dt, \quad x > 0.$$

The domain of definition can be extended to $\mathbf{R} - \{0, -1, -2, \ldots\}$ via the functional equation $\Gamma(x+1) = x\Gamma(x)$, $x > 0$.

(2) *Gauss's representation:*

$$\Gamma(x) = \lim_{n \to \infty} \frac{n^x n!}{x(x+1) \cdots (x+n)}, \quad x \neq 0, -1, -2, \ldots.$$

(3) *Weierstrass's representation:*

$$\Gamma(x) = e^{-\gamma x} \frac{1}{x} \prod_{n=1}^\infty \left(1 + \frac{x}{n}\right)^{-1} e^{\frac{x}{n}}, \quad \text{or}$$

$$\frac{1}{\Gamma(x)} = e^{\gamma x} x \prod_{n=1}^\infty \left(1 + \frac{x}{n}\right) e^{-\frac{x}{n}}, \quad x \neq 0, -1, -2, \ldots.$$

where $\gamma = \lim_{n \to \infty} \left\{\sum_{k=1}^n \frac{1}{k} - \log n\right\} = 0.5772156\ldots$ is *Euler's constant.*

(4) *Partial fractional expansion:*

$$\Gamma(x) = \sum_{n=0}^\infty \frac{(-1)^n}{n!(x+n)} + \int_1^\infty e^{-t} t^{x-1} dt, \quad x \neq 0, -1, -2, \ldots.$$

(5) *Euler's functional equation:*

$$\Gamma(x)\Gamma(1-x) = \frac{\pi}{\sin \pi x}, \quad x \neq 0, \pm 1, \pm 2, \ldots. \tag{5.6.1}$$

It is our purpose in this section to introduce the gamma function $\Gamma(z)$ in the complex variable z as an important application of contents both in Secs. 5.4 and 5.5, and to develop its basic properties. It is amazing to know that each of the afore-mentioned five properties can be adopted as a starting point to define $\Gamma(z)$, just replacing x by z, where $z \neq 0, -1, -2, \ldots$. Of course, convergence should be taken good care in each case.

Section 5.6.1 will discuss in detail the background why we choose (3) as our definition for $\Gamma(z)$ and then, prove its equivalent representations shown in (5.6.1). Other representations will be introduced as we proceed to Sec. 5.6.2.

Basic and characteristic properties for $\Gamma(z)$ will be the core of Sec. 5.6.2.

Using Lindelöf's residue method, Sec. 5.6.3 studies the asymptotic behavior of $\Gamma(z)$, namely, the well-known Stirling's formula.

5.6.1   *Definition and representations*

To start with, we pose the *problem*: Does there exist a meromorphic function $f(z)$ in $\mathbf{C}$, satisfying the following conditions:

1. $f(1) = 1$, and
2. $f(z + 1) = zf(z)$, wherever $f(z)$ is defined in $\mathbf{C}$?   (5.6.1.1)

If such a function does exist, it should be *a meromorphic function with simple poles at $z = 0, -1, -2, \ldots$ and the corresponding residues*

$$\mathrm{Re}\, s(f(z); -n) = \frac{(-1)^n}{n!}, \quad n \geq 0.$$

Since $f(z)$ is analytic at $z = 1$, and $\lim_{z \to 0} f(z) = \lim_{z \to 0} \frac{f(z+1)}{z} = \infty$ and $\lim_{z \to 0} zf(z) = \lim_{z \to 0} f(z+1) = 1$ holds, 0 is a simple pole (see (4.10.2.1)) with residue equal to 1. By inductive process, it is easily seen that $f(z)$ has a simple pole at $z = -n$ $(n \geq 1)$ and

$$f(z + n + 1) = (z + n)(z + n - 1) \cdots (z + 1)zf(z), \quad z \neq 0, -1, \ldots, -n - 1$$

$$\Rightarrow (z + n)f(z) = \frac{f(z + n + 1)}{z(z + 1) \cdots (z + n - 1)} \to \frac{f(1)}{-n(-n + 1) \cdots (-1)}$$

$$= \frac{(-1)^n}{n!} \quad \text{as } z \to -n.$$

Even if (5.6.1.1) has a solution, it is definitely not a unique one. One might choose any meromorphic function $\varphi(z)$ with the properties that $\varphi(1) = 1$ and $\varphi(z + 1) = \varphi(z)$, $z \in \mathbf{C}$. Then $\varphi(z)f(z)$ will be another solution. For instance, $\varphi(z) = \frac{\tan(i+2\pi z)}{\tan i}$ is a possible candidate. But, in this case, $\varphi(z)f(z)$ turns out to have additional zeros (due to $\tan(i + 2\pi z) = 0$) and simple poles (due to $\tan(i + 2\pi z) = \infty$).

The solution is also not unique if we further impose the restriction that the solution has neither zeros nor poles other than $0, -1, -2, \ldots$. This is because all one needs to do is to choose an entire function $\varphi(z)$ satisfying $\varphi(1) = 1$ and $\varphi(z + 1) = \varphi(z)$, $z \in \mathbf{C}$. Then $\varphi(z)f(z)$ will work, too.

Consequently, the best we can do toward the solution of (5.6.1.1) is the following

Claim. *Suppose $f(z)$ is a meromorphic function in $\mathbf{C}$ satisfying:*

(1) $f(1) = 1$;

(2) $f(z+1) = zf(z)$, *wherever $f(z)$ is defined in $\mathbf{C}$, and*

(3) $f(z)$ *does not have zeros and has simple poles only at nonpositive integers.*

Then, $f(z)$ should be of the form

$$f(z) = e^{-g(z)} \frac{1}{z \prod_{n=1}^{\infty} \left(1 + \frac{z}{n}\right) e^{-\frac{z}{n}}},$$

where $g(z)$ is an entire function having the properties:

$$g(1) = \gamma + 2k\pi i;$$

$$g(z+1) - g(z) = \gamma + 2l\pi i$$

in which k and l are integers, and $\gamma = \lim_{n \to \infty} \left(\sum_{k=1}^{n} \frac{1}{k} - \log n\right)$ is Euler's constant. (5.6.1.2)

Sketch of proof. Note that $\frac{1}{f(z)}$ is still a meromorphic function, no poles and having simple zeros $z = 0, -1, -2, \ldots$. According to Example 3(3) in Sec. 5.5.1 and (5.5.2.2), the entire function $z \prod_{n=1}^{\infty} \left(1 + \frac{z}{n}\right) e^{-\frac{z}{n}}$ has simple zeros only at $z = 0, -1, -2, \ldots$. Hence, the entire function

$$\frac{(f(z))^{-1}}{z \prod_{n=1}^{\infty} \left(1 + \frac{z}{n}\right) e^{-\frac{z}{n}}}$$

does not have any zero in $\mathbf{C}$ and thus, can be written as $e^{g(z)}$ (see (5.5.1)), where $g(z)$ is an entire function. The claimed form for $f(z)$ then follows.

Rewrite $f(z)$ as

$$f(z) = \lim_{n \to \infty} f_n(z), \quad \text{where}$$

$$f_n(z) = \frac{e^{-g(z)}}{z \prod_{k=1}^{n} \left(1 + \frac{z}{k}\right) e^{-\frac{z}{k}}} = \frac{n! \exp\left\{-g(z) + \sum_{k=1}^{n} \frac{z}{k}\right\}}{z(z+1) \cdots (z+n)}. \quad (*_1)$$

Then

$$1 = \frac{zf(z)}{f(z+1)} = \lim_{n \to \infty} \frac{zf_n(z)}{f_n(z+1)}$$

$$= \lim_{n \to \infty} (z+n+1) \exp\left\{g(z+1) - g(z) - \sum_{k=1}^{n} \frac{1}{k}\right\}$$

$$= \lim_{n \to \infty} \left(1 + \frac{z+1}{n}\right) \exp\left\{g(z+1) - g(z) - \left[\sum_{k=1}^{n} \frac{1}{k} - \log n\right]\right\}$$

$$= \exp\{g(z+1) - g(z) - \gamma\}, \quad z \in \mathbf{C}$$

$$\Rightarrow g(z+1) - g(z) = \gamma + 2l\pi i, \ l \text{ an integer.}$$

Similarly,

$$1 = \lim_{n \to \infty} f_n(1) = \lim_{n \to \infty} \frac{\exp\left\{-g(1) + \left[\sum_{k=1}^{n} \frac{1}{k} - \log n\right]\right\}}{1 + \frac{1}{n}} = \exp\left\{-g(1) + \gamma\right\}$$

and thus, $g(1) = \gamma + 2k\pi i$ for integer k. $\qquad \Box$

The simplest such function $g(z)$ in (5.6.1.2) is the one given by $g(z) = \gamma z$. The corresponding $f(z)$ is specifically denoted as

$$\Gamma(z) = e^{-\gamma z} \frac{1}{z \prod_{n=1}^{\infty} \left(1 + \frac{z}{n}\right) e^{-\frac{z}{n}}}, \quad z \in \mathbf{C} - \{0, -1, -2, \ldots\} \quad \text{or}$$

$$\frac{1}{\Gamma(z)} = e^{\gamma z} z \prod_{n=1}^{\infty} \left(1 + \frac{z}{n}\right) e^{-\frac{z}{n}}, \quad z \in \mathbf{C}. \tag{5.6.1.3}$$

$\Gamma(z)$ is called the *gamma function* in *Weierstrass's form*. Refer to Ref. [58], Vol. II, p. 308, for the graph of its modulus surface $\mu = |\Gamma(z)|$ or $\frac{1}{|\Gamma(z)|}$.

Now (1), (2), and (4) in (5.6.1) have the following extensions to the complex variable z.

Equivalent representations of $\Gamma(z)$.

(1) *Gauss's representation*: If $z \in \mathbf{C} - \{0, -1, -2, \ldots\}$,

$$\Gamma(z) = \lim_{n \to \infty} \frac{n! n^z}{z(z+1) \cdots (z+n)}.$$

(2) *Euler's integral representation*: In $\mathrm{Re}\, z > 0$,

$$\Gamma(z) = \int_0^{\infty} e^{-t} t^{z-1} dt \quad (t \in \mathbf{R}).$$

The functional equation $\Gamma(z+1) = z\Gamma(z)$ permits $\Gamma(z)$ to be extended to all $z \neq 0, -1, -2, \ldots$.

(3) *Partial fractional representation*: If $z \in \mathbf{C} - \{0, -1, -2, \ldots\}$,

$$\Gamma(z) = \sum_{n=0}^{\infty} \frac{(-1)^n}{n!(z+n)} + \int_1^{\infty} e^{-t} t^{z-1} dt,$$

where the series $\sum_0^{\infty}$ converges absolutely and locally uniformly in $\mathbf{C} - \{0, -1, -2, \ldots\}$ and the integral $\int_1^{\infty}$ represents an entire function.

$$\tag{5.6.1.4}$$

Proof. (1) Returning to $(*_1)$ with $g(z) = \gamma z$, we have

$$\Gamma(z) = \lim_{n\to\infty} \frac{n! \exp\left\{\sum_{k=1}^{n} \frac{z}{k} - \gamma z\right\}}{z(z+1)\cdots(z+n)}$$

$$- \lim_{n\to\infty} \frac{n! \exp\left[\left(\sum_{k=1}^{n} \frac{1}{k} - \log n - \gamma\right)z + z\log n\right\}}{z(z+1)\cdots(z+n)}$$

$$= \lim_{n\to\infty} \frac{n! \exp\{z\log n\}}{z(z+1)\cdots(z+n)} = \lim_{n\to\infty} \frac{n! n^z}{z(z+1)\cdots(z+n)}.$$

Regarding (2) and (3), set temporarily the function

$$f(z) = \int_0^\infty e^{-t} t^{z-1} dt \quad (t \in \mathbf{R},\ z \in \mathbf{C})$$

$$= \int_0^1 e^{-t} t^{z-1} dt + \int_1^\infty e^{-t} t^{z-1} dt, \tag{$*_2$}$$

where the path of integration is along the nonnegative real axis $[0, \infty)$, and $t^{z-1} = e^{(z-1)\log t}$. Since $|e^{-t} t^{z-1}| = e^{-t} t^{\operatorname{Re} z - 1}$ and

$$\lim_{t\to 0}(e^{-t} t^{\operatorname{Re} z - 1}) t^{1-\operatorname{Re} z} = \lim_{t\to 0} e^{-t} = 1, \quad \text{if } \operatorname{Re} z > 0;$$

$$\lim_{t\to\infty} \left(e^{-t} t^{\operatorname{Re} z - 1}\right) t^2 = 0 \quad \text{for any } z \in \mathbf{C},$$

it follows, respectively, that

$$I_1(z) = \int_0^1 e^{-t} t^{z-1} dt \quad \text{absolutely in } \operatorname{Re} z > 0;$$

$$I_2(z) = \int_1^\infty e^{-t} t^{z-1} dt \quad \text{absolutely in } \mathbf{C}. \tag{$*_3$}$$

According to (4.7.4), Exercise B(2) of Sec. 5.3.1 and Exercises B of Sec. 5.3.3 (see also Exercises A(2) and A(3)), $I_1(z)$ is analytic in $\operatorname{Re} z > 0$ while $I_2(z)$ is entire.

Suppose, for the moment, that the function $f(z)$ defined in $(*_2)$ is really identical with $\Gamma(z)$, namely, (2) is proved. Note that

$$e^{-t} = \sum_{n=0}^{\infty} \frac{(-1)^n}{n!} t^n \quad \text{absolutely and uniformly in } [0, 1].$$

$$\Rightarrow I_1(z) = \int_0^1 \left(\sum_{n=0}^{\infty} \frac{(-1)^n}{n!} t^n\right) t^{z-1} dt = \sum_{n=0}^{\infty} \frac{(-1)^n}{n!} \int_0^1 t^{n+z-1} dt$$

$$= \sum_{n=0}^{\infty} \frac{(-1)^n}{n!(z+n)}, \quad \operatorname{Re} z > 0. \tag{$*_4$}$$

But the series in the right converges absolutely and locally uniformly in $\mathbf{C} - \{0, -1, -2, \ldots\}$ and hence, by Weierstrass's theorem (5.3.1.1), represents the analytic function $I_1(z)$ there; moreover, $I_1(z)$ has simple poles at $z = -n$ $(n \geq 0)$ with residues $\frac{(-1)^n}{n!}$. Therefore, by $(*_2)$,

$$\Gamma(z) - I_1(z) = \int_1^\infty e^{-t} t^{z-1} dt, \quad \text{Re } z > 0$$

holds identically. Both functions in the left and the right are entire. They should be coincident everywhere in $\mathbf{C}$ owing to the interior uniqueness property. Hence, (3) holds.

To prove (2), we want to show that, in $(*_2)$,

$$f(z) = \lim_{n \to \infty} f_n(z) = \Gamma(z), \quad \text{Re } z > 0, \quad \text{where}$$

$$f_n(z) = \int_0^n \left(1 - \frac{t}{n}\right)^n t^{z-1} dt \quad \text{for } n \geq 1. \tag{$*_5$}$$

Since $f(z) - \lim_{n \to \infty} \int_0^n e^{-t} t^{z-1} dt$ in $\text{Re } z > 0$, it is sufficient to show that

$$\lim_{n \to \infty} \int_0^n \left[e^{-t} - \left(1 - \frac{t}{n}\right)^{n-1} \right] t^{z-1} dt = 0$$

in order to justify that $f(z) = \lim_{n \to \infty} f_n(z)$. In passing, note that $f_n(z)$ is analytic (entire) for each $n \geq 1$ (see (4.7.4)). Now

$$1 + \frac{t}{n} \leq e^{\frac{t}{n}} \leq \left(1 - \frac{t}{n}\right)^{-1}, \quad |t| \leq n$$

$$\Rightarrow \left(1 + \frac{t}{n}\right)^n \leq e^t \quad \text{and} \quad \left(1 - \frac{t}{n}\right)^n \leq e^{-t}, \quad |t| \leq n$$

$$\Rightarrow 0 \leq e^{-t} - \left(1 - \frac{t}{n}\right)^n = e^{-t} \left[1 - e^t \left(1 - \frac{t}{n}\right)^n\right] \leq e^{-t} \left[1 - \left(1 - \frac{t^2}{n^2}\right)^n\right]$$

$$= e^{-t} \cdot \frac{t^2}{n^2} \left\{ 1 + \left(1 - \frac{t^2}{n^2}\right) + \cdots + \left(1 - \frac{t^2}{n^2}\right)^{n-1} \right\} \leq \frac{t^2 e^{-t}}{n}$$

$$\Rightarrow \left| \int_0^n \left[e^{-t} - \left(1 - \frac{t}{n}\right)^n \right] t^{z-1} dt \right| \leq \frac{1}{n} \int_0^n e^{-t} t^{\text{Re } z+1} dt$$

$$\leq \frac{1}{n} \int_0^\infty e^{-t} t^{\text{Re } z+1} dt \to 0 \quad \text{as } n \to \infty.$$

As a matter of fact, the last step remains true as long as $\text{Re } z > -2$.

Finally, come to the proof of $\lim_{n\to\infty} f_n(z) = \Gamma(z)$ in $\operatorname{Re} z > 0$. Using the change of variables $t = n\zeta$, where ζ is a real variable,

$$f_n(z) = n^z \int_0^1 (1-\zeta)^n \zeta^{z-1} d\zeta = n^z \int_0^1 (1-\zeta)^n d\left\{\frac{\zeta^z}{z}\right\}$$

$$= n^z \left\{ (1-\zeta)^n \frac{\zeta^z}{z} \Big|_{\zeta=0}^1 \right.$$

$$\left. - \int_0^1 \frac{-n}{z}(1-\zeta)^{n-1}\zeta^z d\zeta \right\} \quad \text{(integration by parts)}$$

$$= \frac{n^z \cdot n}{z} \int_0^1 (1-\zeta)^{n-1}\zeta^z d\zeta$$

$$= \cdots \text{(repeated usage of integration by parts)}$$

$$= \frac{n!n^z}{z(z+1)\cdots(z+n-1)} \int_0^1 \zeta^{z+n-1} d\zeta = \frac{n!n^z}{z(z+1)\cdots(z+n)}. \tag{$*_6$}$$

Combining with (1), it follows that $\lim_{n\to\infty} f_n(z) = \Gamma(z)$ in $\operatorname{Re} z > 0$. $\square$

Remark. It is worthy to mention the following relations obtained during the above proof (see (1), $(*_5)$, and $(*_6)$):

$$\int_0^n \left(1-\frac{t}{n}\right)^n t^{z-1} dt = n^z \int_0^1 (1-t)^n t^{z-1} dt$$

$$= \frac{n!z^n}{z(z+1)\cdots(z+n)}, \quad z \neq 0, -1, \ldots, -n;$$

$$\Gamma(z) = \lim_{n\to\infty} \int_0^n \left(1-\frac{t}{n}\right)^n t^{z-1} dt = \lim_{n\to\infty} \int_0^n e^{-t} t^{z-1} dt, \quad \operatorname{Re} z > 0. \tag{5.6.1.5}$$

Recall that $e^{-t} = \lim_{n\to\infty} \left(1-\frac{t}{n}\right)^n$.

In what follows, we try to find some other integral representations for $\Gamma(z)$.

For our purpose and importance by itself, we need the following

Euler's functional equation.

$$\Gamma(z)\Gamma(1-z) = \frac{\pi}{\sin \pi z}, \quad z \neq 0, -1, -2, \ldots.$$

In particular, $\Gamma\left(\frac{1}{2}\right) = \sqrt{\pi}$. $\tag{5.6.1.6}$

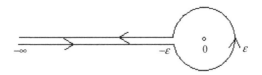

Fig. 5.15

This is easily seen by using (5.6.1.3) and the infinite product for $\sin \pi z$ (see Example 1 in Sec. 5.5.2). Also, refer to Exercises A(4) and A(9).

Here comes

The complex integral representation of $\Gamma(z)$. Fix any $\varepsilon > 0$. Let γ_ε denote the contour composed of the lower side of the cut $(-\infty, 0]$ oriented with the increasing t, followed by the circle $|z| = \varepsilon$ in counterclockwise direction and then the upper side of the cut $(-\infty, 0]$ oriented with the decreasing t. See Fig. 5.15. Then, if $z \neq 0, -1, -2, \ldots$,

$$\Gamma(z) = \frac{1}{2i \sin \pi z} \int_{\gamma_\varepsilon} e^t t^{z-1} dt;$$

$$\frac{1}{\Gamma(z)} = \frac{1}{2\pi i} \int_{\gamma_\varepsilon} e^t t^{-z} dt,$$

where $t^{-z} = e^{-z \operatorname{Log} t}$ is the principal branch in $\mathbf{C} - (-\infty, 0]$: $-\pi < \operatorname{Arg} t \leq \pi$. Moreover, the integrals are independent of the choice of ε.   (5.6.1.7)

Proof. Owing to identity (5.6.1.6), it is sufficient to prove the validity of the first integral for $\Gamma(z)$.

Denote by $\gamma_{\varepsilon,n}$ the part of γ_ε that lies in the closed disk $|z| \leq n$ for $n = 1, 2, \ldots$. Along the lower side of $\gamma_{\varepsilon,n}$, $t = |t| e^{-\pi i}$; while, along the upper side, $t = |t| e^{\pi i}$. Hence,

$$\int_{\gamma_{\varepsilon,n}} e^t t^{-z} dt = \left\{ \int_{-n}^{-\varepsilon} + \int_{|t|=\varepsilon} + \int_{-\varepsilon}^{-n} \right\} e^t t^{-z} dt$$

$$= \int_{-n}^{-\varepsilon} e^t e^{-z(\log |t| - \pi i)} dt + \int_{-\pi}^{\pi} e^{\varepsilon e^{i\theta}} e^{-z(\log \varepsilon + i\theta)} \varepsilon i e^{i\theta} d\theta$$

$$+ \int_{-\varepsilon}^{-n} e^t e^{-z(\log |t| + \pi i)} dt$$

$$= (e^{\pi i z} - e^{-\pi i z}) \int_{\varepsilon}^{n} e^{-t} t^{-z} dt + i \int_{-\pi}^{\pi} e^{\varepsilon e^{i\theta}} \cdot \varepsilon^{1-z} e^{(1-z)\theta i} d\theta.$$

$$(*_7)$$

To estimate the last integral in $(*_7)$, consider $\operatorname{Re} z < 1$ and then

$$\left| \int_{-\pi}^{\pi} e^{\varepsilon e^{i\theta}} \varepsilon^{1-z} e^{(1-z)\theta i} d\theta \right| \leq \int_{-\pi}^{\pi} \varepsilon^{1-\operatorname{Re} z} e^{\varepsilon \cos\theta + \operatorname{Im} z \cdot \theta} d\theta$$

$$\leq 2\pi \varepsilon^{1-\operatorname{Re} z} e^{\varepsilon + \pi |\operatorname{Im} z|} \to 0 \quad \text{as } \varepsilon \to 0.$$

Therefore, if $\operatorname{Re} z < 1$, namely $\operatorname{Re}(1-z) > 0$, let $\varepsilon \to 0$ in $(*_7)$ and we obtain

$$\int_{\gamma_n} e^t t^{-z} dt = 2i \sin \pi z \int_0^n e^{-t} t^{-z} dt,$$

where γ_n denotes the path $[-n, 0]$, oriented from $-n$ to 0 and then from 0 to $-n$. Letting $n \to \infty$ again, then

$$\int_{\gamma} e^t t^{-z} dt = 2i \sin \pi z \int_0^{\infty} e^{-t} t^{-z} dt$$

$$= 2i \sin \pi z \Gamma(1-z), \quad \operatorname{Re}(1-z) > 0, \qquad (*_8)$$

where γ denotes the path $(-\infty, 0]$, oriented both forward and backward.

What remains is to show that, for any $0 < \varepsilon_1 < \varepsilon_2$,

$$\int_{\gamma_{\varepsilon_1}} e^t t^{-z} dt = \int_{\gamma_{\varepsilon_2}} e^t t^{-z} dt \qquad (*_9)$$

holds and hence, this common function is the one represented by $\int_{\gamma} e^t t^{-z} dt$ in $(*_8)$. Let τ denote the contour composed of the circles $|z| = \varepsilon_1$, $|z| = \varepsilon_2$, and the segments $[-\varepsilon_2, -\varepsilon_1]$ oriented twice. See Fig. 5.16. By Cauchy's integral theorem,

$$\int_{\tau} e^t t^{-z} dt = \int_{\gamma_{\varepsilon_1}} e^t t^{-z} dt - \int_{\gamma_{\varepsilon_2}} e^t t^{-z} dt = 0.$$

and $(*_9)$ is thus followed.

Combining $(*_8)$ and $(*_9)$, we obtain, for any $\varepsilon > 0$,

$$\Gamma(1-z) = \frac{1}{2i \sin \pi z} \int_{\gamma_{\varepsilon}} e^t t^{-z} dt, \quad \operatorname{Re}(1-z) > 0$$

$$\Rightarrow \Gamma(z) = \frac{1}{2i \sin \pi z} \int_{\gamma_{\varepsilon}} e^t t^{z-1} dt, \quad \operatorname{Re}(z) > 0.$$

Finally, the recursive identity $\Gamma(z+1) = z\Gamma(z)$ permits us to extend the above integral to $\mathbf{C} - \{0, -1, -2, \ldots\}$.   □

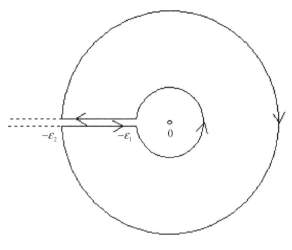

Fig. 5.16

Finally, the

Other two complex integral representations for $\Gamma(z)$.

(1) Let γ_ε be as in (5.6.1.7) (see Fig. 5.15). Then

$$\frac{1}{\Gamma(z+1)} = \frac{1}{2\pi i} \int_{\gamma_\varepsilon} \frac{e^t}{t^{z+1}} dt, \quad z \neq 0, -1, -2, \ldots.$$

(2) (Refer to Sec. 4.14.)

$$\frac{t^z}{\Gamma(z+1)} = \frac{1}{2\pi i} \int_{\alpha-i\infty}^{\alpha+i\infty} \frac{e^{t\zeta}}{\zeta^{z+1}} d\zeta, \quad -1 < \operatorname{Re} z < 0, \quad \alpha > 0, \quad t > 0.$$

$$(5.6.1.8)$$

Recall that we had used (2) in Example 2 of Sec. 4.14.

Proof. (1) The integral can be decomposed as

$$\frac{1}{2\pi i} \int_{-\infty}^{-\varepsilon} \frac{e^t}{t^{z+1}} dt \quad \text{(along the lower side of } (-\infty, -\varepsilon])$$

$$+ \frac{1}{2\pi i} \int_{|t|=\varepsilon} \frac{e^t}{t^{z+1}} dt + \frac{1}{2\pi i} \int_{-\varepsilon}^{-\infty} \frac{e^t}{t^{z+1}} dt$$

$$\text{(along the upper side of } (-\infty, -\varepsilon]). \tag{$*$10}$$

Note that, along the lower side of $(-\infty, 0]$, $t^{z+1} = e^{-(z+1)\pi i}|t|^{z+1}$; along the upper side, $t^{z+1} = e^{(z+1)\pi i}|t|^{z+1}$. Hence, in $(*_{10})$,

the first integral + the third integral

$$= \frac{1}{2\pi i}\int_{\varepsilon}^{\infty}\frac{e^{-t}}{e^{-(z+1)\pi i}t^{z+1}}dt - \frac{1}{2\pi i}\int_{\varepsilon}^{\infty}\frac{e^{-t}}{e^{(z+1)\pi i}t^{z+1}}dt$$

$$= \frac{1}{2\pi i}(e^{(z+1)\pi i} - e^{-(z+1)\pi i})\int_{\varepsilon}^{\infty}e^{-t}t^{-z-1}dt$$

$$= \frac{\sin(z+1)\pi}{\pi}\int_{\varepsilon}^{\infty}e^{-t}t^{-z-1}dt \to \frac{\sin(z+1)\pi}{\pi}\int_{0}^{\infty}e^{-t}t^{-z-1}dt$$

$$= \frac{-\sin z\pi}{\pi}\Gamma(-z) \quad \text{as } \varepsilon \to 0. \tag{$*_{11}$}$$

While, along $|t| = \varepsilon$ or $t = \varepsilon e^{i\theta}$, $0 \le \theta \le 2\pi$,

the second integral in $(*_{10})$

$$= \frac{1}{2\pi i}\int_{-\pi}^{\pi}\frac{e^{\varepsilon e^{i\theta}}}{e^{(z+1)(\log\varepsilon+i\theta)}}\varepsilon i e^{i\theta}d\theta = \frac{1}{2\pi}\int_{-\pi}^{\pi}\frac{e^{\varepsilon e^{i\theta}}\cdot e^{i\theta}}{\varepsilon^z e^{i(z+1)\theta}}d\theta$$

$$\to 0 \quad \text{as } \varepsilon \to 0, \text{ if } \operatorname{Re} z < 0.$$

Putting these results in $(*_{10})$, we have

$$\frac{1}{2\pi i}\int_{\gamma_\varepsilon}\frac{e^t}{t^{z+1}}dt = \frac{-\sin \pi z}{\pi}\Gamma(-z) = \frac{1}{\Gamma(z+1)}, \quad \text{if } \operatorname{Re} z < 0.$$

Note that integral in the left still converges on $\operatorname{Re} z \ge 0$. Hence, via $\Gamma(z+1) = z\Gamma(z)$, the above expression can be extended to $\mathbf{C} - \{0, -1, -2, \ldots\}$.

(2) Let γ be the path shown in Fig. 5.17. Then

$$\frac{1}{2\pi i}\int_{\gamma}\frac{e^{t\zeta}}{\zeta^{z+1}}d\zeta = 0 = \frac{1}{2\pi i}\left\{\int_{\alpha-iR}^{\alpha+iR} + \int_{C_R} + \int_{C_\varepsilon} + \int_{\gamma_{\varepsilon,R}}\right\}\frac{e^{t\zeta}}{\zeta^{z+1}}d\zeta. \tag{$*_{12}$}$$

The integral along $\gamma_{\varepsilon,R}$ can be evaluated as in $(*_{11})$ when $\operatorname{Re} z < 0$ is imposed:

$$\lim_{\substack{\varepsilon \to 0 \\ R \to \infty}}\frac{1}{2\pi i}\int_{\gamma_{\varepsilon,R}}\frac{e^{t\zeta}}{\zeta^{z+1}}d\zeta = \frac{\sin \pi z}{\pi}\int_{0}^{\infty}e^{-t\zeta}\zeta^{-z-1}d\zeta = \frac{\sin \pi z}{\pi}\cdot t^z\Gamma(-z)$$

$$= \frac{-t^z}{\Gamma(1+z)}, \quad t > 0;$$

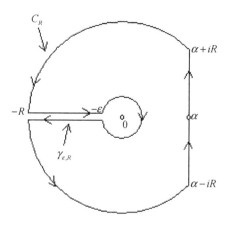

Fig. 5.17

if $-1 < \operatorname{Re} z$ is considered, then

$$\lim_{\varepsilon \to 0} \frac{1}{2\pi i} \int_{C_\varepsilon} \frac{e^{t\zeta}}{\zeta^{z+1}} d\zeta = 0 \quad \text{and} \quad \lim_{R \to \infty} \frac{1}{2\pi i} \int_{C_R} \frac{e^{t\zeta}}{\zeta^{z+1}} d\zeta = 0.$$

Substituting these results in $(*_{12})$, we obtain the claimed integral.   □

Exercises A

(1) Try to use (5.6.1.3) to derive (1) in (5.6.1.4), following the steps:

    (a) Use the definition of Euler's constant γ to show that

$$e^{-\gamma z} = \lim_{n \to \infty} \prod_{k=1}^{n} \left(1 + \frac{1}{k}\right)^{z} e^{-\frac{z}{k}}.$$

    (b) Then, rewrite $\Gamma(z)$ as

$$z\Gamma(z) = \lim_{n \to \infty} \prod_{k=1}^{n} \left(1 + \frac{1}{k}\right)^{z} \left(1 + \frac{z}{k}\right)^{-1} = \lim_{n \to \infty} \frac{n!(n+1)^{z}}{\prod_{k=1}^{n}(z+k)}.$$

(2) Adopt the following steps to show that $\Gamma(z) = \int_0^\infty t^{z-1} e^{-t} dt$ is analytic in $\operatorname{Re} z > 0$.

    (a) For any $0 < a < b < \infty$, show that $\int_a^b e^{-t} t^{z-1} dt$ is entire.

    (b) Rewrite $\Gamma(z) = I_1(z) + I_2(z)$ as in $(*_2)$. Show that

$$I_1(z) = \int_0^1 e^{-t} t^{z-1} dt$$

$$= \sum_{1}^{\infty} u_n(z) \quad \text{absolutely and locally uniformly in } \operatorname{Re} z > 0,$$

where $u_n(z) = \int_{1/(n+1)}^{1/n} e^{-t} t^{z-1} dt$, $n \geq 1$;

$$I_2(z) = \int_1^{\infty} e^{-t} t^{z-1} dt$$

$$= \sum_{n=1}^{\infty} v_n(z) \quad \text{absolutely and locally uniformly in } \mathbf{C},$$

where $v_n(z) = \int_n^{n+1} e^{-t} t^{z-1} dt$, $n \geq 1$.

(c) Then, use Weierstrass's theorem (5.3.1.1).

(3) Use the following steps to define $\Gamma(z)$, $z \neq 0, -1, -2, \ldots$.

  (a) Use Morera's theorem (3.4.2.13) or Integrals of Cauchy type (4.7.1) and Exercise B(2) of Sec. 5.3.1 to show that $\Gamma(z) = \int_0^{\infty} e^{-t} t^{z-1} dt$ is analytic in $\operatorname{Re} z > 0$.

  (b) Show that $\Gamma(z+1) = z\Gamma(z)$ holds in $\operatorname{Re} z > 0$.

  (c) Use $\Gamma(z) = \frac{\Gamma(z+1)}{z}$ to extend $\Gamma(z)$ from $\operatorname{Re} z > 0$ to $\mathbf{C} - \{0, -1, -2, \ldots\}$.

(4) Use $\sin \pi z = \pi z \prod_{-\infty}^{\infty} \left(1 - \frac{z}{n}\right) e^{\frac{z}{n}}$ to define $\Gamma(z)$ in the following steps.

  (a) Note that $\sin \pi z$ is the simplest entire function having all the integers as its only simple zeros. Then, the simplest entire function having all negative integers as its only simples zeros is

$$G(z) = \prod_{n=1}^{\infty} \left(1 + \frac{z}{n}\right) e^{-\frac{z}{n}}.$$

  What is $G(-z)$? Does $zG(z)G(-z) = \frac{\sin \pi z}{\pi}$ hold?

  (b) Show that $G(z-1) = ze^{\gamma(z)}G(z)$ for some entire function $\gamma(z)$. Try to show that $\gamma'(z) \equiv 0$. Hence $\gamma(z) = \gamma$, a constant, and $G(z-1) = ze^{\gamma}G(z)$, $z \in \mathbf{C}$.

  (c) Choose $z = 1$. Show that $\gamma = \prod_{n=1}^{\infty} \left(1 + \frac{1}{n}\right) e^{-\frac{1}{n}} = \lim_{n \to \infty} \left(\sum_{k=1}^{n} \frac{1}{k} - \log n\right)$.

  (d) Let $H(z) = G(z)e^{\gamma z}$. Then $H(z-1) = zH(z)$. Define the *gamma function* by

$$\Gamma(z) = \frac{1}{zH(z)} = e^{-\gamma z} \frac{1}{z} \prod_{n=1}^{\infty} \left(1 + \frac{z}{n}\right)^{-1} e^{\frac{z}{n}}.$$

  Then $\Gamma(z+1) = z\Gamma(z)$ and

$$\Gamma(z)\Gamma(1-z) = \frac{\pi}{\sin \pi z}, \quad z \neq 0, \pm 1, \pm 2, \ldots.$$

(5) Try to use $\Gamma(z)$ to express the following products.

    (a) $\prod_{n=1}^{\infty} \left(1 - \frac{1}{2n}\right) e^{\frac{1}{2n}}$.

    (b) $\prod_{n=1}^{\infty} \left(1 + \frac{z}{2n-1}\right) \left(1 - \frac{z}{2n}\right)$.

    (c) $\prod_{n=1}^{\infty} \left(1 + \frac{(-1)^n}{2n}\right)$.

(6) In $\operatorname{Re} z > 0$, show that

$$\Gamma(z) = 2 \int_0^\infty e^{-t^2} t^{2z-1} dt = \int_0^1 \left(\log \frac{1}{t}\right)^{z-1} dt.$$

(7) Integrate $e^{-\zeta} \zeta^{z-1}$ along the path shown in Fig. 5.18 and show that

$$\int_0^\infty e^{-it} t^{z-1} dt = \Gamma(z) e^{-\frac{\pi i}{2} z}, \quad 0 < \operatorname{Re} z < 1.$$

(8) Integrate $e^{-\zeta} \zeta^{z-1} = e^{-\zeta + (z-1)\operatorname{Log} \zeta}$, $\operatorname{Log} \zeta = \log |\zeta| + i \operatorname{Arg} \zeta$ and $0 \le \operatorname{Arg} \zeta < 2\pi$, along the path γ_ε shown in Fig. 5.19. Show that, if $z \ne 0, -1, -2, \ldots$,

$$\Gamma(z) = \frac{1}{e^{2\pi i z} - 1} \int_{\gamma_\varepsilon} e^{-t} t^{z-1} dt;$$

$$\frac{1}{\Gamma(z)} = -\frac{e^{\pi i z}}{2\pi i} \int_{\gamma_\varepsilon} e^{-t} t^{-z} dt.$$

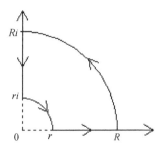

Fig. 5.18

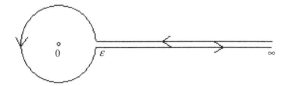

Fig. 5.19

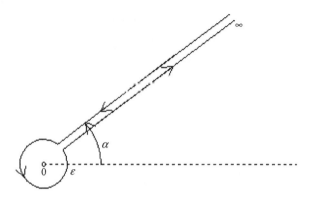

Fig. 5.20

Alternatively, designate $\operatorname{Log} \zeta = \log |\zeta| + i \operatorname{Arg} \zeta$ and $\alpha \leq \operatorname{Arg} \zeta < 2\pi + \alpha$ and integrate $e^{-\zeta} \zeta^{z-1} = e^{-\zeta + (z-1) \operatorname{Log} \zeta}$ along the path γ_ϵ shown in Fig. 5.20. What happens to these two integral expressions?

(9) Adopt Euler's integral $\Gamma(z) = \int_0^\infty e^{-t} t^{z-1} dt$ to prove Euler's functional equation (5.6.1.6). Fill up the gaps in the following process:

$$\Gamma(z)\Gamma(1-z)$$

$$= \int_0^\infty \int_0^\infty e^{-(t+s)} t^{z-1} s^{-z} \, dt \, ds, \quad 0 < \operatorname{Re} z < 1$$

$$= \int_0^\infty \int_0^\infty e^{-x} y^{z-1} \frac{dx \, dy}{1+y} \quad \left(\text{setting } t+s = x \quad \text{and} \quad \frac{t}{s} = y \right)$$

$$= \int_0^\infty e^{-x} dx \cdot \int_0^\infty \frac{y^{z-1}}{1+y} dy.$$

It is advised to use the residue method to justify that $\int_0^\infty \frac{y^{z-1}}{1+y} dy = \frac{\pi}{\sin \pi z}$, $0 < \operatorname{Re} z < 1$.

5.6.2   Basic and characteristic properties

The most basic ones among all, in addition to Euler's equation (5.6.1.6), are the following

Basic properties.

(1) $\Gamma(1) = 1$.

(2) $\Gamma\left(\frac{1}{2}\right) = \int_0^\infty e^{-t} t^{-\frac{1}{2}} dt = 2 \int_0^\infty e^{-x^2} dx = \sqrt{\pi}$.

(3) $\Gamma(z+1) = z\Gamma(z)$, $z \neq 0, -1, -2, \ldots$. Hence,

   1. $\Gamma(z+n) = z(z+1)\cdots(z+n-1)\Gamma(z)$ if $n > 1$ is an integer; in particular,

$$\Gamma\left(n+\frac{1}{2}\right) = \frac{(2n)!}{4^n n!}\sqrt{\pi}, \quad n = 0, 1, 2, \ldots.$$

   2. $\Gamma(n+1) = n!$.

   3. $\lim_{z \to -n} \Gamma(z) = \infty$ and

$$\mathrm{Re}\, s(\Gamma(z), -n) = \frac{(-1)^n}{n!} \quad \text{for } n \geq 0.$$

   4. $\lim_{n \to \infty} \frac{\Gamma(z+n)}{n^z \Gamma(n)} = 1$, $z \neq -n$ for $n = 1, 2, 3, \ldots$.

(4) *Legendre's product formula:*

$$\sqrt{\pi}\Gamma(2z) = 2^{2z-1}\Gamma(z)\Gamma\left(z+\frac{1}{2}\right).$$

(5) *Gauss's product formula:*

$$(2\pi)^{\frac{n-1}{2}}\Gamma(z) = n^{z-\frac{1}{2}}\Gamma\left(\frac{z}{n}\right)\Gamma\left(\frac{z+1}{n}\right)\cdots\Gamma\left(\frac{z+n-1}{n}\right). \quad (5.6.2.1)$$

Proof. Only (3)4, (4), and (5) will be proved.

About (3)4: By Gauss's representation for $\Gamma(z)$,

$$\lim_{n \to \infty} \frac{\Gamma(z+n)}{n^z \Gamma(n)} = \lim_{n \to \infty} \frac{z(z+1)\cdots(z+n-1)}{n^z(n-1)!} \cdot \frac{n! n^z}{z(z+1)\cdots(z+n)}$$

$$= \lim_{n \to \infty} \frac{n}{z+n} = 1.$$

About (4): Adopt Weierstrass's representation for $\Gamma(z)$. Differentiate $\mathrm{Log}\,\Gamma(z)$ logarithmically and we obtain

$$\frac{\Gamma'(z)}{\Gamma(z)} = -\gamma - \frac{1}{z} - \sum_{n=1}^{\infty}\left(\frac{1}{z+n} - \frac{1}{n}\right), \quad (5.6.2.2)$$

absolutely and locally uniformly in $\mathbf{C} - \{0, -1, -2, \ldots\}$. This function is called *Gauss's ψ-function* or *digamma function*. Differentiate (5.6.2.2) again to obtain

$$\frac{d}{dz}\left(\frac{\Gamma'(z)}{\Gamma(z)}\right) = \sum_{n=0}^{\infty} \frac{1}{(z+n)^2}, \quad z \neq 0, -1, -2, \ldots. \quad (*_1)$$

Now, $\Gamma(z)\Gamma(z+\frac{1}{2})$ and $\Gamma(2z)$ obviously have the same set of poles and corresponding residues. Hence, there is an entire function $g(z)$ such that $\Gamma(z)\Gamma\left(z+\frac{1}{2}\right) = e^{g(z)}\Gamma(2z)$ holds. By using $(*_1)$,

$$\frac{d}{dz}\frac{\Gamma'(z)}{\Gamma(z)} + \frac{d}{dz}\frac{\Gamma'\left(z+\frac{1}{2}\right)}{\Gamma\left(z+\frac{1}{2}\right)} = \sum_{n=0}^{\infty}\frac{1}{(z+n)^2} + \sum_{n=0}^{\infty}\frac{1}{\left(z+n+\frac{1}{2}\right)^2}$$

$$= 4\sum_{n=0}^{\infty}\frac{1}{(2z+n)^2} = 2\frac{d}{dz}\frac{\Gamma'(2z)}{\Gamma(2z)}$$

$\Rightarrow g''(z) \equiv 0$

$\Rightarrow g(z) = az + b$ for some constants a and b and hence,

$$\Gamma(z)\Gamma\left(z+\frac{1}{2}\right) = e^{az+b}\Gamma(z).$$

Setting $z = 1$ and $\frac{1}{2}$, respectively, and using $\Gamma(\frac{1}{2}) = \sqrt{\pi}$, $\Gamma(1) = 1$, $\Gamma(\frac{3}{2}) = \frac{1}{2}\sqrt{\pi}$, and $\Gamma(2) = 1$, we find that $a = -2\log 2$ and $b = \frac{1}{2}\log \pi + \log 2$. The result follows.

One can also prove (4) by setting nz for z in (5) (see (5.6.2.1)) and then letting $n = 2$. For other proofs, see Exercises A(2) and A(3).

Yet for another proof, we use Euler's integral for $\Gamma(z)$ as our starting point. Replacing t by x^2, then

$$\Gamma(z) = 2\int_0^{\infty} e^{-x^2}x^{2z-1}dx, \quad \text{Re } z > 0$$

$$\Rightarrow \Gamma\left(z+\frac{1}{2}\right) = 2\int_0^{\infty} e^{-y^2}y^{2z}dy, \quad \text{Re } z > 0$$

$\Rightarrow$ (multiplying both side by side, and then the factor 2^{2z-1})

$$2^{2z-1}\Gamma(z)\Gamma\left(z+\frac{1}{2}\right) = 4\int_0^{\infty}\int_0^{\infty} e^{-(x^2+y^2)}(2xy)^{2z-1}ydxdy;$$

$$2^{2z-1}\Gamma(z)\Gamma\left(z+\frac{1}{2}\right) = 4\int_0^{\infty}\int_0^{\infty} e^{-(x^2+y^2)}(2xy)^{2z-1}xdxdy$$

(interchanging the role of x and y in the first integral)

$\Rightarrow$ (adding up) $2^{2z-1}\Gamma(z)\Gamma\left(z+\frac{1}{2}\right)$

$$= 4\iint_{\Omega} e^{-(x^2+y^2)}(2xy)^{2z-1}(x+y)dxdy,$$

where $\Omega = \{(x,y)|0 \le x < \infty, 0 \le y \le x\}$. Invoke the change of variables: $u = x^2 + y^2$ and $v = 2xy$ with the Jacobian determinant $(4(x^2 - y^2))^{-1}$. Note that $u - v = (x - y)^2 \ge 0$. Hence

$$2^{2z-1}\Gamma(z)\Gamma\left(z + \frac{1}{2}\right) = \int_0^\infty v^{2z-1}dv \int_0^\infty \frac{e^{-u}}{\sqrt{u-v}}du$$

$$= 2\int_0^\infty e^{-v}v^{2z-1}dv \int_0^\infty e^{-t^2}dt \quad (\text{setting } u = v + t^2)$$

$$= \sqrt{\pi}\Gamma(2z).$$

About (5): Use nz to replace z in (5). We just only need to prove

$$n^{\frac{1}{2}-nz}(2\pi)^{\frac{n-1}{2}}\Gamma(nz) = \prod_{k=0}^{n-1}\Gamma\left(z + \frac{k}{n}\right).$$

To see this, set

$$f(z) = \prod_{k=0}^{n-1}\Gamma\left(z + \frac{k}{n}\right).$$

This $f(z)$ has the same simple poles at $-m-\frac{k}{n}$, $m \ge 0$, and $0 \le k \le n-1$, as the function $\Gamma(nz)$ does. Therefore, there is an entire function $g(z)$ so that

$\quad f(z) = e^{g(z)}\Gamma(nz)$

$\Rightarrow$ (by definition (5.6.1.3) for $\Gamma(z)$ and the process similar to the proof of (4)) $g(z)$ is a polynomial $az + b$ of degree ≤ 1 and hence

$$f(z) = ce^{az}\Gamma(nz), \quad c = e^b. \tag{$*_2$}$$

$\Rightarrow$ (replacing z by $z + \frac{1}{n}$) $f(z + \frac{1}{n}) = f(z)\Gamma(z+1) = ce^{az+\frac{a}{n}}\Gamma(nz+1)$

$$\tag{$*_3$}$$

$\Rightarrow$ (using $\Gamma(z+1) = z\Gamma(z)$) $f(z) = ce^{az+\frac{a}{n}} \cdot n\Gamma(nz)$

$\Rightarrow$ (comparing with $(*_2)$) $ne^{\frac{a}{n}} = 1$ and hence, $e^a = n^{-n}$.

Setting $z = 0$ in $(*_3)$, we obtain

$$\prod_{k=1}^{n-1}\Gamma\left(\frac{k}{n}\right) = ce^{\frac{a}{n}} = \frac{c}{n}. \tag{$*_4$}$$

To compute c in $(*_4)$: Observe that

$$\left\{\prod_{k=1}^{n-1}\Gamma\left(\frac{k}{n}\right)\right\}^2 = \prod_{k-1}^{n-1}\left\{\Gamma\left(\frac{k}{n}\right)\Gamma\left(1-\frac{k}{n}\right)\right\} = \prod_{l_0=1}^{n-1}\frac{\pi}{\sin k\pi/n}$$

$$= \frac{\pi^{n-1}}{\left(\prod_{k=1}^{n-1}\sin k\pi/n\right)} = \frac{\pi^{n-1}}{n/2^{n-1}} = \frac{(2\pi)^{n-1}}{n},$$

where we have used the identity $\prod_{k=1}^{n-1}\sin\frac{k\pi}{n} = \frac{n}{2^{n-1}}$ (see Exercise A(12) of Sec. 1.5). It follows that

$$c = n\cdot\prod_{k=1}^{n-1}\Gamma\left(\frac{k}{n}\right) = n\cdot\frac{(2\pi)^{(n-1)/2}}{\sqrt{n}} = \sqrt{n}(2\pi)^{n-1/2}.$$

Substituting these values of c and e^a into $(*_2)$, the result follows. $\square$

For another proof, see Exercise A(4).

(5.6.1.5) indicates that, no matter Weierstrass's infinite product representation or Euler's integral representation being adopted, the gamma function $\Gamma(z)$ can always be represented in Gauss's form and hence, enjoys the property that $\lim_{n\to\infty}\frac{\Gamma(z+n)}{n^z\Gamma(n)} = 1$.

As a matter of fact, we have the following

Characteristic properties of $\Gamma(z)$. There exists a unique analytic function $F(z)$ in $\mathbf{C} - \{0, -1, -2, \ldots\}$, satisfying:

1. $F(1) = 1$;
2. $F(z+1) = zF(z)$, and
3. $\lim_{n\to\infty}\frac{F(z+n)}{n^z F(n)} = 1$.

Indeed, $F(z) = \Gamma(z)$, holds for $z \neq 0, -1, -2, \ldots$ (5.6.2.3)

This is easily seen as follows: Note that $F(z+n)=z(z+1)\cdots (z+n-1)F(z)$, for integers $n \geq 1$ and $z \neq 0, -1, \ldots, -n$, and thus $F(n+1) = n!$, $n \geq 0$. Therefore we have

$$1 = \lim_{n\to\infty}\frac{F(z+n)}{n^z F(n)} = \lim_{n\to\infty}\left\{\frac{z(z+1)\cdots(z+n-1)}{n^z(n-1)!}\right\}F(z) = \frac{F(z)}{\Gamma(z)}$$

for any $z \neq 0, -1, -2, \ldots$.

As two important applications of the gamma function, we will *sketch* the Beta and hypergeometric functions in the following two sections, respectively.

Section (1) The β-function

The function defined by

$$B(z,w) = \int_0^1 t^{z-1}(1-t)^{w-1}dt \quad (t \in \mathbf{R}) \tag{5.6.2.4}$$

is called the *Beta function* of two complex variables z and w, commonly knows as *Euler's second integral*. Since the integral converges absolutely and locally uniformly if $\mathrm{Re}\,z > 0$ and $\mathrm{Re}\,w > 0$, it follows that $B(z,w)$ *is an analytic function on* $\mathrm{Re}\,z > 0$ *and on* $\mathrm{Re}\,w > 0$, *respectively.* This can be justified by using (4.7.4) and Exercise B(2) of Sec. 5.3.1 (refer to $(*_2)$, $(*_3)$ of Sec. 5.6.1 and Exercise A(2) there).

Moreover, we have following

Integral representation of $B(z,w)$. If $\mathrm{Re}\,z > 0$ and $\mathrm{Re}\,w > 0$, then

$$B(z,w) = \frac{\Gamma(z)\Gamma(w)}{\Gamma(z+w)}$$

$$= \int_0^\infty \frac{t^{z-1}}{(1+t)^{z+w}}dt$$

$$= 2\int_0^{\pi/2} (\sin\theta)^{2z-1}(\mathrm{con}\,\theta)^{2w-1}dw$$

$$= \frac{1}{(1-e^{2\pi iz})(1-e^{2\pi iw})}$$

$$\times \int_{\gamma_{0,1}} \zeta^{z-1}(1-\zeta)^{w-1}d\zeta \quad (\text{due to L. Pochhammer (1890)}).$$

The first formula can be used to define $B(z,w)$ in $\mathbf{C}^2 - \{(z,w)|z,w,z+w = 0, -1, -2, \ldots\}$.

In the fourth representation, $\zeta^{p-1}(1-\zeta)^{q-1}$, where $p = \mathrm{Re}\,z > 0$ and $q = \mathrm{Re}\,w > 0$, has a single-valued analytic branch in $\mathbf{C} - (-\infty, 0] - [1, \infty)$ (refer to Example 5 in Sec. 2.7.3) and $\gamma_{o,1}$ is the path (see Fig. 5.21), composed of four Jordan closed curves APA and AQA (both in counterclockwise direction), ARA and ASA (both in clockwise direction), where A is a fixed point in the interval $(0, 1)$. \hfill (5.6.2.5)

Proof. Use $t(1-t)^{-1}$ to replace t in (5.6.2.4) and one obtains the second representation; substitute $t = \sin^2\theta$ (or $\cos^2\theta$) and the third one follows.

As for the first formula: Set $F(z) = B(z,w)\Gamma(z+w)$, $\mathrm{Re}\,z > 0$. It is easy to check that $B(1,w) = \frac{1}{w}$, $w \neq 0$, and hence $F(1) = \Gamma(w)$ holds. In case

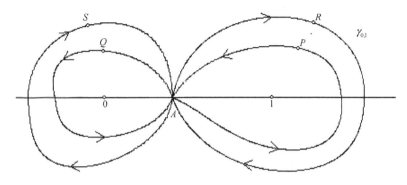

Fig. 5.21

$z \neq 0$ and $w \neq 0$, by using integration by parts (see (2.9.10)),

$$B(z+1, w) = \int_0^1 t^z (1-t)^{w-1} dt = -\frac{1}{w} t^z (1-t)^w \Big|_0^1 + \int_0^1 \frac{z}{w} t^{z-1} (1-t)^w dt$$

$$= \frac{z}{w} B(z, w+1) = \frac{z}{w} [B(z, w) - B(z+1, w)]$$

$$\Rightarrow B(z+1, w) = \frac{z}{z+w} B(z, w), \quad z+w \neq 0, \quad \operatorname{Re} z > -1, \quad \operatorname{Re} w > 0,$$

and $zw \neq 0$

$$\Rightarrow \text{(by induction)} \ B(z+n, w) = \frac{(z+n-1) \cdots z}{(z+w)^n} B(z, w), \quad z+w \neq 0,$$

$\operatorname{Re} z > -n, \quad \operatorname{Re} w > 0, \quad \text{and} \quad zw \neq 0.$

Consequently, $F(z+1) = B(z+1, w) \Gamma(z+1+w) = zF(z)$ and

$$\lim_{n \to \infty} \frac{F(z+n)}{n^z F(n)} = \lim_{n \to \infty} \frac{B(z+n, w) \Gamma(z+n+w)}{n^z B(n, w) \Gamma(n+w)} = 1.$$

Comparing to (5.6.2.3), $F(z) = \Gamma(z) \Gamma(w)$ does hold and hence the formula follows. For other proofs see Exercises A(11), A(12) and A(14).

As for the fourth one: The Jordan closed curve APA can be deformed continuously in $\mathbf{C} - (-\infty, 0]$ into a closed curve γ_P shown in Fig. 5.22. In this case, as the point ζ starts from a point t in $(0, 1)$ and winds around the small circle in the counterclockwise sense and back to the original point t, the values assumed by the function $f(z) = \zeta^p (1-\zeta)^{q-1}$ are changed from

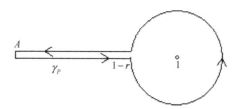

Fig. 5.22

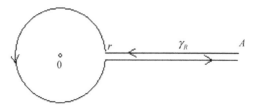

Fig. 5.23

$f(t)$ continuously to $f(t)e^{2\pi iq}$. Hence, by Cauchy's integral theorem,

$$\int_{APA} f(\zeta)d\zeta = \int_{\gamma_P} f(\zeta)d\zeta = \int_a^{1-\gamma} f(t)dt + \int_{C_\gamma} f(\zeta)d\zeta$$

$$+ e^{2\pi iq}\int_{1-\gamma}^a f(t)dt \quad \to \quad \int_a^1 f(t)dt + 0 + e^{2\pi iq}\int_1^a f(t)dt$$

$$= (1 - e^{2\pi iq})\int_a^1 f(t)dt \quad \text{as } r \to 0,$$

where $a + 0i = a$ is the point A. Similarly, when AQA is deformed in $\mathbf{C} - [1, \infty)$ into a closed curve γ_Q shown in Fig. 5.23, the values of $f(\zeta)$ are changed continuously from $f(t)e^{2\pi iq}$ to $f(t)e^{2\pi iq} \cdot e^{2\pi ip}$. Hence

$$\int_{AQA} f(\zeta)d\zeta = \int_{\gamma_Q} f(\zeta)d\zeta \to -e^{2\pi iq}\int_0^a f(t)dt$$

$$+ e^{2\pi i(p+q)}\int_0^a f(t)dt \quad \text{as } r \to 0.$$

When the same process is applying to ARA and ASA, we obtain, as $r \to 0$,

$$\int_{ARA} f(\zeta)d\zeta \to e^{2\pi i(p+q)}\int_a^1 f(t)dt - e^{2\pi i(p+q)} \cdot e^{-2\pi iq}\int_a^1 f(t)dt;$$

$$\int_{ASA} f(\zeta)d\zeta \to -e^{2\pi ip}\int_0^a f(t)dt + e^{2\pi ip} \cdot e^{-2\pi ip}\int_0^a f(t)dt.$$

Combining these results together, we have, as $r \to 0$,

$$\int_{\gamma_{0,1}} \zeta^p (1 - \zeta)^{q-1} d\zeta$$

$$= (1 - e^{2\pi iq} + e^{2\pi i(p+q)} - e^{2\pi ip}) \int_0^1 t^p (1 - t)^{q-1} dt$$

$$= (1 - e^{2\pi ip})(1 - e^{2\pi iq}) B(\operatorname{Re} z, \operatorname{Re} w), \quad p > 0, \quad q > 0. \tag{$*_6$}$$

Replace p by z and q by w, respectively, in $(*_6)$. The result follows because of the interior uniqueness theorem. $\qquad\square$

Section (2) The hypergeometric function

Let a, b, and c be complex constants, all not equal to $0, -1, -2, \ldots$. The function defined by *the hypergeometric series*

$$F(a, b, c, z) = 1 + \sum_{n=1}^{\infty} \frac{a(a+1)\cdots(a+n-1)b(b+1)\cdots(b+n-1)}{n!c(c+1)\cdots(c+n-1)} z^n$$

$$= \frac{\Gamma(c)}{\Gamma(a)\Gamma(b)} \sum_{n=0}^{\infty} \frac{\Gamma(a+n)\Gamma(b+n)}{n!\Gamma(c+n)} z^n \tag{5.6.2.6}$$

is called *Gauss's hypergeometric function*. In case a or b or both is permitted to assume the value 0 or a negative integer, $F(a, b, c, z)$ will then reduce to a polynomial. We will not discuss these cases unless specified. Note the following

Convergence and divergence of the hypergeometric series. Its radius of convergence is equal to 1. Hence,

(1) The series converges absolutely and locally uniformly in $|z| < 1$ to the analytic function $F(a, b, c, z)$.

(2) If $\operatorname{Re}(c - a - b) > 0$, the series converges absolutely and uniformly on $|z| = 1$.

(3) In case a, b, and c are reals,

    1. if $a + b - 1 < c \le a + b$, the series diverges at $z = 1$ and converges conditionally elsewhere on $|z| = 1$;

    2. if $c \le a + b - 1$, the series diverges everywhere on $z = 1$.   (5.6.2.7)

The proof is left to the readers (see also Exercise B(1) of Sec. 3.3.2).

    Many elementary functions are special cases of $F(a, b, c, z)$. Take, for instance:

$$F(a, b, b, z) = (1 - z)^{-a};$$

$$zF(1, 1, 2, -z) = \operatorname{Log}(1 + z);$$

$$zF\left(\frac{1}{2},\frac{1}{2},\frac{3}{2},z^2\right) = \text{Arc}\sin z = \int_0^z \frac{dt}{\sqrt{1-t^2}} = z\sum_{n=0}^{\infty} C_n^{2n} \frac{1}{2n+1}\left(\frac{z}{2}\right)^{2n};$$

$$zF\left(1,\frac{1}{2},\frac{3}{2},-z^2\right) = \text{Arc}\tan z = \int_0^z \frac{dt}{1+t^2} = \sum_{n=0}^{\infty} \frac{(-1)^n}{2n+1}z^{2n+1};$$

$$\lim_{b\to\infty} F(a,b,a,\frac{z}{b}) = e^z;$$

$$\frac{\pi}{2}F\left(\frac{1}{2},\frac{1}{2},1,z\right) = \int_0^{\pi/2} \frac{d\theta}{\sqrt{1-z\sin^2\theta}} \quad \text{(complete elliptic integral)};$$

$$\frac{\pi}{2}F\left(\frac{1}{2},-\frac{1}{2},1,z\right) = \int_0^{\pi/2} \sqrt{1-z\sin^2\theta}\,d\theta \quad \text{(complete elliptic integral)},$$

$$\text{etc.} \tag{5.6.2.8}$$

Here are some

Elementary properties of $F(a,b,c,z)$.

(1) $F(a,b,c,z) = F(b,a,c,z)$.

(2) $(a-b)F(a,b,c,z) = aF(a+1,b,c,z) - bF(a,b+1,c,z)$.

(3) $F(a,b,c,z) = F(a,a+b,c+1,z) - \frac{a(c-b)}{c(c+1)}zF(a+1,b+1,c+2,z)$.

(4) $\frac{d^m}{dz^m}F(a,b,c,z) = \frac{(a)_m(b)_m}{(c)_m}F(a+m,b+m,c+m,z)$, $|z| < 1$, for $m \geq 1$,
where $(a)_m = a(a+1)\cdots(a+m-1)$, etc.

(5)   (Ref. [1], pp. 316–317) $F(a,b,c,z)$ is a solution of *the hypergeometric differential equation*

$$(z - z^2)w'' + [c - (a+b+1)z]w' - abw = 0$$

(see also Exercise B(1) of Sec. 3.3.2).

(6)   (Ref. [1], p. 318) If $\operatorname{Re} c > \operatorname{Re} b > 0$, then

$$F(a,b,c,z) = \frac{\Gamma(c)}{\Gamma(b)\Gamma(c-b)}\int_0^1 t^{b-1}(1-t)^{c-b-1}(1-tz)^{-a}dt, \quad |z| < 1.$$

Also the integral in the right converges to an analytic function absolutely and locally uniformly in $\mathbf{C} - [1,\infty)$, so it extends $F(a,b,c,z)$ analytically from $|z| < 1$ to $\mathbf{C} - [1,\infty)$.

(7)   Suppose $c \neq 0, -1, -2, \ldots$ (and a and b may assume the value 0 or negative integers) and $\operatorname{Re}(c-a-b) > 0$. Then $F(a,b,c,1)$ is analytic in a or b or c, respectively.

(8) In case $\mathrm{Re}\,(c - a - b) > 0$, then

$$F(a, b, c, 1) = \frac{\Gamma(c)\Gamma(c - a - b)}{\Gamma(c - a)\Gamma(c - b)}. \tag{5.6.2.9}$$

Proof. (1)–(4) are left as Exercise A(18).

(5): For simplicity, set $D = z\frac{d}{dz}$, a differential operator, and $w = F(a, b, c, z)$. Since $(D + k)z^n = z \cdot nz^{n-1} + kz^n = (n + k)z^n$ for constants k, direct computation shows that

$$D(D + c - 1)w = \sum_{n=0}^{\infty} \frac{(a)_n (b)_n}{n!(c)_n} D(D - c + 1)z^n$$

$$= \sum_{n=1}^{\infty} \frac{(a)_n (b)_n}{(n - 1)!(c)_{n-1}} z^n$$

$$= z\sum_{n=0}^{\infty} \frac{(a + n)(a)_n \cdot (b + n)(b)_n}{n!(c)_n} z^n$$

$$= z(D + a)(D + b)w$$

$$\Rightarrow [D(D + c - 1) - z(D + a)(D + b)]w = 0.$$

The result follows by noting that $Dw = zw'$, $D(D - 1)w = z^2 w''$.

(6): Using $\Gamma(b + n) = (b)_n \Gamma(b)$ and (5.6.2.5), then

$$\frac{(b)_n}{(c)_n} = \frac{\Gamma(b + n)\Gamma(c)}{\Gamma(b)\Gamma(c + n)} = \frac{\Gamma(c)}{\Gamma(b)\Gamma(c - b)} \cdot \frac{\Gamma(b + n)\Gamma(c - b)}{\Gamma(c + n)}$$

$$= \frac{\Gamma(c)}{\Gamma(n)\Gamma(c - b)} B(b + n, c - b)$$

$$= \frac{\Gamma(c)}{\Gamma(b)\Gamma(c - b)} \int_0^1 t^{b+n-1}(1 - t)^{c-b-1}dt,$$

$$\mathrm{Re}\,c > \mathrm{Re}\,b > 0 \quad \text{for } n \geq 0.$$

$$\Rightarrow F(a, b, c, z) = \frac{\Gamma(c)}{\Gamma(b)\Gamma(c - b)} \sum_{0}^{\infty} \frac{(a)_n}{n!} z^n \cdot \int_0^1 t^{b+n-1}(1 - t)^{c-b-1}dt$$

$$= \frac{\Gamma(c)}{\Gamma(b)\Gamma(c - b)} \int_0^1 t^{b-1}(1 - t)^{c-b-1} \left(\sum_{n=0}^{\infty} \frac{(a)_n}{n!}(zt)^n \right) dt$$

(see the reason below)

$$= \frac{\Gamma(c)}{\Gamma(b)\Gamma(c - b)} \int_0^1 t^{b-1}(1 - t)^{c-b-1}(1 - tz)^{-a}dt,$$

$$|z| < 1. \tag{$*_7$}$$

The interchange of $\sum_{n=0}^{\infty}$ and $\int_0^1$ is valid because the series $\sum_0^{\infty}\frac{(a_n)}{n!}$ $(zt)^n$, $|z| < 1$, converges uniformly to $(1 - tz)^{-a}$ in the t-interval $[0,1]$.

The right integral in $(^*7)$ converges, indeed, locally uniformly in $\mathbf{C} - [1, \infty)$. To see this, choose $(1 - tz)^{-a} = e^{-a\mathrm{Log}(1-tz)}$, the principal branch in $\mathbf{C} - [1, \infty)$. Fix any small $r > 0$, $\varepsilon > 0$, and large $R > 0$. On the compact set $K = \{z | r \le |z - 1| \le R$ and $\varepsilon \le \mathrm{Arg}(z - 1) \le 2\pi - \varepsilon\}$, we have the estimate

$$\left| \int_0^1 t^{b-1}(1 - t)^{c-b-1}(1 - tz)^{-a}dt \right|$$

$$\le \int_0^1 |t^{b-1}(1 - t)^{c-b-1}(1 - tz)^{-a}|dt$$

$$\le \max_{\substack{t \in [0,1] \\ z \in K}} |(1 - tz)^{-a}| \int_0^1 t^{\mathrm{Re}\,b-1}(1 - t)^{\mathrm{Re}\,c-\mathrm{Re}\,b-1}dt < \infty,$$

if $\mathrm{Re}\,c > \mathrm{Re}\,b > 0$.

This shows that the right integral in $(^*7)$ converges uniformly in the compact set K.

(7): Fix $\delta > 0$ so that $\mathrm{Re}\,(c - a - b) \ge 2\delta$ holds. Using Gauss's representation for $\Gamma(z)$,

$$\frac{(a)_n(b)_n}{n!(c)_n} \bigg/ \frac{1}{n^{1+z}}$$

$$= \frac{(a)_{n+1}}{n!n^a} \cdot \frac{(b)_{n+1}}{n!n^b} \cdot \frac{n!n^c}{(c)_{n+1}} \cdot \frac{n(c+n)}{(a+n)(b+n)} \cdot \frac{1}{n^{c-a-b-\delta}}$$

$$\to \frac{\Gamma(c)}{\Gamma(a)\Gamma(b)} \cdot 1 \cdot 0 = 0 \quad \text{as } n \to \infty.$$

Comparison test says that the series $F(a, b, c, 1) = 1 + \sum_{n=1}^{\infty} \frac{(a)_n(b)_n}{n!(c)_n}$ converges locally uniformly in $\mathrm{Re}\,(c - a - b) > 0$. Since each term in the series is an analytic function in a or b or c, the claim follows by Weierstrass's theorem (5.3.1.1).

(8) By (7) and Abel's limit theorem (5.1.3.1), if $z = x\,(0 < x < 1)$ is real, then

$$F(a, b, c, 1) = \lim_{x \to 1^-} F(a, b, c, x)$$

$$= \lim_{x \to 1^-} \frac{\Gamma(c)}{\Gamma(b)\Gamma(c - b)} \int_0^1 t^{b-1}(1 - t)^{c-b-1}(1 - tx)^{-a}dx,$$

under the additional assumption that $\operatorname{Re} c > \operatorname{Re} b > 0$
and then, using (6).

$$= \frac{\Gamma(c)}{\Gamma(b)\Gamma(c-a)} \int_0^1 t^{b-1}(1 \quad t)^{c-b-1} \cdot \lim_{x \to 1^-} (1 - tx)^{-a} dx$$

(see the reason below)

$$= \frac{\Gamma(a)}{\Gamma(b)\Gamma(c-b)} \int_0^1 t^{b-1}(1-t)^{c-b-a-1} dt$$

$$= \frac{\Gamma(c)\Gamma(c-b-a)}{\Gamma(b)\Gamma(c-b)}, \quad \text{if } \operatorname{Re}(c-a-b) > 0$$

and $\operatorname{Re} c > \operatorname{Re} b > 0.$ $\qquad (*8)$

To see that $\lim_{x \to 1^-}$ and $\int_0^1$ is interchangeable, all one needs to do is to show that the integral $\int_0^1$ converges uniformly in the x-interval $[0, 1]$. Since

$$|t^{b-1}(1-t)^{c-b-1}(1-tx)^{-a}|$$

$$\leq \begin{cases} t^{\operatorname{Re} b-1}(1-t)^{\operatorname{Re}(c-a-b)-1}, & \text{if } \operatorname{Re} a \geq 0 \\ t^{\operatorname{Re} b-1}(1-t)^{\operatorname{Re}(c-b)-1}, & \text{if } \operatorname{Re} a < 0 \end{cases}$$

$\Rightarrow$ (under the assumption that $\operatorname{Re}(c-a-b) > 0$ and $\operatorname{Re} c > \operatorname{Re} b > 0$)

$$\int_0^1 t^{\operatorname{Re} b}(1-t)^{\operatorname{Re}(c-a-b)-1} dt = B(\operatorname{Re} b, \operatorname{Re}(c-a-b)), \text{ and}$$

$$\int_0^1 t^{\operatorname{Re} b-1}(1-t)^{\operatorname{Re}(c-b)-1} dt = B(\operatorname{Re} b, \operatorname{Re}(c-b))$$

both exist. Therefore the claim follows.

Finally, by (7), we know that both sides of $(*8)$ are analytic in the domain $\operatorname{Re}(c-a-b) > 0$ and coincident in the subdomain $\operatorname{Re} c > \operatorname{Re} b > 0$. Hence, they coincide throughout $\operatorname{Re}(c-a-b) > 0$. $\qquad \square$

Exercises A

(1) Prove the following identities.

   (a) $\overline{\Gamma(z)} = \Gamma(\bar{z})$, namely, $\Gamma(z)$ is symmetric with respect to the real axis.

   (b) $\Gamma(z)\Gamma(-z) = \frac{\pi}{z \sin \pi z}$; consequently, $|\Gamma(iy)|^2 = \frac{\pi}{y \sinh \pi y}$ for $y = \operatorname{Im} z.$

(c) $\left| \Gamma \left(\frac{1}{2} + iy \right) \right|^2 = \frac{\pi}{\cosh \pi y}$ and hence, $\lim_{y \to \pm\infty} \Gamma \left(\frac{1}{2} + iy \right) = 0$.

(d) $\int_0^1 (1-t)^n t^{z-1} dt = \frac{n!}{z(z+1)\cdots(z+n)}$, $\mathrm{Re}\, z > 0$, for integer $n \geq 1$.

(e) $\prod_{n=1}^{\infty} \left(\frac{2n}{2n+1} \right) e^{\frac{1}{2n}} = \frac{\sqrt{\pi}}{2} e^{\frac{\gamma}{2}}$.

(f) $\Gamma \left(\frac{1}{2} + z \right) \Gamma \left(\frac{1}{2} - z \right) = \frac{\pi}{\cos \pi z}$.

(g) $(1-z)\left(1 + \frac{1}{2}z\right)\left(1 - \frac{1}{3}z\right)\left(1 + \frac{1}{4}z\right)\cdots = \frac{\sqrt{\pi}}{\Gamma\left(1 + \frac{1}{2}z\right)\Gamma\left(1 - \frac{1}{2}z\right)}$.

(h) $\Gamma \left(\frac{1}{6} \right) = 2^{-\frac{1}{3}} \left(\frac{3}{\pi} \right)^{\frac{1}{2}} \left[\Gamma \left(\frac{1}{3} \right) \right]^2$.

(2) Set $f(z) = \frac{2^{2z} \Gamma(z) \Gamma\left(z + \frac{1}{2}\right)}{\Gamma(2z)}$. Adopt the following steps to reprove Legendre's duplication formula.

   (a) $f(z)$ is an entire function satisfying $f(z+1) = f(z)$, $z \in \mathbf{C}$.

   (b) Use $\Gamma(z) = \sqrt{2\pi}(z-1)^{z-\frac{1}{2}} e^{-z+1} \left(1 + O\left(\frac{1}{z} \right) \right)$ for all large $|z|$ (see Example 3 in Sec. 4.14 or Sec. 5.6.3 below) to show that $f(z)$ is bounded in $|z| \geq 1$.

   (c) Use Liouville's theorem to show that $f(z) = \sqrt{2\pi}$ by noting that $\lim_{|z| \to \infty} f(z) = \sqrt{2\pi}$.

(3) Use (5.6.2.2), the series expansion of $\frac{\Gamma'(z)}{\Gamma(z)}$, to show that

$$\frac{2\Gamma'(2z)}{\Gamma(2z)} - \frac{\Gamma'(z)}{\Gamma(z)} - \frac{\Gamma'\left(z + \frac{1}{2}\right)}{\Gamma\left(z + \frac{1}{2}\right)} = 2 \log 2$$

to reprove Legendre's duplication formula. By the way, show that $\Gamma'\left(\frac{1}{2}\right) = -(\gamma + 2\log 2)\sqrt{\pi}$.

(4) Set

$$f(z) = \frac{n^{nz} \Gamma(z) \Gamma\left(z + \frac{1}{n}\right) \cdots \Gamma\left(z + \frac{n-1}{n}\right)}{n\Gamma(nz)}.$$

Imitate the method given in Exercise (2) to reprove Gauss's product formula. Let $z = \frac{1}{n}$ and show that

$$\Gamma \left(\frac{1}{n} \right) \Gamma \left(\frac{2}{n} \right) \cdots \Gamma \left(\frac{n-1}{n} \right) = \frac{(2\pi)^{(n-1)/2}}{\sqrt{\pi}}.$$

(5) *Basic properties of Gauss's ψ-function* (5.6.2.2).

   (a) $\psi(1) = -\gamma$ (hence, Euler's constant γ is the slope of the tangent to the curve $y = -\Gamma(x)$ at $x = 1$).

   (b) $\psi(z+1) = \frac{1}{z} + \psi(z)$ and hence, $\psi(z + n) = \sum_{k=0}^{n-1} \frac{1}{z+k} + \psi(z)$ for integer $n \geq 1$ and $\psi(n+1) = \sum_{k=1}^{n} \frac{1}{k} - \gamma$.

(c) $\psi(z) = \psi(1 - z) - \pi \cot \pi z$.

(d) $\lim_{n\to\infty}[\psi(z+n)-\psi((1+n)] = \lim_{n\to\infty}[\psi(z+n)-\text{Log}\,(\zeta+n)] = 0$
for $z, \zeta \neq 0, -1, -2, \ldots$.

(e) $\psi(z) = \text{Log}\, z + \sum_{n=0}^{\infty}\left[\frac{1}{z+n} - \text{Log}\left(1 + \frac{1}{z+n}\right)\right],\ z \neq 0, -1, -2, \ldots$.

(f) $\int_0^1 \psi(z + t)dt = \text{Log}\, z,\ z \in \mathbf{C} - (-\infty, 0]$.

(6) *Characteristic properties of ψ-function.* There exists a unique analytic function $G(z)$ in $\mathbf{C} - \{0, -1, -2, \ldots\}$ satisfying:

1. $G(1) = -\gamma$;

2. $G(z + 1) - G(z) = \frac{1}{z}$;

3. $\lim_{n\to\infty}[G(z + n) - G(1 + n)] = 0$.

Indeed, $G(z) = \psi(z)$ on $z \in \mathbf{C} - \{0, -1, -2, \ldots\}$.

(7) *Integral representation of $\psi(z)$.* If $\text{Re}\, z > 0$, then (refer to Exercise (8) if necessarily)

$$\psi(z) = \int_0^{\infty}\left(\frac{e^{-t}}{t} - \frac{e^{-zt}}{1 - e^{-t}}\right)dt$$

$$= -\gamma + \int_0^{\infty}\frac{e^{-t} - e^{-zt}}{1 - e^{-t}}dt.$$

Try to use Exercise (6). By which, one needs to show that

$$\int_0^{\infty}\left(\frac{e^{-t}}{t} - \frac{e^{-t}}{1 - e^{-t}}\right)dt = \int_0^{\infty}\frac{e^{-t}(1 - t - e^{-t})}{t(1 - e^{-t})}dt = -\gamma.$$

(8) *Integral representation of $\text{Log}\,\Gamma(z)$.* If $\text{Re}\, z > 0$, then

$$\text{Log}\,\Gamma(z) = \int_0^{\infty}\left\{(z - 1)e^{-t} - \frac{e^{-t} - e^{-zt}}{1 - e^{-t}}\right\}\frac{dt}{t}.$$

Try the following steps: Set $\Gamma_n(z) = \frac{n!n^{z-1}}{z(z+1)\cdots(z+n-1)},\ n \geq 1$. By Gauss's representation, then $\Gamma(z) = \lim_{n\to\infty}\Gamma_n(z)$.

(a) $\text{Log}\,\Gamma_n(z) = (z - 1)\int_0^{\infty}\frac{e^{-t}-e^{-nt}}{t}dt - \sum_{k=1}^{n}\int_0^{\infty}\frac{e^{-kt}-e^{-(z+k-1)t}}{t}dt$
(see Exercises B of Sec. 4.12.4C).

(b) Rewrite $\text{Log}\,\Gamma_n(z)$ as

$$\text{Log}\,\Gamma_n(z) = \int_0^{\infty}\left\{(z - 1)e^{-t} - \frac{e^{-t} - e^{-zt}}{1 - e^{-t}}\right\}\frac{dt}{t} - R_n(z), \quad \text{where}$$

$$R_n(z) = \int_0^{\infty}\left\{(z - 1) - \frac{e^{-t} - e^{-zt}}{1 - e^{-t}}\right\}\frac{e^{-nt}}{t}dt.$$

Observe that $\{\cdots\}\frac{1}{t}$ in $R_n(z) \to \frac{1}{2}(z-1)$ as $t \to 0$; $\to 0$ as $t \to \infty$. Then show that $\lim_{n\to\infty} R_n(z) = 0$ if $\text{Re}\, z > 0$.

(9) Use Exercise (8) to reprove the first integral formula in Exercise (7).

(10) Show that

$$\int_0^\infty \frac{(1-t^a)(1-t^b)}{1-t} \frac{dt}{\log 1/t}$$

$$= \operatorname{Log} \frac{\Gamma(a+b+1)}{\Gamma(a+1)\Gamma(b+1)}, \quad \operatorname{Re}(a+b) > -1, \quad \operatorname{Re} a > -1,$$

$$\operatorname{Re} b > -1.$$

What is

$$\int_0^\infty \frac{(1-t^a)(1-t^b)(1-t^c)}{1-t} \frac{dt}{\log 1/t}, \quad \text{if } \operatorname{Re}(a+b+c) > -1,$$

$$\operatorname{Re}(a+b) > -1, \operatorname{Re}(a+c) > -1 \quad \text{and} \quad \operatorname{Re} a > -1?$$

(11) Set $u = \rho\cos\theta$, $v = \rho\sin\theta\,(0 \le \theta \le \frac{\pi}{2})$ in $\Gamma(z)\Gamma(w) = 4\int_0^\infty \int_0^\infty e^{-(u^2+v^2)}u^{2w-1}v^{2z-1}dudv$ (why?) to reprove $B(z,w) = \frac{\Gamma(z)\Gamma(w)}{\Gamma(z+w)}$, $\operatorname{Re} z > 0$, $\operatorname{Re} w > 0$. Give precise reason for each step.

(12) Set $u = tv$ in $\Gamma(z)\Gamma(w) = \int_0^\infty t^{z-1}e^{-t}dt \cdot \int_0^\infty u^{w-1}e^{-u}du$ and then, change the order of integration (why?) to reprove $B(z,w) = \frac{\Gamma(z)\Gamma(w)}{\Gamma(z+w)}$, $\operatorname{Re} z > 0$, $\operatorname{Re} w > 0$.

(13) *Basic properties of the Beta-function.*

   (a) $B(z,1) = \frac{1}{z}$, $z \ne 0$.

   (b) $B(z,w) = B(w,z)$.

   (c) $B(z,w+1) = \frac{w}{z+w}B(z,w)$, $z+w \ne 0$, $\operatorname{Re} z > -1$, $\operatorname{Re} w > -1$ and $zw \ne 0$.

   (d) $B(z+w,n)B(z,w) = B(z,w+n)B(w,n)$, $\operatorname{Re} z > 0$, $\operatorname{Re} w > -n$, $\operatorname{Re}(z+w) > 0$, $w \ne 0$ and integer $n \ge 1$.

   (e) $B(z,w) = B(z+1,w)+B(z,w+1)$, under the assumptions as (c).

   (f) $zB(z,w+1) = wB(z+1,w)$, under the assumptions as (c).

   (g) $\lim_{y\to\infty} \frac{y^x B(x,y)}{\Gamma(x)} = 1$, $x > 0$, $y > 1$.

(14) Try the following steps to prove $B(z,w) = \frac{\Gamma(z)\Gamma(w)}{\Gamma(z+w)}$, $\operatorname{Re} z > 0$, $\operatorname{Re} w > 0$.

   (a) Use (d) in Exercise (13) to show that $B(x,y) = \frac{B(x,y+n)B(y,n)}{B(x+y,n)}$, $x > 0$, $y > 0$, and $n \ge 1$.

(b) Use (g) in Exercise (13) to show that (due to T. S. Nanjundiah (1969))

$$B(x,y) = \lim_{n \to \infty} \frac{(y+n)^{-x}\Gamma(x) \cdot n^{-y}\Gamma(y)}{n^{-(x+y)}\Gamma(x+y)}$$

$$= \frac{\Gamma(x)\Gamma(y)}{\Gamma(x+y)}, \qquad x > 0, \quad y > 0.$$

(c) Finally, use the interior uniqueness theorem.

(15) Show the following.

(a) $\int_a^b (t-a)^{z-1}(b-t)^{w-1}dt = (b-a)^{z+w-1}B(z,w)$, where a and b are real with $a < b$, $\operatorname{Re} z > 0$, $\operatorname{Re} w > 0$.

(b) $\int_0^1 (1-x^p)^{1/q}dx = \int_0^1 (1-x^q)^{1/p}dx$, $p > 0$, $q > 0$.

(16) Prove (5.6.2.7) in detail.

(17) Prove (5.6.2.8) in detail.

(18) Prove (1)–(4) in (5.6.2.9).

Exercises B

Recall that, in Example 3 of Sec. 4.9.2, the coefficients $J_n(z)$ of the Laurent series $e^{\frac{z}{2}(\zeta-\zeta^{-1})} = \sum_{-\infty}^{\infty} J_n(z)\zeta^n$, $0 < |z| < \infty$, are called the *Bessel functions of the first kind*. $J_n(z)$ can be computed by residues as follow. Setting $\zeta = \frac{2\eta}{z}$,

$$J_n(z) = \frac{1}{2\pi i}\left(\frac{z}{2}\right)^n \int_{|\eta|=\frac{1}{2}|z|} \frac{1}{\eta^{n+1}} e^{\eta - \frac{z^2}{4\eta}} d\eta = \left(\frac{z}{2}\right)^n \operatorname{Re} s\left(\frac{1}{\eta^{n+1}} e^{\eta - \frac{z^2}{4\eta}}; 0\right),$$

where

$$\operatorname{Re} s\left(\frac{1}{\eta^{n+1}} e^{\eta - \frac{z^2}{4\eta}}; 0\right)$$

$$= \text{the coefficient of } \eta^n \text{ in the Laurent series}$$

$$\text{expansion of } e^{\eta - \frac{z^2}{4\eta}} \text{ at } \eta = 0.$$

$$= \text{the coefficient of } \eta^n \text{ in the product series}$$

$$\left\{\sum_{k=0}^{\infty} \frac{1}{k!}\eta^k\right\}\left\{\sum_{k=0}^{\infty} \frac{(-1)^k}{k!}\frac{1}{\eta^k}\left(\frac{z}{2}\right)^{2k}\right\}$$

$$= \sum_{k=0}^{\infty} (-1)^k \frac{1}{(k+n)!k!}\left(\frac{z}{2}\right)^{2k}.$$

Imitating this form of $J_n(z)$, we are able to define the *generalized Bessel function*

$$J_\nu(z) = \left(\frac{z}{2}\right)^\nu \sum_{k=0}^\infty \frac{(-1)^k}{k!\Gamma(k+\nu+1)} \left(\frac{z}{2}\right)^{2k}, \quad \nu \neq 0, -1, -2, \ldots$$

where ν is any presumed complex number other than the nonpositive integers.

(1) (a) Show that, for $\nu \neq 0, -1, -2, \ldots$,

$$J_\nu(z) = \left(\frac{z}{2}\right)^\nu \cdot \frac{1}{2\pi i} \cdot \int_{\gamma_\varepsilon} \frac{1}{\zeta^{\nu+1}} e^{\zeta - \frac{z^2}{4\zeta}} d\zeta$$

where γ_ε is the contour shown in Fig. 5.15.

(b) If $-1 < \operatorname{Re}\nu < 0$, then

$$J_\nu(z) = \left(\frac{z}{2}\right)^\nu \cdot \frac{1}{2\pi i} \cdot \int_{\alpha-i\infty}^{\alpha+i\infty} \frac{1}{\zeta^{\nu+1}} e^{\zeta - \frac{z^2}{4\zeta}} d\zeta$$

where $\alpha > 0$ (refer to Sec. 4.13).

(c) In case $\operatorname{Re} z > 0$ and $\nu \neq 0, -1, -2, \ldots$, then

$$J_\nu(z) = \frac{1}{2\pi} \cdot \int_\gamma e^{iz\sin\zeta - i\nu\zeta} d\zeta$$

where γ is the contour shown in Fig. 5.24.

Note. One may need to use (5.6.1.8) in (a) and (b). The integral in (c) may be written as

$$J_\nu(z) = -\frac{\sin\nu\pi}{\pi} \int_0^\infty e^{-z\sinh t} e^{-\nu t} dt + \frac{1}{\pi} \int_0^\pi \cos(z\sin t - \nu t) dt;$$

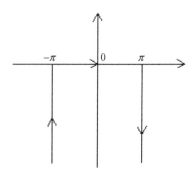

Fig. 5.24

in particular, if ν is chosen to be a positive integer n, then

$$J_n(z) = \frac{1}{n} \int_0^\pi \cos(z \sin t - nt) dt.$$

F. W. Bessel introduced this integral when solving η as a function of θ from Kepler's equation $\theta = \eta - \varepsilon \sin \eta \, (-1 < \varepsilon < 1)$.

Suppose the improper integral $\int_0^1 x^{z-1} f(x) dx$ converges absolutely on the strip $0 < a \le \operatorname{Re} z \le b$. The analytic function

$$(Mf)(z) = \int_0^1 x^{z-1} f(x) dx, \quad a < \operatorname{Re} z < b$$

is called the *Mellin transformation* of the real-valued function $f(x)$. Putting $x = e^{-t}$, it can be treated as a (two-sides) Laplace transform (refer to Sec. 4.13.2). Hence, it has *the inversion formula* (refer to (4.13.2.7))

$$f(x) = \frac{1}{2\pi i} \int_{\alpha-i\infty}^{\alpha+i\infty} (Mf)(z) x^{-z} dz, \quad a < \alpha < b.$$

For details, see Ref. [13], pp. 244–248.

(2) $\Gamma(z) = \int_0^\infty x^{z-1} e^{-x} dx$, $\operatorname{Re} z > 0$ is a special kind of Mellin transformation. The associated inversion formula is

$$e^{-x} = \frac{1}{2\pi i} \int_{a-i\infty}^{a+i\infty} \Gamma(z) x^{-z} dz, \quad a > 0.$$

Try the following steps:

(a) Integrate $\Gamma(z)$ over a rectangle γ with vertices at $a \pm iR$ and $b \pm iR$, where $0 < a < b$ and $R > 0$. Then $\Gamma(z) = \frac{1}{2\pi i} \int_\gamma \frac{\Gamma(z)}{\zeta - z} d\zeta$, $z \in \operatorname{Int} \gamma$. Along the upper side, the absolute value of the integral is $\le \frac{(b-a)\Gamma(b)}{2\pi \{\sqrt{a^2+R^2} - |z|\}} \to 0$ as $R \to \infty$. So does along the lower side. Hence, after letting $R \to \infty$,

$$\Gamma(z) = \frac{1}{2\pi i} \int_{a-i\infty}^{a+i\infty} \frac{\Gamma(\zeta)}{z - \zeta} d\zeta + \frac{1}{2\pi i} \int_{b-i\infty}^{b+i\infty} \frac{\Gamma(\zeta)}{z - \zeta} d\zeta, \quad a < \operatorname{Re} z < b$$

$$= \frac{1}{2\pi i} \int_0^\infty x^{z-1} \left\{ \int_{a-i\infty}^{a+i\infty} \Gamma(\zeta) x^{-\zeta} d\zeta \right\} dx.$$

In the second identity, $\frac{1}{z-\zeta} = \int_0^1 x^{z-\zeta-1} d\zeta = \int_1^\infty x^{z-\zeta-1} d\zeta$ and Cauchy's integral theorem have been used.

(b) To show $e^{-x} = \int_{a-i\infty}^{a+i\infty} \Gamma(\zeta)x^{-\zeta}d\zeta$ (which converges for $x > 0$), integrate $\Gamma(\zeta)x^{-\zeta}$ along a rectangle γ_n with vertices at $-n - \frac{1}{2} \pm iR$ and $a \pm iR$. Show that

$$\frac{1}{2\pi i}\int_{\gamma_n} \Gamma(\zeta)x^{-\zeta}dx = \sum_{k=0}^{n} \operatorname{Re}s(\Gamma(\zeta)x^{-\zeta}; -k) = \sum_{k=0}^{n} \frac{(-1)^k x^k}{k!} \to e^{-x}$$

as $n \to \infty$.

5.6.3   *The asymptotic function of $\Gamma(z)$; Stirling's formula*

Example 3 in Sec. 4.14 had proved Stirling's formula for $\Gamma(z)$:

$$\Gamma(z+1) = \sqrt{2\pi}z^{z+\frac{1}{2}}e^{-z}\left(1 + O\left(\frac{1}{z}\right)\right).$$

An instructive proof using the residue method was due to *E. Lindelöf: Calcul des résidue et ses applications à la théorie des fonctions, Paris 1905, VIII +144P. Reprinted 1947.* We will adopt this method in this section to give an alternative derivation of the formula.

The main ideas are as follows. Starting from derivative of Gauss's ψ-function (see $(*_1)$ in Sec. 5.6.2)

$$\frac{d}{dz}\left(\frac{\Gamma'(z)}{\Gamma(z)}\right) = \sum_{n=0}^{\infty} \frac{1}{(z+n)^2}, \qquad \text{absolutely and locally uniformly in } \mathbf{C}$$

$$-\{0, -1, -2, \ldots\}, \tag{5.6.3.1}$$

the residue method is invoked to rewrite it as

$$\frac{d}{dz}\left(\frac{\Gamma'(z)}{\Gamma(z)}\right) = \frac{1}{2z^2} + \int_0^\infty \coth\pi\eta \cdot \frac{2\eta z}{(\eta^2 + z^2)^2}d\eta$$

$$= \frac{1}{z} + \frac{1}{2z^2} + \int_0^\infty \frac{4\eta z}{(\eta^2 + z^2)^2} \cdot \frac{d\eta}{e^{2\pi\eta} - 1}, \quad z \neq 0, -1, -2, \ldots.$$

$$\tag{5.6.3.2}$$

Integrate the above in $\operatorname{Re} z > 0$ and we obtain

$$\frac{\Gamma'(z)}{\Gamma(z)} = \operatorname{Log} z - \frac{1}{2z} - \int_0^\infty \frac{2\eta}{\eta^2 + z^2} \cdot \frac{d\eta}{e^{2\pi\eta} - 1}, \quad \operatorname{Re} z > 0. \tag{5.6.3.3}$$

The choice of the principal branch $\operatorname{Log} z$ in $\mathbf{C} - (\infty, 0]$ is purposely designed to have the constant of integration become zero. Integrate again and we obtain, after the determination of constant of integration and some computation, the following

Stirling's formula of $\Gamma(z)$.

$$\log \Gamma(z) = \frac{1}{2} \log 2\pi \quad z + \left(z - \frac{1}{2}\right) \log z + J(z), \quad \operatorname{Re} z > 0, \quad \text{or}$$

$$\Gamma(z) = \sqrt{2\pi} z^{z-\frac{1}{2}} e^{-z} e^{J(z)}, \quad \operatorname{Re} z > 0$$

where log means real logarithm in case $z = x > 0$, and

$$J(z) = \frac{1}{\pi} \int_0^\infty \frac{z}{\eta^2 + z^2} \log \frac{1}{1 - e^{-2\pi\eta}} d\eta (\eta \in \mathbf{R}), \quad \operatorname{Re} z > 0$$

$$= \sum_{k=1}^n (-1)^{k-1} \frac{B_{2k}}{(2k-1)2k} \cdot \frac{1}{z^{2k-1}} + J_n(z)$$

$$\sim \sum_{n=1}^\infty (-1)^{n-1} \frac{B_{2n}}{(2n-1)2n} \cdot \frac{1}{z^{2n-1}} \quad \text{as } n \to \infty$$

in which $\lim_{z\to\infty} J(z) = 0$ if $\operatorname{Re} z \geq x_0 > 0$ and B_{2n} are Bernoulli's numbers; moreover,

$$J_n(z) = \frac{(-1)^n}{z^{2n+1}} \cdot \frac{1}{\pi} \int_0^\infty \frac{\eta^{2n}}{1 + (\eta/z)^2} \log \frac{1}{1 - e^{-2\pi\eta}} d\eta, \quad n \geq 1$$

satisfying $\lim_{z\to\infty} z^{2n} J_n(z) = 0$ if $\operatorname{Re} z \geq x_0 > 0$. Consequently,

$$\Gamma(z) \sim \sqrt{2\pi} z^{z-\frac{1}{2}} e^{-z} \left(1 + \frac{1}{12z} + \frac{1}{288z^2} + \cdots\right). \tag{5.6.3.4}$$

The following proof is based on Ref. [1], pp. 201–206.

Proof.

Step 1. To represent (5.6.3.1) as the complex line integral (5.6.3.2).

The appearance of the series in (5.6.3.1) suggests that we need a function having simple poles at integers $k \geq 0$ with the associated residues $\frac{1}{(z+k)^2}$. Since $\pi \cot \pi z$ has simple pole at k with the residue 1 (refer to (4.15.3)),

$$\Phi(\zeta) = \frac{\pi \cot \pi z}{(z + \zeta)^2}$$

does work. Here ζ is the variable while z ($\operatorname{Re} z > 0$) stands as a parameter for the moment.

Fix an integer $n \geq 1$. Integrate $\Phi(z)$ along the path γ_n shown in Fig. 5.25, where γ_n is a rectangle having vertices at $\pm Ri$ and $n + \frac{1}{2} \pm Ri$

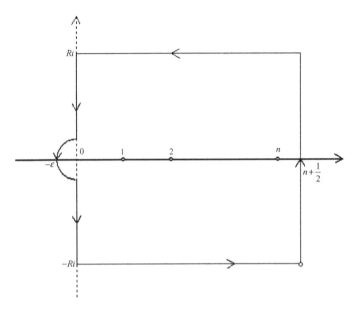

Fig. 5.25

with an outward dented half circle around the point $z = 0$. Then, applying
(4.12.3.3) to the dented half circle, the residue theorem says that

$$\text{P.V.} \frac{1}{2\pi i} \int_{\gamma_n} \Phi(\zeta) d\zeta = \left(\frac{3\pi}{2} - \frac{\pi}{2} \right) \frac{i}{2\pi i} \operatorname{Re} s(\Phi(\zeta); 0) + \sum_{k=1}^{n} \operatorname{Re} s(\Phi; k)$$

$$= \frac{1}{2z^2} + \sum_{k=1}^{n} \frac{1}{(z+k)^2} = -\frac{1}{2z^2} + \sum_{k=0}^{n} \frac{1}{(z+k)^2}. \quad (*_1)$$

We try to let $\varepsilon \to 0$, $R \to \infty$, and $n \to \infty$ in order to calculate the
integral in the left side.

Along the horizontal side $\zeta = \sigma + iR$, $0 \leq \sigma \leq n + \frac{1}{2}$:

$$\cot \pi z = i \frac{e^{2\pi i \sigma} e^{-2\pi R} + 1}{e^{2\pi i \sigma} e^{-2\pi R} - 1} \to i \quad \text{uniformly as } R \to \infty$$

$$\Rightarrow \left| \int_{\substack{\text{upper} \\ \text{side}}} \Phi(\zeta) d\zeta \right| \leq \int_{\substack{\text{upper} \\ \text{side}}} \frac{\pi |\cot \pi z|}{|\zeta|^2 - |z|^2} |d\zeta| \quad \text{(if } R \text{ is large enough)}$$

$$\leq \frac{M\pi \left(n + \frac{1}{2} \right)}{R^2 - |z|^2} \quad (M > 0, \text{ a constant})$$

$$\to 0 \quad \text{as } R \to \infty \quad \text{while } n \text{ is fixed.} \quad (*_2)$$

Since $\lim_{R\to\infty} \cot \pi z = -i$ uniformly, similarly we have

$$\lim_{R\to\infty} \int_{\substack{\text{lower}\\\text{side}}} \Phi(\zeta)d\zeta = 0 \quad \text{where } n \text{ is also kept fixed.} \qquad (*_3)$$

Along the right vertical side $\zeta = n + \frac{1}{2} + i\tau$, $-R < \tau < R$ (even if $-\infty < \tau < \infty$), simply denoted as $[-Ri, Ri]$:

$$\cot \pi \zeta = i \frac{e^{-2\pi\tau} - 1}{e^{-2\pi\tau} + 1}, \quad -\infty < \tau < \infty$$

$$\Rightarrow |\cot \pi z| \le 1 \quad \text{for } n \ge 1 \quad \text{and} \quad -\infty < \tau < \infty$$

$$\Rightarrow \left| \int_{[-Ri,Ri]} \Phi(\zeta)d\zeta \right| \le \pi \int_{[-Ri,Ri]} \frac{|d\zeta|}{|\zeta + z|^2}, \quad n \ge 1. \qquad (*_4)$$

To evaluate the integral in right above, observe that $\bar\zeta = 2n + 1 - \zeta$ along the line $\zeta = n + \frac{1}{2}$. Then, choose n large enough so that $0 < |z| < n + \frac{1}{2}$ and integrate $g(\zeta) = [(\zeta + z)(2n + 1 - \zeta + \bar z)]^{-1}$ along the path γ shown in Fig. 5.26. By the residue theorem,

$$\frac{1}{2\pi i} \int_\gamma g(\zeta)d\zeta = \operatorname{Re} s(g(\zeta); -z) = \frac{1}{2n + 1 + 2\operatorname{Re} z}$$

$$= \frac{1}{2\pi i} \int_{C_{\tilde R}} g(\zeta)d\zeta + \frac{1}{2\pi i} \int_{[-Ri,Ri]} \frac{d\zeta}{|\zeta + z|^2},$$

$$\text{where } R = \sqrt{\tilde R^2 - \left(n + \frac{1}{2}\right)^2}. \qquad (*_5)$$

The circular arc $C_{\tilde R}$ in Fig. 5.26 is given by $z = \tilde R e^{i\theta}$, $-\theta_0 \le \theta \le 2\pi - \theta_0$, and $\theta_0 = \operatorname{Arg}\left(n + \frac{1}{2} + i\sqrt{\tilde R^2 - \left(n + \frac{1}{2}\right)^2}\right)$, and hence

$$\left| \int_{C_{\tilde R}} g(\zeta)d\zeta \right| \le \frac{2\pi\tilde R}{(\tilde R - |z|)(\tilde R - 2n - 1 - |z|)}$$

$$\text{(if } \tilde R \text{ is large enough)} \to 0 \quad \text{as } \tilde R \to \infty.$$

$$\Rightarrow (\text{by}(*_5)) \lim_{R\to\infty} \frac{1}{2\pi i} \int_{[-Ri,Ri]} \frac{d\zeta}{|\zeta + z|^2} = \frac{1}{2\pi i} \int_{\zeta = n + \frac{1}{2}} \frac{d\zeta}{|\zeta + z|^2}$$

$$= \frac{1}{2n + 1 + 2\operatorname{Re} z}.$$

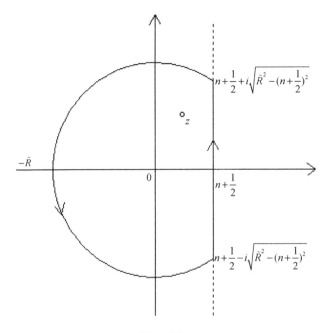

$$n+\frac{1}{2}+i\sqrt{\tilde{R}^2-(n+\frac{1}{2})^2}$$

$$-\tilde{R}$$

$$n+\frac{1}{2}$$

$$n+\frac{1}{2}-i\sqrt{\tilde{R}^2-(n+\frac{1}{2})^2}$$

Fig. 5.26

As a result, letting $R \to \infty$ in $(*_4)$, we obtain

$$\left| \int_{\zeta=n+\frac{1}{2}} \Phi(\zeta)d\zeta \right| \le \frac{2\pi^2}{2n+1+2\mathrm{Re}\, z} \to 0 \quad \text{as } n \to \infty. \qquad (*_6)$$

Finally, along the imaginary axis (as $\varepsilon \to 0$):

$$\mathrm{P.V.} \int_{\infty}^{-\infty} \Phi(\zeta)d\zeta = \int_{\infty}^{-0} \frac{\pi\cot\pi i\eta}{(z+i\eta)^2}id\eta + \int_0^{-\infty} \frac{\pi\cot\pi i\eta}{(z+i\eta)^2}id\eta$$

$$= \int_0^{\infty} \coth\pi\eta \cdot \frac{4\pi i\eta z}{(\eta^2+z^2)^2}d\eta. \qquad (*_7)$$

Now, letting $\varepsilon \to 0$, $R \to \infty$, and $n \to \infty$ in $(*_1)$ and using the results obtained in $(*_2)$, $(*_3)$, $(*_6)$, and $(*_7)$, we have

$$\frac{1}{2\pi i} \int_0^{\infty} \coth\pi\eta \cdot \frac{4\pi i\eta z}{(\eta^2+z^2)^2}d\eta = -\frac{1}{2z^2} + \sum_{k=0}^{\infty} \frac{1}{(z+k)^2}.$$

Note that $\coth\pi\eta = 1 + \frac{2}{e^{2\pi\eta}-1}$ and $\int_0^{\infty} \frac{2\eta z}{(\eta^2+z^2)^2}d\eta = -\frac{z}{\eta^2+z^2}\Big|_0^{\infty} = \frac{1}{z}$ in the above integral. We are now able to rewrite (5.6.3.1) as (5.6.3.2). For a general setting, see Exercises B.

Step 2. To derive (5.6.3.3) from (5.6.3.2).
Consider the right half-plane $\operatorname{Re} z > 0$.

We try to integrate both sides of (5.6.3.2), with respect to z, along any rectifiable curve joining 1 to a point z and lying completely in $\operatorname{Re} z > 0$. In doing so, we need to check that the integral in the right does converges locally uniformly in $\operatorname{Re} z > 0$. To this end, fix any compact set K in $\operatorname{Re} z > 0$. Set $m = \min_{z \in K} |\operatorname{Re} z|$ and $M = \max_{z \in K} |z|$. Then, the estimate

$$\left| \int_0^\infty \frac{4\eta z}{(\eta^2 + z^2)^2} \cdot \frac{d\eta}{e^{2\pi \eta} - 1} \right| \le 4M \int_0^\infty \frac{\eta}{(\eta^2 + m^2)^2} \cdot \frac{d\eta}{e^{2\pi \eta} - 1} < \infty$$

justifies the claim.

After integrating (5.6.3.2) with respect to z, we obtain

$$\frac{\Gamma'(z)}{\Gamma(z)} = c_1 + \operatorname{Log} z - \frac{1}{2z} - \int_0^\infty \frac{2\eta}{\eta^2 + z^2} \cdot \frac{d\eta}{e^{2\pi \eta} - 1}, \qquad \operatorname{Re} z > 0, \qquad (*_8)$$

where c_1 is a constant to be determined in the next step and $\operatorname{Log} z$ is the principal branch in $\mathbf{C} - (-\infty, 0]$. Note that the integral in $(*_8)$ also converges locally uniformly in $\operatorname{Re} z > 0$.

Step 3. Integrate both sides of $(*_8)$ and determine the constants of integration.
Direct integration of both sides of $(*_8)$ will cause a multiple-valued function $\tan^{-1} \frac{z}{\eta}$. To avoid this difficulty, integrate first by parts the integral appeared in the right and we have

$$\int_0^\infty \frac{2\eta}{\eta^2 + z^2} \frac{d\eta}{e^{2\pi \eta} - 1}$$

$$= \frac{1}{2\pi} \int_0^\infty \frac{2\eta}{\eta^2 + z^2} d\{\log(1 - e^{-2\pi \eta})\}$$

$$= \frac{1}{2\pi} \frac{2\eta}{\eta^2 + z^2} \log(1 - e^{-2\pi \eta}) \bigg|_0^\infty - \frac{1}{2\pi} \int_0^\infty \log(1 - e^{-2\pi \eta}) \frac{2(z^2 - \eta^2)}{(\eta^2 + z^2)^2} d\eta$$

$$= -\frac{1}{\pi} \int_0^\infty \frac{z^2 - \eta^2}{(\eta^2 + z^2)^2} \log(1 - e^{-2\pi \eta}) d\eta,$$

where $\log(1 - e^{-2\pi \eta})$ takes the real logarithm.

Under this preparatory work, integrate both sides of $(*_8)$ to obtain

$$\operatorname{Log}\Gamma(z) = c_2 + c_1 z + \left(z - \frac{1}{2}\right)\operatorname{Log} z - z$$

$$+ \frac{1}{\pi}\int_0^1 \frac{z}{\eta^2 + z^2}\log\frac{1}{1 - e^{-2\pi\eta}}d\eta, \quad \operatorname{Re} z > 0, \qquad (*_9)$$

where $\operatorname{Log} z$ and $\operatorname{Log}\Gamma(z)$ are real logarithms in case $z = x > 0$ is real. Therefore, both c_1 and c_2 are real constant. The integral in the right is specifically denoted as

$$J(z) = \frac{1}{\pi}\int_0^\infty \frac{z}{\eta^2 + z^2}\log\frac{1}{1 - e^{-2\pi\eta}}d\eta, \quad \operatorname{Re} z > 0. \qquad (*_{10})$$

In order to determine c_1 and c_2 in $(*_9)$, we need to prove that $\lim_{z\to\infty} J(z) = 0$ in case $\operatorname{Re} z \geq x_0 > 0$. Separate $J(z)$ into two integrals as

$$J(z) = \int_0^{|z|/2} + \int_{|z|/2}^\infty = J_1(z) + J_2(z).$$

On $[0, \frac{|z|}{2}]$, $|\eta^2 + z^2| \geq |z|^2 - |\eta|^2 \geq |z|^2 - \frac{1}{4}|z|^2 = \frac{3}{4}|z|^2$ holds and thus,

$$|J_1(z)| \leq \frac{4}{3\pi|z|}\int_0^\infty \log\frac{1}{1 - e^{-2\pi\eta}}d\eta.$$

While, on $[\frac{|z|}{2}, \infty)$, $|\eta^2 + z^2| = |z - i\eta||z + i\eta| \geq x_0|z|$ (this follows obviously by the relative position of the points z, $\pm i\eta$, the line $\operatorname{Re} z = x_0$ and the set $\eta \geq \frac{|z|}{2}$). Consequently,

$$|J_2(z)| \leq \frac{1}{\pi x_0}\int_{|z|/2}^\infty \log\frac{1}{1 - e^{-2\pi\eta}}d\eta.$$

Since

$$\lim_{\eta\to\infty}\frac{\log(1 - e^{-2\pi\eta})^{-1}}{e^{-2\pi\eta}} = 1 \quad \text{and} \quad \int_0^\infty e^{-2\pi\eta}d\eta < \infty,$$

by comparison, it follows that $\int_0^\infty \log(1 - e^{-2\pi\eta})^{-1}d\eta < \infty$. Hence both $J_1(z) \to 0$ and $J_2(z) \to 0$ hold as $z \to \infty$. And the claim follows.

To prove $c_1 = 0$: Applying $\Gamma(z + 1) = z\Gamma(z)$ to $(*_9)$ and using the fact that $\lim_{z\to\infty}\left(z + \frac{1}{2}\right)$. $\operatorname{Log}\left(1 + \frac{1}{z}\right) = 1$ (here, one may just assume that

$z = x > 0$), then

$$c_2 + c_1(z+1) + \left(z + \frac{1}{2}\right) \text{Log}\,(z+1) - z - 1 + J(z+1)$$

$$- \text{Log}\,z + c_2 + c_1 z + \left(z - \frac{1}{2}\right) \text{Log}\,z - z + J(z)$$

$$\Rightarrow c_1 = -\left(z + \frac{1}{2}\right) \text{Log}\left(1 + \frac{1}{z}\right) + 1 + J(z) + J(z+1) \to -1 + 1 = 0$$

as $z \to \infty$.

To prove $c_2 = \frac{1}{2}\log 2\pi$: Set $z = \frac{1}{2} + iy$ in $\Gamma(z)\Gamma(1-z) = \frac{\pi}{\sin \pi z}$. By $(*_8)$ with $c_1 = 0$, we have

$$2c_2 - 1 + iy\,\text{Log}\left(\frac{1}{2} + iy\right) - iy\,\text{Log}\left(\frac{1}{2} - iy\right) + J\left(\frac{1}{2} + iy\right) + J\left(\frac{1}{2} - iy\right)$$

$$= \log \pi - \log \cosh \pi y + 2k\pi i, \qquad\qquad (*_{11})$$

where k is a constant. Since Log is the principle branch in $\mathbf{C} - (\infty, 0]$, each term in both sides, except $2k\pi i$, turns out to be real if $y = 0$ is assumed. It follows that $k = 0$. On the other hand, if $|y| > \frac{1}{2}$,

$$iy\,\text{Log}\left(\frac{1}{2} + iy\right) - iy\,\text{Log}\left(\frac{1}{2} - iy\right)$$

$$= iy\,\text{Log}\frac{\frac{1}{2} + iy}{\frac{1}{2} - iy} = iy\left\{\pi i + \text{Log}\frac{1 + \frac{1}{2iy}}{1 - \frac{1}{2iy}}\right\}$$

$$= iy\left\{\pi i + \sum_{n=1}^{\infty} \frac{(-1)^{n-1}}{n}\left(\frac{1}{2iy}\right)^n + \sum_{n=1}^{\infty} \frac{1}{n}\left(\frac{1}{2iy}\right)^n\right\}$$

$$= -\pi y + 1 + O\left(\frac{1}{y^3}\right);$$

if $|y| > 0$,

$$\log \cosh \pi y = \log \frac{1 + e^{-2\pi y}}{2e^{-\pi y}} = \pi y - \log 2 + O\left(\frac{1}{e^{2\pi y}}\right).$$

Putting these two results in $(*_{11})$, known that $k = 0$, then

$$2c_2 - 1 - \pi y + 1 + O\left(\frac{1}{y^3}\right) + J\left(\frac{1}{2} + iy\right) + J\left(\frac{1}{2} - iy\right)$$

$$= \log \pi - \pi y + \log 2 + O\left(\frac{1}{e^{2\pi y}}\right).$$

Let $y \to \infty$ in both sides after πy is cancelled and we get $c_2 = \frac{1}{2}\log 2\pi$.

Another proof for $c_2 = \frac{1}{2}\log 2\pi$: Set $z = \frac{1}{2} + iy$ in $\Gamma(z)\Gamma(1-z) = \frac{\pi}{\sin \pi z}$.
Then

$$\Gamma\left(\frac{1}{2} + iy\right)\Gamma\left(\frac{1}{2} - iy\right) = \frac{2\pi e^{-\pi y}}{1 + e^{-2\pi y}}$$

$\Rightarrow$ (observing that $\Gamma\left(\frac{1}{2} - iy\right) = \overline{\Gamma\left(\frac{1}{2} + iy\right)}$

(refer to Exercise A(1)(a) of Sec. 5.6.2))

$$\mathrm{Re}\,\mathrm{Log}\,\Gamma\left(\frac{1}{2} + iy\right) = \log\sqrt{2\pi} - \frac{\pi y}{2} - \frac{1}{2}\log\left(1 + e^{-2\pi y}\right).$$

On the other hand, set $z = \frac{1}{2} + iy$ in (*9) with $c_1 = 0$ and we have

$$\mathrm{Re}\,\Gamma\left(\frac{1}{2} + iy\right) = c_2 - y\tan^{-1} 2y - \frac{1}{2} + \mathrm{Re}\,J\left(\frac{1}{2} + iy\right),$$

where $\tan^{-1}$ denotes the principal branch. Combining the last two expressions, we have

$$c_2 = \log\sqrt{2\pi} + y\left(\tan^{-1} 2y - \frac{\pi}{2}\right) + \frac{1}{2} - \frac{1}{2}\log\left(1 + e^{-2\pi y}\right)$$

$$- \mathrm{Re}\,J\left(\frac{1}{2} + iy\right). \tag{*12}$$

For x and y restricted to $x > 0$ and $y > 0$,

$$\mathrm{Re}\,J(x + iy) = \frac{1}{\pi}\int_0^\infty \frac{x(x^2 + y^2 + \eta^2)}{(x^2 + y^2 + \eta^2 + 2y\eta)[x^2 + (y - \eta)^2]}\log\frac{1}{1 - e^{-2\pi\eta}}d\eta$$

$$< \frac{x}{\left(x^2 + \frac{y^2}{4}\right)}\int_0^\infty \log\frac{1}{1 - e^{-2\pi\eta}}dy$$

$$+ \log\frac{1}{1 - e^{-\pi y}}\int_{\frac{y}{2}}^\infty \frac{x}{x^2 + (y - \eta)^2}d\eta$$

$$< \frac{4x}{\pi y^2}\int_0^\infty \log\frac{1}{1 - e^{-2\pi\eta}}d\eta + \log\frac{1}{1 - e^{-\pi y}}\int_{-\infty}^\infty \frac{x}{x^2 + \eta^2}d\eta$$

$$\Rightarrow \lim_{y\to\infty}\mathrm{Re}\,J\left(\frac{1}{2} + iy\right) = 0.$$

Since $\lim_{y\to\infty} y\left(\tan^{-1} 2y - \frac{\pi}{2}\right) = -\frac{1}{2}$, setting $y \to \infty$ in (*12), it follows that $c_2 = \log\sqrt{2\pi}$.

In conclusion, we have proved that $\log\Gamma(z) = \frac{1}{2}\log 2\pi - z + \left(z - \frac{1}{2}\right)\log z + J(z)$, $\mathrm{Re}\,z > 0$. What remains is to estimate $J(z)$.

Step 4. The estimate of $J(z)$.

Using the identity

$$\frac{z}{\eta^2 + z^2} = \frac{1}{z} \cdot \frac{1}{1 + (\eta/z)^2}$$

$$= \frac{1}{z} \left\{ 1 - \frac{\eta^2}{z^2} + \left(\frac{\eta^2}{z^2}\right)^2 - \cdots + (-1)^{n-1} \left(\frac{\eta^2}{z^2}\right)^{n-1} \right.$$

$$\left. + (-1)^n \frac{(\eta^2/z^2)^n}{1 + (\eta/z)^2} \right\},$$

$J(z)$ in $(*_{10})$ can be rewritten as

$$J(z) = \sum_{k=1}^{n} \frac{C_k}{z^{2k-1}} + J_n(z), \quad \text{where}$$

$$C_k = (-1)^{k-1} \frac{1}{\pi} \int_0^\infty \eta^{2k-2} \cdot \log \frac{1}{1 - e^{-2\pi\eta}} d\eta \quad \text{for } 1 \le k \le n, \quad \text{and}$$

$$J_n(z) = (-1)^n \frac{1}{z^{2n+1}} \cdot \frac{1}{\pi} \int_0^\infty \frac{\eta^{2n}}{1 + (\eta/z)^2} \log \frac{1}{1 - e^{-2\pi\eta}} d\eta, \quad n \ge 1.$$

To evaluate C_k: By invoking the change of the variables $t = 2\pi\eta$, then

$$C_k = (-1)^{k-1} \frac{1}{\pi} \int_0^\infty \left(\frac{t}{2\pi}\right)^{2k-2} \{-\log(1 - e^{-t})\} \cdot \frac{dt}{2\pi}$$

$$= (-1)^k \frac{1}{\pi} \cdot \frac{1}{(2\pi)^{2k-1}} \int_0^\infty t^{2k-2} \log(1 - e^{-t}) dt. \qquad (*_{13})$$

It is well-known that $\log(1 - e^{-t}) = -\sum_{m=0}^\infty \frac{e^{-mt}}{m}$ $(t > 0)$ converges locally uniformly in $(0, \infty)$. Therefore,

$$\int_0^\infty t^{2k-2} \log(1 - e^{-t}) dt$$

$$= -\sum_{m=1}^\infty \frac{1}{m} \int_0^\infty t^{2k-2} e^{-mt} dt$$

$$= -\sum_{m=1}^\infty \frac{1}{m} \int_0^\infty \left(\frac{t}{m}\right)^{2k-2} e^{-t} \frac{dt}{m} = -\sum_{m=1}^\infty \frac{1}{m^{2k}} \int_0^\infty t^{2k-2} e^{-t} dt$$

$$= -\Gamma(2k-1) \sum_{m=1}^\infty \frac{1}{m^{2k}}, \quad 1 \le k \le m.$$

According to Exercise 5 of Sec. 5.4.1, we have $\sum_{m=1}^{\infty} \frac{1}{m^{2k}} = 2^{2k-1} \frac{B^{2k}}{(2k)!} \pi^{2k}$, $k \geq 1$, it follows that

$$C_k = (-1)^k \frac{1}{\pi} \cdot \frac{1}{(2\pi)^{2k-1}} \cdot (-1)(2k-2)! \cdot 2^{2k-1} \cdot \frac{B_{2k}}{(2k)!} \pi^{2k}$$

$$= (-1)^{k-1} \frac{B_{2k}}{(2k-1)(2k)}, \quad k \geq 1.$$

To the last, in case $\operatorname{Re} z \geq x_0 > 0$,

$$|z^{2n} J_n(z)| \leq \frac{1}{|z|} \cdot \frac{1}{\pi} \int_0^{\infty} \frac{\eta^{2n}}{|1+(\eta/z)^2|} \log \frac{1}{1-e^{-2\pi\eta}} d\eta$$

$$= \frac{1}{\pi} \int_0^{\infty} \frac{\eta^{2n}|z|}{|\eta^2 + z^2|} \log \frac{1}{1-e^{-2\pi\eta}} d\eta \to 0 \quad \text{as } z \to \infty,$$

exactly the same as the proof that $J(z) \to 0$, if $\operatorname{Re} z \geq x_0 > 0$, in $(*_{10})$.

$\square$

As a supplement to (5.6.3.4), we want to introduce another form of Stirling's formula and *sketch* as follows.

Set

$$\mu(z) = \operatorname{Log} \Gamma(z+1) - \left(z + \frac{1}{2}\right) \operatorname{Log} z + z - \log \sqrt{2\pi}, \quad z \in \mathbf{C} - (\infty, 0],$$

$$(5.6.3.5)$$

called the *Binet function*. $\mu(z)$ assumes real values in case $z = x > 0$. $\mu(z)$ has some other representations such as $\mu(z) = \int_0^1 \left(t - \frac{1}{2}\right) \psi(z+t) dt$ (where $\psi(z)$ is Gauss's ψ-function $\frac{\Gamma'(z)}{\Gamma(z)}$) $= -\int_0^{\infty} \frac{p(t)}{z+t} dt$ (where $p(t) = t - [t] - \frac{1}{2}$ for real t), both having their own interests (see Exercises A(1) and A(2)). Differentiate now $\mu(z)$ and we obtain

$$\mu'(z) = \psi(z+1) - \operatorname{Log} z + \frac{1}{2z}.$$

Using the first integral representation of $\psi(z)$ in Exercise A(7) of Sec. 5.6.2 and the facts that $\operatorname{Log} z = \int_0^{\infty} \frac{e^{-t} - e^{-tz}}{t} dt$ (see Exercises B of Sec. 4.12.4C) and $\frac{1}{z} = \int_0^{\infty} e^{-zt} dt$, it follows that

$$\mu'(z) = \int_0^{\infty} \left(\frac{1}{t} - \frac{1}{2} - \frac{1}{e^t - 1}\right) e^{-zt} dt, \quad \operatorname{Re} z > 0.$$

Since $\left|\frac{1}{t} - \frac{1}{2} - \frac{1}{e^t-1}\right| < \frac{1}{2}$ if $t > 0$ and $\lim_{t \to 0^+} \left(\frac{1}{t} - \frac{1}{2} - \frac{1}{e^t-1}\right) = 0$, the above integral converges locally uniformly in $\operatorname{Re} z > 0$. Then, integrate $\mu'(z)$ along the line segment joining 1 to z ($\operatorname{Re} z > 0$) and change the order of integration. As a result, we have

$$\mu(z) - \mu(1) = \int_0^\infty \left(\frac{1}{e^t - 1} + \frac{1}{2} - \frac{1}{t}\right) \frac{e^{-zt}}{t} dt$$
$$- \int_0^\infty \left(\frac{1}{e^t - 1} + \frac{1}{2} - \frac{1}{t}\right) \frac{e^{-t}}{t} dt. \qquad (*14)$$

Applying the known expansion (see Exercise A(4)(h) of Sec. 5.4.2)

$$\frac{1}{e^t - 1} + \frac{1}{2} - \frac{1}{t} = 2t \sum_{n=1}^\infty \frac{1}{t^2 + 4n^2\pi^2}$$

$$\Rightarrow \left(\frac{1}{e^t - 1} + \frac{1}{2} - \frac{1}{t}\right) \frac{1}{t} \le \frac{2}{4\pi^2} \sum_{n=1}^\infty \frac{1}{n^2} = \frac{1}{2\pi^2} \cdot \frac{\pi^2}{6} = \frac{1}{12}$$

to the first integral in $(*14)$ above, we obtain the estimate

$$\left| \int_0^\infty \left(\frac{1}{e^t - 1} + \frac{1}{2} - \frac{1}{t}\right) \frac{e^{-zt}}{t} dt \right| \le \frac{1}{12} \int_0^\infty e^{-(\mathrm{Re}\, z)t} dt = \frac{1}{12\mathrm{Re}\, z}, \quad \mathrm{Re}\, z > 0.$$

While, direct computation (see Exercise A(4)) shows that

$$\mu(1) = 1 - \log\sqrt{2\pi} = \int_0^\infty \left(\frac{1}{e^t - 1} + \frac{1}{2} - \frac{1}{t}\right) \frac{e^{-t}}{t} dt = -\int_0^\infty \frac{p(t)}{1+t} dt$$

which can also be justified by showing that $\mu(z) \to 0$ in $(*14)$ as $\mathrm{Re}\, z > 0$ and $z \to 0$ (see Exercise A(3)). Consequently, $(*14)$ reduces to

$$\mu(z) = \int_0^\infty \left(\frac{1}{e^t - 1} + \frac{1}{2} - \frac{1}{t}\right) \frac{e^{-zt}}{t} dt, \quad \mathrm{Re}\, z > 0 \qquad (*15)$$

$$\Rightarrow |\mu(z)| \le \frac{1}{12\mathrm{Re}\, z}, \quad \mathrm{Re}\, z > 0.$$

We summarize the above as the

Stirling's formula in Binet form.

$$\Gamma(z+1) = \left(\frac{z}{e}\right)^z \sqrt{2\pi z}\, e^{\mu(z)}, \quad z \in \mathbf{C} - (\infty, 0],$$

where the Binet function $\mu(z)$ (see (5.6.3.5)) satisfies $|\mu(z)| \le \frac{1}{12\mathrm{Re}\, z}$, $\mathrm{Re}\, z > 0$. In particular, for real $x > 0$,

$$\Gamma(x) = \sqrt{2\pi} x^{x-\frac{1}{2}} e^{-x} e^{\frac{\theta(x)}{12x}}$$

with $0 < \theta(x) < 1$.   (5.6.3.6)

There are some other ways to derive this remarkable formula (see Exercise A(6)).

Exercises A

(1) Use Exercise A(5)(f) of Sec. 5.6.2 to derive the following *Raabe integral*

$$\int_0^1 \operatorname{Log}\Gamma(z+t)dt = z\operatorname{Log} z - z + \log\sqrt{2\pi}, \quad z \in \mathbf{C} - (\infty, 0].$$

Then, integrating by parts, one obtains an integral representation

$$\mu(z) = \int_0^1 \left(t - \frac{1}{2}\right)\psi(z+t)dt, \quad z \in \mathbf{C} - (\infty, 0]$$

for the Binet function (5.6.3.5).

(2) Let $P(t) = t - [t] - \frac{1}{2}$ for real t, where $[t]$ denotes Gauss's integer part of t. Try to show

$$\mu(z) = -\int_0^\infty \frac{P(t)}{z+t}dt, \quad z \in \mathbf{C} - (\infty, 0]$$

by the following steps.

(a) By noting that $P(t) + [t] = t - \frac{1}{2}$, show that

$$\mu(z) = \int_0^1 P(t)\psi(z+t)dt.$$

(b) Using the fact that $P(t+1) = P(t)$ and Exercise A(5)(b) of Sec. 5.6.2, show that

$$\mu(z) = -\int_0^n \frac{P(t)}{z+t}dt + \int_n^{n+1} P(t)\psi(z+t)dt, \quad n \geq 1 \quad \text{integers}.$$

(c) Show that $\lim_{n\to\infty} \int_n^{n+1} P(t)\psi(z+t)dt = 0$ by using that $\int_0^1 P(t)dt = 0$, $|P(t)| \leq \frac{1}{2}$, and Exercise A(5)(d) of Sec. 5.6.2.

(3) Fix $0 < \delta < \pi$. Show that, when restricted $|\operatorname{Arg} z| \leq \pi - \delta$,

$$\lim_{z\to\infty} \mu(z) = 0.$$

Try the following steps.

(a) Set $L(x) = \int_0^x P(t)dt$, $x > 0$. Show that $L(0+) = 0$ and $|L(x)| \leq \frac{1}{8}$ for $x > 0$.

(b) Use integration by parts to show that $\int_0^\infty \frac{P(t)}{z+t}dt = \int_0^\infty \frac{L(t)}{(z+t)^2}dt$.

(c) In case $z = re^{i\theta}$, $|z+t|^2 \geq (r+t)^2 - (r+t)^2\sin^2\frac{\theta}{2} = (r+t)^2\cos^2\frac{\theta}{2}$. Hence,

$$|\mu(z)| \leq \frac{1}{8\cos^2\frac{\theta}{2}}\int_0^\infty \frac{dt}{(r+t)^2} = \frac{1}{8r}\sec^2\frac{\theta}{2}.$$

(4) Show that

$$\int_0^\infty \frac{P(t)}{1+t}\,dt = \frac{1}{2}\log 2\pi - 1$$

by the following steps.

(a) Show first that $\int_0^\infty \frac{P(t)}{1+t}\,dt$ does converge by noting that $\int_0^x P(t)dt = \frac{1}{2}P(x)^2 - \frac{1}{8}$ for $x \geq 0$.

(b) Use Gauss's representation for $\Gamma(z)$ in (5.6.1.4) by setting $z = \frac{1}{2}$ and $\Gamma\left(\frac{1}{2}\right) = \sqrt{\pi}$ to show that

$$\sqrt{\pi} = \lim_{n \to \infty} \frac{2^{2n+1}[(n+1)!]^2(n+1)^{-3/2}}{(2n+1)!}.$$

(c) Use (b) and the continuity of $\mathrm{Log}\,z$ along $z = x > 0$ to show that

$$\frac{1}{2}\log \pi = \lim_{n \to \infty} \left\{ (2n+1)\log 2 + 2\sum_{k=0}^{n}\log(1+k) - \frac{3}{2}\log(1+n) \right.$$

$$\left. - \sum_{k=0}^{2n}\log(1+k) \right\}.$$

(d) Suppose $f(t)$ is continuously differentiable (namely, $f'(t)$ exists and are continuous) in $[0,\infty)$. Performing integration by parts to $P(t)f'(t)$ over each $[k, k+1]$, $0 \leq k \leq n-1$, we have

$$\sum_{k=0}^{n} f(k) = \int_0^n f(t)dt + \frac{1}{2}[f(0) + f(n)] + \int_0^n f'(t)P(t)dt, \quad n \geq 1.$$

Setting $f(t) = \mathrm{Log}\,(z+t)$, show that

$$\sum_{k=0}^{n} \mathrm{Log}\,(z+k) = \left(z + n + \frac{1}{2} \right)\mathrm{Log}\,(z+n) - n - \left(z - \frac{1}{2} \right)\mathrm{Log}\,z$$

$$+ \int_0^n \frac{P(t)}{z+t}\,dt.$$

Then, let $z = 1$ and use $2n$ to replace n in the resulted identity to show that

$$\frac{1}{2}\log \pi = \lim_{n \to \infty} \left\{ (2n+1)\log 2 + \left(2n + \frac{3}{2} \right)\log\left(\frac{1+n}{1+2n} \right) \right.$$

$$\left. + 2\int_0^n \frac{P(t)}{1+t}\,dt - \int_0^{2n} \frac{P(t)}{1+t}\,dt \right\}.$$

(e) Finally, show that

$$\lim_{n\to\infty}\left\{(2n+1)\log 2+\left(2n+\frac{3}{2}\right)\log\left(\frac{1+n}{1+2n}\right)\right\}=1-\frac{1}{2}\log 2;$$

$$\lim_{n\to\infty}\int_0^n\frac{P(t)}{1+t}dt=\lim_{n\to\infty}\int_0^{2n}\frac{P(t)}{1+t}dt=\int_0^\infty\frac{P(t)}{1+t}dt.$$

(5) (Ref. [1], pp. 205–206) Use Stirling's formula (5.6.3.4) or (5.6.3.6) to prove Euler's representation: $\Gamma(z)=\int_0^\infty e^{-t}t^{z-1}dt$, $\operatorname{Re}z>0$. Try the following steps: Set $F(z)=\int_0^\infty e^{-t}t^{z-1}dt$, $\operatorname{Re}z>0$, temporarily.

(a) $F(z+1)=zF(z)$ can be used to extend $F(z)$ analytically into $\mathbf{C}-\{0,-1,-2,\dots\}$. Hence, $\frac{F(z)}{\Gamma(z)}$ is analytic with period 1 in $\operatorname{Re}z>0$.

(b) $|F(z)|\le\int_0^\infty e^{-t}e^{\operatorname{Re}z-1}dt=F(\operatorname{Re}z)$, $1\le\operatorname{Re}z\le 2$, shows that $F(z)$ is bounded in the strip. In case $1\le\operatorname{Re}z\le 2$ and $|\operatorname{Im}z|$ is large, Stirling's formula implies that

$$\log|\Gamma(z)|=\frac{1}{2}\log 2\pi-\operatorname{Re}z+\left(\operatorname{Re}z-\frac{1}{2}\right)\log|z|$$
$$-\operatorname{Im}z\cdot\operatorname{Arg}z+\operatorname{Re}J(z)$$

$$\Rightarrow\quad\lim_{|\operatorname{Im}z|\to\infty}\frac{\log|\Gamma(z)|}{(\pi/2)|\operatorname{Im}z|}=-\lim_{|\operatorname{Im}z|\to\infty}\frac{\operatorname{Im}z}{|\operatorname{Im}z|}\cdot\frac{\operatorname{Arg}z}{\pi/2}\ge-1$$

$$\Rightarrow\quad\left|\frac{F(z)}{\Gamma(z)}\right|\sim Me^{\pi|\operatorname{Im}z|/2},\quad 1\le\operatorname{Re}z\le 2,$$

for some constant $M>0$.

(c) By (a) and Exercises B of Sec. 4.9.2, $\frac{F(z)}{\Gamma(z)}$ can be expressed as an analytic function of $\zeta=e^{2\pi iz}$ with isolated singularities at $z=0$ and ∞.

(d) By (b), $\left|\frac{F(z)}{\Gamma(z)}\right|\sim|\zeta|^{-\frac{1}{4}}$ as $\zeta\to 0$ and $\frac{F(z)}{\Gamma(z)}\sim|\zeta|^{\frac{1}{4}}$ as $\zeta\to\infty$. Hence 0 and ∞ are removable singularities and hence $\frac{F(z)}{\Gamma(z)}$ is a constant. Why is $F(z)\equiv\Gamma(z)$ in $\operatorname{Re}z>0$?

(6) Set $u(x)=(x+\frac{1}{2})\log(1+\frac{1}{x})-1$, $x>0$. Show that

1. $0<u(x)<\frac{1}{12x(x+1)}$;

2. $0<J(x)=-\int_0^\infty\frac{P(x)}{t+x}dt<\int_0^\infty\frac{[t]-t+\frac{1}{2}}{t+x}dt<\frac{1}{2x}$, $x>0$, and then

3. $\Gamma(x)=\sqrt{2\pi}x^{x-\frac{1}{2}}e^{-x}e^{\frac{\theta(x)}{12x}}$, $x>0$, $0<\theta(x)<1$.

Exercises B

The method used in Step 1 in the proof of (5.6.3.4) can be developed as a general setting. We state as

A summation formula for $\sum_{n=0}^{\infty} f(n)$. Suppose $f(z)$ is analytic in $\operatorname{Re} z \geq 0$ and satisfies the conditions: $z = x + iy$,

1. $\lim_{y \to \infty} |f(x \pm iy)| e^{\mp 2\pi y} = 0$ locally uniformly in $x \in [0, \infty)$;
2. $\int_0^{\infty} |f(x \pm iy)| \frac{1}{e^{2\pi y}-1} dy$ converges for $x \geq 0$ and approaches 0 as $x \to \infty$.

If one of the following

$$\sum_{n=0}^{\infty} f(n) \quad \text{and} \quad \int_0^{\infty} f(x)dx$$

converges, then the other converges, too, and

$$\sum_{n=0}^{\infty} f(n) = \frac{1}{2}f(0) + \int_0^{\infty} f(x)dx + i \int_0^{\infty} \frac{f(iy) - f(-iy)}{e^{2\pi y} - 1} dy.$$

Note that $\lim_{y \to 0} \frac{f(iy)-f(-iy)}{e^{2\pi y}-1} = \frac{i}{\pi} f'(0)$, hence the second integral in the right does converges at $y = 0$.   (5.6.3.7)

Sketch of proof. Adopt the path γ_n in Fig. 5.25. Then, as usual, we have

$$\frac{1}{2\pi i} \int_{\gamma_n} f(z) \cdot \pi \cot \pi z \, dz = \frac{1}{2}f(0) + \sum_{k=1}^{n} f(k).$$

Denote by $\gamma_n^{(1)}$ and $\gamma_n^{(2)}$, respectively, the part of γ_n that lies in the upper and the lower half-plane. Since $\cot \pi z = -i[1 + 2(e^{-2\pi i z} - 1)^{-1}]$, so (as $\varepsilon \to 0$)

$$\frac{1}{2\pi i} \int_{\gamma_n^{(1)}} f(z) \cdot \pi \cot \pi z \, dz$$

$$= -\frac{1}{2} \int_{\gamma_n^{(1)}} f(z) \left[1 + \frac{2}{e^{-2\pi i z} + 1} \right] dz$$

$$= \frac{1}{2} \int_0^{n+\frac{1}{2}} f(x)dx \quad \left(\text{integrating along the real interval } \left[0, n + \frac{1}{2}\right] \right)$$

$$- \int_{\gamma_n^{(1)}} \frac{f(z)}{e^{-2\pi i z} - 1} dz.$$

And so does a similar formula for integration along $\gamma_n^{(2)}$. Adding together, we have

$$\int_0^{n+\frac{1}{2}} f(x)dx + \int_{\gamma_n^{(1)}} \frac{f(z)}{e^{-2\pi i z} - 1} dz + \int_{\gamma_n^{(2)}} \frac{f(z)}{e^{2\pi i z} - 1} dz = \frac{1}{2}f(0) + \sum_{k=1}^{n} f(k).$$

$$(*_{16})$$

Finally, the restricted condition will impose that the second and the third integrals in the left approach 0 as $n \to \infty$. $\square$

(1) Choose $f(z+t) = \frac{1}{(z+t)^2}$, Re $z > 0$, and use (5.6.3.7) to derive (5.6.3.2).

(2) Apply (5.6.3.7) to $f(z) = \frac{1}{(z+1)^2}$ and show that

$$\sum_{n=1}^{\infty} \frac{1}{n^3} = 1 + \int_0^\infty \frac{6y - 2y^3}{(1+y^2)^3} \cdot \frac{1}{e^{2\pi y} - 1} dy.$$

(3) Let $\gamma = \lim_{n\to\infty}\{1 + \frac{1}{2} + \cdots + \frac{1}{n} - \log n\}$ be Euler's constant. Show that, by applying $(*16)$ to $f(z) = \frac{1}{z+1}$,

$$\gamma = \frac{1}{2} + 2 \int_0^\infty \frac{y}{1+y^2} \cdot \frac{1}{e^{2\pi y} - 1} dy.$$

5.7   The Riemann zeta Function $\zeta(z)$

Let $z = x + iy$ and $n \geq 1$ be integers.

In this section, $\log n$ in $n^{-z} = e^{-z \log n}$ is always understood as the real logarithm.

Since $|n^{-z}| = n^{-x}$ and $\sum_1^\infty \frac{1}{n^x} \leq \sum_1^\infty \frac{1}{n^\delta}$ for $x \geq \delta > 1$, hence $\sum_1^\infty \frac{1}{n^z}$ converges uniformly in Re $z = x \geq \delta > 1$. It follows that (refer to Example 3 in Sec. 5.1.1)

The Riemann zeta function. The series

$$\zeta(z) = \sum_{n=1}^{\infty} \frac{1}{n^z}$$

converges absolutely in Re $z > 1$ and uniformly in every closed half-plane Re $z \geq \delta > 1$ (and, hence, locally uniformly in Re $z > 1$) to an analytic function $\zeta(z)$, called *the Riemann zeta function*, simply denoted as *the ζ-series*.   (5.7.1)

This function originated from number theory. It plays an important role in the study of distribution of prime numbers.

Let $2, 3, 5, 7, 11, 13, 17, 19, 23, \ldots, p_n, \ldots$ be the sequence of all the prime numbers (suppose there are infinitely many of them and are arranged in increasing magnitude).

Absolute convergence of the ζ-series in Re $z > 1$ permits us to write

$$\zeta(z)(1 - 2^{-z}) = \sum_{n=1}^{\infty} \frac{1}{n^z} - \sum_{n=1}^{\infty} \frac{1}{(2n)^z} = \sum_{m_2 \geq 1} \frac{1}{m_2^z},$$

where m_2 runs through all the odd positive integers ≥ 1. Since

$$\sum_{m_2 \geq 1} \left| \frac{1}{m_2^z} \right| \leq \sum_{m_2 > 1} \frac{1}{m_2^x} \leq \sum_{n=1}^{\infty} \frac{1}{n^x},$$

the absolute convergence of $\sum_{m_2 \geq 1} \frac{1}{m_2^z}$ implies that

$$\zeta(z)(1 - 2^{-z})(1 - 3^{-z}) = \sum_{m_2 \geq 1} \frac{1}{m_2^z} - \sum_{m_2 \geq 1} \frac{1}{(3m_2)^z} = \sum_{m_3 \geq 1} \frac{1}{m_3^z},$$

where m_3 runs through all positive integers that do not contain factors 2 and 3. Continue this process and we have, in general,

$$\zeta(z)(1 - 2^{-z})(1 - 3^{-z}) \cdots (1 - p_n^{-z}) = \sum_{m_n \geq 1} \frac{1}{m_n^z}, \qquad (*_1)$$

where m_n assumes these positive integers that contain none of the prime factors $2, 3, \ldots, p_n$. Remind that the first term in the series $(*_1)$ is 1. Furthermore, since

$$\left| \sum_{m_n \geq 1} \frac{1}{m_n^z} \right| \leq 1 + \sum_{m_n \geq p_{n+1}} \frac{1}{m_n^x}$$

the series in $(*_1)$ converges absolutely in $\operatorname{Re} z > 1$ and uniformly in $\operatorname{Re} z \geq \delta > 1$, and approaches 1 in $\operatorname{Re} z > 1$ as $n \to \infty$. Therefore, setting $n \to \infty$ in $(*_1)$, we obtain

$$\zeta(z) \lim_{n \to \infty} \prod_{k=1}^{n} (1 - p_k^{-z}) = 1 \qquad (*_2)$$

absolutely in $\operatorname{Re} z > 1$ and uniformly in $\operatorname{Re} z \geq \delta > 1$. In conclusion, summarize the above as

The infinite product representation of the ζ-function. Let $p_1 = 2$, $p_3 = 3, \ldots, p_n, \ldots$, be the sequence of all prime numbers in increasing magnitude. Then,

$$\zeta(z) = \prod_{n=1}^{\infty} (1 - p_n^{-z})^{-1}, \quad \operatorname{Re} z > 1 \quad \text{or}$$

$$\frac{1}{\zeta(z)} = \prod_{n=1}^{\infty} \left(1 - \frac{1}{p_n^z} \right), \quad \operatorname{Re} z > 1$$

absolutely in $\operatorname{Re} z > 1$ and uniformly in each closed half-plane $\operatorname{Re} z \geq \delta > 1$.

(5.7.2)

From $(*_2)$, it is easily seen that $\zeta(z) \neq 0$ in $\operatorname{Re} z > 1$. Note that, in the above process lending to $(*_2)$, it is presumed that there are infinitely many primes. In case it is not and p_N is the largest prime, then we would stop at

$(*_1)$ which turns out to be $\zeta(z)(1 - 2^{-z}) \cdots (1 - p_N^{-z}) = 1$. This will imply that $\zeta(z)$ has a finite limit as $z \to 1$, contradicting to $\sum_1^\infty \frac{1}{n} = \infty$. This argument, due to Euler, obtains purely algebraic result through analytic method.

The following is divided into two subsections.

Section (1) Complex line integral representation of the ζ-function and its analytic extension

Use nt to replace t in $\Gamma(z) = \int_0^\infty e^{-t} t^{z-1} dt$, $\mathrm{Re}\, z > 0$, $n \geq 1$, and we have

$$n^{-z}\Gamma(z) = \int_0^\infty t^{z-1} e^{-nt} dt, \quad \mathrm{Re}\, z > 0, \quad n \geq 1.$$

When restricted to $\mathrm{Re}\, z \geq 1 + x_0$ $(x_0 > 0)$:

$$\Gamma(z)\zeta(z) = \Gamma(z) \lim_{n \to \infty} \sum_{k=1}^n k^{-z} = \lim_{n \to \infty} \int_0^\infty t^{z-1} \sum_{k=1}^n e^{-kt} dt$$

$$= \lim_{n \to \infty} \int_0^\infty t^{z-1} \frac{e^{-t} - e^{-(n+1)t}}{1 - e^{-t}} dt$$

$$= \int_0^\infty \frac{t^{z-1}}{e^t - 1} dt - \lim_{n \to \infty} \int_0^\infty t^{z-1} \frac{e^{-(n+1)t}}{1 - e^{-t}} dt, \qquad (*_3)$$

where $\int_0^\infty \frac{t^{z-1}}{e^t - 1} dt$ converges absolutely and uniformly in $\mathrm{Re}\, z \geq 1 + x_0$ (why?). While, on using $1 + t \leq e^t$, $t \geq 0$,

$$\left| \int_0^\infty t^{z-1} \frac{e^{-(n+1)t}}{1 - e^{-t}} dt \right| \leq \int_0^\infty t^{\mathrm{Re}\, z - 2} e^{-nt} dt$$

$$= n^{1 - \mathrm{Re}\, z} \Gamma(\mathrm{Re}\, z - 1) \to 0 \quad \text{as } n \to \infty.$$

Letting $n \to \infty$ in $(*_3)$, we obtain

A relation between Γ-function and ζ-function.

$$\zeta(z) = \frac{1}{\Gamma(z)} \int_0^\infty \frac{t^{z-1}}{e^t - 1} dt, \quad \mathrm{Re}\, z > 1$$

converges absolutely and locally uniformly in $\mathrm{Re}\, z > 1$. $\hspace{2cm}$ (5.7.3)

Note that, in $t^{z-1} = e^{(z-1)\log t}$, $\log t$ means real logarithm and the path of integration is taken along the positive real axis $(0, \infty)$.

Consider the integrand in (5.7.3)

$$F(t) = \frac{t^{z-1}}{e^t - 1}, \qquad (*_4)$$

where t now acts as a *complex variable* and z a fixed complex constant. Since $e^t = 1$ at $t = 2n\pi i$ $(n = 0, \pm 1, \pm 2, \ldots)$, $F(t)$ has a simple pole at $z = 0$.

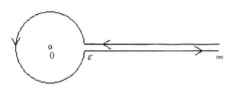

Fig. 5.27

Choose $0 < \varepsilon < 2\pi$. Construct a path γ_ε shown in Fig. 5.27, starting at ∞ and decreasing to ε along the upper side of $(0, \infty)$, then circulating around the circle $|t| = \varepsilon$ in the counterclockwise direction and finally, returning to ∞ along the lower side of $(0, \infty)$.

Log t in $t^{z-1} = e^{(z-1)\mathrm{Log}\, t}$ is the principle branch defined in $\mathbf{C} - [0, \infty)$; hence, $\mathrm{Log}\, t = \log t$ is the real logarithm for real $t > 0$ along the upper side of $(0, \infty)$ and $\mathrm{Log}\, t = \log t + 2\pi i$ along the lower side of $(0, \infty)$. Integrate $F(t)$ along γ_ε and we have

$$\int_{\gamma_\varepsilon} \frac{t^{z-1}}{e^t - 1} dt = \int_\infty^\varepsilon \frac{t^{z-1}}{e^t - 1} dt + \int_{|z|=\varepsilon} \frac{t^{z-1}}{e^t - 1} dt + \int_\varepsilon^\infty \frac{t^{z-1} e^{2\pi i(z-1)}}{e^t - 1} dt$$

$$= (e^{2\pi i z} - 1) \int_\varepsilon^\infty \frac{t^{z-1}}{e^t - 1} dt + \int_{|z|=\varepsilon} \frac{t^{z-1}}{e^t - 1} dt. \qquad (*_5)$$

Owing to Cauchy's integral theorem, $\int_{|t|=\varepsilon_1} = \int_{|t|=\varepsilon_2}$ holds for any $0 < \varepsilon_1 \le \varepsilon_2 < 2\pi$. This indicates that the integral $\int_{|t|=\varepsilon}$ in $(*_5)$ is a constant, independent of the choice of such ε. It can be evaluated as follows: Along $|t| = \varepsilon$, if $z = x + iy$,

$$|t^{z-1}| = e^{(x-1)\log \varepsilon - y \mathrm{Arg}\, t} \le \varepsilon^{x-1} e^{2\pi |y|},$$

$$|e^t - 1| \ge M\varepsilon \quad (M > 0, \text{ a constant})$$

$$\Rightarrow \left| \int_{|t|=\varepsilon} \frac{t^{z-1}}{e^t - 1} dt \right| \le \int_{|t|=\varepsilon} \frac{\varepsilon^{x-1} e^{2\pi |y|}}{M\varepsilon} |dt|$$

$$= 2\pi M^{-1} \varepsilon^{x-1} e^{2\pi |y|} \to 0 \quad \text{as } \varepsilon \to 0 \text{ if } \mathrm{Re}\, z = x > 1.$$

On letting $\varepsilon \to 0$ in the right side in $(*_5)$ and using the above result, we have, by (5.7.3).

$$\int_{\gamma_\varepsilon} \frac{t^{z-1}}{e^t - 1} dt = (e^{2\pi i z} - 1) \int_0^\infty \frac{t^{z-1}}{e^t - 1} dt = (e^{2\pi i z} - 1)\zeta(z)\Gamma(z), \quad \mathrm{Re}\, z > 1.$$

Summarize as

The complex line integral representation of the ζ-function. Let γ_ε $(0 < \varepsilon < 2\pi)$ be as in Fig. 5.27. Then, if $\operatorname{Re} z > 1$,

$$\zeta(z) = \frac{1}{(e^{2\pi i z} - 1)\Gamma(z)} \int_{\gamma_\varepsilon} \frac{t^{z-1}}{e^t - 1} dt = \frac{e^{-\pi i z}\Gamma(1 - z)}{2\pi i} \int_{\gamma_\varepsilon} \frac{t^{z-1}}{e^t - 1} dt$$

$$= -\frac{\Gamma(1 - z)}{2\pi i} \int_{\gamma_\varepsilon} \frac{(-t)^{z-1}}{e^t - 1} dt,$$

where t acts as a complex variable, and $t^{z-1} = e^{(z-1)\operatorname{Log} t}$ and $\operatorname{Log} t$ is the principal branch in $\mathbf{C} - [0, \infty)$ (and hence, designate $\operatorname{Log}(-t) = \log|-t| + i\operatorname{Arg}(-t)$ with $-\pi < \operatorname{Arg}(-t) = \operatorname{Arg} t - \pi < \pi$ on the third integral).
$$(5.7.4)$$

The integral in (5.7.4)

$$\int_{\gamma_\varepsilon} \frac{t^{z-1}}{e^t - 1} dt, \quad z \in \mathbf{C} \tag{$*_6$}$$

defines an *entire function*. This is so according to (4.7.4), or, more accurately, according to Exercise B(2) of Sec. 5.3.3. Therefore, the meromorphic function

$$\frac{e^{-\pi i z}\Gamma(1 - z)}{2\pi i} \int_{\gamma_\varepsilon} \frac{t^{z-1}}{e^t - 1} dt, \quad z \in \mathbf{C} - \{1, 2, \ldots\} \tag{$*_7$}$$

has possible simple poles at $z = 1, 2, \ldots$. Since $\zeta(z)$ is already known to be analytic in $\operatorname{Re} z > 1$, the formula in (5.7.4) shows that the poles of $(*_7)$ at $z = 2, 3, \ldots$ must cancel out the zeros of $(*_6)$. So, the function in $(*_7)$ has only one simple pole at $z = 1$ with the associated residue given by

$$\left(\frac{e^{-\pi i z}}{2\pi i} \int_{\gamma_\varepsilon} \frac{t^{z-1}}{e^t - 1} dt \quad \text{evaluated at } z = 1\right) \cdot \operatorname{Re} s(\Gamma(1 - z); 1)$$

$$= e^{-\pi i}\operatorname{Re} s\left(\frac{1}{e^t - 1}; 0\right) \operatorname{Re} s(\Gamma(1 - z); 1) = (-1) \cdot 1 \cdot (-1) = 1.$$

In summary, the importance of (5.7.4) lies in the following fact.

The analytic continuation of the ζ-function form $\operatorname{Re} z > 1$ to $\mathbf{C} - \{1\}$. Define

$$\zeta(z) = \frac{e^{-\pi i z}\Gamma(1 - z)}{2\pi i} \int_{\gamma_\varepsilon} \frac{t^{z-1}}{e^t - 1} dt, \quad z \in \mathbf{C} - \{1\}.$$

Then $\zeta(z)$ is meromorphic in $\mathbf{C}-\{1\}$ with only one simple pole at $z = 1$ and

$$\operatorname{Re} s(\zeta(z); 1) = 1.$$

Known that $\frac{t}{e^t-1} = 1 - \frac{t}{2} + \sum_{n=1}^{\infty} \frac{B_{2n}}{(2n)!} t^{2n}$, $|t| < 2\pi$ (see Example 5 of Sec. 4.8), it follows immediately that

$$\zeta(0) = -\frac{1}{2};$$

$$\zeta(-2n) = 0, \quad n \geq 1;$$

$$\zeta(1 - 2n) = \frac{-B_{2n}}{2n}, \quad n \geq 1.$$

The points $-2n(n \geq 1)$ are called the *trivial zeros* of the ζ-function.

(5.7.5)

Some books state $\frac{t}{e^t-1} = 1 - \frac{t}{2} + \sum_{n=1}^{\infty} (-1)^{n-1} \frac{B_{2n}}{(2n)!} t^{2n}$. So, in this case, $\zeta(1 - 2n) = (-1)^n \frac{B_{2n}}{2n}$, $n \geq 1$.

Section (2) Riemann's functional equation about the ζ-function

Riemann recognized the following rather simple relation between $\zeta(1 - z)$, $\operatorname{Re} z < 0$ and $\zeta(z)$, $\operatorname{Re} z > 1$, via the gamma function.

The functional equation.

$$\zeta(z) = 2^z \pi^{z-1} \sin \frac{\pi z}{2} \Gamma(1 - z)\zeta(1 - z), \quad z \in \mathbf{C} - \{1\}, \quad \text{or}$$

$$\zeta(1 - z) = 2^{1-z} \pi^{-z} \cos \frac{\pi z}{2} \Gamma(z)\zeta(z), \quad z \in \mathbf{C} - \{0\}.$$

In particular,

$$\zeta(2n) = 2^{2n-1} \pi^{2n} (-1)^n \frac{B_n}{(2n)!}, \quad n \geq 1;$$

$$\zeta'(0) = -\frac{1}{2} \log 2\pi. \tag{5.7.6}$$

The series representation (5.7.1) shows that $|\zeta(z)|$ is dominated by $\zeta(\operatorname{Re} z)$, namely, $|\zeta(z)| \leq \zeta(\operatorname{Re} z)$, in the half-plane $\operatorname{Re} z > 1$. As a consequence of this functional equation, the behavior of $\zeta(z)$ can also be controlled in $\operatorname{Re} z < 0$. For instance, since $\zeta(z) \neq 0$ in $\operatorname{Re} z > 1$ (see $(*_2)$ or (5.7.2)), it follows that the only zeros of $\zeta(z)$ in $\operatorname{Re} z < 0$ are the trivial ones $-2n$ for $n \geq 1$ (see (5.7.5)). Therefore, all its possible nontrivial zeros should lie in the so-called *critical strip* $0 \leq \operatorname{Re} z \leq 1$. It can be shown that $\zeta(z)$ does not have zeros along the vertical lines $\operatorname{Re} z = 0$ and $\operatorname{Re} z = 1$. And its *nonreal* zeros thus lie in $0 < \operatorname{Re} z < 1$ and appears in conjugate pairs and are symmetric with respect to the *critical line* $\operatorname{Re} z = \frac{1}{2}$. Riemann *conjectured* that *all the nonreal zeros of* $\zeta(z)$ *lie in the line* Re $z = \frac{1}{2}$. This is neither proved nor disproved yet. G. H. Hardy showed that there are infinitely many zeros lying on Re $z = \frac{1}{2}$; N. Levinson proved in 1975 that asymptotically more than one-third of the zeros lie on Re $z = \frac{1}{2}$; P. Brent showed in 1978 that there are 81×10^6 zeros on the line; and all zeros found by computer work lie on the line, too.

The following proof illustrates another less trivial usefulness of the residue method as Lindelöf's proof of Stirling's formula (see Sec. 5.6.3) had shown.

Proof. Recall the integrand $F(t)$ in $(*_4)$ has simple poles at $\pm 2n\pi i$ for $n \geq 0$. Choose a path γ_n shown in Fig. 5.28, composed of the following curves: the upper side of the cut $[(2n + 1)\pi, \infty)$, descending from ∞ to $(2n + 1)\pi$; then followed by a square with vertices at $(2n + 1)\pi(\pm 1 \pm i)$, in counterclockwise direction; and finally, along the lower side of $[(2n+1)\pi, \infty)$ form $(2n + 1)\pi$ to ∞. Let $\gamma_{\varepsilon,n}$ be the circle $|t| = \varepsilon$, where $0 < \varepsilon < 2\pi$. Note that $\gamma_n - \gamma_{\varepsilon,n}$ approaches γ_ε in Fig. 5.27 as $n \to \infty$.

Now, for $1 \leq k \leq n$,

$$\operatorname{Re} s(F(t); 2k\pi i) + \operatorname{Re} s(F(t); -2k\pi i)$$

$$= (2k\pi)^{z-1} e^{\pi i(z-1)} \{ e^{\pi i(z-1)/2} + e^{-\pi i(z-1)/2} \}$$

$$= -2(2k\pi)^{z-1} e^{\pi i z} \sin \frac{1}{2}\pi z.$$

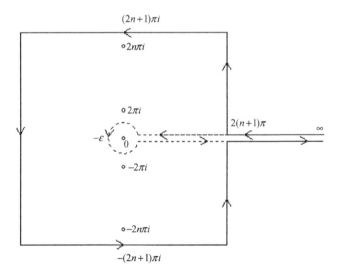

Fig. 5.28

By the residue theorem,

$$\int_{\gamma_n - \gamma_{\varepsilon,n}} \frac{t^{z-1}}{e^t - 1} dt = 2\pi i \sum_{k=1}^{n} \{\operatorname{Re} s(F(t); 2k\pi i) + \operatorname{Re} s(F(t); -2k\pi i)\}$$

$$\Rightarrow \int_{\gamma_{\varepsilon,n}} \frac{t^{z-1}}{e^t - 1} dt = \int_{\gamma_n} \frac{t^{z-1}}{e^t - 1} dt + 4\pi i e^{\pi i z} \sin \frac{\pi z}{2} \sum_{k=1}^{n} (2k\pi)^{z-1}. \qquad (*_8)$$

What remains is to evaluate the integral $\int_{\gamma_n}$ as $n \to \infty$.

Divide γ_n into $\gamma_n^{(1)} + \gamma_n^{(1)}$ where $\gamma_n^{(1)}$ is the part on the square and $\gamma_n^{(2)}$ that outside the square, namely, the upper and lower side of the cut $[(2n+1)\pi, \infty)$. Along $\gamma_n^{(1)}$:

$$|e^t - 1| \geq \min_{\gamma_n^{(1)}} |e^t - 1| = m > 0$$

$$|t^{z-1}| = e^{(\operatorname{Re} z - 1) \log |t| - (\operatorname{Im} z) \operatorname{Arg} t}$$

$$\leq A e^{(\operatorname{Re} z - 1) \log (2n+1)\sqrt{2}} (A > 0, \text{ a constant})$$

$$= A[(2n+1)\sqrt{2}]^{\operatorname{Re} z - 1} = O(n^{\operatorname{Re} z - 1}).$$

$$\Rightarrow \left| \int_{\gamma_n^{(1)}} \frac{t^{z-1}}{e^t - 1} dt \right| \leq \frac{O(n^{\operatorname{Re} z - 1})}{m} (16n + 8)$$

$$= O(n^{\operatorname{Re} z}) \to 0 \quad \text{as } n \to \infty, \quad \text{if } \operatorname{Re} z < 0.$$

Along the upper side of $\gamma_n^{(2)}$:

$$\left| \frac{t^{z-1}}{e^t - 1} \right| \leq \frac{O(t^{\operatorname{Re} z - 1})}{e^\pi - 1}, \quad \text{if } \operatorname{Re} z < 0;$$

while, along the lower side of $\gamma_n^{(2)}$:

$$\left| \frac{t^{z-1}}{e^t - 1} \right| \leq \frac{e^{(\operatorname{Re} z - 1) \log t - (\operatorname{Im} z) \cdot 2\pi}}{e^\pi - 1} \leq \frac{O(t^{\operatorname{Re} z - 1})}{e^\pi - 1}, \quad \text{if } \operatorname{Re} z < 0.$$

Therefore, there is a constant $M > 0$ such that, if $\operatorname{Re} z < 0$,

$$\left| \int_{\gamma_n^{(2)}} \frac{t^{z-1}}{e^t - 1} dt \right| \leq \frac{2M}{e^\pi - 1} \int_{2n+1}^{\infty} t^{\operatorname{Re} z - 1} dt$$

$$= \frac{-2M}{(e^\pi - 1)\operatorname{Re} z} (2n+1)^{\operatorname{Re} z} \to 0 \quad \text{as } n \to \infty.$$

On using these estimate and letting $n \to \infty$ in $(*_8)$ under the assumption that $\operatorname{Re} z < 0$, we obtain

$$\int_{\gamma_\varepsilon} \frac{t^{z-1}}{e^t - 1} dt = 4\pi i e^{\pi i z} \sin \frac{\pi z}{2} \sum_{k=1}^{\infty} (2k\pi)^{z-1}$$

$$= 4\pi i e^{\pi i z} (2\pi)^{z-1} \sin \frac{\pi z}{2} \zeta(1 - z), \quad \operatorname{Re} z < 0.$$

$$\Rightarrow \text{ (by (5.7.5)) } \zeta(z) = 2^z \pi^{z-1} \sin \frac{\pi z}{2} \Gamma(1 - z) \zeta(1 - z), \quad \operatorname{Re} z < 0. \quad (*_9)$$

Remind, as we have already shown, that $\zeta(z)$ is analytic in $\mathbf{C} - \{1\}$ and hence, $\zeta(1-z)$ is analytic in $\mathbf{C} - \{0\}$, and $\Gamma(1-z)$ is analytic in $\mathbf{C} - \{1, 2, \ldots\}$. Hence, the function in the right in $(*_9)$ is analytic in $\mathbf{C} - \{1, 2, \ldots\}$. Wherein, the simple zeros $2n$ $(n \geq 1)$ of $\sin \frac{\pi z}{2}$ cancel against the simple poles $2n$ of $\Gamma(1 - z)$, and the simple zeros $2n + 1 (n \geq 1)$ of $\zeta(1 - z)$ cancel against the simple poles $2n + 1$ of $\Gamma(1 - z)$, too. Consequently, this function is indeed analytic in $\mathbf{C} - \{1\}$ and agrees with $\zeta(z)$, by $(*_9)$, on the open set $\operatorname{Re} z < 0$. Hence, $(*_9)$ is true for all $z \neq 1$.

By the way, to compute $\zeta'(0)$: Differentiate logarithmically the equivalent equation obtained by replacing z by $1 - z$ in $(*_9)$ and we obtain

$$-\frac{\zeta'(1 - z)}{\zeta(1 - z)} = -\log 2\pi - \frac{\pi}{2} \tan \frac{\pi z}{2} + \frac{\Gamma'(z)}{\Gamma(z)} + \frac{\zeta'(z)}{\zeta(z)}. \quad (*_{10})$$

The meromorphic functions in the right have the Laurent expansions at $z = 1$ as:

$$\frac{\pi}{2} \tan \frac{\pi z}{2} = -\frac{1}{z - 1} + O(|z - 1|)$$

(see Exercise A(5) of Sec. 4.8 and Exercise A(2) of Sec. 4.9.2);

$$\frac{\Gamma'(z)}{\Gamma(z)} = \frac{\Gamma'(1)}{\Gamma(1)} + \cdots = -\gamma + \text{analytic terms in } (z - 1) \text{ (see (5.6.2.2))};$$

$$\frac{\zeta'(z)}{\zeta(z)} = \frac{-(z - 1)^{-2} + c + \cdots}{(z - 1)^{-1} + \gamma + c(z - 1) + \cdots}$$

$$= -\frac{1}{z - 1} + \gamma + \cdots \text{ (by using (5.1.1.4))},$$

where γ is Euler's constant. Letting $z \to 1$ in $(*_{10})$ and noting that $\zeta(0) = -\frac{1}{2}$ (see (5.7.5)), it comes $\zeta'(0) = -\frac{1}{2} \log 2\pi$. $\qquad \square$

As for another form of the functional equation, we use the factor $1 - z$ to cancel out the simple pole of $\zeta(z)$ at $z = 1$ in $(1 - z)\zeta(z)$, and the simple poles of $\Gamma\left(\frac{z}{2}\right)$ to cancel against the simple zeros $-2n\,(n \geq 1)$ of $\zeta(z)$ in $\Gamma\left(\frac{z}{2}\right)\zeta(z)$. And one more factor z is needed in order to remove the simple pole of $\Gamma\left(\frac{z}{2}\right)$ at 0. Consequently, we formulate

Another form of the ζ-function. The function

$$\xi(z) = \frac{1}{2}z(1-z)\pi^{-z/2}\Gamma\left(\frac{z}{2}\right)\zeta(z)$$

is entire of order 1 and satisfies the relation:

$$\xi(z) = \xi(1-z), \quad z \in \mathbf{C} - \{0,1\}, \quad \text{or}$$

$$\xi\left(\frac{1}{2}+iz\right) = \xi\left(\frac{1}{2}-iz\right), \quad z \in \mathbf{C} - \left\{\pm\frac{1}{2}i\right\},$$

which is equivalent to Legendre's duplication formula (see (4) in (5.6.2.1)):

$$\sqrt{\pi}\,\Gamma(2z) = 2^{2z-1}\Gamma(z)\Gamma\left(z+\frac{1}{2}\right). \tag{5.7.7}$$

Sketch of proof. In case $z \neq 0,1$,

$$\xi(z) = \xi(1-z) \quad \text{holds.}$$

$\Leftrightarrow$ (by definition of $\xi(z)$ and (5.7.6))

$$\pi^{-z/2}\Gamma\left(\frac{z}{2}\right)\zeta(z) = \pi^{(z-1)/2}\Gamma\left(\frac{1-z}{2}\right)\zeta(1-z)$$

$$= 2^{1-z}\pi^{-(z+1)/2}\Gamma(z)\Gamma\left(\frac{1-z}{2}\right)\cos\frac{\pi z}{2}$$

$$\Leftrightarrow \quad \cos\frac{\pi z}{2}\Gamma(z)\Gamma\left(\frac{1-z}{2}\right) = 2^{z-1}\pi^{1/2}\Gamma\left(\frac{z}{2}\right)$$

$$\Leftrightarrow \quad \left(\text{on using } \Gamma\left(\frac{1-z}{2}\right)\Gamma\left(\frac{1+z}{2}\right) = \frac{\pi}{\cos\pi z/2}\right)\sqrt{\pi}\,\Gamma(z)$$

$$= 2^{z-1}\Gamma\left(\frac{z}{2}\right)\Gamma\left(\frac{1+z}{2}\right).$$

Set $2z$ in place of z in this last identity and we get Legendre's formula.

As for the order of $\xi(z)$, it is sufficient to estimate the growth of $|\zeta(z)|$ for $\operatorname{Re}z \geq \frac{1}{2}$ because of $\xi(z) = \xi(1-z)$. Owing to (5.5.3.4) and Exercise A(1) there, it is thus sufficient to estimate the growth of $\Gamma\left(\frac{z}{2}\right)$ and $\zeta(z)$, respectively.

The growth of $\Gamma\left(\frac{z}{2}\right)$: By Stirling's formula (5.6.3.4),

$$\Gamma\left(\frac{z}{2}\right) = \sqrt{2\pi}\left(\frac{z}{2}\right)^{(z-1)/2} e^{-z/2} e^{J(z/2)}$$

$$\Rightarrow \log\left|\Gamma\left(\frac{z}{2}\right)\right| = \frac{1}{2}\log 2\pi + \mathrm{Re}\left(\frac{z-1}{2}\right)\log\left|\frac{z}{2}\right|$$

$$- \mathrm{Im}\left(\frac{z-1}{2}\right)\mathrm{Arg}\frac{z}{2} - \frac{1}{2}\mathrm{Re}\,z + \mathrm{Re}\,J\left(\frac{z}{2}\right)$$

$$\Rightarrow \log\left|\Gamma\left(\frac{z}{2}\right)\right| \leq M|z|\log|z| \quad \text{for some constant } M > 0 \text{ and large } |z|$$

and $\mathrm{Re}\,z > 0$. $(*_{11})$

Moreover, by Stirling's formula in Binet form (5.6.3.6), this estimate is precise for real $z = x$.

The growth of $\zeta(z)$: Denote by $[x]$ the Gaussian integer for real x. It can be shown that, if $\mathrm{Re}\,z > 1$,

$$\zeta(z) = z\int_1^\infty \frac{[x]}{x^{z+1}}dx$$

$$= \sum_{k=1}^N \frac{1}{k^z} + \frac{N^{1-z}}{z-1} - z\int_N^\infty \frac{x-[x]}{x^{z+1}}dx \qquad (*_{12})$$

for any positive integer $N \geq 1$. The advantage of the second expression in $(*_{12})$ over the first one lies on the fact that the integral $\int_N^\infty \frac{x-[x]}{x^{z+1}}dx$ converges for $\mathrm{Re}\,z > 0$. Therefore, $(*_{12})$ is still valid for $\mathrm{Re}\,z > 0$. When restricted to $\mathrm{Re}\,z \geq \frac{1}{2}$,

$$\left|\int_N^\infty \frac{x-[x]}{x^{z+1}}dx\right| \leq \int_N^\infty \frac{1}{x^{\mathrm{Re}\,z+1}}dx = \frac{N^{-\mathrm{Re}\,z}}{\mathrm{Re}\,z} \leq 2N^{-\frac{1}{2}}$$

$$\Rightarrow |\zeta(z)| \leq N + \frac{N^{\frac{1}{2}}}{|z|-1} + 2N^{-\frac{1}{2}}|z|$$

$$\leq M|z|^{\frac{2}{3}}, \quad \text{where } M \text{ is a constant,} \qquad (*_{13})$$

if $|z|$ is large and N is chosen as the integer closest to $|z|^{\frac{2}{3}}$.

Combining $(*_{11})$ and $(*_{13})$, and the very definition (5.5.3.3) of the order, it follows that $\xi(z)$ is of order 1. $\square$

The infinite product representation (5.7.2) and the fact that $\zeta(z)$ is analytic in $\{\mathrm{Re}\,z \geq 1\} - \{1\}$ so that $(z-1)\zeta(z)$ is analytic and zero-free in $\mathrm{Re}\,z \geq 1$ can be used to prove the following remarkable

Prime number theorem. *Let $\pi(N)$ denote the number of the primes not larger than the positive integer N. Then,*

$$\lim_{N \to \infty} \frac{\pi(N) \log N}{N} = 1,$$

namely, $\pi(N) \sim N/\log N$. (5.7.8)

One may refer to Ref. [71] for a proof.

Remark (A millennium prize problem). Let C be an elliptic curve over $\mathbf{Q}$ (the rational field). An affine model for the curve in Weierstrass form is given by

$$y^2 = x^3 + ax + b, \quad a, b \in \mathbf{Z} \text{ (the set of integers)}.$$

Set

$\triangle =$ the discriminant of the cubic;

$N_p =$ the cardinal number of the set {the solutions of $y^2 = x^3 + ax + b$ mod p}, and

$a_p = p - N_p$

The incomplete L-series of C is defined as

$$L(C, s) = \prod_{p \nmid 2\triangle} (1 - a_p p^{-s} + p^{1-2s})^{-1},$$

considered as a function of the complex variables s (compare to Exercise A(10) of Sec. 5.5.2). It has been proved in 1995, 2001, that $L(C, s)$ can be analytically continued from $\operatorname{Re} s > \frac{3}{2}$ to the whole complex plane. Hence comes the

Conjectures (B. Birch and H. Swinnerton-Dyer, 1965). *The Taylor series expansion of $L(C, s)$ at $s = 1$ has the form*

$$L(C, s) = c(s - 1)r + higher \ order \ terms$$

with $c \neq 0$ and $r = \operatorname{rank}(C(\mathbf{Q}))$.

For details, read *Andrew Wiles: The Birch and Swinnerton–Dyer Conjecture, Clay Math Institute Millennium Problems, www.claymath.org/ millennium.*

5.8   Normal Families of Analytic (Meromorphic) Functions

The concept of normal family was initiated by P. Montel in 1907.

Let Ω be a domain in $\mathbf{C}^*$ (see Remark 2 in Sec. 2.4) and $\mathfrak{F}$ be a family of analytic or meromorphic function in Ω. If every sequence in $\mathfrak{F}$ contains a subsequence that

1. converges locally uniformly in Ω to an analytic or a meromorphic function, or
2. converges locally uniformly in Ω to ∞ (namely, for each compact set K in Ω and each positive number R, there is an integer $N = N(K, R) \geq 1$ so that $|f_n(z)| \geq R$ for all $n \geq N$ and $z \in K$),   (5.8.1)

then $\mathfrak{F}$ is said to be *normal* in Ω. In case Ω is a domain in the finite complex plane $\mathbf{C}$ and $\mathfrak{F}$ contains only analytic functions and Case 2 does not happen, the convergence is in the Euclidean metric; otherwise, in spherical chord metric (see Sec. 1.6). This is the classical definition for the normal family and is the version for functions of Bolzano–Weierstrass's theorem for bounded infinite point set or sequence (see Sec. 1.9).

Section 5.8.1 will establish criteria for normality. Here, in Exercises B, we will formulate the concept of normal family in terms of metric space with Ascoli–Arzela's theorem as our main concern.

Lots of concrete examples are listed in Sec. 5.8.2. Important ones among all are families of uniformly bounded analytic functions, of univalent functions in the unit disk, etc.

The main theme in Sec. 5.8.3 is to prove *Montel's normality criterion*, obtained by Montel in 1912, which relates to the values taken by members of the family. There are several proofs for it. The one we used is an application of the elliptic modular function (see (5.8.3.5)) and the monodromy theorem (see (5.2.2.1) and (5.2.2.2)). This criterion will, in turn, be adopted to prove *Picard's first* and *second theorems* as we have mentioned in (4.10.3.3) and even more such as Julia direction, covering problems, Landau and Schottky theorems, etc., all beyond the scope of this book. Also, Schwarz–Ahlfors's Lemma in Sec. 5.8.4 can be used to prove Picard's and Montel's theorems through the method of Gauss's curvature. See Refs. [2, 51] for details.

As a meaningful application of the concept of the normal family and Schwarz–Pick's theorem, and as a continuation of Sec. 5.4.3, Sec. 5.8.5 presents some results in complex dynamical systems which are of current research.

5.8.1   *Criteria for normality*

Section (1) Family of analytic functions

We had shown in (5.3.3.10) that the local uniform boundedness is a sufficient condition for a sequence of analytic functions to be normal in a domain. It is indeed a necessary condition, too. To extend this result to a family $\mathfrak{F}$ of analytic functions defined in a domain Ω, some terminologies about $\mathfrak{F}$ are introduced as follows:

1. $\mathfrak{F}$ is *locally uniformly bounded in Ω if*, for each compact set K *in* Ω,

$$\sup_{f \in \mathfrak{F}} \sup_{z \in K} |f(z)| < \infty \qquad (5.8.1.1)$$

   holds. This is equivalent to say that $\mathfrak{F}$ is uniformly bounded in a neighborhood of each point in Ω.

2. $\mathfrak{F}$ is *locally (uniformly) equicontinuous if*, for each compact set K in Ω and $\varepsilon > 0$, there is a $\delta = \delta(K, \varepsilon) > 0$ such that

$$\sup_{f \in \mathfrak{F}} \sup_{\substack{z,\, z' \in K \\ |z - z'| < \delta}} |f(z) - f(z')| < \varepsilon \qquad (5.8.1.2)$$

   holds. Equivalently, for each point $z_0 \in \Omega$ and $\varepsilon > 0$, there is a $\delta = \delta(z_0, \varepsilon) > 0$ so that $|f(z) - f(z_0)| < \varepsilon$ wherever $|z - z_0| < \delta$, simultaneously for all $f \in \mathfrak{F}$.

It should be clear to the readers what it means that $\mathfrak{F}$ is *uniformly bounded* or (uniformly) *equicontinuous* on the whole set Ω.

Our main result is the

Criteria for normality of a family $\mathfrak{F}$ of analytic functions in a domain Ω in **C**. The following are equivalent.

(1) $\mathfrak{F}$ is normal, namely, every sequence in $\mathfrak{F}$ contains a subsequence converging locally uniformly (to an analytic function in Ω).

(2) $\mathfrak{F}$ is *local normal* in Ω, namely, $\mathfrak{F}$ is normal in a neighborhood of each point in Ω.

(3) $\mathfrak{F}$ is locally uniformly bounded in Ω.

(4) $\mathfrak{F}$ possesses the following properties:

   1. $\mathfrak{F}$ is locally equicontinuous in Ω, and
   2. there is a point $z_0 \in \Omega$ so that set $\{f(z_0) | f \in \mathfrak{F}\}$ is bounded.

   Consequently, the family $\{f' | f \in \mathfrak{F}\}$ is locally uniformly bounded in Ω and hence, is normal in Ω. \qquad (5.8.1.3)

Proof. (1) $\Rightarrow$ (2): This is obvious.

(2) $\Rightarrow$ (3): Suppose, on the contrary, that $\mathfrak{F}$ fails to be locally uniformly bounded in Ω. Then, there is a compact subset K of Ω such that, for each integer $n \geq 1$, there are a function $f_n \in \mathfrak{F}$ and a point $z_n \in K$ so that $|f_n(z_n)| \geq n$. We may suppose the sequence z_n itself converging to a point $z_0 \in K$.

By assumption, $\mathfrak{F}$ is normal in a neighborhood O_{z_0} of z_0. We may, as well, assume that the sequence f_n itself converges locally uniformly to an analytic function f in O_{z_0}. In particular, $f_n(z_n) \to f(z_0)$ as $n \to \infty$ (why? see Exercises B(4) and (5) of Sec. 5.3.1). This obviously contradicts to the fact that $|f_n(z_n)| \geq n$ for $n \geq 1$.

(3) $\Rightarrow$ (4): Only the local equicontinuity is needed to be proved.

Fix any point $\zeta_0 \in \Omega$. Choose a closed disk $|z - \zeta_0| \leq r$, contained in Ω. For any two points z and z' in $|\zeta - \zeta_0| < r$, we have by Cauchy's integral theorem,

$$f(z) - f(z') = \frac{1}{2\pi i} \int_{|\zeta - \zeta_0| = r} \left(\frac{1}{\zeta - z} - \frac{1}{\zeta - z'} \right) f(\zeta) d\zeta$$

$$= \frac{z - z'}{2\pi i} \int_{|\zeta - \zeta_0| = r} \frac{f(\zeta)}{(\zeta - z)(\zeta - z')} d\zeta$$

$$\Rightarrow |f(z) - f(z')|$$

$$\leq \frac{|z - z'|}{2\pi} \cdot 2\pi r \cdot \max_{|\zeta - \zeta_0| = r} |f(\zeta)| \cdot \frac{1}{(r - |z - \zeta_0|)(r - |z' - \zeta_0|)}$$

$$\leq \frac{4}{r} \max_{|\zeta - \zeta_0| = r} |f(\zeta)| \cdot |z - z'| \text{ if } |z - \zeta_0| \leq \frac{r}{2} \text{ and } |z' - \zeta_0| \leq \frac{r}{2}. \qquad (*_1)$$

$(*_1)$ holds for any $f \in \mathfrak{F}$. Since $\sup_{f \in \mathfrak{F}} \max_{|\zeta - \zeta_0| = r} |f(\zeta)| < \infty$, this proves the equicontinuity on the disk $|\zeta - \zeta_0| \leq \frac{r}{2}$.

Let K be a compact set in Ω. The open disks $|\zeta - \zeta_0| < \frac{r}{4}$, where $\zeta_0 \in K$ and $|\zeta - \zeta_0| \leq r$ is contained in Ω, form an open covering of K. Select a finite subcovering $|\zeta - \zeta_j| < \frac{r_j}{4}$ for $1 \leq j \leq n$. Set $M_j = \sup_{f \in \mathfrak{F}} \sup_{|\zeta - \zeta_j| = r_j} |f(\zeta)|$ for $1 \leq j \leq n$, and $r = \min_{1 \leq j \leq n} r_j$ and $M = \max_{1 \leq j \leq n} M_j$. For a given $\varepsilon > 0$, let $\delta = \min \left\{ \frac{r}{4}, \frac{\varepsilon r}{4M} \right\}$. Choose any $z, z' \in K$ and $|z - z'| < \delta$, then $|z - \zeta_k| < \frac{r_k}{4}$ for some k and it follows that

$$|z' - \zeta_k| \leq |z - z'| + |z - \zeta_k| < \delta + \frac{r_k}{4} < \frac{r}{4} + \frac{r_k}{4} \leq \frac{r_k}{2}$$

$$\Rightarrow (\text{by}(*1)) |f(z) - f(z')| \leq \frac{4}{r_k} M_k |z - z'| < \frac{4}{r} M \delta \leq \varepsilon.$$

This finishes the proof of the equicontinuity on K.

(4) $\Rightarrow$ (3): Let $|f(z_0)| \leq M$ for each $f \in \mathfrak{F}$.

Choose any point $z^* \in \Omega$. Let γ be a rectifiable curve in Ω connecting z_0 to z^*. As a point set, γ is compact. For $\varepsilon = 1$, there is a $\delta > 0$ so that $|f(z) - f(z')| < 1$ whenever $z, z' \in \gamma$ with $|z - z'| < \delta$ and $f \in \mathfrak{F}$. Select a positive integer n, large enough. Let $z_1 = z_0, z_2, \ldots, z_n = z^*$ be n points on γ (in this ordering) so that each of their consecutive distance is less than δ; then,

$$|f(z_j) - f(z_{j-1})| < 1, \quad 1 \leq j \leq n,$$

$$\Rightarrow |f(z^*)| = |f(z_n)| \leq \sum_{j=1}^{n} |f(z_j) - f(z_{j-1})| + |f(z_0)| \leq n + M, \quad f \in \mathfrak{F}.$$

$$(*_2)$$

This proves that $\{f(z) | f \in \mathfrak{F}\}$ is bounded for *each* point $z \in \Omega$.

Fix any point $\zeta_0 \in \Omega$ and a closed disk $|z - \zeta_0| \leq r$, contained in Ω. Let $|f(\zeta_0)| \leq M$ for each $f \in \mathfrak{F}$. By the equicontinuity of $\mathfrak{F}$ on $|z - \zeta_0| \leq r$, there is a $0 < \delta \leq r$ such that $|f(z) - f(\zeta_0)| < 1$ if $|z - \zeta_0| \leq \delta$ and $f \in \mathfrak{F}$. As a consequence,

$$|f(z)| \leq 1 + |f(\zeta_0)| \leq M + 1, \quad |z - \zeta_0| \leq \delta \quad \text{and} \quad f \in \mathfrak{F}.$$

Hence, $\mathfrak{F}$ is uniformly bounded on the disk $|z - \zeta_0| \leq \delta$. This, in turn, means or implies that $\mathfrak{F}$ is locally uniformly bounded in Ω.

(4) $\Rightarrow$ (1): Former part of the proof is exactly the same as we proved (1) $\Rightarrow$ (2) in (5.3.3.10). Just choose a countable *dense* set $E = \{z_1, z_2, \ldots\}$ in Ω. By the argument leading to $(*_2)$, $\{f_n(z_k)\}_{n=1}^{\infty}$ is bounded for each $k \geq 1$ if f_n is any presumed sequence from $\mathfrak{F}$. The Cantor diagonal process shows that f_n has a subsequence f_{n_j}, converging at each point z_k, $k \geq 1$.

To show that f_{n_j} converges locally uniformly in Ω, consider a fixed compact set K in Ω. On assuming that $\mathfrak{F}$ is equicontinuous on K, for $\varepsilon > 0$, there is a $\delta > 0$ as that $|f(z) - f(z')| < \frac{\varepsilon}{3}$ whenever $z, z' \in K$ with $|z - z'| < \delta$ and $f \in \mathfrak{F}$. Cover K by a finite number of open disks with radius $\frac{\delta}{2}$, say $O_1, \ldots, O_m$. Select a point $z_k \in E \cap O_k$ for $1 \leq k \leq m$. Since $\{f_{n_j}(z_k)\}$ is Cauchy for each k $(1 \leq k \leq m)$, there is a positive integer j_0 so that $|f_{n_i}(z_k) - f_{n_j}(z_k)| < \frac{\varepsilon}{3}$ if $i, j \geq j_0$, and $1 \leq k \leq m$.

Now, for each fixed $z \in K$, there is some k $(1 \leq k \leq m)$ so that $|z - z_k| < \frac{\delta}{2}$. In this case,

$$|f_{n_i}(z) - f_{n_i}(z_k)| < \frac{\varepsilon}{3} \quad \text{and} \quad |f_{n_j}(z) - f_{n_j}(z_k)| < \frac{\varepsilon}{3}$$

$$\Rightarrow |f_{n_i}(z) - f_{n_j}(z)|$$

$$\leq |f_{n_i}(z) - f_{n_i}(z_k)| + |f_{n_i}(z_k) - f_{n_j}(z_k)| + |f_{n_j}(z_k) - f_{n_j}(z)|$$

$$< \frac{\varepsilon}{3} + \frac{\varepsilon}{3} + \frac{\varepsilon}{3} = \varepsilon \quad \text{if } i, j \geq j_0 \quad \text{and} \quad z \in K.$$

By completeness of the complex field $\mathbf{C}$, this means that $f_{n_j}(z)$ converges for each $z \in K$ and, even more, converges uniformly on K. Hence $\mathfrak{F}$ is normal in Ω.

Finally, to show that the family $\{f' | f \in \mathfrak{F}\}$ is locally uniformly bounded and, at the same time, to give $(3) \Rightarrow (4)$ *another proof*: Here, we assume (3) holds in advance.

Choose a point $z_0 \in \Omega$. Let $r = \text{dist}(z_0, \partial\Omega)$. For any point $z \in \Omega$ with $|z - z_0| \leq \frac{r}{2} < \rho < r$, we obtain by Cauchy's integral formula

$$f'(z) = \frac{1}{2\pi i} \int_{|\zeta - z_0| = \rho} \frac{f(\zeta)}{(\zeta - z)^2} d\zeta$$

$$\Rightarrow |f'(z)| \leq \frac{\rho}{(\rho - \frac{r}{2})^2} \sup_{f \in \mathfrak{F}} \sup_{|\zeta - z_0| = \rho} |f(\zeta)|, \quad f \in \mathfrak{F}.$$

This proves the first claim.

On the other hand, for z, z' in $|\zeta - z_0| \leq \rho < r$, we have

$$f(z) - f(z') = \int_z^{z'} f'(\zeta) d\zeta \quad \text{(integrating along the segment } \overline{zz'}\text{)}, \quad f \in \mathfrak{F}$$

$$\Rightarrow |f(z) - f(z')| \leq \sup_{f \in \mathfrak{F}} \sup_{|\zeta - z_0| \leq \rho} |f'(\zeta)| \cdot |z - z'|, \quad f \in \mathfrak{F}.$$

This shows that $\mathfrak{F}$ is locally equicontinuous on Ω.   □

Section (2) Family of meromorphic functions

A meromorphic function assumes the value ∞ at its poles. In order to handle this case, we will adopt the spherical chord distance $d(,)$ as our metric in this subsection. Recall that (see $(1.6.8)$)

$$d(z_1, z_2) = \begin{cases} \dfrac{2|z_1 - z_2|}{\sqrt{(1 + |z_1|^2)(1 + |z_2|^2)}}, & z_1, z_2 \in \mathbf{C} \\ \dfrac{2}{\sqrt{1 + |z_1|^2}}, & z_1 \in \mathbf{C} \quad \text{and} \quad z_2 = \infty \end{cases} \tag{5.8.1.4}$$

and $(\mathbf{C}^*, d(,))$ turns out to be a complete metric space (see Exercises B of Sec. 1.9). Moreover, when restricted to compact sets in $\mathbf{C}$ the classical Euclidean metric $||$ is *equivalent* to $d(,)$, namely, $m|z_1 - z_2| \leq d(z_1, z_2) \leq M|z_1 - z_2|$ for some constants $m > 0$ and $M > 0$.

When we say that $\lim_{n\to\infty} f_n(z) = f(z)$ in the spherical distance $d(,)$, it means that $\lim_{n\to\infty} d(f_n(z), f(z)) = 0$ and vice versa (see (2.1.4)).

Let $f_n(z)$, $n \geq 1$, be a sequence of meromorphic (including analytic) functions defined on a domain Ω in $\mathbf{C}$. Suppose $f_n(z)$ converges in the sense $d(,)$, locally uniformly, in Ω. The *question* is: What might the limit function $f(z)$ look like?

To answer this question, two cases are described.

Case (1) Suppose all of $f_n(z)$, $n \geq 1$, are meromorphic in Ω.

The limit function $f(z)$ is continuous in Ω in the sense $d(,)$ owing to the local uniform convergence (refer to (2.2.2) and (2.3.3)). Fix a point $z_0 \in \Omega$.

If $f(z_0) \neq \infty$, there is an open neighborhood O_{z_0} of z_0 in Ω on which $f(z)$ is bounded; hence, $f_n(z)$ is bounded in O_{z_0} for all sufficiently large n and, in particular, is analytic in O_{z_0}. By Weierstrass's theorem (5.3.1.1), $f(z)$ will then be analytic in O_{z_0}.

In case $f(z_0) = \infty$: Since $d(f_n(z), f(z)) = d\left(\frac{1}{f_n(z)}, \frac{1}{f(z)}\right)$, it follows that $\frac{1}{f_n(z)}$ converges to $\frac{1}{f(z)}$ locally uniformly in Ω. Applying the last paragraph to $\frac{1}{f(z_0)} = 0$, we obtain that $\frac{1}{f(z)}$ is analytic in a neighborhood of z_0 and has a zero at z_0. Hence, $f(z)$ is meromorphic in such a neighborhood of z_0 or identically equal to ∞ there.

In conclusion, $f(z)$ is either meromorphic in Ω or identically equal to ∞.

Case (2) Suppose all of $f_n(z)$, $n \geq 1$, are analytic in Ω. Fix a point $z_0 \in \Omega$.

In case $f(z_0) \neq \infty$, then $f(z)$ is analytic in a neighborhood of z_0 as we already proved.

If $f(z_0) = \infty$ happens, there is an open neighborhood O_{z_0} of z_0 on which $f_n(z) \neq 0$ for all sufficiently large n. As before, $\frac{1}{f(z)}$ is analytic in O_{z_0} and has a zero at z_0; as a consequence, $\frac{1}{f(z)}$ is identically equal to 0 in O_{z_0} according to Hurwitz's theorem (5.3.2.2). And this means that $f(z) \equiv \infty$ in O_{z_0}.

In conclusion, $f(z)$ is either analytic in Ω or identically equal to ∞.

We summarize the above as the

Generalized Weierstrass's and Hurwitz's theorems. *Let $f_n(z)$, $n \geq 1$, be a sequence of meromorphic (analytic included) functions defined in a domain Ω in $\mathbf{C}$. Suppose $f_n(z)$ converges in the spherical sense $d(,)$, locally uniformly, to the limit function $f(z)$ in Ω.*

(1) *If all of $f_n(z)$, $n \geq 1$, are meromorphic, then $f(z)$ is either meromorphic or ∞ identically.*

(2) *If all of $f_n(z)$, $n \geq 1$, are analytic, then $f(z)$ is either analytic or ∞ identically.*

$$(5.8.1.5)$$

After this preparatory work, we try to extend (5.8.1.3) to a family $\mathfrak{F}$ of meromorphic functions defined in a domain Ω. Wherein, (1) and (2) are trivially fulfilled in the sense $d(,)$. And condition 2 in (4) becomes unnecessary because $\mathbf{C}^*$, endowed with the bounded metric $d(,)$, is a compact metric space. What is less trivial is (3).

A subset of $\mathbf{C}^*$ that contains ∞ is unbounded in $\mathbf{C}$, but it is always bounded in the sense $d(,)$. Even though $d(,)$ and $||$ are equivalent on bounded subsets of $\mathbf{C}$, as we noted before, the concept of "local uniform boundedness" in $\mathbf{C}$ will not have direct extension to $\mathbf{C}^*$ under $d(,)$.

We are able to overcome this difficult by considering infinitesimally. Well, (5.8.1.4) suggests that the arc length element on the Riemann sphere S^2 under the stereographic projection $\Phi : \mathbf{C}^* \to S^2$ is given by

$$\frac{2|dw|}{1 + |w|^2}, \quad w \in \mathbf{C}^*$$

(see Exercise B(1) of Sec. 1.6). Suppose a curve γ in the z-plane is mapped into a curve $f(\gamma)$ in $\mathbf{C}^*$ under a mapping $w = f(z)$ and then into image curve $\Phi(f(\gamma))$ on S^2 under Φ. Then (see (2.4.5) and (3.5.1.5))

$$\text{the length of } \Phi(f(\gamma)) = \int_\gamma \rho(f)|dz|, \quad (5.8.1.6)$$

where

$$\rho(f)(z) = \frac{2|f'(z)|}{1 + |f(z)|^2} \quad (5.8.1.7)$$

and $\rho(f)|dz|$ is the arc length element of $\Phi(f(\gamma))$.

In what follows, we try to prove (5.8.1.8) below.

Now, suppose that $\rho(f)$, $f \in \mathfrak{F}$, are locally uniformly bounded in the Euclidean sense in Ω. Choose a closed disk $|z - z_0| \leq r$, contained in Ω. Set $\sup_{f \in \mathfrak{F}} \sup_{|z-z_0| \leq r} |\rho(f)(z)| = M$. Then $0 \leq M < \infty$. For any two points z_1, z_2 in $|z - z_0| \leq r$, by (5.8.1.6), we have

$$d(f(z_1), f(z_2))$$

$$\leq \text{ the length of the image curve } \Phi(f(\overline{z_1 z_2})) \text{ of the segment } \overline{z_1 z_2}$$

$$= \int_{\overline{z_1 z_2}} \rho(f)|dz| \leq M \int_{\overline{z_1 z_2}} |dz| = M|z_1 - z_2|.$$

This immediately shows that $\mathfrak{F}$ is equicontinuous on $|z - z_0| \leq r$ and hence, locally equicontinuous in Ω. The same argument as in the proof

(4) $\Rightarrow$ (1) for (5.8.1.3) then, in turn, implies that $\mathfrak{F}$ is normal in the sense $d(,)$.

Conversely, suppose that $\mathfrak{F}$ is normal in the sense $d(,)$. We try to prove that $\{\rho(f)|f \in \mathfrak{F}\}$ is locally uniformly bounded. If, on the contrary, there is a compact set K in Ω so that $\sup_{f \in \mathfrak{F}} \sup_{z \in K} \rho(f(z)) = \infty$. This implies that there is a sequence f_n in $\mathfrak{F}$ such that

$$\lim_{n \to \infty} \sup_{z \in K} \rho(f_n(z)) = \infty. \tag{$*_3$}$$

On choosing a subsequence if necessary, we may assume that the sequence $f_n(z)$ itself converges in Ω locally uniformly to a function $f(z)$. $f(z)$ is either meromorphic or identically equal to ∞ in Ω, owing to (5.8.1.5).

Fix a point $z_0 \in K$. Choose a closed disk $\overline{D_{z_0}} : |z - z_0| \leq r$, contained in Ω. In case f is meromorphic, r is so chosen that $f(z)$ has only one zero or one pole in $|z - z_0| \leq r$ but not both. Then $f(z)$ or $\frac{1}{f(z)}$ is analytic in $\overline{D_{z_0}}$.

In case $f(z)$ is analytic in $\overline{D_{z_0}}$, it is bounded there. Uniform continuity of $f(z)$ on $\overline{D_{z_0}}$ then implies that $f_n(z)$ are also bounded in $\overline{D_{z_0}}$ for all large n; in particular, they are analytic in $\overline{D_{z_0}}$. Weierstrass's theorem (5.3.1.1) says that both $f_n(z)$ and $f_n'(z)$ converge, respectively, to $f(z)$ and $f'(z)$ uniformly on a slightly smaller open disk; and hence, $\rho(f_n(z)) \to \rho(f(z))$ uniformly on the smaller disk as $n \to \infty$. Since $\rho(f(z))$ is bounded in $\overline{D_{z_0}}$, it follows that $\rho(f_n(z))$, $n \geq 1$, are also bounded on the smaller disk.

In case $\frac{1}{f(z)}$ is analytic in $\overline{D_{z_0}}$, apply the same argument to $\rho(\frac{1}{f_n(z)}) = \rho(f_n(z))$ and we obtain that $\rho(f_n(z))$, $n \geq 1$, are bounded on a smaller disk.

Since the compact set K can be covered by a finite number of such smaller disks, both cases imply that $\rho(f_n(z))$, $n \geq 1$, are bounded on K: This contradicts to $(*_3)$. Hence, $\{\rho(f)|f \in \mathfrak{F}\}$ should be locally uniformly bounded if $\mathfrak{F}$ is normal.

In conclusion, we summarize as the

Criteria for normality of a family $\mathfrak{F}$ of meromorphic functions in a domain Ω in C. The following are equivalent.

(1) $\mathfrak{F}$ is normal in the spherical chord distance $d(,)$ (see the definition (5.8.1)) in Ω.

(2) $\mathfrak{F}$ is locally normal in the sense $d(,)$ in Ω.

(3) (F. Marty, 1931) The family

$$\left\{ \frac{2|f'(z)|}{1 + |f(z)|^2} \Big| f \in \mathfrak{F}, \quad z \in \Omega \right\}$$

is locally uniformly bounded in the Euclidean sense in Ω.

(4) $\mathfrak{F}$ is locally equicontinuous in the sense $d(,)$ in Ω. \hfill (5.8.1.8)

In contract to (5.8.1.3), the family $\{f'|f \in \mathfrak{F}\}$ is not necessarily normal in this case. For instance, $\mathfrak{F} = \{n(z^2 - n)|n \geq 1\}$ is normal in $\mathbf{C}$ since $f_n(z) = n(z^2 - n) \to \infty$ uniformly on each compact subset of $\mathbf{C}$. Yet $\{2nz|n \geq 1\}$ is not so. Concrete examples are illustrated in Sec. 5.8.2.

Exercises B

The method of proofs and the results obtained in the context can be put in a general yet abstract setting after a careful examination of the whole process. At this moment, go back to the Remark in Sec. 5.3.1 and we will be free to adopt notations and results stated there in the following exercises.

(1) Let $\mathfrak{F}$ be a nonempty set in $(F(\Omega, X), \rho(,))$. Then, the following are equivalent:

  1. $\mathfrak{F}$ is *relatively compact*, i.e., every sequence in $\mathfrak{F}$ contains a convergent subsequence.
  2. The closure $\bar{\mathfrak{F}}$ is *compact*, i.e., every sequence in $\mathfrak{F}$ contains a subsequence converging to a point in $\bar{\mathfrak{F}}$.
  3. *In case* (X, d) *is a complete metric space*, $\mathfrak{F}$ *is totally bounded*, i.e., for each $\varepsilon > 0$, there are finitely many points $f_1, \ldots, f_n \in \mathfrak{F}$ so that, for each $f \in \mathfrak{F}$, there is some f_j satisfying $\rho(f, f_j) < \varepsilon$. In other words,

$$\mathfrak{F} \subseteq \bigcup_{j=1}^{n} B_\varepsilon(f_j),$$

  where $B_\varepsilon(f_j) = \{f \in F(\Omega, X)|d(f, f_j) < \varepsilon\}$ for $1 \leq j \leq n$.

  Try to interpret these statement using the language of local uniform convergence in Ω and the metric space (X, d).

(2) *G. Ascoli* (1843–1896, Italy)–*C. Arzelà* (1847–1912, Italy) *Theorem.* Let $\mathfrak{F}$ be a family of *continuous* function form a domain Ω in the plane to a metric space (X, d). Then

  1. $\mathfrak{F}$ is *normal*, i.e., the closure $\bar{\mathfrak{F}}$ is compact in $(F(\Omega, X), \rho(,))$.
  $\Leftrightarrow$ 2. $\mathfrak{F}$ is equicontinuous on every compact set in Ω and, for each $z \in \Omega$, the closure of the set $\{f(z)|f \in \mathfrak{F}\}$ is compact in X.

  In case (X, d) *is a compact (and hence, complete) metric space*, the condition that $\overline{\{f(z)|f \in \mathfrak{F}\}}$ is compact in X may be deleted from 2.

5.8.2 *Examples*

According to (5.8.1.3) or (5.8.1.8), a family $\mathfrak{F}$ of analytic function is normal if it is locally uniformly bounded in the classical sense $\|$ or $\{\frac{2|f'(z)|}{1+|f(z)|^2}|f \in \mathfrak{F}\}$

is locally uniformly bounded in the sense $\|\cdot\|$. In the latter case, a sequence $f_n(z)$ in $\mathfrak{F}$ is permitted to converge to ∞ locally uniformly.

For instance, the family

$$\mathfrak{F} = \{f : \Omega \to \mathbf{C} \text{ is analytic. } |\text{Re } f(z) \geq 0 \text{ in the domain } \Omega\} \qquad (5.8.2.1)$$

is normal in Ω because $\{e^{-f}|f \in \mathfrak{F}\}$ is uniformly bounded in Ω. Let $R > 0$ be a constant and $w_0 \in \mathbf{C}$ be a fixed point. Then the family

$$\mathfrak{F} = \{f : \Omega \to \mathbf{C} \text{ is analytic. } \||f(z) - w_0| \geq R \quad \text{for all } z \text{ in the domain } \Omega\}$$
$$(5.8.2.2)$$

is normal in Ω, too. This is because $\left\{\frac{1}{f(z)-w_0}|f \in \mathfrak{F}\right\}$ is uniformly bounded in Ω.

Suppose $b_n \geq 0$ for $n \geq 0$ and the series $\sum_0^\infty b_n z^n$ converges in $|z| < 1$. Then the family.

$$\mathfrak{F} = \left\{\sum_0^\infty a_n z^n \||a_n| \leq b_n \quad \text{for} \quad n \geq 0\right\}$$

is normal in $|z| < 1$. This is so because, for $0 < r < 1$, $|\sum_0^\infty a_n z^n| \leq \sum_0^\infty b_n r^n < \infty$ on $|z| \leq r$. In particular, $\{\sum_0^\infty a_n z^n \||a_n| \leq M, n \geq 0\}$ is normal in $|z| < 1$ wherein $M > 0$ is a constant.

A less trivial one is the following

Example 1. Let Ω be a domain and $M > 0$ be a constant. Show that

$$\mathfrak{F} = \left\{f : \Omega \to \mathbf{C} \text{ is analytic. } \left|\iint_\Omega |f(z)|^2 dxdy \leq M < \infty\right.\right\}$$

is normal.

Solution. Choose a closed disk $|z - z_0| \leq r$, contained in Ω. For any $f \in \mathfrak{F}$,

$$(f(z))^2 = \frac{1}{2\pi i}\int_{|\zeta-z_0|=r}\frac{(f(\zeta))^2}{\zeta - z}d\zeta, \quad |z - z_0| \leq \frac{r}{2}.$$

$$\Rightarrow |f(z)|^2 \leq \frac{2}{r} \cdot \frac{1}{2\pi}\int_0^{2\pi}|f(z_0 + re^{i\theta})|^2 rd\theta, \quad |z - z_0| \leq \frac{r}{2}$$

$$\Rightarrow \int_0^r |f(z)|^2 rdr = \frac{r^2}{2}|f(z)|^2 \leq \frac{1}{\pi}\iint_{|\zeta-z_0|\leq r}|f(z_0 + re^{i\theta})|^2 dxdy$$

$$\Rightarrow |f(z)|^2 \leq \frac{2}{\pi r^2}\iint_\Omega |f(z)|^2 dxdy \leq \frac{2M}{\pi r^2}$$

$$\Rightarrow |f(z)| \leq \sqrt{\frac{2M}{\pi r^2}}, \quad |z - z_0| \leq \frac{r}{2} \quad \text{and} \quad f \in \mathfrak{F}.$$

This shows that $\mathfrak{F}$ is locally uniformly bounded in Ω and hence, is normal in Ω.

An alternative way is as follows. According to (2) in (5.8.1.3), it is harmless to assume that Ω is the unit disk $|z| < 1$. Suppose $f(z) = \sum_0^\infty a_n z^n \in \mathfrak{F}$. By assumption,

$$
\iint_\Omega |f(z)|^2 dx dy = \int_0^{2\pi} \int_0^1 |f(re^{i\theta})|^2 r dr d\theta
$$

$$
= \int_0^{2\pi} \int_0^1 \left\{ \sum_0^\infty a_n r^n e^{in\theta} \right\} \left\{ \sum_0^\infty \overline{a_n} r^n e^{-in\theta} \right\} r dr d\theta
$$

$$
= \pi \sum_0^\infty \frac{|a_n|^2}{n+1} \le M
$$

$\Rightarrow$ (by using Cauchy–Schwarz's inequality)

$$
|f(z)| \le \sum_0^\infty |a_n| r^n, \quad \text{if } |z| \le r < 1.
$$

$$
\le \left\{ \frac{1}{r} \sum_0^\infty |a_n|^2 \frac{r^{n+1}}{n+1} \right\}^{1/2} \left\{ \sum_{n=0}^\infty (n+1) r^n \right\}^{1/2}
$$

$$
\le \left(\frac{M}{\pi} \right)^{1/2} \frac{1}{1-r}, \quad |z| \le r < 1 \quad \text{and} \quad f \in \mathfrak{F}.
$$

Hence $\mathfrak{F}$ is locally uniformly bounded in $|z| < 1$ and then is normal.

Example 2. Find the domain(s) of normality for each of the following families.

(1) $\mathfrak{F} = \{z^n | n \ge 1\}$.

(2) $\mathfrak{F} = \{\sin nz | n \ge 1\}$.

(3) $\mathfrak{F} = \{f(z) = az + b | a, b \in \mathbf{C} \text{ and } a \ne 0\}$.

(4) $\mathfrak{F} = \{e^{az} \sin az | a \in \mathbf{C}\}$.

Solution. (1) $\mathfrak{F}$ is locally uniformly bounded in $|z| < 1$. In case $|z| \ge R > 1$, for any $M > 0$, choose $n \ge \frac{\log M}{\log R}$ and we have $|z^n| \ge R^n \ge M$. Hence, any sequence in $\mathfrak{F}$ converges to ∞ locally uniformly in $|z| > 1$. Fix any point z_0 with $|z_0| = 1$. In any neighborhood O_{z_0} of z_0, $z^n \to 0$ in $O_{z_0} \cap \{|z| < 1\}$ while $z^n \to \infty$ in $O_{z_0} \cap \{|z| > 1\}$. In conclusion, F is normal in $|z| < 1$ and $|z| > 1$, respectively.

On the other hand, using (3) in (5.8.1.8), the family

$$\left\{ \frac{2n|z|^{n-1}}{1+|z|^{2n}} \,\middle|\, n \geq 1 \right\} \text{ is locally uniformly bounded in a domain } \Omega.$$

$\Leftrightarrow \Omega$ is contained either in $|z| < 1$ or in $|z| > 1$.

And we obtain the same conclusion.

(2) Note that $|\sin nz| = \sqrt{\sin^2 nx + \sinh^2 ny}$, where $z = x + iy$. Hence, $|\sin nz| \to \infty$ as $n \to \infty$ in case $\mathrm{Im}\, z \neq 0$. $\mathfrak{F}$ is not normal in $\mathrm{Im}\, z > 0$ and $\mathrm{Im}\, z < 0$, respectively, in the Euclidean metric; but it is in the spherical metric because every sequence in $\mathfrak{F}$ converges to ∞ uniformly there.

Or, on using (3) in (5.8.1.8), the family

$$\frac{2n|\cos nz|}{1+|\sin nz|^2} = \frac{2n\sqrt{\cos^2 nx + \sinh^2 ny}}{1+(\sin^2 nx + \sinh^2 ny)} \left(\leq \frac{2n}{|\cosh ny|} \right), \quad n \geq 1$$

is locally bounded either in $\mathrm{Im}\, z > 0$ or in $\mathrm{Im}\, z < 0$, but not so on domains with nonempty intersection with the real axis.

(3) $\mathfrak{F}$ cannot be uniformly bounded in any domain in $\mathbf{C}$ and hence, is not normal there in the Euclidean sense. While, the family

$$\frac{2|a|}{1+|az+b|^2}, \quad a, b \in \mathbf{C} \text{ and } a \neq 0$$

is not bounded at $z = 0$. This shows that $\mathfrak{F}$ is not normal in $\mathbf{C}$ in the spherical sense.

(4) Set $f_n(z) = e^{(2n+\frac{1}{2})z} \sin(2n + \frac{1}{2})z$, $n \geq 1$. Then $\lim_{n \to \infty} f_n(\pi) = \infty$ while $\lim_{n \to \infty} f_n(2\pi) = 0$. Consequently, $f_n(z)$ cannot have a subsequence converging locally uniformly to either an analytic function or ∞ in $|z - \pi| < \pi + 1$. Then $\mathfrak{F}$ is not normal in $|z - \pi| < \pi + 1$ in the spherical metric.

As a prelude to the next example, consider the sequence $f_n(z) = \frac{1}{n}(e^{nz} - 1)$ for $n \geq 1$. Observe that $\lim_{n \to \infty} f_n(0) = 0$ and $\lim_{n \to \infty} f_n(x) = \infty$, where x is a fixed real number between 0 and 1 (excluded). Hence, $f_n(z)$, $n \geq 1$, contains no subsequence locally uniformly convergent in $|z| < 1$ to an analytic function or to ∞. And then, the family of analytic functions $f(z)$ on $|z| < 1$ that satisfy $f(0) = 0$ and $f'(0) = 1$ is not normal. Refer to Exercise A(3) of Sec. 5.8.3. Yet we have the following important

Normality of the family of normalized univalent (analytic) functions on the open disk. The family

$$\mathcal{S} = \{f(z) \text{ is univalent on } |z| < 1. \, |f(0) = 0 \text{ and } f'(0) = 1\}$$

is normal in $|z| < 1$ and is compact in the sense that the limit function of any locally uniformly convergent sequence from $\mathcal{S}$ is still in $\mathcal{S}$. Moreover,

(1) (Koebe's $\frac{1}{4}$-*theorem*, 1907; Bieberbach and Faber, 1916) For each $f \in \mathcal{S}$, the image $f(\{|z| < 1\})$ always contains the open disk $|w| < \frac{1}{4}$.

(2) $\{f' | f \in \mathcal{S}\}$ is normal in $|z| < 1$.

(3) $\{\frac{1}{f'} | f \in \mathcal{S}\}$ is normal in $|z| < 1$.

Consequently, letting Ω be a domain in $\mathbf{C}$ and $z_0 \in \Omega$ a fixed point, then

$$\mathcal{S}(\Omega; z_0) = \{f : \Omega \to \mathbf{C} \text{ is univalent. } \|f(z_0)\| \leq M \text{ (a constant) and}$$

$$|f'(z_0)| \leq M\}$$

is both normal and compact in Ω.

Note. A one-to-one analytic function will always be simply called an *univalent function*. (5.8.2.3)

Recall that (1) is the one mentioned in Exercise B(4) of Sec. 4.8 and the constant $\frac{1}{4}$ is sharp as the univalent function $\frac{z}{(1-z)^2}$, $|z| < 1$, having its image the domain $\mathbf{C} - (\infty, \frac{1}{4}]$ (see Example 1 in Sec. 3.5.7).

For a proof, we adopt the one appeared in Ref. [63], pp. 215–217.

Proof. Let f be univalent in $|z| < 1$. Then

$$f\left(\frac{z + \zeta}{1 + \bar{\zeta}z}\right) \text{ is univalent and analytic in } |z| < 1 \text{ for each fixed } \zeta, |\zeta| < 1.$$

$\Rightarrow$ The normalized

$$g(z) = \frac{f(\frac{z+\zeta}{1+\bar{\zeta}z}) - f(\zeta)}{f'(\zeta)(1 - |\zeta|^2)} \in \mathcal{S}.$$

$\Rightarrow$ The second coefficient of the Taylor series expansion of $g(z)$ at $z = 0$ is

$$\frac{g''(0)}{2!} = \frac{1}{2}\left\{\frac{f''(\zeta)(1 - |\zeta|^2)}{f'(\zeta)} - 2\bar{\zeta}\right\}.$$

$\Rightarrow$ (by using Exercise B(3) of Sec. 4.8 that says $|\frac{g''(0)}{2}| \leq 2$, and replacing ζ by z)

$$\left|\frac{zf''(z)}{f'(z)} - \frac{2|z|^2}{1 - |z|^2}\right| \leq \frac{4|z|}{1 - |z|^2}. \tag{$*_1$}$$

Here, we should remind the readers that $f'(z) \neq 0$ throughout $|z| < 1$ owing to the univalence (see (3.5.1.9)). Consequently, $\log f'(z)$ has single-valued branches in $|z| < 1$ (see (3.3.1.8) or (4.4.2)). Fix such a branch. According to Exercise A(6) of Sec. 3.2.2,

$$\frac{d}{dz} \log f'(z) = \frac{r}{z} \left(\frac{\partial}{\partial r} \log |f'(z)| + i \frac{\partial}{\partial r} \arg f'(z) \right), \quad |z| = r < 1$$

$$\Rightarrow \operatorname{Re} \frac{z f''(z)}{f'(z)} = r \frac{\partial}{\partial r} \log |f'(z)|, \quad |z| = r < 1.$$

Then, $(*_1)$ reduces to

$$\frac{2r - 4}{1 - r^2} \leq \frac{\partial}{\partial r} \log |f'(z)| \leq \frac{4 + 2r}{1 - r^2}, \quad |z| = r < 1.$$

$\Rightarrow$ (integrating these inequalities from 0 to $|z|$ and assuming that

$f'(0) = 1$, and then by taking exponentials)

$$\frac{1 - |z|}{(1 + |z|)^3} \leq |f'(z)| \leq \frac{1 + |z|}{(1 - |z|)^3}, \quad |z| < 1. \tag{5.8.2.4}$$

This is a *local distortion* theorem for $|f'(z)|$ of the univalent mapping $w = f(z)$ with $f'(0) = 1$. Both the lower and the upper bounds in (5.8.2.4) are sharp as the function $\frac{1+z}{(1-z)^3}$ shows.

On integrating the second inequality in (5.8.2.4) along the segment connecting 0 to z and by assuming that $f(0) = 0$, we obtain

$$|f(z)| = \left| \int_0^z f'(z) dz \right| \leq \int_0^r |f'(z)| dr \leq \int_0^r \frac{1 + r}{(1 - r)^3} dr = \frac{r}{(1 - r)^2},$$

$$|z| = r < 1, \tag{$*_2$}$$

an upper bound for $|f(z)|$. To obtain a lower bounded for $|f(z)|$, consider first the case that $|f(z)| < \frac{1}{4}$. By (1), Koebe's quarter theorem, the line segment connecting 0 to $f(z)$ in the w-plane lies entirely in the range $f(\{|z| < 1\})$. Let γ be the arc in $|z| < 1$, which is mapped by $w = f(z)$ onto this line segment. Then $dw = f'(z) dz > 0$ along γ and then, we have

$$|f(z)| = \left| \int_\gamma f'(z) dz \right| = \int_\gamma |f'(z)| dr \geq \int_0^r \frac{1 - r}{(1 + r)^3} dr = \frac{r}{(1 + r)^2}. \tag{$*_3$}$$

Since $\frac{r}{(1+r)^2} \leq \frac{1}{4}$ if $0 \leq r < 1$, $(*_3)$ therefore holds, too, in case $|f(z)| \geq \frac{1}{4}$. Combining $(*_2)$ and $(*_3)$, we have a *local distortion* theorem for $|f(z)|$ if $f \in \mathcal{S}$:

$$\frac{|z|}{(1 + |z|)^2} \leq |f(z)| \leq \frac{|z|}{(1 - |z|)^2}, \quad |z| < 1. \tag{5.8.2.5}$$

For another form of distortion theorem for $\mathcal{S}$, see Exercise B(4).

The second inequality in (5.8.2.5) plus the maximum modulus principle show that $\mathcal{S}$ is locally uniformly bounded in $|z| < 1$ and hence, is normal by (5.8.1.3).

Suppose $f_n \in \mathcal{S}$ converges in $|z| < 1$ locally uniformly to an analytic function f. Since $f_n(0) = 0$ and $f'_n(0) = 1$ for each $n \geq 1$, Weierstrass's theorem (5.3.1.1) says that $f(0) = 0$ and $f'(0) = 1$ hold, too. That $f'(0) = 1$ shows that f is not a constant function. Hurwitz's theorem (5.3.2.2), in turn, implies that f is univalent in $|z| < 1$. Hence $f \in \mathcal{S}$ and $\mathcal{S}$ is compact.

(2) is a consequence of (5.8.1.3).

As for (3): Suppose on the contrary that $\{\frac{1}{f'} | f \in \mathcal{S}\}$ is not normal in $|z| < 1$. Then there are a compact set K in $|z| < 1$, and functions $f_n \in \mathcal{S}$ and points $z_n \in K$ for $n \geq 1$, such that

$$\left| \frac{1}{f'_n(z_n)} \right| \geq n, \quad n \geq 1$$

$$\Rightarrow |f'_n(z_n)| \leq \frac{1}{n}, \quad n \geq 1.$$

After adjusting to subsequences, we may assume that z_n converges to a point $z_0 \in K$ and the sequence f_n itself converges in $|z| < 1$ locally uniformly to an analytic function $f \in \mathcal{S}$. Since f is univalent, so $f'(z_0) \neq 0$. On the other hand, since f_n converges uniformly to f in K, therefore (refer to Exercises B(4) and (5)(b) of Sec. 5.3.1) $\lim_{n\to\infty} f'_n(z_n) = f'(z_0) = 0$ holds, a contradiction. Thus, the claim follows.

To prove the last statement, we may just assume that Ω is an open disk $|z - z_0| < R$ owing to (2) in (5.8.1.3). Set

$$g(\zeta) = \frac{f(z_0 + R\zeta) - f(z_0)}{Rf'(z_0)}, \quad |\zeta| < 1 \quad \text{and} \quad f \in \mathcal{S}(\Omega; z_0).$$

Then $\{g(\zeta) | f \in \mathcal{S}(\Omega; z_0)\}$ is the family $\mathcal{S}$ on $|\zeta| < 1$ and is thus compact. Let $g_n(\zeta)$ be a sequence converging locally uniformly to $g(\zeta)$ in $|\zeta| < 1$. Note that $f_n(z_0 + R\zeta) = Rf'_n(z_0)g_n(\zeta) + f_n(z_0), n \geq 1, |\zeta| < 1$. Since $|f_n(z_0)| \leq M$ and $|f'_n(z_0)| \leq M$ for $n \geq 1$, there are convergent subsequences $f_{n_j}(z_0)$ and $f'_{n_j}(z_0)$ for $j \geq 1$. Under this circumstance, $f_n(z)$ has a subsequence $f_{n_j}(z)$ converging locally uniformly to an analytic function in $|z - z_0| < R$ and this limit function is still in $\mathcal{S}(\Omega; z_0)$. □

Sketch of another proof for (5.8.2.3). Since each $f \in \mathcal{S}$ satisfies $f(0) = 0$ and $f'(0) = 1$, it is certain that 0 is an interior point of the image $f(D)$, where D denotes the unit disk $|z| < 1$ in what follows. If we are able to show that *each* $f \in \mathcal{S}$, after some adjustment, will not assume points in the *same* open disk, then (5.8.2.2) says that $\mathscr{S}$ is thus normal. Ideas here will be helpful in proving the Riemann mapping theorem (see (6.1.1)).

First, note that, for each $f \in \mathscr{S}$,

$$\alpha_f = \text{dist}(0, \mathbf{C} - f(D)) \text{ satisfies } 0 < \alpha_f \le 1. \tag{5.8.2.6}$$

$\alpha_f > 0$ is obvious owing to the openness of f. The disk $|w| < \alpha_f$ is contained in $f(D)$. The analytic function $f^{-1}(\alpha_f \zeta)$ form $|\zeta| < 1$ to $|z| < 1$ maps 0 to 0 and hence, by Schwarz's lemma, $|f^{-1}(\alpha_f \zeta)| \le |\zeta|$ on $|\zeta| < 1$. Setting $f^{-1}(\alpha_f \zeta) = z$, we obtain $\alpha_f \zeta = f(z)$ and then

$$|z| \le \frac{|f(z)|}{\alpha_f}, \quad |z| < 1$$

$$\Rightarrow \alpha_f \le \frac{|f(z)|}{|z|}, \quad |z| < 1$$

$$\Rightarrow \text{(on letting } z \to 0) \; \alpha_f \le |f'(0)| = 1.$$

This verifies (5.8.2.6). As a consequence, the open disk $|w| < \alpha_f$ is contained in $f(D)$ for each $f \in \mathscr{S}$. See Fig. 5.29.

We do hope that *the open unit disk $|z| < 1$ is contained in $f(D)$* for each $f \in \mathscr{S}$. This is easily seen once we replace $f(z)$ by $\frac{f(z)}{\alpha_f}$, without changing the univalence.

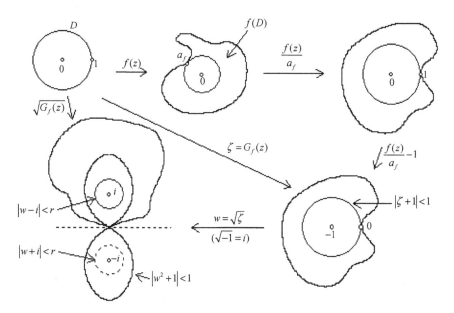

Fig. 5.29

It is well-known already that $\sqrt{\zeta}$ has single-valued branches on a simply connected domain, not containing 0 (see (2.7.2.6) or (3.3.1.5) or (4.4.2)), and maps it onto a simply connected domain in a half-plane. To take advantage of this fact, just shift each image domain of $\frac{f(z)}{\alpha_f}$ to the left one unit, namely, set

$$G_f(z) = \frac{f(z)}{\alpha_f} - 1, \quad |z| < 1.$$

Then $0 \notin G_f(D)$ and $|\zeta + 1| < 1$ is contained in $G_f(D)$ for each $f \in \mathscr{S}$ (see Fig. 5.29).

Now, the image of $|\zeta+1| < 1$ under $w = \sqrt{\zeta}$ is the lemniscate $|w^2+1| < 1$ (see Example 4 in Sec. 1.4.3); hence, it is possible to choose a branch of $\sqrt{G_f}(z)$ on $G_f(D)$ so that i situates as an interior point of $\sqrt{G_f}(D)$. As a consequence, there is a number r, $0 < r < 1$, independent of $f \in \mathscr{S}$, such that $|w - i| < r$ is contained in $\sqrt{G_f}(\{|\zeta + 1| < 1\}) \subseteq \sqrt{G_f}(D)$ for each $f \in \mathscr{S}$; therefore, $|w + i| < r$ is not contained in $\sqrt{G_f}(D)$ for each $f \in \mathscr{S}$. It follows that $\{\sqrt{G_f(z)}|f \in \mathscr{S}\}$ is normal in D and so is the original family $\mathscr{S}$. $\square$

For a third proof, see Exercise B(1) of Sec. 5.8.3.

Remark (Extremal problems and extremal functions). We had touched problems like this in (3.4.5.4), (3.4.5.5), Exercise B(2) of Sec. 3.4.5 and Exercise B(2) of Sec. 4.8. At the present, we formulate the following

The extremal problem for the coefficients of the Taylor series expansions of functions in $\mathscr{S}$. For each fixed integer $k \geq 1$, define the linear functional $L_k : \mathscr{S} \to \mathbf{C}$ by

$$L_k(f) = \text{the } k\text{-th coefficient } a_k \text{ in } \sum_{n=0}^{\infty} a_n z^n \text{ of the Taylor series expansion}$$

of $f \in \mathscr{S}$ at $z = 0$.

Then L_k is continuous, namely $L_k(f_n) \to L_k(f)$ if $f_n \in \mathscr{S} \to f \in \mathscr{S}$ locally uniformly in $|z| < 1$. Moreover, there is a function $f_M \in \mathscr{S}$, called an *extremal function*, so that

$$|L_k(f_M)| = \sup_{f \in \mathscr{S}} |L_k(f)|,$$

namely, $|a_k(f)| \leq |L_k(f_M)|$ holds for all $f \in \mathscr{S}$.   (5.8.2.7)

That L_k is continuous is proved by using Cauchy's inequality or (5.3.1.2). It can be shown that $|a_2| \leq 2$, $|a_3| \leq 3$, and $|a_4| \leq 4$ (refer to Exercise B(3) of Sec. 4.8 and Refs. [2, 26]) with the same extremal function $f_M(z) = \frac{z}{(1-z)^2} = \sum_{n=1}^{\infty} nz^n, |z| < 1$. These facts lead naturally to the conjecture that $|a_n| \leq n$ for $n \geq 1$ with $f_M(z)$ as the extremal function for each $n \geq 1$. This is the *Bieberbach conjecture* (1935) which was completely solved until 1985 by Prof. L. de Branges [21].

As an application of (5.8.2.3), we have an

Example 3. Let Ω be a domain and $z_0 \in \Omega$ be a fixed point. Then the family

$$\mathfrak{F} = \{f \text{ is univalent in } \Omega. | |f(z_0)| \leq 1 \quad \text{and} \quad |f'(z_0)| \leq 1\}$$

and the family $\mathfrak{F}' = \{f' | f \in \mathfrak{F}\}$ are normal in Ω. Compare to $\mathcal{S}(\Omega; z_0)$ in (5.8.2.3).

Sketch of proof. Suppose ζ is a point in Ω so that $M(\zeta) = \sup_{f \in F}\{|f(\zeta)| + |f'(\zeta)|\} < \infty$ holds. Choose $r > 0$ so that $|z - \zeta| < r$ is contained in Ω. Set

$$F(z) = \frac{f(\zeta + rz) - f(\zeta)}{rf'(\zeta)}, \quad |z| < 1 \quad \text{and} \quad f \in \mathfrak{F}$$

Then $\{F | f \in \mathfrak{F}\}$ is normal in $|z| < 1$, by (5.8.2.3). And hence, there is a constant $M_0 > 0$ so that

$$|F(z)| \leq M_0, \quad |z| \leq \frac{1}{2} \quad \text{and} \quad f \in \mathfrak{F}$$

$$\Rightarrow |f(z)| \leq r|f'(\zeta)|M_0 + |f(\zeta)| \leq rM(\zeta)M_0 + M(\zeta),$$

$$|z - \zeta| \leq \frac{1}{2} \quad \text{and} \quad f \in \mathfrak{F}.$$

On applying Cauchy's integral formula to $f'(z)$ on $|z - \zeta| = \frac{1}{2}$,

$$|f'(z)| \leq \frac{8}{r}(rM(\zeta)M_0 + M(\zeta)), \quad |z - \zeta| \leq \frac{r}{4} \quad \text{and} \quad f \in \mathfrak{F}.$$

Set $\Omega_0 = \{\zeta \in \Omega | \text{ both } \mathfrak{F} \text{ and } \mathfrak{F}' \text{ are (locally) uniformly bounded in a neighborhood of } \zeta\}$. By assumption and the result obtained in the last paragraph, $z_0 \in \Omega_0$ holds and thus, Ω_0 is a nonempty open set in Ω. Ω_0 is obviously closed. $\Omega_0 = \Omega$ is therefore true and the claim follows.
□

To the end, we present (refer to the proof of (5.8.3.12) and Exercise A (9) there)

Example 4. (Ref. [1], p. 227). Let f be an entire function. Then

the family $\mathfrak{F} = \{f(kz) | k \in \mathbf{C} \text{ is a constant}\}$ is normal in

a ring domain $0 < r_1 < |z| < r_2$.

$\Leftrightarrow$ f is a polynomial.

Proof. ($\Rightarrow$) Set $f_n(z) = f(nz)$ on $r_1 < |z| < r_2$ for $n \geq 1$.

Then f_n has a subsequence f_{n_k} converging locally uniformly in $r_1 < |z| < r_2$ to an analytic function f or to ∞. Fix any compact set $\rho_1 \leq |z| \leq \rho_2$, where $r_1 < \rho_1 < \rho_2 < r_2$.

In case $f_{n_k} \to f$ uniformly in $\rho_1 \leq |z| \leq \rho_2$: The boundedness of f on $\rho_1 \leq |z| \leq \rho_2$ shows that f_{n_k}, $k \geq 1$, are uniformly bounded there. Say $M > 0$ so that

$$|f_{n_k}(z)| \leq M, \quad \rho_1 \leq |z| \leq \rho_2, \quad k \geq 1.$$

$$\Rightarrow |f(z)| \leq M, \quad n_k\rho_1 \leq |z| \leq n_k\rho_2, \quad k \geq 1.$$

For any $k \geq 1$, then either

1. $n_{k+1}\rho_1 \leq n_k\rho_2$, then $|f(z)| \leq M$ on $n_k\rho_1 \leq |z| \leq n_{k+1}\rho_2$, or
2. $n_k\rho_2 < n_{k+1}\rho_1$, applying the maximum modulus principle to the boundary $|z| = n_k\rho_2$ and $|z| = n_{k+1}\rho_1$ on which $|f(z)| \leq M$ is known, then $|f(z)| \leq M$ on $n_k\rho_2 \leq |z| \leq n_{k+1}\rho_1$.

Combining together, then $|f(z)| \leq M$ on $|z| \geq n_1\rho_1$. This shows that ∞ is a removable singularity of the entire function f. f is thus a constant by Liouville's theorem.

In case $f_{n_k} \to \infty$ uniformly in $\rho_1 \leq |z| \leq \rho_2$: For any $M > 0$, there is $k_0 \geq 1$ so that $|f_{n_k}(z)| \geq M$ on $\rho_1 \leq |z| \leq \rho_2$ for $k \geq k_0$; thus, $|f(z)| \geq M$ on $n_k\rho_1 \leq |z| \leq n_k\rho_2$ for $k \geq k_0$. Similar argument shows that $|f(z)| \geq M$ on $|z| \geq n_{k_0}\rho_1$; therefore, $\lim_{z \to \infty} f(z) = \infty$ and ∞ is a pole of f. f turns out to be a polynomial.

($\Leftarrow$) Suppose $f(z) = a_n z^n + \cdots + a_1 z + a_0$, where $n \geq 0$ and $a_n \neq 0$. Try to use (5.8.1.8) to show that $\mathfrak{F}$ is normal.

Fix a constant $k \in \mathbf{C}$ and $k \neq 0$. Consider

$$\frac{2|kf'(kz)|}{1 + |f(kz)|^2} = \frac{2\left|\frac{na_n}{k^n}z^{n-1} + \frac{(n-1)a_{n-1}}{k^{n+1}}z^{n-2} + \cdots + \frac{a_1}{k^{2n-1}}\right|}{\frac{1}{|k|^{2n}} + \left|a_n z^n + \frac{a_{n-1}}{k}z^{n-1} + \cdots + \frac{a_1}{k^{n-1}}z + \frac{a_0}{k^n}\right|^2}. \quad (*4)$$

Choose a compact set $\rho_1 \leq |z| \leq \rho_2$, where $r_1 < \rho_1 < \rho_2 < r_2$ and $|k_0| \geq 1$ large enough so that

$$\left|\frac{(n-l)a_{n-l}}{k_0^{n+l}}\right| \leq 1 \quad \text{for} \quad 0 \leq l \leq n-1, \quad \text{and}$$

$$\left|a_n z^n + \frac{a_{n-1}}{k}z^{n-1} + \cdots + \frac{a_1}{k^{n-1}}z + \frac{a_0}{k^n}\right|$$

$$\geq |a_n||z|^n - \left|\frac{a_{n-1}}{k}\right||z|^{n-1} - \cdots - \left|\frac{a_1}{k^{n-1}}\right||z| - \left|\frac{a_0}{k^n}\right|$$

$$\geq |a_n||z|^n - \left|\frac{a_{n-1}}{k_0}\right||z|^{n-1} - \cdots - \left|\frac{a_1}{k_0^{n-1}}\right||z| - \left|\frac{a_0}{k_0^n}\right|, \quad \text{if } |k| \geq |k_0|$$

$$\geq |a_n|\rho_1^n - \left|\frac{a_{n-1}}{k_0}\right|\rho_2^{n-1} - \cdots - \left|\frac{a_1}{k_0^{n-1}}\right|\rho_2 - \left|\frac{a_0}{k_0^n}\right| \underset{\text{(def.)}}{=} \nu > 0.$$

Under this circumstance, by $(*_4)$, we obtain

$$\frac{2|kf'(kz)|}{1+|f(kz)|^2} \leq \frac{2(1+\rho_2+\cdots+\rho_2^{n-1})}{\alpha} \underset{\text{(def.)}}{=} M_1, \quad \rho_1 \leq |z| \leq \rho_2$$

$$\text{and} \quad |k| \geq |k_0|. \tag{$*_5$}$$

Note that M_1 depends only on ρ_1, ρ_2, and $|k_0|$ (disregarding the coefficients $a_n, a_{n-1}, \ldots, a_0$).

On the other hand, the continuous function $\frac{2|kf'(kz)|}{[1+f(kz)^2]}$ is bounded on the compact set $\{\rho_1 \leq |z| \leq \rho_2\} \times \{|k| \leq |k_0|\}$, say

$$\frac{2|kf'(kz)|}{1+|f(kz)|^2} \leq M_2, \quad (z,k) \in \{\rho_1 \leq |z| \leq \rho_2\} \times \{|k| \leq |k_0|\}. \tag{$*_6$}$$

$(*_5)$ and $(*_6)$ together say that the family

$$\left\{\frac{2|kf'(kz)|}{1+|f(kz)|^2}\,\middle|\, k \in \mathbf{C} \text{ and } r_1 < |z| < r_2\right\}$$

is locally uniformly bounded in $r_1 < |z| < r_2$.

Therefore, by (5.8.1.8), $\mathfrak{F}$ is normal.

Exercises A

(1) Let $\mathfrak{F}$ be a family of analytic functions defined on a domain Ω and missed the values of nonnegative real numbers. Show that F is normal in Ω.

(2) Let Ω be a domain and z_0 be a fixed point in Ω. Define $\mathfrak{F} = \{f$ is univalent and analytic in $\Omega\,|\,|f(z_0)| \leq 1$ and $|f'(z_0)| \geq 1\}$. Show that $\{\frac{1}{f'}|f \in \mathfrak{F}\}$ is normal in Ω.

(3) Let $\mathfrak{F}$ be a family of functions, analytic in $|z| < 1$, continuous on $|z| \leq 1$, and satisfying the inequality $\int_{|z|=1} |f(z)||dz| \leq 1$. Then $\mathfrak{F}$ is normal in $|z| < 1$.

(4) Suppose $\mathfrak{F}$ is a normal family of analytic functions in a domain Ω and g is an entire function. In case $\mathfrak{F}$ is also compact, then the family

$$g(\mathfrak{F}) = \{g \circ f \,|\, f \in \mathfrak{F}\}$$

is normal and compact in Ω. Note that, if $\mathfrak{F}$ fails to be compact, the result may cease to be true. For instance, $\mathfrak{F} = \{az|a \in \mathbf{C}\}$ together with $g(z) = e^z \sin z$ is such a case in $|z| > 1$ (see Example 2(3)).

(5) Suppose $\mathfrak{F}$ is a family of analytic functions from $|z| < 1$ into itself. For each $f(z) = \sum_0^\infty a_n(f)z^n \in \mathfrak{F}$, set $K_n = \sup_{f \in \mathfrak{F}} |a_n(f)|$ for $n \geq 0$. Show that $\mathfrak{F}$ is normal in $|z| < 1$ if and only if $0 \leq K_n < \infty$ for $n \geq 0$ and

$$\varlimsup_{n \to \infty} K_n^{\frac{1}{n}} = K \leq 1$$

hold.

(6) Let $\mathfrak{F}$ be a family of analytic functions $f(z)$ from $|z| < 1$ into itself and satisfying the condition that $f(0) = f'(0) = f''(0) = 1$. Show that there is a unique function f_M in $\mathfrak{F}$ so that $f_M(\frac{1}{2}) = \sup_{f \in \mathfrak{F}} |f(\frac{1}{2})|$.

(7) Let $0 < a_k < 1$ for $k = 1, 2, \ldots, n$. Suppose $\mathfrak{F}$ is a family of analytic functions $f(z)$ from $|z| < 1$ into itself and satisfying that $f(0) = 0$ and $f(a_k) = 0$ for $1 \leq k \leq n$. Show that

$$\sup_{f \in \mathfrak{F}} |f'(0)| = \prod_{k=1}^{n} a_k.$$

What is a corresponding extremal function?

(8) Let $\mathfrak{F}$ be the family of analytic functions $f(z)$ from $|z| < 1$ into itself. Show that $f_M(z) = z$ is an extremal function so that

$$\left| \frac{\partial \mathrm{Re}\, f_M}{\partial x}(0) \right| = \sup_{f \in \mathfrak{F}} \left| \frac{\partial \mathrm{Re}\, f}{\partial x}(0) \right| = 1.$$

Exercises B

(1) If f is univalent in a domain Ω, and $z_0 \in \Omega$, then

$$\frac{1}{4}|f'(z_0)|\mathrm{dist}(z_0, \partial\Omega) \leq \mathrm{dist}(f(z_0), \partial f(\Omega)) \leq 4|f'(z_0)|\mathrm{dist}(z_0, \partial\Omega).$$

In particular, for $f \in \mathcal{S}$ (see (5.8.2.3)),

$$\frac{1}{4} \leq \mathrm{dist}(0, \partial f(D)) \leq 1,$$

where $\Omega = D$ is the unit open disk $|z| < 1$.

(2) (*A distortion theorem*) If $f \in \mathcal{S}$, then

$$\mathrm{dist}(f(z), \partial f(D)) \geq \frac{1}{16}(1 - |z|^2), \quad z \in D = \{|z| < 1\}.$$

(3) If $f(z) = z + \sum_{n=2}^{\infty} a_n z^n \in \mathcal{S}$, then $|a_n| \leq en^2$ for $n \geq 2$.

(4) (*Koebe–Faber distortion theorem*) If $f \in \mathcal{S}$, then

(a)
$$\frac{\text{dist}(z, \partial f(D))}{1 - |z|^2} \leq |f'(z)|$$

$$\leq 4 \frac{\text{dist}(z, \partial f(D))}{1 - |z|^2}, \quad z \in D = \{|z| < 1\}, \text{ and}$$

(b) $\max_{|z|=r} |f(z)| \leq \frac{4r}{(1-r)^2}$, $|z| = r < 1$.

Note that the inequality in (b) provides yet another proof of the normality for $\mathcal{S}$. In case f is just univalent in $|z| < 1$, then (b) implies that

$$\max_{|z|=r} |f(z)| \leq |f(0)| + \frac{4r|f'(0)|}{(1-r)^2}, \quad |z| = r < 1;$$

if, in addition, $f(0) = 0$ and $f(z)$ does not assume the value 1 in $|z| < 1$, then the second inequality in (a) implies that

$$|f'(0)| \leq 4.$$

Try the following steps to give a proof:

1. Fix a point z, $|z| < 1$. Let $w = f(z)$. Choose $0 < \delta < \text{dist}(w, \partial f(D))$. Apply Schwarz–Pick's lemma (see (3.4.5.2)) to $f^{-1}(w + \delta\eta)$ on $\eta < 1$ to obtain the lower estimate in (a).

2. Show that $h(\eta) = f(\frac{\eta+z}{1+\bar{z}\eta})$, where z is fixed, $|z| < 1$, and $|\eta| < 1$, has its image containing an open disk with center at $h(0) = f(z)$ and radius $\frac{1}{4}|h'(0)|$, and then, obtain the upper estimate in (a).

3. Let $w = f(z)$, $|z| < 1$. Then $\text{dist}(w, \partial f(D)) \leq |w - 0| + \text{dist}(0, \partial f(D)) \leq 1 + |w|$ (why?). Since $|dw| = |f'(z)||dz|$, try to use (a) to show that $\frac{|dw|}{1+|w|} \leq \frac{4|dz|}{(1-|z|^2)}$, $|z| < 1$. In case γ is the line segment connecting 0 to a fixed point $z_0 \in D$, then

$$\int_{f(\gamma)} \frac{|dw|}{1 + |w|} \leq 4 \int_\gamma \frac{|dz|}{1 - |z|^2} = 2 \log \frac{1 + |z_0|}{1 - |z_0|}$$

(refer to Exercise B(5) of Sec. 3.4.5);

$$\geq \int_0^{|w_o|} \frac{dt}{1 + t} = \log(1 + |w_o|), \text{ where } w_o = f(z_0).$$

(5) (*A distortion theorem* (Koebe, 1909, 1910)) Let Ω be a domain. For each compact set K in Ω, there is a constant $c = c(\Omega, K)$, depending only on Ω and K, so that

$$\frac{1}{c} \leq \left| \frac{f'(z)}{f'(\zeta)} \right| \leq c, \quad z, \zeta \in K$$

holds for any univalent function f on Ω.

(6) (Montel, 1925) Let Ω be a domain and $z_0 \in \Omega$. Let f_n be a sequence of univalent functions in Ω, satisfying

   1. $f_n(z_0)$ converges; and
   2. $f'_n(z_0)$ converges to 0.

   Then f_n converges to a constant function locally uniformly in Ω.

(7) Let Ω be a domain and E be an infinite set in Ω, having a limit point in Ω.

   (a) Let $\mathfrak{F}$ be a family of univalent functions $f(z)$ in Ω, satisfying the property that $\{f(a)|f \in \mathfrak{F}\}$ is bounded for each $a \in E$. Then $\mathfrak{F}$ is normal in Ω.

   (b) Suppose $f_n \in \mathfrak{F}$, $n \geq 1$, and $f_n(a)$, $n \geq 1$, converges for each $a \in E$. Then f_n converges locally uniformly in Ω.

   (c) Suppose Ω is the disk $|z| < 1$ and $E = \{a_1, a_2, \ldots\}$ is a set in Ω, composed of distinct points and satisfying $\prod_{n=1}^{\infty}(1 - a_n) = \infty$. Then the results in (a) and (b) are still valid.

(8) Let Ω be a domain. Denote by $\Lambda(\Omega)$ the set of all analytic functions defined on Ω. Set $\|f\|_{\infty} = \sup_{z \in \Omega} |f(z)|$ for $f \in A(\Omega)$.

   (a) Show that $(A(\Omega), \|\|_{\infty})$ is a *Banach space* (i.e., a complete normed linear space).

   (b) Show that the unit ball $\{f \in A(\Omega) \mid \|f\|_{\infty} < 1\}$ in $A(\Omega)$ is normal in Ω.

(9) Let f be analytic in $|z| < 1$. If $0 \leq r < 1$, define

$$M_p(f;r) = \begin{cases} \left[\dfrac{1}{2\pi} \displaystyle\int_0^{2\pi} |f(re^{i\theta})|^p d\theta \right]^{\frac{1}{p}}, & 1 \leq p < \infty \\[2mm] \sup_{|z|=r} |f(z)|, & p = \infty \end{cases},$$

and

$$\|f\|_p = \sup_{0 \leq r < 1} M_p(f;r), \quad 1 \leq p \leq \infty.$$

The set $H^p = \{f \text{ is analytic in } |z| < 1 | \|f\|_p < \infty\}$ is called the *Hardy p-space*.

   (a) In case $1 \leq p < \infty$, $(H^p, \|\|_p)$ is a Banach space.

   (b) If $1 \leq p < \infty$, the unit ball $\{f \in H^p | \|f\|_p < 1\}$ in H^p is normal in $|z| < 1$.

5.8.3   The elliptic modular function: Montel's normality criterion and Picard's theorems

P. Montel: *Sur les familles de fonctions analytiques, qui admettent des valeurs exceptionnelles dans un domaine, Ann.Sci. École Norm. Sup. (3)29 (19,12), 487-535, FM43,509* related the normality to the values assumed by members of the family and realized that the modular function could be used to derive a criterion for normality. We state formally as

Montel's normality criterion. Let Ω be a domain in $\mathbf{C}$ (or $\mathbf{C}^*$).

(1) *Family of analytic functions.* Let $\mathfrak{F}$ be a family of analytic functions on Ω that *every f in $\mathfrak{F}$ omits two fixed distinct* numbers in $\mathbf{C}$. Then $\mathfrak{F}$ is normal in Ω.

(2) *Family of meromorphic functions.* Let $\mathfrak{F}$ be a family of meromorphic functions on Ω that *every f in $\mathfrak{F}$ omits three fixed distinct* numbers in $\mathbf{C}^*$ (∞ included). Then $\mathfrak{F}$ is normal in Ω.   (5.8.3.1)

Three subsections are divided.

Even though there are many alternative proofs of this criterion (see the descriptive remarks in Sec. 5.8.4), the classical proof (see Section (2)) using the elliptic modular function (see Section (1)) is still of great value and beauty in argument, and has many implications by itself. Section (3) will be devoted to some applications of this theorem, in particular, in the proofs of Picard's theorems.

Section (1)   The elliptic modular function

Consider the open set

$$\Delta = \left\{ w \,\Big|\, 0 < \operatorname{Re} w < 1 \quad \text{and} \quad \left| w - \frac{1}{2} \right| > \frac{1}{2}, \operatorname{Im} w > 0 \right\} \qquad (*_1)$$

in the $w = u + iv$ plane. It is a simply connected domain and its boundary $\partial\Delta$ is a curvilinear triangle with vertices at 0, 1, and ∞. Number the sides of $\partial\Delta$ as: I, the circular arc on the circle $|w - \frac{1}{2}| = \frac{1}{2}$; II, the vertical half line $\operatorname{Re} w = 1$, $\operatorname{Im} w > 0$, and III, the vertical half line, $\operatorname{Re} w = 0$, $\operatorname{Im} w > 0$. See Fig. 5.30.

According to the Riemann mapping theorem (see (6.1.1)) or, more precisely, the Schwarz–Christoffel formula (see Sec. 6.2), there is a conformal mapping $z = \varphi(w)$ (namely, one-to-one analytic or univalent function) that maps Δ onto the open half-plane

$$\mathrm{H} : \operatorname{Im} z > 0. \qquad (*_2)$$

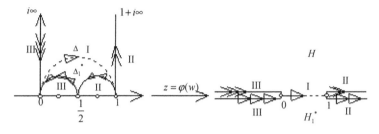

Fig. 5.30

By (6.1.2), this $z = \varphi(w)$ can be continuously extended to a homeomorphism mapping $\bar{\Delta}$ (the closure in $\mathbf{C}^*$) onto $\bar{H}$ (the closure in $\mathbf{C}^*$) so that

The vertices 0, 1, and ∞ are mapped onto the points 0, 1, and ∞, respectively, and:

The side I$\rightarrow$ the segment I: $(0, 1)$;

The side II$\rightarrow$ the segment II: $(1, \infty)$, and

The side III$\rightarrow$ the segment III: $(\infty, 0)$, all along the real axis. $(*_3)$

See Fig. 5.30. Such a $z = \varphi(w)$ is therefore uniquely defined (why? Refer to (2.5.4.8) and Example 3 there or (6.1.2.2)).

Let the symmetry principle (3.4.6.2) or (6.1.2.3) come into playing the role that extends the domain Δ of $z = \varphi(w)$ to the whole half strip $0 < \operatorname{Re} w < 1$, $\operatorname{Im} w > 0$, and then to the whole upper half-plane $\operatorname{Im} w > 0$ in the following manner.

Continue $z = \varphi(w)$ analytically across the side I from Δ to the curvilinear triangle Δ_1^*, with circular arcs lying on the circles $|w - \frac{1}{2}| = \frac{1}{2}$, $|w - \frac{3}{4}| = \frac{1}{4}$, and $|w - \frac{1}{4}| = \frac{1}{4}$. At the same time, the range of the extended $z = \varphi(w)$ goes across the side I from H into the lower half-plane H_1^*, with the corresponding boundaries indicated by arrows shown in Fig. 5.30. It is cautioned that $z = \varphi(w)$ maps $\Delta \cup I \cup \Delta_1^*$ univalently onto $H \cup I \cup H_1^*$, and $0, 1, \frac{1}{2}$ onto $0, 1, \infty$, respectively.

Similarly, continue the resulted $z = \varphi(w)$ across the circular arc on $|w - \frac{1}{4}| = \frac{1}{4}$ from Δ_1^* to the curvilinear triangle Δ_{11}^*; across the circular arc on $|w - \frac{3}{4}| = \frac{1}{4}$ from Δ_1^* to the curvilinear triangle Δ_{12}^*. See Fig. 5.31.

Explanation. The twice-extended function is still denoted by $z = \varphi(w)$.

1. $z = \varphi(w)$ maps

$$\Delta_{11}^* = \left\{ w \,\Big|\, \left|w - \frac{1}{4}\right| < \frac{1}{4}, \left|w - \frac{1}{6}\right| > \frac{1}{6}, \left|w - \frac{5}{12}\right| > \frac{1}{12} \text{ and } \operatorname{Im} w > 0 \right\}$$

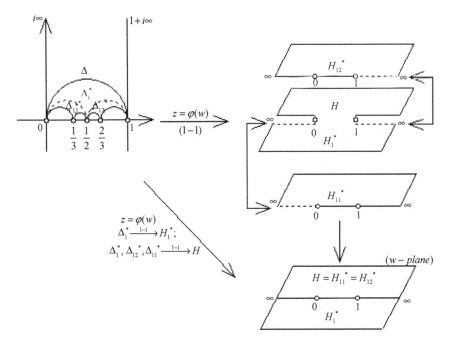

Fig. 5.31

univalently onto the upper half-plane H_{11}^* : $\operatorname{Im} z > 0$, pasted with H_1^* along the lower side of the cut III : $(\infty, 0)$. Note that $\varphi(0) = 0$, $\varphi(\tfrac{1}{3}) = 1$, and $\varphi(\tfrac{1}{2}) = \infty$.

2. $z = \varphi(w)$ maps

$$\Delta_{12}^* = \left\{ w \,\middle|\, \left|w - \frac{3}{4}\right| < \frac{1}{4}, \left|w - \frac{5}{6}\right| > \frac{1}{6}, \left|w - \frac{7}{12}\right| > \frac{1}{12} \text{ and } \operatorname{Im} w > 0 \right\}$$

univalently onto the upper half-plane H_{12}^* : $\operatorname{Im} z > 0$, pasted with H_1^* along the lower side of the cut II : $(1, \infty)$. Note that $\varphi(\tfrac{2}{3}) = 0$, $\varphi(1) = 1$, and $(\tfrac{1}{2}) = \infty$.

In Fig. 5.31, the curvilinear triangles Δ, Δ_1^*, Δ_{11}^*, and Δ_{12}^* are non-Euclidean ones (refer to Exercise B(5) of Sec. 3.4.5 or Appendix B). Moreover, $z = \varphi(w)$ maps $\Delta \cup I \cup \Delta_1^* \cup \mathrm{II} \cup \Delta_{11}^* \cup \mathrm{III} \cup \Delta_{12}^*$ onto $\mathbf{C}^* - \{0, 1, \infty\}$, covers the open upper half-plane three times (namely, $H = H_{11}^* = H_{12}^*$) and covers the open lower half-plane one time (namely, H_1^*), and covers once each of the line segments $(\infty, 0), (0, 1)$, and $(1, \infty)$.

Continue this process and we can extend the domain of definition of $z = \varphi(w)$ to the whole *open half strip*

$$0 < \operatorname{Re} w < 1 \quad \text{and} \quad \operatorname{Im} w > 0, \tag{5.8.3.2}$$

and the corresponding range *the punctured sphere*

$$\mathbf{C}^*(0, 1, \infty) \underset{\text{(def.)}}{=} \mathbf{C}^* - \{0, 1, \infty\} = \mathbf{C} - \{0, 1\}, \text{ simply denoted as}$$

$$\mathbf{C}(0, 1) \quad \text{or} \quad \mathbf{C}_{0,1} \tag{5.8.3.3}$$

with the three distinct points 0, 1, and ∞ removed from $\mathbf{C}^*$. In this case, $\mathbf{C}^*(0, 1, \infty)$ is, of course, covered countably infinitely many times by the half strip, under the map $z = \varphi(w)$.

Still by repeated use of the symmetry principle with respect to the vertical half lines $\operatorname{Re} w = n$, $\operatorname{Im} w > 0$ for $n = 0, \pm 1, \pm 2, \ldots$, the domain of definition of $z = \varphi(w)$ can be further extended to *the open half plane* $\operatorname{Im} w > 0$ with the same *range* $\mathbf{C}^*(0, 1, \infty)$. See Fig. 5.32.

Take three equidistant points A, B, and C on the unit circle $|\zeta| = 1$ in the ζ-plane. There is a unique bilinear transformation $w = \tau(\zeta)$, mapping $|\zeta| = 1$ onto $\operatorname{Im} w = 0$, the open disk $|\zeta| < 1$ onto $\operatorname{Im} w > 0$, and the points

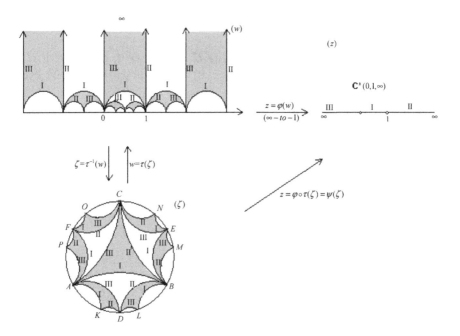

Fig. 5.32

A, B, C onto $0, 1, \infty$, respectively. In particular, the curvilinear triangle $\triangle \widehat{ABC}$ whose sides $\widehat{AB}$, $\widehat{BC}$, and $\widehat{CA}$ are circular arcs perpendicular to $|\zeta| = 1$ is mapped onto Δ in $(*_1)$. See Fig. 5.32, where each shaded triangle is mapped by $z = \varphi(w)$ or

$$z = \psi(\zeta) = \varphi \circ \tau(\zeta) \tag{5.8.3.4}$$

univalently onto the open upper half-plane $\operatorname{Im} z > 0$, while the unshaded ones onto the open lower half-plane $\operatorname{Im} z < 0$. Readers should realize how the conformality of $w = \tau(\zeta)$ plays between the two configurations constructed on $\operatorname{Im} w > 0$ and $|\zeta| < 1$, respectively. One of the advantages of the configuration on $|\zeta| < 1$ over that on $\operatorname{Im} w > 0$ lies on the fact that the former is a bounded set. And hence, it is easier to visualize physically what such curvilinear polygon ADBECF might look like. How can one pinpoint the corresponding polygon of that on $\operatorname{Im} w > 0$? In doing this, one needs to know beforehand to which point among 0, 1, and ∞ the vertices $A, B, \ldots, F$ will correspond. As a matter of fact, we can start from the very beginning to construct $z = \psi(\zeta)$ between $|\zeta| < 1$ and $\mathbf{C}^*(0, 1, \infty)$ instead of $z = \varphi(w)$: Reflect $\triangle \widehat{ABC}$ across its sides $\widehat{AB}$, $\widehat{BC}$, and $\widehat{CA}$ to obtain $\triangle \widehat{ABD}$, $\triangle \widehat{BCE}$, and $\triangle \widehat{CAF}$ and hence, the polygon ADBECF. And, at the same time, the range of $z = \psi(\zeta)$ will be extended to the lower half-plane from $\operatorname{Im} z > 0$ across the line segments $(0, 1), (1, \infty)$, and $(\infty, 0)$, respectively. And so forth.

In conclusion, we have

The elliptic modular function $z = \psi(\zeta)$ *(or* $z = \varphi(w)$): The function $z = \psi(\zeta)$ defined by (5.8.3.4) is a single-valued analytic function mapping the unit open disk $|\zeta| < 1$ onto the punctured sphere $\mathbf{C}^*(0, 1, \infty)$. Moreover,

(1) Every point on $|\zeta| = 1$ is either a vertex (countable infinite of them in total) of a curvilinear polygon or a limit point of these vertices. Hence $|\zeta| = 1$ is the *natural boundary* of $z = \psi(\zeta)$ in the sense that $z = \psi(\zeta)$ cannot be analytically continued across it into $|\zeta| > 1$.

(2) $\mathbf{C}^*(0, 1, \infty)$ is covered countably infinitely many time by $|\zeta| < 1$ under $z = \psi(\zeta)$. More precisely, each point $z_0 \in \mathbf{C}^*(0, 1, \infty)$ has a neighborhood $O(z_0)$ so that $\psi^{-1}(O_{z_0})$ is a countable union of pairwise disjoint components, each of them is mapped onto O_{z_0} both conformally and topologically by $z = \psi(\zeta)$. $\tag{5.8.3.5}$

As for (1) is concerned, suppose on the contrary that there exists a point on $|\zeta| = 1$ which is not a limit point of such vertices. Then there is a sequence σ_n of circular arcs on $|\zeta| = 1$ such that

1. $\sigma_{n+1} \subseteq \sigma_n$ for $n \geq 1$;
2. the end points of each σ_n are some two consecutive vertices of a curvi-linear polygon γ_n;
3. as $n \to \infty$, σ_n approaches a circular arc σ on $|\zeta| = 1$, whose interior points are not limit points of vertices.

Construct, in $|\zeta| < 1$, a circular arc passing the end points of σ and per-pendicular to $|\zeta| = 1$. Reflect A, B, and C with respect to σ and then at least one of their image points A', B', and C' would be an interior point of σ. If n is large enough, the symmetric points of A, B, and C with respect to σ_n will be sufficiently close to A', B', and C', respectively. This will lead to a contradiction and the former part of (1) is therefore true. That $z = \psi(\zeta)$ is single-valued is obvious from its constructive definition or by the mon-odromy theorem (5.2.2.2). For the latter part of (1), let ζ_0 be a fixed point on $|\zeta| = 1$ and D_{ζ_0} an open disk neighborhood of ζ_0. Note that $z = \psi(\zeta)$ assumes only the values $0, 1$, and ∞ at so many vertices lying along the arc $D_{\zeta_0} \cap \{|\zeta| = 1\}$. But, $z = \psi(\zeta)$ assumes every value in $\mathbf{C}^*(0, 1, \infty)$ on the open set $D_{\zeta_0} \cap \{|\zeta| < 1\}$. This indicates that $z = \psi(\zeta)$ cannot be continued extended to the closed set $\bar{D}_{\zeta_0} \cap \{|\zeta| \leq 1\}$, not to mention the analytic continuation across the arc $D_{\zeta_0} \cap \{|\zeta| = 1\}$ into $|\zeta| > 1$.

How about the inverse function $\zeta = \psi^{-1}(z)$?

According to (2) in (5.8.3.5), $\zeta = \psi^{-1}(z)$ is multiple-valued. If restricted the range of $\zeta = \psi^{-1}(z)$ to a presumed non-Euclidean curvilinear triangle, then the corresponding domain of definition of it will be either the upper or the lower half-plane. Under this circumstance, we obtain a *single-valued univalent branch* of $\zeta = \psi^{-1}(z)$.

In particular, when restricted to the *closed* $\triangle \widehat{ABC}$ (the interior plus the three sides disregarding the vertices A, B, and C), and the *open* $\triangle \widehat{ABD}$ (only the interior is considered), then we have the *principal branch*

$$\zeta = \psi^{-1}(z) : \mathbf{C}^*(0, 1, \infty) \to \text{closed } \triangle \widehat{ABC} \bigcup \text{open } \triangle \widehat{ABD} \qquad (5.8.3.6)$$

It is single-valued, univalent, and onto.

Continue analytically any fixed branch of $\zeta = \psi^{-1}(z)$, with domain an open set in $\mathbf{C}^*(0, 1, \infty)$, along all rectifiable curves so that their range will fill up the entire disk $|\zeta| < 1$. The resulted *complete* analytic configuration (see Ref. [75], or complete analytic function) is nothing else but the multiple-valued $\zeta = \psi^{-1}(z)$ itself.

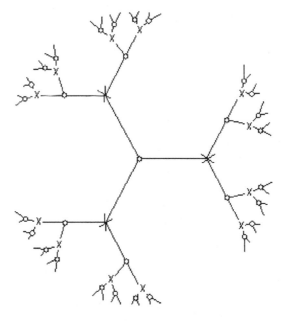

Fig. 5.33

The Riemann surface of $\zeta = \psi^{-1}(z)$ is composed of countably infinitely many copies of the z-plane, pasted properly (see the right figure in Fig. 5.31) together along the upper and lower sides of the cuts $(0,1)$, $(1,\infty)$, and $(0,\infty)$. The line complex is as in Fig. 5.33. Note that the points $0, 1$, and ∞ are not on the Riemann surface. $\zeta = \psi^{-1}(z)$ turns out to be single-valued on the surface.

There are corresponding words for $z = \varphi(w)$ and the upper half-plane $\operatorname{Im} w > 0$. The readers are supposed to be able to speak them out.

Remark (The modular group and Poincaré's metric on $\mathbf{C}^*(0, 1, \infty)$). Let w be a fixed point in the half strip (5.8.3.2). The symmetric point of w with respect to the line $\operatorname{Re} w = 1$ is the point $w^* = 2 - \bar{w}$ which, in turn, has the symmetric point $w^{**} = 4 - \overline{(2 - \bar{w})} = 2 + w$ with respect to $\operatorname{Re} w = 2$. By the construction of $z = \varphi(w)$,

$$\varphi(w^*) = \varphi(2 - \bar{w}) = \overline{\varphi(w)}$$
$$\Rightarrow \varphi(w^{**}) = \varphi(2 + w) = \overline{\varphi(w^*)} = \varphi(w).$$

Hence, $z = \varphi(w)$ is of *period* 2 and

$$\varphi(w + 2n) = \varphi(w), \quad \operatorname{Im} w > 0 \quad \text{and} \quad n = 0, \pm 1, \pm 2, \ldots. \tag{5.8.3.7}$$

On the other hand, the symmetric point of w with respect to $|w - \frac{1}{2}| = \frac{1}{2}$ is the point $\frac{1}{2} + \frac{1}{4\bar{w} - 2}$. Hence,

$$\varphi\left(\frac{1}{2} + \frac{1}{4\bar{w} - 2}\right) = \overline{\varphi(w)}$$

$$\Rightarrow \varphi(2 - w) = \varphi(-w) = \overline{\varphi(\bar{w})} = \varphi\left(\frac{1}{2} + \frac{1}{4w - 2}\right) = \varphi\left(\frac{w}{2w - 1}\right)$$

$$\Rightarrow \varphi(w) = \varphi\left(\frac{w}{2w + 1}\right), \quad \operatorname{Im} w > 0. \tag{5.8.3.8}$$

(5.8.3.7) and (5.8.3.8) together show that $z = \varphi(w)$ is *invariant* under the translation $w \to w + 2$, the transformation $w \to \frac{w}{2w+1}$, and their compositions. In other words, $z = \varphi(w)$ remains unchanged under the bilinear transformations

$$w \to \frac{aw + b}{cw + d}, \quad \text{where} \quad \begin{bmatrix} a & b \\ c & d \end{bmatrix} = \begin{bmatrix} 1 & 0 \\ 0 & 1 \end{bmatrix} \pmod{2}. \tag{5.8.3.9}$$

The set of all such transformations forms a group under the composite operation and is called the *modular or automorphic group* and its elements the *modular transformations*. This group is useful in number theory.

Distinct single-valued branches of $w = \varphi^{-1}(z)$ are related by the modular transformations (5.8.3.9). It is well-known that Poincaré's metric $\frac{dw}{\operatorname{Im} w}$ is invariant under the bilinear transformations preserving the upper half-plane (refer to Exercise B(5) of Sec. 3.4.5 or Appendix B); consequently,

$$\eta(z) = \frac{|(\varphi^{-1})'(z)|}{\operatorname{Im} \varphi^{-1}(z)}, \quad z \in \mathbf{C}^*(0, 1, \infty) \tag{5.8.3.10}$$

is well-defined, independent of the branches chosen. Therefore, we define

$$ds = \eta(z)|dz| \tag{5.8.3.11}$$

or simply $\eta(z)$ as *Poincaré's metric* on the punctured sphere $\mathbf{C}^*(0, 1, \infty)$. This is the metric induced on $\mathbf{C}^*(0, 1, \infty)$ from that on the upper half-plane $\operatorname{Im} w > 0$ by $z = \varphi(w)$. It can be used to derive the dilatation of the moduli under quasiconformal mappings. See Ref. [55], Chap. 4. See Sec. 5.8.4 for further explanation. □

Section (2) The proof of Montel's criterion (5.8.3.1)

Given first that $\mathfrak{F}$ is a family of analytic function in a domain Ω. According to (2) in (5.8.1.3), one may just suppose that Ω is the open unit disk. If a and b are the two finite complex numbers not assumed in Ω by every

$f \in \mathfrak{F}$, replace f by $F(z) = \frac{f(z)-a}{b-a}$, $z \in \Omega$. Then $F(z) \neq 0, 1$ on Ω, and the normality of $\{F | f \in \mathfrak{F}\}$ is equivalent to the one for $\mathfrak{F}$. Hence it is harmless to assume that the exceptional values are 0 and 1, for simplicity.

Choose any sequence f_n in $\mathfrak{F}$.

Fix any point $z_0 \in \Omega$. The sequence $f_n(z_0)$, $n \geq 1$, has a limit point α in $\mathbf{C}^*$. Four cases are divided according to $\alpha \in \mathbf{C}^*(0, 1, \infty)$ or $\alpha = 0, 1, \infty$, respectively.

Case 1. $\alpha \in \mathbf{C}^*(0, 1, \infty)$.
On replacing by a subsequence if necessary, we may assume that $\lim_{n\to\infty} f_n(z_0) = \alpha$ holds. Choose an open (disk) neighborhood O_α of α in $\mathbf{C}^*(0, 1, \infty)$. There is an integer $n_0 \geq 1$ so that $f_n(z_0) \in O_\alpha$ if $n \geq n_0$; hence, $f_n(\Omega) \cap O_\alpha$ is nonempty and open for $n \geq n_0$. Denote by $\zeta = \psi^{-1}(w)$ the principle branch (5.8.3.6) of the inverse function of the modular function (note that z there has been replaced by w here). Let U_n be the component of the open set $f_n^{-1}(f_n(\Omega) \cap O_\alpha)$ that contains the point z_0; consequently, the composite functions $\psi^{-1}(f_n(z))$, $n \geq n_0$, are well-defined and analytic on U_n. Since f_n is defined on the whole disk Ω and has its range $f_n(\Omega)$ containing in $\mathbf{C}^*(0, 1, \infty)$, the function element $(\psi^{-1}(f_n(z)), U_n)$ can be analytically continued throughout Ω, and the resulted function, still denoted by $\psi^{-1}(f_n(z))$, is single-valued and analytic in Ω, according to the monodromy theorem (5.2.2.2).

The functions $\psi^{-1}(f_n(z))$, $n \geq n_0$, $z \in \Omega$ all have their ranges in the open disk $|\zeta| < 1$ and $\lim_{n\to\infty} \psi^{-1}(f_n(z_0)) = \beta$, the principal value of $\psi^{-1}(\alpha)$. By using (3) in (5.8.1.3), a subsequence $g_j(z) = \psi^{-1}(f_{n_j}(z)), j \geq 1$, of the sequence $\psi^{-1}(f_n(z)), n \geq n_0$, converges to an analytic function $g(z)$ in Ω locally uniformly. Hence, $g(z_0) = \beta$ holds. Since $|\beta| < 1$, $g(z)$ is not a constant function of absolute equal to 1; as a matter of fact, $|g(z)| < 1$ holds throughout Ω. To see this, observing that $|g_j(z)| < 1, z \in \Omega$, and $j \geq 1$, implies that $|g(z)| \leq 1$ on Ω and then the maximum principle justifies that $|g(z)| < 1$ should hold for all $z \in \Omega$. Consequently $\psi(g(z))$ is well-defined and analytic in Ω.

Set $M = \max_{z \in K} |g(z)|$, where K is a compact set in Ω. Note that $M < 1$. The uniform convergence indicates that, for some $M < M' < 1$ and $j_0 \geq 1$,

$|g_j(z) - g(z)| \leq M' - M$, $z \in K$ and $j \geq j_0$

$\Rightarrow |g_j(z)| \leq M' - M + |g(z)| < M'$, $z \in K$ and $j \geq j_0$, and hence,

$g_j(z)$, $j \geq j_0$, are locally uniformly bounded by M' (in absolute value) on K.

$\Rightarrow$ (Since $w = \psi(\zeta)$ is bounded on $|\zeta| \le M' < 1$, and recall that $g_j(z) = \psi^{-1}(f_{n_j}(z))$)

$$|f_{n_j}(z)| \le \max_{|\zeta| \le M'} |\psi(\zeta)| < \infty, \quad z \in K \quad \text{and} \quad j \ge j_0.$$

Hence $f_{n_j}(z)$, $j \ge j_0$ (or just $j \ge 1$), are locally uniformly bounded in Ω. Therefore, $f_{n_j}(z)$, $j \ge 1$, has a subsequence converging locally uniformly in Ω. So does the original sequence $f_n(z)$, $n \ge 1$.

Case 2. $\alpha = 1$.

We still suppose that $\lim_{n \to \infty} f_n(z_0) = 1$ holds. Choose $n_0 \ge 1$ so that $|f_n(z_0) - 1| < 1$ if $n \ge n_0$. Let U_n be the component of $f_n^{-1}(f_n(\Omega) \cap \{|w - 1| < 1\})$ that contains z_0. Denote by $\text{Log } w$ the principal branch of $\log w$ in $|w - 1| < 1$, determined by $\log 1 = 0$ (see (1) in (2.7.3.2)). The composite function $\text{Log } f_n(z)$, $n \ge n_0$, is well-defined in U_n, and is single-valued and analytic there. Continue the function element $(\text{Log } f_n(z), U_n)$, $n \ge n_0$, analytically throughout the whole disk Ω and still denote the resulted single-valued analytic function by $\text{Log } f_n(z)$. Since $f_n(z) \ne 0, 1$ in Ω, hence $\text{Log } f_n(z) \ne \infty, \pm 2\pi i$ in Ω. Set

$$g_n(z) = \frac{1}{4\pi i}(\text{Log } f_n(z) + 2\pi i), \quad z \in \Omega, \quad n \ge n_0.$$

Observe that $g_n(z)$ is analytic in Ω, and $g_n(z) \ne 0, 1, \infty$ there; moreover, $\lim_{n \to \infty} g_n(z_0) = \frac{1}{2}$ because $\lim_{n \to \infty} \text{Log } f_n(z_0) = 0$.

By invoking Case 1 to the sequence $g_n(z)$, it has a subsequence $g_{n_j}(z)$, $j \ge 1$, converging to an analytic function $g(z)$ locally uniformly in Ω; in particular, $\lim_{j \to \infty} g_{n_j}(z_0) = g(z_0) = \frac{1}{2}$. Yet $g_{n_j}(z) \ne \frac{1}{2}$ everywhere in Ω since $f_n(z) \ne 1$ in Ω. Using Hurwitz's theorem (see (1) in (5.3.2.2)), it follows that $g(z) = \frac{1}{2}$ in Ω which, in turn, implies that the subsequence $f_{n_j}(z)$, $j \ge 1$, converges to 1 locally uniformly in Ω.

Case 3. $\alpha = 0$.

Suppose that $\lim_{n \to \infty} f_n(z_0) = 0$ holds. Therefore the function $1 - f_n(z) \ne 0, 1, \infty$ on Ω for each $n \ge 1$, and $\lim_{n \to \infty}(1 - f_n(z_0)) = 1$ holds. Applying Case 2 to the sequence $1 - f_n(z)$, $n \ge 1$, we obtain that the sequence $f_n(z)$ has a subsequence $f_{n_j}(z)$ converging to 0 locally uniformly in Ω.

Case 4. $\alpha = \infty$.

Suppose again that $\lim_{n \to \infty} f_n(z_0) = \infty$ holds. Then $1 - \frac{1}{f_n(z)} \ne 0, 1, \infty$ on Ω for $n \ge 1$, and $\lim_{n \to \infty}(1 - \frac{1}{f_n(z_0)}) = 1$ holds. Applying Case 2 to the sequence $1 - \frac{1}{f_n(z)}, n \geq 1$, then $f_n(z)$ has a subsequence $f_{n_j}(z)$ converging to ∞ locally uniformly in Ω.

This ends up the normality proof for the analytic family $\mathfrak{F}$.

Suppose $\mathfrak{F}$ is a family of meromorphic functions in Ω such that each member of it does not assume three distinct points $a, b, c \in \mathbf{C}^*$ in Ω.

In case one of a, b, and c is the point ∞, then each function in $\mathfrak{F}$ is analytic in Ω and $\mathfrak{F}$ is a family of analytic functions in Ω. That $\mathfrak{F}$ is thus normal is just proved.

In case a, b, and c are finite complex numbers, consider the family

$$\left\{ \frac{1}{f(z) - a} \,\middle|\, f \in \mathfrak{F} \right\}$$

composed of analytic functions in Ω. Each member of this family omits two distinct values $\frac{1}{b-a}$ and $\frac{1}{c-a}$ in Ω. Hence it is normal in Ω. So is the original $\mathfrak{F}$. $\qquad\qquad\qquad\qquad\qquad\qquad\qquad\qquad\qquad\qquad\square$

Section (3) Picard's theorems

An entire function omitting the values in a given disk $|w - w_0| < r$ must then reduce to a constant, in view of Liouville's theorem. So is an entire function which does not assume values in the segment $[-1, 1]$, with the help of the inverse function $\eta = w - \sqrt{w^2 - 1}$ ($\eta(\infty) = 0$) of the Joukouski function $w = \frac{1}{2}(\eta + \frac{1}{\eta})$.

The well-known entire functions $\sin z$ and $\cos z$ assume any presumed finite complex number, even at infinitely many points (see (2.6.2.6), (2.7.4.1), and (2.7.4.5)). While, e^z does not assume the value 0, but assumes each nonzero complex number at infinitely many points, too.

Does there exist a nonconstant entire function omitting two distinct finite complex values? *E. Picard: Mémoire sur les fonctions entières, Ann École Norm.,t.9, 1880* gave the answer negative. It is the

Picard's first (or little) theorem (refer to (4.10.3.3)).

(1) *A nonconstant entire function assumes each finite complex number, except at most one finite number (called the exceptional value).*

(2) *A transcendental entire function assumes each finite complex number at infinitely many points, except at most one finite number.*

$$(5.8.3.11)$$

(2) means that, for any presumed $w \in \mathbf{C}$, with at most one exception, the equation $f(z) = w$ has infinitely many solution if $f(z)$ is transcendental.

One may suppose that the entire function $f(z)$ omits the values 0 and 1, namely, its range $f(\mathbf{C}) \subseteq \mathbf{C}(0, 1)$, and try to show that $f(z)$ thus reduces to a constant. Recall that $\mathbf{C}(0,1)$ is the domain of the *multiple-valued* inverse function $\zeta = \psi^{-1}(w)$, of the elliptic modular function $w = \psi(\zeta)$, whose range fills up the unit disk $|\zeta| < 1$. It is tempted to think that $\psi^{-1}(f(z))$ is bounded and entire, and then reduces to a constant; since $\zeta = \psi^{-1}(w)$ is not a constant, $f(z)$ will turn out to be constant, too. Unfortunately $\psi^{-1}(f(z))$ is *not single-valued*, and all one needs to do is to *uniformize* $\psi^{-1}(f(z))$ *locally* to a single-valued branch and then, to use *analytic continuation* to extend the local branch to a *bounded entire function* in $\mathbf{C}$. This is exactly the method adopted in proving Case 1 for Montel's criterion.

Proof. Assume that the entire function $f(z)$ omits the values 0 and 1. Choose a single-valued branch of $\zeta = \psi^{-1}(w)$ on an open neighborhood O of the point $f(0)$ in the domain $\mathbf{C}(0,1)$. Fix an open disk $D : |z| < r$ so that $f(D) \subseteq O$ holds. Since $f(z)$ is entire, the function element $(\psi^{-1}(f(z)), D)$ can be thus analytically continued into the whole plane; and the monodromy theorem (5.2.2.2) says that the extended function, still denoted by $\psi^{-1}(f(z))$, is single-valued in $\mathbf{C}$ and hence, is a bounded entire function. Therefore $\psi^{-1}(f(z))$ is a constant function in $\mathbf{C}$. Since $\zeta = \psi^{-1}(w)$ is locally univalent, it follows that $f(z)$ should be constant in $\mathbf{C}$. This proves (1).

As for (2), suppose that $f(z)$ is transcendental (and then ∞ is an essential singularity). Suppose on the contrary that $f(z)$ assumes two distinct finite complex numbers a and b only *finitely* many times. Then there is an $R > 0$ so that $f(z) \neq a, b$ on $|z| > R$, contradicting to Picard's second theorem (see (5.8.3.12) below). This proves (2).   □

Casorati–Weierstrass's theorem (see (4.10.3.1)) is completed as

Picard's second (or great) theorem. *Suppose $f(z)$ is analytic in $0 < |z - z_0| < \rho$ and has z_0 as an essential singularity. Then $f(z)$ assumes each finite complex number in $0 < |z - z_0| < \rho$, except at most one finite complex number.*   (5.8.3.12)

In case $z_0 = \infty$ is the essential singularity, $0 < |z - z_0| < \rho$ should be replaced by $\rho < |z| < \infty$ in the above statement.

Proof. By performing the transformation $z \to z - z_0$ or $z \to \frac{1}{z}$, we may suppose that $z_0 = 0$, and that $f(z)$ does not assume the values $a, b \in \mathbf{C}$, where $a \neq b$, in $0 < |z| < \rho$. Construct a sequence of ring domains:

$$\Omega_n : \frac{\rho}{2^{n+1}} < |z| < \frac{\rho}{2^n} \quad \text{for } n \geq 0,$$

and consider the associated sequence of analytic functions:

$$f_n(z) = f\left(\frac{z}{2^n}\right), \quad z \in \Omega_0 \quad \text{for } n \geq 1.$$

The sequence $f_n(z)$, $n \geq 1$, forms a normal family of analytic functions in Ω_0, by (5.8.3.1), because $f_n(z) \neq a$, b for each $n \geq 1$. Hence we may suppose the sequence $f_n(z)$ itself converges to an analytic function $g(z)$ or to ∞ locally uniformly in Ω_0.

In case $g(z)$ is an analytic function: Fix any r, $\frac{\varrho}{2} < r < \rho$. $g(z)$ is bounded on $|z| = r$. The uniform convergence on $|z| = r$ implies that

$\qquad |f_n(z)| \leq M$ on $|z| = r$ for $n \geq 1$ and some constant $M > 0$.
$\Rightarrow$ (by the maximum principle) $|f(z)| \leq M$ on $0 < |z| \leq r$.
$\Rightarrow z_0 = 0$ is a removable singularity of f (see (4.10.1.1)).

This is a contradiction.

In case $g(z) \equiv \infty$ in Ω_0: On applying the former case to the sequence $\frac{1}{f_n(z)}$, $n \geq 1$, in Ω_0, we obtain that $\frac{1}{f(z)}$ is analytic at $z_0 = 0$. Hence, $z_0 = 0$ is a removable singularity or a pole of $f(z)$, a contradiction, too. $\square$

Let $f(z)$ be a nonconstant meromorphic function in $\mathbf{C}$ and $f(z)$ do not assume three distinct values a, b, and c in $\mathbf{C}^*$.

In case $c = \infty$: Then $f(z)$ is a nonconstant entire function, not assuming the values a and b. This contradicts to Picard's first theorem.

In case a, b, and c are in $\mathbf{C}$: Suppose $c \neq 0$. Then the entire function $\frac{1}{f(z)-c}$ does not take the values $\frac{1}{a-c}$ and $\frac{1}{b-c}$ in $\mathbf{C}$, a contradiction, too.

Let ∞ be an essential singularity of f. In case $f(z)$ assumes three distinct values a, b, and c in $\mathbf{C}$ only at finitely many points, then $f(z)$ does not assume a, b, and c in $\rho \leq |z| < \infty$ for some $\rho > 0$. This contradicts to what we obtained in the last two paragraphs.

Combining together, we have

The value distribution of a nonconstant meromorphic function $f(z)$ on $\mathbf{C}$.

(1) $f(z)$ assumes all values (including ∞) in $\mathbf{C}^*$, except at most two distinct values (in $\mathbf{C}^*$).

(2) If ∞ is an essential singularity of $f(z)$, then $f(z)$ assumes each finite complex number at infinitely many points, except at most two distinct values (in $\mathbf{C}$).   (5.8.3.13)

To the end, we remind the readers that, in Sec. 4.10.3, we had raised examples using Picard's theorems and Casorati–Weierstrass's theorem.

Some further applications of these theorems are continued in the exercises there.

Exercises A

(1) Let $f_n(z) = e^{nz}$, $n \geq 1$. Show that $\{f_n(z)|n \geq 1\}$ is not normal in **C**. This example shows that Montel's criterion is no more true if only one exceptional value is permitted.

(2) Let $f_n(z)$, $n \geq 1$, be analytic in a domain Ω so that any subsequence of it never converges locally uniformly in Ω.

  (a) Show that the set $\overline{\lim}_{n \to \infty} f_n(\Omega) = \bigcap_{n=1}^{\infty} \bigcup_{m \geq n} f_m(\Omega)$ is either **C** or **C** with a point deleted.

  (b) Show that there is a point z_0 in Ω such that, for any open neighborhood O of z_0, the set $\bigcap_{n=1}^{\infty} \bigcup_{m \geq n} f_m(O)$ is either **C** or **C** with a point deleted.

(3) Let $\mathfrak{F} = \{f(z)$ be analytic in $|z| < 1$ so that $f(0) = 0$, $f'(0) = 1$, and $f(z) \neq 0$ in $0 < |z| < 1\}$. Show that $\mathfrak{F}$ is normal and compact.

(4) Let $\mathfrak{F}$ be as in Exercise (3). Show that there is a constant $\alpha > 0$ such that, for each $f \in \mathfrak{F}$, the range $f(\{|z| < 1\})$ always contains the open disk $|w| < \alpha$.

(5) Let a be a fixed constant. Show that the family $\mathfrak{F} = \{f(z)$ is univalent in a domain but $f(z) \neq a$ on $\Omega\}$ is normal.

(6) Suppose a nonconstant entire function $f(z)$ assumes real value if z is real. Show that $f(z)$ assumes all nonreal complex values.

(7) A periodic entire function has a fixed point. Prove it.

(8) Let $p(z)$ be a nonzero polynomial. Show that $e^z - p(z) = 0$ has infinitely many solutions.

(9) Let $f(z)$ be analytic in $0 < |z| < \rho$ with the point 0 being an essential singularity. Set $f_n(z) = f(\frac{z}{2^n})$, $0 < |z| < \rho$, for $n \geq 1$.

  (a) Show that $\{f_n|n \geq 1\}$ is not normal in $\frac{\rho}{2^4} < |z| < \rho$ (and hence, in $0 < |z| < \rho$).

  (b) Suppose $0 < 2r < \rho$. Show that $\{f_n|n \geq 1\}$ is not normal in $\frac{r}{4} < |z| < r$.

Exercises B

(1) Let $\mathscr{S}$ be the family of the normalized univalent analytic functions in $|z| < 1$, defined in (5.8.2.3). Try to use Montel's criterion (5.8.3.1) to prove that $\mathscr{S}$ is normal and compact in $|z| < 1$, and that there is a

constant $\alpha > 0$ so that

$$\{|w| < \alpha\} = \bigcap_{f \in \mathcal{S}} f(\{|z| < 1\})$$

with $\frac{z}{(1+e^{i\theta}z)^2}$ $(\theta \in \mathbf{R}.)$ as an extremal function whose range is the domain obtained from $\mathbf{C}$ by deleting the ray $\mathrm{Arg}(w - \frac{1}{4}e^{i\theta}) = \theta$. α, which is equal to $\frac{1}{4}$, is called *Koebe's constant*. Observe the following step.

(a) Let α_f be as in (5.8.2.6). Choose $a_f \in \partial f(\{|z| < 1\})$ so that $|a_f| = \alpha_f$. Denote by $g_f(z)$ the branch of $\sqrt{1 - \frac{f(z)}{a_f}}$ $(f \in \mathcal{S})$ in $|z| < 1$, uniquely determined by $\sqrt{1} = 1$. Show that such a $g_f(z)$ omits the value 0 and -1 in $|z| < 1$. Hence $\{g_f | f \in \mathcal{S}\}$ is normal in $|z| < 1$.

(b) Let $f_n \in \mathcal{S}$ be a sequence. Denote $g_n = g_{f_n}$, $n \geq 1$. Let a subsequence g_{n_j}, $j \geq 1$, converge to g locally uniformly in $|z| < 1$. Show that g is analytic owing to $\lim_{j \to \infty} g_{n_j}(0) = 1$. Hence $\frac{f_{n_j}}{a_{n_j}}$, where $a_{n_j} = a_{f_{n_j}}$ for $j \geq 1$, converges locally uniformly in $|z| < 1$ to an analytic function and $\lim_{j \to \infty} a_{n_j} = a$ holds with $0 < |a| \leq 1$, by Weierstrass's theorem (5.3.1.1) and $f'_{n_j}(0) = 1$ for $j \geq 1$. Consequently, $f_{n_j} \to f$ locally uniformly in $|z| < 1$. And Hurwitz's theorem (5.3.2.2) is needed to justify that f is univalent.

(c) If no such α exists, then there is a sequence $f_n \in \mathcal{S}$ with $|a_{f_n}| < \frac{1}{n}$ for $n \geq 1$. This will contradict to $0 < |a| \leq 1$.

(2) (G. Julia, 1919) Suppose $f(z)$ is analytic in $0 < |z| < \rho$ and has 0 as an essential singularity. Then, there is a real number θ_0 such that, for each ε with $0 < \varepsilon < \frac{\pi}{2}$, $f(z)$ assumes each finite complex number infinitely many times in the sector domain

$$S_\varepsilon, 0 < |z| < \rho, \quad |\mathrm{Arg}\, z - \theta_0| < \varepsilon$$

with at most one exception (indeed, $f(z)$ assumes the same number at each point of a sequence converging to 0). $\mathrm{Arg}\, z = \theta_0$ is called *Julia direction* of $f(z)$ at 0. Try the following steps.

(a) Let f_n be as in Exercise A(9). Then $\{f_n | n \geq 1\}$ is not normal in $\frac{\rho}{2^4} < |z| < \rho$. Hence, there is a point z_0 so that $\{f_n | n \geq 1\}$ fails to be normal in any open neighborhood of z_0 (see(3) in (5.8.1.3)). Set $\theta_0 = \mathrm{Arg}\, z_0$.

(b) In particular, $\{f_n | n \geq 1\}$ fails to be normal in S_ε. By Montel's criterion, for any $w \in \mathbf{C}$ with at most one exception, there are a sequence n_k, $k \geq 1$, of positive integers and a sequence ζ_{n_k}, $k \geq 1$, of points in S_ε so that $f_{n_k}(\zeta_{n_k}) = f(\frac{\zeta_{n_k}}{2^{n_k}}) = w$, $k \geq 1$.

Note. This theorem initiated the study of *value distribution* in an angular domain, somewhat different from the one in the whole plane. Interested readers might refer to Refs. [73, 82, 84].

(3) Show that the positive and the negative imaginary axes are the only two Julia directions of e^z.

(4) Show that e^{e^z} has infinitely many Julia directions (refer to Exercise B(2)(0) of Sec. 2.7.3 and Example 5(4) of Sec. 3.3.1).

(5) Suppose f is analytic in a simply connected domain Ω on which f does not assume the values 0 and 1.

  (a) Show that there is an analytic functions $g(z)$ in Ω, not assuming the values $0, \pm 1, \pm 2, \ldots$, such that $f(z) = e^{2\pi i g(z)}$ in Ω.

  (b) Show that there are analytic functions $h(z)$ and $p(z)$ in Ω such that $g(z) = h(z)^2$ and $g(z) - 1 = p(z)^2$ in Ω, respectively.

  (c) Show that there is an analytic function $F(z)$ in Ω such that $h(z) - p(z) = e^{F(z)}$.

  (d) Show that

$$f(z) = -e^{\pi i \cosh 2F(z)}, \quad z \in \Omega.$$

and deduce that $F(z)$ does not assume the values $\pm \log(\sqrt{m} + \sqrt{m-1}) + \frac{n\pi i}{2}$, where $m \geq 1$ are integers and $n = 0, \pm 1, \pm 2, \ldots$. Denote by S the set of these exceptional values.

  (e) Fix any point $z_0 \in \mathbf{C}$. Show that the open disk $|z - z_0| < 1$ always has nonempty intersection with the set S, defined in (d). Consequently, the image $F(\Omega)$ does not contain any open disk of radius 1.

(6) Suppose $f(z)$ is analytic in $|z| < 1$ and $f'(0) = 1$. Then the range $f(\{|z| < 1\})$ contains an open disk of radius $\frac{1}{16}$, called a *Landau constant*. Try the following steps.

  (a) Consider $f_r(z) = \frac{1}{r}f(rz)$, $|z| \leq 1$, where $0 < r < 1$ is fixed. Suppose the result is known to be valid in case $f(z)$ is analytic in the *closed* disk $|z| \leq 1$. Then $f_r(z)$ is analytic in $|z| \leq 1$ and hence, the image of $|z| < 1$ under $f(rz)$ contains an open disk of radius $\frac{r}{16}$. Set $r \to 1$.

  (b) Suppose $f(z)$ is analytic in $|z| \leq 1$ from now one. Set $M(r) = \max_{|z|=r} |f'(z)|$, $0 \leq r \leq 1$. Then $\varphi(r) = (1-r)M(r)$ is continuous on $0 \leq r \leq 1$ (see Exercise A(6) of Sec 3.4.4), and $\varphi(0) = 1$ and

$\varphi(1) = 0$. Let $r_0 = \sup_{0 \le r \le 1} \varphi(r)$. Note that $0 \le r_0 < 1$ and $\varphi(r_0) = 1$; also, $\varphi(r) < 1$ in case $r_0 < r \le 1$.

(c) Choose z_0, $|z_0| = r_0$, so that $\varphi(r_0) = (1 - r_0)|f'(z_0)| = 1$. Hence, $|f'(z_0)| = \frac{1}{1-r_0}$. Set $g(z) = f(z + z_0) - f(z_0)$. Then $g(z)$ is analytic in $|z| \le 1 - r_0$, $g(0) = 0$, and $|g'(0)| = \frac{1}{1-r_0}$. It is thus sufficient to show that $g(z)$ has its range containing an open disk of radius $\frac{1}{16}$.

(d) Show that $|g'(z)| \le \frac{2}{1-r_0}$ if $|z| = \frac{1-r_0}{2}$ and hence $|g(z)| = |\int_0^z g'(\zeta)d\zeta| \le 1$ if $|z| \le \frac{1-r_0}{2}$. Fix a point $w \notin g(\{|z| \le \frac{1-r_0}{2}\})$. Then $w \ne 0$. Choose a branch $h(z) = \sqrt{1 - \frac{g(z)}{w}} = 1 + a_1 z + a_2 z^2 + \cdots$ in $|z| < \frac{1-r_0}{2}$, where $a_1 = -\frac{g'(0)}{2w}$. By Passeval's identity (see Exercise B(2) of Sec. 3.4.2 or (5.1.2.9)),

$$1 + |a_1|^2 \left(\frac{1 - r_0}{2}\right)^2 \le \sum_0^\infty |a_n|^2 \left(\frac{1 - r_0}{2}\right)^{2n}$$

$$= \frac{1}{2\pi} \int_0^{2\pi} \left| h\left(\frac{1 - r_0}{2}e^{i\theta}\right) \right|^2 d\theta \le 1 + \frac{1}{|w|}$$

which implies that $|w| \ge \frac{1}{16}$.

Note. Suppose f is analytic in $|z| < 1$ and satisfies the condition that $|f'(0)| = 1$. Then its range always contains an open disk of radius $\frac{1}{16}$, called a *Bloch* or *Landau constant* (refer to Exercise B(6) of Sec. 4.8). In general setting, let f be analytic in $|z| < 1$ and $f'(0) = 1$. Set $\beta_f = \sup\{r|f \text{ is univalent in an open disk } U \text{ contained in } |z| < 1 \text{ such that the image } f(U) \text{ contains an open disk of radius } r\}$ and β be the infimum of all such β_f. β is called the *Bloch (1893–1948) constant*. It can be shown that $0.43 \le \beta \le 0.47$ yet the precise value is still unknown. See Ref. [2] or [12]. However, this fact provides *another proof of Picard's little theorem*. To see this, let f be a nonconstant entire function, omitting the values 0 and 1. Then, $f(z) = -e^{\pi i \cosh 2F(z)}$, $|z| < 1$, according to Exercise 5(d) above. Choose $z_0 \in \mathbf{C}$ so that $F'(z_0) \ne 0$. Construct the auxiliary function $G(z) = \frac{1}{16}F(\frac{16z}{F'(z_0)} + z_0)$, $|z| < 1$. Then $G(z)$ is entire and the image $G(\{|z| < 1\})$ contains an open disk of radius $\frac{1}{16}$; consequently, $F(z)$ maps $|z - z_0| < \frac{16}{|F''(z_0)|}$ onto a domain, containing an open disk of radius 1 , which contradicts to Exercise 5(e) above.

(7) See Ref. [83] for yet another proof of Montel's criterion.

5.8.4   Remarks on Schottky's theorems and Schwarz–Ahlfors' Lemma: Other proofs of Montel's criterion and Picard's theorems

In this section, we try to sketch descriptively theorems of Schottky type and Schwarz–Ahlfors' Lemma. Both are, directly or indirectly, influenced by the elliptic modular function, and provide other proofs for Montel's and Picard's theorems.

Three sections are divided.

Section (1) Theorems of Schottky type

Let $f(z)$ be analytic in $|z| \leq 1$, and omit the values 0 and 1.

In Exercise B(5) of Sec. 5.8.3, choose $g(z)$ and $F(z)$ so that $0 \leq \operatorname{Im} g(0)$, $\operatorname{Im} F(0) < 2\pi$.

In case $\frac{1}{2} \leq |f(0)| \leq \alpha$ (we may suppose that $2 \leq \alpha < \infty$, a constant): Then

$$|g(0)| \leq \frac{1}{2\pi} \log \alpha + 1 \underset{\text{(def.)}}{=} C_0(\alpha);$$

$$|\sqrt{g(0)} \pm \sqrt{g(0) - 1}| \leq C_0(\alpha)^{\frac{1}{2}} + [C_0(\alpha) + 1]^{\frac{1}{2}} \underset{\text{(def.)}}{=} C_1(\alpha)$$

$\Rightarrow$ (by considering the cases that $|\sqrt{g(0)}| \geq 1$ or < 1, respectively)

$$|F(0)| \leq \log C_1(\alpha) + 2\pi \underset{\text{(def.)}}{=} C_2(\alpha).$$

According to Exercise B(6) of Sec. 5.8.3, for fixed z, $|z| < 1$, the image of $|\zeta| < 1 - |z|$ under $F(z + \zeta)$ contains an open disk of radius $\frac{1}{16}(1 - |z|)|F'(z)|$; while Exercise B(5)(e) implies that this radius is strictly less than 1. Hence

$$|F'(z)| \leq \frac{16}{1 - |z|}, \quad |z| < 1$$

$$\Rightarrow |F(z)| \leq |F(0)| + |F(z) - F(0)| \leq C_2(\alpha) + \frac{16|z|}{1 - |z|} \leq C_2(\alpha) + \frac{16r}{1 - r}$$

$$\underset{\text{(def.)}}{=} C_3(\alpha, r) \quad \text{if } |z| \leq r < 1.$$

$$\Rightarrow |f(z)| \leq e^{\pi |\cosh 2F(z)|} \leq e^{\pi e^{2C_3(\alpha, r)}} \underset{\text{(def.)}}{=} C_4(\alpha, r) \quad \text{if } |z| \leq r < 1. \quad (*_1)$$

In case $0 < |f(0)| < \frac{1}{2}$: Then $\frac{1}{2} < |1 - f(0)| \leq 1 + |f(0)| < \frac{3}{2} < 2$ holds. Applying $(*_1)$ to $1 - f(z)$, we obtain that $|f(z)| \leq 1 + C_4(2, r)$ if $|z| \leq r < 1$.

Setting $C(\alpha, r) = \max(C_4(\alpha, r), 1 + C_4(2, r))$, we have a

Schottky's theorem (I). *Suppose $f(z)$ is analytic in $|z| < 1$ and omits the values 0 and 1. Suppose also that $|f(0)| \leq \alpha$. Then for each r, $0 \leq r < 1$, there is a constant $C(\alpha, r)$ depending only on α and r, so that*

$$|f(z)| \leq C(\alpha, r), \quad |z| < r. \tag{5.8.4.1}$$

The essence of Schottky's theorem(s) lies on the fact that $f(z)$ is bounded on $|z| \leq r$ by a constant depending only on $f(0)$ and r ($0 \leq r < 1$). Schottky published this result in *Sitzungsber. Preuss. Akad. Wiss. Berlin, Phys. Math.* (1904), where the bound $C(\alpha, r)$ was not explicitly given by a numerical quantity. It was A. Ostrowski who, in the same journal published in 1925, numericalized the upper bound via an elementary method as

$$\sup_{|z|=1-r} |f(z)| \leq e^{169\frac{d}{r} \log \frac{e}{r}}, \quad 0 < r < 1, \tag{5.8.4.2}$$

where $d = \max\{1, \log |f(0)|\}$. Another remarkable upper bound obtained by Ahlfors in 1938 via differential geometric method is stated in (5.8.4.21). Ref. [84] obtained the estimates

$$\log |f(z)| < e^{70}(4 + \log^+ |f(0)|)^2 \frac{1}{(1 - |z|)^4}, \quad |z| < 1. \tag{5.8.4.3}$$

All these estimates can be used to provide Montel's and Picard's theorems another proofs. Say, for instance, (5.8.4.1) does work for Montel's criterion: Let $\mathfrak{F}$ be as in (1) of (5.8.3.1) in case Ω is the disk $|z| < 1$. Set $A = \{f \in \mathfrak{F} \,|\, |f(0)| \leq 1\}$ and $B = \{f \in \mathfrak{F} \,|\, |f(0)| > 1\}$. By (5.8.4.1), A is obviously normal in $|z| < 1$ owing to local uniform boundedness. So is the family $\{\frac{1}{f} \,|\, f \in B\}$. Hurwitz's theorem (5.3.2.2) says that any locally uniformly convergent sequence from B either converges to ∞ or to an analytic function. Hence B is normal, too. Readers are urged to try more cases.

An analytic function f, defined in $|z| < 1$ and omitting the values 0 and 1, has its range as an open set in the punctured sphere $\mathbf{C}^*(0, 1, \infty) = \mathbf{C}(0, 1)$ which is the domain for the inverse function $\zeta = \psi^{-1}(w)$ of the elliptic modular function $w = \psi(\zeta)$, $|\zeta| < 1$. It seems naturally that $w = \psi(\zeta)$ can be used to obtain another type of Schottky's theorem. We sketch as follows.

Fix a point z where $|z| < 1$ and set $w = \psi(z)$. Choose $\delta > 0$ so that the open disk $|\eta - w| < \delta$ is contained in $\mathbf{C}(0, 1)$. Then (see step 1 in Exercise B(4) of Sec. 5.8.2)

$$\frac{|\psi'(z)|}{\delta} \geq \frac{1}{1 - |z|^2}, \quad |z| < 1. \tag{$*_2$}$$

Let γ_w be a piecewise differentiable curve in $\mathbf{C}(0,1)$ connecting $w_0 = \psi(0)$ to w, so that the distance from it to either 0 or 1 is larger than δ. The branch of $\zeta = \psi^{-1}(w)$, uniquely determined by $\psi^{-1}(w_0) = 0$, can be analytically continued along γ_w (refer to Sec. 5.2.1) and the resulted analytic function produces a piecewise differentiable curve γ_z in $|\zeta| < 1$, connecting 0 to z. Therefore, $(*_2)$ also holds for all $\zeta \in \gamma_z$; and then the *hyperbolic distance* from 0 to z (refer to Exercise B(5) of Sec. 3.4.5 or Appendix B) is

$$d(0, z) = \frac{1}{2} \log \frac{1 + |z|}{1 - |z|} \le \int_{\gamma_z} \frac{|d\zeta|}{1 - |\zeta|^2} \le \frac{1}{\delta} \int_{\gamma_z} |\psi'(\zeta)||d\zeta|$$

$$= \frac{1}{\delta} \int_{\gamma_w} |dw|. \tag{$*_3$}$$

Now, exhaust $\mathbf{C}(0,1)$ by an increasing sequence of domains $C^n(0,1)$, define by

$$C^n(0, 1) = \Big\{ w \in \mathbf{C}(0,1) | w \text{ can be connected to } w_0 \text{ by a piecewise}$$
$$\text{differentiable curve } \gamma_w \text{ in } \mathbf{C}(0,1) \text{ whose length is } < n$$
$$\text{and whose distance to either 0 or 1 is } > \frac{1}{n} \Big\} \tag{$*_4$}$$

for each integer $n \ge 1$. Observe that $\mathbf{C}(0,1) = \bigcup_{n=1}^{\infty} C^n(0,1)$. Then $(*_3)$ implies that

$$d(0, z) < n^2 \quad \text{if } |z| < 1 \quad \text{and} \quad w = \psi(z) \in C^n(0,1).$$

This relation can be restated as, for each $n \ge 1$,

$$C^n(0, 1) \subseteq \psi\left(\left\{ |\zeta| < \frac{e^{2n^2} - 1}{e^{2n^2} + 1} \right\} \right) = \psi(\{d(0, \zeta) < n^2\}). \tag{$*_5$}$$

Geometrically, this shows how concentric disks, either euclidean or noneuclidean, in $|\zeta| < 1$, can be used to cover the whole punctured sphere $\mathbf{C}(0,1)$ through the modular function $w = \psi(\zeta)$.

Denote temporarily by D_r the open disk $|\zeta| < r$ for $0 < r < 1$. For $\varepsilon > 0$, set

$$\Omega_\varepsilon = \left\{ z \in \mathbf{C}(0,1) | \varepsilon < |z| < \frac{1}{\varepsilon} \quad \text{and} \quad |z - 1| > \varepsilon \right\}. \tag{$*_6$}$$

See Fig. 5.34. For a given $\varepsilon > 0$, one can find $n \ge 2$ and $0 < r < 1$ such that $\Omega_\varepsilon \subseteq C^n(0,1) \subseteq \psi(D_r)$. Hence, for $0 < \varepsilon < \frac{1}{2}$ and $0 < r < s < 1$, it follows that

$$\Omega_\varepsilon \subseteq \psi(D_r) \subseteq \psi(D_s) \subseteq \Omega_\delta \tag{$*_7$}$$

if $\delta > 0$ is sufficiently small.

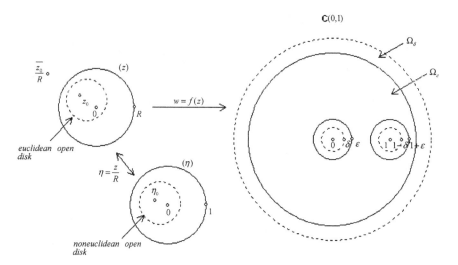

Fig. 5.34

Finally, let such $f(z)$ come into play.

Set $f(0) = a_0$ for a moment. Recall that $a_0 \neq 0, 1$. There is a $z_0, |z_0| < 1$ so that $\psi(z_0) = a_0$. Pick up any fixed single-valued branch $g(w)$ of $\zeta = \psi^{-1}(w)$ in an open neighborhood V_0 of a_0 in $\mathbf{C}(0,1)$ so that $g(a_0) = z_0$. Choose an open neighborhood U_0 of 0 in $|z| < 1$ so that $f(U_0) \subseteq V_0$. Just as in the proof of Montel's criterion, the function element $(g \circ f, U_0)$ can be analytically continued into a single-valued analytic function in $|z| < 1$, owing to the monodromy theorem (5.2.2.2). Still denote this extended function by

$$h(z) = g \circ f(z) : \{|z| < 1\} \to \{|\zeta| < 1\}. \tag{$*_8$}$$

Then Schwarz–Pick's Lemma (3.4.5.2) and Exercise B(5) (or Appendix B) there imply that

$$d(h(z_1), h(z_2)) \leq d(z_1, z_2), \quad |z_1| < 1, \quad |z_2| < 1. \tag{$*_9$}$$

This shows, in particular, that if $h(z_1) \in D_r$ and $d(z_1, z_2) < |s - r|$, then $h(z_2) \in D_s$ should hold.

Suppose $|z_1| < R$ and $f(z_1) \in \Omega_\varepsilon$. Then $(*_7)$ implies that $h(z_1) = g \circ f(z_1) \in D_r$. In case $|z_2| < R$ and $d(\frac{z_1}{R}, \frac{z_2}{R}) < |s - r|$, then $(*_9)$ says that $h(z_2) \in D_s$: consequently, still by $(*_7)$, $f(z_2) = \psi \circ h(z_2) \in D_\delta$ for some $\delta > 0$.

In conclusion, we have another

Schottky's theorem (II). *Suppose $f(z)$ is analytic in $|z| < R$ and omits the values 0 and 1. Then, for any $0 < \varepsilon < \frac{1}{2}$ and $M > 0$, there is a $\delta = \delta(\varepsilon, M) > 0$ satisfying the property: for each z, $|z| < R$, and $f(z) \in \Omega_\varepsilon$ (see $(*_6)$ and Fig. 5.34), then*

$$f(\zeta) \in \Omega_\delta$$

whenever $|\zeta| < R$ with the hyperbolic distance $d(\frac{z}{R}, \frac{\zeta}{R}) < M$. $\qquad$ (5.8.4.4)

Fix a point z_0, $|z_0| < R$, note that

$$\text{the hyperbolic open disk: } d(\eta, \eta_0) < M, \quad \eta = \frac{z}{R}(|z| < R), \quad \eta_0 = \frac{z_0}{R}$$

$$\Leftrightarrow \text{ the Euclidean open disk: } \frac{R|z - z_0|}{|R^2 - \overline{z_0}z|} < \alpha, \quad \alpha = \frac{e^{2M} - 1}{e^{2M} + 1}. \qquad (*_{10})$$

See Fig. 5.34. In particular, when $z_0 = 0$, the disk $d(\eta, 0) < M$ is the Euclidean disk $|z| < R\alpha$.

Schottky's theorem means geometrically that

$$f(\text{hyperbolic disk in } (*_{10})) = f(\text{Euclidean disk in}(*_{10})) \subseteq \Omega_\delta, \qquad (5.8.4.4)'$$

where $\delta = \delta(\varepsilon, M)$ depends only on ε and M, in which ε determines the position of $f(z_0) \in \Omega_\varepsilon$ and M decides the magnitude of the radius of the open disk. In case $z_0 = 0$, δ depends only on $f(0)$ and M, and this is the classical form of Schottky's theorem.

The last paragraph leading to (5.8.4.4) also indicates that $f(\zeta) \notin \Omega_\varepsilon$ if $f(z) \notin \Omega_\delta$ and $\zeta \in O$, where O denotes the open disk in $(*_{10})$. As a domain in $\mathbf{C}(0,1)$, $f(O)$ should lie in one of the three components of the set $\mathbf{C} - \Omega_\varepsilon$. These considerations lead to get another

Schottky's theorem (III). *Let $f(z)$, R, ε, M, and δ be as in (5.8.4.4). Then, for each z, $|z| < R$, and for all ζ, $|\zeta| < R$, satisfying the hyperbolic distance $d(\frac{\zeta}{R}, \frac{z}{R}) < M$:*

1. *if $|f(z)| \le \delta$, then $|f(\zeta)| \le \varepsilon$;*
2. *if $|f(z) - 1| \le \delta$, then $|f(\zeta) - 1| \le \varepsilon$, and*
3. *if $|f(z)| \ge \frac{1}{\delta}$, then $|f(\zeta)| \ge \frac{1}{\varepsilon}$.* $\qquad$ (5.8.4.5)

This type of Schottky's theorem can be adopted to prove Montel's criterion. See Exercise A(4).

Many mathematicians improved or extended Schottky's theorems and Montel's criterion.

A remarkable one is the following

Miranda's theorem (1935). *Suppose $f(z)$ is analytic in $|z| < 1$ and omits the value 0, and there is a nonnegative integer n so that $f^{(n)}(z) - 1$ does not assume the value 0 in $|z| < 1$. Then there is a positive constant M_n, independent of $f(z)$, such that*

$$\log M(r, f) \leq \frac{1}{1 - r} \left[64 \log^+ |f(0)| + \frac{M_n}{(1 - r)^4} \right], \quad 0 \leq r < 1,$$

where $M(r, f) = \max_{|z|=r} |f(z)|, 0 \leq r < 1$.   (5.8.4.6)

In case $n = 0$, another *type of Schottky's theorem* is stated as

$$\log M(r, f) \leq \frac{1}{1 - r} \left[12 \log^+ |f(0)| + C_0 \log \frac{2}{1 - r} \right], \quad 0 \leq r < 1,$$
(5.8.4.7)

where C_0 is another positive constant different from M_0. For proofs, see Ref. [12], p. 426 or Ref. [82], p. 119. Hiong King-Loi improved (5.8.4.6) in *Scientia Sinica 7 (1958)* (reappeared in *Chinese Math. 9 (1967), 390–399 (1968)*) as

$$\log M(r, f) \leq \frac{1}{1 - r} \left[H_n |\log |f(0)|| + K_n \log \frac{2}{1 - r} \right], \quad 0 \leq r < 1,$$
(5.8.4.8)

where H_n and K_n are two positive constants. W. K. Hayman improved (5.8.4.7) in 1947 as

$$\log M(r, f) \leq [\pi + \log^+ |f(0)|] \frac{1 + r}{1 - r}, \quad 0 \leq r < 1.$$ (5.8.4.9)

Miranda's theorem can be used to prove Picard's theorem. See Exercise A(5).

Hayman, in 1959, improved the so-called *Nevanlinna's second fundamental theorem* and conjectured a criterion for normality. Gu Yong-Xing confirmed this conjecture. We state as

Gu's normality criterion (1979). *Let Ω be a domain, k a positive integer and $a, b \in \mathbf{C}$ with $b \neq 0$. Then the family*

$$\mathfrak{F} = \{f(z) \text{ is meromorphic in } \Omega | f(z) \neq a \text{ and } f^{(k)}(z) \neq b \text{ hold in } \Omega\}$$

is normal in Ω.   (5.8.4.10)

In case $k = 0$, $a \neq b$, and $\mathfrak{F}$ composed of analytic functions, this result reduces to Montel's normality criterion. Miranda's theorem can be used to prove the case that $\mathfrak{F}$ is such a family of analytic functions. While, for the meromorphic case, see Ref. [82], pp. 135–149. On p. 161 in this book, one can also find the proof of the following

Bloch–Valiron's normality criterion. *Let $\mathfrak{F}$ be a family of meromorphic functions in a domain Ω. Suppose there are n distinct complex numbers $a_1, \ldots, a_n$ (one of them is allowed to be ∞) so that, for each $f \in \mathfrak{F}$, each of the equations*

$$f(z) = a_j, \quad 1 \leq j \leq n$$

has at least m_j (included) solutions in Ω, where $m_j \geq 1$ for $1 \leq j \leq n$. If

$$\sum_{j=1}^{n} \left(1 - \frac{1}{m_j}\right) > 2$$

holds, then $\mathfrak{F}$ is normal in Ω. $\hspace{2cm}$ (5.8.4.11)

Section (2) Schwarz–Ahlfors's Lemma

For general reference, we start with some terminologies:

A nonnegative real-valued C^2 function $\rho(z)$ on a domain Ω is said to define or to be a *metric* on Ω, and $ds_\rho - \rho(z)|dz|$ a *Riemannian metric* on Ω, acting as the *arc length element*. Some isolated zeros of $\rho(z)$ are permitted in Ω, and are called the *isolated singularity* of $\rho(z)$ or ds_ρ. In case $\rho(z) > 0$ throughout Ω, ds_ρ and the Euclidean metric $|dz|$ are conformal.

The *ρ-distance* between two distinct points z_1 and z_2 in Ω is defined by

$$d_\rho(z_1, z_2) = \inf_{\gamma} \int_{\gamma} \rho(z)|dz|, \hspace{2cm} (*_{11})$$

where inf is taken over all rectifiable curves γ in Ω, connecting z_1 to z_2. An extremal curve is called a *geodesic* connecting z_1 and z_2. The *curvature* of $\rho(z)$ at a point $z \in \Omega$ is defined by

$$K(z, \rho) = -\frac{\Delta \log \rho(z)}{\rho(z)^2}, \hspace{2cm} (5.8.4.12)$$

where $\Delta = \frac{\partial^2}{\partial x^2} + \frac{\partial^2}{\partial y^2} = 4\frac{\partial^2}{\partial z \partial \bar{z}} = \frac{\partial^2}{\partial r^2} + \frac{1}{r}\frac{\partial}{\partial r} + \frac{1}{r^2}\frac{\partial^2}{\partial \theta^2}$ ($z = x + iy = re^{i\theta}$) is the *Laplacian operator*. This is the Gaussian curvature in the Riemannian space (Ω, ds_ρ).

So far in this book, we have mentioned the following metrics, except the usual *Euclidean* (or *parabolic*) metric $|dz|$ (see Appendix B):

1. *The spherical or elliptic metric* (see Exercise B(2) of Sec. 1.6 or Appendix B)

$$ds = \frac{2|dz|}{1 + |z|^2}, \quad z \in \mathbf{C} \quad \text{with } K(z, \sigma) = 1 \quad \text{where } \sigma(z) = \frac{2}{1 + |z|^2}.$$

2. *The non-Euclidean or hyperbolic metric* (see Exercise B(5) of Sec. 3.4.5 or Appendix B)

$$ds = \frac{2|dz|}{1 - |z|^2}, \quad |z| < 1 \text{ with } K(z, \lambda) = -1 \text{ where } \lambda(z) = \frac{2}{1 - |z|^2}; \text{ or}$$

$$ds = \frac{|dz|}{\text{Im } z}, \quad \text{Im } z > 0 \text{ with } K(z, \lambda) = -1 \text{ where } \lambda(z) = \frac{1}{\text{Im } z}$$

They are also called *Poincaré's metrics* on $|z| < 1$ and $\text{Im } z > 0$, respectively.

3. *Poincaré's metric* on the punctured sphere $\mathbf{C}^*(0, 1, \infty)$ (see (5.8.3.10))

$$ds = \frac{|(\varphi^{-1})'(z)||dz|}{\text{Im } \varphi^{-1}(z)}, \quad \text{where } z = \varphi(w) : \{\text{Im } z > 0\} \to \mathbf{C}^*(0, 1, \infty); \text{ or}$$

$$ds = \frac{2|(\psi^{-1})'(z)| \, |dz|}{1 - |\psi^{-1}(z)|^2}, \quad \text{where } z = \psi(\zeta) : \{|\zeta| < 1\} \to \mathbf{C}^*(0, 1, \infty).$$

$$(*_{12})$$

In general, in case Ω is a simply connected domain whose boundary contains at least two distinct points, Riemann's mapping theorem (see (6.1.1)) says that there is a univalent (analytic) function $f(z)$ mapping Ω onto $|w| < 1$. In this case, call

$$(f^*\lambda)(z) \underset{(\text{def.})}{=} \lambda(f(z))|f'(z)| = \frac{2|f'(z)|}{1 - |f(z)|^2} \text{ or } (f^*\lambda)(z)|dz|, \quad z \in \Omega,$$

$$(*_{13})$$

Poincaré's metric on Ω *induced* by $f(z)$. It is invariant under conformal self mappings of Ω.

Quite generally, an induced metric, such as $(*_{13})$, can be loosely defined as follows.

Let $f(z) : \Omega_1$ (a domain) $\to \Omega_2$ (a domain) be a nonconstant analytic function (not necessarily univalent) and $\rho(w)$ be a metric on Ω_2. Then

$$(f^*\rho)(z) \underset{(\text{def.})}{=} \rho(f(z))|f'(z)|, \quad z \in \Omega_1, \tag{5.8.4.13}$$

is a metric on Ω_1, *induced* from ρ by $f(z)$. A vivid example is provided by (5.8.1.7). Moreover, $K(z, f^*\rho) = K(f(z), \rho)$ at points z where both $f'(z) \neq 0$ and $\rho(f(z)) \neq 0$ hold; in particular, the equality is valid wherever $\rho(w) > 0$ in case $f(z)$ is univalent. Hence the curvature is a *conformal invariant*.

Let us go back to the Schwarz's lemma (3.4.5.1).

It was G. Pick who, in 1915, firstly realized the geometric aspect of this lemma that a nonconstant analytic function $f(z)$ from $|z| < 1$ into itself decreases the non-Euclidean length of two points (see (3.4.5.2)), namely,

$$\frac{|f(z_1) - f(z_2)|}{|1 - \overline{f(z_2)}f(z_1)|} \leq \frac{|z_1 - z_2|}{|1 - \overline{z_2}z_1|}, \quad |z_1| < 1, \quad |z_2| < 1$$

$\Rightarrow$ (in terms of $(*_{13})$) $(f^*\lambda)(z) \leq \lambda(z) = \dfrac{2}{1 - |z|^2}, \quad |z| < 1$ or

$$ds_{f^*\lambda} \leq ds_\lambda \tag{$*_{14}$}$$

with equality only if $f(z)$ is a bilinear transformation of $|z| < 1$ onto itself. Via the concept of curvature introduced in (5.8.4.12), we have

$$\lambda(z)^2 = -K(z, \lambda)^{-1}\Delta \log \lambda(z) = \Delta \log \lambda(z);$$

$$(f^*\lambda)(z)^2 = -K(z, f^*\lambda)^{-1}\Delta \log (f^*\lambda)(z) = \Delta \log (f^*\lambda)(z),$$

and hence, $(*_{14})$ is equivalent to the validity of the inequality

$$\Delta[\log \lambda(z) - \log (f^*\lambda)(z)] \geq 0, \quad |z| < 1. \tag{$*_{15}$}$$

Owing to the fact that $\lim_{|z| \to 1} \log \lambda(z) = \infty$, *if it is possible to keep* $\log (f^*\lambda)(z)$ *remain bounded in* $|z| < 1$, then $\lim_{|z| \to 1}[\log \lambda(z) - \log (f^*\lambda)(z)] = \infty$ will imply that the C^2 function $\log \lambda(z) - \log (f^*\lambda)(z)$ would have a minimum at some point z_0 in $|z| < 1$. This, in turn, implies that $\Delta[\log \lambda(z_0) - \log(f^*\lambda)(z_0)] \geq 0$ holds (why? the trace of the positive semidefinite Hessian matrix at z_0). Then $(*_{15})$ holds throughout $|z| < 1$.

This observation seems reasonably lead to the following

Schwarz–Ahlfors's Lemma (1938). *Suppose* $f(z):\{|z| < 1\} \to \Omega$ *(a domain) is analytic and there is a metric* $\rho(w)$ *on* Ω *such that its curvature* $K(w, \rho) \leq -1$ *holds throughout* Ω. *Then the induced metric*

$$(f^*\rho)(z) = \rho(f(z))|f'(z)| \leq \lambda(z) = \frac{2}{1 - |z|^2}, \quad |z| < 1$$

holds everywhere in $|z| < 1$; *in short,*

$$ds_{f^*\rho} \leq ds_\lambda.$$

Hence, an analytic function on $|z| < 1$ *will decrease Poincaré's metric.*

(5.8.4.14)

The classical Schwarz's Lemma (3.4.5.1) and Schwarz–Pick's Lemma (3.4.5.2) are indeed special cases of this one. See Exercise A(9). *Some historical comments*: This theorem appeared in Ref. [5] which initiated the study of geometric behaviors of analytic functions by means of differential geometry. Based on this point of view, R. M. Robinson [69] proved Picard's theorems without recourse to the traditional method using the elliptic modular function. Some mathematicians, like H. Grauert and H. Reckziegel [37], Z. Kobayashi [49], D. Minda and G. Schober [59], made contributions in this direction. And theorems of this type appeared at first time in the books by Ahlfors [2] and S. G. Krantz [51]. It is highly recommended to read the article [59] and the books [2, 51] for detailed account about materials within the present Section (2).

Sketch of proof (Ref. [2]). Fix $0 < r < 1$. Construct Poincaré's metric

$$\lambda_r(z) = \frac{2r}{r^2 - |z|^2}, \quad |z| < r$$

on $|z| < r$ induced by $h(z) = \frac{z}{r}$ from $|z| < r$ onto $|w| < 1$ (see $(*_{13})$). The process leading to $(*_{15})$ reads now as (with $f^*\rho$ replacing $f^*\lambda$)

$$\Delta[\log(f^*\rho)(z) - \log\lambda_r(z)] \geq (f^*\rho)(z)^2 - \lambda_r(z)^2, \qquad (*_{16})$$

owing to $K(z, f^*\rho) = K(f(z), \rho) \leq -1$. The set $E = \{z \mid |z| < r$ and $\log(f^*\rho)(z) > \log\lambda_r(z)\}$ is open in $|z| < r$ and does not contain the isolated singularities of $f^*\rho$. $F(z) = \log(f^*\rho)(z) - \log\lambda_r(z)$ is positive in E and $(*_{16})$ shows that it is subharmonic there (see (6.4.1)). Hence $F(z)$ does not assume local maximum except it is a constant. In case $E \neq \emptyset$, then the boundary $\partial E \neq \emptyset$. ∂E has empty intersection with the circle $|z| = r$, because $F(z) = 0$ along ∂E by the continuity of $F(z)$, while $\lim_{|z|\to r} \log\lambda_r(z) = +\infty$ and $\lim_{|z|\to r} \log(f^*\rho)(z) \geq -\infty$. Consequently, $F(z)$ attains a local maximum on the compact set $\bar{E}$ which is contained in $|z| < r$; and hence $F(z) \equiv 0$ on $|z| < r$, a contradiction. Thus $E = \emptyset$ holds and

$$\log(f^*\rho)(z) \leq \log\lambda_r(z), \quad |z| < r$$

$$\Rightarrow (f^*\rho)(z) \leq \lambda_r(z), \quad |z| < r$$

$$\Rightarrow (\text{on letting } r \to 1^-)(f^*\rho)(z) \leq \lambda(z), \quad |z| < 1.$$

This finishes the proof.

While, Minda and Schober [59] improved the above argument as follows. Consider

$$v(z) = \frac{(f^*\rho)(z)}{\lambda_r(z)}, \quad |z| < r.$$

$v(z)$ is continuous on $|z| \leq r$ and ≥ 0; $v(z) \to 0$ as $|z| \to r$ because $(f^*\rho)(z)$ is bounded in $|z| \leq r$ and $\lambda_r(z) \to \infty$ as $|z| \to r$. Hence $v(z)$ assumes its maximum at some z_0, $|z_0| < 1$. It is then sufficient to show that $v(z_0) \leq 1$. In case $f^*(\rho)(z_0) = 0$, it follows that $f^*(\rho)(z) \equiv 0$ in $|z| \leq r$. If $f^*(\rho)(z_0) > 0$, then $\log v(z)$ assumes its maximum at z_0. Hence

$$0 \geq \Delta \log v(z_0) = \Delta \log(f^*\rho)(z_0) - \Delta \log \lambda_r(z_0)$$
$$= -K(z_0, f^*\rho)(f^*\rho)(z_0)^2 + K(z_0, \lambda)\lambda_r(z_0)^2$$
$$\geq (f^*\rho)(z_0)^2 - \lambda_r(z_0)^2$$

$$\Rightarrow v(z_0) \leq 1.$$

What remains is the same as near the end of the last paragraph. $\qquad\square$

A slight generalization of (5.8.4.14) is at hand.

Observe that the metric $c\lambda(z)$, where $c > 0$ is a constant, has the curvature $-\frac{1}{c^2}$. As a consequence, for a fixed constant $A > 0$,

$$\lambda_R^A(z) = \frac{2R}{\sqrt{A}(R^2 - |z|^2)}, \quad |z| < R \qquad (5.8.4.15)$$

is a metric on $|z| < R$ with the *constant* curvature $K(z, \lambda_R^A) = -A$. On applying (5.8.4.14) to Poincaré's metric $\lambda(\zeta) = \sqrt{A}R\lambda_R^A(R\zeta)$, $|\zeta| < 1$, and the metric $\sqrt{B}\rho(w)$ on Ω, where $B > 0$ is a constant, we have

$$\sqrt{B}\rho(f(R\zeta))R|f'(R\zeta)| \leq \sqrt{A}R\lambda_R^A(R\zeta), \quad |\zeta| < 1.$$

This leads to

A generalization of Schwarz–Ahlfors's Lemma. *Suppose $f(z) : \{|z| < R\} \to \Omega$ (a domain) is analytic and Ω carries a metric $\rho(w)$ with curvature $K(w, \rho) < -B$ throughout Ω, where $B > 0$ is a constant. Then*

$$(f^*\rho)(z) = \rho(f(z))|f'(z)| \leq \frac{\sqrt{A}}{\sqrt{B}}\lambda_R^A(z), \quad |z| < R,$$

where $\lambda_R^A(z)$ is given by (5.8.4.15). $\qquad (5.8.4.16)$

Remark (Ultrahyperbolic metrics and Schottky's theorem). A careful examination of Ahlfors's proof for (5.8.4.14) shows that the central idea behind it lies on the fact that the continuous function $\log \lambda_r(z) - \log (f^*\rho)(z)$ attains its (local) minimum in $|z| < 1$. In Ref. [2], Ahlfors formally introduced the notion of an *ultrahyperbolic metric* in a domain Ω: It is a metric $\rho|dz|$ in Ω, satisfying the following two properties:

1. ρ is upper semicontinuous (namely, $\overline{\lim}_{z \to z_0}\rho(z) \leq \rho(z_0)$ for each $z_0 \in \Omega$).
2. At every point $z_0 \in \Omega$ with $\rho(z_0) > 0$, there is a C^2 metric ρ_0 in a neighborhood O_{z_0} of z_0 in Ω such that $K(z, \rho_0) \leq -1$ and $\rho \geq \rho_0$ in O_{z_0} and $\rho(z_0) = \rho_0(z_0)$. $\qquad (5.8.4.17)$

In this case, $\log \lambda(z) - \log \rho(z)$ is lower semicontinuous in Ω and the local minimum of $\log \lambda(z) - \log \rho_0(z)$ is still assured. Localize the proof for (5.8.4.14) at each point z_0 where $\rho(z_0) > 0$ and we obtain

A generalization of Schwarz–Ahlfors's Lemma. *Let $f(z)$ be analytic from $|z| < 1$ into a domain Ω on which there is given an ultrahyperbolic metric $\rho(z)$. Then*

$$(f^*\rho)(z) = \rho(f(z))|f'(z)| \le \lambda(z) = \frac{2}{1 - |z|^2}, \quad |z| < 1. \qquad (5.8.4.18)$$

This result remains valid even if Ω is a Riemann surface. The notion of an ultrahyperbolic metric has recently found many new applications in the theory of several complex variables.

There is no ultrahyperbolic metric either in the whole plane $\mathbf{C}$ or in a punctured plane $\mathbf{C} - \{z_0\}$, where $z_0 \in \mathbf{C}$ is a fixed point. $(*_{12})$ shows that we do have one in the punctured plane $\mathbf{C}(0,1)$ via the elliptic modular function. Indeed, we do have

The existence and uniqueness of an ultrahyperbolic metric. Let Ω be a domain in $\mathbf{C}$ whose complement (in $\mathbf{C}$) contains at least two distinct points.

(1) There is a metric $\rho(z)$ in Ω whose curvature $K(z, \rho) \le -A$, where $A > 0$ is a constant.

(2) There exists a unique *maximal* ultrahyperbolic metric λ_Ω in Ω, such that

$$K(z, \lambda_\Omega) = -1, \quad z \in \Omega.$$

Here "maximal" means that, for any ultrahyperbolic metric $\rho(z)$ in Ω, $\rho(z) \le \lambda_\Omega(z)$ always holds. Such a $\lambda_\Omega(z)$ is called *Poincaré's metric* in Ω. (5.8.4.19)

(1) is left as Exercise A(11). The existence proof in (2) is concerned with the *uniformization of the Riemann surface* (see Chap. 7; in particular, the last paragraph on the last page) and one may refer to Ref. [2], p. 151 or Ref. [51], Chap. 3.

Analytic expression such as the one appeared in $(*_{12})$ is not of great value. What we need at the present are the following

Bounds for Poincaré's metric (Ahlfors, 1938).

(1) In case $\Omega_1 \subseteq \Omega_2$, then $\lambda_{\Omega_2} \le \lambda_{\Omega_1}$ in Ω_1.
(2) For each $z \in \Omega$,

$$\lambda_\Omega(z) \le \frac{2}{\operatorname{dist}(z, \partial\Omega)}$$

always holds. In case z is the center of an open disk Ω, the equality holds.

(3) Let $\lambda_{0,1}(z)$ denote Poincaré's metric of the three-points punctured sphere $\mathbf{C}^*(0,1,\infty) = \mathbf{C}(0,1)$. Then

$$\lambda_{0,1}(z) \le \left(|z| \log \frac{1}{|z|}\right)^{-1}, \quad |z| \le 1;$$

$$\lambda_{0,1}(z) \ge \left|\frac{\zeta'(z)}{\zeta(z)}\right| [4 - \log |\zeta(z)|]^{-1}, \quad |z| \le 1, \quad |z| \le |z-1|,$$

where $\zeta(z) = \frac{\sqrt{1-z}-1}{\sqrt{1-z}+1} (\mathrm{Re}\sqrt{1-z} > 0)$ maps $\mathbf{C} - [1,\infty)$ conformally onto $|w| < 1$, $\zeta(0) = 0$, and preserves the symmetry with respect to the real axis.   (5.8.4.20)

(1) and (2) are obvious. Note that $0 < |z| < 1$ is contained in $\mathbf{C}(0,1)$. $w = \log z$ maps the universal covering surface (the infinitely many-sheeted "unit open disk" $|z| < 1$ on the Riemann surface $R_{\log z}$ (see Fig. 2.72 and refer to basic example (4) in Sec. 7.5.5) of the punctured disk $0 < |z| < 1$ univalently onto the left half-plane $\mathrm{Re}\, w < 0$. $(*_{12})$ and $(*_{13})$ together show that the Poincaré's metric of $0 < |z| < 1$ is given by

$$\frac{|dw|}{|\mathrm{Re}\, w|} = \frac{|dz|}{|z| \log \frac{1}{|z|}}, \quad 0 < |z| < 1.$$

The upper bound is thus obtained by using (1). The estimate for lower bound is complicated and better refer to Ref. [2], pp. 17–18.

Based on (5.8.4.14) and (3) in (5.8.4.20), comes the remarkable

Picard–Schottky's theorem (IV) (Ahlfors, 1938). *Suppose $f(z)$ is analytic in $|z| < 1$ and omits the values 0 and 1. Then*

$$\log |f(z)| \le [7 + \log^+ |f(0)|] \frac{1+|z|}{1-|z|}, \quad |z| < 1. \tag{5.8.4.21}$$

As claimed by Ahlfors in his paper, this estimate was the best numerical upper bound obtained so far in 1938. Jenkins improved this explicit bound in 1955 by using the elliptic modular function. However, Picard's theorems now turns out to be easily proved by these estimates. For detail, refer to Ref. [2], pp. 19–21.

Section (3) Some applications

(5.8.4.16) can be adopted to provide easy proofs for Liouville's, Picard's, and Montel's theorems.

Let $f(z)$ be an entire function, bounded by a constant $M > 0$ throughout **C**. Fix any $R > 0$. Then $f(z)$ maps $|z| < R$ into $|w| < M$. Compare Poincaré's metric $\lambda_R(z)$ on $|z| < R$ and the metric $f^*\lambda_M(z)$ induced on $|z| < R$ by Poincaré's metric $\lambda_M(w)$ on $|w| < M$ (see (5.8.4.15) with $A = 1$). Then

$$(f^*\lambda_M)(z) = \frac{2M|f'(z)|}{M^2 - |f(z)|^2} \leq \frac{2R}{R^2 - |z|^2}, \quad |z| < R \qquad (*_{17})$$

$$\Rightarrow \text{ (letting } R \to \infty) \ f'(z) = 0, \quad z \in \mathbf{C}.$$

Hence $f(z)$ is a constant. This is *Liouville's theorem*.

The essence of this differential geometric proof lies on the fact that it is possible to construct a metric $\rho(w)$, namely Poincaré's in this case, in the range Ω with the curvature $\leq -B$, where $B > 0$ is a constant. Suppose this is the case. Then $(*_{17})$ is replaced by

$$(f^*\rho)(z) = \rho[f(z)]|f'(z)| \leq \frac{1}{\sqrt{B}} \frac{2R}{R^2 - |z|^2}, \quad |z| < R$$

$$\Rightarrow \text{ (letting } R \to \infty) \ \ \rho(f(z))|f'(z)| = 0, \quad z \in \mathbf{C}.$$

Since the singularities of $\rho(w)$ are isolated, $f'(z) \equiv 0$ and hence $f(z)$ is a constant. We obtain

An extension of Liouville's theorem. *Suppose an entire function $f(z)$ has its range in a domain Ω on which it carries a metric $\rho(w)$ whose curvature $K(w, \rho) \leq -B$ throughout Ω, where $B > 0$ is a constant. Then $f(z)$ is a constant.* (5.8.4.22)

Combining (1) in (5.8.4.19) (see Exercise A(11)) and (5.8.4.22), it follows easily *Picard's little theorem* (5.8.3.11).

Finally, come to *Montel's normality criterion* (5.8.3.1).

We still use $\sigma(z) = \frac{2}{1+|z|^2}$ to denote the spherical metric. Then (3) in (5.8.1.8) now reads as: *A family $\mathfrak{F}$ of analytic (or meromorphic) functions in a domain is normal if and only if the family $\{f^*\sigma | f \in \mathfrak{F}\}$ of the induced metrics is locally uniformly bounded in Ω.*

Suppose $\mathfrak{F}$ is a family of analytic functions in a domain Ω and each member of which omits the values 0 and 1, namely, its range lies in $\mathbf{C}(0, 1)$. Let $\rho(w)$ be the metric defined on $\mathbf{C}(0, 1)$ as in Exercise A(11). Then the curvature $k(w, \rho) \leq -A$, $w \in \mathbf{C}(0, 1)$, where $A > 0$ is a constant.

Fix an open disk $|z - z_0| < r$ in Ω. Choose Poincaré's metric $\lambda_R(z) = \frac{2r}{r^2 - |z - z_0|^2}$ on $|z - z_0| < r$. By (5.8.4.16), we have

$$(f^*\rho)(z) = \rho(f(z))|f'(z)| \leq \frac{1}{\sqrt{A}}\lambda_R(z) = \frac{2r}{\sqrt{A}(r^2 - |z - z_0|^2)},$$

$$|z - z_0| < r, \ f \in \mathfrak{F}. \qquad (*_{18})$$

Compare $\rho(w)$ and the spherical metric $\sigma(w)$ in $\mathbf{C}(0, 1)$ as follows:

$$\frac{\sigma(w)}{\rho(w)} = \frac{2|w|^{5/6}|w - 1|^{5/6}}{(1 + |w|^2)(1 + |w|^{1/3})^{1/2}(1 + |w - 1|^{1/3})^{1/2}}, \quad w \in \mathbf{C}(0, 1)$$
$$\rightarrow 0,$$

as $w \to 0$ or 1 or ∞. Hence, there are constants $M_1 > 0$ and $\delta > 0$ so that

$$\frac{\sigma(w)}{\rho(w)} \leq M_1 \ \text{if} \ |w| < \delta \ \text{or} \ |w - 1| < \delta \ \text{or} \ |w| > \frac{1}{\delta}.$$

Since $\frac{\sigma(w)}{\rho(w)}$ is continuous on the compact set $K = \mathbf{C}^* - \{|w| < \delta\} - \{|w-1| < \delta\} - \{|w| > \frac{1}{\delta}\}$, there is another constant $M_2 > 0$ so that

$$\frac{\sigma(w)}{\rho(w)} \leq M_2 \quad \text{if} \ w \in K.$$

Consequently,

$$\frac{\sigma(w)}{\rho(w)} \leq M = \max\{M_1, M_2\}, \quad w \in \mathbf{C}(0, 1). \qquad (*_{19})$$

Combining $(*_{18})$ and $(*_{19})$, we have

$$(f^*\sigma)(z) = \sigma(f(z))|f'(z)| \leq M\rho(f(z))|f'(z)|$$

$$\leq \frac{2Mr}{\sqrt{A}(r^2 - |z - z_0|^2)}, \quad |z - z_0| < r \quad \text{and} \quad f \in \mathfrak{F}.$$

This shows that $\{f^*\sigma | f \in \mathfrak{F}\}$ is uniformly bounded on $|z - z_0| < r$ and hence is locally so in Ω. The normality of $\mathfrak{F}$ is thus proved. $\qquad \square$

Exercises A

(1) (Landau) Suppose $f(z)$ is analytic in $|z| < R$, $f'(0) \neq 0$, and $f(z) \neq 0$, 1 throughout. Then

$$R \leq R(a_0, a_1) = \frac{1 - |g(a_0)|}{|g'(a_0)||a_1|}, \quad a_0 = f(0) \quad \text{and} \quad a_1 = f'(0),$$

where $g(w)$ is any fixed single-valued branch at a_0 of the inverse function $\zeta = \psi^{-1}(w)$ of the elliptic modular function $w = \psi(\zeta) : \{|z| < 1\} \to \mathbf{C}(0, 1)$ and $R(a_0, a_1)$ is a constant, dependent only on a_0 and

a_1, independent of the choice of $g(w)$. The equality $R = R(a_0, a_1)$ holds if and only if $f(z) = \psi(h(z))$, $|z| < R$, where $\zeta = h(z) : \{|z| < R\} \to \{|\zeta| < 1\}$ is a linear transformation. Moreover, for any fixed $a_0 \neq 0, 1$, and $a_1 \neq 0$, there exists *such* a function $f(z)$, satisfying $f(0) = a_0$, $f'(0) = a_1$, so that $R = R(a_0, a_1)$ holds.

Try the following steps.

(a) Apply Schwarz–Pick's Lemma to the function $h(z)$ in $(*_8)$ and compute $h(0)$, $h'(0)$.

(b) In case $\tilde{g}(w)$ is another such a branch, then $\tilde{g} = T \circ g$ where $T : \{|\zeta| < 1\} \to \{|\zeta| < 1\}$ is a modular transformation (see (5.8.3.9)). Observe that $\frac{|T'(z_0)|}{1-|T(z_0)|^2} = \frac{1}{1-|z_0|^2}$ (why?).

(c) Set $R = \frac{1-|g(a_0)|}{|g'(a_0)||a_1|}$ for given a_0 and $a_1 \neq 0$. Then $h(z) = \frac{z+Rg(a_0)}{R+g(a_0)z}$ maps $|z| < R$ onto $|\zeta| < 1$. Let $f_0(z) = \psi(h(z))$, $|z| < R$. Choose θ so that $e^{i\theta}f'(0) = a_1$. Does $f(z) = f_0(e^{i\theta}z)$ work?

(2) Use Exercise (1) to prove Picard's little theorem.

(3) Suppose $f(z)$ is analytic in $|z| < R$ and $f(z) \neq 0, 1$ there. For any fixed $0 < K < \infty$ and $0 < M < \infty$, then there is an $N(0 \leq N < \infty)$ so that $|f(\zeta)| \leq N$ whenever $|z| < R$ and $|f(z)| \leq K$, and for all ζ, $|\zeta| < R$, with the hyperbolic distance $d(\frac{\zeta}{R}, \frac{z}{R}) < M$.

(4) Let $f_n(z)$, $n \geq 1$, be a sequence of analytic functions in $|z| < R$ so that $f_n(z) \neq 0, 1$ there for each $n \geq 1$.

(a) If there is a point z_0, $|z_0| < R$, so that $\lim_{n\to\infty} f_n(z_0) = 0$ or 1 or ∞, then $f_n(z)$ converges locally uniformly to 0 or 1 or ∞ in $|z| < R$, respectively. Try to use (5.8.4.5).

(b) In case $\lim_{n\to\infty} f_n(z) = f(z)$ exists for each z, $|z| < R$, then $f_n(z)$ converges to $f(z)$ locally uniformly in $|z| < R$. Try to use (5.8.4.4), (5.8.3.1), and Exercise A(3) of Sec. 5.3.3.

(c) Use (a) and (b) to give Montel's normality criterion another proof.

(5) Let $f(z)$ be an entire function that omits the value 0, and that $f^{(k)}(z)$ does not assume a nonzero finite complex number for some nonnegative integer k. Then $f(z)$ is a nonzero constant function. Use this fact to prove Picard's theorems. Try to apply Miranda's theorem to $f(z) = \frac{f(Rz)}{R^k}$ for any fixed $R \geq 1$, under the assumption that $f^{(k)}(z) \neq 1$, $z \in \mathbf{C}$.

(6) (Pólya–Saxer) Let $f(z)$ be an entire function such that $f(z)$, $f'(z)$, and $f''(z)$ never vanish in $\mathbf{C}$. Then $f(z) = e^{az+b}$ for some constants a and b. Try to use Exercise (5).

(7) Suppose $f(z)$ is an entire function, not of the form $z + b$ (b, a constant). Then $f \circ f(z)$ must have a fixed point. *Hint*: Suppose on the contrary and consider $\frac{f(f(z)) - z}{f(z) - z}$ and use Picard's little theorem.

(8) Show that the curvature defined in (5.8.4.13) is indeed a conformal invariant.

(9) Prove that Schwarz's Lemma (3.4.5.1) and Schwarz–Pick's Lemma (3.4.5.2) are special cases of Schwarz–Ahlfors's Lemma (5.8.4.14).

(10) Show that there is no ultrahyperbolic metric in $\mathbf{C}$ or in $\mathbf{C} - \{z_0\}$, where $z_0 \in \mathbf{C}$ is a fixed point.

(11) Let Ω be as in (5.8.4.19). Let $a, b \in \mathbf{C} - \Omega$ where $a \neq b$. After invoking a transformation $\frac{z-a}{b-a}$, we may suppose that $a = 0$ and $b = 1$, and $\Omega = \mathbf{C}(0, 1)$. Set

$$\rho(z) = \frac{(1 + |z|^{1/3})^{1/2}}{|z|^{5/6}} \cdot \frac{(1 + |z - 1|^{1/3})^{1/2}}{|z - 1|^{5/6}}, \quad z \in \mathbf{C}(0, 1).$$

Compute $K(z, \rho)$, $z \in \mathbf{C}(0, 1)$. Observing that $K(z, \rho) < 0$ for $z \in \mathbf{C}(0, 1)$ and the limits of $K(z, \rho)$ as $z \to 0, 1, \infty$, respectively, show that $K(z, \rho) \leq -A$ for some constant $A > 0$.

Exercises B

Read Ref. [2], pp. 12–21 for a complete understanding of the Remark.

*5.8.5 *An application: Some results in complex dynamical system*

As an important application of both Montel's normality criterion (5.8.3.1) and Schwarz–Pick's Lemma (3.4.5.2), or its generalized form (5.8.5.6) below, this short section is going to prove a remarkable result in modern complex dynamical system due to Sullivan in early 1980s (see (5.8.5.11)). The method of argument is quite similar to that given in Sec. 5.4.3; in particular, in Sections (3) and (4) there.

Let Ω be a domain in $\mathbf{C}^*$ and $f : \Omega \to \Omega$ be an analytic (self-) mapping. The iterate sequence f^n, $n \geq 1$, is called a *complex dynamical system* on Ω. Point sequences $f^n(z)$, $n \geq 1$, determined by a point z, is called the *orbit* of z. The questions are:

1. What is the behavior of $f^n(z)$ if n is large enough?
2. How does the behavior depend on the initial point z and does the limit exist as $n \to \infty$?

Fatou and Julia studied iterate sequences systematically before World War I. The usage of computer work draws out wonderful and complicated figures of iterate functions, which provide fruitful examples for fractal geometry. This revives the study of complex dynamical system after 1980.

Recall that (see Exercise A(5) of Sec. 5.3.4):

Fatou set $F(f) = \{z \in \Omega | f^n, n \geq 1,$ is normal in a neighborhood of $z\}$;
Julia set $\Im(f) = \Omega - F(f) = \{z \in \Omega | f^n(z), n \geq 1,$ is not normal$\}$.
$$(5.8.5.1)$$

Note that $F(f)$ is open, while $\Im(f)$ is closed.

Both Fatou and Julia sets are invariant under conjugation (see (5.3.4.2)). More precisely, if $\Phi \colon \Omega \to \Omega$ is a conjugation between f and g, then

$$F(f) = \Phi(F(g)) \quad \text{and} \quad \Im(f) = \Phi(\Im(g)). \qquad (5.8.5.2)$$

Take, for instance, Exercise A(5) of Sec. 5.3.4 as an example: Let $f(z) = z^2 - 2 \colon \mathbf{C}^* \to \mathbf{C}^*$. Then $\Phi(\zeta) = \zeta + \frac{1}{\zeta}, |\zeta| > 1$ conjugates g to f, namely $f = \Phi \circ g \circ \Phi^{-1}$. It is easy to find that $F(g) = \{|z| < 1\} \cup \{|z| > 1\}$ and $\Im(g) = \{|z| = 1\}$; as a consequence, $F(f) = \mathbf{C}^* - [-2, 2]$ and $\Im(f) = [-2, -2]$. Readers are asked to do Exercises A(6) and A(7) there for more practice.

In case the initial point z lies in the Fatou set, then its orbit is stable in the following sense: It might be convergent, or periodic after n is large enough, or has a convergent subsequence in its worst situation. If $\Im(f) = \phi$, then $F(f) = \Omega$ and the iterate sequence turns out to be simple.

The Julia set, for most functions, are complicated. It might be a totally disconnected perfect set (like Cantor ternary set), or a curve of Hausdorff dimension greater than one, or even the whole plane $\mathbf{C}^*$.

When does the Julia set become empty?

The simplest case is that Ω is a bounded domain. Then $f^n(z)$, $n \geq 1$, is uniformly bounded in Ω and hence, by (5.8.1.3), it is normal on Ω. In this case, $F(f) = \Omega$ and $\Im(f) = \phi$.

If $\Omega = \mathbf{C}^*$ minus three distinct points or $\Omega = \mathbf{C}$ minus two distinct points, then by Montel's normality criterion (5.8.3.1), $f^n(z)$, $n \geq 1$, is normal in Ω. Consequently $F(f) = \Omega$ and $\Im(f) = \phi$, too.

It follows that only if Ω is one of the following three domains:

1. $\mathbf{C}^*$;
2. $\mathbf{C}$;
3. $\mathbf{C} - \{a\}, a \in \mathbf{C}$, $\qquad\qquad\qquad\qquad\qquad\qquad (5.8.5.3)$

the Julia set will be nonempty.

Two subsections are divided.

Section (1) Some terminologies and generalized Schwarz–Pick's Lemma

To extend (5.3.4.1). For a point $z_0 \in \Omega$, suppose n is the smallest positive integer so that $f^n(z_0) = z_0$, i.e., z_0 is a fixed point of f^n. In this case z_0 is called a *periodic point* and n its *period*; and

$$\lambda = (f^n)'(z_0) \qquad (5.8.5.4)$$

the *multiplier* of f at z_0. z_0 is classified as

1. *attractive* if $|\lambda| < 1$; *superattractive* if $\lambda = 0$;
2. *repellent* or *repulsive* if $|\lambda| > 1$;
3. *neutral* if $|\lambda| = 1$. $\qquad (5.8.5.5)$

An attractive or superattractive point belongs to the Fatou set (see (5.3.4.7)), while a repulsive point is in the Julia set. A neutral point could be in the Fatou set or in the Julia set.

Suppose $g\colon \Omega$ (a domain in $\mathbf{C}^*$) $\to \{|w| < 1\}$ is a univalent, onto (conformal) mapping. Recall that (see $(*_{13})$ and (5.8.4.13) in Sec. 5.8.4)

$$ds = \lambda(z)|dz|, \quad \text{where } \lambda(z) = \frac{2|g'(z)|}{1 - |g(z)|^2}$$

the *Poincaré's metric* on Ω induced by g from the hyperbolic metric $\lambda(w) = \frac{2}{1-|w|^2}$ on $|w| < 1$.

Then Schwarz–Pick's Lemma (3.4.5.2) can be easily extended to

Generalized Schwarz–Pick's Lemma. *Suppose domains Ω_1 and Ω_2 have Poincaré's metrics $ds_1 = \lambda_1(z)|dz|$ and $ds_2 = \lambda_2(w)|dw|$, respectively. In case $f : \Omega_1 \to \Omega_2$ is analytic, then*

$$\lambda_2(f(z))|f'(z)| \le \lambda_1(z), \quad z \in \Omega_1$$

with equality if and only if f is univalently from Ω_1 onto Ω_2. $\qquad (5.8.5.6)$

Compare to Schwarz–Ahlfor's Lemma (5.8.4.14).

Proof. The composite map $F = g_2 \circ f \circ g_1^{-1}\colon\{|\zeta| < 1\} \to \{|\eta| < 1\}$ is analytic, and, by the classic Schwarz–Pick's Lemma,

$$\frac{|F'(\zeta)|}{1 - |F(\zeta)|^2} \le \frac{1}{1 - |\zeta|^2}$$

$$\Rightarrow \text{ (by setting } z = g_1^{-1}(\zeta)) \quad \frac{|g_2'(f(z))|\,|f'(z)|}{1 - |g_2(f(z))|^2} \le \frac{|g_1'(z)|}{1 - |g_1(z)|^2}, \quad z \in \Omega_1.$$

The result follows, with equality if and only if F is a linear transformation, which in turn implies that $f = g_2^{-1} \circ F \circ g_1 : \Omega_1 \to \Omega_2$ is univalent and onto.

□

Section (2) Sullivan's Theorem

Choose $\Omega = \mathbf{C}^*$ and let $f : \mathbf{C}^* \to \mathbf{C}^*$ be an onto self-mapping. This means that f is a rational function of degree greater or equal to one (see Example 6 in Sec. 4.10.2).

If f is of degree 1, then $f(z)$ is a linear transformation. Refer to $(*_1)$–$(*_6)$ in Sec. 5.3.4 for its dynamical system.

Henceforth, we suppose that f is of degree at least equal to 2. In this case, the Julia set is nonempty and perfect.

Let G be a component of the Fatou set $F(f)$. Then $f^n(G) \subseteq F(f)$ for each $n \geq 1$. Fatou conjectured that the sequence

$$G, G_1 = f(G), \ldots, G_n = f(G_{n-1}) = f^n(G), \ldots \qquad (5.8.5.7)$$

will eventually be *periodic*, namely, there are two positive integers m and k so that

$$G_{n+k} = G_n, \quad n \geq m.$$

Sullivan confirmed this conjecture in early 1980s.

Without loss of its generality, we may suppose that the component G itself is periodic, i.e., $G_k = G$ or $f^k(G) = G$ holds for some $k \geq 1$. Then we have the following

Lemma 5.1. *Let f be a rational function of degree > 1. Suppose G is a component of the Fatou set $F(f)$ so that $f^k(G) = G$ holds and the restriction $f|_G : G \to G$ is not a homeomorphism (i.e., fails to be univalent or onto). Then:*

(1) *If f^k has a fixed point z_0 in G, then z_0 is attractive or superattractive and, as $n \to \infty$,*

$$f^{kn}(z) \to z_0, \quad z \in G,$$

locally uniformly on G.

(2) *If f^k does not have a fixed point in G, then as $n \to \infty$*

$$f^{kn}(z) \to \zeta \in \partial G \text{ (a fixed boundary point)}, \quad z \in G,$$

locally uniformly on G. $\qquad (5.8.5.8)$

The component G in (1) is called an *attractive (stability) domain* or a *superattractive (stability) domain* if z_0 is attractive or superattractive, respectively. While G in (2) is usually called a *parabolic domain*. Readers should refer to (5.3.4.11) and (5.3.4.12) and their proofs.

Before we start to prove, note that

$$|\lambda| = |(f^k)'(z_0)| \le 1 \qquad (5.8.5.9)$$

holds. Since $f^k(z_0) = z_0$, this follows easily by invoking (5.3.4.6) to f^k. But the generalized Schwarz–Pick's Lemma (5.8.5.6) provides a quick and simpler proof for this inequality. To see this, observe that $G \subseteq \mathbf{C}^* - \Im(f)$ and $\Im(f)$ is a nonempty perfect set. Hence $\mathbf{C}^* - G$ contains at least two distinct points and then G has a Poincaré's metric, say $\lambda(z)|dz|$. Recalling $f|_G : G \to G$ is a self-mapping of G, it follows that

$$\lambda(f^k(z))|(f^k)'(z)| \le \lambda(z), \quad z \in G$$

$\Rightarrow$ (setting $z = z_0$ and noting that $f^k(z_0) = z_0$)

$$\lambda(z_0)|(f^k)'(z_0)| \le \lambda(z_0)$$

$\Rightarrow$ (since $\lambda(z_0) > 0$) $\quad |(f^k)'(z_0)| \le 1.$

This proves (5.8.5.9).

Proof of (5.8.5.8). (1) By assumption that $f|_G : G \to G$ fails to be univalent or onto, the multiplier λ must satisfy $|\lambda| < 1$ (see (5.3.4.11)). Therefore z_0 is attractive or superattractive.

Since $|\lambda| < 1$, there is an open neighborhood O of z_0 and a constant $0 < \alpha < 1$ so that (see $(*_8)$ in Sec. 5.3.4)

$$|f^k(z) - z_0| \le \alpha|z - z_0|, \quad z \in O$$

$$\Rightarrow |f^{kn}(z) - z_0| \le \alpha^n|z - z_0|, \quad z \in O.$$

Hence $f^{kn}(z)$ converges to z_0 uniformly on O, as $n \to \infty$.

We *claim* that $f^{kn}(z)$ will then converge to z_0 locally uniformly on the whole set G. *Geometrically* this means that for any fixed point $z \in G$, the orbit of z is always attracted to that of z_0; more precisely, $f^{kn+j}(z) \to f^j(z_0)$, as $n \to \infty$, for all $z \in G$ and $0 \le j \le k - 1$.

To prove the claim, we adopt contradictory argument (as we did in Sec. 5.3.4). Suppose that there is a point ζ in G and a subsequence $f^{kn_j}(z)$, $j \ge 1$, so that $f^{kn_j}(\zeta)$ does not converge to z_0. Since f^m, $m \ge 1$, is normal, a subsequence of $f^{kn_j}(z)$, say itself, will converge locally uniformly on G to an analytic function g (see (5.3.1.1)). Consequently, $g(z) \equiv z_0$ on O and, in turn, on the whole set G. This is contradictory and the claim follows.

(2) As we stated in (1) of (5.3.4.12), we first try to prove that any convergent subsequence $f^{kn_j}(z)$, $j \geq 1$, will always converge locally uniformly on G to a *constant* function h. Rewrite

$$f^{kn_{j+1}}(z) = f^{k(n_{j+1}-n_j)}(z) \circ f^{n_j}$$

$\Rightarrow h = g \circ h$, where g is the limit function of $f^{k(n_{j+1}-n_j)}$, $j \geq 1$.

In case h is not a constant, it must be univalent in a neighborhood of some point; in this case, g should be an identity on that neighborhood and hence on the whole set G. Consequently

$$f^{k(n_{j+1}-n_j)} \to \text{the identity map on } G, \text{ locally uniformly on } G,$$

contradicting to the assumption that $f^k|_G$ is not homeomorphic. Compare to $(*_{13})$ and $(*_{14})$ in Sec. 5.3.4.

Now, suppose

$$f^{kn_j}(z) \to \zeta \text{ (a point in } \bar{G}), \text{ locally uniformly on } G.$$

$$\Rightarrow f^k(\zeta) = \lim_{j\to\infty} f^k \circ f^{kn_j}(z) = \lim_{j\to\infty} f^{kn_j} \circ f^k(z) = \zeta.$$

Hence ζ is a *fixed point* of f^k. Since f^k is assumed to have no fixed point in G, it follows that $\zeta \in \partial G$ should hold.

Finally, we have to show that the sequence

$$f^{kn}(z) \to \zeta \quad \text{locally uniformly on } G.$$

As we had shown that any convergent subsequence of f^{kn}, $n \geq 1$, always converges locally uniformly on G to a fixed point of f^k, all we need to do is to prove that f^k has only one fixed point. Suppose on the contrary that f^k has fixed points $\zeta_1, \ldots, \zeta_l$ where $l > 1$ (why does f^k have only finitely many fixed point?). Choose disjoint open neighborhoods O_j of ζ_j, $1 \leq j \leq l$, so that

$$f^k(O_j) \bigcap \bigcup_{p \neq j} O_p = \phi, \quad 1 \leq j \leq l$$

$\Rightarrow$ there exists infinitely many m_j so that

$$f^{km_j}(z) \notin \bigcup_{p=1}^{l} O_p, \quad z \in G.$$

This indicates that the limit point of a convergent subsequence of f^{km_j}, $j \geq 1$, is not one of the fixed points $\zeta_1, \ldots, \zeta_l$, a contradiction. Hence $l = 1$ holds.

  This ends the proof of (5.8.5.8). $\square$

  Similar argument leads to

Lemma 5.2. *Let f and G be as in Lemma 5.1. But now suppose $f^k|_G : G \to G$ is homeomorphic (namely, univalent and onto).*

(1) *In case f^k has a fixed point z_0 in G, then z_0 is neutral and G is called a Siegel (stable) domain.*

(2) *In case f^k does not have fixed point in G, then G is doubly-connected and is called a Herman (stable) ring domain.*   (5.8.5.10)

  f^k acts as a rotation on a Siegel or Herman domain. For detail, refer to Ref. [15] and the references within.

  Combining Lemmas 5.1 and 5.2, we have the remarkable

Sullivan Theorem. *Every periodic component of the Fatou set of a rational function of degree > 1 can only be one of the following five types of stable domains:*

1. *Superattractive domain.*
2. *Attractive domain.*
3. *Parabolic domain.*
4. *Siegel domain.*
5. *Herman ring domain.*   (5.8.5.11)

CHAPTER 6

Conformal Mapping and Dirichlet's Problems

Introduction

In Secs. 2.5.2, 2.5.4, 2.5.5, and 3.5.7, we have practiced quite a few examples of mapping specified simply connected domains *one-to-one* and *analytically onto* another simply connected domains, in particular, the open disks or the half-planes. As a continuation of the geometric properties of analytic functions studied in Sec. 3.5, this chapter is going to study the *global* geometric mapping properties and the related Dirichlet's problems. By a *conformal* or a *univalent* mapping, we always mean a *one-to-one analytic* function mapping a plane domain *onto* another plane domain, unless otherwise stated.

Sketch of the Content

Riemann initiated the global geometric study of analytic functions by formulating that, any two simply connected domains Ω_1 and Ω_2, each of whose boundaries contain more than one point, are *conformally equivalent*. Section 6.1 devotes to the proof of this *Riemann mapping theorem*, using R. Koebe's method, and to the *boundary correspondence*, saying that the conformal mapping can be homeomorphically extended from $\overline{\Omega_1}$ onto $\overline{\Omega_2}$ if Ω_1 and Ω_2 are Jordan domains.

In case Ω is a polygon, the *Schwarz–Christoffel formula* gives explicitly the conformal mapping that maps Ω onto the open disk $|w| < 1$ or the half-plane $\operatorname{Im} w > 0$. Section 6.2 also contains lots of concrete examples, especially if Ω is a triangle or a rectangle. The application to the *elliptic function* is only sketched in Exercise B of Sec. 6.2.2.

The next main theme of this chapter is *Dirichlet's problem*, of finding a harmonic function in a domain with a presumed boundary values. Section 6.3 solves this problem for a disk which is usually called *Schwarz's theorem*, by using *Poisson's integral* for harmonic functions. Basic properties of *Green's function* of a domain with respect to a point (pole) are also studied for later usage in Sec. 6.6. Included are the *symmetry principle* and *Harnack's principle.*

The extension of harmonic functions to the *subharmonic* ones is in Sec. 6.4. Basic properties are proved for the need of Sec. 6.5.

Section 6.5 adopts *Perron's method,* concerning a family of subharmonic functions with specified boundary properties, to solve *Dirichlet's problem for a class of general domains* Ω. The best result that we obtained is the following: *Dirichlet's problem is solvable for any domain whose complement in* **C** *is such that no component reduces to a point* (see (6.5.7)); in particular, a simply connected domain whose boundary contains more than one point and a domain bounded by a finite number of Jordan curves are in this class. *These results are used in Sec. 6.6.*

The last section (Sec. 6.6) relies mainly on the concepts of *harmonic measure* (see Sec. 6.6.1) and Green's function (see Sec. 6.3.1) to determine the *canonical mapping* that maps a finitely-connected domain univalently onto a *canonical domain.* The canonical domains introduced are:

1. Annulus with concentric circular slits ((6.6.2.1); Exercise B(5) of Sec. 6.6.2).
2. Domain with spiral slits (Exercise B(1) of Sec. 6.6.2).
3. Disk with concentric circular slits (Exercise B(3) of Sec. 6.6.2).
4. Domain with vertical slits (6.6.3.3).
5. Domain with horizontal slits (6.6.3.5).
6. Domain with parallel slits (6.6.3.6).
7. Horizontal strip with parallel segment slits (6.6.4.2).
8. n-Sheeted open strip (6.6.4.3).

The *extremal problem method* (as the one in proving the Riemann mapping theorem) is also touched in the exercises in finding the canonical domains and mappings mentioned above.

6.1   The Riemann Mapping Theorem

In Secs. 2.5 (about elementary rational functions) and Sec. 2.6 (about elementary transcendental functions), we experienced many examples showing how some particularly-shaped simply connected domains, whose boundaries compose mostly of line segments and circular arcs, are mapped univalently and analytically onto another simply connected domains. The highlights in this direction culminate in Sec. 3.5.7, including Secs. 2.7.2–2.7.4 wherein Riemann surfaces of elementary multiple-valued functions are constructed for descriptive purpose.

The general setting is the *problem*: Does there always exist a univalent analytic function mapping a given domain Ω_1 onto another domain Ω_2? If it does, Ω_1 and Ω_2 are said to be *conformally equivalent*. Riemann asserted that the answer is in the affirmative if both are simply connected domains, each of whose boundaries contains at least two distinct points. Refer to Remark 3 in Sec. 2.4 for related explanation.

As a preparatory work, recall that a nonconstant analytic function maps a domain onto a domain (see (3.4.4.4) and (3.5.1.12)); if, in addition, it is univalent, then it maps a simply connected domain onto a simply connected one (see (2.4.13), (3.5.1.9), and (3.5.5.1)). Conformal equivalence among domains is an equivalent relation (namely, reflexive, symmetric, and transitive). Hence, we may choose the open unit disk as one of the two simply connected domains preassigned; Liouville's theorem suggests that the other one cannot be the whole plane $\mathbf{C}$ or the plane $\mathbf{C}$ with a point removed.

Riemann formulated

The Riemann mapping theorem. *Let Ω be a simply connected domain in $\mathbf{C}^*$ where boundary (in $\mathbf{C}^*$) contains at least two distinct points. Fix a point $z_0 \in \Omega$ and a real number θ ($0 \leq \theta < 2\pi$). Then there exists a unique univalent analytic function $f(z)$ mapping Ω onto the open unit disk $|w| < 1$, satisfying the conditions:*

1. $f(z_0) = 0$;
2. $\mathrm{Arg}\, f'(z_0) = \theta$. $\hspace{5cm}$ (6.1.1)

This is an existence and uniqueness theorem. The explicit analytic expression for such an $f(z)$ cannot be found, in general, except in very restricted cases (see Sec. 6.2). However, the geometric properties of the domains that are being mapped lead to analytic properties of the mapping function (see, for instance, Sec. 6.1.2). This theorem, initiated the study of the geometric function theory, does not have its counterpart in the theory of function of several complex variables (*Poincaré's theorem*: There is no biholomorphic map between the open unit ball $\{z = (z_1, \ldots, z_n) \in \mathbf{C}^n | |z_1|^2 + \cdots + |z_n|^2 < 1\}$ and the open polycylinder $\{z \in \mathbf{C}^n | |z_1| < 1, \ldots, |z_n| < 1\}$, where $n \geq 2$).

For pedagogical (not for logical) reason, we have cited this theorem previously in quite a few places, such as in Secs. 3.4.5, 3.5.7, 5.8.2, 5.8.3, and 5.8.4, etc.

Note that only condition 1 cannot single out such an $f(z)$ uniquely because $e^{i\theta}f(z)$ is also such a function for each real θ if $f(z)$ merely satisfies $f(z_0) = 0$. It is condition 2 that does so (see the beginning of Sec. 6.1.1). Condition 2 means geometrically that $w = f(z)$ maps a segment, passing z_0 and parallel to the positive real axis, onto a curve passing 0 and having an inclination angle θ to the positive real axis (refer to (3.5.1.1)); consequently, the tangents to all curves passing through the point z_0 experience a rotation of angle θ when subjected to the transformation $w = f(z)$ mapping them into curves passing through the center 0 of the unit disk $|w| < 1$.

Two sections are divided.

Section 6.1.1 is devoted fully to the proof of the theorem.

Section 6.1.2 extends the function $f(z)$ homeomorphically to the closure $\overline{\Omega}$, mapping onto $|w| \leq 1$, in case Ω is a Jordan domain. By a *Jordan domain* Ω in $\mathbf{C}^*$, we mean the one whose boundary is a closed Jordan curve γ in $\mathbf{C}^*$, namely, $\Omega = \text{Int}\,\gamma$ (see (2.4.11) and Remark 2 after it). We formally state as

The boundary correspondence theorem. *Let $f(z)$ be a univalent analytic function mapping a Jordan domain Ω_1 onto a Jordan domain Ω_2. Then $f(z)$ can be continuously extended as a homeomorphism from $\overline{\Omega_1}$ onto $\overline{\Omega_2}$ so that $w = f(z)$ sets up a homeomorphism between the boundaries $\partial\Omega_1$ and $\partial\Omega_2$.* (6.1.2)

For detailed account, refer to Ref. [58], Vol. III, Chap. 2; for a little up-to-date treatment, refer to Ref. [66], Sec. 4 of Chap. IX, and Ref. [9], Sec. 8 of Chap. 4.

For simplicity, *a univalent function will always mean an analytic function that is univalent.* Recall that a univalent function is conformal (see (3.5.1.9)) but not conversely.

6.1.1   *Proof*

P. Koebe gave the theorem its first successful proof. Later proofs of it are variants of Koebe's original proof.

Uniqueness:

In case $g : \Omega \to \{|w| < 1\}$ is another such function, then the composite map $g \circ f^{-1} : \{|w| < 1\} \to \{|w| < 1\}$ is univalent, onto and maps 0 into 0. According to Example 1 in Sec. 3.4.5, there is a real α so that $g(f^{-1}(w)) = e^{i\alpha}w$, $|w| < 1$. Since $f(z_0) = 0$, then $g'(z_0)(f^{-1})'(0) = g'(z_0)(f'(z_0))^{-1} = e^{i\alpha}$ and $\text{Arg}\,f'(z_0) = \text{Arg}\,g'(z_0) = \theta$ together show that $\alpha = 0$ should hold. Hence $g(z) = f(z)$ throughout Ω.

Existence:

If $f(z)$ is known to be the required mapping, then the function $g(z) = e^{-i\theta} f(z)$ will map Ω univalently onto $|w| < 1$, satisfying the properties that $g(z_0) = 0$ and $g'(z_0) > 0$. Hence, we may prove the theorem in the case that $\theta = 0$.

Consider the family $\mathfrak{F}$ of all the functions $g(z)$ satisfying the conditions:

1. $g(z)$ is univalent (and analytic) in Ω;

2. $|g(z)| \le 1$ in Ω;

3. $g(z_0) = 0$ and $g'(z_0) > 0$. $\hspace{2cm} (*_1)$

Then we try to show that $\mathfrak{F}$ is *nonempty* and hence is a *normal* family. Finally, try to solve the *extremal problem* $\sup_{g \in \mathfrak{F}} g'(z_0)$ and show that the *extremal function* $f(z)$ is the required one (refer to the Remark in Sec. 5.8.2 and (5.8.2.7) there).

Step 1. The nonemptyness of $\mathfrak{F}$.

There is a point $a \ne \infty$ not in Ω, since, otherwise, $\Omega = \mathbf{C}$ or $\mathbf{C}^*$ will contradict to our assumption that $\partial\Omega$ contains at least two distinct points. Since $z - a \ne 0$ on Ω, $\sqrt{z-a}$ has two distinct single-valued branches (see (3.3.1.5) or (4.4.2), etc.). Fix one of them as $h(z)$. h *is univalent on* Ω for, if $h(z_1) = h(z_2)$, then

$$h(z_1)^2 = h(z_2)^2, \quad \text{namely } z_1 - a = z_2 - a \text{ or } z_1 = z_2.$$

Moreover, $h(z_1) \ne -h(z_2)$ *for any* $z_1, z_2 \in \Omega$ for, if so, then $h(z_1) = -h(z_2)$ implies that $h(z_1)^2 = h(z_2)^2$, i.e., $z_1 = z_2$ which, in turn, implies that $h(z_1) = 0$, contradicting to the fact that $h(z) \ne 0$ throughout Ω.

$w = h(z)$ is open and hence the image $h(\Omega)$ contains an open disk $|w - h(z_0)| < \rho$ for some $\rho > 0$. Since $-h(z) \notin h(\Omega)$ for all $z \in \Omega$, it follows that $|-h(z) - h(z_0)| = |h(z) + h(z_0)| \ge \rho$ comes true on Ω; in particular, $2|h(z_0)| \ge \rho$ holds. Set

$$g_0(z) = k \frac{h(z) - h(z_0)}{h(z) + h(z_0)}, \quad z \in \Omega,$$

where k is a constant to be determined later on so that this g_0 will be in $\mathfrak{F}$. First

$$|g_0(z)| = |k||h(z_0)| \left| \frac{1}{h(z_0)} - \frac{2}{h(z) + h(z_0)} \right| \le |k||h(z_0)| \left(\frac{2}{\rho} + \frac{2}{\rho} \right)$$

$$= \frac{4|k||h(z_0)|}{\rho} = 1,$$

if, in the last equality, $|k| = \frac{\rho}{4|h(z_0)|}$ is chosen. Set $\theta_0 = \operatorname{Arg} k$ temporarily. Hence

$$g_0(z) = \frac{\rho e^{i\theta_0}}{4|h(z_0)|} \cdot \frac{h(z) - h(z_0)}{h(z) + h(z_0)}, \quad z \in \Omega, \qquad (*_2)$$

is univalent in Ω, $g_0(z_0) = 0$, and $|g_0(z)| \leq 1$ there. On the other hand,

$$g_0'(z_0) = k \frac{h'(z_0)}{2h(z_0)}$$

$$= |k| e^{i\theta_0} \frac{|h'(z_0)| e^{i\varphi_0}}{2|h(z_0)| e^{i\psi_0}}, \quad \text{where } \theta_0 = \operatorname{Arg} k, \quad \varphi_0 = \operatorname{Arg} h'(z_0), \quad \text{and}$$

$$\psi_0 = \operatorname{Arg} h(z_0)$$

$$= \frac{\rho |h'(z_0)|}{8|h(z_0)|^2} e^{i\theta_0} e^{i(\varphi_0 - \psi_0)}$$

$$> 0 \quad \text{if } \theta_0 = \psi_0 - \varphi_0 \quad \text{is chosen.}$$

$$\Rightarrow e^{i\theta_0} = e^{-i(\varphi_0 - \psi_0)} = \frac{h(z_0)}{|h(z_0)|} \cdot \frac{|h'(z_0)|}{h'(z_0)}.$$

In conclusion, rewrite $(*_2)$ in its final form as

$$g_0(z) = \frac{\rho}{4} \frac{|h'(z_0)|}{h'(z_0)} \frac{h(z_0)}{|h(z_0)|^2} \frac{h(z) - h(z_0)}{h(z) + h(z_0)}, \quad z \in \Omega. \qquad (*_3)$$

This $g_0(z)$ is indeed in $\mathfrak{F}$.

Step 2. An extremal problem.
According to (3) in (5.8.1.3), $\mathfrak{F}$ is normal (compare to (5.8.2.3)). Let

$$M = \sup_{g \in \mathfrak{F}} g'(z_0). \qquad (*_4)$$

At present, M could be $+\infty$. However, $\mathfrak{F}$ has a sequence $g_n(z)$, $n \geq 1$, satisfying $g_n'(z_0) \to M$ as $n \to \infty$. The sequence $g_n(z)$ has a subsequence, say $g_n(z)$ itself, converging to an analytic function $f(z)$ locally uniformly in Ω. Observe that $g_n(z_0) = 0$, $g_n'(z_0) > 0$, and $|g_n(z)| \leq 1$ for each $n \geq 1$. These facts lead to $f(z_0) = 0$, $f'(z_0) = M > 0$ (also $M < \infty$), and $|f(z)| \leq 1$ on Ω. Hurwitz's theorem (5.3.2.2) guarantees that $f(z)$ is univalent in Ω because $f(z)$ is not a constant.

We conclude that $f(z)$ is in $\mathfrak{F}$ and is the extremal function solving $(*_4)$.

Step 3. The extremal function is required.
What remains is to show that $f(z)$ maps Ω *onto* the whole disk $|w| < 1$.

Suppose on the opposite that there is a point w_0 in $|w| < 1$ which does not lie in the domain $f(\Omega)$. We try to construct a function F in $\mathfrak{F}$ with $F'(z_0) > M$ in order to contradict to $(*_4)$. Once this is done, as a domain, $f(\Omega) = \{|w| < 1\}$ should hold because $f(z)$ cannot assume values lying on the circle $|w| = 1$.

Consider the following three auxiliary maps:

$$\zeta = f_1(w) = \frac{w - w_0}{1 - \overline{w_0}w} : \{|w| < 1\} \to \{|\zeta| < 1\};$$

$$\eta = f_2(\zeta) = \sqrt{\zeta} : \{|\zeta| < 1\} \to \{|\eta| < 1 \text{ and } \operatorname{Im}\eta > 0\},$$

a branch defined by $\sqrt{-1} = i$;

$$\sigma = f_3(\eta) = e^{i\theta_0}\frac{\eta - \eta_0}{1 - \overline{\eta_0}\eta} : \{|\eta| < 1 \text{ and } \operatorname{Im}\eta > 0\} \to \{|\sigma| < 1\},$$

$$\eta_0 = \sqrt{-w_0},$$

where θ_0 is a real constant to be determined yet. *Explain* as follows: Since $f(z) \neq w_0$ on Ω, the simply connected domain $f_1 \circ f(\Omega)$ does not contain the point 0 and this, in turn, implies that a single-valued branch of $G(z) = f_2 \circ f_1 \circ f(z)$ can be defined on $f_1 \circ f(\Omega)$ (see (3.3.1.5)). It is understood that we fix such a branch in the following argument.

The composite map

$$F(z) = f_3 \circ f_2 \circ f_1 \circ f(z) = f_3 \circ G(z) = e^{i\theta_0}\frac{G(z) - G(z_0)}{1 - \overline{G(z_0)}G(z)}, \quad \text{where}$$

$$G(z) = \sqrt{\frac{f(z) - w_0}{1 - \overline{w_0}f(z)}} \quad \text{and} \quad z \in \Omega, \tag{$*_5$}$$

is univalent and $F(z_0) = f_3 \circ G(z_0) = f_3(\eta_0) = 0$. To determine θ_0 so that $F'(z_0) > 0$: A direct computation shows that

$$F'(z_0) = f_3'(G(z_0))G'(z_0) = e^{i\theta_0}\frac{G'(z_0)}{1 - |\eta_0|^2}, \quad \text{where}$$

$$G'(z_0) = \frac{1}{2\eta_0} \cdot f'(z_0)(1 - |w_0|^2).$$

On choosing θ_0 so that $G'(z_0) = e^{-i\theta_0}|G'(z_0)|$ and recalling that $f'(z_0) = M$ and $\eta_0 = \sqrt{-w_0}$, we have

$$F'(z_0) = \frac{|G'(z_0)|}{1 - |\eta_0|^2} = \frac{1 - |w_0|^2}{1 - |w_0|} \cdot \frac{M}{2|w_0|^{1/2}} = \frac{1 + |w_0|}{2|w_0|^{1/2}}M > M.$$

contradicting to $(*_4)$.

This finishes the proof of the Riemann mapping theorem.

Remark 1 (On variants of the proof). Ω is always assumed to be a simply connected domain whose boundary in $\mathbf{C}^*$ contains more than one point.

About Step 1:
In case Ω is bounded, say, Ω is contained in the disk $|z| < R$. For any fixed point $z_0 \in \Omega$, the function $g(z) = \frac{1}{2R}(z - z_0)$ is univalent in Ω, $g(z_0) = 0$, $g'(z_0) > 0$, and $|g(z)| < 1$ on Ω. While the function $\frac{g(z)}{g'(z_0)} = h(z)$ satisfies $h(z_0) = 0$, $h'(z_0) = 1$, and $h(z)$ remains bounded on Ω.

In case Ω is unbounded but has an exterior point, say $\zeta_0 \in \mathbf{C}$. Suppose the disk $|z - \zeta_0| < r$ is contained in the exterior of Ω. Then $g(z) = \frac{1}{z - \zeta_0}$ is bounded by $\frac{1}{r}$ on Ω and univalent; moreover, $g(z)$ maps Ω onto a bounded simply connected domain $g(\Omega)$.

In case Ω is unbounded with empty exterior, then the boundary $\partial\Omega$ is a continuum (namely, a compact connect set in $\mathbf{C}^*$, containing infinitely many points). Fix two distinct points a and b along $\partial\Omega$. Then $\frac{z-a}{z-b}$ is univalent in Ω and never assumes the values 0 and ∞ there; consequently, $\sqrt{\frac{z-a}{z-b}}$ has two distinct single-valued univalent branches, say $g_1(z)$ and $g_2(z)$, on Ω (see (3.3.1.5)). Note that, $g_2(z) = -g_1(z)$ for all $z \in \Omega$. Therefore $g_1(\Omega)$ and $g_2(\Omega)$ are pairwise disjoint simply connected domains. Either of $g_1(\Omega)$ and $g_2(\Omega)$ is contained in the exterior of the other. According to the last paragraph, there are bounded univalent functions on $g_1(\Omega)$ or $g_2(\Omega)$, and so are on the original Ω.

In conclusion, there are (in fact, infinitely many) bounded univalent functions on Ω. If $g(z)$ is such a function and z_0 is a fixed point in Ω, then $h(z) = g(z) - g(z_0)$ is bounded, univalent and $h(z_0) = 0$ and $h'(z_0) \neq 0$; and $e^{i\theta}h(z)$ has a positive derivative at z_0 for some real θ; and $\frac{h(z)}{h'(z_0)}$ has its derivative equal to 1 at z_0.

About Step 2:
It is harmless, in $(*_1)$, to drop the condition 3 and, instead, consider the family $\mathfrak{F} = \{g(z)$ is univalent in Ω, $g(z_0) = 0$, and $|g(z)| \leq 1$ (even if $|g(z)| < 1$) there$\}$. In this case, $(*_4)$ is changed to $M = \sup_{g \in \mathfrak{F}} |g'(z_0)|$. That $M < \infty$ can be proved in advance. To see this, choose any $0 < r < 1$ such that the disk $|z - z_0| \leq r$ is contained in Ω. Via Cauchy's inequality, $|g'(z_0)| \leq \frac{1}{r}$ for any $g \in \mathfrak{F}$; it follows that $M \leq \frac{1}{r}$. In case $f(z)$ is the corresponding extremal function, after multiplying $f(z)$ by a constant of modulus 1, the resulted $f(z)$ will then satisfy $f'(z_0) = M > 0$.

Of course, there are some other choices of the family $\mathfrak{F}$.

In the formulation of $g_0(z)$ in $(*_3)$, $\rho > 0$ can be chosen smaller so that the closed disk $|w - h(z_0)| \leq \rho$ lies entirely in $h(\Omega)$. Then, $2|h(z_0)| > \rho$ holds; consequently, $|g_0(z)| < 1$ holds on Ω. Suppose this is the case in what follows. Hence $g_0(z)$ maps Ω univalently onto a domain $g_0(\Omega)$ in $|\zeta| < 1$. Note that $g_0(z_0) = 0 \subset g_0(\Omega)$.

Consider the family

$$\mathfrak{F} = \{\varphi(\zeta) \,|\, \varphi(\zeta) \text{ is univalent on } g_0(\Omega), \varphi(0) = 0, \varphi'(0) > 0, \text{ and }$$

$$|\varphi(z)| < 1 \text{ on } g_0(\Omega)\}. \tag{$*_1$}'$$

$\mathfrak{F}$ is not empty because $r\zeta (0 < r \leq 1)$ is such a function. $\mathfrak{F}$ is thus normal and, by Cauchy's inequality or Schwarz's Lemma, there is some $r > 0 (r \leq 1)$ so that $0 < \varphi'(0) \leq \frac{1}{r}$ for all $\varphi \in \mathfrak{F}$. Instead of $(*_4)$, set

$$M = \sup_{\varphi \in \mathfrak{F}} \varphi'(0). \tag{$*_4$}'$$

$M \geq 1$ since $\varphi(\zeta) = \zeta \in \mathfrak{F}$, and also, $M < \infty$ holds. The normality of $\mathfrak{F}$ shows that there is a $\varphi_0 \in \mathfrak{F}$ so that $\varphi_0'(0) = M$ holds. Also, $|\varphi_0(\zeta)| \leq 1$ on $g_0(\Omega)$, and, by the maximum principle, indeed $|\varphi_0(\zeta)| < 1$ throughout $g_0(\Omega)$ because $\varphi_0(\zeta)$ is not a constant.

Exactly the same technique as Step 3 shows that this $\varphi_0(\zeta)$ maps $g_0(\Omega)$ univalently *onto* the whole disk $|w| < 1$. By the uniqueness, the composite map $f(z) = \varphi_0(g_0(z))$ is the required map.

Yet another choice of $\mathfrak{F}$ is given by an

$$\mathfrak{F} = \{g(z) \text{ is a } bounded \text{ univalent function on } \Omega \text{ satisfying } g(z_0) = 0$$

$$\text{and } g'(z_0) = 1\}. \tag{$*_1$}''$$

$\mathfrak{F}$ is nonempty. Instead of $(*_4)$, let

$$R_{z_0} = \inf_{g \in \mathfrak{F}} \sup_{z \in \Omega} |g(z)|. \tag{$*_4$}''$$

Note that $0 \leq R_{z_0} < \infty$. There is a sequence $g_n(z)$ in $\mathfrak{F}$ such that $\sup_{z \in \Omega} |g_n(z)| \to R_{z_0}$ as $n \to \infty$. $g_n(z)$, $n \geq 1$, are then uniformly bounded on Ω, and hence is normal as a family of functions. A subsequence of it, say itself, converges to an analytic function $f(z)$ locally uniformly in Ω. $f(z_0) = 0$ and $f'(z_0) = 1$, in turn, imply that $f(z)$ is univalent on Ω by Hurwitz's theorem. For any fixed $\zeta \in \Omega$, $|g_n(\zeta)| \leq \sup_{z \in \Omega} |g_n(z)|$ for $n \geq 1$ indicate, after letting $n \to \infty$, that $|f(\zeta)| \leq R_{z_0}$ for all $\zeta \in \Omega$. Hence $f \in \mathfrak{F}$. This implies that $\sup_{z \in \Omega} |f(z)| \geq R_{z_0}$. Hence, $\sup_{z \in \Omega} |f(z)| = R_{z_0}$ holds, namely, $|f(z)| \leq R_{z_0}$ throughout Ω. That $|f(z)| = R_{z_0}$ for some $z \in \Omega$ is not possible (why?) shows that $|f(z)| < R_{z_0}$ for all $z \in \Omega$. Step 3 is still applicable here to guarantee that $f(z)$ indeed maps Ω *onto* the whole open disk $|w| < R_{z_0}$. This $f(z)$ is the required one. $\qquad\qquad\square$

Remark 2 (The conformal radius of the domain Ω relative to the point $z_0 \in \Omega$). Let $w = f(z)$ be the unique univalent function mapping Ω onto the open unit disk $|w| < 1$ and satisfying the extra conditions:

$$f(z_0) = 0 \quad \text{and} \quad f'(z_0) > 0, \qquad\qquad (6.1.1.1)$$

where Ω denotes a simply connected domain with more than one boundary point and $z_0 \in \Omega$ is a preassigned point. Then the function

$$F(z) = \frac{f(z)}{f'(z_0)}$$

maps Ω univalently onto $|w| < \frac{1}{f'(z_0)}$ and satisfies the conditions:

$$F(z_0) = 0 \quad \text{and} \quad F'(z_0) = 1.$$

Such an $F(z)$ is unique. The number (see also $(*_4)''$)

$$R_{z_0} = \frac{1}{f'(z_0)} \qquad\qquad (6.1.1.2)$$

is well-defined and is called the *conformal radius* of Ω relative to the point z_0. Let us read back to $(*_1)''$ and $(*_4)''$: R_{z_0} can also be characterized geometrically by

$$R_{z_0} = \frac{1}{\pi} \inf_{g \in \mathfrak{F}} \iint_\Omega |g'(z)|^2 dx dy \quad (z = x + iy). \qquad\qquad (6.1.1.3)$$

For reference, we list some

Basic properties of the conformal radius. Let Ω and Ω^* be simply connected domains whose boundaries each contain more than one point.

(1) Suppose $\Omega \subseteq \Omega^*$ and $z_0 \in \Omega$, a fixed point. Then $R_{z_0} \le R^*_{z_0}$, the conformal radius of Ω^* relative to z_0.

(2) (A variant of (6.1.1.3).) Suppose $0 \in \Omega$. Suppose the univalent function $f(z) = z + a_2 z^2 + \cdots + a_n z^n + \cdots$ maps $|z| < 1$ onto Ω. Then Ω has

$$\text{the } inner\ area \underset{\text{(def.)}}{=} \lim_{r \to 1^-} \iint_{|z| \le r} |f'(z)|^2 dx dy$$

$$= \lim_{r \to 1^-} \pi \left(r^2 + \sum_{n=2}^{\infty} n|a_n|^2 r^{2n} \right)$$

$$= \pi \left(1 + \sum_{n=2}^{\infty} n|a_n|^2 \right) \ge \pi$$

with equality if and only if $f(z) = z$, the identity.

(3) Suppose $\Omega \subseteq \mathbf{C}$ and $z_0 \in \Omega$, a fixed point. Then (refer to Exercise B(4) of Sec. 5.8.2)

$$d(z_0, \partial\Omega) \le R_{z_0} \le \min(\delta, 4d(z_0, \partial\Omega)),$$

where $d(z_0, \partial\Omega)$ is the distance from z_0 to $\partial\Omega$ and $\delta = \sup_{z\in\partial\Omega} |z - z_0|$. The first equality holds if and only if Ω is $|z - z_0| < R_{z_0}$.   (6.1.1.4)

Sketch of proof for (2) and (3). Set $z = re^{i\theta}, 0 \le r < 1$, and $0 \le \theta < 2\pi$. Then

$$f'(z) = f'(re^{i\theta}) = 1 + \sum_{n=2}^{\infty} na_n r^{n-1} e^{i(n-1)\theta}$$

$$\Rightarrow \iint_{|x|\le r} |f'(z)|^2 dxdy = \iint_{|x|\le r} f'(\rho e^{i\theta})\overline{f'(\rho e^{i\theta})}\rho d\rho d\theta$$

$$= \pi\left(r^2 + \sum_{n=2}^{\infty} n|a_n|^2 r^{2n}\right) \le \pi\left(1 + \sum_{n=2}^{\infty} n|a_n|^2\right).$$

And the result in (2) follows. Note that $1 + \sum_{n=2}^{\infty} n|a_n|^2$ could be $+\infty$.

As for (3): After a translation, we may suppose that $z_0 = 0 \in \Omega$. Let $F(w)$ be the unique univalent function mapping Ω onto $|z| < R_{z_0}$ that satisfies $F(0) = 0$ and $F'(0) = 1$. Set $f(z) = F^{-1}(R_{z_0}z), |z| < 1$. Note that $f'(0) = R_{z_0}$. On applying (2) to $\frac{f(z)}{R_{z_0}}$, the first equality holds if and only if $f(z) = R_{z_0}z, |z| < 1$.

On the other hand, let $z = F(w) = w + b_1 w^2 + \cdots$. Then $w = F^{-1}(z) = z + c_2 z^2 + \cdots, |z| < R_{z_0}$. On applying Exercise B(4) of Sec. 5.8.2 to $F^{-1}(R_{z_0}z)$ at $z_0 = 0$, where $|z| < 1$, the result follows (even including the first inequality).   □

Exercises A

In this Exercise, Ω always denotes a simply connected domain with more than one boundary point, unless otherwise stated; while, a univalent function always means the one that is analytic at the same time.

(1) Can $0 < |z| < \infty$ be mapped univalently onto $0 < |z| < 1$?

(2) Let $f(z)$ be an entire function taking no values in a continuum E whose complement $\mathbf{C}^* - E$ is connected. Show that $f(z)$ is a constant.

(3) Suppose Ω is symmetric with respect to the real axis and $z_0 \in \Omega$ lies on the real axis. Let $f(z)$ be the mapping in (6.1.1) with $\theta_0 = 0$. Show that $f(z)$ preserves symmetry, namely, $f(\bar{z}) = \overline{f(z)}$. What happens if Ω is also symmetric with respect to z_0?

(4) Suppose $\Omega \neq \mathbf{C}$, $z_0 \in \Omega$, and $f(z)$ is the mapping in (6.1.1) with $\theta_0 = 0$. Let $g(z)$ be any univalent function mapping Ω onto $|w| < 1$. Try to represent $g(z)$ in terms of $f(z)$.

(5) Find the conformal radius for each of the following domains Ω at the point z_0.

   (a) $\Omega: |z| < R$, at z_0 $(z_0 < R)$.
   (b) $\Omega: \operatorname{Im} z > 0$, at $z_0 = hi$ $(h > 0)$.

(6) Suppose that $f(z)$ maps $|z| < 1$ univalently onto a domain Ω. Find the conformal radius of Ω at $a = f(z_0)$, where $|z_0| < 1$.

(7) Show that

$$\lambda_\Omega(z) = \frac{2|f'(z)|}{1 - |f(z)|^2}$$

has the same value for every univalent function $w = f(z)$ mapping Ω onto $|w| < 1$. This is *Poincaré's metric* defined in $(*_{13})$ of Sec. 5.8.4 (see also (5.8.4.19)).

(8) Let Ω^* be a simply connected domain with more than one boundary point. Let $f(z) : \Omega \to \Omega^*$ be analytic. Then

$$\lambda_{\Omega^*}(f'(z))|f'(z)| \leq \lambda_\Omega(z), \quad z \in \Omega.$$

This is a rather special case of Schwarz–Ahlfor's Lemma (5.8.4.14) and, in turn, Schwarz–Pick's Lemma (3.4.5.2) is a special case of this inequality, too.

(9) Let $f(z)$ be analytic in $|z| < 1$, and suppose $f(z)$ assumes no nonpositive values on $|z| < 1$. Show that

$$\frac{|f'(z)|}{|f(z)|} \leq \frac{4}{1 - |z|^2}, \quad \text{and}$$

$$\left(\frac{1 - |z|}{1 + |z|} \right)^2 \leq \left| \frac{f(z)}{f(0)} \right| \leq \left(\frac{1 + |z|}{1 - |z|} \right)^2, \quad |z| < 1.$$

In case $f(0) = 1$ holds, then $f(z) = \left(\frac{1+z}{1-z} \right)^2$ gives both equalities for $z = |z|$ and $z = -|z|$, respectively. Try to prove similar estimates for analytic function $g(z)$ in $|z| < 1$ with positive real part, i.e., $\operatorname{Re} g(z) > 0$.

6.1.2   The boundary correspondence

To prove (6.1.2), it is sufficient to suppose that one of the Jordan domains Ω_1 and Ω_2 is the open unit disk $|z| < 1$. Now, suppose $w = f(z)$ is an univalent (analytic) function, mapping a Jordan domain Ω, bounded by a closed Jordan curve γ, onto $|w| < 1$. Note that $\Omega = \text{Int}\,\gamma$ is a *bounded simply connected domain with its boundary* $\partial\Omega = \gamma$. We will prove (6.1.2) in this case.

A rigorous and complete proof is quite intricate and lasts long. The main difficulty one might encounter is to prove a characteristic property of a Jordan closed curve, saying that *every point of it is simple and accessible from within*. Roughly speaking, this means that, *for each point $\zeta \in \gamma$, it is always possible to construct a continuous curve Γ connecting ζ to a point in Ω such that Γ lies completely in Ω except its end point ζ.* We will take this fact for grant in the following proof. For this and a detailed proof for (6.1.2) and its extensions, read Ref. [58], Vol. III, Chap. 2; in particular, Sec. 8 within.

Sketch of proof for (6.1.2). First, try to show that $f(z)$ is uniformly continuous on the bounded domain Ω and hence, $f(z)$ has a continuous extension to the closure $\bar{\Omega} = \overline{\text{Int}\,\gamma}$. When the same process is adopted to the inverse function $z = f^{-1}(w):\{|w < 1|\} \to \Omega$, the claimed result will follow.

The uniform continuity of $f(z)$ on Ω:

Suppose on the opposite that $f(z)$ fails to be uniformly continuous on Ω. Then there are a positive number ε_0 and two sequence z_n, z_n', $n \geq 1$, of points in Ω satisfying

$$|z_n - z_n'| < \frac{1}{n}, \quad n \geq 1, \quad \text{and}$$

$$|f(z_n) - f(z_n')| \geq \varepsilon_0, \quad n \geq 1. \tag{$*_1$}$$

As a bounded sequence, z_n has a subsequence z_{n_k} converging to a point $\zeta \in \bar{\Omega}$. Since $|z_{n_k} - z_{n_k}'| \to 0$ as $k \to \infty$, so the subsequence z_{n_k}' also converges to the same point ζ. For simplicity, we may assume the original sequences $z_n \to \zeta$ and $z_n' \to \zeta$ hold. We *claim* that the limit point $\zeta \in \partial\Omega = \gamma$ should hold. If this is not the case, then $\zeta \in \Omega$ and, by continuity of $f(z)$, $|f(z_n) - f(z_n')| \to |f(\zeta) - f(\zeta)| = 0$ contradicting to the fact that $|f(z_n) - f(z_n')| \geq \varepsilon_0$ for each $n \geq 1$.

Set $w_n = f(z_n)$ and $w_n' = f(z_n')$, $n \geq 1$. Since both sequences w_n, $n \geq 1$, and w_n', $n \geq 1$, are bounded, we may assume, as in the last paragraph, that $\lim_{n\to\infty} w_n = \eta$ and $\lim_{n\to\infty} w_n' = \eta'$. Hence, $|\eta| \leq 1$ and $|\eta'| \leq 1$, and

$$|\eta - \eta'| \geq \varepsilon_0 \tag{$*_2$}$$

hold. We *claim* that $|\eta| = |\eta'| = 1$. For otherwise, suppose that $|\eta| < 1$ is true. Then $z_n = f^{-1}(w_n) \to \zeta = f^{-1}(\eta) \in \partial\Omega$, as $n \to \infty$, and this contradicts to the fact that $f^{-1}(\eta) \in \Omega$ since $f(z)$ is univalent from Ω onto $|w| < 1$. Moreover, $(*_2)$ indicates that

$$|w_n - w_n'| \geq \frac{\varepsilon_0}{2}$$

for all sufficiently large n. Hence, it is possible to choose two points w_{01}, w_{01}' in $|w| < 1$ so that the line segments $\overline{w_{01}w_n}$ and $\overline{w_{01}'w_n'}$ have distance between them larger than $\frac{\varepsilon_0}{2}$.

Since γ is a Jordan closed curve, we can construct a continuous curve Γ connecting ζ to a point z_0 in Ω so that Γ lies entirely in Ω except the endpoint ζ. Let C_r be a circle with center at ζ and radius $r > 0$ such that z_0 lies outside it. Let z^* be the first point that the curve Γ meets the circle C_r when a point is starting from ζ and moving along Γ. Also, let ζ_1 and ζ_2 be the first two points that the circle C_r meets the curve γ if a point is starting from z^* and moving along C_r in both directions. Note that the circular arc $\overset{\frown}{\zeta_1\zeta_2}$ lies in Ω except at the endpoints ζ_1 and ζ_2. See Fig. 6.1. Denote by Ω_r the subdomain of Ω bounded by $\overset{\frown}{\zeta_1\zeta_2}$ and the part of γ that contains the point ζ. Let σ and σ' be the preimages in Ω of $\overline{w_{01}w_n}$ and $\overline{w_{01}'w_n'}$ under $w = f(z)$, respectively, so that $z_{01} = f^{-1}(w_{01})$ and $z_{01}' = f^{-1}(w_{01}')$ are their respective endpoints. Choose r small enough so that both z_{01} and z_{01}' lie outside Ω_r, while both z_n and z_n' are in Ω_r if n is large enough. Under these circumstances, let τ_1 and τ_2 be the respective points of intersection of σ and σ' with C_r. Then

$$\frac{\varepsilon_0}{2} < |f(\tau_2) - f(\tau_1)| = \left| \int_{\tau_1}^{\tau_2} f'(z)dz \right| = \left| \int_{\theta_1}^{\theta_2} f'(\zeta + re^{i\theta})rie^{i\theta}d\theta \right|$$

$$(0 \leq \theta_1 < \theta_2 \leq 2\pi) \leq \int_{\theta_1}^{\theta_2} |f'(\zeta + re^{i\theta})|rd\theta$$

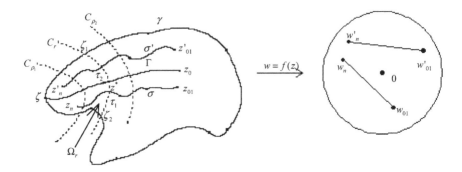

Fig. 6.1

⇒ (by Cauchy–Schwarz's inequality)

$$\frac{\varepsilon_0^2}{4} \leq \int_{\theta_1}^{\theta_2} 1^2 d\theta \cdot \int_{\theta_1}^{\theta_2} |f'(\zeta + re^{i\theta})|^2 r^2 d\theta \leq 2\pi \int_{\theta_1}^{\theta_2} |f'(\zeta + re^{i\theta})|^2 r^2 d\theta$$

⇒ $$\frac{\varepsilon_0^2}{4r} \leq 2\pi \int_{\theta_1}^{\theta_2} |f'(\zeta + re^{i\theta})|^2 r d\theta$$

⇒ (integrating both sides with respect to r from ρ_1 to ρ_2, where
$0 < \rho_1 \leq r \leq \rho_2$)

$$\frac{\varepsilon_0^2}{4} \log \frac{\rho_2}{\rho_1} \leq 2\pi \int_{\rho_1}^{\rho_2} \int_{\theta_1}^{\theta_2} |f'(\zeta + re^{i\theta})|^2 r dr d\theta \leq \iint_\Omega |f'(z)|^2 dx dy$$

$$\leq 2\pi \cdot \text{the area of } f(\Omega) \leq 2\pi^2.$$

On letting $\rho_1 \to 0^+$ and observing that $\log \dfrac{\rho_2}{\rho_1} \to \infty$, we get a contradiction.
Hence, $f(z)$ is uniformly continuous on Ω.

The continuous extension of $f(z)$ from Ω to the bounded closure $\bar{\Omega}$:
For any fixed $\varepsilon > 0$, let $\delta = \delta(\varepsilon) > 0$ be such that, for any two points z_1, $z_2 \in \Omega$,

$$|f(z_1) - f(z_2)| < \varepsilon \qquad (*_3)$$

whenever $|z_1 - z_2| < \delta$.

Choose any point $\zeta \in \partial\Omega$. Let $z_n \in \Omega$ for $n \geq 1$ and $z_n \to \zeta$ as $n \to \infty$.
Then the sequence z_n, $n \geq 1$, is Cauchy. There is an integer $N = N(\delta) > 0$
so that $|z_n - z_m| < \delta$ if $n \geq m \geq N$. According to $(*_3)$, this means that
$|f(z_n) - f(z_m)| < \varepsilon$ if $n \geq m \geq N$. That $f(z_n)$, $n \geq 1$, is Cauchy shows
that $f(z_n)$ does converge in $\mathbf{C}$.

Suppose z_n', $n \geq 1$, is another sequence of points in Ω, converging to the
same boundary point ζ. Then $f(z_n')$, $n \geq 1$, also converges and, as a matter
of fact, converges to the *same limit* as $f(z_n)$ does. To see this, choose an
integer $N > 0$ so that $|z_n - \zeta| < \frac{\delta}{2}$ and $|z_n' - \zeta| < \frac{\delta}{2}$ whenever $n \geq N$. Hence
$|z_n - z_n'| < \delta$ if $n \geq N$ and, then, $(*_3)$ indicates that $|f(z_n) - f(z_n')| < \varepsilon$ if
$n \geq N$. The claimed result thus follows.

In conclusion, we define

$$f(\zeta) = \lim_{\substack{z \in \Omega \\ z \to \zeta}} f(z), \quad \zeta \in \partial\Omega. \qquad (*_4)$$

This very definition indicates that $f(\zeta)$ is continuous from within the
domain Ω.

Let $\varepsilon > 0$ and $\delta > 0$ be as in $(*_3)$. For any two points $\zeta_1, \zeta_2 \in \partial\Omega$ with $|\zeta_1 - \zeta_2| < \frac{\delta}{2}$. There are points $z_1, z_2 \in \Omega$ such that

$$|\zeta_1 - z_1| < \frac{\delta}{4} \quad \text{and} \quad |f(\zeta_1) - f(z_1)| < \frac{\varepsilon}{2};$$

$$|\zeta_2 - z_2| < \frac{\delta}{4} \quad \text{and} \quad |f(\zeta_2) - f(z_2)| < \frac{\varepsilon}{2}$$

$$\Rightarrow |z_1 - z_2| \le |z_1 - \zeta_1| + |\zeta_1 - \zeta_2| + |\zeta_2 - z_1| < \delta, \text{ and hence (by}(*_3)),$$

$$|f(z_1) - f(z_2)| < \varepsilon$$

$$\Rightarrow |f(\zeta_1) - f(\zeta_2)| \le |f(\zeta_1) - f(z_1)| + |f(z_1) - f(z_2)|$$

$$+ |f(z_2) - f(\zeta_2)| < 2\varepsilon$$

whenever $\zeta_1, \zeta_2 \in \partial\Omega$ with $|\zeta_1 - \zeta_2| < \frac{\delta}{2}$. This shows that $f(z)$ is continuous along $\partial\Omega$.

Combining together, we obtain that the extended $f(z)$ is (uniformly) continuous on $\bar{\Omega}$. Moreover if $\zeta \in \partial\Omega$ is a boundary point, then it is necessary that $|f(\zeta)| = 1$. For otherwise, $|f(\zeta)| < 1$ will imply that $f^{-1}(f(\zeta)) = \zeta \in \Omega$, an interior point since $w = f(z)$ maps Ω univalently onto $|w| < 1$.

The homeomorphism of $w = f(z)$ from $\bar{\Omega}$ onto $|w| < 1$:
On applying the above known result to the inverse univalent function $z = f^{-1}(w)$, mapping $|w| < 1$ onto Ω, $z = f^{-1}(w)$ can be (uniformly) continuously extended from $|w| < 1$ to the closure $|w| \le 1$. Also, whenever η is a point on $|w| = 1$, then the extended $z = f^{-1}(w)$ has its value at η a point on the boundary $\partial\Omega$.

Suppose that there are points $\zeta_1, \zeta_2 \in \partial\Omega$, $\zeta_1 \ne \zeta_2$ so that $f(\zeta_1) = f(\zeta_2) = \eta$ lies on $|w| = 1$. Choose a sequence $z_n \in \Omega \to \zeta_1$ and a sequence $z'_n \in \Omega \to \zeta_2$. Then $f(z_n)$ and $f(z'_n)$ are in $|w| < 1$ for each $n \ge 1$. Set

$$w_n = \begin{cases} f(z_m) \text{ if } n = 2m - 1 \text{ for } m = 1, 2, \ldots \\ f(z'_m) \text{ if } n = 2m \text{ for } m = 1, 2, \ldots \end{cases}$$

Then $|w_n| < 1$ for $n \ge 1$ and $w_n \to \eta$ as $n \to \infty$. Yet $f^{-1}(w_n) = z_m$ or z'_m depending on $n = 2m - 1$ or $n = 2m$, respectively, and $f^{-1}(w_n)$ does not converge. This is a contradiction. Hence $w = f(z)$ is one-to-one on $\partial\Omega$ and, then, is one-to-one on $\bar{\Omega}$.

This finishes the proof for (6.1.2) in case $\Omega_1 = \Omega$ and Ω_2 is the open disk $|w| < 1$. $\qquad\square$

The result in (6.1.2) can be suitably extended to a well-behaved sub-curve of the boundary $\partial\Omega$ of a simply connected domain Ω. A Jordan curve γ, contained in the boundary of $\partial\Omega$ as a relative open subset, is said to be an *accessible arc* of $\partial\Omega$ if there is a Jordan curve γ' with the same endpoints as γ, contained in Ω except its end points, such that the interior $\mathrm{Int}(\gamma \cup \gamma')$ is contained in Ω. And the interior $\mathrm{Int}(\gamma \cup \gamma')$ is said to be *an adjacent domain* to γ. It is simply connected. Note that an accessible arc might possess more than one adjacent domain. Take, for instance, the domain $\Omega = \{|z| < 1\} - [0, 1)$. Then the open interval $(0, 1)$ is an accessible arc of $\partial\Omega$ with two disjoint adjacent domains; intuitively speaking, this means that $(0, 1)$ is two-sided. Refer to Remark 1 of Sec. 3.4.6 for related concepts.

Then we have

An extension of (6.1.2). Let Ω be a simply connected domain with more than one boundary point and $w = f(z)$ be a univalent function mapping Ω onto $|w| < 1$. If γ is an accessible arc of $\partial\Omega$ and $D \subseteq \Omega$ is a fixed adjacent domain to γ, then $w = f(z)$ sets up a homeomorphism between γ and an (open) arc σ on $|w| < 1$ such that $w = f(z)$ maps $\Omega \cup \gamma$ homeomorphically onto $\{|w| < 1\} \cup \sigma$.   (6.1.2.1)

This follows easily by applying (6.1.2) to the adjacent domain D.

However, (6.1.2) fails in the case of general simply connected domain. Such an example is provided by the simply connected domain Ω whose boundary $\partial\Omega = \{|z| = 1\} \cup \{|z| = 2\} \cup \{\text{the spiral } r = \frac{3}{2} + \frac{1}{\pi}\tan\theta\}$ (see Fig. 2.10(d)). Then any function, mapping Ω univalently onto $|w| < 1$, cannot be continuously extended to $\bar\Omega$. For details, see Ref. [58], Vol. III, p. 91.

In what follows, we prove two important applications of the symmetry principle (see Sec. 3.4.6). First, we have

A uniqueness mapping theorem. *Let γ_1 and γ_2 be two rectifiable Jordan closed curves.*

(1) *Suppose $w = f(z)$ maps the Jordan domain $\mathrm{Int}\,\gamma_1$ univalently on the Jordan domain $\mathrm{Int}\,\gamma_2$. Then $f(z)$ preserves the orientation of γ_1 and γ_2.*

(2) *Fix three distinct points $z_1, z_2, z_3 \in \gamma_1$ and three distinct points $w_1, w_2, w_3, \in \gamma_2$, such that they induce the same orientation on both γ_1 and γ_2. Then there exists a unique function $w = f(z)$ mapping $\mathrm{Int}\,\gamma_1$ univalently onto $\mathrm{Int}\,\gamma_2$ so that $f(z_k) = w_k$ for $k = 1, 2, 3$.*   (6.1.2.2)

The existence of such an f in (2) is left as an exercise to the readers.

Proof. By (6.1.2), $w = f(z)$ is a homeomorphism between $\overline{\mathrm{Int}\,\gamma_1}$ and $\overline{\mathrm{Int}\,\gamma_2}$; in particular, between γ_1 and γ_2. On applying the generalized Cauchy integral theorem $(3.4.2.4)'$ to the Argument principle $(3.5.3.4)$, we have, for $w_0 \in \text{Int}\,\gamma_2$,

$$\frac{1}{2\pi i}\int_{\gamma_2}\frac{dw}{w-w_0} = \frac{1}{2\pi i}\int_{\gamma_1}\frac{f'(z)dz}{f(z)-w_0},$$

which is equal to 1 or -1 (see Exercises B(1) of Sec. 2.7.1 and (5) in $(3.5.2.4)$) because γ_2 is a Jordan closed curve. Therefore, $w = f(z)$ traverses γ_2 in the same direction as z traverses γ_1. This is (1).

As for (2), consider first the cases where γ_1 and γ_2 are circles $|z| < 1$ and $|w| < 1$, respectively. Suppose there are two such functions $w = f(z)$ and $w = g(z)$, mapping $|z| < 1$ univalently onto $|w| < 1$ and transforming z_k into w_k for $k = 1, 2, 3$. According to $(6.1.2)$, both $f(z)$ and $g(z)$ are homeomorphic between $|z| \le 1$ and $|w| \le 1$; and by the symmetry principle $(3.4.6.2)$, both can be analytically continued across the unit circle $|z| = 1$ into $|z| > 1$. Therefore, the extended $f(z)$ and $g(z)$ are univalent meomorphic functions in $\mathbf{C}^*$. According to Example 6 and Exercise A(6) of Sec. 4.10.2, they must be linear fractional transformations. Since $f(z_k) = g(z_k)$ for $k = 1, 2, 3$, it follows that $f(z) \equiv g(z)$ throughout $\mathbf{C}$ (see $(2.5.4.13)$).

Returning to the general cases, let $f(z)$ and $g(z)$ be two such functions, mapping $\text{Int}\,\gamma_1$ univalently onto $\text{Int}\,\gamma_2$ and satisfying $f(z_k) = g(z_k) = w_k$ for $k = 1, 2, 3$. Also, let $\zeta = \varphi(z)$ and $\eta = \psi(w)$ be two functions mapping $\text{Int}\,\gamma_1$ and $\text{Int}\,\gamma_2$ univalently onto $|\zeta| < 1$ and $|\eta| < 1$, respectively. Set $\zeta_k = \varphi(z_k)$ and $\eta_k = \psi(w_k)$ for $k = 1, 2, 3$. Then

$$\eta = F(\zeta) = \psi \circ f \circ \varphi^{-1}(\zeta) \quad \text{and} \quad \eta = G(\zeta) = \psi \circ g \circ \varphi^{-1}(\zeta)$$

both map $|z| < 1$ univalently onto $|\eta| < 1$ and carry ζ_k into η_k for $k = 1, 2, 3$. By what we have just obtained in the last paragraph, $F(\zeta) \equiv G(\zeta)$ on $|\zeta| < 1$ which, in turn, implies that $f(z) \equiv g(z)$ on $\text{Int}\,\gamma_1$. The proof is finished. $\qquad\square$

Remark 1 (An extension of (6.1.2.2)). Suppose γ_1 and γ_2 are unbounded locally rectifiable Jordan closed curves (i.e., passing the point at infinity). Let Ω_k be either of the simply connected domains with boundary γ_k for $k = 1, 2$. Then $(6.1.2.2)$ remains valid for a univalent function $w = f(z)$ mapping Ω_1 onto Ω_2 and carrying three distinct points along γ_1 into three distinct points along γ_2.

This is easily seen by invoking preliminary linear fractional transformations carrying Ω_1 and Ω_2 onto bounded Jordan domains. $\qquad\square$

Second, we come back to (3.4.6.3) and restate it as

A generalization of the symmetry principle. Let Ω be a simply connected domain whose boundary contains a regular, free one-sided, analytic Jordan arc γ. Let $w = f(z)$ be a function mapping Ω univalently onto a disk or a half-plane. Then

(1) $f(z)$ has an analytic extension to $\Omega \cup \gamma$, namely, can be continued analytically across γ, and

(2) γ is mapped one-to-one and onto an (open) arc of the circle or an (open) segment of the line.   (6.1.2.3)

Readers should review Remark 1 in Sec. 3.4.6 for terminologies concerned. This result remains valid even if γ is a regular, free two-sided analytic arc or an accessible regular analytic arc.

(6.1.2.3) is an obvious consequence of (6.1.2.1) and the symmetry principle (3.4.6.2), and, of course, the definition of an analytic curve.

However, we try to give a direct proof based on (3.4.6.2).

Direct proof. Consider the special case that γ is an open interval on a certain line. Because rotations are unimportant, we may as well suppose that γ is the interval $a < x < b$ on the real axis and Ω lies in the upper half-plane. Let $z_0 \in \Omega$ be the point such that $f(z_0) = 0$ holds (implicitly, this means $f(z)$ maps Ω onto $|w| < 1$).

For each point $x \in \gamma$, consider a disk $D(x)$ with center at x so small that $z_0 \notin D(x)$. Set $D^+(x) = D(x) \cap \Omega$. Then $f(z) \neq 0$ on $D^+(x)$ and then choose a single-valued branch $\log f(z) = F_x(z)$ on $D^+(x)$. Since $\operatorname{Re} F_x(z) = \log|f(z)| \to 0$ as $z \to x$ for $|f(z)| \to 1$, the symmetry principle (3.4.6.2) implies that there is an analytic extension, still denoted by $F_x(z)$, to all of $D(x)$. Define

$$g_x(z) = e^{F_x(z)}, \quad z \in D(x).$$

Then $g_x(z)$ is analytic in $D(x)$ and $g_x(z) = f(z)$ in $D^+(x)$.

Set $D = \Omega \cup \bigcup_{x \in \gamma} D(x)$. Then D is a domain. Define

$$g(z) = \begin{cases} f(z), & \text{if } z \in \Omega - \bigcup_{x \in \gamma} D(x) \\ g_x(z), & \text{if } z \in \bigcup_{x \in \gamma} D(x) \quad \text{and} \quad z \in D(x) \end{cases}. \qquad (*_5)$$

This $g(z)$ is well-defined. For, if $z \in D^+(x) \cap D^+(x')$, then $g_x(z) = f(z) = g_{x'}(z)$; by the interior uniqueness theorem (see (4) in (3.4.2.9)), $g_x(z) = g_{x'}(z)$ throughout $D(x) \cap D(x')$. Hence $g(z)$ is single-valued and analytic in D. This proves (1).

Note that, by (1) of (3.4.6.5) or (6.1.2), $|g(x)| = 1$ if $x \in \gamma$. This implies that $g'(x) \neq 0$ for each $x \in \gamma$. Otherwise, if $g'(x_0) = 0$ for some $x_0 \in \gamma$, then $g(x_0)$ is a multiple value and the two subinterval of γ that are divided by x_0 will be mapped onto two arcs on $|w| = 1$ that form an angle $n\pi$ with $n \geq 2$ (see Section Two of Sec. 3.5.1; in particular, the geometric meaning of (3.5.1.6)). This is clearly impossible.

In $\bigcup_{x \in \gamma} D(x)$, $g(z) \neq 0$ always holds and hence, $\log g(z)$ has single-valued branches. Fix such a branch, still denoted by $g(z)$, and set $u(z) = \log |g(z)|$ and $v(z) = \operatorname{Arg} g(z)$. In this case, if $x \in \gamma$,

$$\frac{\partial u}{\partial x}(x) = \lim_{h \to 0} \frac{u(x+h) - u(x)}{h} = \lim_{h \to 0} \frac{0 - 0}{h} = 0 = \frac{\partial v}{\partial y}(x);$$

$$\frac{\partial u}{\partial y}(x) = \lim_{y \to 0^+} \frac{u(x+iy) - u(x)}{y} \leq 0 \quad \text{(recall that } \Omega \text{ lies in } \operatorname{Im} z > 0\text{).}$$

$$\Rightarrow \frac{\partial u}{\partial y}(x) = -\frac{\partial v}{\partial x}(x) \leq 0 \quad \text{along } \gamma.$$

If $\frac{\partial v}{\partial x}(x_0) = 0$ for some $x_0 \in \gamma$, then

$$\frac{d}{dz} \log g(z)|_{z=x_0} = \frac{\partial u}{\partial x}(x_0) + i\frac{\partial v}{\partial x}(x_0) = 0 = \frac{g'(x_0)}{g(x_0)}$$

$$\Rightarrow g'(x_0) = 0,$$

a contradiction. Consequently,

$$\frac{\partial v}{\partial x}(x) = \frac{\partial}{\partial x}\operatorname{Arg} g(x) > 0 \quad \text{along } \gamma. \tag{$*_6$}$$

This means $\operatorname{Arg} g(x)$ is strictly increasing along γ (and hence moves constantly in the same direction). Thus $w = g(z)$ maps γ one-to-one and onto an open arc on the circle $|w| = 1$. This proves (2).

Return to the general case that γ is a regular, free one-sided, analytic Jordan curve. Recall (see Remark 1 in Sec. 3.4.6) that there is a function $\varphi(t) : (a, b) \to \mathbf{C}$ satisfying:

1. For each $t_0 \in (a, b)$, there is an open interval $N(t_0) = (t_0 - \rho, t_0 + \rho) \subseteq (a, b)$ such that $\varphi(t) = \sum_{n=0}^{\infty} a_n(t - t_0)^n$ converges in $N(t_0)$, i.e., $\varphi(t)$ is *real analytic*.
2. $\varphi'(t) \neq 0$ for all $t \in (a, b)$, i.e., $\varphi(t)$ is *regular*.
3. $\varphi(t_1) = \varphi(t_2)$ if and only if $t_1 = t_2 \in (a, b)$, i.e., $\varphi(t)$ is *Jordanian* or *simple*.

Indeed, $\varphi(z)$ can be defined as the sum of the convergent series $\sum_{n=0}^{\infty} a_n(z - t_0)^n$ in the open disk $D(t_0) = \{|z - t_0| < \rho\}$ (see (3.3.2.4)). Hence, $\varphi(z)$ is analytic in $D(t_0)$ for each $t_0 \in (a, b)$. Set

$$D = \bigcup_{t_0 \in (a,b)} D(t_0).$$

Then D is a domain, symmetric with respect to the real axis. By exactly the same argument as in $(*_5)$, $\varphi(z)$ is well-defined and analytic in D. Under these circumstances, $\varphi(t)$ is said to determine a *regular analytic Jordan curve* γ whose locus is the set $\varphi(a, b)$.

Since γ is free and one-sided as part of the boundary $\partial\Omega$ of Ω, we may suppose that $\varphi(z) \in \Omega$ if $z \in D^+ = D \cap \{\operatorname{Im} z > 0\}$, while $\varphi(z) \notin \Omega$ if $z \in D^- = D \cap \{\operatorname{Im} z < 0\}$. Moreover, we may suppose that $\varphi(z) \neq z_0$ for all $z \in \Omega$, where $z_0 \in \Omega$ is the point that $f(z_0) = 0$ holds.

Now, $f \circ \varphi(z) \neq 0$ on D^+. Fix a single-valued branch $\log f \circ \varphi(z)$ on D^+ and analytically continue it across (a, b) into the lower half D^-, by using the symmetry principle (3.4.6.2). Denote by $G(z)$ the extended function on D. In case $t \in (a, b)$, $\varphi'(t) \neq 0$, and therefore, φ has an analytic inverse φ^{-1} in a neighborhood $N(\varphi(t))$ (see (3.4.7.1) or (3.5.1.4)).

Define

$$F(z) = e^G \circ \varphi^{-1}(z), \quad z \in N(\varphi(t)) \quad \text{for each } t \in (a, b).$$

Then $F(z)$ is well-defined and analytic in $\bigcup_{t \in (a,b)} N(\varphi(t))$, a domain containing γ. When $z \in \Omega \cap N(\varphi(t))$ is restricted, then $\varphi^{-1}(z) \in D^+$ holds and

$$F(z) = e^G \circ \varphi^{-1}(z) = f \circ \varphi(z) \circ \varphi^{-1}(z) = f(z).$$

Thus $F(z)$ is indeed an analytic extension of $f(z)$ from Ω to $\Omega \cup \gamma$.

The same argument as in the special case (see $(*_5)$) will show that $w = F(z)$ maps γ one-to-one and onto an open circular arc of the unit circle $|w| = 1$.   □

Right at this moment, readers are asked to be familiar with the content of Example 3 in Sec. 3.5.5 for the mere purpose of the following

Example. Two rectangles R_1 and R_2 are conformally equivalent so that the two set of four vertices correspond to each other if and only if their respective ratios of the altitude to the base are equal, namely, R_1 and R_2 are similar.

Solution. Let R be a rectangle with vertices at $-a$, a, $a + bi$, $-a + bi$ where $a > 0$ and $b > 0$. Let R' be another rectangle with vertices at $-a'$, a',

$a' + b'i$, $-a' + b'i$ where $a' > 0$ and $b' > 0$. These assumptions are permitted because translation and rotation does not alter the length of segments.

The necessity: Suppose $w = f(z)$ maps R onto R' univalently such that the vertices in that ordering corresponds to each other. According to (6.1.2.3), $w = f(z)$ can be analytically continued into the whole plane as a univalent entire function. Note that $w = f(z)$ has removable singularities at points $ma + nbi$ for $m, n = 0, \pm 1, \pm 2, \ldots$ (why?). ∞ is a simple pole of $f(z)$ owing to the univalence and hence $f(z)$ is of the form $\alpha z + \beta$, where $\alpha \neq 0$ and β are constants. $f(a) = a'$ and $f(-a) = -a'$ shows that $\beta = 0$. While, $f(a + bi) = \alpha(a + bi) = a' + b'i$ shows that $\frac{b}{2a} = \frac{b'}{2a'}$ holds. Hence R and R' are similar.

The sufficiency: Suppose $\frac{b}{2a} = \frac{b'}{2a'} = \alpha$ holds. Then $w = \frac{a'}{a}z$ maps R univalently onto R'.

If R and R' are situated in their general positions, then a similarity transformation $w = az + b$, $|a| = 1$, will work.

Remark 2. Let Ω be a simply connected domain whose boundary $\partial\Omega$ contains an accessible arc (or four distinct accessible points). Then Ω can be mapped univalently onto the upper half-plane in such a way that the four given accessible boundary points go into four distinct points a_1, a_2, a_3, a_4 on the real axis. According to Exercise B(2) of Sec. 2.5.4, there is a linear fractional transformation mapping the upper half-plane onto itself so that a_1, a_2, a_3, a_4 are mapped into $-1, k, k, 1$ respectively, where k depends on these four points. Finally, Example 3 of Sec. 3.5.5 shows that

$$w = f(z) = \int_0^z \frac{d\zeta}{\sqrt{(1 - \zeta^2)(1 - k^2\zeta^2)}} \quad (0 < k < 1)$$

maps the upper half-plane univalently onto the rectangle with vertices $\pm K$, $\pm K + iK'$ in such a way that $-\frac{1}{k}, -1, 1, \frac{1}{k}$ are mapped into $-K + iK', -K, K, K + iK'$, respectively.

In conclusion, *such* a simply connected domain Ω can always be mapped univalently onto a rectangle with four preassigned accessible boundary points going into the vertices. □

6.2   Conformal Mapping of Polygons: The Schwarz–Christoffel Formulas

A *polygon* is a simply connected domain in $\mathbf{C}^*$ whose boundary is a closed polygonal curve without self-intersection. The Riemann mapping theorem (6.1.1) says that a polygon can always be mapped univalently onto an open disk or a half-plane.

The Schwarz–Christoffel formula gives the mapping function needed an explicit analytic expression (Sec. 6.2.1). Example 3 in Sec. 3.5.5 shows how the upper half-plane is mapped univalently onto a rectangle. More illustrative examples will be presented in Sec. 6.2.2, including the triangle functions of Schwarz. The final Sec. 6.2.3 will devote briefly to *unbounded* polygon whose boundary is a "generalized polygonal curve" in $\mathbf{C}^*$, namely, it sides may be traversed more than once and are allowed to pass the point at infinity. Section 6.2.1 will also discuss the case for *unbounded* polygon with ∞ as an interior point or with ∞ acting as a vertex.

There is a corresponding theory for curvilinear polygons whose sides compose of circular arcs. Unfortunately, considerations of space do not permit us to include them here. Interested reader should refer to Refs. [63], pp. 198–209, [35] or [44], Vol. III, pp. 375–384.

6.2.1   *The Schwarz–Christoffel formulas for polygons*

Section (1) A bounded polygon with finite vertices

Let Ω be a *bounded* polygon with vertices $w_1, \ldots, w_n$ in the w-plane. The consecutive vertices $w_1, \ldots, w_n$ (with $w_{n+1} = w_1$) is arranged in the *positive* cyclic order so that the induced orientation is consistent with the *positive orientation* of the boundary $\partial\Omega$. The latter means that, if walking along $\partial\Omega$, the domain Ω is always situated to the left side. The (inner) *angle* at w_k is defined as the value $\arg \frac{w_{k-1} - w_k}{w_{k+1} - w_k}$ *between* 0 *and* 2π and is denoted by $\alpha_k \pi$, $0 < \alpha_k < 2$, for $1 \le k \le n$. The number $\beta_k \pi = (1 - \alpha_k)\pi$, $-1 < \beta_k < 1$, is called the *outer angle* at w_k, for convenience. See Fig. 6.2. Note that $\alpha_1 + \cdots + \alpha_n = n - 2$, while $\beta_1 + \cdots + \beta_n = 2$. And the polygon is *convex* if and only if $\beta_k > 0$ for $1 \le k \le n$.

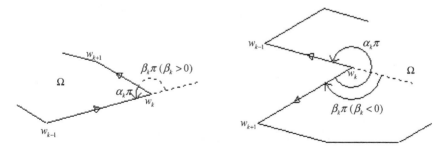

Fig. 6.2

The open unit disk $|z| < 1$ and the open half-plane $\mathrm{Im}\, z > 0$ are well known to be conformally equivalent via Linear fractional transformation (see Example 1 in Sec. 2.5.4). And such a transformation map $|z| \le 1$ homeomorphically onto $\mathrm{Im}\, z \ge 0$ so that the circle $|z| = 1$ is homeomorphic to the extended real axis (refer to (6.1.2.2) or (6.1.2.3)). Hence we will focus our attention between the half-plane $\mathrm{Im}\, z > 0$ in the z-plane and the (bounded) polygon Ω in the w-plane in what follows until (6.2.1.5) where we restate formulas obtained in terms of $|z| < 1$ and Ω.

Our main *theme* is to formulate *explicitly* a function $w = f(z)$ mapping $\mathrm{Im}\, z > 0$ univalently onto Ω that sets up a homeomorphism between $\mathrm{Im}\, z \ge 0$ and $\bar{\Omega}$. The existence of such a function is provided by (6.1.1) and (6.1.2) or (6.1.2.3).

Let $a_1, \ldots, a_n$ be the distinct points, none of them is the point ∞, on the real axis that are mapped under $w = f(z)$ onto the vertices $w_1, \ldots, w_n$, respectively. Note that $-\infty < a_1 < \cdots < a_n < +\infty$ owing to consistence of orientation on $\mathrm{Im}\, z = 0$ and $\partial \Omega$. Denote by

$\gamma_k = (a_k, a_{k+1})$, $1 \le k \le n - 1$, the open interval along the real axis;

$\gamma_n = (a_n, a_1) = (a_n, +\infty) \cup \{\infty\} \cup (-\infty, a_1)$, the open interval, with

endpoints a_n and a_1, that passes through the point ∞, and

$\Gamma_k = \overline{w_k w_{k+1}}$, $1 \le k \le n$, the open side component of $\partial \Omega$ that

connects w_k to w_{k+1}. $(*_1)$

Be aware again that $w_{n+1} = w_1$. $w = f(z)$ maps γ_k one-to-one and onto Γ_k. And, by the symmetry principle (see (6.1.2.3)), $w = f(z)$ can be continued analytically across γ_k into the lower half-plane. Thus, for each $k(1 \le k \le n)$, we obtain a function $F_k(z)$ satisfying the following properties:

1. $F_k(z)$ is analytic in $\{\mathrm{Im}\, z > 0\} \cup \gamma_k \cup \{\mathrm{Im}\, z < 0\}$;
2. $F_k(z) = f(z)$ if $\mathrm{Im}\, z > 0$, and $F_k(z)$ is the symmetric point of $f(\bar{z})$ with respect to Γ_k if $\mathrm{Im}\, z < 0$; and
3. $F_k' \ne 0$ in $\{\mathrm{Im}\, z > 0\} \cup \gamma_k \cup \{\mathrm{Im}\, z < 0\}$. $(*_2)$

Property 2 means that $w = F_k(z)$ preserves the symmetry. As for 3, since $F_k(z)$ is also univalent in $\mathrm{Im}\, z < 0$, $F_k' \ne 0$ holds there; while, in a neighborhood of each point on γ_k, $F_k(z)$ is univalent and hence $F_k' \ne 0$ along γ_k. It is possible that $F_k(z)$ is not more univalent in $\{\mathrm{Im}\, z > 0\} \cup \gamma_k \cup \{\mathrm{Im}\, z < 0\}$. This phenomenon does really happen as indicated in Fig. 3.41.

Combining together, we have n extended functions $F_1(z), \ldots, F_n(z)$, in total. They all agree with the original $f(z)$ in $\mathrm{Im}\, z > 0$; yet, distinct in

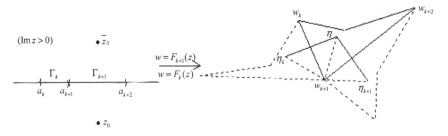

Fig. 6.3

general in $\operatorname{Im} z < 0$. As a matter of fact, any two consecutive of them (in this ordering) differ only by an entire linear transformation. To see this, fix any point z_0 in $\operatorname{Im} z < 0$. Then $\overline{z_0}$ is in $\operatorname{Im} z > 0$. Set

$$\eta = F_k(\overline{z_0}) = F_{k+1}(\overline{z_0}) = f(\overline{z_0}), \quad \eta_k = F_k(z_0), \quad \text{and} \quad \eta_{k+1} = F_{k+1}(z_0)$$

for each $k(1 \leq k \leq n)$. According to property 2 in $(*_2)$, η and η_k are symmetric with respect to the side Γ_k, while η and η_{k+1} are symmetric with respect to Γ_{k+1}. See Fig. 6.3. Elementary geometric consideration shows that $\eta_k - w_{k+1} = e^{2\pi i \alpha_{k+1}}(\eta_{k+1} - w_{k+1})$, namely,

$$F_k(z) - w_{k+1} = e^{2\pi i \alpha_{k+1}}(F_{k+1}(z) - w_{k+1}) \quad \text{for all } z \text{ in } \operatorname{Im} z < 0$$

$$\Rightarrow F_k(z) = A_k F_{k+1}(z) + B_k, \quad \operatorname{Im} z < 0, \tag{$*_3$}$$

where A_k and B_k are constants, depending on $k(1 \leq k \leq n)$. On differentiating both sides of $(*_3)$, we have

$$F'_k(z) = A_k F'_{k+1}(z), \; F''_k(z) = A_k F''_{k+1}(z), \quad \operatorname{Im} z < 0$$

$$\Rightarrow \frac{F''_k(z)}{F'_k(z)} = \begin{cases} \dfrac{F''_{k+1}(z)}{F'_{k+1}(z)}, & \operatorname{Im} z < 0 \text{ for } 1 \leq k \leq n \quad (*_4) \\[2ex] \dfrac{f''(z)}{f'(z)}, & \operatorname{Im} z > 0 \quad\quad\quad\quad\quad (*_5) \end{cases}$$

Note that $(*_3)$ and, hence, $(*_4)$ are still valid on γ_k for $1 \leq k \leq n$, owing to the fact that $f(z)$, $f'(z)$, and $f''(z)$ are simultaneously analytically continuable (see (5.2.1.3)). In conclusion, we obtain a function $g(z)$ satisfying the conditions:

1. $g(z)$ is analytic in $\mathbf{C}$ except at isolated singularities $a_1, \ldots, a_n$;
2. $g(z) = \frac{f''(z)}{f'(z)}$, $\operatorname{Im} z > 0$ (remember that $g(z)$ is given by $(*_4)$ in case $\operatorname{Im} z < 0$). $\tag{$*_6$}$

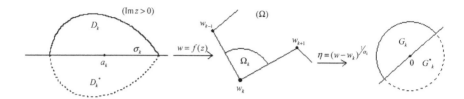

Fig. 6.4

What is next is to *show* that each of the points $a_1, \ldots, a_n$ is a simple pole of $g(z)$ and that ∞ is a removable singularity of it.

Fix a vertex w_k ($1 \le k \le n$). Consider a sector domain Ω_k with vertex at w_k and two sides along Γ_{k-1} and Γ_k, respectively. Set $D_k = f^{-1}(\Omega_k)$. Then D_k is a simply connected domain in $\operatorname{Im} z > 0$ whose boundary contains an open interval σ_k, along the real axis and passing the point a_k. See Fig. 6.4. Note that $w = f(z)$ maps $\overline{D_k}$ homeomorphically onto $\overline{\Omega_k}$; in particular, it maps σ_k one-to-one and onto the two sides of Ω_k. A fixed single-valued branch of $\eta = (w - w_k)^{1/\alpha_k}$ (see (2.7.3.4)) maps Ω_k univalently onto an open half disk G_k with center at 0, and homeomorphically from $\overline{\Omega_k}$ onto $\overline{G_k}$. The composite map

$$\eta = \varphi_k(z) = (f(z) - w_k)^{1/\alpha_k} \tag{$*_7$}$$

maps D_k univalently onto G_k, $\overline{D_k}$ homeomorphically onto $\overline{G_k}$ and segment σ_k onto the diameter of G_k. Note that $\varphi_k(a_k) = 0$. By the symmetry principle (3.4.6.3) or using (6.1.2.3) directly, $\eta = \varphi_k(z)$ can be analytically continued from D_k across σ_k into its symmetric domain D_k^* in $\operatorname{Im} z < 0$. Still denote the extended function by $\eta = \varphi_k(z)$ which is univalent in $D_k \cup \sigma_k \cup D_k^*$. Hence in an open neighborhood O of a_k, $\eta = \varphi_k(z)$ can be factored as

$$\varphi_k(z) = (z - a_k)\psi_k(z),$$

where $\psi_k(z)$ is analytic in O and $\psi_k(z) \ne 0$ throughout O (see (3.4.2.9), etc.). Via ($*_7$), we thus obtain, in O,

$$f(z) = w_k + (z - a_k)^{\alpha_k} h_k(z), \quad h_k(z) = \psi_k(z)^{\alpha_k} \quad \text{(a branch in } O)$$

$$\Rightarrow g(z) = \frac{f''(z)}{f'(z)} = \frac{\alpha_k - 1}{z - a_k} + \frac{h'(z)}{h(z)}, \quad z \in O - \{a_k\}. \tag{$*_8$}$$

This means that a_k is a simple pole of $g(z)$ with residue $\operatorname{Res}(g; a_k) = \alpha_k - 1$; moreover, $\frac{\alpha_k - 1}{z - a_k}$ is the principle part of the Laurent series expansion of $g(z)$ at a_k because $\frac{h'(z)}{h(z)}$ is analytic at a_k.

To prove that ∞ is a removable singularity of $g(z)$, note first that $w = f(z)$ is a bounded function in $\operatorname{Im} z > 0$. Hence the function $w = F_n(z)$ (see $(*_2)$) is bounded in $\{\operatorname{Im} z > 0\} \cup \gamma_n \cup \{\operatorname{Im} z < 0\}$; in particular, ∞ is thus a removable singularity of $w = F_n(z)$. In an open neighborhood O of ∞, since $F_n(z)$ is univalent,

$$F_n(z) = a_0 + \frac{a_{-1}}{z} + \cdots, \quad a_{-1} \neq 0$$

$$\Rightarrow F_n'(z) = -\frac{1}{z^2} h(z), \quad \text{where } h(z) \text{ is analytic in } O \text{ and } h(\infty) = a_{-1} \neq 0.$$

$$\Rightarrow \frac{F_n''(z)}{F_n'(z)} = -\frac{2}{z} + \frac{h'(z)}{h(z)},$$

where $\dfrac{h'(z)}{h(z)}$ is analytic in a neighborhood of ∞.

$$\Rightarrow (\text{see } (*_5) \text{ and } (*_6)) \ \lim_{z \to \infty} g(z) = 0 \ (\text{recall that } h'(\infty) = 0). \qquad (*_9)$$

Hence ∞ is a removable singularity of $g(z)$ (see (4.10.1.2)).

Finally, consider the entire function

$$G(z) = g(z) - \sum_{k=1}^{n} \frac{\alpha_k - 1}{z - a_k}.$$

Since $G(\infty) = 0$ (see $(*_9)$), $G(z)$ is bounded and hence, is the constant zero function. Consequently,

$$g(z) = \sum_{k=1}^{n} \frac{\alpha_k - 1}{z - a_k}, \quad z \in \mathbf{C}^* - \{a_1, \ldots, a_n\}$$

$$\Rightarrow (\text{see } (*_6)) \frac{f''(z)}{f'(z)} = \sum_{k=1}^{n} \frac{\alpha_k - 1}{z - a_k}$$

$$\text{which is analytic everywhere in } \operatorname{Im} z > 0. \qquad (*_{10})$$

Fix any point z_0 in $\operatorname{Im} z \geq 0$ and integrate the above expression along any rectifiable curve in $\operatorname{Im} z > 0$, connecting z_0 to a point z ($\operatorname{Im} z > 0$). We obtain

$$\log f'(z) = \sum_{k=1}^{n} (\alpha_k - 1) \log(z - a_k) + a,$$

where a is a constant, and the $\log(z - a_k)$, $1 \leq k \leq n$, are some fixed branches. Therefore,

$$f'(z) = c_1 \prod_{k=1}^{n} (z - a_k)^{\alpha_k - 1}, \quad c_1 = e^a$$

and another integration yields the required representation (6.2.1.1) below.

In conclusion, we obtain

The Schwarz–Christoffel formula for a bounded polygon with finite vertices.
Let Ω be a bounded polygon with finite vertices $w_1, \ldots, w_n$ which, in this
ordering, induce the positive orientation of the boundary $\partial\Omega$. Let $\alpha_k\pi$ ($0 <$
$\alpha < 2$) be the interior angle at the vertex w_k for $1 \leq k \leq n$ so that
$\beta_k\pi = (1 - \alpha_k)\pi$ ($-1 < \beta_k < 1$) is the corresponding exterior angle (see
Fig. 6.2). Then the functions $w = f(z)$ which map $\operatorname{Im} z > 0$ univalently
onto Ω are of the form

$$w = f(z) = c_1 \int_{z_0}^{z} \prod_{k=1}^{n} (\zeta - a_k)^{\alpha_k - 1} d\zeta + c_2, \quad \operatorname{Im} z > 0,$$

where z_0 is any fixed point in $\operatorname{Im} z \geq 0$, $c_1 \neq 0$, and c_2 are complex constants, and $a_1, \ldots, a_n$ are points on the real axis, with increasing ordering
$-\infty < a_1 < \ldots < a_n < +\infty$, and $w_k = f(a_k)$, $1 \leq k \leq n$. Recall that the
sum of the interior angles is

$$(\alpha_1 + \alpha_2 + \cdots + \alpha_n)\pi = (n - 2)\pi. \tag{6.2.1.1}$$

The constants c_1 and c_2 are determined by the size and position of the
polygon Ω in the plane $\mathbf{C}$ (subject to a similarity transformation). Of the
n points $a_1, \ldots, a_n$, three can be chosen arbitrarily to correspond to three
of the *given* vertices $w_1, \ldots, w_n$ provided that they have the same ordering as the corresponding vertices. When this is the case, such a mapping
is uniquely determined according to (6.1.2.2): the other vertices are then
uniquely determined by Ω; in particular, by the following $(n - 2)$ independent conditions

$$|c_1| \left| \int_{a_j}^{a_{j+1}} \prod_{k=1}^{n} (t - a_k)^{a_k - 1} dt \right|$$

$$= \text{the length the side } \Gamma_j \quad (\text{see } (*_1)), \quad 1 \leq j \leq n - 2. \tag{6.2.1.2}$$

For an immediate illustration, see Example 3 in Sec. 3.5.5; for more, see
Sec. 6.2.2.

Section (2) An unbounded polygon with finite vertices and having ∞ as an interior point

Let Ω, $w_1, \ldots, w_n$, and $a_1, \ldots, a_n$ be as in (6.2.1.1) except that Ω is
unbounded now with ∞ as an interior point. *Note* that, in this case, the
exterior $\operatorname{Ext}\Omega$ is a bounded polygon with the same vertices $w_1, \ldots, w_n$.
These vertices are in clockwise ordering when viewed from $\operatorname{Ext}\Omega$.

As before, $g(z)$ has a simple pole with residue $\alpha_k - 1$ at every a_k for $1 \le k \le n$ (see $(*_8)$). But now $g(z)$ has two extra poles at $b = f^{-1}(\infty)$ and its conjugate $\bar{b}$ (refer to $(*_6)$). The univalence of $w = f(z)$ shows that b is necessary a simple pole. Hence

$$f(z) = \frac{a_{-1}}{z - b} + \cdots \quad \text{with } a_{-1} \neq 0, \quad \text{in a neighborhood of } b.$$

$$\Rightarrow g(z) = \frac{f''(z)}{f'(z)} = -\frac{2}{z - b} + \text{regular part, in a neighborhood of } b. \quad (*_{11})$$

Hence $g(z)$ has a simple pole at b with residue -2. On the other hand, any $F_k(z)$ in $(*_2)$ also has a simple pole at $\bar{b}$ and it follows that

$$g(z) = \frac{F_k''(z)}{F_k'(z)} = -\frac{2}{z - \bar{b}} + \text{regular part, in a neighborhood of } \bar{b}. \quad (*_{11})'$$

This means that $g(z)$ also has a simple pole at $\bar{b}$ with residue -2. Similar argument leading to $(*_{10})$ shows that

$$\frac{f''(z)}{f'(z)} = \sum_{k=1}^{n} \frac{\alpha_k - 1}{z - \alpha_k} - \frac{2}{z - b} - \frac{2}{z - \bar{b}}, \quad \text{Im } z > 0. \quad (*_{12})$$

Integrating this expression twice, we obtain

The Schwarz–Christoffel formula for an unbounded polygon with finite vertices and having ∞ as an interior point. Let Ω be such a polygon, and $w_1, \ldots, w_n$ and $a_1, \ldots, a_n$ be as in (6.2.1.1). Then any function $w = f(z)$ mapping $\text{Im } z > 0$ univalently onto Ω is of the form

$$w = f(z) = c_1 \int_{z_0}^{z} \frac{\prod_{k=1}^{n} (\zeta - a_k)^{\alpha_k - 1}}{(\zeta - b)^2 (\zeta - \bar{b})^2} \, d\zeta + c_2,$$

where z_0 in $\text{Im } z \ge 0$ is fixed point and $b = f^{-1}(\infty)$ is the inverse image in $\text{Im } z > 0$ of the point ∞ under $w = f(z)$. In this case, the sum of the interior angles is

$$(\alpha_1 + \cdots + \alpha_n)\pi = (n + 2)\pi. \quad (6.2.1.3)$$

The last identity follows from the fact that the sum of *all* the residues is equal zero (see (4.11.1.7)). In this case, by $(*_8)$, $(*_9)$, $(*_{11})$, and $(*_{11})'$, we have

$$\sum_{k=1}^{n} \text{Res}\left(\frac{f''}{f'}; a_k\right) + \text{Res}\left(\frac{f''}{f'}; \infty\right) + \text{Res}\left(\frac{f''}{f'}; b\right) + \text{Res}\left(\frac{f''}{f'}; \bar{b}\right)$$

$$= \sum_{k=1}^{n} (\alpha_k - 1) - (-2) - 2 - 2 = 0. \quad (*_{13})$$

Section (3) A polygon with ∞ corresponding to one of its vertices

Now, suppose $a_n = \infty$ in (6.2.1.1).

Choose any fixed point $a < a_1$. The transformation

$$\zeta = \frac{1}{a - z} \qquad (*_{14})$$

maps $\operatorname{Im} z > 0$ univalently onto $\operatorname{Im} \zeta > 0$, the real axis one-to-one and onto the real axis, and the points $a_1, \ldots, a_{n-1}, \infty$ into the points $b_1, \ldots, b_{n-1}, b_n = 0$, respectively, where $-\infty < b_1 < \cdots < b_{n-1} < b_n = 0$. Hence, on applying (6.2.1.1) to $\eta = F(\zeta) = f(a - \frac{1}{\zeta})$, we have

$$\eta = F(\zeta) = c_1 \int_{\zeta_0}^{\zeta} \prod_{k=1}^{n-1} (\tau - b_k)^{\alpha_k - 1} \cdot \tau^{\alpha_n - 1} d\tau + c_2$$

mapping $\operatorname{Im} \zeta > 0$ univalently onto Ω. Equivalently,

$$w = f(z) = F\left(\frac{1}{a - z}\right)$$

$$= c_1 \int_{z_0}^{z} \prod_{k=1}^{n-1} \left(\frac{1}{a - t} - b_k\right)^{\alpha_k - 1} \cdot \frac{1}{(a - t)^{\alpha_n - 1}} \cdot \frac{1}{(a - t)^2} dt + c_2$$

$$= c_1' \int_{z_0}^{z} \prod_{k=1}^{n-1} (t - a_k)^{\alpha_k - 1} \cdot (a - t)^{-\sum_{k=1}^{n} (\alpha_k - 1) - 2} dt + c_2$$

$$\left(\text{recall that } b_k = \frac{1}{a - a_k}, \, 1 \leq k \leq n\right)$$

$$= c_1' \int_{z_0}^{z} \prod_{k=1}^{n-1} (t - a_k)^{\alpha_k - 1} dt + c_2 \qquad \left(\text{recall that } \sum_{k=1}^{n} \alpha_k = n - 2\right).$$

Alternatively, we may keep a_n finite and adopt the transformation $\zeta = \frac{1}{a_n - z}$ and finally let $a_n \to \infty$ in the last few steps.

We summarize as

The Schwarz–Christoffel formula for a polygon with ∞ corresponding to one of its vertices.

(1) Let $\Omega, w_1, \ldots, w_n$, and $-\infty < a_1 < \cdots < a_{n-1} < a_n = \infty$ be as in (6.2.1.1) so that the point $a_n = \infty$ corresponds to the n-th vertex w_n. Then the mapping is given by

$$w = f(z) = c_1 \int_{z_0}^{z} \prod_{k=1}^{n-1} (\zeta - a_k)^{\alpha_k - 1} d\zeta + c_2.$$

(2) Let $\Omega, w_1, \ldots, w_n$, and $-\infty < a_1 < \cdots < a_{n-1} < a_n = \infty$ be as in (6.2.1.3) so that the point $a_n = \infty$ corresponds to w_n and that ∞ is an interior point of Ω. Then the mapping is given by

$$w = f(z) = c_1 \int_{z_0}^{z} \frac{\prod_{k=1}^{n-1}(\zeta - a_k)^{\alpha_k - 1}}{(z - b)^{\Omega}(\zeta - \bar{b})^{2}} \, d\zeta + o_2.$$

Namely, the factor $(z - a_n)^{\alpha_n - 1}$ disappears from the product $\prod_{k=1}^{n}(\zeta - a_k)^{\alpha_k - 1}$ in both cases.   (6.2.1.4)

The derivation in (2) is left as an exercise by noting that, in this case, $\sum_{k=1}^{n} \alpha_k = n + 2$.

Section (4) The mapping of the unit disk onto a polygon

In the case of (6.2.1.1):

The transformation

$$\zeta = \frac{z - \sigma}{z - \bar{\sigma}}, \quad \text{where } \sigma \text{ is a fixed point in } \operatorname{Im} z > 0, \qquad (*15)$$

maps $\operatorname{Im} z > 0$ univalently onto $|\zeta| < 1$, the real axis $\operatorname{Im} z = 0$ onto the circle $|\zeta| = 1$ and σ into the center 0. The inverse mapping is $z = \frac{\sigma - \bar{\sigma}\zeta}{1 - \zeta}$. On applying (6.2.1.1) to $F(z) = f\left(\frac{z-\sigma}{z-\bar{\sigma}}\right)$, $\operatorname{Im} z > 0$, we have

$$F(z) = c_1 \int_{z_0}^{z} \prod_{k=1}^{n}(\tau - a_k)^{\alpha_k - 1} d\tau + c_2$$

mapping $\operatorname{Im} z > 0$ univalently onto Ω. Setting $\zeta_k = \frac{a_k - \sigma}{a_k - \bar{\sigma}}$ for $1 \leq k \leq n$, then

$$w = f(z) = F\left(\frac{\sigma - \bar{\sigma}\zeta}{1 - \zeta}\right)$$

$$= c_1 \int_{(\sigma - \bar{\sigma}\zeta_0)/(1-\zeta_0)}^{(\sigma - \bar{\sigma}\zeta)/(1-\zeta)} \prod_{k=1}^{n}(\tau - a_k)^{\alpha_k - 1} d\tau + c_2$$

$$= c_1 \int_{\zeta_0}^{\zeta} \prod_{k=1}^{n}\left(\frac{\sigma - \bar{\sigma}t}{1 - t} - \frac{\sigma - \bar{\sigma}\zeta_k}{1 - \zeta_k}\right)^{\alpha_k - 1} \cdot \frac{\sigma - \bar{\sigma}}{(1 - t)^2} dt + c_2$$

$$= c_1' \int_{\zeta_0}^{\zeta} \prod_{k=1}^{n}(t - \zeta_k)^{\alpha_k - 1} dt + c_2 \quad \left(\text{noting that } \sum_{k=1}^{n} \alpha_k = n - 2\right),$$

where ζ_0 and ζ are in $|\zeta| < 1$. This form coincides with that in (6.2.1.1) except notations for variables and constants concerned.

In the case of (6.2.1.3):

We try to map the center $\zeta = 0$ of the open disk $|\zeta| < 1$ into the point at infinity ∞. For this purpose, choose σ equal to b in $(*_{15})$, the preimage of ∞ under the univalent mapping in (6.2.1.3). Using the formula $\sum_{k=1}^{n} \alpha_k = n + 2$ in this case. Similar process will give the required

$$
w = f(z) = c_1' \int_{z_0}^{z} \prod_{k=1}^{n} (t - \zeta_k)^{\alpha_k - 1} \cdot \frac{dt}{t^2} + c_2.
$$

We summarize as

The Schwarz–Christoffel formula that maps the open unit disk onto a polygon. Let Ω *be a polygon with finite vertices* $w_1, \ldots, w_n$ *and corresponding interior angles* $\alpha_1 \pi, \ldots, \alpha_n \pi$.

(1) In case Ω is a bounded polygon: Any function $w = f(z)$ mapping $|z| < 1$ univalently onto Ω is of the form

$$
w = f(z) = c_1 \int_{z_0}^{z} \prod_{k=1}^{n} (\zeta - a_k)^{\alpha_k - 1} d\zeta + c_2, \quad |z| < 1,
$$

where z_0 is a fixed point in $|z| < 1$, c_1 and c_2 are constants, and $a_k = f^{-1}(w_k) = e^{i\theta_k}$ $(0 \leq \theta_k < 2\pi)$, $1 \leq k \leq n$, are points on $|z| = 1$ arranged in the ordering $\theta_1 < \cdots < \theta_n$.

(2) In case ∞ is an interior point of Ω:

$$
w = f(z) = c_1 \int_{z_0}^{z} \prod_{k=1}^{n} (\zeta - a_k)^{\alpha_k - 1} \frac{d\zeta}{\zeta^2} + c_2, \quad |z| < 1,
$$

where $f^{-1}(\infty)$ is designated to be the center 0.   (6.2.1.5)

Section (5) The converse problem

Given a transformation (function) defined by

$$
w = f(z) = c \int_{0}^{z} (\zeta - a_1)^{\alpha_1 - 1} \cdots (\zeta - a_n)^{\alpha_n - 1} d\zeta, \quad\quad (6.2.1.6)
$$

where $c > 0$ is a constant, $a_1, \ldots, a_n$ are *distinct* real numbers (say, $a_1 < \cdots < a_n$) and $\alpha_1, \ldots, \alpha_n$ are real numbers but not necessarily distinct. The *question* is: *When does the transformation map the upper half-plane univalently onto a bounded polygon* Ω *so that the points* $a_1, \ldots, a_n$ *are mapped into the vertices?*

The way we treated Examples 2 and 3 of Sec. 3.5.5 can be definitely and unambiguously imitated and generalized to handle this problem.

The (absolute) convergence of the integral at a_k $(1 \leq k \leq n)$ and at ∞: To converge at a_k, one needs to have $\alpha_k - 1 > -1$, namely, $\alpha_k > 0$ for $1 \leq k \leq n$. If $|\zeta|$ is large, the *integrand*

$$\varphi(\zeta) = \prod_{k=1}^{n} (\zeta - a_k)^{\alpha_k - 1} = \zeta^{\alpha_1 + \cdots + \alpha_k - n} \prod_{k=1}^{n} \left(1 - \frac{a_k}{\zeta}\right)^{\alpha_k - 1}$$

is dominated by $\zeta^{\alpha_1 + \cdots + \alpha_k - n}$. Hence the integral converges at ∞ only if $\alpha_1 + \cdots + \alpha_n - n < -1$, namely, $\alpha_1 + \cdots + \alpha_n < n - 1$ holds.

The single-valued branch of the integrand: For each k $(1 \leq k \leq n)$, $(\zeta - a_k)^{\alpha_k - 1} = e^{(\alpha_k - 1)\log(\zeta - a_k)}$ has a single-valued branch in $\operatorname{Im}\zeta > 0$ (see (3.3.1.5)); then so does the integrand $\varphi(\zeta)$ in (6.2.1.6).

We choose the following branch, for certainty:

In case $-\infty < \zeta < a_1$: Then $\zeta - a_k < 0$ for $1 \leq k \leq n$. We *designate*

$$\operatorname{Arg}(\zeta - a_k) = \pi, \quad 1 \leq k \leq \pi$$

$$\Rightarrow \operatorname{Arg}\varphi(\zeta) = (\alpha_1 - 1)\pi + \cdots + (\alpha_n - 1)\pi \quad \text{(which is less than } -\pi\text{)}.$$

$$(*_{16})$$

In case $a_1 < \zeta < a_2$: When ζ starts from the left side of a_1 and winds around a_1 along an upper half circle C_ε with center at a_1 and reaches the right side of a_1 (see Fig. 6.5), we have (refer to (1)b in (2.7.1.4) and Example 2(4) of Sec. 2.7.2, etc.)

$$\Delta_{C_\varepsilon}\operatorname{Arg}(\zeta - a_1) = 0 - \pi = -\pi;$$

$$\Delta_{C_\varepsilon}\operatorname{Arg}(\zeta - a_k) = \pi - \pi = 0, \quad 2 \leq k \leq n$$

$$\Rightarrow \operatorname{Arg}\varphi(\zeta)$$

$$= \operatorname{Arg} e^{(\alpha_1 - 1)i[\Delta_{C_\varepsilon}\operatorname{Arg}(\zeta - a_1)]} + (\alpha_1 - 1)\pi + (\alpha_2 - 1)\pi + \cdots + (\alpha_n - 1)\pi$$

$$= (\alpha_2 - 1)\pi + \cdots + (\alpha_n - 1)\pi.$$

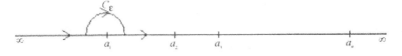

Fig. 6.5

Proceed in this manner. In general, in case, $a_k < \zeta < a_{k+1}$:

$$\Delta_{C_\varepsilon}\operatorname{Arg}(\zeta - a_1) = 0 - 0 = 0, \quad 1 \le l \le k - 1;$$

$$\Delta_{C_\varepsilon}\operatorname{Arg}(\zeta - a_k) = 0 - \pi = -\pi;$$

$$\Delta_{C_\varepsilon}\operatorname{Arg}(\zeta - a_1) = \pi - \pi = 0, \quad k + 1 \le l \le n$$

$$\Rightarrow \operatorname{Arg}\varphi(\zeta) = -(\alpha_k - 1)\pi + (\alpha_k - 1)\pi + (\alpha_{k+1} - \pi) + \cdots + (\alpha_k - 1)\pi$$

$$= (\alpha_{k+1} - 1)\pi + \cdots + (\alpha_n - 1)\pi, \quad 1 \le k \le n - 1. \qquad (*_{17})$$

Consequently, if $a_n < \zeta < +\infty$,

$$\operatorname{Arg}\varphi(z) = 0. \qquad (*_{18})$$

Designate $a_0 = -\infty$ and $a_{n+1} = \infty$, for convenience. Note that $a_0 = a_{n+1} = \infty$ in the extended number system $\mathbf{C}^*$. Then $(*_{17})$ is valid for $0 \le k \le n$; namely, it includes $(*_{16})$ if $k = 0$ and $(*_{18})$ if $k = n$ when the right side of $(*_{17})$ is designated to be zero in this case.

   The appearance of the vertices, the sides, and then the polygon:
   Set

$$w_k = c \int_0^{a_k} (\zeta - a_1)^{\alpha_1 - 1} \cdots (\zeta - a_n)^{\alpha_n - 1} d\zeta, \quad 1 \le k \le n. \qquad (*_{19})$$

Note that $w_0 = w_{n+1}$. By $(*_{17})$,

$$w = w_{k-1} + c \int_{a_{k-1}}^z \varphi(\zeta)d\zeta$$

$$= w_{k-1} + c \exp\left\{\sum_{l=k}^n (\alpha_l - 1)\pi i\right\} \int_{a_{k-1}}^z |\varphi(\zeta)|d\zeta,$$

$$z \in (a_{k-1}, a_k), \quad 1 \le k \le n. \qquad (*_{20})$$

Wherein $\sum_{l=k}^n (\alpha_k - 1)\pi i$ remains a constant as long as $z \in (a_{k-1}, a_k)$; yet $\int_{a_{k-1}}^z |\varphi(\zeta)|d\zeta$ changes continuously from 0 to

$$c \int_{a_{k-1}}^{a_k} |\varphi(\zeta)|d\zeta = l_k, \quad 1 \le k \le n \qquad (*_{21})$$

as z increases from a_{k-1} steadily to a_k. More precisely, as z increases from a_{k-1} to a_k, $w = w(z)$ starts from w_{k-1} and traces out the segment $\overline{w_{k-1}w_k}$ until w_k, with its length equal to l_k (see $(*_{21})$) and its inclination the angle $\sum_{l=1}^n (\alpha_k - 1)\pi$ with the positive real axis. See Fig. 6.6. In particular, in case $k = 1$, we have the segment $\overline{w_0w_1}$, where w_0 is the one obtained in $(*_{19})$ by setting $k = 0$; in case $k = n+1$, we have the segment $\overline{w_nw_{n+1}}$ (see $(*_{18})$), where $w_{n+1} = w_0$.

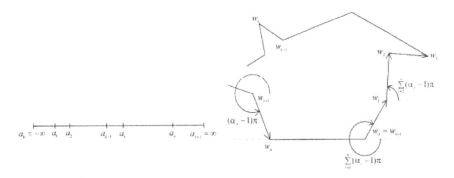

Fig. 6.6

In conclusion, $w = w(z)$ maps the real axis onto the boundary of a polygon with vertices $w_0, w_1, \ldots, w_n$. The boundary segments could be self-intersecting at a point which is not a vertex; and the mapping then fails to be one-to-one.

A further assumption: Suppose now that the boundary $w_0 w_1 \ldots w_n$ is not self-intersecting; namely, it constitutes a rectifiable Jordan closed curve. Under this circumstance, $w = w(z)$ maps $\operatorname{Im} z > 0$ univalently onto a polygon Ω with vertices $w_0, w_1, \ldots, w_n$ (see (3.5.5.1)).

The conformation of the interior and exterior angles at a vertex:
Since $-\infty < a_1 < a_2 < \cdots < a_n < \infty$, the vertices $w_0, w_1, \ldots, w_n$ are arranged along the boundary in the counterclockwise direction. According to Fig. 6.7, the *interior angle* at w_k $(1 \le k \le n)$ is equal to

$$\pi + \sum_{l=k}^{n} (\alpha_l - 1)\pi - \sum_{l=k+1}^{n} (\alpha_l - 1)\pi = \alpha_k \pi, \quad 1 \le k \le n. \qquad (*_{22})$$

See Fig. 6.7.

As a matter of fact, in addition to $\alpha_k > 0$, α_k has to satisfy

$$\alpha_k < 2, \quad 1 \le k \le n. \qquad (*_{23})$$

To see this, fix any k $(1 \le k \le n)$. In case $|\zeta - a_k| < r < \min(a_k - a_{k-1}, a_{k+1} - a_k)$ (in case $k = 1$, choose $r < a_2 - a_1$; in case $k = n$, choose $r < a_n - a_{n-1}$), the function $g(\zeta) = \prod_{\substack{l=1 \\ l \ne k}}^{n} (\zeta - a_l)^{\alpha_l - 1}$ has single-valued branches which are all different from zero. Fix such a branch and develop it into its Taylor series at a_k as

$$g(\zeta) = \sum_{n=0}^{\infty} b_n^{(k)} (\zeta - a_k)^n, \quad |\zeta - a_k| < r,$$

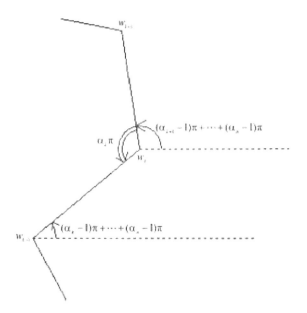

Fig. 6.7

where $b_0^{(k)} = g(a_k) \neq 0$. Then, in $|\zeta - a_k| < r$,

$$w = w_k + c \int_{a_k}^{z} f(\zeta)d\zeta = w_k + c \int_{a_k}^{z} g(\zeta)(\zeta - a_k)^{\alpha_k - 1}d\zeta$$

$$= w_k + c\frac{b_0^{(k)}}{\alpha_k}(z - a_k)^{\alpha_k}\left\{1 + \frac{\alpha_k}{\alpha_k + 1}\frac{b_1^{(k)}}{b_0^{(k)}}(z - a_k) + \cdots\right\}.$$

Choose $r > 0$ small enough such that, if $|z - a_k| < r$, then

$$\left|\frac{\alpha_k}{\alpha_k + 1}\frac{b_1^{(k)}}{b_0^{(k)}}(z - a_k) + \cdots\right| < 1$$

$\Rightarrow$ (since $c > 0$ and $\alpha_k > 0$)

$$\arg(w - w_k) = \arg b_0^{(k)} + \alpha_k\arg(z - a_k)$$

$$+ \arg\left\{1 + \frac{\alpha_k}{\alpha_k + 1}\frac{b_1^{(k)}}{b_0^{(k)}}(z - a_k) + \cdots\right\}$$

$$\to \arg b_0^{(k)} + \alpha_k\arg(z - a_k) \quad \text{as } z \to a_k.$$

This indicates the fact that, if z approaches a_k along a ray starting at a_k and pointing toward the upper half-plane and with the inclination angle

θ with the positive real axis, then $w = w(z)$ will approach w_k along a curve whose tangent at w_k has the inclination angle $\operatorname{Arg} b_0^{(k)} + \alpha_k \theta$ with the positive real axis. As a consequence, when z starts form $a_k + r$ and winds along the upper half circle $\gamma : |z - a_k| = r$, $\operatorname{Im} z > 0$, in the anticlockwise direction and terminates at $a_k - r$, the variation of $\operatorname{Arg}(z - a_k)$ is given by

$$\Delta_\gamma \operatorname{Arg}(z - a_k) = \pi - 0$$

$$\Rightarrow \Delta_\gamma \operatorname{Arg}(w - w_k) = (\operatorname{Arg} b_0^{(k)} + \alpha_k \pi) - (\operatorname{Arg} b_0^{(k)} - 0) = \alpha_k \pi, \quad 1 \leq k \leq n.$$

Owing to the univalence, the image curve $w(\gamma)$ is a Jordan curve in the polygon Ω, starting at a point on the side $\overline{w_k w_{k+1}}$ and ending at point on the side $\overline{w_{k-1} w_k}$; it never winds around the vertex w_k. Hence, $\alpha_k \pi$ as the interior angle of the polygon at w_k, $\alpha_k \pi < 2\pi$ does holds; namely, $\alpha_k < 2$ for $1 \leq k \leq n$.

What is the *interior angle* at $w_0 = w_{n+1}$? It is

$$-\left\{ \sum_{l=1}^{n} (\alpha_l - 1)\pi - (-\pi) \right\} = (n-1)\pi - \sum_{l=1}^{n} \alpha_k \pi. \qquad (*24)$$

See Fig. 6.8 (also, Fig. 6.6). Since Ω has $(n+1)$ vertices, the sum of its interior angle is $[(n+1) - 2]\pi = (n-1)\pi$. Hence, $(*24)$ is exactly what we want; moreover,

$$(n-1)\pi - \sum_{l=1}^{n} \alpha_l \pi < 2\pi$$

$$\Rightarrow n - 3 < \alpha_1 + \cdots + \alpha_n. \qquad (*25)$$

We summarize the above as

The necessary conditions that the upper half-plane is mapped univalently onto a bounded polygon. Suppose $a_1, \ldots, a_n$ are n distinct real number

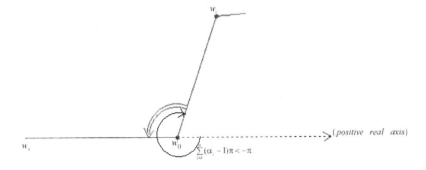

Fig. 6.8

arranged in the ordering $a_1 < \cdots < a_n$; $c > 0$ is a constant, and $\alpha_1, \ldots, \alpha_n$ are real numbers (not necessarily unequal). If the *Schwarz–Christoffel transformation*

$$w = w(z) = c \int_0^z (\zeta - a_1)^{\alpha_1 - 1} \cdots (\zeta - a_n)^{\alpha_n - 1} d\zeta$$

maps the upper half-plane $\operatorname{Im} z > 0$ univalently onto a bounded polygon Ω with $(n+1)$ *vertices* $w_0, w_1, \ldots, w_n$, where

$$w_k = c \int_0^{a_k} (\zeta - a_1)^{\alpha_1 - 1} \cdots (\zeta - a_n)^{\alpha_n - 1} d\zeta$$

$$(a_0 = -\infty, \quad a_{n+1} = +\infty \quad \text{and} \quad w_0 = w_{n+1})$$

and with the corresponding *interior angles*

$$[(n-1) - (\alpha_1 + \cdots + \alpha_n)]\pi, \alpha_1 \pi, \ldots, \alpha_n \pi,$$

then it is necessary that

1. $0 < \alpha_k < 2$, $1 \le k \le n$;
2. $n - 3 < \alpha_1 + \cdots + \alpha_n < n - 1$.

In this case, $w = w(z)$ maps the real interval $[a_{k-1}, a_k]$, $2 \le k \le n$, one-to-one and onto the side segment $\overline{w_{k-1} w_k}$ whose

(a) length is equal to $c \int_{a_{k-1}}^{a_k} |(\zeta - a_1)^{\alpha_1 - 1} \cdots (\zeta - a_n)^{\alpha_n - 1}| d\zeta$, and whose
(b) inclination angle with the positive real axis is equal to $(\alpha_k - 1)\pi + \cdots + (\alpha_n - 1)\pi$.

In particular, $w = w(z)$ maps $[a_n, a_{n+1}]$ and $[a_0, a_1]$, respectively, one-to-one and onto the sides $\overline{w_n w_0}$ (lying on the real axis) and $\overline{w_0 w_1}$ (with the inclination angle $(\alpha_1 - 1)\pi + \cdots + (\alpha_n - 1)\pi$ with the positive real axis). (6.2.1.7)

If the condition

$$\alpha_1 + \cdots + \alpha_n = n - 2 \tag{6.2.1.8}$$

holds, then the interior angle (see $(*_{24})$) at w_0 is equal to π. In other words, w_0 is no more a vertex of the polygon but is an *interior point* of the side $\overline{w_n w_1}$. In this case, $w = w(z)$ maps $\operatorname{Im} z > 0$ univalently onto a bounded polygon with n *vertices* $w_1, \ldots, w_n$ only.

Remark. For each formula from (6.2.1.3) to (6.2.1.5) there is a similar counterpart as (6.2.1.7) and (6.2.1.8). We will feel free to use them if necessary. □

6.2.2 *Examples*

Let us make a general remark before we start to illustrate examples.

Given a Schwarz–Christoffel transformation $w = w(z)$, mapping $\operatorname{Im} z > 0$ univalently onto a polygon Ω with vertices $w_1, \ldots, w_n$. Set $a_k = f^{-1}(w_k)$ for $1 \le k \le n$. Let the segments γ_k and Γ_k $(1 \le k \le n)$ be as in $(*_1)$ of Sec. 6.2.1. The repeated usage of the symmetry principle to γ_k and Γ_k will continue $w = f(z)$ analytically to a *function*, still denoted by $w = f(z)$ for simplicity, which maps a *covering surface* R_z of the z-plane univalently onto a *covering surface* R_w of the w-plane. Both R_z and R_w are infinitely-sheeted, in general; and their *branch points* lie over the points a_k and the vertices w_k, respectively, with orders finite or infinite. See Fig. 6.9 for the readers to convince their understanding and imagination. As a matter of fact, we had used this process explicitly in the construction of a nonconstant elliptic function (see Note (2) in Example 3 of Sec. 3.5.5) and, implicitly, in $(*_6)$ of Sec. 6.2.1. Another such an example is provided by the elliptic modular function (see Sec. 5.8.3).

This process is particularly interesting when R_w is a simply connected domain, say, the entire w-plane. In other words, the reflected images of the polygon Ω fill out the plane and form a *tessellation*, without overlapping and without gaps. In this case, the extended $w = f(z)$ maps R_z univalently onto the w-plane; equivalently, this means that $w = f(z)$, as a function of the variable z in the conventional "owe-sheeted" plane, is *multiple-valued* (indeed, one-to-infinite) and R_z acts as the *Riemann surface* for it. Note that the inverse function $z = f^{-1}(w) : \mathbf{C}^* \to \mathbf{C}^*$ is then *single-valued* but not univalent. Please refer to Figs. 3.72–3.77 to convince oneself firmly or see the examples in the following.

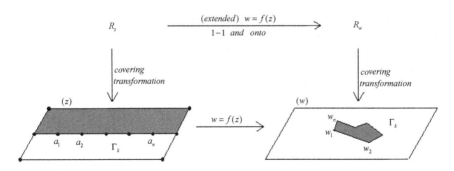

Fig. 6.9

Section (1) Triangles

Let Ω be a triangle with *finite* vertices w_1, w_2, w_3, and the corresponding interior angles $\alpha_1 \pi, \alpha_2 \pi, \alpha_3 \pi$, where $\alpha_1 + \alpha_2 + \alpha_3 = 1$. Choose $a_1 = 0$, $a_2 = 1$, and $a_3 = \infty$, to be mapped into w_1, w_2, and w_3, respectively. By (6.2.1.4), the function mapping $\operatorname{Im} z > 0$ univalently onto the triangle $\Delta w_1 w_2 w_3$ is given by

$$w = f(z) = c_1 \int_0^z \zeta^{\alpha_1 - 1}(1 - \zeta)^{\alpha_2 - 1} d\zeta + c_2. \qquad (*_1)$$

Since $f(0) = w_1$ and $f(1) = w_2$,

$$c_2 = w_1;$$

$$c_1 = \frac{w_2 - w_1}{\int_0^1 x^{\alpha_1 - 1}(1 - x)^{\alpha_2 - 1} dx}$$

$$= \frac{w_2 - w_1}{B(\alpha_1, \alpha_2)} \quad (B(\alpha_1, \alpha_2), \quad \text{the beta function (see (5.6.2.4))}$$

$$= \frac{\pi(w_2 - w_1)}{\Gamma(\alpha_1)\Gamma(\alpha_2)\Gamma(\alpha_3)\sin \pi \alpha_3} \quad (\text{using } \alpha_1 + \alpha_2 + \alpha_3 = 1 \text{ and } (5.6.1.6)).$$

Recall: A reflection with respect to a line segment (or circular arc) preserves the magnitude of an angle but reverses its orientation. Hence an even number of reflection preserves both the magnitude and the orientation of an angle, and it results in a conformal mapping.

As repeated images of $\Delta w_1 w_2 w_3$ form a tessellation of the plane, it is thus necessary that repeated reflections across sides with a common vertex should ultimately lead back to the original triangle after *even number* of steps. This will be the case only if the angles must be of the form $\frac{\pi}{n_1}, \frac{\pi}{n_2}, \frac{\pi}{n_3}$ with integral denominators. The condition $\frac{1}{n_1} + \frac{1}{n_2} + \frac{1}{n_3} = 1$ is fulfilled only by the triples $(3, 3, 3)$, $(2, 4, 4)$, and $(2, 3, 6)$. They correspond to an equilateral triangle, an isosceles right triangle, and half an equilateral triangle, respectively. The corresponding functions $w = f(z)$ are known as the *Schwarz triangle functions*.

The case $\alpha_1 = \alpha_2 = \alpha_3 = \frac{1}{3}$:

Designate the vertices $w_1 = 0$, $w_2 = a$ $(a > 0)$, and then $w_3 = \frac{1 + \sqrt{3}i}{2}a$. Then $(*_1)$ reduces to

$$w = c_1 \int_0^z \zeta^{-\frac{2}{3}}(1 - \zeta)^{-\frac{2}{3}} d\zeta, \quad \text{where}$$

$$c_1 = \frac{a\pi}{\left[\Gamma\left(\frac{1}{3}\right)\right]^3 \sin \frac{\pi}{3}} = \frac{2\pi a}{\sqrt{3} \left[\Gamma\left(\frac{1}{3}\right)\right]^3}. \qquad (*_2)$$

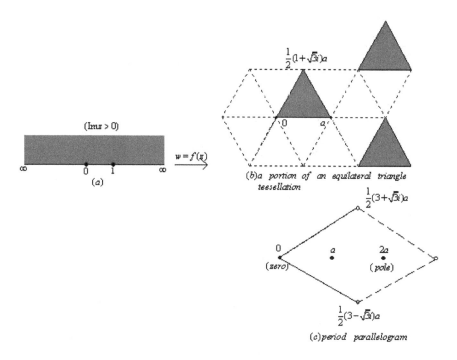

Fig. 6.10

When z travels along the real axis in the ordering $0 \to 1 \to \infty \to 0$, $w = f(z)$ winds around the triangle $\triangle w_1 w_2 w_3$ in the counterclockwise direction $w_1 \to w_2 \to w_3 \to w_1$. The triangle net (i.e., tessellation) can be easily drawn. Part of it is shown in Fig. 6.10(a). Furthermore, we can derive from it the following information:

1. The extended $w = f(z)$ is a one-to-infinite function; for each point z in $\operatorname{Im} z > 0$, there associates exactly only one point $f(z)$ inside a triangle, so does for each point in $\operatorname{Im} z < 0$.
2. The inverse function $z = f^{-1}(w)$ is a single-valued meromorphic function in the w-plane. It is doubly periodic, with

$$\omega_1 = \frac{1}{2}(3 + \sqrt{3}i)a \quad \text{and} \quad \omega_2 = \frac{1}{2}(3 - \sqrt{3}i)a = \overline{\omega_1}$$

as two *fundamental periods* whose ratio $\frac{\omega_1}{\omega_2} = \frac{1+\sqrt{3}i}{2}$, a nonreal constant. A *period parallelogram* is shown in Fig. 6.10(c), where the vertices $\frac{1}{2}(3 \pm \sqrt{3}i)a$ and the dotted sides are dismissed. There are six triangles contained in the parallelogram.

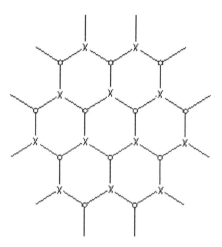

Fig. 6.11

3. Consequently, $z = f^{-1}(w)$ has the zeros-set $\{m_1\omega_1 + m_2\omega_1 | m_1, m_2 = 0, \pm 1, \ldots\}$ and the poles-set $\{\frac{1+\sqrt{3}i}{2}a + m_1\omega_1 + m_2\omega_2 | m_1, m_2 = 0, \pm 1, \ldots\}$. Only *one* zero 0 and *one* pole $2a$ lie in the period parallelogram, both of the same order 3, called the *order* of $z = f^{-1}(w)$.

This $z = f^{-1}(w)$ is an *elliptic function* of *order* 3. Also, Fig. 6.10(b) shows that the *Riemann surface* R_z for $w = f(z)$ has its *line complex* as in Fig. 6.11. The branch points lie over 0, 1, and ∞. Each branch point is of the order 2.

The case $\alpha_1 = \alpha_3 = \frac{1}{4}$ *and* $\alpha_2 = \frac{1}{2}$:

Let the vertices of the isosceles right triangle be $w_1 = 0$, $w_2 = a$ $(a > 0)$, and $w_3 = (1 + i)a$. In this case, $(*_1)$ becomes

$$w = f(z) = c_1 \int_0^z \zeta^{-\frac{3}{4}}(1 - \zeta)^{-\frac{1}{2}} d\zeta, \quad \text{where} \tag{$*_3$}$$

$$c_1 = \frac{a\pi}{\left[\Gamma\left(\frac{1}{4}\right)\right]^2 \Gamma\left(\frac{1}{2}\right) \sin\frac{\pi}{4}} = \frac{a\sqrt{2\pi}}{\left[\Gamma\left(\frac{1}{4}\right)\right]^2}.$$

The tessellation of the isosceles right triangle net is shown in Fig. 6.12(b).

1. The extended $w = f(z)$ is one-to-infinite; it maps either $\operatorname{Im} z \geq 0$ or $\operatorname{Im} z \leq 0$ univalently onto the closure of a simple isosceles right triangle.

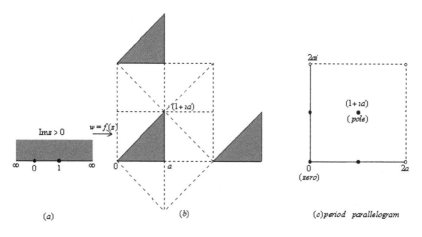

Fig. 6.12

2. The inverse function $z = f^{-1}(w)$ is a single-valued meromorphic function in the w-plane. It is doubly-periodic, with

$$\omega_1 = 2a \text{ and } \omega_2 = 2ai$$

as two *fundamental periods* whose ratio is $\frac{\omega_1}{\omega_2} = -i$. A *period parallelogram*, as shown in Fig. 6.12(c), contains eight congruent isosceles right triangles.

3. $z = f^{-1}(w)$ has the zeros-set $\{m_1\omega_1 + m_2\omega_2 | m_1, m_2 = 0, \pm1, \ldots\}$ and the poles-set $\{(1 + i)a + m_1\omega_1 + m_2\omega_1 | m_1, m_1 = 0, \pm1, \ldots\}$. Both zero and pole are of order 4. A period parallelogram contains only one zero and only one pole, yet it contains 2 incongruent one-point (namely, a and ai).

$z = f^{-1}(w)$ is an elliptic function of order 4. Meanwhile Fig. 6.12(b) shows that the *Riemann surface* R_z for $w = f(z)$ has its *line complex* as in Fig. 6.13. There are infinitely many branch points over each of the points 0, 1, and ∞. These over 0 and ∞ are of order 3; that over 1 is of order 1.

The case $\alpha_1 = \frac{1}{6}$, $\alpha_2 = \frac{1}{2}$, and $\alpha_3 = \frac{1}{3}$:
Let the vertices of the corresponding right triangle be $w_1 = 0$, $w_2 = a$ $(a > 0)$, and $w_3 = (1 + \frac{i}{\sqrt{3}})a$. In this case, $(*_1)$ reduces to

$$w = f(z) = c_1 \int_0^z \zeta^{-\frac{5}{6}}(1 - \zeta)^{-\frac{1}{2}} d\zeta, \quad \text{where} \tag{$*_4$}$$

$$c_1 = \frac{a\pi}{\Gamma\left(\frac{1}{6}\right)\Gamma\left(\frac{1}{2}\right)\Gamma\left(\frac{1}{3}\right)\sin\frac{\pi}{3}} = \frac{2a\sqrt{\pi}}{\sqrt{3}\Gamma\left(\frac{1}{6}\right)\Gamma\left(\frac{1}{3}\right)}.$$

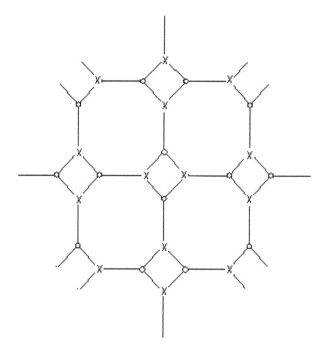

Fig. 6.13

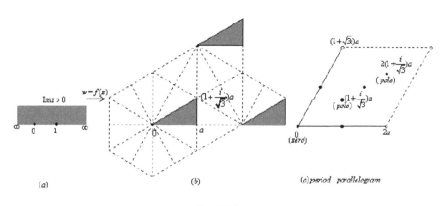

Fig. 6.14

The tessellation of the right triangle net is shown in Fig. 6.14(b). The inverse function $z = f^{-1}(w)$ is single-valued and meromorphic in the w-plane. It has two *fundamental periods*

$$\omega_1 = 2a \quad \text{and} \quad \omega_2 = (1 + \sqrt{3}i)a$$

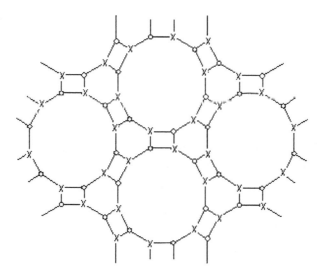

Fig. 6.15

whose ratio $\frac{\omega_1}{\omega_2} = -1 + \sqrt{3}i$ is not real. A period parallelogram is shown in
Fig. 6.14(c). The *parallelogram* contains 12 congruent right triangles; one
zero 0 of order 6; two poles $(1 + \frac{i}{\sqrt{3}})a$ and $2(1 + \frac{i}{\sqrt{3}})a$, each of order 3,
and three one-point, each of multiplicity 2. This $z = f^{-1}(w)$ is an *elliptic
function* of order 6. The *line complex* of its *Riemann surface* R_z, as suggested by Fig. 6.14(b), is shown in Fig. 6.15: R_z is infinitely-sheeted and
has branch points over 0, 1, and ∞, with the order 5, 1, and 2, respectively.
Note that, if instead we designate $\alpha_1 = \frac{1}{3}$, $\alpha_2 = \frac{1}{2}$, $\alpha_3 = \frac{1}{6}$, and $w_1 = 0$,
$w_2 = a$, $w_3 = (1 + \sqrt{3}i)a$ where $a > 0$, then the resulted $z = f^{-1}(w)$ will
have the periods $3a \pm \sqrt{3}ai$. Readers are asked to give out the details.

Next, we suppose Ω is a triangle with a vertex at ∞.

Choose $a_1 = -1$, $a_2 = 1$, and $a_3 = \infty$ such that $w_k = f(a_k)$ for
$k = 1, 2, 3$, and $w_3 = \infty$. The case $\alpha_1 = \frac{\pi}{2}$, $\alpha_2 = \frac{\pi}{2}$, and $\alpha_3 = 0$ is the only
possibility in order to have the triangle $\triangle w_1 w_2 w_3$ to generate a tessellation
of the whole w-plane. It is reasonable to have the interior angle at $w_3 = \infty$
equal to 0, as indicated in Fig. 6.16 where a moving P is approaching ∞ in
the vertical strip. For simplicity, we choose $w_1 = -\frac{\pi}{2}$ and $w_2 = \frac{\pi}{2}$.

In this case, the corresponding Schwarz–Christoffel transformation is

$$w = \int_0^z \frac{d\zeta}{\sqrt{1 - \zeta^2}} = \sin^{-1} z, \quad \text{Im } z > 0 \qquad (*_5)$$

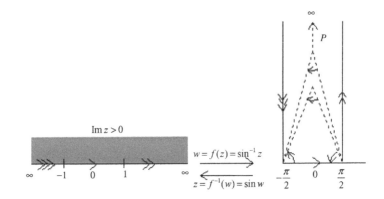

Fig. 6.16

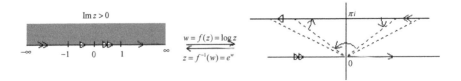

Fig. 6.17

the restriction to the upper half-plane $\operatorname{Im} z > 0$ of the principle branch of the inverse arcsin function (see (2) in (2.7.4.5)). See Figs. 2.86 and 2.87 for its Riemann surface and line complex.

Finally, let Ω be a triangle with two vertices at ∞.

In this degenerated case (for detail, see Sec. 6.2.3), choose $a_1 = -\infty$, $a_2 = 1$, and $a_3 = +\infty$ such that $w_k = f(a_k)$ for $k = 1, 2, 3$, and $w_2 = 0$, $w_1 = w_3 = \infty$. Figure 6.17 shows why we designate the interior angle $\alpha_1 = \alpha_3 = 0$. As a matter of fact, we better treat Ω as a degenerated triangle with two sides only. Then $\alpha_2 = \pi$.

The associated Schwarz–Christoffel transformation is

$$ w = \int_1^z \frac{d\zeta}{\zeta} = \log z, \quad \operatorname{Im} z > 0, \tag{$*_6$} $$

the restriction to $\operatorname{Im} z > 0$ of the principle branch of the logarithm function (see (3) in (2.7.3.2)). See Figs. 2.73 and 2.74 for its Riemann surface and line complex.

Section (2) Polygon with n vertices ($n \geq 4$)

Let $w = f(z) : \{\operatorname{Im} z > 0\} \to \Omega$ (polygon) be a Schwarz–Christoffel transformation.

Then $n = 4$ is the only possibility in order to have the inverse function $z = f^{-1}(w)$ single-valued in the whole w-plane; in this case, Ω turns out to be a rectangle. Choose four points $-\frac{1}{k}$, -1, 1, and $\frac{1}{k}$ ($0 < k < 1$) along the real axis. The Schwarz–Christoffel transformation mapping $\operatorname{Im} z > 0$ univalently onto the rectangle is thus of the form

$$w = f(z) = c_1 \int_0^z \frac{d\zeta}{\sqrt{1 - \zeta^2}\sqrt{1 - k^2 \zeta^2}} + c_2, \quad \text{where} \tag{$*7$}$$

$c_2 = f(0) = 0$;

$$c_1 = \frac{K}{\int_0^1 \frac{1}{\sqrt{(1-x^2)(1-k^2 x^2)}}\, dx} \quad \text{in which } K = f(1) \text{ is a vertex of the rectangle.}$$

In this case, the height K' of the rectangle is determined by $f(\frac{1}{k}) = K + iK'$; namely,

$$K' = c_1 \int_1^{1/k} \frac{dx}{\sqrt{(x^2 - 1)(1 - k^2 x^2)}} = c_1 \int_0^1 \frac{dx}{\sqrt{(1 - x^2)[1 - (1 - k^2)x^2]}}$$

(setting $x = [1 - (1 - k^2)t^2]^{-1/2}$ and then changing t back to x).

We may fix k ($0 < k < 1$) in advance and choose $c_1 = 1$. Then the mapping reduces to

$$w = f(z) = \int_0^z \frac{d\zeta}{\sqrt{1 - \zeta^2}\sqrt{1 - k^2 \zeta^2}}, \quad \operatorname{Im} z > 0.$$

This is exactly the one we studied in Example 3 of Sec. 3.5.5. The inverse $z = f^{-1}(w)$ is an elliptic function of order 2 (the simplest among all), having periods $4K$ and $2K'i$. The period parallelogram with vertices at $-K \pm iK'$ and $3K \pm iK'$ (see Fig. 3.77) contains one zero 0 of order 2 and one pole $-iK'$ of order 2. See Figs. 3.80 and 3.81 for the Riemann surface R_z for $w = f(z)$ and its line complex.

As an application of (6.2.1.7), we list the following two examples.

Example 1. Show that

$$w = \int_0^z \frac{dz}{(1 - z^n)^{2/n}}, \quad |z| < 1$$

maps $|z| < 1$ univalently onto a regular n-gon with side length

$$\frac{1}{n} 2^{1 - \frac{4}{n}} \frac{\left(\Gamma\left(\frac{1}{2} - \frac{1}{n}\right)\right)^2}{\Gamma\left(1 - \frac{2}{n}\right)}, \quad n \geq 3.$$

Solution. The equation $1 - z^n = 0$ has the roots $1, \omega, \omega^2, \ldots, \omega^{n-1}$, where $\omega = e^{2\pi i/n}$. Set $\alpha_k = 1 - \frac{2}{n}$ and $\beta_k = \frac{2}{n}$ for $1 \leq k \leq n$. Note that

$\alpha_1 + \cdots + \alpha_n = n - 2$ (see (6.2.1.8)). According to (6.2.1.5) and (6.2.1.7), this $w = f(z)$ maps $|z| < 1$ univalently onto a polygon, each of its n vertices having the interior angle $(1 - \frac{2}{n})\pi$ and the exterior angle $\frac{2\pi}{n}$. So it is a regular n-gon.

Note that $w(0) = 0$ (and note that $w'(0) = 1$). The vertices are given by

$$w_1 = w(1) = \int_0^1 \frac{dz}{(1 - z^n)^{2/n}} = \frac{1}{n} \int_0^1 t^{\frac{1}{n} - 1}(1 - t)^{1 - \frac{2}{n} - 1} dt$$

$$= \frac{1}{n} \cdot \frac{\Gamma\left(\frac{1}{n}\right) \Gamma\left(1 - \frac{2}{n}\right)}{\Gamma\left(1 - \frac{1}{n}\right)} \quad \text{(see (5.6.2.4) and (5.6.2.5))};$$

$$w_k = w(e^{2k\pi i/n}) = \int_0^{2k\pi i/n} \frac{dz}{(1 - z^n)^{2/n}} = \int_0^1 \frac{e^{2k\pi i/n} dz}{(1 - e^{2k\pi i} z^n)^{2/n}}$$

$$= e^{2k\pi i/n} \int_0^1 \frac{dz}{(1 - z^n)^{2/n}} = e^{2k\pi i/n} w_1, \quad 1 \le k \le n - 1.$$

Hence the side length is given by

$$2w_1 \sin \frac{\pi}{n} = \frac{2}{n} \cdot \frac{\Gamma\left(\frac{1}{n}\right) \Gamma\left(1 - \frac{2}{n}\right)}{\Gamma\left(1 - \frac{1}{n}\right)} \cdot \frac{\pi}{\Gamma\left(\frac{1}{n}\right) \Gamma\left(1 - \frac{1}{n}\right)} \quad \text{(see (5.6.1.6))}$$

$$= \frac{2\pi}{n} \cdot \frac{\Gamma\left(1 - \frac{2}{n}\right)}{\pi \left(\Gamma\left(1 - \frac{2}{n}\right)\right)^2} \cdot 2^{-\frac{4}{n}} \left(\Gamma\left(\frac{1}{2} - \frac{1}{n}\right)\right)^2$$

$$= \frac{1}{n} \cdot 2^{1 - \frac{4}{n}} \frac{\left(\Gamma\left(\frac{1}{2} - \frac{1}{n}\right)\right)^2}{\Gamma\left(1 - \frac{2}{n}\right)},$$

where, in the second equality, we have used Legendre's product formula $\sqrt{\pi} \Gamma(2z) = 2^{2z-1} \Gamma(z) \Gamma(z + \frac{1}{2})$ (see (4) in (5.6.2.1)) by setting $z = \frac{1}{2} - \frac{1}{n}$.

Example 2. Show that

$$w = \int_0^z \frac{(1 - z^5)^{2/5}}{(1 + z^5)^{4/5}} dz, \quad |z| < 1$$

maps $|z| < 1$ univalently onto the convex starlike domain shown in Fig. 6.18, whose circumscribed circle has the radius

$$R = \int_{-1}^0 (1 - t^5)^{2/5}(1 + t^2)^{-4/5} dt = \frac{\Gamma\left(\frac{1}{10}\right) \Gamma\left(\frac{1}{5}\right)}{2^{2/5} \cdot 5\Gamma\left(\frac{3}{10}\right)},$$

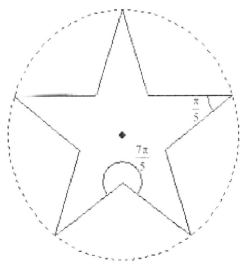

Fig. 6.18

and the inscribed circle has the radius

$$r = \int_0^1 (1-t^2)^{2/5}(1+t^5)^{-4/5}dt = \frac{\Gamma\left(\frac{7}{10}\right)\Gamma\left(\frac{2}{5}\right)}{2^{2/5}\cdot 5\Gamma\left(\frac{9}{10}\right)}.$$

Solution. The roots $z_k = e^{2k\pi i/5}$ $(0 \le k \le 4)$ of $z^5 - 1 = 0$ are mapped into vertices of interior angle $\pi - \frac{2}{5}\pi = \frac{7}{5}\pi$ each; while the roots $\zeta_k = e^{(\pi+2k\pi)i/5}$ $(0 \le k \le 4)$ of $z^5 + 1 = 0$ are mapped into vertices of interior angle $\pi - \frac{4}{5}\pi = \frac{1}{5}\pi$ each.

According to (6.2.1.5) and (6.2.1.7), $w = w(z)$ maps $|z| < 1$ univalently onto a starlike domain shown in Fig. 6.18. z_k $(0 \le k \le 4)$ are mapped into

$$w_0 = w(z_0) = \int_0^1 \frac{(1-t^5)^{2/5}}{(1+t^5)^{4/5}}dt$$

$$= \int_1^0 \left(\frac{2t^{1/2}}{1+t^{1/2}}\right)^{2/5}\left(\frac{1+t^{1/2}}{2}\right)^{4/5}\cdot\frac{1}{5}\left(\frac{1+t^{1/2}}{1-t^{1/2}}\right)^{4/5}\frac{-t^{-1/2}}{(1+t^{1/2})^2}dt$$

$$\left(\text{by changing the variable } \left(\frac{1-t^5}{1+t^5}\right)^2 \to t\right)$$

$$= \frac{1}{5 \cdot 2^{2/5}} \int_0^1 t^{-3/10}(1 - t)^{-4/5} dt$$

$$= \frac{1}{5 \cdot 2^{2/5}} \cdot \frac{\Gamma\left(\frac{7}{10}\right)\Gamma\left(\frac{1}{5}\right)}{\Gamma\left(\frac{9}{10}\right)},$$

$$w_k = w(z_k) = \int_0^{e^{2k\pi i/5}} \frac{(1 - z^5)^{2/5}}{(1 + z^5)^{4/5}} dz = \cdots = e^{2k\pi i/5} w_0, \quad 1 \le k \le 4;$$

$\zeta_k \ (0 \le k \le 4)$ are mapped into

$$\widetilde{w_0} = w(\zeta_0) = \int_0^{e^{\pi i/5}} \frac{(1 - z^5)^{2/5}}{(1 + z^5)^{2/5}} dz = e^{\pi i/5} \int_0^1 \frac{(1 + t^5)^{2/5}}{(1 - t)^{4/5}} dt = \cdots$$

$$= e^{\pi i/5} \cdot \frac{1}{5 \cdot 2^{2/5}} \int_0^1 t^{-9/10}(1 - t)^{-4/5} dt$$

$$= e^{\pi i/5} \cdot \frac{1}{5 \cdot 2^{2/5}} \frac{\Gamma\left(\frac{1}{10}\right)\Gamma\left(\frac{1}{5}\right)}{\Gamma\left(\frac{3}{10}\right)},$$

$$\widetilde{w_k} = w(\zeta_k) = \cdots = e^{2k\pi i/5}\widetilde{w_0}, \quad 1 \le k \le 4.$$

*Section (3) Curvilinear triangle with circular arcs as sides

Suppose now that Ω is a simply connected domain bounded by a finite number of circular arcs, which will be just referred to as a *curvilinear polygon*. The Riemann mapping theorem still guarantees that there is a function $w = f(z)$ mapping $\operatorname{Im} z > 0$ univalently onto Ω such that it induces a homeomorphism between $\operatorname{Im} z \ge 0$ and $\bar{\Omega}$. The essence leading to the Schwarz–Christoffel formula (6.2.1.1) can be similarly generalized to obtain a corresponding formula, even though much more complicated, in the present case. Instead of $\frac{w''}{w'}$ (see $(*_3)$ and $(*_4)$ in Sec. 6.2.1), but now the so-called *Schwarzian derivative* of $w = f(z)$

$$\{w, z\} = \left(\frac{w''}{w'}\right)' - \frac{1}{2}\left(\frac{w''}{w'}\right)^2 = \frac{w'''}{w'} - \frac{3}{2}\left(\frac{w''}{w'}\right)^2 \tag{$*_4$}'$$

is *invariant* under the linear fractional transformations which carry circular arcs into circular arcs. At a point a_k, on the real axis, which is mapped into a vertex w_k of angle $\pi\alpha_k$ (see $(*_8)$ in Sec. 6.2.1), instead we have now

$$\{w, z\} = \frac{1 - \alpha_k^2}{2(z - a_k)^2} + \frac{\gamma_k}{z - a_k} + h_k(z), \quad \text{where}$$

$$\gamma_k = \frac{1 - \alpha_k^2}{\alpha_k}\frac{h_k'(a_k)}{h_k(a_k)} \quad \text{for } 1 \le k \le n. \tag{$*_8$}'$$

Note $h_k(z)$ is a real function if z is real, and hence, γ_k is real. Applying the Argument principle (see Sec. 3.5.3) instead of Liouville's theorem, $(*10)$ becomes

$$\{w, z\} = \frac{1}{2} \sum_{k=1}^{n} \frac{1 - \alpha_k^2}{(z - a_k)^2} + \sum_{k=1}^{n} \frac{\gamma_k}{z - a_k} + \gamma, \qquad (*10)'$$

where γ is a real constant. If none of the points $a_1, \ldots, a_k$ is the point at infinity, then $w = f(z)$ will be analytic at ∞. The substitution of the expansion $f(z) = b_0 + \sum_{m=1}^{\infty} \frac{b_m}{z^m}$ near ∞ into $\{w, z\}$ in $(*8)'$ will yield the following *four* conditions:

$$\gamma = 0;$$

$$\sum_{k=1}^{n} \gamma_k = 0;$$

$$\sum_{k=1}^{n} \left(2a_k\gamma_k + 1 - \alpha_k^2\right) = 0;$$

$$\sum_{k=1}^{n} [\gamma_k a_k^2 + a_k(1 - \alpha_k^2)] = 0. \qquad (*9)'$$

The last three relations should be satisfied by the n real constants $\gamma_1, \ldots, \gamma_n$. Finally, a consideration of the linearly independent solutions of the linear differential equation $u''(z) + p(z)u(z) = 0$ will lead to

The Schwarz–Christoffel formula for mapping $\operatorname{Im} z > 0$ *univalently onto a curvilinear polygon with n circular arc sides.* Let the points $a_1, \ldots, a_n$ on the real axis be mapped respectively into the vertices $w_1, \ldots, w_n$ of interior angles $\alpha_1\pi, \ldots, \alpha_n\pi$. Then such a function is of the form

$$w = f(z) = \frac{u_1(z)}{u_2(z)},$$

where $u_1(z)$ and $u_2(z)$ are two linearly independent solutions of the linear differential equation

$$u''(z) + \left[\frac{1}{4} \sum_{k=1}^{n} \frac{1 - \alpha_k^2}{(z - a_k)^2} + \frac{1}{2} \sum_{k=1}^{n} \frac{\gamma_k}{z - a_k} \right] u(z) = 0$$

and the real constants $\gamma_1, \ldots, \gamma_n$ are subject to the last three relations in $(*9)'$. \hfill (6.2.2.1)

According to (2) in (6.1.2.2), $(n-3)$ points out of $a_1, \ldots, a_n$ and hence, $(n-3)$ constants out of $\gamma_1, \ldots, \gamma_n$ can be chosen independently. This raises difficulty in practice.

In case $n = 3$, the complete solution is given by

The Schwarz–Christoffel formula for mapping $\operatorname{Im} z > 0$ *univalently onto a curvilinear triangle* $\Delta w_1 w_2 w_3$. Choose $a_1 = 0$, $a_2 = \infty$, and $a_3 = 1$ so that they are mapped into w_1, w_2, and w_3, respectively. Then such a function is of the form

$$w = f(z) = \frac{u_1(z)}{u_2(z)},$$

where $u_1(z)$ and $u_2(z)$ are two linearly independent solutions of the *hypergeometric equation* (see Exercise B(1) of Sec. 3.3.2 and (5) in (5.6.2.9))

$$z(1 - z)u'' + [c - (a + b + 1)z]u' - abu = 0,$$

where $a = \frac{1}{2}(1 + \alpha_2 - \alpha_1 - \alpha_3)$, $b = \frac{1}{2}(1 - \alpha_1 - \alpha_2 - \alpha_3)$, $c = 1 - \alpha_1$, and $\alpha_1\pi$, $\alpha_2\pi$, $\alpha_2\pi$ are the interior angles at w_1, w_2, w_3, respectively.

$$(6.2.2.2)$$

Owing to the choice of a_1, a_2, and a_3, caution that now the vertices w_1, w_2, and w_3 are arranged in the clockwise direction and $\alpha_1\pi$, $\alpha_2\pi$, and $\alpha_3\pi$ are considered to be positive angles, even in the clockwise sense.

Z. Nehari [63], pp. 198–208, has presented a detailed account on (6.2.2.1) and (6.2.2.2), to which the readers are referred. See also pp. 308–316 in the book for the following sketch about the *Schwarzian triangle functions* $w = f(z)$ defined in (6.2.2.2).

We will concentrate on these functions $w = f(z)$ whose inverse functions $z = f^{-1}(w)$ are single-valued in the w-plane. Exactly like the explanation followed $(*_1)$, $\frac{1}{\alpha_1}$, $\frac{1}{\alpha_2}$, and $\frac{1}{\alpha_3}$ all should be positive integers or $+\infty$ in order to have $z = f^{-1}(w)$ single-valued. $z = f^{-1}(w)$ is called *automorphic* if there is a subgroup S, called an *automorphic group*, of the group of the linear fractional transformations (see (1) in (2.5.4.13)) such that

$$f^{-1}(Tw) = f^{-1}(w) \text{ for all } T \in S \quad \text{and} \quad w. \qquad (6.2.2.3)$$

In other words, $z = f^{-1}(w)$ is reproduced if w is subject to each member of the group S.

The case $\alpha_1 + \alpha_2 + \alpha_3 > 1$. These are five possibilities.

Case 1. $\alpha_1 = 1$ and $\alpha_2 = \alpha_3 = \frac{1}{n}$ $(n \geq 2)$.
This is a degenerated polygon with two sides. See Fig. 6.19. The inverse function $z = f^{-1}(w)$ is of the form $z = w^n$, for simplicity. The Riemann surface R_z and its line complex are shown in Figs. 2.48 and 2.50, respectively. The point a_1 corresponding to $\alpha_1 = 1$ is a removable singularity, while the points a_2 and a_3 corresponding to $\alpha_2 = \alpha_3 = \frac{1}{n}$ are branch points

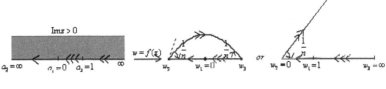

Fig. 6.19

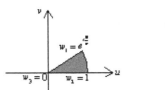

Fig. 6.20

of order $n - 1$ each. The associated automorphic group is the cyclic group $\{e^{2k\pi i/n}w|k = 0, 1, 2, \ldots, n - 1\}$, generated by $e^{2\pi i/n}$.

Case 2. $\alpha_1 = \alpha_2 = \frac{1}{2}$ and $\alpha_3 = \frac{1}{n}$ $(n \geq 1)$ (Ref. [50], pp. 53–54).
The triangle $\Delta w_1 w_2 w_3$ is the shaded one in Fig. 6.20. The consecutive reflections of $\Delta w_1 w_2 w_3$ with respect to its three sides will result in $4n$ conformally congruent triangles which constitute a tessellation of the extended plane $\mathbf{C}^*$. Since two consecutive triangles together cover the extended z-plane $\mathbf{C}^*$ once, the inverse function is a rational function of degree $\frac{4n}{2} = 2n$ (see Sec. 2.5.3 and Example 6 in Sec. 4.10.2). This can also be justified by observing that $z = f^{-1}(w)$ has n double poles $w_2 e^{2k\pi i/n}$ for $0 \leq k \leq n-1$. $z = f^{-1}(w)$ is automorphic. The Riemann surface for $w = f(z)$ is $2n$-sheeted. The branch points corresponding to $\alpha_1 = \alpha_2 = \frac{1}{2}$ are of order 1 each, while that corresponding to $\alpha_3 = \frac{1}{n}$ is of order $n-1$. The line complex is shown in the right figure in Fig. 6.20.

Case 3. $\alpha_1 = \frac{1}{2}$ and $\alpha_2 = \alpha_3 = \frac{1}{3}$ (Ref. [50], p. 55).
The triangle $\Delta w_1 w_2 w_3$ is one of the shaded parts appeared in Fig. 6.21. The images of the three vertices, under the stereographic projection (see Sec. 1.6), on the Riemann sphere are the vertices of a face of a regular tetrahedron inscribed to the sphere. The whole sphere can be tessellated by 24 conformally congruent triangles. The configuration is drawn out in the extended w-plane in the left figure in Fig. 6.21. The inverse function $z = f^{-1}(w)$ is therefore a rational function of order 12. It is automorphic. The Riemann surface R_z for $w = f(z)$ is 12-sheeted, the branch points

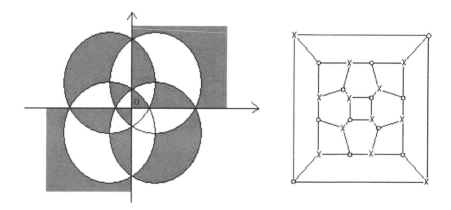

Fig. 6.21

corresponding to $\alpha_2 = \alpha_3 = \frac{1}{3}$ are of order 2 each, and that corresponding to $\alpha_1 = \frac{1}{2}$ is of order 1. The line complex is shown in the right side of Fig. 6.21.

Case 4. $\alpha_1 = \frac{1}{2}$, $\alpha_2 = \frac{1}{3}$, and $\alpha_3 = \frac{1}{4}$ (Ref. [50], p. 56).

In this case, the images of the three vertices of $\Delta w_1 w_2 w_3$ on the Riemann sphere are the vertices of a face of a regular octahedron inscribed to the sphere. And hence 48 such conformally congruent triangles form a tessellation of the whole sphere. The inverse function $z = f^{-1}(w)$ is a rational function of order 24 and is also an automorphic function. The line complex of the 24-sheeted Riemann surface R_z for $w = f(z)$ is shown in Fig. 6.22. The branch points corresponding to $\alpha_1 = \frac{1}{2}$ are of order 1 each, those corresponding to $\alpha_2 = \frac{1}{3}$ are of order 2 each, and those corresponding to $\alpha_3 = \frac{1}{4}$ are of order 3 each.

Case 5. $\alpha_1 = \frac{1}{2}$, $\alpha_2 = \frac{1}{3}$, and $\alpha_3 = \frac{1}{5}$.

The image of the vertices of $\Delta w_1 w_2 w_3$ on the Riemann sphere form the vertices of a face of a regular 20-gon inscribed to the sphere. The whole sphere is tessellated by 120 conformally congruent triangles. The inverse function is thus a rational function of order 60. It is automorphic, too. The Riemann surface for $w = f(z)$ is 60-sheeted, and the branch points corresponding to $\alpha_1 = \frac{1}{2}$, $\alpha_2 = \frac{1}{3}$, and $\alpha_3 = \frac{1}{5}$ are of the orders 1, 2, and 4 each, respectively. The portion of the associated line complex that lies in the first quadrant is shown in Fig. 6.23, while these in the other three quadrants are obtained by performing the reflection with respect to the axes but paying attention to the exchange of the positions of 0 and $\times$.

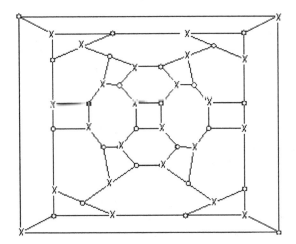

Fig. 6.22

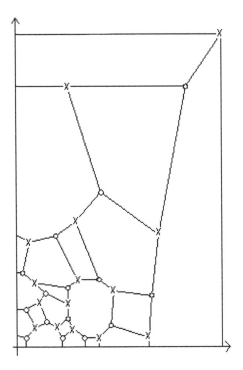

Fig. 6.23

The case $\alpha_1 + \alpha_2 + \alpha_3 = 1$.

By invoking a linear fraction transformation, $\Delta w_1 w_2 w_3$ is transformed into a triangle whose sides are rectilinear and then the mapping function can be furnished by the Schwarz–Christoffel formula as discussed in Section (1).

The case $\alpha_1 + \alpha_2 + \alpha_3 < 1$.

In order to have the inverse function $z = f^{-1}(w)$ to be single-valued, it is necessarily possible to construct a circle containing the triangle $\Delta w_1 w_2 w_3$ inside and intersecting the three circles making up the triangle at right angles. Such a circle is called the *orthogonal circle* of the triangle.

Then the symmetry principle permits us to analytically continue $z = f^{-1}(w)$ throughout the interior of the fixed orthogonal circle Γ. Monodromy theorem (see (5.2.2.2)) says that the extended $z = f^{-1}(w)$ is still single-valued in the disk Int Γ and Γ itself is the natural boundary for it. It is, of course, automorphic. See Ref. [63], pp. 308–316 for a detailed account.

The Riemann surface R_z for $w = f(z)$ is infinitely-sheeted. The branch points corresponding to α_1, α_2, and α_3 are of infinite order each. The tessellation of Int Γ by the conformally congruent triangles for the case $\alpha_1 = \frac{1}{7}$, $\alpha_2 = \frac{1}{2}$, and $\alpha_3 = \frac{1}{3}$ is nicely drawn in Fig. 41 on Nehari's book [63], p. 312. As an exercise, the readers are asked to draw the corresponding line complex in this case.

An extremal case happens if $\alpha_1 = \alpha_2 = \alpha_3 = 0$. Such a concrete example is the *elliptic modular function* we introduced in Section (1) of Sec. 5.8.3 for the main purpose of proving Montel's normality criterion. Go back and read carefully if you missed it previously. Also, refer to the latter part of Exercise B for further explanation.

Exercises A

(1) Map the upper half-plane $\operatorname{Im} z > 0$ univalently onto the rhombus with 0, a $(a > 0)$, and $a(1 + e^{i\alpha\pi})$ as three of its vertices in such a way that 0, 1, and ∞ are mapped into these three vertices, respectively.

(2) Map the open disk $|z| < 1$ univalently onto the rhombus, with side length l and having an obtuse angle $\alpha\pi$.

(3) Map $|z| < 1$ univalently onto the strip domain $-\frac{1}{4}\pi < \operatorname{Re} w < \frac{1}{4}\pi$ such that i and $-i$ are mapped into ∞.

Note. This exercise together with $(*_5)$ and $(*_6)$ shows that the mappings of a disk or a half-plane onto a parallel strip, or onto a half strip with two right angles, are special cases of the Schwarz–Christoffel formula.

(4) Show that

$$w = f(z) = \int_0^z \frac{dz}{\sqrt{z(1-z^2)}}$$

maps Im $z > 0$ univalently onto a square with side length $\frac{1}{2\sqrt{2\pi}} \left(\Gamma\left(\frac{1}{4}\right)\right)^2$. Find the vertices of the square. Show that the inverse function $z = f^{-1}(w)$ is a single-valued doubly-periodic function. Construct the Riemann surface R_z for the extended $w = f(z)$ and draw its line complex.

(5) Show that

$$w = f(z) = \int_0^z \frac{dz}{\sqrt{1-z^4}}$$

maps $|z| < 1$ univalently onto a square whose diagonal is of length equal to $\frac{1}{2\sqrt{2\pi}} \left(\Gamma\left(\frac{1}{4}\right)\right)^2$. What is its side length? Where are the vertices?

(6) Show that

$$w = \int_0^z \frac{(1+t^n)^\lambda}{(1-t^2)^{2/n+\lambda}} dt, \quad -1 < \lambda < 1 - \frac{2}{n}$$

maps $|z| < 1$ univalently onto a starlike domain.

(7) Show that a univalent function $w = w(z)$, mapping $|z| < 1$ onto a convex n-polygon and satisfying the condition $w(0) = 0$ and $w'(0) = 1$, is of the form

$$w(z) = \int_0^z \frac{dz}{\prod_{k=1}^n (1 - e^{i\theta_k} z)^{\beta_k}}, \quad \beta_k > 0, \quad 0 \le \theta_k < 2\pi$$
$$\text{for} \quad 1 \le k \le n.$$

Also show that, on $|z| = r < 1$,

$$|w'(z)| \le \frac{1}{1-r^2} \quad \text{and} \quad |w(z)| \le \frac{r}{1-r}.$$

In particular,

$$w(z) = \int_0^z \frac{dz}{(1-z^n)^{2/n}}$$

maps $|z| < 1$ onto a regular n-gon. Find the vertices $w_k = w(e^{2k\pi i/n})$, $0 \le k \le n-1$, and the image points $w(e^{(2k+1)\pi i/n})$, $0 \le k \le n-1$.

(8) Show that

$$w = \int_z^\infty \frac{dz}{(z^3 - z)^{2/3}}$$

maps Im $z > 0$ univalently onto an equilateral triangle. Pinpoint the vertices.

(9) Show that

$$w = \int_0^z \frac{dz}{z^{1/2}(1 - z)^{2/3}(1 + z)^{5/6}}$$

maps Im $z > 0$ univalently onto a right triangle with interior angle $\frac{\pi}{2}$, $\frac{\pi}{3}$, and $\frac{\pi}{6}$. Find the vertices of the triangle.

(10) Map the interior of a rectangle univalently onto the exterior of another rectangle.

Exercises B

(1) Read Nehari's book [63] to convince oneself the validity of (6.2.2.1) and (6.2.2.2).

(2) Find the corresponding automorphic group for each of these automorphic functions mentioned from Cases 2–5 in Section (3).

In what follows, we give a sketch about general properties of elliptic functions. For details about elliptic functions, refer to Refs. [58], Vol. III; [44], Vol. II; [71] and the bibliographs listed inside; for a quick outlook, see Ref. [1], Chap. 7. A function $f(z) : \mathbf{C} \to \mathbf{C}$ is said to be *doubly periodic* with two periods ω_1 and ω_2 if $f(z + \omega_1) = f(z + \omega_2) = f(z)$ for all z and the ratio Im$\frac{\omega_2}{\omega_1} > 0$. A doubly periodic function $f(z)$ is called *elliptic* if it is a meromorphic function. The parallelogram Δ with vertices 0, ω_1, $\omega_1 + \omega_2$, and ω_2 is called the (fundamental) *period parallelogram*. It is tacitly understood that Δ does not contain the two closed sides sharing the vertex $\omega_1 + \omega_2$. Two points z and z' are said to be *congruent* with respect to ω_1 and ω_2 if $z - z' = m_1\omega_1 + m_2\omega_2$ for some integers m_1 and m_2. Congruence is an equivalent relation. Each equivalent class has exactly only one representative in Δ. The action of translation on Δ will tessellate the entire plane. And, for convenience, we will consider the translated set $\Delta_a = \{a + z | z \in \Delta\}$ as a period parallelogram, so that $f(z)$ has no poles on the boundary of Δ_a. The behavior of an elliptic function on $\mathbf{C}$ is therefore narrowly visualized on a period parallelogram. Do the following problems.

(3) An elliptic function without poles is a constant.

Note. An elliptic function can only have finitely many poles in a period parallelogram. When an elliptic function is mentioned, it will always mean a nonconstant one.

(4) The sum of the residues of the poles in a period parallelogram of an elliptic function is zero. Hence, a nonconstant elliptic function has equally many poles as it has zeros.

Note. The sum of the orders of the poles of an elliptic function in a period parallelogram is called the *order* of the function. Therefore the order of an elliptic function is ≥ 2.

(5) An elliptic function of order $n \geq 2$ assumes every value in $\mathbf{C}^* n$ times, counting the multiplicities, in a period parallelogram.

(6) The zeros $a_1, \ldots, a_m$ and the poles $b_1, \ldots, b_m$ of an elliptic function satisfy the property that $\sum_{k=1}^m a_k$ and $\sum_{k=1}^m b_k$ are congruent with respect to the periods.

The elliptic functions of order 2, the simplest among all: The one with two simple poles and opposite residues was due to Jacobi; the other with a double pole and residue zero was due to Weierstrass, some 30 years behind that of Jacobi's, and was called the *Weierstrass $\wp$-function.* Fix two nonzero complex constants ω_1 and ω_2 with $\mathrm{Im}\frac{\omega_2}{\omega_1} > 0$. The $\wp$-function is defined by

$$\wp(z) = \frac{1}{z^2} + \sum_{\omega \neq 0} \left[\frac{1}{(z-\omega)^2} - \frac{1}{\omega^2} \right], \qquad (6.2.2.4)$$

where the sum ranges over all the periods $\omega = 2n_1\omega_1 + 2n_2\omega_2$, $n_1, n_2 = \pm 1, \pm 2, \ldots$ except 0. The series on the right converges absolutely and locally uniformly in $\mathbf{C} - \{2n_1\omega_1 + 2n_2\omega_2 | n_1, n_2 = 0, \pm 1, \pm 2, \ldots\}$ (refer to (5.4.1.6)). Then $\wp(z)$ is an *even* elliptic function of *order* 2, with fundamental periods $2\omega_1$ and $2\omega_2$. $\wp(z)$ has a double pole at 0 and hence it covers $\mathbf{C}^*$ twice when z ranges over the fundamental period parallelogram Δ_0 with vertices at 0, $2\omega_1$, $2(\omega_1 + \omega_2)$, and $2\omega_2$. Note that the derivative

$$\wp'(z) = -2 \sum_{\omega} \frac{1}{(z-\omega)^3} \quad \text{(including } \omega = 0\text{)} \qquad (6.2.2.5)$$

is an odd elliptic function of order 3. While the normalized primitive

$$\zeta(z) = \frac{1}{z} + \sum_{\omega \neq 0} \left[\frac{1}{z-\omega} + \frac{1}{\omega} + \frac{z}{\omega^2} \right] \qquad (6.2.2.6)$$

satisfies the conditions that $\zeta(z + 2\omega_1) = \zeta(z) + \eta_1$, $\zeta(z + 2\omega_2) = \zeta(z) + \eta_2$, where η_1 and η_2 are constants satisfying *Legendre's relation* $\eta_1\omega_2 - \eta_2\omega_1 = \pi i$. The Laurent series of $\zeta(z)$ at $z = 0$ will give rise to that of $\wp(z) = -\zeta'(z)$ at $z = 0$, and will eventually lead to the differential equation

$$\wp'(z)^2 = 4\wp(z)^3 - g_2\wp(z) - g_3, \quad g_2 = 60\sum_{\omega \neq 0}\frac{1}{\omega^4} \quad \text{and} \quad g_3 = 140\sum_{\omega \neq 0}\frac{1}{\omega^6}$$

satisfied by $\wp(z)$. Setting $w = \wp(z)$ the above expression is rewritten as

$$\left(\frac{dw}{dz}\right)^2 = 4w^3 - g_2w - g_3. \tag{6.2.2.7}$$

Since $\wp(2\omega_1 - z) = \wp(z)$, $\wp'(2\omega_1 - z) = -\wp'(z)$ implies that $\wp'(\omega_1) = 0$. Similarly, $\wp'(\omega_2) = \wp'(\omega_1 + \omega_2) = 0$. The points ω_1, ω_2, and $\omega_1 + \omega_2$ all lie in the parallelogram Δ_0 (and hence are mutually incongruent); they are therefore precisely the three distinct simple zeros of $\wp'(z)$ in Δ_0. Set

$$e_1 = \wp(\omega_1), \quad e_2 = \wp(\omega_2), \quad \text{and} \quad e_3 = \wp(\omega_1 + \omega_2). \tag{6.2.2.8}$$

Note that then $4w^3 - g_2w - g_3 = 4(w - e_1)(w - e_2)(w - e_3)$. Moreover, (6.2.2.7) implies that

$$w' = \sqrt{4w^3 - g_2w - g_3}, \quad w = \wp(z)$$

in which the sign of the square root must be chosen so that it actually equals to the already known *single-valued* function $\wp'(z)$. Consequently,

$$z - z_0 = \int_\gamma \frac{dw}{\sqrt{4w^3 - g_2w - g_3}}, \quad w = \wp(z) \tag{6.2.2.9}$$

shows that $w = \wp(z)$ is the *inverse* of an elliptic integral of the first kind, where the path γ of integration is the image under $\wp(z)$ of a path from z_0 to z that avoids the zeros and poles of $\wp'(z)$. In particular, letting $z_0 \to 0$ (and then $w = \wp(z_0) \to \infty$), we obtain

$$z = \int_\infty^w \frac{d\zeta}{\sqrt{4\zeta^3 - g_2\zeta - g_3}}, \quad w = \wp(z), \tag{6.2.2.10}$$

where the square root is chosen to agree with $\wp'(z)$. On the other hand, if the period ω_1 and ω_2 are multiplied by t simultaneously, then the e_k

are multiplied by t^{-2}. From (6.2.2.8), we then conclude that

$$\lambda(\tau) = \frac{e_3 - e_1}{e_2 - e_1}, \quad \tau = \frac{\omega_2}{\omega_1} \tag{6.2.2.11}$$

depends only on the ratio τ. $\lambda(\tau)$ is entire and does not assume the values 0 and 1 in the τ-plane. It can be shown that $\lambda(\tau)$ maps the domain

$$\left\{ \tau \in \mathbf{C} \middle| 0 < \operatorname{Re}\tau < 1 \text{ and } \left| \tau - \frac{1}{2} \right| > \frac{1}{2}, \operatorname{Im}\tau > 0 \right\}$$

univalently onto the upper half-plane, and can be extended continuously to the boundary in such a way that $\tau = 0, 1, \infty$ correspond to $\lambda = 0, 1, \infty$, respectively. This $\lambda(\tau)$ is nothing new but the *elliptic modular function* $z = \varphi(w)$ mentioned in Section (1) of Sec. 5.8.3.

(7) (Refs. [1], p. 282; [44], Vol. II, p. 169; [50], pp. 197–200) Show that the function

$$J(\tau) = 1 - \frac{1}{27} \left[-(1 + \lambda) \left(1 + \frac{1}{1 - \lambda} \right) \left(1 + \frac{\lambda - 1}{\lambda} \right) \left(1 + \frac{1}{\lambda} \right) \right.$$

$$\left. \times \left(1 + \frac{\lambda}{\lambda - 1} \right) (1 + 1 - \lambda) \right] = \frac{4}{27} \frac{(1 - \lambda + \lambda^2)^3}{\lambda^2 (1 - \lambda)^2}$$

$$= \frac{-4(e_1 e_2 + e_2 e_3 + e_3 e_1)^3}{(e_1 - e_2)^2 (e_2 - e_3)^2 (e_3 - e_1)^2}$$

is automorphic with respect to the *full modular group* (see also (5.8.3.9))

$$\left\{ \begin{bmatrix} \omega_2' \\ \omega_1' \end{bmatrix} = \begin{bmatrix} a & b \\ c & d \end{bmatrix} \begin{bmatrix} \omega_2 \\ \omega_1 \end{bmatrix} \middle| a, b, c, d \text{ are integers and } ad - bc = \pm 1 \right\}.$$

Show that $J(\tau)$ maps the domain $\{ \tau \in \mathbf{C} | 0 < \operatorname{Re}\tau < \frac{1}{2} \text{ and } |\tau| > 1, \operatorname{Im}\tau > 0 \}$ univalently onto the upper half-plane, and can be extended continuously to the boundary. Decide where it takes the values 0, 1, and ∞, and with what multiplicities. Construct the Riemann surface for the inverse $\tau = J^{-1}(\zeta)$ of $\zeta = J(\tau)$ and its line complex.

6.2.3   The Schwarz–Christoffel formula for the generalized polygon

A simply connected domain Ω in $\mathbf{C}^*$ is called a *generalized polygon* if its boundary $\partial\Omega$ is a closed polygonal curve in $\mathbf{C}^*$ whose sides may be traversed more than twice and are allowed to have infinite length. We adopt the following convention:

1. The boundary $\partial\Omega$ is oriented in the counterclockwise direction such that Ω always lies on the left side when traveling along it.
2. The *interior angle* $\alpha\pi$ at a *finite* vertex lies between 0 and 2π (included).
3. In case $\partial\Omega$ passes ∞, ∞ is always regarded as a vertex of the polygon and is assigned a *negative interior angle* $\alpha\pi$ between -2π and 0 (included).   (6.2.3.1)

Then, we have

The Schwarz–Christoffel formula for a generalized polygon with ∞ a boundary point (or vertex). Suppose the polygonal boundary $\partial\Omega$ of such a polygon Ω has the interior angles $\alpha_1\pi, \ldots, \alpha_n\pi$ ($0 \le \alpha_k \le 2$ for $1 \le k \le n$) at its finite vertices $w_1, \ldots, w_n$ and the interior angles in the interval $[-2\pi, 0]$ at infinite vertices. Then the functions $w = f(z)$ that map $\mathrm{Im}\, z > 0$ univalently onto Ω are of the form

$$w = f(z) = c_1 \int_{z_0}^{z} \prod_{k=1}^{n} (\zeta - a_k)^{\alpha_k - 1} d\zeta + c_2,$$

exactly as in (6.2.1.1), where $a_k = f^{-1}(w_k)$, $1 \le k \le n$, are arranged as $-\infty < a_1 < \cdots < a_n < \infty$. In this case, the sum of the interior angles (included these at infinite) equals $(n - 2)\pi$.   (6.2.3.2)

Moreover, in case $a_n = \infty$ and $w_n = f(a_n)$ holds, then

$$w = f(z) = c_1 \int_{z_0}^{z} \prod_{k=1}^{n-1} (\zeta - a_k)^{\alpha_k - 1} d\zeta + c_2 \qquad (6.2.3.3)$$

as in (1) of (6.2.1.4); in case $a_n = \infty$ corresponds to w_n, a finite vertex, yet ∞ is an interior point of Ω, then

$$w = f(z) = c_1 \int_{z_0}^{z} \frac{\prod_{k=1}^{n-1} (\zeta - a_k)^{\alpha_k - 1}}{(\zeta - b)^2 (\zeta - \bar{b})^2} d\zeta + c_2, \quad b = f^{-1}(\infty), \quad (6.2.3.4)$$

as in (2) of (6.2.1.4). Of course, (1) in (6.2.1.5) is still valid under the assumption in (6.2.3.2).

This section has been divided into two sub-sections.

Section (1) The proof of (6.2.3.2)

The readers are required to recall the process of proving (6.2.1.1). Now we must check $(*_8)$ in Sec. 6.2.1 for a number of new possibilities.

Case 1. $\alpha_k \pi = 2\pi$ (i.e., $\alpha_k = 2$) at a finite vertex w_k.

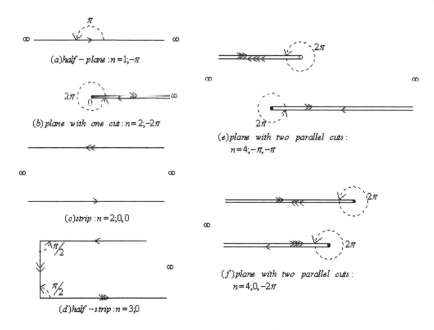

Fig. 6.24

Then $(*_8)$ turns out to be $f(z) = w_k + (z - a_k)^2 h_k(z)$ which is analytic at a_k and the result remains valid, namely, a_k is a simple pole of $\frac{f''(z)}{f'(z)}$ with residue $\alpha_k - 1 = 2 - 1 = 1$.

Case 2. The vertex at ∞ has an interior angle $\alpha\pi$, where $-2 < \alpha < 0$. This means the sides at ∞ are nonparallel lines or rays. The composite of $w \to w^{1/\alpha}$ and $w = f(z)$ (see $(*_7)$) turns $(*_8)$ into $f(z) = z^\alpha h(z)$ which is analytic at ∞; and the result remains valid.

The remaining cases are these in each of which ∞ is a vertex whose sides lie on the same line or on distinct parallel lines or rays. See Fig. 6.24 for such polygons, where n denotes the number of sides and the interior angles at ∞ are shown after it. Readers should convince themselves why the vertices at ∞ possess the interior angles indicated.

Case 3. The two sides of the vertices at ∞ lie on the same line or ray, such as in Figs. 6.24(a) and 6.24(b).

Then the angle $\alpha\pi$ at ∞ is either $-\pi$ or -2π. Let $f(a_k) = \infty$. In case $\alpha = -1$, $(*_8)$ becomes $f(z)^\alpha = \frac{1}{f(z)} = (z - a_k)h(z)$ which is analytic at a_k and $\frac{f''(z)}{f'(z)}$ has a simple pole at a_k with residue $\alpha - 1 = -1 - 1 = -2$. In case

$\alpha = -2$, ($*_8$) becomes $f(z)^{\frac{1}{\alpha}} = f(z)^{-\frac{1}{2}}$ (a fixed branch) $= (z - a_k)h(z)$, still analytic at a_k. Then

$$f(z) = \frac{\varphi(z)}{(z - a_k)^2}, \quad \varphi(z) \text{ is analytic at } a_k \text{ and } \varphi(a_k) \neq 0.$$

$$\Rightarrow \frac{f''(z)}{f'(z)} = \frac{-3}{z - a_k} + \text{regular part, in a neighborhood of } a_k.$$

The residue of $\frac{f''(z)}{f'(z)}$ at the simple pole a_k is then equal to $\alpha - 1 = -2 - 1 = -3$.

Case 4. The two sides of the vertex at ∞ lie on distinct parallel lines or rays.

See Figs. 6.24(c)–6.24(f) in which the angle $\alpha_k \pi$ at ∞ could be 0 or $-\pi$ or -2π. Let $f(a_k) = \infty$. We further divide into three subcases.

If $\alpha_k = 0$ as in Figs. 6.24(c), 6.24(d), and 6.24(f): By invoking an entire linear fractional transformation, we may assume that the sides Γ_{k-1} and Γ_k lie on the two parallel lines $\operatorname{Im} w = \pm h$ ($h > 0$). Draw a line segment, perpendicular to both Γ_{k-1} and Γ_k, to form a half-strip neighborhood Ω_k of $w_k = \infty$. The function $\eta = e^{bw}$ ($b = -\frac{2\pi}{h}$) maps Ω_k univalently onto a half-disk G_k with center at 0 (image of ∞). See Fig. 6.25. Then ($*_7$) turns out to be $\eta = \varphi_k(z) = e^{bf(z)}$, mapping $f^{-1}(\Omega_k)$ univalently onto G_k and $\overline{f^{-1}(\Omega_k)}$ homeomorphically onto $\overline{G_k}$. Routine application of the symmetry principle shows that $\eta = \varphi_k(z)$ is univalently analytic in a neighborhood of a_k. Under this circumstance, ($*_8$) becomes

$$e^{bf(z)} = (z - a_k)h_k(z), \quad h_k(a_k) \neq 0$$

$$\Rightarrow \frac{f''(z)}{f'(z)} = \frac{-1}{z - a_k} + \text{regular part, in a neighborhood of } a_k.$$

$$\Rightarrow \operatorname{Res}\left(\frac{f''(z)}{f'(z)}; a_k\right) = -1 = 0 - 1 = \alpha_k - 1.$$

If $\alpha_k = -1$ as in Fig. 6.24(e): Suppose the two sides Γ_{k-1} and Γ_k lie on the lines $\operatorname{Im} w = 0$ and $\operatorname{Im} w = \pi$, respectively. Consider a neighborhood

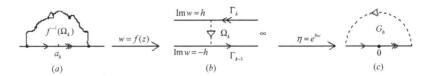

Fig. 6.25

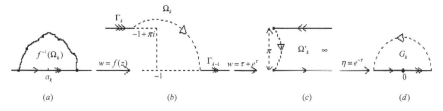

Fig. 6.26

domain Ω_k of ∞, bounded by Γ_{k-1}, Γ_k and a circular arc shown in Fig. 6.26(b). The function $w = \tau + e^\tau$ maps Ω_k univalently on to a domain Ω'_k shown in Fig. 6.26(c) (see Exercise A(7) of Sec. 3.4.1 or Example 2 below) which, in turn, is mapped univalently onto a half-disk G_k with center at 0 under the function $\eta = e^{-\tau}$. In this case, $(\ast_7)$ and $(\ast_8)$ become:

$f(z) = \tau + e^\tau$ and $\eta = e^{-\tau}$ which define η implicitly as a univalent function $\varphi_k(z)$ of z in a neighborhood of a_k, namely, $\eta = \varphi_k(z) = (z - a_k)\psi_k(z)$, $\psi_k(a_k) \neq 0$.

$$\Rightarrow f(z) = \frac{1}{\varphi_k(z)} - \log \varphi_k(z), \quad \text{in a neighborhood of } a_k.$$

$$\Rightarrow \frac{f''(z)}{f'(z)} = \frac{-2}{z - a_k} + \text{regular part, in a neighborhood of } a_k.$$

Hence, a_k is a simple pole of $\frac{f''(z)}{f'(z)}$ with the residue $\alpha_k - 1 = -1 - 1 = -2$.

If $\alpha = -2$ as in Fig. 6.24(f): We may suppose that the two sides Γ_{k-1} and Γ_k lie on the parallel lines $\operatorname{Im} w = -\pi$ and $\operatorname{Im} w = \pi$, respectively. Consider a neighborhood domain Ω_k of ∞, bounded by Γ_{k-1}, Γ_k and a circular arc shown in Fig. 6.27(b). Then, map Ω_k consecutively and univalently onto Ω'_k by $w = \tau + e^\tau$, and then onto a half-disk G_k by $\eta = e^{-\tau/2}$. Then, $(\ast_7)$

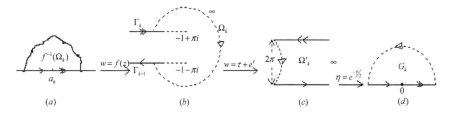

Fig. 6.27

and (∗8) become:

$f(z) = \tau + e^{\tau}$ and $\eta = e^{-\tau/2}$ define $\eta = \varphi_k(z) = (z - a_k)\varphi_k(z)$, $\varphi_k(a_k) \neq$ 0 in a neighborhood of a_k.

$$\Rightarrow f(z) = \frac{1}{(\varphi_k(z))^2} - 2\log \varphi_k(z), \text{ in a neighborhood of } a_k.$$

$$\Rightarrow \frac{f''(z)}{f'(z)} = \frac{-3}{z - a_k} + \text{regular part, in a neighborhood of } a_k.$$

This shows that a_k is a simple pole of $\frac{f''(z)}{f'(z)}$ with residue $\alpha_k - 1 = -2 - 1 = -3$.

In conclusion, in each possible case, $a_k = f^{-1}(w_k)$ is always a simple pole of $\frac{f''(z)}{f'(z)}$ with residue $\alpha_k - 1$. Therefore, (∗10) in Sec. 6.2.1 is still valid under the present situation and the result (6.2.3.1) follows by exactly the same argument as before.

Recall that the residue of $\frac{f''(z)}{f'(z)}$ at ∞ is equal to 2 (see (∗9) in Sec. 6.2.1) if ∞ is not an interior of the polygon Ω. According to (4.11.1.7),

$$\sum_{k=1}^{n} \text{Res}\left(\frac{f''}{f'}; a_k\right) + \text{Res}\left(\frac{f''}{f'}; \infty\right) = \sum_{k=1}^{n}(\alpha_k - 1) + 2 = 0$$

$$\Rightarrow \sum_{k=1}^{n} \alpha_k = n - 2.$$

This finishes the proof of (6.2.3.1).   □

Section (2) Examples

Example 1. Map the upper half-plane univalently onto the generalized polygons shown in Fig. 6.24.

Solution.

(a) Choose $a_1 = \infty$. Then

$$w = f(z) = C_1 \int_0^z dz + C_2 = C_1 z + C_2$$

are possible candidates, where C_1 and C_2 are constants. As z traverses the real axis from left to right, so does $w = f(z)$; and it follows that $C_1 > 0$ and C_2 is real. Therefore, $w = C_1 z + C_2$ will work for any choice of positive C_1 and real C_2.

(b) Choose $a_1 = 0$ and $a_2 = \infty$ so that $w_1 = f(a_1)$ has the interior angle 2π and $w_2 = f(a_2) = \infty$ has the interior angle -2π. Then

$$w = f(z) = C_1 \int_0^z (\zeta - 0)^{2-1} d\zeta + C_2 = \frac{C_1}{2} z^2 + C_2.$$

Since $f(a_1) = f(0) = C_2$, so $C_2 = w_1$. That $f(1) - f(0) = \frac{C_1}{2} + C_2 - C_2 = \frac{C_1}{2}$ is real and positive shows that $C_1 > 0$. Therefore, $w = \frac{C_1}{2} z^2 + w_1$ is a required one if $C_1 > 0$.

(c) Choose $a_1 = 0$ and $a_2 = \infty$. Since the vertices $\infty = f(a_k)$ have the interior angles equal to 0 for $k = 1, 2$, then

$$w = f(z) = C_1 \int_1^z (\zeta - 0)^{0-1} d\zeta + C_2 = C_1 \log z + C_2,$$

where $\log z$ is a fixed branch of the logarithmic function in the upper half-plane, say the principal branch $\mathrm{Log}\, z$ is determined by $\log 1 = 0$. Choose $1 < t_1 < t_2 < \infty$ and set $b_k = f(t_k)$ for $k = 1, 2$. Then

$$b_1 = C_1 \,\mathrm{Log}\, t_1 + C_2 \quad \text{and} \quad b_2 = C_1 \,\mathrm{Log}\, t_2 + C_2$$

$$\Rightarrow C_1 = \frac{b_2 - b_1}{\mathrm{Log}\, t_2 - \mathrm{Log}\, t_1} \quad \text{and} \quad C_2 = \frac{b_1 \,\mathrm{Log}\, t_2 - b_2 \,\mathrm{Log}\, t_1}{\mathrm{Log}\, t_2 - \mathrm{Log}\, t_1}.$$

(d) Choose $a_1 = \infty$, $a_2 = -1$, and $a_3 = 1$ so that $w(a_1) = \infty$, $w(a_2) = ih$ ($h > 0$), and $w(a_3) = 0$. Then

$$w = f(z) = C_1 \int_0^z (\zeta + 1)^{\frac{1}{2} - 1} (\zeta - 1)^{\frac{1}{2} - 1} d\zeta + C_2$$

$$= C_1' \int_0^z \frac{d\zeta}{\sqrt{1 - \zeta^2}} + C_2 = C_1' \sin^{-1} z + C_2,$$

where $\sin^{-1} z$ is the branch that maps $\mathrm{Im}\, z > 0$ univalently onto $-\frac{1}{2}\pi < \mathrm{Re}\, w < \frac{1}{2}\pi$, $\mathrm{Im}\, w > 0$ (see (2.7.4.5)), and the constants C_1' and C_2 are univalently determined by the following system of equations

$$C_1' \sin^{-1}(-1) + C_2 = ih \quad \text{and} \quad C_1' \sin^{-1}(1) + C_2 = 0.$$

Since $\sin^{-1}(-1) = -\frac{\pi}{2}$ and $\sin^{-1}(1) = \frac{\pi}{2}$, the solved $C_1' = -\frac{ih}{\pi}$ and $C_2 = \frac{ih}{2}$. Hence the required solution is $w = ih(\frac{-1}{\pi} \sin^{-1} z + \frac{1}{2})$. Refer the readers to Example 2 of Sec. 3.5.5 for further explanation.

(e) Suppose the boundary rays are given by $w = u_1 + v$, $v \geq 0$, and $w = u_2 + ih - v$ with $h \neq 0$ and $v \geq 0$. The four vertices of the generalized polygon are ∞, u_1, ∞, $u_2 + ih$ (in this ordering) and the corresponding interior angles are $-\pi$, 2π, $-\pi$, 2π, respectively. Let $a_1 = 0$, $a_2 = 1$,

$a_3 = \infty$, $a_4 = a$ ($a < 0$) be points on the real axis that are mapped into the vertices. Then

$$w = f(z) = C_1 \int_0^z \zeta^{-2}(\zeta - 1)(\zeta - a)d\zeta + C_2 = C_1 z - (a+1)C_1 \text{Log}\, z$$

$$-aC_1 z^{-1} + C_2,$$

where $\text{Log}\, 1 = 0$. To compute C_1 and C_2, note that

$$f(1) = u_1, \text{ i.e., } C_1 - aC_1 + C_2 = u_1;$$

$$f(a) = u_2 + ih, \text{ i.e., } C_1 a - (a+1)C_1 \text{Log}\, a - C_1 + C_2 = u_2 + ih.$$

$$\Rightarrow C_1 = -\frac{h}{\pi(1+a)}; \quad C_2 = u_1 + \frac{h}{\pi}\frac{1-a}{1+a}, \quad \text{and}$$

$$2\frac{h}{\pi}\frac{1-a}{1+a} + \frac{h}{\pi}\log|a| = u_2 - u_1.$$

Solve the last equation for a and then determine C_1 and C_2. Note that $h \neq 0$ implies that $a \neq -1$.

(f) Suppose the boundary rays have the equations $w = u_1 - t$ ($t \geq 0$) and $w = u_2 + ih - t$ ($h \neq 0, t \geq 0$), respectively. In this case, the vertices are u_1, ∞, $u_2 + ih$, ∞ and the corresponding interior angles are 2π, -2π, 2π, 0. Let the inverse images of the vertices on the real axis be given by a ($a > 0$), ∞, -1, and 0 (in this ordering). Then

$$w = f(z) = C_1 \int_1^z (\zeta - a)(\zeta + 1)\zeta^{-1} d\zeta + C_2$$

$$= C_1 \left(\frac{1}{2}z^2 - (a-1)z - a\, \text{Log}\, z\right) + C_2,$$

where $\text{Log}\, 1 = 0$. To compute C_1 and C_2, use the equations

$$f(a) = u_1, \text{ i.e., } C\left(\frac{1}{2}a^2 - (a-1)a - a\log a\right) + C_2 = u_1;$$

$$f(-1) = u_2 + ih, \text{i.e., } C\left(\frac{1}{2} + (a-1) - a\,\text{Log}\,(-1)\right) + C_2 = u_2 + ih.$$

$$\Rightarrow C_1 = -\frac{h}{a\pi}; \quad C_2 = u_2 + \frac{h(2a-1)}{2a\pi}, \quad \text{and}$$

$$\frac{ah}{2\pi} - \frac{h}{2a\pi} + \frac{h}{\pi}\log a = u_1 - u_2.$$

Solve the last equation for a and then determine the constants C_1 and C_2. Note that $a \neq 1$ since $u_1 \neq u_2$.

Example 2. Show that $w = f(z) = z + e^z$ maps the horizontal strip $0 < \operatorname{Im} z < \pi$ univalently onto the domain Ω obtained by deleting the slit $-\infty < \operatorname{Re} w \leq -1$, $\operatorname{Im} w = \pi$ from the upper half-plane $\operatorname{Im} w > 0$. Find the images of the strips $(2n - 1)\pi < \operatorname{Im} z < (2n + 1)\pi$, $n = 0, \pm 1, \ldots$, under $w - f(z)$ and construct the Riemann surface R_m for the inverse function $z = f^{-1}(w)$. Refer to Exercise A(7) of Sec. 3.4.1 for a preliminary discussion.

Solution. $\zeta = e^z$ maps $0 < \operatorname{Im} z < \pi$ univalently onto the upper half-plane $\operatorname{Im} \zeta > 0$ (see (2.6.1.3)). Ω can be treated as a generalized 3-gon with vertices $w_1 = \infty$, $w_2 = -1 + \pi i$, and $w_3 = \infty$. Let $a_1 = \infty$, $a_2 = -1$, and $a_3 = 0$ be points (in order) on the real axis that are mapped into w_1, w_2, and w_3. See Fig. 6.28. By (6.2.3.3),

$$w = \int_1^\zeta (\zeta + 1)^{2-1} \zeta^{-1-0} d\zeta = \int_1^\zeta \frac{\zeta + 1}{\zeta} d\zeta = \zeta + \operatorname{Log} \zeta \quad (\operatorname{Log} 1 = 0)$$

maps $\operatorname{Im} \zeta > 0$ univalently onto Ω. The composite map $w = z + e^z$ therefore has the required property.

The univalence and ontoness can be proved directly without recourse to the Schwarz–Christoffel formula. It is worth to do so because the importance of the map $w = z + e^z$ by itself in the flow pattern about fluid (refer to Sec. 3.2.4) passing an obstacle.

Set $z = x + iy$ and $w = u + iv$. Rewrite $w = z + e^z$ as

$$u = x + e^x \cos y,$$

$$v = y + e^x \sin y.$$

Then $w = f(z)$ maps the line segment $z = x_0 + iy$ into the planar u-curve

$$u = x_0 + e^{x_0} \cos y, \quad v = y + e^{x_0} \sin y$$

$$\Rightarrow v = \cos^{-1} \frac{u - x_0}{e^{x_0}} + \{e^{2x_0} - (u - x_0)^2\}^{1/2}.$$

See the dotted curves in Fig. 6.29. In particular, if $x_0 = 0$, then the u-curve is $v = \cos^{-1} u + (1 - u^2)^{1/2}$ passing the point $(\frac{\pi}{2} + 1)i$. These

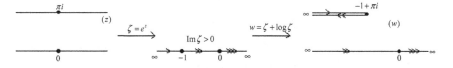

Fig. 6.28

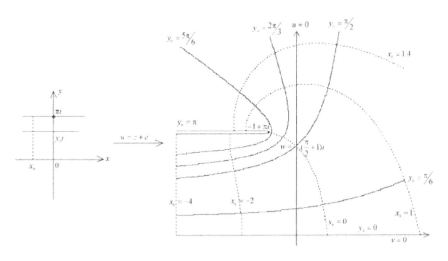

Fig. 6.29

are *equipotentials* of the complex potential $w = f(z)$. On the other hand, $w = f(z)$ maps the line $z = x + iy_0$ into the v-curve

$$u = x + e^x \cos y_0, \quad v = y_0 + e^x \sin y_0$$

$$\Rightarrow u = (v - y_0) \cot y_0 + \log \frac{v - y_0}{\sin y_0} \quad (y_0 \neq n\pi \quad \text{for} \quad n = 0, \pm 1, \ldots).$$

See the black curves in Fig. 6.29. In particular, $w = f(z)$ maps $y = 2n\pi$ $(n = 0, \pm 1, \ldots)$ *univalently* onto the curves $u = x + e^x$, $v = 2n\pi$, namely, $\operatorname{Im} w = 2n\pi$; maps $y = (2n+1)\pi$ $(n = 0, \pm 1, \ldots)$ *univalently* onto $u = x - e^x$, $v = (2n + 1)\pi$, namely, $\operatorname{Im} w = (2n + 1)\pi$, $-\infty < \operatorname{Re} w \leq -1$, in *two-to-one* manner. These curves are *streamlines* of $w = f(z)$. When y_0 increases steadily from 0 to π, the corresponding v-curves vary form $v = 0$ upward until bend into the horizontal slit $\operatorname{Im} w = \pi$, $-\infty < \operatorname{Re} w \leq -1$. The equipotentials are orthogonal to the streamlines.

It is easy to see, by using (3.4.1.1), that $w = f(z)$ is univalent in $0 < \operatorname{Im} z < 1$. An alternate way is to use (3.5.5.1) and we are able to prove simultaneously both the univalence and the ontoness. Moreover, the concept of winding number (Sec. 3.5.2) and the method of the argument principle (Sec. 3.5.3) can help achieve these purposes, too. Even the purely real method as used in calculus can be adopted to prove the univalence.

Owing to $f(z + 2n\pi i) = f(z) + 2n\pi i$, $n = 0, \pm 1, \pm 2, \ldots$, the symmetry principle helps us to justify the following fact: $w = f(z)$ maps the strip domains, for $n = 0, \pm 1, \ldots$,

$$D_n = \{(2n-1)\pi < \operatorname{Im} z < (2n+1)\pi\} \text{ univalently onto the domains}$$

$$\Omega_n = \mathbf{C} - \{\operatorname{Im} w = (2n-1)\pi, \ -\infty < \operatorname{Re} w \le -1\} - \{\operatorname{Im} w = (2n+1)\pi,$$
$$-\infty < \operatorname{Re} w \le -1\}.$$

Since $f'(z) = 1 + e^z = 0 \Leftrightarrow e^z = -1 \Leftrightarrow z = (2n+1)\pi$, $n = 0, \pm 1, \ldots$, and $f''((2n+1)\pi) = -1 \neq 0$, it follows that $f((2n+1)\pi) = -1 + (2n+1)\pi i$, $n = 0, \pm 1, \ldots$, are algebraic branch point of order 1 of the Riemann surface R_w for $z = f^{-1}(w)$; and ∞ is a logarithmic branch point, but it does not lie on R_w. See Fig. 6.30 for R_w. What does its line complex look like?

Example 3. Map $\operatorname{Im} z > 0$ univalently onto each of the following generalized polygons shown in Fig. 6.31. (These domains are modeled after those appeared in the book by Kober [50].) If possible, try to draw up the equipotentials and the streamlines for the complex potentials $w = f(z)$ obtained. How about the Riemann surfaces for the inverse functions $z = f^{-1}(w)$ or $w = f(z)$?

Solution. Let Ω denote the generalized polygon with vertices w_k and the associated interior angles $\alpha_k \pi$ for $1 \le k \le n$, and a_k the preimages on the real axis under $w = f(z)$.

(a) $n = 4$, $w_1 = \infty$, $w_2 = 0$, $w_3 = \infty$, $w_4 = \pi(1 + \sqrt{a^2 - 1}i)$ and $\alpha_1 = 0$, $\alpha_2 = \frac{1}{2}$, $\alpha_3 = 0$, $\alpha_4 = \frac{3}{2}$. Choose $a_1 = \infty$, $a_2 = 0$, $a_3 = 1$, and $a_4 = a > 1$. According to (6.2.3.3),

$$w = f(z) = C \int_0^z (\zeta - 0)^{\frac{1}{2}-1}(\zeta - 1)^{0-1}(\zeta - 1)^{\frac{3}{2}-1}d\zeta$$

$$= C \int_0^z \frac{(\zeta - a)^{1/2}}{(\zeta - 1)\zeta^{1/2}}d\zeta$$

$$= C' \left\{ \cos^{-1}\frac{a - 2z}{a} + \sqrt{a-1}\cosh^{-1}\frac{az + a - 2z}{a(1 - z)} \right\},$$

where $\cos^{-1}$ and $\cosh^{-1}$ denote the branches defined uniquely by the conditions that $\cos^{-1}(-1) = \pi$ and $\cosh^{-1}(-1) = \pi i$ (see (2.7.4.2) and (3.3.1.9)), respectively. Then $f(a_4) = f(a) = C'(\pi + \sqrt{a-1}\pi i) = \pi(1 + \sqrt{a-1}i)$ shows that $C' = 1$. The required map is thus given by

$$w = f(z) = \cos^{-1}\frac{a - 2z}{a} + \sqrt{a-1}\cosh^{-1}\frac{az + a - 2z}{a(1 - z)}.$$

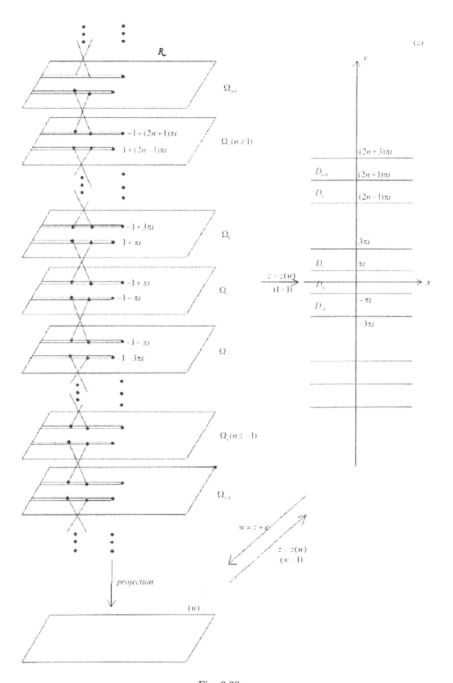

Fig. 6.30

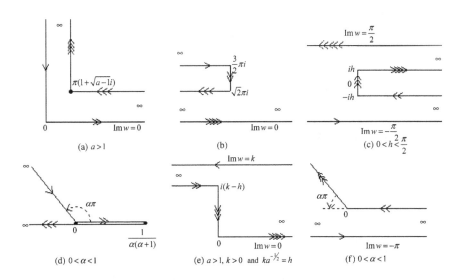

Fig. 6.31

(b) $n = 4$, $w_1 = \infty$, $w_2 = \frac{3}{2}\pi i$, $w_3 = \sqrt{2}\pi i$, $w_4 = \infty$ and $\alpha_1 = -\pi$, $\alpha_2 = \alpha_3 = \frac{3}{2}\pi$, $\alpha_4 = 0$. Choose $a_1 = \infty$, $a_2 = -1$, $a_3 = 0$, and $a_4 = 1$. By (6.2.3.3), we have

$$w = f(z) = C_1 \int_0^z (\zeta + 1)^{3/2-1}(\zeta - 0)^{3/2-1}(\zeta - 1)^{0-1}d\zeta + C_2$$

$$= C_1 \left\{ \sqrt{z}\sqrt{z+1} + \frac{3}{2}\cosh^{-1}(2z+1) - \sqrt{2}\cosh^{-1}\frac{3z+1}{z-1} \right\} + C_2,$$

where $\sqrt{\ }$ is the branch defined by $\sqrt{-1} = i$ and $\cosh^{-1}$ is the branch defined by $\cosh^{-1}(-1) = \pi i$. To compute C_1 and C_2: Set

$$f(-1) = \frac{3}{2}\pi i \Rightarrow \frac{3}{2}\pi i C_1 + C_2 = \frac{3}{2}\pi i;$$

$$f(0) = \sqrt{2}\pi i \Rightarrow \sqrt{2}\pi i C_1 + C_2 = \sqrt{2}\pi i.$$

Hence $C_1 = 1$ and $C_2 = 0$. The required map is then

$$w = f(z) = \sqrt{z}\sqrt{z+1} + \frac{3}{2}\cosh^{-1}(2z+1) - \sqrt{2}\cosh^{-1}\frac{3z+1}{z-1}.$$

(c) $n = 5$, $w_1 = \infty$, $w_2 = \infty$, $w_3 = -ih$, $w_4 = ih$, $w_5 = \infty$, and $\alpha_1 = 0$, $\alpha_2 = 0$, $\alpha_3 = \alpha_4 = \frac{3}{2}\pi$, $\alpha_5 = 0$. Choose $a_1 = \infty$, $a_2 = -1$, $a_3 = -a$ $(0 < a < 1)$, $a_4 = a$, and $a_5 = 1$. By (6.2.3.3),

$$w = f(z) = C \int_0^z (\zeta + 1)^{0-1}(\zeta + a)^{3/2-1}(\zeta - a)^{3/2-1}(\zeta - 1)^{0-1}d\zeta$$

$$= C \int_0^z \frac{\sqrt{\zeta^2 - a^2}}{\zeta^2 - 1}d\zeta$$

$$= C\left\{\frac{\pi}{2}i - \cosh^{-1}\frac{z}{a} + \frac{\sqrt{1 - a^2}}{2}\cosh^{-1}\frac{z^2(2 - a^2) - a^2}{a^2(z^2 - 1)}\right\},$$

where $\cosh^{-1}(-1) = \pi i$. The conditions $f(a) = -f(-a) = hi$ will give rise to $C = 1$ and

$$h = \frac{1}{2}\pi - \frac{1}{2}\pi\sqrt{1 - a^2}$$

$$\Rightarrow a = \frac{2}{\pi}\sqrt{\pi h - h^2}.$$

(d) $n = 3$, $w_1 = \infty$, $w_2 = 0$, $w_3 = \{a(\alpha + 1)\}^{-1}$, and $\alpha_1 = -(1 + \alpha)\pi$, $\alpha_2 = \alpha\pi$, $\alpha_3 = 2\pi$. Choose $a_1 = \infty$, $a_2 = -1$, and $a_3 = 0$. Then, by (6.2.3.3),

$$w = f(z) = C \int_{-1}^z (\zeta + 1)^{\alpha-1}(\zeta - 0)^{2-1}d\zeta = C \int_{-1}^z (\zeta + 1)^{\alpha-1}\zeta d\zeta$$

$$= C\left\{-\frac{1}{1+\alpha}(z + 1)^{1+\alpha} + \frac{1}{\alpha}(z + 1)^\alpha\right\},$$

where $(z+1)^{1+\alpha}$ is the branch defined by $\mathrm{Log}\,1 = 0$ (namely, $1^{1+\alpha} = 1$). The condition $f(0) = w_3$ shows that $C = 1$.

(e) $n = 4$, $w_1 = \infty$, $w_2 = \infty$, $w_3 = i(k - h)$, $w_4 = 0$, and $\alpha_1 = 0$, $\alpha_2 = 0$, $\alpha_3 = \frac{3}{2}\pi$, $\alpha_4 = \frac{\pi}{2}$. Choose $a_1 = \infty$, $a_2 = 0$, $a_3 = 1$, and $a_4 = a > 1$. Then

$$w = f(z) = C \int_a^z (\zeta - 0)^{0-1}(\zeta - 1)^{3/2-1}(\zeta - a)^{1/2-1}d\zeta$$

$$= C \int_a^z \frac{(\zeta - 1)^{1/2}}{\zeta(\zeta - a)^{1/2}}d\zeta$$

$$= C\left\{\cosh^{-1}\frac{2z - a - 1}{a - 1} - \frac{1}{\sqrt{a}}\cosh^{-1}\frac{(a + 1)z - 2a}{(a - 1)z}\right\},$$

where $\cosh^{-1}(-1) = \pi i$. The condition $f(0) = i(k-h)$ shows the $C = \frac{k}{\pi}$ if $ka^{-1/2} = h$ is preassigned.

(f) $n = 3$, $w_1 = \infty$, $w_2 = \infty$, $w_3 = 0$, and $\alpha_1 = -\alpha$, $\alpha_2 = 0$, $\alpha_3 = \alpha + 1$.
Choose $a_1 = \infty$, $a_2 = 0$, and $a_3 = 1$. Then

$$w = f(z) = C \int_1^z (\zeta - 0)^{0-1}(\zeta - 1)^{\alpha+1-1} d\zeta = C \int_1^z \frac{(\zeta - 1)^\alpha}{\zeta} d\zeta$$

$$= C' \int_1^z \frac{(1 - \zeta)^\alpha}{\zeta} d\zeta = C'' \int_z^1 \frac{(1 - \zeta)^\alpha}{\zeta} d\zeta,$$

where $1^\alpha = 1$. As z traces the real segment $[0, 1]$, the corresponding
$w = f(z)$ assumes positive real values. Consequently, the constant C''
should be positive. In case $\zeta = t > 1$, then $(1 - \zeta)^\alpha = e^{-\alpha\pi i}(t - 1)^\alpha$
(why $-\alpha\pi i$? why not $\alpha\pi i$? Observe that as ζ, in the upper half-plane,
approaches a point $t > 1$, $\mathrm{Arg}(1 - \zeta)$ keeps less then 0 and approaches
$\mathrm{Arg}(1 - t) = -\pi$ by continuity). Hence, for fixed $z \in (1, \infty)$,

$$\int_z^1 \frac{(1 - \zeta)^\alpha}{\zeta} d\zeta = e^{-\alpha\pi i} \int_z^1 \frac{(t - 1)^\alpha}{t} dt = e^{(1-\alpha)\pi i} \int_1^z \frac{(t - 1)^\alpha}{t} dt.$$

Therefore $w = f(z)$ traces the ray $\mathrm{Arg}\, w = (1 - \alpha)\pi$ as z goes from 1
to ∞ along the real axis. Recall that the interior angle at ∞ is $-\alpha\pi$;
hence, the exterior angle at ∞ is $(\alpha - 1)\pi = -(1 - \alpha)\pi$.

In particular, suppose $\alpha = \frac{p}{q}$ where p and q are positive integers with
$p < q$. By changing the variables from ζ to η, where $\eta = (1 - \zeta)^{1/q}$ or
$\zeta = 1 - \eta^q$, then

$$w = f(z) = \int_{(1-z)^{1/q}}^0 \frac{\eta^p}{1 - \eta^q} \cdot (-q\eta^{q-1}) d\eta = q \int_0^{(1-z)^{1/q}} \frac{\eta^{p+q-1}}{1 - \eta^q} d\eta.$$

For instance, if $\alpha = \frac{1}{3}$ ($p = 1, q = 3$), then

$$w = f(z) = -3(1 - z)^{1/3} - \frac{3}{2} \log[1 - (1 - z)^{1/3}] + \frac{1}{2} \log z$$

$$+ \sqrt{3} \tan^{-1} \frac{\sqrt{3}}{1 + 2(1 - z)^{-1/3}}$$

where $\sqrt[3]{1} = 1$, $\log 1 = 0$, and $\tan^{-1} \sqrt{3} = \frac{\pi}{3}$.

Example 4. Map the open unit disk $|z| < 1$ univalently onto the gen-
eralized regular 4n-gon, obtained by deleting the 2n-segments joining
the origin 0 to the points $1, e^{\pi i/n}, \ldots, e^{(2n-1)\pi i/n}$, respectively, from the
plane $\mathbf{C}^*$. See Fig. 6.32. Then try to construct the Riemann surface for the
mapping $w = f(z)$.

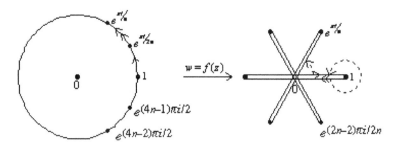

Fig. 6.32

Solution. The vertices of the $4n$-gon are $1, 0, e^{(2n-1)\pi i/n}, 0, \ldots, 0, e^{\pi i/n}, 0$ with the interior angles $2\pi, \frac{\pi}{n}, 2\pi, \ldots, 2\pi, \frac{\pi}{n}$, respectively. Note that the ordering of these vertices is in the clockwise direction so that (the interior of) the polygon lies on the left side when traveling along the boundary. Such a mapping $w = f(z)$ does exist owing to the Riemann mapping theorem. Suppose $w = f(z)$ maps the circular sector $-\frac{\pi}{2n} < \operatorname{Arg} z < 0$, $|z| < 1$ onto the sector domain $0 < \operatorname{Arg} w < \frac{\pi}{2n}$ so that $1, 0, e^{-\pi i/2n}$, go into $1, \infty, 0$, respectively; more precisely, $[0, 1]$ to $[1, \infty]$, $[0, e^{-\pi i/2n}]$ to $[\infty, 0]$ along the ray $\operatorname{Arg} w = \frac{\pi}{2n}$, and the circular arc $\overline{1e^{-\pi i/2n}}$ to the segment $[0, 1]$. Applying the symmetry principle successively to the radii $\operatorname{Arg} z = \frac{k\pi}{2n}$ $(0 \le k \le 4n-1)$, one can realize that the inverse images of these $4n$ vertices on $|w| = 1$, under the extended $w = f(z)$, must be the vertices of a regular $4n$-gon,

$$1, e^{\pi i/2n}, e^{2\pi i/2n}, \ldots, e^{(2n-1)\pi i/n}, e^{(4n-1)\pi i/2n}$$

in this ordering (and in the counterclockwise sense). Using (6.2.1.5) now (see the words after (6.2.3.2)) and recalling that $f^{-1}(\infty) = 0$, we have

$$w = f(z) = C_1 \int_1^z (\zeta - 1)^{2-1} (\zeta - e^{\pi i/2n})^{1/n-1} (\zeta - e^{2\pi i/2n})^{2-1} \cdots$$

$$(\zeta - e^{(4n-2)\pi i/2n})^{2-1} \cdot (\zeta - e^{(4n-1)\pi i/2n})^{1/n-1} \frac{1}{\zeta^2} d\zeta + C_2$$

$$= C_1 \int_1^z (\zeta^{2n} - 1)(\zeta^{2n} + 1)^{1/n-1} \frac{1}{\zeta^2} d\zeta + C_2$$

$$= C_1 \int_1^z (\zeta^n + \zeta^{-n})^{1/n-1} (\zeta^{n-1} - \zeta^{-n-1}) d\zeta + C_2$$

$$= C_1'(z^n + z^{-n})^{1/n} + C_2'.$$

Setting $z = 1$ and $z = e^{\pi i/2n}$, respectively, then

$$f(1) = 1 = 2^{1/n}C_1' + C_2' \quad \text{and} \quad f(e^{\pi i/2n}) = 0 = C_1'(i - i)^{1/n} + C_2' = C_2'$$

$$\Rightarrow \quad C_2' = 0 \quad \text{and} \quad C_1' = 2^{-1/n}.$$

A required map is then given by

$$w = f(z) = \left[\frac{1}{2}(z^n + z^{-n}) \right]^{1/n}.$$

Note that $f\left(\frac{1}{z}\right) = f(z)$. Therefore, the same function $w = f(z)$ maps $|z| > 1$ univalently onto the same generalized polygon. Also refer to Exercise A(11) of Sec. 3.5.7.

Some elementary geometric mapping properties of $w = f(z)$ can be described as follows. Decompose $\eta = \frac{1}{2}(z^n + z^{-n})$ as

$$z \to \zeta = z^n \to \eta = \frac{1}{2}(\zeta + \zeta^{-1}).$$

Then $w = f(z) = \eta^{\frac{1}{n}}$. The mapping properties and Riemann surfaces for $\zeta = z^n$ and $\eta = \frac{1}{2}(\zeta + \zeta^{-1})$ or their inverses (see Secs. 2.5.2, 2.5.5, and 2.7.2) will lead naturally to those for $w = f(z)$.

Recall that $\eta = \eta(z)$ has poles of order n at $z = 0$ and $z = \infty$. It follows that the inverse $z = z(\eta)$ has two algebraic branch points of order $n - 1$ at $\eta = 0$ and $\eta = \infty$. Moreover

$$\eta' = \frac{1}{2}\left(nz^{n-1} - \frac{n}{z^{n+1}}\right) = 0$$

$$\Rightarrow \quad z_k = e^{2k\pi i/2n} = e^{k\pi i/n}, \quad 0 \le k \le 2n - 1 \quad \text{and} \quad \eta''(z_k) \ne 0.$$

This shows that the inverse $z = z(\eta)$ has $2n$ algebraic points of order 1 at the points $\eta_k = \eta(z_k) = \pm 1$ for $0 \le k \le 2n - 1$ ($\eta(z_k) = 1$ if k is even; $= -1$ if k is odd).

Via $\zeta = z^n$, the following sector domains

$$\Omega_k : \frac{k\pi}{n} < \operatorname{Arg} z < \frac{(k+1)\pi}{n}, \quad 0 \le k \le 2n - 1$$

are fundamental domains (see (2.7.4)) of $\eta = \eta(z)$. See Fig. 6.33 for the images of Ω_0 and Ω_1 under $\eta = \eta(z)$. Combining Fig. 6.33 together for $\Omega_0, \Omega_1, \ldots, \Omega_{2n-1}$, we have the Riemann surface for $z = z(\eta)$ as in Fig. 6.34 and its line complex in Fig. 6.35 ($n = 6$). The case $n = 1$ is the Joukowski function and its Riemann surface (see Figs. 2.61 and 2.62). Note that these univalent functions that preserve $\eta = \eta(z)$ unchanged form a group, generated by

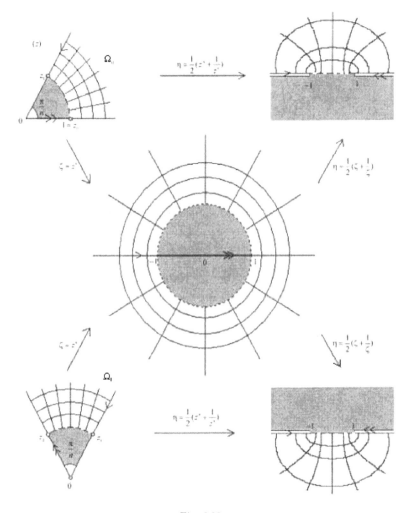

Fig. 6.33

1. rotation: $\eta = e^{2\pi i/n} z$, and
2. reflection: $\eta = \frac{1}{z}$.

Elements in this group give rise to self-homeomorphisms (conformal mappings) of the Riemann surface $z = z(\eta)$ onto itself so that a point and its images lie over the same point in the classical η-plane. This group is usually called the *covering transformation group* of the Riemann surface (refer to Sec. 7.5.4).

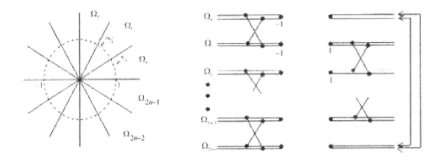

Fig. 6.34

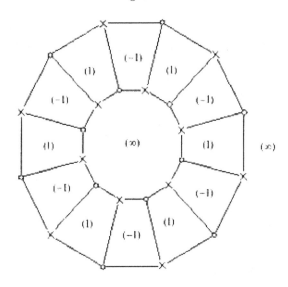

Fig. 6.35

Let us go back to original

$$w = f(z) : z \to \zeta = z^n \to \eta = \frac{1}{2}\left(\zeta + \frac{1}{\zeta}\right) \to w = \eta^{\frac{1}{n}}.$$

The *first* component map $\zeta = z^n$ maps the z-plane univalently onto the n-sheeted Riemann surface R_ζ and the inscribed $4n$-gon onto an n-sheeted domain D_1 with center at 0 (the circular arcs in Ω_0 and Ω_1 of Fig. 6.34 are mapped to form a complete ordinary circle in the first sheet, etc). The *second* map $\eta = \frac{1}{2}(\zeta + \frac{1}{\zeta})$ maps R_ζ univalently onto the $2n$-sheeted Riemann surface R_η and D_1 onto a $2n$-sheeted domain D_2 whose boundary consists of $4n$ line segments of unit length whose projections onto the real axis are alternatively the segments $[0,1]$ and $[-1,0]$ (try to draw it in the right figure of Fig. 6.34). Finally the *third* map $w = \eta^{\frac{1}{n}}$ maps R_η univalently onto the classical w-plane and D_2 onto the given regular $4n$-gon.

Example 5. Map the upper half-plane $\text{Im}\, z > 0$ univalently onto the genera-
lized $2n$-polygon, obtained by deleting the n-segments joining the origin 0 to the points $e^{2k\pi i/n}$, $0 \le k \le n-1$, respectively, from $\mathbf{C}^*$ (see Fig. 6.36). Con-struct the Riemann surface for the function $z = f^{-1}(w)$, where $w = f(z)$ is the mapping acquired.

Solution. The generalized polygon

$$\Omega = \mathbf{C}^* - \bigcup_{k=0}^{n-1} \{\text{the segment connecting to 0 to } e^{2k\pi i/n}\}$$

is a regular $2n$-gon with the vertices (in clockwise ordering)

$$1, 0, e^{2(n-1)\pi i/n}, 0, \ldots, 0, e^{2\pi i/n}, 0$$

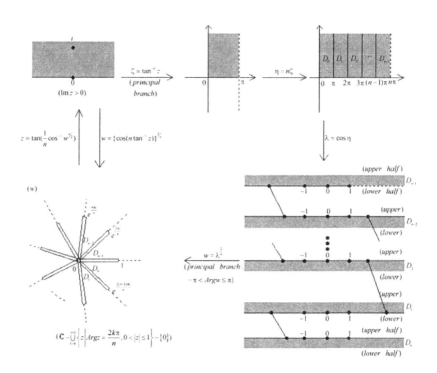

Fig. 6.36

and the corresponding vertex angles $2\pi, \frac{2\pi}{n}, 2\pi, \frac{2\pi}{n}, \ldots, \frac{2\pi}{n}, 2\pi, \frac{2\pi}{n}$, respectively. For simplicity, choose their preimages on the real axis as

$$1 = \tan\frac{0\pi}{n}, \tan\frac{\pi}{2n}, \tan\frac{\pi}{n}, \tan\frac{3\pi}{2n}, \ldots, \tan\frac{(2n-3)\pi}{2n}, \tan\frac{(2n-2)\pi}{2n},$$

$$\tan\frac{(2n-1)\pi}{2n};$$

in other words $f^{-1}(0) = \tan\frac{(2k+1)\pi}{2n}$ for $0 \le k \le n-1$ and $f^{-1}(e^{-2k\pi i/n}) = \tan\frac{k\pi}{n}$ for $0 \le k \le n-1$. According to (6.2.3.2), we have

$$w = f(z)$$

$$= C_1 \int_0^z \prod_{k=0}^{n-1}\left(t - \tan\frac{(2k+1)\pi}{2n}\right)^{\frac{2}{n}-1} \cdot \prod_{k=0}^{n-1}\left(t - \tan\frac{k\pi}{n}\right)^{2-1} dt + C_2$$

$$= C_1 \int_0^z \left\{\prod_{k=0}^{n-1}\left(t - \tan\frac{(2k+1)\pi}{2n}\right)\right\}^{\frac{2}{n}-1} \cdot \left\{\prod_{k=0}^{n-1}\left(t - \tan\frac{k\pi}{n}\right)\right\} dt + C_2.$$

Recall that $\cos n\zeta$ is a polynomial in $z = \tan\zeta$ of degree n (see Exercise B(2) of Sec. 1.5) whose zeros are $\tan\frac{(2k+1)\pi}{2n}$ for $0 \le k \le n-1$; and $\sin n\zeta$ is also a polynomial in $z = \tan\zeta$ of degree n whose zeros are $\tan\frac{k\pi}{n}$ for $0 \le k \le n-1$. Hence, after adjusting constants C_1 and C_2, we obtain an acquired mapping

$$w = f(z) = (\cos(n\tan^{-1}z))^{2/n},$$

where $\tan^{-1}$ denotes the principal branch of Arctan function (see (2.7.4.7)). It is an algebraic function. For instance,

$$n = 1: \ w = \frac{1}{1+z^2};$$

$$n = 2: \ w = \frac{1-z^2}{1+z^2};$$

$$n = 3: \ w = \frac{(1-3z^2)^{2/3}}{1+z^2};$$

$$n = 4: \ w = \frac{(1-6z^2+z^4)^{1/2}}{1+z^2}, \ \text{etc.}$$

It will be within our expectation that the Riemann surfaces for $z = f^{-1}(w)$ or $w = f(z)$ are compact ones and have only algebraic branch points (see (3.5.6.10)).

Conversely, the mapping $w = (\cos(n\tan^{-1}z))^{2/n}$ can be decomposed as $z \to \zeta = \tan^{-1}z \to \eta = n\zeta \to \lambda = \cos\eta \to w = \lambda^{2/n}$. See Fig. 6.36.

For $n = 1$, the Riemann surface R_w for $z = \sqrt{(1-w)/w}$ is as in Fig. 6.37. It has two algebraic branch points of order 1 over $w = 0$ and $w = 1$. For $n = 2$, the Riemann surface R_w for $z = \sqrt{(1-w)/(1+w)}$, as in Fig. 6.38, has two algebraic branch points of order 1 over $w = -1$ and $w = 1$. Note that, in this case, $w = 0$ is an ordinary point since $w(\pm 1) = 0$ but $w'(\pm 1) \neq 0$. Both Riemann surfaces have the same line complex as that of $z = \sqrt{w}$ (see $n = 2$ in Fig. 2.49). The readers are asked to construct the Riemann surface for the case $n = 3$: It is 3-sheeted obtained by pasting crosswise along the cuts $[0, 1]$, $[0, e^{2\pi i/3}]$, and $[0, e^{4\pi i/3}]$, respectively, and has an algebraic branch point of order 2 over $w = 0$, three algebraic branch points of order 1 over $w = 1$, $e^{2\pi i/3}$, and $e^{4\pi i/3}$ each. The line complex is as in Fig. 6.39. Guess what the Riemann surface and its line complex look like for the general n?

Exercises A

(1) Map $|z| < 1$ univalently onto the domain $\Omega = \mathbf{C}^* - [\infty, -\frac{1}{4}]$, viewed as a 2-gon obtained by deleting the ray from $-\frac{1}{4}$ to ∞ along the negative real axis. The map $w = f(z)$ is required to satisfy $f(-1) = -\frac{1}{4}$, $f(1) = \infty$, and $f(0) = 0$. Refer to Example 1 of Sec. 3.5.7, Exercise B(4) of Sec. 4.8, (5.8.2.3), and (5.8.2.6)).

(2) Find the polygons in the w-plane onto which

$$w = f(z) = \int_0^z \frac{\zeta - \lambda}{\sqrt{\zeta(\zeta - 1)}} d\zeta = \sqrt{z(z-1)} + (1 - 2\lambda)$$

$$\left[\mathrm{Log}(\sqrt{z} + \sqrt{z - 1}) - \frac{\pi i}{2} \right]$$

maps univalently the upper half-plane $\mathrm{Im}\, z > 0$ if

(a) $\lambda < 0$;
(b) $0 < \lambda < \frac{1}{2}$;
(c) $\lambda = \frac{1}{2}$;
(d) $\frac{1}{2} < \lambda < 1$;
(e) $\lambda > 1$.

(3) Find $w = f(z)$ mapping $|z| < 1$ univalently onto the exterior of the regular n-gon with center at 0 and one vertex at $w = 1$, given that $f(0) = \infty$ and $f(x) > 0$ if $0 < x < 1$.

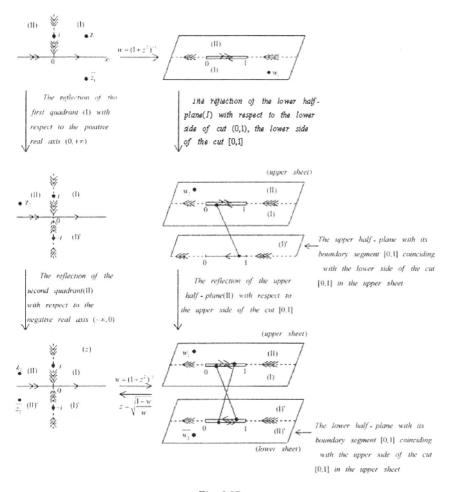

Fig. 6.37

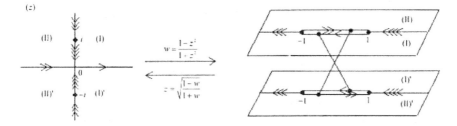

Fig. 6.38

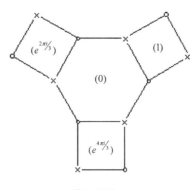

Fig. 6.39

(4) Show that

$$w = f(z) = \int_0^z \frac{dz}{\sqrt{1 - z^4}(1 + z^4)}$$

maps $|z| < 1$ univalently onto the cruciform polygon shown in Fig. 6.40. Prove that the width

$$a = \frac{1}{\sqrt{2}} \int_{-1}^1 \frac{dx}{(1 + x^4)\sqrt{1 - x^4}} = \frac{1}{8}\left[\sqrt{2\pi} + \pi^{-1/2}\left(\Gamma\left(\frac{1}{4}\right)\right)^2\right].$$

(5) Show that

$$w = f(z) = \int_0^z \frac{1 - 2z^2 \cos 2\alpha + z^4}{(1 - z^2)(1 + z^2)^2}\,dz$$

$$= \int_0^z \left\{\frac{\sin^2 \alpha}{1 - z^2} + \frac{(1 - z^2)\cos^2 \alpha}{(1 + z^2)^2}\right\}\,dz$$

$$= \frac{1}{2}\sin^2 \alpha \mathrm{Log}\left(\frac{1 + z}{1 - z}\right) + \frac{2\cos^2 \alpha}{1 + z^2}$$

maps $|z| < 1$ univalently onto $\mathbf{C}^*$ with four horizontal rays deleted, as shown in Fig. 6.41, where a and b are given by

$$a + bi = w(e^{i\alpha}) = \frac{1}{2}\sin^2 \alpha \mathrm{Log}\cot \frac{1}{2}\alpha + \frac{1}{2}\cos \frac{1}{2}\alpha + \frac{1}{2}\pi i \sin^2 \alpha.$$

Exercises B

See Ref. [50] for more concrete examples. Also, refer to *Werner von Koppenfels und Friedmamn Stallmamn: Praxis Der Konformen Abbildung, Springer-Verlag, 1959*, for further information.

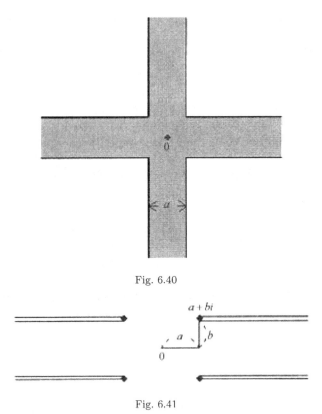

Fig. 6.40

Fig. 6.41

6.3 Harmonic Function and the Dirichlet Problem for a Disk

Let $v(z) = v(x, y)$ be a continuous real-valued function on the boundary $\partial\Omega$ of a domain Ω. The problem of finding a harmonic function $u(z) = u(x, y)$ on Ω (see Sec. 3.4.3), continuous on $\bar{\Omega}$, such that $u(z) = v(z)$ at every point of $\partial\Omega$ is known as the *Dirichlet problem*, of great importance in mathematical physics and in function theory itself. The solution $u(z)$, if exists, is unique owing to the maximum and minimum principle (see (3.4.3.7)).

Readers are suppose to be familiar with all the materials presented in Sec. 3.4.3.

Two subsections are divided.

Section 6.3.1 continues the content of Sec. 3.4.3 and investigates some further properties of harmonic function. In particular, the roles played by the *conjugate differential* $^*du$ of the differential du of a harmonic function $u(z)$ on a domain are studied: The influence of $\int_\gamma {}^*du = 0$ (γ is a cycle homologous to zero in Ω) on the existence of a single-valued conjugate harmonic function, on the arithmetic mean over concentric circles, on the isolated singularities of a harmonic function and on the integral representation of a harmonic function through Green's function, etc. Moreover, basic properties of Green's function of a domain with respect to a point (or pole) are studied, in addition to its connection to the Riemann mapping theorem (see (6.3.1.12)).

The integral representation of a harmonic $u(z)$ in $|z| < 1$ (and continuous on $|z| \le 1$) through its boundary values along $|z| = 1$ is *Poisson's formula (integral)*. This enables us to represent an analytic function $f(z)$ in $|z| < R$ (and continuous on $|z| \le R$) in terms of its real part, the so-called *Schwarz's formula*. Besides these, the core of Sec. 6.3.2 is to solve the most fundamental *Dirichlet problem for a disk*: For any function $u(z)$ continuous on the circle $|z| = R$, there is a unique harmonic function, specifically denoted by $P_u(z)$, in $|z| < R$, whose boundary values along $|z| = R$ coincide with the given $u(z)$, namely,

$$\lim_{z \to Re^{i\theta}} P_u(z) = u(Re^{i\theta}), \quad 0 \le \theta \le 2\pi.$$

6.3.1   *Some further properties of harmonic functions: Green's function*

Let $u(z)$ be harmonic in a domain Ω. We introduce two differentials (see Sec. 2.8):

$$\text{the differential of } u: \quad du = \frac{\partial u}{\partial x}dx + \frac{\partial u}{\partial y}dy;$$

$$\text{the conjugate differential of } du: \quad {}^*du = -\frac{\partial u}{\partial y}dx + \frac{\partial u}{\partial x}dy. \quad (6.3.1.1)$$

If $u(z)$ has a single-valued harmonic conjugate function $v(z)$ in Ω, then the Cauchy–Riemann equations say that $^*du = dv$, the total differential of v. (6.3.1.1) has remarkable, both geometrically and physically, interpretations as follows: Let γ be a regular curve in Ω. The tangent vector at a point on γ is given by $dx + idy = (dx, dy)$; while the outward normal which points to the right of the tangent is given by $dy - idx = (dy, -dx)$. Then (referring to Fig. 3.8 and (3.2.4.4), with $u(z) + iv(z)$ there replaced by) the complex potential $f(z) = \frac{\partial u}{\partial x} - i\frac{\partial u}{\partial y}$ (see (1) in (3.4.3.4)) has

its component along the tangent $= \langle \overline{f(z)}, (dx, dy) \rangle = du;$

its component along the normal $= \langle \overline{f(z)}, (dy, -dx) \rangle = {}^*du.$ (6.3.1.2)

The classical notation for ${}^*du$ is

$$\frac{\partial u}{\partial n}|dz|, \quad \text{where} \frac{\partial u}{\partial n} = \frac{\partial u}{\partial x}\frac{dy}{ds} - \frac{\partial u}{\partial y}\frac{dx}{ds} \quad \text{and} \quad ds = |dz|. \qquad (*_1)$$

$\frac{\partial u}{\partial n}$ means the directional derivative of u in the outward normal and is called the *outward normal derivative* of u with respect to the curve γ.

Form the complex differential

$$f(z)dz = \left(\frac{\partial u}{\partial x} - i\frac{\partial u}{\partial y} \right)(dx + idy) = du + i{}^*du. \qquad (*_2)$$

Since $f(z)$ is analytic in Ω and du is an exact differential, Cauchy's integral theorem (4.3.3.1) says that, for any cycle γ in Ω which is homologous to zero in Ω,

$$\int_\gamma f(z)dz = \int_\gamma du + i\int_\gamma {}^*du = i\int_\gamma {}^*du = 0$$

$$\Rightarrow \int_\gamma {}^*du = 0. \qquad (*_3)$$

Physically, this means that the flux of the flow $f(z)$ through γ is identically equal to zero (see (3.2.4.5)).

$(*_3)$ can be still generalized to a pair $u_1(z)$ and $u_2(z)$ of harmonic functions in Ω. According to (4.3.4.1) and (4.3.4.2), it is sufficient to assume that $\gamma = \partial R$, where R is a rectangle in Ω, with sides parallel to the coordinate axes. Since R is simply connected, both $u_1(z)$ and $u_2(z)$ have single-valued harmonic conjugate functions $v_1(z)$ and $v_2(z)$, respectively. Then

$$u_1{}^*du_2 - u_2{}^*du_1 = u_1dv_2 - u_2dv_1 = u_1dv_2 + v_1du_2 - (u_2dv_1 + v_1du_2)$$

$$= \operatorname{Im}(u_1 + iv_1)d(u_2 + iv_2) - d(u_2v_1)$$

$\Rightarrow$ for $\gamma = \partial R$ and hence, any cycle γ homologous to zero in Ω, by Cauchy's integral theorem,

$$\int_\gamma u_1{}^*du_2 - u_2{}^*du_1 = \int_\gamma \operatorname{Im}(u_1 + iv_1)d(u_2 + iv_2) - \int_\gamma d(u_2v_1)$$

$$= \operatorname{Im}\int_\gamma (u_1 + iv_1)d(u_2 + iv_2) = 0. \qquad (*_4)$$

An alternate way to prove this formula is to use Exercise B(1) of Sec. 2.9. Try it.

In conclusion, we have

Some basic properties of harmonic functions in terms of differentials. Let Ω be a domain.

(1) Let $u(z)$ be harmonic in Ω. Then

$$\int_\gamma {}^*du = 0$$

for every cycle γ in Ω *which is homologous to zero in* Ω.

(2) Let $u(z)$ be harmonic in Ω. Then (refer to(3.4.3.4))

$u(z)$ has a single-valued harmonic conjugate $v(z)$, unique up to an additive constant.

$$\Leftrightarrow \int_\gamma {}^*du = 0 \quad \text{for every cycle } \gamma \quad \text{in } \Omega.$$

This will be the case if Ω is simply connected.

(3) Let $u_1(z)$ and $u_2(z)$ be harmonic in Ω. Then

$$\int_\gamma u_1{}^*du_2 - u_2{}^*du_1 = 0$$

for every cycle γ homologous to zero in Ω.   (6.3.1.3)

$(*_1)$ looks simpler if γ is a circle $|z| = r$. In this case, $x = r\cos\theta$, $y = r\sin\theta$ and then, $dx = -r\sin\theta d\theta$, $dy = r\cos\theta d\theta$, $ds = |dz| = rd\theta$; hence

$$\frac{\partial u}{\partial n} = \left(\frac{\partial u}{\partial r}\frac{\partial r}{\partial x} + \frac{\partial u}{\partial \theta}\frac{\partial \theta}{\partial x}\right)\frac{dy}{ds} - \left(\frac{\partial u}{\partial r}\frac{\partial r}{\partial y} + \frac{\partial u}{\partial \theta}\frac{\partial \theta}{\partial y}\right)\frac{dx}{ds}$$

$$= \left[\frac{\partial u}{\partial r}\cos\theta + \frac{\partial u}{\partial \theta}\left(-\frac{1}{r}\sin\theta\right)\right]\cos\theta - \left[\frac{\partial u}{\partial r}\sin\theta + \frac{\partial u}{\partial \theta}\cdot\frac{1}{r}\cos\theta\right](-\sin\theta)$$

$$= \frac{\partial u}{\partial r}$$

$$\Rightarrow {}^*du = \frac{\partial u}{\partial n}|dz| = r\frac{\partial u}{\partial r}d\theta \quad \text{along} \quad |z| = r. \tag{6.3.1.4}$$

Choose $u_1(z) = \log|z|$ and $u_2(z) = u(z)$ which are harmonic in $0 < |z| < \rho$, and the cycle $\gamma = C_1 - C_2$, where C_k is a circle $|z| = r_k < \rho$, $k = 1, 2$, oriented in the counterclockwise sense. Then $(*_4)$ yields, since γ

is homologous to zero in $0 < |z| < \rho$,

$$\int_{\gamma} \log r \cdot r \frac{\partial u}{\partial r} d\theta - u \cdot r \frac{1}{r} d\theta = 0$$

$$\Rightarrow \log r_1 \int_{C_1} r_1 \frac{\partial u}{\partial r} d\theta - \int_{U_1} u d\theta - \log r_2 \int_{C_2} r_2 \frac{\partial u}{\partial r} d\theta - \int_{C_2} u d\theta.$$

While ($*_3$), applying to $u_1(z)$ here, yields similarly

$$\int_{C_1} r_1 \frac{\partial u}{\partial r} d\theta = \int_{C_2} r_2 \frac{\partial u}{\partial r} d\theta.$$

These two identities together imply that

$$\int_{|z|=r} r \frac{\partial u}{\partial r} d\theta = \text{constant}, \quad \text{if } 0 < r < \rho, \quad \text{and}$$

$$= 0 \quad \text{if } u(z) \text{ is harmonic in } |z| < \rho;$$

$$\log r \int_{|z|=r} r \frac{\partial u}{\partial r} d\theta - \int_{|z|=r} u d\theta = \text{constant}, \quad \text{if } 0 < r < \rho. \qquad (*_5)$$

Hence, we have proved

The arithmetic mean of a harmonic function over concentric circles. Suppose $u(z)$ is harmonic in $0 < |z| < \rho$. Then

$$\frac{1}{2\pi} \int_{|z|=r} u(z) d\theta = \alpha \log r + \beta, \quad 0 < r < \rho$$

is a linear function of $\log r$, where α and β are constants. Moreover, if $u(z)$ is harmonic in $|z| < \rho$, then

$$\frac{1}{2\pi} \int_{|z|=r} {}^* du = \frac{1}{2\pi} \int_{|z|=r} r \frac{\partial u}{\partial r} d\theta = \alpha = 0 \quad \text{for } 0 \le r < \rho;$$

in this case, $\frac{1}{2\pi} \int_{|z|=r} u(z) d\theta = \beta = u(0)$ or

$$u(0) = \frac{1}{2\pi} \int_0^{2\pi} u(re^{i\theta}) d\theta \quad \text{for } 0 \le r < \rho,$$

called the *mean-value property* of $u(z)$ at 0. $\qquad (6.3.1.5)$

In general, if $u(z)$ is harmonic in $|z - z_0| < \rho$, the *mean-value property* reads as

$$u(z_0) = \frac{1}{2\pi} \int_0^{2\pi} u(z_0 + re^{i\theta}) d\theta, \quad 0 \le r < \rho. \qquad (6.3.1.6)$$

This is exactly the one obtained in (3) of (3.4.3.5), directly via Cauchy's integral formula. The most importance of all is that it mainly relies on the mean-value property of a harmonic function that harmonic(and hence analytic) functions enjoy the maximum–minimum principle. See(3.4.3.7) and its proof.

In what follows, two subsections are divided.

Section (1) The isolated singularities of a harmonic function

We turn to the study of the isolated singularities of a harmonic function.

A point z_0 is called an *isolated singularity* of a harmonic function $u(z)$ if $u(z)$ is harmonic in a deleted neighborhood $0 < |z - z_0| < \rho$ of z_0.

The most natural way to investigate the behavior of $u(z)$ at z_0 is to use the single-valued analytic function $f(z) = \frac{\partial u}{\partial x} - i \frac{\partial u}{\partial y}$ (see (1) in (3.4.3.4)). For simplicity, suppose $z_0 = 0$. This $f(z)$ has the Laurent series expansion at $z_0 = 0$

$$f(z) = \sum_{n=-\infty}^{\infty} a_n z^n$$

$$\Rightarrow F(z) = C + a_{-1} \log z$$

$$+ \sum_{\substack{n = -\infty \\ n \neq -1}}^{\infty} \frac{a_n z^{n+1}}{n + 1}, \quad \text{multiple-valued in } 0 < |z| < \rho$$

$$\text{unless } a_{-1} = 0.$$

$$\Rightarrow u(z) = \operatorname{Re} F(z)$$

$$= \alpha \log r + \sum_{n=-\infty}^{\infty} (\alpha_n \cos n\theta + \beta_n \sin n\theta) r^n, \quad 0 < |z| = r < \rho,$$

$$(*6)$$

where $\alpha = \operatorname{Re} a_{-1}$ (note that $\operatorname{Im} a_{-1} = 0$ since $u(z)$ is single-valued), $\alpha_0 = \operatorname{Re} C$ and $z = r e^{i\theta}$, $0 \le r \le 2\pi$. This means that $u(z)$ is the *real part* of the multiple-valued analytic function

$$F(z) = \alpha \log z + \sum_{-\infty}^{\infty} (\alpha_n - i\beta_n) z^n, \quad 0 < |z| < \rho, \qquad (*7)$$

which, in turn, indicates that 0 is an isolated singularity of $F(z)$. Meanwhile, 0 could be a removable singularity, a pole or an essential singularity of $\sum_{-\infty}^{\infty} (\alpha_n - i\beta_n) z^n$. Note that $(*6)$ also implies the arithmetic mean

$$\frac{1}{2\pi} \int_{|z|=r} u(z) d\theta = \alpha \log r + \alpha_0, \quad \text{with } \alpha_0 = \beta \quad \text{in (6.3.1.5)}. \qquad (*8)$$

There are three cases to be considered.

The removable singularity of a harmonic function. Let $u(z)$ be harmonic in $0 < |z| < \rho$. Then the following are equivalent.

(1) $u(z)$ can be defined at $z = 0$ so that $u(z)$ is harmonic in $|z| < \rho$.

(2) $u(z)$ is bounded in $0 < |z| < \rho$.

(3) $\int_{|u|=r} {}^*du = \int_{|z|=r} r \frac{\partial u}{\partial r}\, dr = 0, \quad 0 \le r < \rho.$

(4) In $(*_6)$, $u(z) = \sum_{n=0}^{\infty}(\alpha_n \cos n\theta + \beta_n \sin n\theta)r^n$ locally uniformly in $|z| < \rho$, where $z = re^{i\theta}$, $0 \le r < \rho$.

Then, z_0 is called a *removable* singularity of $u(z)$.   (6.3.1.7)

Proof. (1) $\Leftrightarrow$ (4) follows from (3) in (3.4.3.4) and the Taylor series representation of an analytic function at 0. (1) $\Rightarrow$ (2) is obvious. (2) $\Leftrightarrow$ (3) is a consequence of the arithmetic mean of $u(z)$ (see $(*_5)$, (6.3.1.5) or $(*_6)$, and $(*_8)$).

To prove (2) $\Rightarrow$ (1) (Ref. [1], p. 166): (1) in (6.3.1.3) and (3) above together indicate that

$$\int_\gamma {}^*du = 0$$

for every rectifiable closed curve in $0 < |z| < \rho$. According to (2) in (3.4.3.4), $u(z)$ has a conjugate harmonic function $v(z)$ in $0 < |z| < \rho$. The bounded assumption of $u(z)$ shows that the single-valued analytic function e^{u+iv} is also bounded in $0 < |z| < \rho$. Hence 0 is a removable singularity of e^{u+iv} (see (3.4.2.17) or (4.10.1.1)); this, in turn, implies that 0 is a removable singularity of $u + iv$ (refer to Example 7 of Sec. 4.10.2), and then, is of u.   □

The second case is

The logarithmic pole of a harmonic function. Let $u(z)$ be harmonic in $0 < |z| < \rho$. Then the following are equivalent.

(1) There is a constant $\alpha \ne 0$ so that $u(z) - \alpha \log|z|$ has a removable singularity at 0.

(2) $u(z)$ can be expressed as $u(z) = \alpha \log|z| + h(z)$ in $0 < |z| < \rho$, where $\alpha \ne 0$ is a constant and $h(z)$ is harmonic in $|z| < \rho$.

(3) In $(*_6)$ and $(*_7)$, $u(z) = \alpha \log r + \sum_{n=0}^{\infty}(\alpha_n \cos n\theta + \beta_n \sin n\theta)r^n$ in $0 < |z| < \rho$, where $\alpha \ne 0$ and $z = re^{i\theta}$, $0 < r < 1$, $0 < r < \rho$.

Then $z = 0$ is called a *logarithmic pole* of $u(z)$ and $\alpha \log|z|$ the *principle part* of $u(z)$ at the logarithmic pole 0.   (6.3.1.8)

The last case is

The essential singularity of a harmonic function. Let $u(z)$ be harmonic in $0 < |z| < \rho$. In $(*_6)$, there is at least one nonzero coefficient α_{-n}, β_{-n} $(n \geq 1)$; equivalently, in $(*_7)$, $\sum_{n=-\infty}^{\infty} (\alpha_n - i\beta_n)z^n$ has a pole or an essential singularity at 0. Then $u(z)$ takes all real values in every deleted neighborhood of $z = 0$. In this case, 0 is called an *essential singularity* of $u(z)$.

$$(6.3.1.9)$$

Proof. Suppose $\alpha_{-m} \neq 0$ for some $m \geq 1$. Since $u(z)$ is continuous on the connected set $0 < |z| < \rho$, it is sufficient, via the intermediate-value property (see (2.2.6)), to show that $u(z)$ takes value of arbitrarily large absolute value. Suppose on the contrary that $u(z)$ is bounded above by M, namely, $u(z) \leq M$ on $0 < |z| < \rho$. Observe that, for $0 < r < \rho$,

$$\int_0^{2\pi} u(z) \cos m\theta d\theta = \pi(\alpha_m r^m + \alpha_{-m} r^{-m}), \text{ and, by } (*_6),$$

$$\int_0^{2\pi} u(z)d\theta = 2\pi(\alpha_0 + \alpha \log r)$$

$$\Rightarrow 2\pi(\alpha_0 + \alpha \log r) \pm \pi(\alpha_m r^m + \alpha_{-m} r^{-m}) = \int_0^{2\pi} u(z)(1 \pm \cos m\theta)$$

$$\leq 4\pi M. \tag{$*_9$}$$

Choose the sign ± 1 so that $|\alpha_{-m}| = \alpha_{-m}$ or $-\alpha_{-m}$. Then

$$|\alpha_{-m}|r^{-m} + 2\alpha \log r + (2\alpha_0 \pm \alpha_m r^m) \leq 4M, \quad 0 < r < \rho.$$

This contradicts to the fact that $|\alpha_{-m}|r^{-m} + 2\alpha \log r \to +\infty$ as $r \to 0^+$. The claim follows. $\qquad\square$

Section (2) Green's function of a domain with respect to a point

The function $u(z) = -\log|z|$ is harmonic and positive in $0 < |z| < 1$, and has the properties that $\lim_{z \to 0} u(z) = +\infty$ and $\lim_{|z| \to 1} u(z) = 0$. This $u(z)$ has a logarithmic pole at 0.

A little more general is the function

$$u(z) = -\log\left|\frac{z - a}{1 - \bar{a}z}\right|, \quad |z| < 1 \quad \text{but } z \neq a \quad \text{(a fixed point in } |z| < 1\text{)}.$$

$$(*_{10})$$

It is harmonic in $|z| < 1$ except at $z = a$, which is a logarithmic pole, and has the boundary values zero along $|z| = 1$.

One more step further is the case where Ω is a simply connected domain with more than one boundary point. Fix any point $z_0 \in \Omega$. By the Riemann mapping theorem (6.1.1), there exists a unique function $w = f(z)$ mapping Ω univalently onto $|w| < 1$ and satisfying the conditions that $f(z_0) = 0$ and $f'(z_0) > 0$. Set the harmonic function

$$g(z, z_0) = \log \frac{1}{|f(z)|}, \quad z \in \Omega - \{z_0\}. \tag{$*_{11}$}$$

Rewrite $f(z) = (z - z_0)\varphi(z)$, where $\varphi(z)$ is analytic and nonzero throughout Ω. Then

$$g(z, z_0) = \log \frac{1}{|z - z_0|} + \log \frac{1}{|\varphi(z)|}, \quad z \in \Omega - \{z_0\}. \tag{6.3.1.10}$$

This $g(z, z_0)$ enjoys the same properties as $u(z)$ in $(*_{10})$ did, namely, $g(z, z_0)$ has a logarithmic pole at z_0 with the principal part $-\log|z - z_0|$, and it is positive in $\Omega - \{z_0\}$ and approaches zero as z approaches the boundary of Ω (since then $w = f(z)$ approaches $|w| = 1$, see (3.4.6.5)). $g(z, z_0)$ is called *Green's function* of Ω with respect to z_0.

In general, let Ω be a domain and $z_0 \in \Omega$ be a fixed point. Suppose there is a real-valued function $g(z, z_0)$ satisfying the following properties:

1. $g(z, z_0)$ is harmonic in $\Omega - \{z_0\}$, and has a logarithmic pole at z_0 with principal part $-\log|z - z_0|$;

2. $g(z, z_0) > 0$ in $\Omega - \{z_0\}$ and $\lim_{z \to \zeta} g(z, z_0) = 0$ for any $\zeta \in \partial\Omega$.

$$\tag{6.3.1.11}$$

Such a function $g(z, z_0)$ is called *Green's function* with *pole* z_0 of the domain Ω with respect to the point $z_0 \in \Omega$. It is *uniquely determined* if it does exist. To see this, observe first that $g(z, z_0) + \log(z - z_0)$ is harmonic throughout Ω (see (6.3.1.8)). If $h(z, z_0)$ is another such function, then $g(z, z_0) - h(z, z_0)$ is harmonic in Ω and vanishes on the boundary $\partial\Omega$. By the maximum principle (see (3.4.3.7) and refer to (3.4.4.5), (3.4.4.6)) it follows that $g(z, z_0) \equiv h(z, z_0)$ in Ω.

The existence of Green's function in a general domain under some constrained boundary condition belongs to the scope of the Dirichlet problem with the boundary value $\log|\zeta - z_0|$, to be fully developed in Sec. 6.5; in particular, in (6.5.7). This function is of fundamental importance both in the theory of harmonic functions and in the theory of conformal mappings.

Take, for instance, the case $(*_{11})$ or (6.3.1.10), to convince oneself the above point of view. We state the following

Equivalent statement of the Riemann mapping theorem (6.1.1). Let Ω be a Jordan domain whose boundary is a Jordan closed rectifiable curve γ. Fix a point $z_0 \in \Omega$. Then

(1) There exists a unique function $w = f(z)$ mapping Ω univalently onto $|w| < 1$ so that $f(z_0) = 0$ and $f'(z_0) > 0$.

$\Leftrightarrow$ (2) There exists the (unique) Green's function $g(z, z_0)$ for the domain Ω with pole at z_0.   (6.3.1.12)

It is precisely this idea that can be generalized to obtain the canonical domains for multiply connected domains (see Sec. 6.6.3). See also the explanation after (6.5.7).

Proof. Only (2) $\Rightarrow$ (1) needs to be proved.

Set $g(z, z_0) = -\log|z - z_0| + G(z, z_0)$, where $G(z, z_0)$ is harmonic in Ω. Since Ω is simply connected, G has a harmonic conjugate on Ω (see (3.4.3.4)); namely, there is an analytic function $F(z)$ in Ω such that $\operatorname{Re} F(z) = G(z, z_0)$. Let

$$f(z) = (z - z_0)e^{-F(z)}.$$

Then $f(z)$ is analytic in Ω and $f(z_0) = 0$. Since

$$-\log|f(z)| = -\log|z - z_0| + G(z, z_0) = g(z, z_0) \to 0 \quad \text{as} \quad z \to \partial\Omega,$$

it follows that $|f(z)| \to 1$ as $z \to \partial\Omega$. Just as in the proof for (6.1.2) in Sec. 6.1.2, $f(z)$ can be continuously extended to $\bar{\Omega}$ (even analytically extended to $\bar{\Omega}$ if γ is an analytic curve, see (6.1.2.3) and its proof). By the very definition (6.3.1.11) or by the maximum principle, $g(z, z_0) > 0$ in $\Omega - \{z_0\}$ and hence, $-\log|f(z)| > 0$ in $\Omega - \{z_0\}$, which, in turn, implies that $|f(z)| < 1$ throughout Ω. Moreover,

$$\frac{f'(z)}{f(z)} = \frac{1}{z - z_0} - F'(z), \quad z \in \Omega - \{z_0\}$$

$$\Rightarrow \frac{1}{2\pi i} \int_\gamma \frac{f'(z)}{f(z)} dz = \frac{1}{2\pi i} \int_\gamma \frac{1}{z - z_0} dz = n(\gamma; z_0) = n(0; f(\gamma)) = 1$$

$\Rightarrow$ (By the Argument principle, see also (4) and (5) in (3.5.2.4).)

$$n(w_0, f(\gamma)) = 1 \quad \text{for any point} \quad w_0 \quad \text{in } |w| < 1.$$

Thus $w = f(z)$ assumes every value w in $|w| < 1$ just once. This proves the univalence of $f(z)$. $\square$

We have another

Two basic properties of a Green's function. Suppose $g(z, z_0)$ is Green's function of a domain Ω with respect to a point $z_0 \in \Omega$.

(1) $g(z, z_0)$ is *invariant* under conformal mappings. Namely, if $z = z(\zeta)$ maps a domain D univalently onto the domain Ω such that $z_0 = z(\zeta_0)$ for some $\zeta_0 \in D$, then $g(z(\zeta), z(\zeta_0))$ is Green's function of D with respect to ζ_0.

(2) In case Ω is a domain of finite connectivity, then $g(z, z_0)$ is *symmetric* as a function of z and z_0 in Ω. Namely,

$$g(z_1, z_2) = g(z_2, z_1), \quad z_1, \quad z_2 \in \Omega, \quad z_1 \neq z_2.$$

And hence, $g(z_1, z_2)$ is simultaneously continuous in both variables, for $z_1 \neq z_2$. $\qquad (6.3.1.13)$

Proof. According to (3.4.6.5), $z = z(\zeta)$ approaches the boundary of Ω as ζ tends to a boundary point of D; hence, $g(z(\zeta), z(\zeta_0))$ has the boundary values zero. Observe that

$$g(z(\zeta), z(\zeta_0)) - \log|\zeta - \zeta_0| = (g(z(\zeta), z(\zeta_0)) - \log|z(\zeta) - z(\zeta_0)|)$$
$$+ (\log|z(\zeta) - z(\zeta_0)| - \log|\zeta - \zeta_0|).$$

The function inside the first bracket in the right is harmonic in $z(\zeta)$, and, in turn, is a harmonic function in $\zeta \in D$ (see (2) in (3.4.3.5)). While, since $z_0 - z(\zeta_0)$ is a simple zero of $z(\zeta) - z(\zeta_0)$, the function in the second bracket is also harmonic in D. By the uniqueness, the claim follows. This proves (1).

As for (2), set $g_1(z) = g(z, z_1)$ and $g_2(z) = g(z, z_2)$ for the moment.

By performing preliminary conformal mappings, we may assume that Ω is bounded by analytic curves $\gamma_1, \ldots, \gamma_n$ (refer to (6.6.8) if necessary). Let C_1 and C_2 be two small circles about z_1 and z_2, respectively, described in the positive sense, so that $\overline{\text{Int } C_1}$ and $\overline{\text{Int } C_2}$ have empty intersection and are contained in Ω. Then the cycle $\gamma_1 + \cdots + \gamma_n - C_1 - C_2 = \gamma$ is homologous to zero in $\Omega - \{z_1, z_2\}$ (see Sec. 4.3.2). Since $g_1(z)$ and $g_2(z)$ vanish on $\gamma_1 + \cdots + \gamma_n$, we have, by (3) in (6.3.1.3),

$$\int_\gamma g_1{}^*dg_2 - g_2{}^*dg_1 = -\int_{C_1+C_2} g_1{}^*dg_2 - g_2{}^*dg_1 = 0. \qquad (*_{12})$$

Let $g_1(z) = -\log|z - z_1| + G_1(z)$, where $G_1(z)$ is harmonic in Ω. Then

$$^*dg_1 = -d\,\text{Arg}(z - z_1) + {}^*dG_1$$

$$\Rightarrow \int_{C_1} g_1{}^* dg_2 - g_2{}^* dg_1 = - \int_{C_1} \log |z - z_1|{}^* dg_2 + \int_{C_1} (G_1{}^* dg_2 - g_2{}^* dG_1)$$

$$+ \int_{C_1} g_2 d \operatorname{Arg}(z - z_1),$$

where

$$\int_{C_1} (G_1{}^* dg_2 - g_2{}^* dG_1) = 0$$

(since both G_1 and g_2 are harmonic in $\overline{\operatorname{Int} C_1}$);

$$\int_{C_1} \log |z - z_1|{}^* dg_2 = 0$$

(since $|z - z_1|$ is constant on C_1 and $^* dg_2$ is exact in $\overline{\operatorname{Int} C_1}$);

$$\int_{C_1} g_2 d \operatorname{Arg}(z - z_1) = 2\pi g_2(z_1) \quad \text{(by the mean-value property)}.$$

$$\Rightarrow \int_{C_1} g_1{}^* dg_2 - g_2{}^* dg_1 = 2\pi g_2(z_1). \tag{$*13$}$$

Similarly,

$$\int_{C_2} g_1{}^* dg_2 - g_2{}^* dg_1 = -2\pi g_1(z_2).$$

Then $(*12)$ says that $g_2(z_1) = g_1(z_2)$ should hold.

The proof of the last statement in (2) is left as Exercise A(9).

In case $n = 1$, namely, Ω is a simply connected domain, see Exercise A(1) for a direct proof.   □

As an another application of (3) in (6.3.1.3) and, in particular, of $(*13)$, we have the following

Representation theorems of harmonic function in terms of Green's function. Let Ω be a domain and $z_0 \in \Omega$ be a fixed point such that Green's function $g(z, z_0)$ with pole z_0 exists.

(1) Let the closed disk $|z - z_0| \le \rho$ ($\rho > 0$) be contained in Ω. If $u(z)$ is harmonic in $|z - z_0| \le \rho$, then

$$\int_{|z - z_0| = \rho} g^* du - u^* dg = 2\pi u(z_0).$$

(2) Suppose Ω is a bounded domain whose boundary consists of n Jordan closed analytic (or rectifiable) curves $\gamma_1, \dots, \gamma_n$ ($n \ge 1$). If $u(z)$ is

harmonic in the closed domain $\bar{\Omega}$, then

$$u(z_0) = -\frac{1}{2\pi} \int_\gamma u^* dg = -\frac{1}{2\pi} \int_\gamma u \frac{\partial g}{\partial n} |dz|,$$

where $\gamma = \gamma_1 + \cdots + \gamma_n$ is in the positive direction (with respect to Ω).

(3) Let Ω be an n-connected domain as stated in (4.3.2.5), where $n \geq 2$, and $\gamma_1, \ldots, \gamma_{n-1}$ be a homology basis for Ω. The multiple-valued conjugate function of a harmonic function $u(z)$ on $\bar{\Omega}$ has *periods*

$$\int_{\gamma_k} {}^* du = \int_{\gamma_k} \frac{\partial u}{\partial n} |dz|, \quad 1 \leq k \leq n - 1.$$

Hence, for any cycle $\gamma \sim \sum_{k=1}^{n-1} C_k \gamma_k \ (\mathrm{mod}\,\Omega)$,

$$\int_\gamma {}^* du = \sum_{k=1}^{n-1} C_k \int_{\gamma_k} {}^* du. \tag{6.3.1.14}$$

In (2) and (3) above, what is

$$\frac{1}{2\pi} \int_{\gamma_k} {}^* dg? \tag{6.3.1.15}$$

It is the harmonic measure of γ_k with respect to Ω, to be formally introduced in Sec. 6.6.1. See (6.6.1.5).

Proof.

(1) Let C be the circle $|z - z_0| = \rho$. Applying $(*_{12})$ to the integration of $g^* du - u^* dg$ along C, the same computation leading to $(*_{13})$ will acquire the result.

(2) Choose a small circle $C : |z - z_0| = \rho$ so that $|z - z_0| \leq \rho$ is contained in Ω. Then the cycle $\gamma - C$ is homologous to zero in $\Omega - \{z_0\}$. Observing that $g(z, z_0)$ vanishes along $\gamma = \gamma_1 + \cdots + \gamma_n$, then, by (1),

$$\int_{\gamma - C} u^* dg - g^* du = \int_\gamma u^* dg - \int_C (u^* dg - g^* du)$$

$$= \int_\gamma u^* dg - (-2\pi u(z_0)) = 0$$

yields the result.

(3) Just follows by noting that $\int_\sigma {}^* du = 0$, where $\sigma = \gamma - \sum_{k=1}^{n-1} C_k \gamma_k \sim 0$ in Ω. $\qquad\square$

Exercises A

(1) Let $g(z, z_0)$ be defined as in $(*_{11})$. Show that $g(z_1, z_0) = g(z_0, z_1)$ for any $z_1 \in \Omega$ with $z_1 \neq z_0$, by using the uniqueness of the mapping function and Green's function.

(2) If $a \neq 0$, show that

$$\frac{1}{2\pi} \int_0^{2\pi} \log |\rho e^{i\theta} - a| d\theta = \begin{cases} \log |a|, & \rho = 0 \\ \log \rho, & 0 < |a| < \rho \\ \log |a|, & \rho > |a| \end{cases}.$$

Note that, in case $a = \rho e^{i\theta_0}$ for some θ_0, try to use $|\rho e^{i\theta} - a| = 2\rho \sin \frac{|\theta - \theta_0|}{2} \geq \frac{2}{\pi}\rho|\theta - \theta_0|$ and $|\log |\rho e^{i\theta} - a|| \leq |\log (\frac{2}{\pi}\rho|\theta - \theta_0|)|$, if $|\rho e^{i\theta} - a| < 1$, to show that the improper integral is absolutely convergent. Refer to Example 4 in Sec. 4.12.1 or $(*_5)$ in Sec. 5.5.3.

(3) Suppose $f(z)$ is analytic in $|z| < R$ and has a zero of order λ ($\lambda \geq 0$) at $z = 0$. Also suppose that $a_1, \ldots, a_n, \ldots$ be the sequence of all the zeros (counting multiplicities) of $f(z)$, other than 0, in $|z| < R$, satisfying the property that it does not have a limit point in $|z| < R$. For $0 \leq \rho < R$, let $\nu(\rho)$ be the number of these zeros that are contained in the *closed* disk $|z| \leq \rho$. Assume that $0 < |a_1| \leq |a_2| \leq \cdots \leq |a_n| \leq \cdots$.

(a) Set $\varphi(z) = \frac{f(z)}{z^\lambda} \prod_{k=1}^{\nu(\rho)} \left(\frac{\rho^2 - \bar{a}_k z}{\rho(z - a_k)}\right)$ if $|a_k| < \rho$ for $1 \leq k \leq \nu(\rho)$ and apply the mean-value property to $\log |\varphi(z)|$ over the circle $|z| < \rho$. Show that

$$\frac{1}{2\pi} \int_0^{2\pi} \log |f(\rho e^{i\theta})| d\theta = \log \left|\frac{f^{(\lambda)}(0)}{\lambda!}\right| + \log \left(\rho^\lambda \prod_{k=1}^{\nu(\rho)} \frac{\rho}{|a_k|}\right)$$

and hence deduce that the arithmetic mean of $\log |f(z)|$ is an increasing function of ρ ($0 \leq \rho \leq R$). Under what conditions will it be strictly increasing?

(b) Suppose the zeros a_k, for $m \leq k \leq \nu(\rho)$, lie on $|z| = \rho$ and all the previous zeros lie in $|z| < \rho$. Then consider $\varphi(z) = \frac{f(z)}{z^\lambda} \prod_{k=1}^{m-1} \left(\frac{\rho^2 - \bar{a}_k z}{\rho(z - a_k)}\right) \cdot \prod_{k=m}^{\nu(\rho)} \left(\frac{1}{z - a_k}\right)$ instead and show that the result in (a) remains valid. Refer to (5.5.3.5).

(4) (continued from (3)) Set $M(\rho) = \max_{0 \leq \theta \leq 2\pi} |f(\rho e^{i\theta})|$ for $0 < \rho < R$. Show that

(a) *Jensen's inequality:* For $0 < \rho < R$,

$$\frac{\rho^{\nu(\rho)}}{|a_1| \cdots |a_{\nu(\rho)}|} \leq \frac{M(\rho)}{\left|\frac{f^{(\lambda)}(0)}{\lambda!}\right| e^\lambda};$$

(b) $\int_0^\rho \frac{\nu(r)}{r} dr \leq \log \frac{M(\rho)}{\left|\frac{f^{(\lambda)}(0)}{\lambda!}\right|}$ (refer to Exercise B(1) of Sec. 3.4.4).

(5) Suppose $f(z)$ is analytic in $|z| < R < \infty$, not identically equal to zero, and has infinitely many zeros $a_n \neq 0$ for $n \geq 1$. Show that the following are equivalent:

1. $\lim_{\rho \to R} \frac{1}{2\pi} \int_0^{2\pi} \log |f(\rho e^{i\theta})| d\theta < \infty$;
2. the infinite product $\prod_{n=1}^{\infty} \frac{R}{|a_n|}$ converges;
3. the infinite series $\sum_{n=1}^{\infty} (R - |a_n|)$ converges.

As a consequence, if $f(z)$ is analytic in $|z| < R$ and vanishes at all points of a sequence b_n, $n \geq 1$, such that the series $\sum_{n=1}^{\infty} (R - |b_n|)$ diverges, then $f(z) \equiv 0$ in $|z| < R$ (compare to the interior uniqueness theorem, (4) in (3.4.2.9) and refer Exercise A(13) of Sec. 5.5.1).

Note. An analytic function satisfying the condition 1 is called a function of *bounded logarithmic mean*.

(6) Suppose $f(z)$ is an analytic function of bounded logarithmic mean in $|z| < R < \infty$ and all its zeros are given by $0, \ldots, 0$ (λ times) and $a_1, a_2, \ldots, a_n, \ldots$, where $0 < |a_1| \leq |a_2| \leq \cdots \leq |a_n| \leq \cdots$. Show that $f(z)$ is of the form $f(z) = F(z)B(z)$, where $F(z)$ is analytic and nonvanishing in $|z| < R$, and

$$B(z) = \left(\frac{z}{R}\right)^\lambda \prod_{n=1}^{\infty} \frac{R|a_n|}{a_n} \frac{a_n - z}{R^2 - \bar{a}_n z},$$

called the *Blaschke product* corresponding to the zeros a_n of $f(z)$, is analytic and bounded by 1in $|z| < R$.

(7)(a) Let a_n, $n \geq 1$, be a sequence of points in $|z| < R < \infty$ such that $0 < |a_1| \leq \cdots \leq |a_n| \leq \cdots$. If $\sum_{n=1}^{\infty} (R - |a_n|) < \infty$, show that $\lim_{\rho \to R} n(\rho) \log \frac{\rho}{R} = 0$ where $n(\rho)$ denotes the number a_n which satisfies $|a_n| < \rho < R$.

(b) Use (a) to show that the Blaschke product $B(z)$ in Exercise (6) satisfies

$$\lim_{\rho \to R} \frac{1}{2\pi} \int_0^{2\pi} \log |B(\rho e^{i\theta})| d\theta = 0,$$

namely, $B(z)$ has zero logarithmic mean.

(8) (converse to Exercise (7)(b)) Suppose $f(z)$ is analytic and bounded by 1 in $|z| < R < \infty$, and has zero logarithmic mean. Then $f(z)$ can be written in the form

$$f(z) = e^{i\alpha} B(z),$$

where α is a real constant and $B(z)$ is the Blaschke product corresponding to the zeros of $f(z)$.

(9) Prove the last statement in (2) of (6.3.1.13). One might apply the maxi–
mini principle to $g(z, z_0) + \log|z - z_0| = G(z, z_0)$, which is known to
be symmetric, and harmonic in each variable, and has the boundary
values $\log|\zeta - z_0|$ as a function of z.

6.3.2 *Poisson's formula and integral: The Dirichlet problem for a disk; Harnack's principle*

A function $u(z)$, harmonic in $|z| < R$ and continuous on $|z| \le R$, can be
represented by the so-called *Poisson's formula* through its values along the
boundary circle $|z| = R$ and hence, is uniquely determined by its boundary
values. A byproduct is that both its harmonic conjugate $v(z)$ in $|z| < R$
and the resulted analytic function $u(z) + iv(z)$ can be expressed through
its boundary values, too. This is *Schwarz's formula*. The most prominent
perspective, along all, is the introduction of *Poisson's integral*, based on
the idea implicitly provided by Poisson's formula, that solves the *Dirichlet
problem* for a disk, due to H. A. Schwarz. The Riemann mapping theorem
then permits us to extend some of the results obtained to simply connected
domains.

This section has been divided into four.

Section (1) Poisson's and Schwarz's formulas

In fact, we obtained these two formulas in Exercise B(4) of Sec. 3.4.2
through the integral formula presented in Example 4 there. It is worthy
to prove them by using the mean-value property (6.3.1.6) because they are
the generalized versions of the latter.

To start with, suppose that $u(z)$ is harmonic on the closed disk $|z| \le R$.
Fix any point a in $|z| < R$ and consider the bilinear transformation $w = \frac{R(z-a)}{R^2 - \bar{a}z}$ that maps $|z| \le R$ onto $|w| \le 1$ with $z = a$ corresponding to
$w = 0$. The composite $u(z(w))$ of $u(z)$ followed by the inverse function
$z = z(w) = \frac{R(Rw+a)}{R + \bar{a}w}$ is harmonic in $|w| \le 1$ (see (2) in (3.4.3.5)), and, by
(6.3.1.6), we have

$$u(a) = u(w(0)) = \frac{1}{2\pi} \int_{|w|=1} u(z(w)) d\mathrm{Arg}\, w. \qquad (*_1)$$

Since $w = e^{i\theta}$, where $\theta = \mathrm{Arg}\, w \in [0, 2\pi]$, $dw = ie^{i\theta} d\theta = iw d\theta$ holds and
hence,

$$d\mathrm{Arg}\, w = -\frac{i}{w} dw = -i \left(\frac{1}{z-a} + \frac{\bar{a}}{R^2 - \bar{a}z} \right) dz$$

(by logarithmic differentiation)

$$= \left(\frac{z}{z-a} + \frac{\bar{a}z}{R^2 - \bar{a}z} \right) d\theta$$

(since $z = Re^{i\theta}$ and $dz = Rie^{i\theta}d\theta = izd\theta$)

$$= \frac{R^2 - |a|^2}{|z-a|^2} d\theta$$

$$\left(\text{by using } R^2 = z\bar{z} \text{ and } \frac{z}{z-a} + \frac{\bar{a}}{\bar{z}-\bar{a}} = \frac{R^2 - |a|^2}{|z-a|^2} \right)$$

$$= \mathrm{Re} \frac{z+a}{z-a} d\theta \quad \left(\text{since } \frac{\bar{z}}{\bar{z}-\bar{a}} + \frac{a}{z-a} = \frac{R^2 - |a|^2}{|z-a|^2} \right)$$

$$= \frac{R^2 - r^2}{R^2 - 2rR\cos(\theta - \varphi) + r^2} d\theta, \quad \text{where}$$

$$z = Re^{i\theta} \text{ and } a = re^{i\varphi}. \tag{$*_2$}$$

Substituting this result into $(*_1)$, we obtain Poisson's formula stated in (6.3.2.1) below.

In case $u(z)$ is harmonic in $|z| < R$ and continuous on $|z| \le R$, just apply the known result to the harmonic function $u(rz)$ on $|z| \le R$, where $0 < r < 1$, to obtain

$$u(ra) = \frac{1}{2\pi} \int_0^{2\pi} \frac{R^2 - |a|^2}{|z-a|^2} u(rz)d\theta. \tag{$*_3$}$$

Since $u(z)$ is uniformly continuous on $|z| \le R$, it follows that $u(rz) \to u(z)$ uniformly on $|z| = R$ as $r \to 1$. On letting $r \to 1$ in $(*_3)$ now, we still obtain the same formula.

We summarize as the following

Poisson's formula for harmonic function on a disk. Suppose that $u(z)$ is harmonic in $|z| < R$ and continuous on $|z| \le R$. Then, for all $|a| < R$,

$$u(a) = \frac{1}{2\pi} \int_{|z|=R} \frac{R^2 - |a|^2}{|z-a|^2} u(z)d\mathrm{Arg}\,z$$

$$= \frac{1}{2\pi} \int_{|z|=R} \mathrm{Re} \frac{z+a}{z-a} u(z)d\mathrm{Arg}\,z$$

$$= \frac{1}{2\pi} \int_0^{2\pi} \frac{R^2 - r^2}{R^2 - 2Rr\cos(\theta - \varphi) + r^2} u(Re^{i\theta})d\theta, \quad a = re^{i\varphi},$$

where $\frac{R^2 - |a|^2}{|z-a|^2}$ (see $(*_2)$) is called *Poisson's kernel* at point a. (6.3.2.1)

In short, the value $u(a)$ of the harmonic function $u(z)$ at an interior point a is recaptured by the arithmetic mean of its boundary values along

$|z| = R$ times Poisson's kernel at that point a. In particular, setting $u(z) \equiv 1$ in (6.3.2.1), we obtain

$$\int_{|z|=R} \frac{R^2 - |a|^2}{|z - a|^2} d\text{Arg}\, z = 2\pi. \tag{6.3.2.2}$$

According to (4.7.1) and noting that $d\text{Arg}\, z = \frac{dz}{iz}$ in case $z = e^{i\text{Arg}\, z}$,

$$\frac{1}{2\pi i} \int_{|\zeta|=R} \frac{\zeta + z}{\zeta - z} u(\zeta) \frac{d\zeta}{\zeta} \tag{$*_4$}$$

is an analytic function of z in $|z| < R$ or $|z| > R$. As a consequence of this and (6.3.2.1), we obtain the following

Schwarz's formula for an analytic function on a disk in terms of its real part. Let $f(z)$ be analytic in $|z| < R$ and continuous on $|z| \leq R$, and $u(z) = \text{Re}\, f(z)$ be its real part. Then

$$f(z) = \frac{1}{2\pi i} \int_{|\zeta|=R} \frac{\zeta + z}{\zeta - z} u(\zeta) \frac{d\zeta}{\zeta} + i\text{Im}\, f(0), \quad |z| < R$$

and hence

$$u(z) = \text{Re}\, f(z) = \frac{1}{2\pi} \int_{|\zeta|=R} \text{Re} \frac{\zeta + z}{\zeta - z} \cdot \frac{u(\zeta)}{i\zeta} d\zeta;$$

$$v(z) = \text{Im}\, f(z) = \frac{1}{2\pi} \int_{|\zeta|=R} \text{Im} \frac{\zeta + z}{\zeta - z} \cdot \frac{u(\zeta)}{i\zeta} d\zeta + \text{Im}\, f(0)$$

$$= \frac{1}{2\pi} \int_0^{2\pi} \frac{2Rr \sin(\theta - \varphi)}{R^2 - 2Rr \cos(\theta - \varphi) + r^2} \cdot u(Re^{i\theta}) d\theta, \quad z = re^{i\varphi}.$$

$$\tag{6.3.2.3}$$

A generalization of (6.3.2.1) is the following

Poisson's formula for harmonic function on a simply connected domain with more than one boundary point. Let $u(z)$ be harmonic on such a domain Ω. Let z_0 be an arbitrarily fixed point of Ω and $w = f(z)$ be the unique function mapping Ω univalently onto $|w| < 1$ that satisfies $f(z_0) = 0$ and $f'(z_0) > 0$. For $0 < \rho < 1$, denote by $\gamma_\rho = f^{-1}(|w| = \rho)$ the regular Jordan closed curve, the inverse image of the circle $|w| = \rho$ under $w = f(z)$. Then

$$u(z) = \frac{1}{2\pi} \int_{\gamma_\rho} u(\zeta) \frac{|f(\zeta)|^2 - |f(z)|^2}{|f(\zeta) - f(z)|^2} \cdot \frac{|f'(\zeta)|}{|f(\zeta)|} |d\zeta|, \quad z \in \text{Int } \gamma_\rho.$$

In particular (refer to (2) in (6.3.1.14)),

$$u(z_0) = \frac{1}{2\pi} \int_{\gamma_\rho} u(\zeta) \frac{|f'(\zeta)|}{|f(\zeta)|} |d\zeta|$$

$$= \frac{-1}{2\pi} \int_{\gamma_\rho} u(\zeta) \frac{\partial}{\partial n} \log \frac{1}{|f(\zeta)|} |d\zeta| = \frac{-1}{2\pi} \int_{\gamma_\rho} u(\zeta)^* dg(\zeta, z_0),$$

where $g(z, z_0) = -\log |f(z)|$ is Green's function for Ω with the pole z_0 (see
$(*_{11})$ or (6.3.1.10) in Sec. 6.3.1) and $\frac{\partial}{\partial n}$ is the outward normal differentiation
with respect to γ_ρ (see $(*_1)$ in Sec. 6.3.1).   (6.3.2.4)

Proof. $u(f^{-1}(w))$ is harmonic on the open disk $|w| < 1$ (see (2) in
(3.4.3.5)). Fix $0 < \rho < 1$. Then $u(f^{-1}(w))$ is harmonic on $|w| \le \rho$ and,
by (6.3.2.1), can be expressed as

$$u(z) = \frac{1}{2\pi} \int_{|w|=\rho} \frac{|w|^2 - |f(z)|^2}{|w - f(z)|^2} u(f^{-1}(w)) d\text{Arg}\, w, \quad w = f(\zeta), \quad \zeta \in \gamma_\rho.$$

Recalling $(*_2)$ and observing that $|dw| = \rho d\text{Arg}\, w$ is the arc length element
along $|w| = \rho$, it follows that

$$d\text{Arg}\, w = \frac{|dw|}{\rho} = \frac{|dw|}{|w|} = \frac{|f'(\zeta)|}{|f(\zeta)|} |d\zeta|, \quad w = f(\zeta) \quad \text{and} \quad \zeta \in \gamma_\rho.$$

Substituting this relation into the above integral expression, we obtain
the generalized Poisson's formula. In particular, if setting $z = z_0$, then
it reduces to

$$u(z_0) = \frac{1}{2\pi} \int_{\gamma_\rho} u(\zeta) \frac{|f'(\zeta)|}{|f(\zeta)|} |d\zeta|. \tag{$*_5$}$$

On the other hand, Green's function for $|w| < 1$ with pole at $w = 0$ is given
by $g(w, 0) = -\log |w|$; and the corresponding Green's function for Ω with
pole at z_0 is given by $g(z, z_0) = -\log |f(z)|$. According to (6.3.1.4),

$$^*dg(w, 0) = \frac{\partial}{\partial n} g(w, 0) |dw| = \frac{\partial}{\partial \rho} \log \frac{1}{\rho} |dw| = -\frac{1}{\rho} |dw| \text{ along } |w| = \rho,$$

$$\text{where} \quad w = f(\zeta) \text{ for } \zeta \in \gamma_\rho$$

$$\Rightarrow {}^* dg(\zeta, z_0) = -\frac{|f'(\zeta)|}{|f(\zeta)|} |d\zeta| = \frac{\partial g(\zeta, z_0)}{\partial n} |d\zeta| \text{ along } \gamma_\rho.$$

This ends the proof.   □

Two illustrative examples are provided as follows.

Example 1. (Ref. [1], p. 171).

(a) If $f(z)$ is entire and if $\lim_{z \to \infty} z^{-1} \operatorname{Re} f(z) = 0$ holds, then $f(z)$ is a constant.

(b) If ∞ is an isolated singularity of $f(z)$ and if $\lim_{z \to \infty} z^{-1} \operatorname{Re} f(z) = 0$ holds, then ∞ is a removable singularity.

Proof. (a) Choose a strictly decreasing sequence ε_n, $n \geq 1$, so that $\lim_{n \to \infty} \varepsilon_n = 0$. For each $\varepsilon_n > 0$, there is a constant $R_n > 0$ so that $|\zeta^{-1} \operatorname{Re} f(\zeta)| \leq \varepsilon_n$ if $|\zeta| \geq R_n$. For any fixed $z \in \mathbf{C}$, choose $R > 0$ large enough such that $|z| < R < \frac{1}{\varepsilon_n}$ and $R \geq R_n$ for some n (it is assumed that $\lim_{n \to \infty} R_n = +\infty$). Under this circumstance, by Schwarz's formula,

$$|f(z)| \leq \frac{1}{2\pi} \int_{|\zeta|=2R} \frac{|\zeta| + |z|}{|\zeta| - |z|} |\zeta^{-1} u(\zeta)| |d\zeta| + |\operatorname{Im} f(0)|$$

$$\leq \frac{1}{2\pi} \cdot \frac{2R + R}{2R - R} \cdot \varepsilon_n \cdot 4\pi R + |\operatorname{Im} f(0)| = 6\varepsilon_n R + |\operatorname{Im} f(0)|$$

$$< 6 + |\operatorname{Im} f(0)|.$$

Hence $f(z)$ is a bounded entire function and then, by Liouville's theorem, is a constant.

Readers might use Exercise A(1) to give another proof.

(b) According to (4.9.1.3), $f(z)$ can be expressed as $f(z) = f_1(z) + f_2(z)$, where $\lim_{z \to \infty} f_1(z) = 0$ and $f_2(z)$ is an entire function. The assumption that $\lim_{z \to \infty} z^{-1} \operatorname{Re} f(z) = 0$ implies that $\lim_{z \to \infty} z^{-1} \operatorname{Re} f_2(z) = 0$ holds, which, by (a), indicates that $f_2(z)$ turns out to be a constant. The claim then follows.   □

Example 2. (Ref. [1], p. 172). If $u(z)$ is harmonic in $0 < |z| < \rho$ and $\lim_{z \to 0} z u(z) = 0$, then $u(z)$ can be written in the form $u(z) = \alpha \log |z| + u_0(z)$, where $\alpha = \frac{1}{2\pi} \int_{|z|=r} {}^* du$ (see (6.3.1.5)) is a constant $(0 < r < \rho)$ and $u_0(z)$ is harmonic in $|z| < \rho$.

Proof. This can be easily seen by using $(*_6)$, (6.3.1.7), (6.3.1.8), and (6.3.1.9) of Sec. 6.3.1. In the content of this section, set

$$u_0(z) = u(z) - \alpha \log |z|.$$

Then $u_0(z)$ is harmonic in $0 < |z| < \rho$. For any fixed $0 < r < \rho$, observe that

$$\int_{|z|=r} {}^*du_0 = \int_{|z|=r} {}^*du - \alpha \int_{|z|=r} {}^*d\log|z| = 2\pi\alpha - 2\pi\alpha = 0.$$

Therefore, $u_0(z)$ has a single-valued harmonic conjugate $v_0(z)$ in $0 < |z| < \rho$ (see (2) in (6.3.1.3)). Then $f_0(z) = u_0(z) + iv_0(z)$ is analytic in $0 < |z| < \rho$ and satisfies the condition that $\lim_{z\to 0} z\operatorname{Re} f_0(z) = \lim_{z\to 0}(zu(z) - \alpha z \log|z|) = 0$. By example 1, 0 is a removable singularity of $f(z)$ and hence, of $u_0(z)$, too. As a matter of fact, we can cite (3) in (6.3.1.7) directly to conclude that 0 is a removable singularity of $u_0(z)$.   □

Section (2) Poisson's integral and the Dirichlet problem for a disk

The converse problem to (6.3.2.1) is as follows. Suppose $u(z)$ is a piecewise continuous real-valued function on the circle $|z| = R$, namely, $u(\operatorname{Re}^{i\theta})$ is a bounded continuous function in $\theta \in [0, 2\pi]$ except at finite many points. *Poisson's integral* of $u(z)$ in $|z| < R$ is defined by

$$P_u(z) = \frac{1}{2\pi}\int_0^{2\pi} \operatorname{Re}\frac{\operatorname{Re}^{i\theta} + z}{\operatorname{Re}^{i\theta} - z}u(\operatorname{Re}^{i\theta})d\theta, \quad |z| < R. \tag{6.3.2.5}$$

$P_u(z)$, as a function of z, is harmonic in $|z| < R$ because it is the real part of the analytic function defined in $(*_4)$. But $P_u(z)$ is a function of u as well in the following sense:

(1) *linearity:* $P_{u+v} = P_u + P_v$; $P_{\alpha u} = \alpha P_u$ for constant α.
(2) *positivity:* $P_u(z) \geq 0$ on $|z| < R$ if $u \geq 0$ on $|z| = R$.
(3) *constancy:* $P_\alpha(z) = \alpha$ for any constant α. $\qquad\qquad$ (6.3.2.6)

(3) follows directly from (6.3.2.2). P_u is called a *positive linear functional* in u owing to properties 1 and 2. Consequently, any inequality $m \leq u(z) \leq M$ on $|z| = R$ always implies $m \leq P_u(z) \leq M$ in $|z| < R$.

$\qquad$H. A. Schwarz gave Poisson's integral $P_u(z)$ the following *geometrical interpretation:* Fix a point z, $|z| < R$, and a point $\operatorname{Re}^{i\theta}$ on the circle $|z| = R$. Choose another point $\operatorname{Re}^{i\theta^*}$ on $|z| = R$ so that $\operatorname{Re}^{i\theta}$, z, and $\operatorname{Re}^{i\theta^*}$ are collinear. See Fig. 6.42. The similarity of the two triangles shows that

$$(R + |z|)(R - |z|) = |\operatorname{Re}^{i\theta} - z||\operatorname{Re}^{i\theta^*} - z|$$

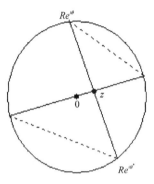

Fig. 6.42

$$\Rightarrow R^2 - |z|^2 = -(Re^{i\theta} - z)(Re^{-i\theta^*} - \bar{z}) \quad \left(\text{since } \frac{Re^{i\theta} - z}{Re^{i\theta^*} - z} < 0\right)$$

$$\Rightarrow \frac{Re^{i\theta}}{Re^{i\theta} - z} = \frac{Re^{-i\theta^*}}{Re^{-i\theta^*} - \bar{z}} \frac{d\theta^*}{d\theta}$$

(treat θ^* as a function of θ; while z is a constant)

$$\Rightarrow \frac{d\theta^*}{d\theta} = \frac{|Re^{i\theta^*} - z|}{|Re^{i\theta} - z|} = \frac{R^2 - |z|^2}{|Re^{i\theta} - z|^2} = \text{Re} \frac{Re^{i\theta} + z}{Re^{i\theta} - z}$$

(taking absolute values of both sides)

$$\Rightarrow P_u(z) = \frac{1}{2\pi} \int_0^{2\pi} u(Re^{i\theta}) d\theta^* = \frac{1}{2\pi} \int_0^{2\pi} u(Re^{i\theta^*}) d\theta. \qquad (6.3.2.7)$$

This means that, to find $P_u(z)$, just replace each value of $u(Re^{i\theta})$ by the value $u(Re^{i\theta^*})$ at the point opposite to z and then take the arithmetic mean over the circle $|z| = R$.

The most fundamental boundary value problem is settled by

Schwarz's theorem (the Dirichlet problem for a disk). Suppose $u(z)$ is a piecewise continuous real-valued function on the circle $|z| = R$. Then Poisson's integral $P_u(z)$ defined by (6.3.2.5) is harmonic in $|z| < R$ and has the following boundary behavior:

(1) $\lim_{z \to Re^{i\theta_0}} P_u(z) = u(Re^{i\theta_0})$ if $u(Re^{i\theta})$ is continuous at $\theta_0 \in [0, 2\pi]$. Moreover, if $u(Re^{i\theta})$ is continuous on $\alpha \le \theta \le \beta$, then $\lim_{z \to Re^{i\theta}} f(z) = u(Re^{i\theta})$ uniformly on $\alpha \le \theta \le \beta$.

(2) Suppose $u(\theta) = u(Re^{i\theta})$ is discontinuous at θ_0 but both $u(\theta_0-)$ and $u(\theta_0+)$ exist. Let $L_\varphi(Re^{i\theta_0})$ be a segment passing the point $Re^{i\theta_0}$ and

making an angle $\varphi(0 < \varphi < \pi)$ with the positive tangent of $|z| = R$ at $Re^{i\theta_0}$. Then, as $z \to Re^{i\theta_0}$ along $L_\varphi(Re^{i\theta_0})$,

$$P_u(z) \to u(\theta_0+) + \frac{\varphi}{\pi}(u(\theta_0-) - u(\theta_0+)). \qquad (6.3.2.8)$$

The restriction on the boundary function $u(z)$ can be relaxed to the Riemann integrability on $[0, 2\pi]$, and even to the Lebesgue integrability; in both cases, $\lim_{z \to Re^{i\theta}} f(z) = u(Re^{i\theta})$ uniformly as $z \to Re^{i\theta}$ in any fixed Stolz domain with vertex at $Re^{i\theta}$ (see Fig. 1.12), for almost all $0 \in [0, 2\pi]$. See Ref. [78], Chap. IV, for a detailed treatment about Poisson's integral.

Proof. For simplicity, we may suppose that $R = 1$ and set $u(\theta) = u(e^{i\theta})$, $0 \le \theta \le 2\pi$.

As for (1): Suppose $u(\theta)$ is continuous at θ_0. Since $P_{u-u(\theta_0)} = P_u - P_{u(\theta_0)} = P_u - u(\theta_0)$ (see (6.3.2.6)), we may assume, without loss of generality, that $u(\theta_0) = 0$.

Given $\varepsilon > 0$, there is an open arc $C_2 = \{e^{i\theta} | \theta_0 - \delta < \theta < \theta_0 + \delta\}$ for some $\delta > 0$, on $|z| = 1$ such that $|u(\theta)| < \frac{\varepsilon}{2}$ if $e^{i\theta} \in C_2$. Set $C_1 = \{|z| = 1\} - C_2$, the complementary arc of C_2 in $|z| = 1$. Let

$$u_1(\theta) = \begin{cases} u(\theta), & e^{i\theta} \in C_1 \\ 0, & e^{i\theta} \in C_2 \end{cases}, \quad \text{and}$$

$$u_2(\theta) = u(\theta) - u_1(\theta).$$

Then (see $(*_4)$)

$$P_{u_j}(z) = \text{Re}\left\{ \frac{1}{2\pi i} \int_{\zeta \in C_j} \frac{\zeta + z}{\zeta - z} u_j(\zeta) \frac{d\zeta}{\zeta} \right\}, \quad j = 1, 2$$

is the real part of an analytic function $f_j(z)$ in $\mathbf{C} - C_j$, and hence, is harmonic in $\mathbf{C} - C_j$. Since $\text{Re}\frac{\zeta+z}{\zeta-z} = \frac{1-|z|^2}{|e^{i\theta}-z|^2} = 0$ on C_2 for all $e^{i\theta} \in C_1$, it follows that $P_{u_1}(z) = 0$ along C_2; and, by continuity, $\lim_{z \to e^{i\theta_0}} P_{u_1}(z) = P_{u_1}(e^{i\theta_0}) = 0$. Hence, there is a $\delta_1 > 0$ so that $|P_{u_1}(z)| < \frac{\varepsilon}{2}$, $|z - e^{i\theta_0}| < \delta_1$, and $|z| < 1$. Note that $|u_2(\theta)| < \frac{\varepsilon}{2}$ along C_2 implies that $|P_{u_2}(z)| < \frac{\varepsilon}{2}$ on $|z| < 1$. Combining together, we have

$$|P_u(z)| = |P_{u_1}(z) + P_{u_2}(z)| \le |P_{u_1}(z)| + |P_{u_2}(z)| < \frac{\varepsilon}{2} + \frac{\varepsilon}{2} = \varepsilon$$

if $|z| < 1$ and $|z - e^{i\theta_0}| < \delta_1$.

The second part is proved similarly by using the uniform continuity of both $u_2(\theta) = u(\theta)$ and $P_{u_1}(\theta)$ on $[\alpha, \beta]$.

A slight variation of this proof, even simpler, is given in Exercise A(3) if $z \to e^{i\theta_0}$ in a Stolz domain with vertex at $e^{i\theta_0}$.

As for (2): We may assume that $\theta_0 = 0$. Set $U(\theta) = u(\theta) - \frac{\theta}{2\pi}[u(2\pi-) - u(0+)]$. Since $U(0+) = U(2\pi-) = u(0+)$, redefine $U(0)$ by $u(0+)$ so that $U(\theta)$ turns out to be continuous at $\theta = 0$. On applying the known result in 1 to Poisson's integral

$$P_U(z) = \frac{1}{2\pi} \int_{|\zeta|=1} \operatorname{Re} \frac{\zeta+z}{\zeta-z} U(\zeta) d\zeta$$

$$\Rightarrow \lim_{z \to 1} P_U(z) = U(0) = u(0+)$$

$$= \lim_{z \to 1} P_u(z) - \frac{1}{2\pi}[u(2\pi-) - u(0+)] \lim_{z \to 1} P_\theta(z),$$

where

$$P_\theta(z) = \frac{1}{2\pi} \int_{|\zeta|=1} \operatorname{Re} \frac{\zeta+z}{\zeta-z} \theta d\zeta = \frac{1}{2\pi} \int_0^{2\pi} \frac{1-r^2}{1-2r\cos(\theta-\psi)+r^2} \theta d\theta,$$

$$z = re^{i\psi} \text{ and } |z| < 1. \tag{$*_6$}$$

It can be shown that $P_\theta(z) \to 2\varphi$ as z approaches 1 along the segment $L_\varphi(1)$ (see Exercise A(4)). Therefore, as soon as $z \to 1$ along $L_\varphi(1)$, then

$$P_u(z) \to u(0+) + \frac{1}{2\pi}[u(2\pi-) - u(0+)] \cdot 2\varphi = u(0+) + \frac{\varphi}{\pi}[u(2\pi-) - u(0+)].$$

This finishes the proof of (6.3.2.8).   □

Based on (6.1.2), we are able to extend (6.3.2.8) to

The Dirichlet problem for a Jordan domain. Let $u(z)$ be a piecewise continuous real function on the boundary $\partial\Omega$ of a Jordan domain Ω, and let $w = f(z)$ be a function mapping Ω univalently onto the disk $|w| < 1$. Then the harmonic function

$$P_u(z) = \frac{1}{2\pi} \int_0^{2\pi} \frac{1-|f(z)|^2}{|e^{i\theta} - f(z)|^2} u(f^{-1}(e^{i\theta})) d\theta, \quad z \in \Omega$$

solves the Dirichlet problem for Ω. In case the boundary $\partial\Omega$ is rectifiable and $w = f(z)$ has a continuous derivative on the closed domain $\bar{\Omega}$, then $P_u(z)$ can be expressed as

$$P_u(z) = \frac{1}{2\pi} \int_{\partial\Omega} \frac{1-|f(z)|^2}{|f(\zeta) - f(z)|^2} u(\zeta) |f'(\zeta)| |d\zeta|, \quad z \in \Omega;$$

and, in particular,

$$P_u(f^{-1}(0)) = \frac{-1}{2\pi} \int_{\partial\Omega} u(\zeta) \frac{\partial}{\partial n} \log \frac{1}{|f(\zeta)|} |d\zeta| = \frac{-1}{2\pi} \int_{\partial\Omega} u(\zeta)^* dg(\zeta, f^{-1}(0)).$$

$$(6.3.2.9)$$

A sufficient condition for $f(z)$ to have a continuous derivative on $\bar{\Omega}$ is known as *Kellog's theorem* (see Ref. [78], p. 361 or Ref. [35], p. 426): If represented by $z = z(s)$ with s being the length parameter, $\partial\Omega$ has a tangent at every point, varied continuously and satisfying $|z'(s_1) - z'(s_2)| \le c|s_1 - s_2|^\alpha, c > 0$ is a constant and $0 < \alpha < 1$. Then $f'(z) \ne 0$ in $\bar{\Omega}$ and $|f'(\zeta_1) - f'(\zeta_2)| \le c'|\zeta_1 - \zeta_2|^\alpha$ for $c' > 0$ a constant and $\zeta_1, \zeta_2 \in \bar{\Omega}$.

If Ω is the upper half-plane and $u(\zeta)$ is a real-valued function along the real axis, we try to derive integral representation analogous to Poisson's integral for $u(z)$ in Im $z > 0$. To this end, consider the mapping $w = \frac{z-i}{z+i}$ which maps Im $z > 0$ onto $|w| < 1$. Note that $w(\infty) = 1$ and $u(\zeta)$ is not defined as $\zeta \to +\infty$ or $-\infty$. For the moment, we suppose that

$$\frac{1}{2\pi} \int_0^{2\pi} \frac{1 - |w|^2}{|e^{i\theta} - w|^2} u\left(\frac{-i(e^{i\theta} + 1)}{e^{i\theta} - 1}\right) d\theta, \quad \text{where}$$

$$e^{i\theta} = \frac{\zeta - i}{\zeta + i} \quad \text{and} \quad \zeta \in \mathbb{R} \cup \{\infty\}$$

is meaningful. Changing the variable w back to z, we obtain the following integral expression in z

$$\frac{1}{2\pi} \int_{-\infty}^{\infty} \frac{4\text{Re}\,i\bar{z}}{|z+i|^2} \cdot \frac{|\zeta + i|^2 |z + i|^2}{|2i|^2 |\zeta - z|^2} \cdot \frac{|2i|}{|\zeta + i|^2} u(\zeta) d\zeta$$

$$= \frac{1}{\pi} \int_{-\infty}^{\infty} \frac{y}{(x - \zeta)^2 + y^2} u(\zeta) d\zeta, \quad z = x + iy \quad \text{with} \quad \text{Im } z = y > 0,$$

$$(6.3.2.10)$$

which is called *Poisson's integral* for $u(\zeta)$ in Im $z > 0$ if the improper integral in the right is convergent. The kernel $\frac{y}{(x-\zeta)^2+y^2}$ is dominated by $\frac{1}{\zeta^2}$ if $|\zeta|$ is large and $z = x + iy$ is fixed in the open upper half-plane. It follows that the integral will converge locally uniformly in Im $z > 0$ in case $u(\zeta)$ *is piecewise continuous and bounded for all real ζ.*

Under this circumstance, we formulate the following analogues of both (6.3.2.1) and (6.3.2.8) as

Poisson's integral and formula for the upper half-plane (Ref. [1], p. 171)

(1) Suppose $u(\zeta)$ is real-valued, piecewise continuous and bounded for all real ζ. Then *Poisson's integral*

$$P_u(z) = \frac{1}{\pi} \int_{-\infty}^{\infty} \frac{y}{(\zeta - x)^2 + y^2} u(\zeta) d\zeta, \quad z = x + iy \text{ with } \mathrm{Im}\, z = y > 0$$

is harmonic in $\mathrm{Im}\, z > 0$ and solves Dirichlet problem for $\mathrm{Im}\, z > 0$: At ζ_0 where $u(\zeta)$ is continuous,

$$\lim_{z \to \zeta_0} P_u(z) = u(\zeta_0);$$

at point ζ_0 where $u(\zeta_0-)$ and $u(\zeta_0+)$ exist, as $z \to \zeta_0$ along a line segment $L_\varphi(\zeta_0)$ which passes ζ_0 and makes an angle φ $(0 < \varphi < \pi)$ with the positive real axis, then

$$P_u(z) \to u(\zeta_0+) + \frac{\varphi}{\pi}[u(\zeta_0-) - u(\zeta_0+)]$$

namely, $P_u(z) - [u(\zeta_0+) + \frac{1}{\pi}(u(\zeta_0-) - u(\zeta_0+))\mathrm{Arg}(z - \zeta_0)] \to 0$ as $z \to \zeta_0$ along $L_\varphi(\zeta_0)$.

(2) Suppose $u(z)$ is harmonic and bounded in $\mathrm{Im}\, z > 0$, and continuous on $\mathrm{Im}\, z \geq 0$. Then $u(z)$ can be represented as the *Poisson's formula*

$$u(z) = \frac{1}{\pi} \int_{-\infty}^{\infty} \frac{y}{(\zeta - z)^2 + y^2} u(\zeta) d\zeta, \quad z = x + iy \text{ with } \mathrm{Im}\, z = y > 0.$$

$$(6.3.2.11)$$

Proof. That for (1) is left as Exercise A(5).

As for (2), construct Poisson's integral $P_u(z)$ for the boundary value function $u(\zeta)$ of $u(z)$ in the upper half-plane $\mathrm{Im}\, z = y > 0$. Then (1) says that $\lim_{z \to \zeta} P_u(z) = u(\zeta)$ for all real ζ (even locally uniformly); hence $\lim_{z \to \zeta}(u(z) - P_u(z)) = 0$ for all real ζ. Since all the lines pass through the point at infinity ∞ (see (1.6.10)), we need to control the behavior of $u(z) - P_u(z)$ as $z \to \infty$ ($\mathrm{Im}\, z > 0$) before we invoke the maximum–minimum principle to infer that $u(z) - P_u(z) \leq 0$ or $u(z) - P_u(z) \geq 0$ throughout $\mathrm{Im}\, z > 0$. For this purpose in mind, consider the harmonic function

$$v(z) = u(z) - P_u(z) - \varepsilon \mathrm{Im}(\sqrt{iz}), \quad \mathrm{Im}\, z > 0,$$

where $\varepsilon > 0$ is a fixed constant and (see Exercise B(1) of Sec. 1.3)

$$\mathrm{Im}\, \sqrt{iz} = \begin{cases} \sqrt{\dfrac{y + \sqrt{x^2 + y^2}}{2}}, & x \neq 0 \\ \sqrt{y}, & x = 0 \end{cases} \quad \text{with } z = x + iy \text{ and } y > 0$$

(why not use Im $\sqrt{z}$?). Then

$$\lim_{z \to \zeta} v(z) = \begin{cases} -\sqrt{\dfrac{|\zeta|}{2}}\, \varepsilon < 0, & \text{if } \zeta \neq 0 \\ 0, & \text{if } \zeta = 0 \\ -\infty, & \text{if } \zeta = \infty \\ & \text{(since both } u(z) \text{ and } P_u(z) \text{ are bounded)} \end{cases}$$

$\Rightarrow v(z) \leq 0$ on Im $z > 0$.

$\Rightarrow$ (letting $\varepsilon \to 0$) $u(z) - P_u(z) \leq 0$, namely, $u(z) \leq P_u(z)$ on Im $z > 0$.

Similarly, $P_u(z) \leq u(z)$ holds on Im $z > 0$. Hence $u(z) = P_u(z)$ throughout Im $z > 0$. $\square$

Section (3) Some applications of the Schwarz's theorem (6.3.2.8)

Schwarz's theorem permits us to take a closer look at the harmonic functions.

Characteristic properties of a harmonic function. Suppose $u(z)$ is a real-valued function on a domain Ω. Then the following are equivalent.

(1) $u(z)$ is harmonic on Ω (recall: $u \in C^2(\Omega)$ and satisfies the Laplacian equation $\Delta u = \frac{\partial^2 u}{\partial x^2} + \frac{\partial^2 u}{\partial y^2} = 0$).

(2) $u(z)$ is continuous on Ω and has the *mean-value property* locally, namely, for each $z \in \Omega$ and for all sufficiently small $r > 0$ (so that $|\zeta - z| \leq r$ is contained in Ω),

$$u(z) = \frac{1}{2\pi} \int_0^{2\pi} u(z + re^{i\theta})d\theta$$

always holds.

(3) $u(z)$ is continuous on Ω, and the partial derivatives $\frac{\partial^2 u}{\partial x^2}, \frac{\partial^2 u}{\partial y^2}$ exist and satisfy $\Delta u = 0$ on Ω.   (6.3.2.12)

Proof. Only (2) $\Rightarrow$ (1) and (3) $\Rightarrow$ (1) are needed to be proved.

(2) $\Rightarrow$ (1): Fix $z_0 \in \Omega$ and consider only these $r > 0$ so that the closed disks $|z - z_0| \leq r$ are contained in Ω and on each of which the mean-value property holds. Choose such an r. Construct Poisson's integral $P_u(z)$ in $|z - z_0| < r$ with the boundary function $u(\zeta)$ on $|\zeta - z_0| = r$. Then $P_u(z)$ is harmonic in $|z - z_0| < r$ and $\lim_{z \to \zeta} P_u(z) = u(\zeta)$ holds for each boundary point ζ. By assumption, the continuous function $P_u(z) - u(z)$ enjoys the mean-value property over every circle $|z - z_0| < \rho \leq r$; and hence, it possesses the

maximum–minimum principle on $|z - z_0| \leq r$ (see the proof of (3.4.3.7)). Since $P_u(\zeta) - u(\zeta) \equiv 0$ along $|\zeta - z_0| = r$, it follows that $P_u(z) = u(z)$ throughout the whole disk $|z - z_0| < r$, and consequently $u(z)$ is harmonic in $|z - z_0| < r$.

(3) $\Rightarrow$ (1) (due to Carathéodory): Adopt the same notation as above. For each $\varepsilon > 0$, consider the function, for $z = x + iy$ and $z_0 = x_0 + iy_0$,

$$v(z) = u(z) - P_u(z) + \varepsilon(x - x_0)^2.$$

This $v(z)$ obeys the maximum principle on $|z - z_0| \leq r$. To see this, observe that $\Delta v = \Delta u - \Delta P_u + 2\varepsilon = 2\varepsilon > 0$. In case $v(z)$ assumes a local maximum at an interior point ζ_0 of $|z - z_0| < r$, then the Hessian matrix at ζ_0

$$\begin{bmatrix} \dfrac{\partial^2 v}{\partial x^2} & \dfrac{\partial^2 v}{\partial x \partial y} \\[2mm] \dfrac{\partial^2 v}{\partial x \partial y} & \dfrac{\partial^2 v}{\partial y^2} \end{bmatrix}$$

is negative semidefinite; in particular, $\frac{\partial^2 v}{\partial x^2} \leq 0$, $\frac{\partial^2 v}{\partial y^2} \leq 0$, and the trace $\Delta v = \frac{\partial^2 v}{\partial x^2} + \frac{\partial^2 v}{\partial y^2} \leq 0$ at ζ_0, a contradiction. Hence $\lim_{z \to \zeta} v(z) = \lim_{z \to \zeta} \varepsilon(x - x_0)^2 \leq \varepsilon r^2$ on $|\zeta - z_0| = r$; and thus $v(z) \leq \varepsilon r^2$ on $|z - z_0| \leq r$. On letting $\varepsilon \to 0$, we obtain that $u(z) - P_u(z) \leq 0$ or $u(z) \leq P_u(z)$ on $|z - z_0| < r$. Similar argument applying to $P_u(z) - u(z) + \varepsilon(x - x_0)^2$ will lead to $P_u(z) \leq u(z)$ on $|z - z_0| < r$. Therefore $u(z) = P_u(z)$ on $|z - z_0| < r$ and $u(z)$ is harmonic there. □

As a second application, Schwarz's theorem can be used to prove the symmetry principle for harmonic functions stated in (3.4.6.4) and, in turn, the symmetry principle for analytic functions (for details, see Sec. 3.4.6). In order to have more practice about the usage of Poisson's integral, we restate it and prove it in the following.

The symmetry (or reflection) principle for harmonic and analytic functions. Let Ω be a symmetry domain with respect to the real axis (see (3.4.1.5) and Fig. 3.15) and $\Omega^+ = \Omega \cap \{\operatorname{Im} z > 0\}$, $\sigma = \Omega \cap \{\operatorname{Im} z = 0\}$. Suppose

1. $v(z)$ is harmonic in Ω^+;
2. $v(z)$ is continuous on $\Omega^+ \cup \sigma$; and
3. $v(z) = 0$ along σ.

Then $v(z)$ has a harmonic extension to Ω which satisfies the symmetric relation

$$v(\bar{z}) = -v(z).$$

In case $v(z)$ is the imaginary part of an analytic function $f(z)$ in Ω^+, then $f(z)$ has an analytic extension to Ω which satisfies $f(\bar{z}) = \overline{f(z)}$.

$$(6.3.2.13)$$

Proof. Construct the function

$$h(z) = \begin{cases} v(z), & z \in \Omega^+ \\ 0, & z \in \sigma \\ -v(\bar{z}), & z \in \Omega^- = \Omega \cap \{\operatorname{Im} z < 0\} \end{cases}.$$

$h(z)$ is continuous on Ω, and harmonic in Ω^+ and Ω^-. What remains is to show that $h(z)$ is harmonic along σ. Fix a point $x_0 \in \sigma$. Consider a disk $|z - x_0| \leq r$, contained in Ω, and construct Poisson's integral $P_h(z)$ of $h(z)$ in $|z - x_0| < r$ (see (6.3.2.5)). In case x is real and $|x - x_0| < r$,

$$P_h(x) = \frac{1}{2\pi} \int_0^{2\pi} \operatorname{Re} \frac{re^{i\theta} + x}{re^{i\theta} - x} h(re^{i\theta}) d\theta = \frac{1}{2\pi} \int_0^{\pi} \cdots d\theta + \frac{1}{2\pi} \int_{\pi}^{2\pi} \cdots d\theta$$

$$= \frac{1}{2\pi} \int_0^{\pi} \operatorname{Re} \frac{re^{i\theta} + x}{re^{i\theta} - x} v(re^{i\theta}) d\theta + \frac{1}{2\pi} \int_{\pi}^{0} \operatorname{Re} \frac{re^{i(2\pi-\theta)} + x}{re^{i(2\pi-\theta)} - x}$$

$$[-v(re^{-i(2\pi-\theta)})] d(2\pi - \theta)$$

$$= \frac{1}{2\pi} \int_0^{\pi} \operatorname{Re} \frac{re^{i\theta} + x}{re^{i\theta} - x} v(re^{i\theta}) d\theta - \frac{1}{2\pi} \int_0^{\pi} \operatorname{Re} \frac{re^{-i\theta} + x}{re^{-i\theta} - x} v(re^{i\theta}) d\theta = 0;$$

hence, $P_h(z)$ vanishes along σ. The function $h(z) - P_h(z)$ is harmonic in the upper half of the disk $|z - x_0| < r$, vanishes on the upper half circle, by (6.3.2.8), and also along the diameter. By the maximum–minimum principle (3.4.3.7), $h(z) = P_h(z)$ in the closed upper half disk, and by the same process, in the closed lower half disk. In conclusion, $h(z) = P_h(z)$ is harmonic in $|z - x_0| < r$, and in particular at $x_0 \in \sigma$.

Next, suppose $v(z) = \operatorname{Im} f(z)$ where $f(z)$ is analytic in Ω^+.

We still use $v(z)$, instead of $h(z)$, to denote the extended harmonic function in Ω.

For fixed $x_0 \in \sigma$ and the closed disk $|z - x_0| \leq r$ contained in Ω, let $-u(z)$ be the harmonic conjugate, unique up to a real constant, of the extended $v(z)$ in $|z - x_0| < r$ so that $-if(z) = v(z) - iu(z)$ or $f(z) = u(z) + iv(z)$ in the upper half disk. Consider

$$g(z) = u(z) - u(\bar{z}), \quad |z - x_0| < r.$$

$g(z)$ is harmonic. Note that $\frac{\partial g}{\partial x} = 0$ along the real diameter; while

$$\frac{\partial g}{\partial y}(x, 0) = \lim_{y \to 0} \frac{g(x, y) - g(x, 0)}{y} = 2\frac{\partial u}{\partial y}(x, 0) = -2\frac{\partial v}{\partial x}(x, 0) = 0.$$

We conclude that the analytic function $\frac{\partial g}{\partial x} - i\frac{\partial g}{\partial y}$ (see (1) in (3.4.3.4)) is identically equal to zero along real diameter and hence in the whole disk $|z - x_0| < r$. Therefore $g(z)$ is a constant which is zero. Hence

$$u(z) = u(\bar{z}) \quad \text{in} \quad |z - x_0| < r.$$

Now extend $f(z)$ analytically from the upper half circle to the whole $|z - x_0| < r$ by defining $f(z) = u(z) + iv(z)$ throughout it; then $f(\bar{z}) = \overline{f(z)}$ holds in $|z - x_0| < r$.

The above process can be repeated in all such disks. Suppose $|z - x_1| < r_1$ and $|z - x_2| < r_2$ are two overlapped disks, and $u_1(z)$ and $u_2(z)$ are the associated harmonic functions obtained as in the last paragraph. Then, $u_1(z) = u_2(z) = \operatorname{Re} f(z)$ in the upper half of their common part, and hence in the whole common part. This means that the extended $u(z)$ is well-defined.

Finally, note that $f(z) = u(z) + iv(z)$ in Ω^+. Define

$$F(z) = \begin{cases} f(z), & z \in \Omega^+ \cup \sigma \\ \overline{f(\bar{z})}, & z \in \Omega^- \end{cases}.$$

Then $F(z)$ is analytic in Ω and its restriction to Ω^+ is the original $f(z)$, and $F(\bar{z}) = \overline{F(z)}$ holds in Ω. $\square$

Section (4) Harnack's inequality and principle

Suppose $u(z)$ is harmonic in $|z| < R$ and continuous on $|z| \leq R$. By applying the inequalities

$$\frac{R-r}{R+r} \leq \frac{R^2 - |z|^2}{|Re^{i\theta} - z|^2} \leq \frac{R+r}{R-r} \quad \text{for } 0 \leq |z| \leq r < R \text{ and } 0 \leq \theta \leq 2\pi \quad (*_7)$$

to Poisson's integral (6.3.2.1),

$$u(z) = \frac{1}{2\pi} \int_0^{2\pi} \frac{R^2 - |z|^2}{|Re^{i\theta} - z|^2} u(Re^{i\theta})d\theta, \quad |z| < R$$

$$\Rightarrow |u(z)| \leq \frac{1}{2\pi} \frac{R+r}{R-r} \int_0^{2\pi} |u(Re^{i\theta})|d\theta, \quad |z| \leq r < R; \quad (*_8)$$

in case $u(z) \geq 0$ throughout $|z| < R$, then

$$\frac{1}{2\pi} \cdot \frac{R-r}{R+r} \int_0^{2\pi} u(Re^{i\theta})d\theta \leq u(z) \leq \frac{1}{2\pi} \cdot \frac{R+r}{R-r} \int_0^{2\pi} u(Re^{i\theta})d\theta$$

$\Rightarrow$ (by the mean-value property)

$$\frac{R-r}{R+r} u(0) \leq u(z) \leq \frac{R+r}{R-r} u(0), \quad |z| \leq r < R \quad (6.3.2.14)$$

called *Harnack's inequality*, valid only for the positive harmonic functions.

We give an instant application of this inequality.

Example 3.

(1) If $u(z)$ is harmonic in **C** and ≥ 0 everywhere, then $u(z)$ is a constant (see Exercise B(6) of Sec. 3.4.3)
(2) (Ref. [1], p. 244) Let K be a compact set in a domain Ω. Show that there is a constant M, depending only on Ω and K, such that $u(z') \leq Mu(z)$ for any positive harmonic function $u(z)$ in Ω and for any two points z, $z' \in K$.

Solution. (1) In (6.3.2.14), letting $R \to \infty$ while keeping r fixed, we have $u(0) \leq u(z) \leq u(0)$ for $z \in \mathbf{C}$. Hence $u(z) = u(0)$, a constant, for $z \in \mathbf{C}$. Readers are urged to use Liouville's theorem to give another proof.

(2) Fix any point $z \in \Omega$. Consider the closed disk $|\zeta - z| \leq R$, contained in Ω. Choose $r = \frac{R}{2}$ in (6.3.2.14). Then, for any positive harmonic function $u(z)$ in Ω,

$$\frac{R - \frac{R}{2}}{R + \frac{R}{2}}u(z) \leq u(\zeta) \leq \frac{R + \frac{R}{2}}{R - \frac{R}{2}}u(z), \quad |\zeta - z| \leq \frac{R}{2}; \quad \text{or}$$

$$\frac{1}{3}u(z) \leq u(\zeta) \leq 3u(z), \quad |\zeta - z| \leq \frac{R}{2}.$$

$$\Rightarrow \frac{1}{3^2}u(\zeta_2) \leq u(\zeta_1) \leq 3^2 u(\zeta_2), \quad |\zeta_1 - z| \leq \frac{R}{2}, \quad \text{and} \quad |\zeta_2 - z| \leq \frac{R}{2}. \quad (*_9)$$

The compactness of K then implies that there are finitely many points $z_1, \ldots, z_n$ in K such that the total of the disks $|\zeta - z_l| \leq \frac{R_l}{2}$, $1 \leq l \leq n$, covers K, where each larger disk $|\zeta - z_l| \leq R_l$ is still contained in Ω. Then $(*_9)$ indicates that $M = 3^{2n}$ will work. $\qquad \square$

The main application of (6.3.2.14) is used to prove the following powerful

Harnack's principle. Suppose $u_n(z)$, $n \geq 1$, are harmonic in a domain Ω and $u_n(z) \leq u_{n+1}(z)$ on Ω for $n \geq 1$. Then either

(1) $u_n(z)$ converges to $+\infty$ locally uniformly in Ω, or
(2) $u_n(z)$ converges locally uniformly in Ω to a harmonic function $u(z)$.

$$(6.3.2.15)$$

A slight *generalization* is as follows: Each $u_n(z)$ is harmonic in a domain Ω_n. Let Ω be a domain such that every point in Ω has a neighborhood, contained in all but finitely many Ω_n, in which $u_n(z) \leq u_{n+1}(z)$ holds if n is large enough. Then (1) and (2) remain valid in Ω. For such a domain, see, for instance, (5.3.2).

Proof. Suppose $\lim_{n \to \infty} u_n(z_0) = +\infty$ for some point $z_0 \in \Omega$. Choose a closed disk $|z - z_0| \le R$, contained in Ω. Apply the left inequality in (6.3.2.14) to $u_n(z) - u_1(z) \ge 0$ for $|z - z_0| \le r = \frac{R}{2}$, and we obtain

$$\frac{1}{3}[u_n(z_0) - u_1(z_0)] \le u_n(z) - u_1(z), \quad |z - z_0| \le r$$

$$\Rightarrow u_n(z) \ge \frac{1}{3}u_n(z_0) + \left[u_1(z) - \frac{1}{3}u_1(z_0)\right], \quad |z - z_0| \le r$$

$$\Rightarrow u_n(z) \ge \frac{1}{3}u_n(z_0) - K, \quad \text{where } K = \max_{|z-z_0| \le r} \left|u_1(z) - \frac{1}{3}u_1(z_0)\right|,$$

$$|z - z_0| \le r$$

$$\Rightarrow \lim_{n \to \infty} u_n(z) = +\infty \text{ uniformly on } |z - z_0| \le r.$$

On the other hand, suppose $\lim_{n \to \infty} u_n(z_0) < +\infty$ for some point $z_0 \in \Omega$. We use the right inequality in (6.3.2.14) instead and we obtain

$$0 \le u_n(z) - u_1(z) \le 3[u_n(z_0) - u_1(z_0)], \quad |z - z_0| \le r = \frac{R}{2}$$

$$\Rightarrow 0 \le \lim_{n \to \infty} u_n(z) - u_1(z) \le 3[\lim_{n \to \infty} u_n(z_0) - u_1(z_0)] \quad \text{(uniformly)},$$

$$|z - z_0| \le r$$

$$\Rightarrow \lim_{n \to \infty} u_n(z) < +\infty \quad \text{(uniformly)}, \ |z - z_0| \le r.$$

Let $S_1 = \{z \in \Omega | u_n(z) \to +\infty\}$ and $S_2 = \{z \in \Omega | \lim_{n \to \infty} u_n(z) < +\infty\}$. Then both S_1 and S_2 are open as shown in the last paragraph. Also $\Omega = S_1 \cup S_2$. Since Ω is connected, so either $S_1 \ne \emptyset$ or $S_2 \ne \emptyset$.

In case $S_2 = \emptyset$, then $S_1 = \Omega$ and it has been shown that $u_n(z) \to +\infty$ locally uniformly in Ω.

In case $S_1 = \emptyset$, then $S_2 = \Omega$. For $n \ge m$, use the right inequality of (6.3.2.14) again and we have

$$0 \le u_n(z) - u_m(z) \le 3[u_n(z_0) - u_m(z_0)], \quad |z - z_0| \le r = \frac{R}{2}$$

$$\Rightarrow \lim_{n \to \infty} u_n(z) = u(z) \text{ uniformly on } |z - z_0| \le r \text{ and hence, locally}$$

uniformly in Ω.

This $u(z)$ is continuous on Ω. To prove the harmonicity of $u(z)$ on Ω, fix any point $z_0 \in \Omega$ and choose $\rho > 0$ so that $|z - z_0| \le \rho$ is contained in Ω.

Then

$$u_n(z_0) = \frac{1}{2\pi} \int_0^{2\pi} u_n(z_0 + re^{i\theta})d\theta, \quad 0 \le r \le \rho, \quad n \ge 1$$

$$\Rightarrow \text{ (by uniform convergence) } u(z_0) = \frac{1}{2\pi} \int_0^{2\pi} u(z_0 + re^{i\theta})d\theta, \quad 0 < r \le \rho.$$

By (2) in (6.3.2.12), it follows that $u(z)$ is harmonic in Ω.   □

As a supplement to (6.3.2.15) and a counterpart of Montel's theorem (see (2) in (5.3.3.10) and (3) in (5.8.1.3)), we have a

Criterion for normality of a sequence of harmonic functions. Let $u_n(z)$, $n \ge 1$, be a sequence of harmonic functions on a domain Ω. If $u_n(z)$, $n \ge 1$, are locally uniformly bounded in Ω, then it has a subsequence converges to a harmonic function locally uniformly in Ω.   (6.3.2.16)

This can be proved by imitating the proof of Montel's theorem but now Cauchy's integral formula should be replaced by Poisson's formula. An easier way is to use Montel's theorem directly and Schwarz's formula (6.3.2.3).

But a direct extension of Vitali's theorem (see (1) in (5.3.3.10)) to harmonic functions is not more true. For instance, set

$$u_{2n-1}(z) = 0, \quad u_{2n}(z) = \text{Im } z, \quad n \ge 1.$$

Then $u_n(z)$ does converge along the real axis, yet it diverges, elsewhere. However, we have a

Vitali's version for harmonic functions. Let $u_n(z)$, $n \ge 1$, be harmonic in a domain Ω. Suppose

1. $u_n(z)$, $n \ge 1$, are locally uniformly bounded in Ω, and
2. $u_n(z)$ converges pointwise on a *subdomain* D of Ω.

Then $u_n(z)$ converges locally uniformly in Ω.   (6.3.2.17)

Sketch of proof. Suppose $u_n(z)$ does not converge pointwise in Ω. By (6.3.2.16), $u_n(z)$, $n \ge 1$, has two subsequences converging locally uniformly in Ω to harmonic functions $v_1(z)$ and $v_2(z)$, respectively, such that $v_1(z) \equiv v_2(z)$ on D and $v_1(z_0) \ne v_2(z_0)$ for some point $z_0 \in \Omega - D$. Set $V(z) = v_1(z) - v_2(z)$, $z \in \Omega$.

Suppose $z_0 \ne \infty$. Choose a simply connected subdomain A containing z_0 and $D \cap A \ne \phi$. Then $V(z)$ has a conjugate harmonic function $U(z)$ in

A so that $F(z) = V(z) + iU(z)$ is analytic in A. Since $\operatorname{Re} f(z) = 0$ in $D \cap A$, $f(z) \equiv 0$ throughout A. Hence $f(z_0) = 0$ implies that $v_1(z_0) = v_1(z_0)$, a contradiction.

Suppose $z_0 = \infty$, just apply the last paragraph to $u_n(\zeta) = u_n(\frac{1}{\zeta})$.

As long as we have proved that $u_n(z)$, $n \geq 1$, converge pointwise in Ω, just imitate the proof of the original Vitali's theorem to show that it does converge locally uniformly in Ω (see the proof of (5.3.3.10)).   □

Exercises A

(1) Prove the following version of Schwarz's formula (6.3.2.3):

$$f(z) = \frac{1}{\pi i} \int_{|\zeta|=R} \frac{u(\zeta)}{\zeta - z} d\zeta - \overline{f(0)}, \quad |z| < R.$$

(2) Suppose $f(z)$ is analytic in $R \leq |z| \leq \infty$ and $u(z) = \operatorname{Re} f(z)$. Derive Schwarz's formula for $f(z)$ and Poisson's formula for $u(z)$ in terms of the boundary values of $u(\zeta)$ on the circle $|\zeta| = R$.

(3) Still set $R = 1$ and $\theta_0 = 0$. For $\varepsilon > 0$, choose $\delta > 0$ so that $|u(e^{i\theta}) - u(1)| < \varepsilon$ in $|\theta| \leq \delta$. Then

$$|P_u(z) - u(1)| \leq \frac{1}{2\pi} \int_{-\pi}^{\pi} \frac{1 - |z|^2}{|e^{i\theta} - z|^2} |u(e^{i\theta}) - u(1)| d\theta$$

$$= \frac{1}{2\pi} \int_{|\theta| \leq \delta} \cdots + \frac{1}{2\pi} \int_{|\theta| \geq \delta} \cdots = \mathrm{I} + \mathrm{II},$$

where $\mathrm{I} \leq \varepsilon$. In case $|\psi| \leq \frac{\delta}{2}$, where $z = re^{i\psi}$, then $|\theta - \psi| \geq \frac{\delta}{2}$ and $|e^{i\theta} - z|^2 \geq \sin^2(\frac{\delta}{2})$; hence,

$$\mathrm{II} \leq \frac{1 - r^2}{2\pi \sin^2\left(\frac{\delta}{2}\right)} \int_{-\pi}^{\pi} (|u(e^{i\theta})| + |u(1)|) d\theta \leq \text{a constant} \cdot (1 - r).$$

Try to finish the proof for (1) in (6.3.2.8).

(4) In $(*_6)$, try to show that

$$\frac{1 - r^2}{1 - 2r\cos(\theta - \psi) + r^2} = \operatorname{Re} \frac{e^{i\theta} + re^{i\psi}}{e^{i\theta} - re^{i\psi}}$$

$$= 1 + 2 \sum_{n=1}^{\infty} r^n \cos n(\theta - \psi), \quad z = re^{i\psi}$$

$$\Rightarrow P_\theta(z) = \pi - 2 \sum_{n=1}^{\infty} \frac{r^n \sin n\psi}{n} = \pi + 2\operatorname{Im}\log(1 - z),$$

$$z = re^{i\psi} \quad \text{and} \quad |z| < 1.$$

Note that $\operatorname{Im}\log(1 - z) = \operatorname{Arg}(1 - z) = \varphi - \frac{\pi}{2}$ if $z \in L_\varphi(1)$.

(5) Prove (1) in (6.3.2.11) in detail.

(6) Suppose $u(z)$ is harmonic in $\mathbf{C}$ and bounded below (or above). Show that $u(z)$ is a constant by Harnack's inequality (6.3.2.14) and by Liouville's theorem, independently.

(7) Let $f(z)$ be analytic in Im $z > 0$ and continuous on Im $z \geq 0$. Derive representation analogous to Schwarz's formula for $f(z)$ in terms of the boundary values of its real part Re $f(z) = u(z)$ along the real axis.

(8) Let C_1 and C_2 be complementary arcs on the unit circle. Set $u(e^{i\theta}) = 1$ on C_1, $u(e^{i\theta}) = 0$ on C_2. Find $P_u(z)$ explicitly and show that $2\pi P_u(z)$ equals the length of the arc, opposite to C_1, cut off by the chords through z and the end points of C_1.

(9) Let $-\infty < a_1 < a_2 < \cdots < a_n < \infty$. Use (6.3.2.11) to find a harmonic function $u(z)$ in Im $z > 0$ with constant boundary values $u_0, u_1, \ldots, u_n$ on the intervals $(-\infty, a_1), (a_1, a_2), \ldots, (a_n, +\infty)$, respectively. And, then, use the result to deduce the Schwarz–Christoffel transformation given in (6.2.1.7).

(10) Suppose $f_n(z)$, $n \geq 1$, are analytic in a domain Ω. If $f_n(z_0)$ converges for some point $z_0 \in \Omega$ and Re $f_n(z)$ converges locally uniformly in Ω, show that $f_n(z)$ converges locally uniformly in Ω.

(11) Suppose a sequence of harmonic functions $u_n(z)$ in a domain Ω converges locally uniformly in Ω to a harmonic function $u(z)$. Set $z = x + iy$.

(a) Show that the partial derivatives $\dfrac{\partial^{k+l} u(x,y)}{\partial x^k \partial y^l}$ are also harmonic in Ω for all $k, l = 1, 2, \ldots$.

(b) For fixed $k, l = 1, 2, \ldots$, show that $\dfrac{\partial^{k+l} u_n(x,y)}{\partial x^k \partial y^l}$ converges to $\dfrac{\partial^{k+l} u(x,y)}{\partial x^k \partial y^l}$ locally uniformly in Ω.

(12) (*Poisson–Jensen formula for harmonic function*) Let $u(z)$ be harmonic in $|z| < R$, except a logarithmic pole at 0 with principal part $\alpha_0 \log |z|$, nonzero logarithmic poles at $z_1, z_2, \ldots$, arranged in order of increasing absolute values, with principal parts $\alpha_k \log |z - z_k|$, $k = 1, 2, \ldots$. Set $z_0 = 0$. Suppose $z_1, \ldots, z_{v(\rho)}$ are these poles that are contained in the open disk $|z| < \rho < R$. Then, for $|z| < \rho$,

$$\frac{1}{2\pi} \int_0^{2\pi} \frac{\rho^2 - |z|^2}{|\rho e^{i\theta} - z^2|} u(\rho e^{i\theta}) d\theta = u(z) - \sum_{k=0}^{v(\rho)} \alpha_k \log \left| \frac{\rho(z - z_k)}{\rho^2 - \overline{z_k} z} \right|.$$

This formula is still valid even if $z_1, \ldots, z_{v(\rho)}$ are these poles lying in the *closed* disk $|z| \leq \rho$ (see Ref. [58], Vol. II, pp. 224–229 for detailed account; also, refer to (5.5.3.5) and Exercise A(3) of Sec. 6.3.1).

(13) Let $f(z)$ be analytic in $|z| < R \le +\infty$ except a pole at 0 of order λ, nonzero zeros at $a_1, a_2, \ldots$ of orders $\alpha_1, \alpha_2, \ldots$, and nonzero poles at $b_1, b_2, \ldots$ of orders $\beta_1, \beta_2, \ldots$, respectively. Suppose $|a_1| \le |a_2| \le \cdots$ and $|b_1| \le |b_2| \le \cdots$. Let the zeros $a_1, \ldots, a_{n(\rho)}$ and the poles $b_1, \ldots, b_{m(\rho)}$ be contained in the closed disk $|z| \le \rho < R$. Then, for $|z| \le r < \rho$,

$$\frac{1}{2\pi} \int_0^{2\pi} \frac{\rho^2 - |z|^2}{|\rho e^{i\theta} - z|^2} \log |f(\rho e^{i\theta})| d\theta = -\lambda \log \rho - \sum_{k=1}^{n(\rho)} \alpha_k \log \left| \frac{\rho(z - a_k)}{\rho^2 - \overline{a_k} z} \right|$$

$$+ \sum_{l=1}^{m(\rho)} \beta_l \log \left| \frac{\rho(z - b_l)}{\rho^2 - \overline{b_l} z} \right|.$$

In case $z = 0$ is a zero of order λ, $-\lambda \log \rho$ should be replaced by $\log \frac{|f(z)|}{|z|^\lambda}$; if $z = 0$ is neither a zero nor a pole, set $\lambda = 0$ in either case. Refer to (5.5.3.5) and Exercise A(3) of Sec. 6.3.1, too.

(14) Prove (6.3.2.16) and (6.3.2.17) in detail.

Exercises B

Let $C^1(\Omega)$ be the vector space of continuously differentiable real functions on a Jordan domain Ω (refer to (2.8.5)). For $u, v \in C^1(\Omega)$, define

$$D(u, v) = \iint_\Omega \left(\frac{\partial u}{\partial x} \frac{\partial v}{\partial x} + \frac{\partial u}{\partial y} \frac{\partial v}{\partial y} \right) dx dy.$$

Then $C^1(\Omega)$, endowed with $D(,)$ is an *inner product space*, namely,

1. $D(u) = D(u, u) \ge 0$; and $= 0 \Leftrightarrow u(z)$ is a constant.
2. $D(u, v) = D(v, u)$.
3. $D(\lambda_1 u_1 + \lambda_2 u_2, v) = \lambda_1 D(u_1, v) + \lambda_2 D(u_2, v)$ for constants $\lambda_1, \lambda_2 \in \mathbf{R}$.

The usual Cauchy–Schwarz inequality now reads as $D(u, v)^2 \le D(u)D(v)$ with equality if and only if $\frac{u}{v} = \lambda$, a real constant. The square of the length of u

$$D(u) = \iint_\Omega \left[\left(\frac{\partial u}{\partial x} \right)^2 + \left(\frac{\partial u}{\partial y} \right)^2 \right] dx dy$$

$$= \iint_\Omega \left[\left(\frac{\partial u}{\partial r} \right)^2 + \frac{1}{r^2} \left(\frac{\partial u}{\partial \theta} \right)^2 \right] r dr d\theta \quad \text{(in polar coordinates)}$$

is known as *Dirichlet integral* of u relative to Ω, denoted as $D_\Omega(u)$ in order to emphasize its dependence on Ω.

(1) Prove that $D(u)$ is a conformal invariant, namely, if $w = f(z)$ maps Ω onto Ω^* univalently and if $u^*(w) = u(f^{-1}(w))$, then

$$D_\Omega(u) = D_{\Omega^*}(u^*).$$

(2) Let $\mu(z)$ be a continuous real function on the boundary $\partial\Omega$. Let $C^1_\mu(\Omega)$ be the family of these functions in $C^1(\Omega)$ which are continuous on $\bar{\Omega}$ and coincide with $\mu(z)$ on $\partial\Omega$. According to (6.3.2.9), $C^1_\mu(\Omega)$ contains a unique harmonic function $u(z)$. Try to use Exercise B(1) of Sec. 2.9 to show that

$$D(u) \le D(h) \quad \text{for all } h \in C^1_\mu(\Omega),$$

known as *Dirichlet principle* for Ω.

(3) In (2), show that, if $h(z) \not\equiv u(z)$, then $D(u) < D(h)$ unless $D(h) = +\infty$ for all $h \in C^1_\mu(\Omega)$. Set Ω the unit disk $|z| < 1$ and

$$\mu(\theta) = \sum_{n=1}^{\infty} \frac{\cos n^4\theta}{n^2}, \quad 0 \le \theta \le 2\pi.$$

Then show that $D(h) = \infty$ for all $h \in C^1_\mu(\Omega)$ does happen.

Note. Refer to Ref. [78], Chap. 1, for a detailed account about Dirichlet principle.

6.4   Subharmonic Functions

The one-dimensional Laplacian equation $\Delta u = \frac{d^2u}{dx^2} = 0$ has all its solutions provided by the linear functions $u(x) = ax + b$. A function $v(x)$ is called *convex* on (a, b) if, geometrically, in any closed subinterval $[c, d]$, it is not greater than the linear function $u(x)$ with the same values as $v(x)$ at c and d, namely, $v(tc + (1 - t)d) \le tv(c) + (1 - t)v(d)$, $0 \le t \le 1$. The two-dimensional version of these concepts leads to the relations between the harmonic functions and the subharmonic functions, to be formally introduced as follows.

Based on (6.3.2.12), we formulate the following

Definition of a (continuous) subharmonic function in a domain. Let $v(z)$ be a real continuous function in a domain Ω. The following are equivalent:

(1) For any fixed $z_0 \in \Omega$, there is a $\rho > 0$ so that $|z - z_0| \le \rho$ is contained in Ω and that

$$v(z_0) \le \frac{1}{2\pi} \int_0^{2\pi} v(z_0 + re^{i\theta})d\theta, \quad 0 \le r < \rho \quad (\text{even } \le \rho).$$

Hence, such a $v(z)$ enjoys *the maximum principle* in Ω:$v(z)$ cannot have a local maximum in Ω except being a constant.

(2) For any harmonic function $u(z)$ in a *closed* disk (or subdomain) $\bar{D}$ which is contained in Ω, $v(z) \leq u(z)$ on the boundary ∂D always implies that $v(z) \leq u(z)$ in the interior D.

(3) For any harmonic function $u(z)$ in an *open* disk (or subdomain) D contained in Ω, the difference $v(z) - u(z)$ satisfies the maximum principle in D.

(4) (In case $v \in C^2(\Omega)$) $\Delta v \geq 0$ in Ω.

Under any of these circumstances, $v(z)$ is said to be *subharmonic* in Ω.

$$(6.4.1)$$

A continuous function $v(z)$ is said to be *subharmonic at a point z_0* if it is subharmonic in a neighborhood of z_0; and it is subharmonic in a domain Ω if and only if it is at each point of Ω. This definition for subharmonicity is not in its acceptable generality (refer to Exercise B below) yet it is suitable for our main purpose in Sec. 6.5 when solving the Dirichlet problem for quite a general class of domains. By the way, (3) implies implicitly (just choosing $u(z) \equiv 0$) that such a $v(z)$ possesses the maximum principle in Ω.

Proof. Let $|z - z_0| \leq \rho$ be contained in Ω such that

$$v(z_0) \leq \frac{1}{2\pi} \int_0^{2\pi} v(z_0 + re^{i\theta})d\theta, \quad 0 \leq r < \rho.$$

Suppose on the contrary that $v(z_0) = \max_{|z-z_0|\leq\rho} v(z)$ and $v(z)$ is not a constant there. Then there is $0 < r < \rho$ so that $v(z_0+re^{i\theta_0}) < v(z_0)$ for some $\theta_0 \in [0, 2\pi]$ and hence, by continuity, for all θ in an open neighborhood of θ_0. This will lead to $v(z_0) < v(z_0)$ (see the proof for (3.4.3.7)), a contradiction. This shows that such a $v(z)$ enjoys the maximum principle in Ω.

(1) $\Rightarrow$ (2): We may suppose that $\bar{D}$ is the closed disk $|z - z_0| \leq \rho$ and that $v(z_0) > u(z_0)$ holds. Then

$$v(z_0) > u(z_0) = \frac{1}{2\pi} \int_0^{2\pi} u(z_0 + \rho e^{i\theta})d\theta \geq \frac{1}{2\pi} \int_0^{2\pi} v(z_0 + \rho e^{i\theta})d\theta \geq v(z_0),$$

a contradiction. Alternatively, the function $v(z) - u(z)$ has the property that $v(z_0) - u(z_0) \leq \frac{1}{2\pi} \int_0^{2\pi} [v(z_0 + re^{i\theta}) - u(z_0 + re^{i\theta})]$, $0 \leq r < \rho$, and $v(z) - u(z) \leq 0$ on $|z - z_0| = \rho$. The result follows from the maximum principle just proved.

(2) $\Rightarrow$ (3): Fix such a subdomain D and any harmonic function $u(z)$ on it. Let $|z - z_0| \leq \rho$ be a closed disk contained in D. Let $P_v(z)$ be Poisson's

integral of $v(z)$ on $|z - z_0| = r$ for any $0 < r \leq \rho$. Then $P_v(z)$ is harmonic in $|z - z_0| < r$ and has the boundary value $v(z)$ along $|z - z_0| = r$. By assumption, $v(z) \leq P_v(z)$ in $|z - z_0| \leq r$; in particular,

$$v(z_0) \leq P_v(z_0) = \frac{1}{2\pi} \int_0^{2\pi} P_v(z_0 + re^{i\theta})d\theta = \frac{1}{2\pi} \int_0^{2\pi} v(z_0 + re^{i\theta})d\theta,$$

$$0 \leq r \leq \rho.$$

This means that $v(z)$ enjoys the property stated in (1); and so does the difference $v(z) - u(z)$ in D. The claimed result then follows.

(3) $\Rightarrow$ (1): Pick $|z - z_0| \leq \rho$, contained in Ω. Construct $P_v(z)$ of $v(z)$ on $|z - z_0| = \rho$. Then $P_v(z) = v(z)$ on $|z - z_0| = \rho$ implies that $v(z) \leq P_v(z)$, by the maximum principle assumption, which, in turn, implies that $v(z)$ has the property stated in (1).

(1) $\Rightarrow$ (4): Assume $\Delta v(z_0) < 0$ for some point $z_0 \in \Omega$; and then, by continuity, it follows that $\Delta v(z) < 0$ in some open disk $|z - z_0| < \rho$, contained in Ω. Equivalently, $\Delta(-v)(z) > 0$ in $|z - z_0| < \rho$. In case $u(z)$ is harmonic in $|z - z_0| < \rho$ and $-v(z) - u(z)$ has a local maximum in $|z - z_0| < \rho$, then $\Delta(-v - u) = \Delta(-v) \leq 0$ for some point there, which is a contradiction. According to (3) above, $-v(z)$ is subharmonic in $|z - z_0| < \rho$.

Therefore, both $v(z)$ and $-v(z)$ are subharmonic in $|z - z_0| < \rho$; and then, for each fixed z there,

$$-v(z) \leq \frac{-1}{2\pi} \int_0^{2\pi} v(z + re^{i\theta})d\theta, \quad \text{and}$$

$$v(z) \leq \frac{1}{2\pi} \int_0^{2\pi} v(z + re^{i\theta})d\theta \text{ for all sufficiently small } r > 0$$

$$\Rightarrow v(z) = \frac{1}{2\pi} \int_0^{2\pi} v(z + re^{i\theta})d\theta, \quad r > 0 \text{ small}$$

$$\Rightarrow v(z) \text{ is harmonic in } |z - z_0| < \rho, \text{ namely, } \Delta v(z) = 0 \text{ holds there.}$$

This contradictory result justifies that $\Delta v(z) \geq 0$ throughout Ω.

(4) $\Rightarrow$ (1): Fix $\varepsilon > 0$ and consider the function $g(z) = v(z) + \varepsilon x^2$, where $z = x + iy$. Then $\Delta g(z) = \Delta v(z) + 2\varepsilon \geq 2\varepsilon > 0$ in Ω. Let u be a harmonic function in any fixed subdomain D in Ω. If the function $h(z) = g(z) - u(z)$ has a local maximum in D, then $\frac{\partial^2 h}{\partial x^2} \leq 0$, $\frac{\partial^2 h}{\partial y^2} \leq 0$, and $\Delta h \leq 0$ hold at some point in D, a contradiction. Hence $h(z)$ enjoys the maximum principle in D; and hence, by (3), it turns out that $g(z)$ is subharmonic in Ω.

Choose $\varepsilon = \frac{1}{n}$ for $n \geq 1$. The functions $v(z) + \frac{1}{n}x^2$, $n \geq 1$, are subharmonic in Ω and, as a sequence, converges to $v(z)$ locally uniformly in Ω. Therefore $v(z)$ is subharmonic in Ω (why? This follows easily by using (1) or refer to (6) in (6.4.3) below).   □

Where are sources of subharmonic function? Here are some of them:

One is: If $u(z)$ is harmonic, then the absolute function $|u|(z) = |u(z)|$ is automatically subharmonic (by (1) in (6.4.1)).

The other is: Analytic functions provide the most important source of all. Let $f(z)$ be analytic in a domain Ω. Then

1. $|f(z)|^p$, $p > 0$;
2. $|z|^q|f(z)|^p$, $p > 0$, $q > 0$ ($q < 0$ is permitted if $0 \notin \Omega$); and
3. $\log(1 + |f(z)|^2)$,   (6.4.2)

are subharmonic in Ω. The proofs for these claims are left as Exercise A(1). As a reminiscence, it is $\log|f(z)|$ ($f(z) \neq 0$ in Ω) and $\arg f(z)$ (if single-valued branch exists in Ω) that are harmonic in Ω.

To the end, we list some

Elementary properties of subharmonic functions. Operations of more than one subharmonic functions are supposed to be permitted on their common domain of definitions.

(1) v is subharmonic and $c > 0$ is a constant $\Rightarrow cv$ is subharmonic.
(2) v_1 and v_2 are subharmonic $\Rightarrow v_1 + v_2$ is subharmonic.
(3) v_1 and v_2 are subharmonic $\Rightarrow v = \max(v_1, v_2)$ (namely, $v(z) = \max\{v_1(z), v_2(z)\}$) is subharmonic.
(4) Let v be subharmonic in a domain Ω and D be an open disk so that the closure $\bar{D} \subseteq \Omega$. Let $P_v(z)$ be the Poisson's integral of $v(z)$ along ∂D. The function $v_D(z)$ defined by

$$v_D(z) = \begin{cases} P_v(z), & z \in \bar{D} \\ v(z), & \Omega - D \end{cases}$$

is subharmonic in Ω and hence, $v(z) \leq v_D(z)$ throughout Ω.
(5) Subharmonicity is a *conformal invariant*: If $v(z)$ is subharmonic in a domain Ω and $w = f(z)$ maps Ω univalently onto a domain Ω', then $v(f^{-1}(w))$ is subharmonic in Ω'.
(6) If a sequence $v_n(z)$ of subharmonic functions in a domain Ω converges to $v(z)$ locally uniformly in Ω or nonincreasingly in Ω (namely, $v_n(z) \geq$

$v_{n+1}(z)$, $n \geq 1$, on Ω and $v_n(z) \to v(z)$), then $v(z)$ is also subharmonic in Ω.   (6.4.3)

To prove (4), note first that, by (2) or (3) in (6.4.1), $v(z) \leq P_v(z)$ in Ω. The detail of it and the proofs of the whole (6.4.3) are left as Exercise A(2).

A continuous real function $v(z)$ in a domain Ω is said to be *super-harmonic* if the negative $-v(z)$ of it is subharmonic in Ω. And it is harmonic if and only if it is both subharmonic and superharmonic. All the properties developed so far for the subharmonic functions in (6.4.1) and (6.4.3) can be easily transferred to the corresponding ones for the superharmonic functions. We will feel no hesitation to cite them if needed.

Exercises A

(1) Prove in detail the subharmonicity of these functions stated in (6.4.2).
(2) Prove (6.4.3) in detail.
(3) State and prove all the properties mentioned in (6.4.1) and (6.4.3) for the superharmonic functions.
(4) Let $u(z)$ be a continuous real function in a domain Ω.

   (a) Show that $u(z)$ is harmonic in Ω if and only if

   $$u(z_0) = \frac{1}{\pi r^2} \iint_{|z-z_0| \leq r} u(z) dx dy$$

   for each point $z_0 \in \Omega$ and all sufficiently small $r > 0$ (such that $|z - z_0| \leq r$ is contained in Ω).
   (b) Try to state and prove a similar result as (a) for a subharmonic function. One might refer to Exercise B below for general setting in this direction.

Exercises B (Ref. [1], p. 247)

Let $v(z)$ be a real function defined on a plane set A. $v(z)$ is said to be *upper semicontinuous* (u.s.c.) at a point $z_0 \in A$ if $-\infty \leq v(z_0) < \infty$, and

$$\overline{\lim}_{z \in A \to z_0} v(z) \leq v(z_0) \text{ (namely, for each } \varepsilon > 0, \text{ there is a } \delta = \delta(z_0, \varepsilon) > 0$$

$$\text{so that } v(z) \leq v(z_0) + \varepsilon \text{ for all } z \in A \text{ satisfying } |z - z_0| < \delta).$$

$u(z)$ is upper semicontinuous on A if $-\infty \leq v(z) < \infty$ on A and $v(z)$ is u.s.c. at every point of A. $v(z)$ is said to be *lower semicontinuous* (l.s.c.)

on A if $-v(z)$ is u.s.c. on A.

(1) Let $v(z)$ be u.s.c. on a compact set K and $v(z) \not\equiv -\infty$ there.

   (a) Show that $v(z)$ attains its (finite) maximum on K.
   (b) (R. Baire) Suppose $v(z) \leq M$ on K. Then there exists a sequence of continuous function $\varphi_n(z)$ on $\mathbf{C}$ such that

   1. $M \geq \varphi_n(z) \geq \varphi_{n+1}(z)$ on $\mathbf{C}$ for $n \geq 1$, and
   2. $\lim_{n\to\infty} \varphi_n(z) = v(z)$ on K.

   Note. $\varphi_n(z)$ may be defined as $\sup_{\zeta \in K}[v(\zeta) - n|\zeta - z|]$ for $n \geq 1$ where z is a point in $\mathbf{C}$.

(2) Let $v(z)$ be u.s.c. in a domain Ω and $v(z) \not\equiv -\infty$ there. Under this assumption, show that (2) $\Leftrightarrow$ (3) $\Leftrightarrow$ (4) in (6.4.1) are still valid.

In what follows, a *subharmonic* function $v(z)$ in a domain Ω will always mean the one defined in Exercise (2) above. So does a superharmonic function. As a matter of fact, (1) *in* (6.4.1) *remains true* in the Lebesgue integral sense under this generally accepted definition, and it is Harnack's principle (6.3.2.15) that makes this possible. We *sketch* as follows: Choose a closed disk $|z - z_0| \leq \rho$, contained in Ω. According to Exercise (1), there is a nonincreasing sequence of continuous functions $\varphi_n(z)$ on $|z - z_0| \leq \rho$ that converges to $v(z)$ there. Construct Poisson's integral $P_n(z) = P_{\varphi_n}(z)$ of $\varphi_n(z)$ along $|z - z_0| = \rho$ for $n \geq 1$. Then

$$P_1(z) \geq P_2(z) \geq \cdots \geq P_n(z) \geq \cdots \geq v(z) \quad \text{in } |z - z_0| \leq \rho$$

$$\Rightarrow \text{(by Harnack's principle) } \lim_{n\to\infty} P_n(z) = P(z) \geq v(z) \text{ in } |z - z_0| \leq \rho.$$

Note that $v(z) \not\equiv -\infty$ is used here to guarantee that the limit function $P(z)$ is harmonic in $|z - z_0| < \rho$ without reducing to $-\infty$ identically. Observing that $P_1(z) - P_n(z)$ nondecreases in n and $P_1(z) - P_n(z) \to P_1(z) - v(z) \geq 0$ along $|z - z_0| = r(0 < r < \rho)$, by Lebesgue's dominated convergence theorem (check real analysis book!), it follows that, as $n \to \infty$,

$$\int_0^{2\pi} [P_1(z) - P_n(z)]d\theta \to \int_0^{2\pi} [P_1(z) - v(z)]d\theta,$$

$$\text{where } z = z_0 + re^{i\theta}, \quad 0 \leq \theta \leq 2\pi. \quad (*_1)$$

Meanwhile, if $v(z_0) > -\infty$,

$$\frac{1}{2\pi} \int_0^{2\pi} [P_1(z) - P_n(z)]d\theta = P_1(z_0) - P_n(z_0) \le P_1(z_0) - v(z_0), \quad n \ge 1$$

$$\Rightarrow \frac{1}{2\pi} \int_0^{2\pi} [P_1(z) - v(z)]d\theta \le P_1(z_0) - v(z_0) < \infty$$

$\Rightarrow P_1(z) - v(z)$ is integrable on $|z - z_0| = r$.

$\Rightarrow v(z)$ is integrable on $|z - z_0| = r$,

and, by $(*_1)$ and the fact that $v(z_0) \le P_n(z_0) = \frac{1}{2\pi} \int_0^{2\pi} P_n(z_0 + re^{i\theta})d\theta$ for $n \ge 1$,

$$-\infty < v(z_0) \le \frac{1}{2\pi} \int_0^{2\pi} v(z_0 + re^{i\theta})d\theta = \lim_{n \to \infty} \frac{1}{2\pi} \int_0^{2\pi} P_n(z_0 + re^{i\theta})d\theta.$$

$$(6.4.4)$$

Similarly, via (see Exercise A(4))

$$v(z_0) \le P_n(z_0) = \frac{1}{\pi r^2} \iint_{|z-z_0| \le r} P_n(z)dxdy, \quad n \ge 1, \quad z = x + iy$$

$\Rightarrow v(z)$ is integrable on $|z - z_0| \le r$, and, by (6.4.4),

$$-\infty < v(z_0) \le \frac{1}{\pi r^2} \iint_{|z-z_0| \le r} v(z)dxdy$$

$$= \lim_{n \to \infty} \frac{1}{\pi r^2} \iint_{|z-z_0| \le r} P_n(z)dxdy. \qquad (6.4.5)$$

This justifies the validity of (1) in (6.4.1) and even its equivalence with (2) in (6.4.1). Moreover, Exercise A(4) is still valid in this new definition. Unfortunately, considerations of space do not permit us to go into more details concerned, and it is urged to refer to Ref. [78], pp. 37–52, for information far beyond this point.

6.5   Perron's Method: Dirichlet's Problem for a Class of General Domains

The first to use subharmonic functions for solving Dirichlet's problem was due to O. Perron: *Eine neue Behandlung der ersten Randwertaufgabe für* $\Delta u = 0$. *Math. Zeits. 18 (1923)*, and, later, simplified by Radó, T. and Riesz, F.: *Über die erste Randwertaufgabe für* $\Delta u = 0$. *Math. Zeits. 22 (1925)*. This method is characterized by extreme generality even though it is completely elementary.

Let Ω be a (bounded) domain in $\mathbf{C}$. A general point in Ω will always be denoted by z while a boundary point of Ω will be denoted by ζ, for simplicity and distinction, in what follows.

This section has been divided in two subsections.

Section (1) Perron's method

Our main result is the following

Perron's existence theorem. *Let Ω be a bounded domain in $\mathbf{C}$ and $f(\zeta)$ be a given bounded real function on the boundary $\partial\Omega$. Let $\mathfrak{V}(f)$ be the family of functions $v(z)$ satisfying:*

1. *$v(z)$ is subharmonic in Ω;*
2. *$\overline{\lim}_{z\to\zeta}v(z) \leq f(\zeta)$ for each $\zeta \in \partial\Omega$.*

Define

$$u(z) = \sup_{v\in\mathfrak{V}(f)} v(z), \quad z \in \Omega.$$

Then $u(z)$ is harmonic in Ω, and is called a solution of the Dirichlet problem for Ω with the boundary value f (even though it is still unknown that $u(z)$ has the boundary values $f(z)$).   (6.5.1)

Note that the subharmonicity in 1 is that defined in (6.4.1), while 2 means that given any $\varepsilon > 0$ and a point $\zeta \in \partial\Omega$, there is a $\delta = \delta(\varepsilon, \zeta) > 0$ such that $v(z) < f(\zeta)+\varepsilon$ whenever $z \in N_\delta(\zeta)\cap\Omega$, where $N_\delta(\zeta)$ is the open disk $|z - \zeta| < \delta$ (or a certain neighborhood of ζ). Also, $\mathfrak{V}(f)$ is nonempty, for it contains all the constants $\leq -M$ if $|f(z)| \leq M$ along $\partial\Omega$. Be aware that, if the subharmonicity mentioned in Exercise B of Sec. 6.4 is adopted here, the final result remains valid since the proof is only needed to subject to minor modifications.

Proof. Let $|f(\zeta)| \leq M$ on $\partial\Omega$. Recall that $\mathfrak{V}(f) \neq \phi$ since it contains all the constants $\leq -M$.

Step 1. To show that $v(z) \leq M$ on Ω for each $v \in \mathfrak{V}(f)$.
Let $\varepsilon > 0$. Set $E = \{z \in \Omega | v(z) \geq M + \varepsilon\}$. Then E is a bounded subset of Ω. To see that E is a closed set, choose any sequence $z_n \in E$ converging to a point $z_0 \in \Omega$. Then $v(z_n) \geq M + \varepsilon$ for $n \geq 1$ and, by continuity of $v(z)$, it follows that $v(z_0) \geq M + \varepsilon$. Hence $z_0 \in E$ and E is closed.

Note. If the subharmonicity in Exercise B of Sec. 6.4 is adopted, the above proof can be modified as follows. To see $\mathbf{C} - E$ is open: Since $\operatorname{Ext}\Omega$ is open, we only need to consider points ζ in $\mathbf{C} - E$ which lie on $\partial\Omega$ and points

z in $\mathbf{C} - E$ which lie in Ω. For $\zeta \in \partial\Omega$, there is a neighborhood $N_\delta(\zeta)$ of ζ such that, by condition 2, $v(z) < f(z) + \varepsilon \leq M + \varepsilon$ for all z in $N_\delta(\zeta) \cap \Omega$; hence, $N_\delta(\zeta) \subseteq \mathbf{C} - E$. For $z \in \Omega$, then $\overline{\lim}_{\eta \to z} v(\eta) \leq v(z) < M + \varepsilon$ implies that there is some neighborhood $N_\delta(z)$ on which $v(\eta) < M + \varepsilon$ holds throughout; and then, $N_\delta(z) \subseteq \mathbf{C} - E$, too. This justifies that $\mathbf{C} - E$ is open and, thus, E is closed.

Now, E is a compact subset of Ω. If $E \neq \phi$, $v(z)$ has a maximum on E and this would be a local maximum in Ω. By (1) in (6.4.1), $v(z)$ is a constant $> M$ which is impossible owing to condition 2.

Therefore, $E = \phi$ for every $\varepsilon > 0$, and it follows that $v(z) \leq M$ on Ω.

Step 2. To minorize $u(z)$ locally by harmonic function.
Fix any point $z_0 \in \Omega$ and consider the closed disk $\bar{D}: |z - z_0| \leq \delta$, contained in Ω.

Since $u(z_0) = \sup_{v \in \mathfrak{V}(f)} v(z_0)$, there is a sequence of $v_n(z) \in \mathfrak{V}(f)$ such that $v_n(z_0) \to u(z_0)$ as $n \to \infty$. Set, for $n \geq 1$,

$$V_n(z) = \max(v_1(z), \ldots, v_n(z)), \quad z \in \Omega.$$

According to (3) in (6.4.3), $V_n(z)$ is subharmonic in Ω. For each $\zeta \in \partial\Omega$, since $\overline{\lim}_{z \to \zeta} v_k(z) \leq f(\zeta)$ for $1 \leq k \leq n$, it follows that $\overline{\lim}_{z \to \zeta} V_n(z) \leq f(\zeta)$ holds, too. Hence $V_n(z) \in \mathfrak{V}(f)$ for $n \geq 1$; moreover, $V_n(z) \leq V_{n+1}(z)$ for $n \geq 1$ on Ω (This is the main advantage of $V_n(z)$ over $v_n(z)$)

Construct, for each $n \geq 1$, a function $u_n(z)$ by

$$u_n(z) = \begin{cases} P_{V_n}(z) \text{ (Poisson's integral of } V_n(z) \text{ for } D), & \text{if } z \in \bar{D} \\ V_n(z), & \text{if } z \in \Omega - \bar{D} \end{cases}.$$

This $u_n(z)$ harmonizes $V_n(z)$ locally in D. These $u_n(z)$, $n \geq 1$, enjoy the following properties:

1. $u_n(z)$ are subharmonic in Ω and $V_n(z) \leq u_n(z)$ in Ω (see (4) in (6.4.3)).
2. $u_n \in \mathfrak{V}(f)$.
3. $u_n(z) \leq u_{n+1}(z)$, $n \geq 1$, on Ω (since $V_n(z) \leq V_{n+1}(z)$ on ∂D implies that $u_n(z) \leq u_{n+1}(z)$ on $\bar{D}$, by maximum principle).
4. In particular, $v_n(z_0) \leq V_n(z_0) \leq u_n(z_0) \leq u(z_0)$. (This last inequality is owing to 2 above and the definition of $u(z)$.)

For each fixed $z \in \Omega$, by Step 1, $u_n(z)$ forms a bounded nondecreasing sequence; hence, $\lim_{n \to \infty} u_n(z) = U(z)$ exists as a finite real number. Property 4 then implies that $u(z_0) \leq U(z_0) \leq u(z_0)$, namely,

$$U(z_0) = u(z_0). \tag{$*_1$}$$

Moreover, property 3 and Harnack's principle (6.3.2.15) together imply that $U(z)$ is harmonic in D and, since $u_n(z) \le u(z)$ for $n \ge 1$,

$$U(z) \le u(z) \quad \text{in } D. \tag{$*_2$}$$

Step 3. To show that $u(z) = U(z)$ for all $z \in D$; hence, $u(z)$ is harmonic in D and then in Ω.

Fix any $z_1 \in D$. Pick $\widehat{w_n}(z) \in \mathfrak{V}(f)$ such that $\widehat{w_n}(z_1) \to u(z_1)$ as $n \to \infty$. Set

$$w_n^+(z) = \max\{v_n(z), \widehat{w_n}(z)\}, \quad z \in \Omega;$$

$$W_n(z) = \max\{w_1^+(z), \dots, w_n^+(z)\}, \quad z \in \Omega;$$

$$w_n(z) = \begin{cases} P_{W_n}(z), & \text{if } z \in \bar{D} \\ W_n(z), & \text{if } z \in \Omega - \bar{D} \end{cases}.$$

As in Step 2, $w_n(z) \to W(z)$ in Ω as $n \to \infty$ and $W(z_1) = u(z_1)$ (see $(*_1)$) and $U(z) \le W(z) \le u(z)$ in D (see $(*_2)$, observing that $v_n \le w_n^+ \Rightarrow V_n \le W_n \Rightarrow u_n \le w_n$ in D); in particular, $U(z_0) = W(z_0) = u(z_0)$ holds.

Now $U(z)$ and $W(z)$ are harmonic in D and $U(z) - W(z) \le 0$ in D. Since $U(z_0) - W(z_0) = 0$, it follows that $U(z) = W(z)$ throughout D; in particular, $u(z_1) = U(z_1)$ holds.

This shows that $u(z) = U(z)$ for all $z \in D$. $\qquad \square$

Instead of the family $\mathfrak{V}(f)$ in (6.5.1), we may as well consider the family $\mathfrak{U}(f)$ of functions $v(z)$ satisfying:

1. $v(z)$ is superharmonic in D;
2. $\underline{\lim}_{z \to \zeta} v(z) \ge f(\zeta)$ for each $\zeta \in \partial D$.

Then, by completely similar argument, the function defined by

$$u(z) = \inf_{v \in \mathfrak{U}(f)} v(z), \quad z \in \Omega \tag{6.5.2}$$

is *harmonic in Ω and is called a solution of the Dirichlet problem for Ω with the boundary value f, too.*

Are the two harmonic functions, defined in (6.5.1) and (6.5.2) coincident on Ω? Here comes

The uniqueness theorem. *Let Ω be a bounded domain in $\mathbf{C}$ and $f(z)$ be a bounded continuous real function on the boundary $\partial \Omega$. If $U(z)$ is a solution of the Dirichlet problem for Ω with the boundary value $f(z)$ (namely,*

$\lim_{z \to \zeta} U(z) = f(\zeta)$ *for each* $\zeta \in \partial\Omega$), *then*

$$U(z) = u(z), \quad z \in \Omega$$

where $u(z)$ is the one in (6.5.1), constructed by Perron's method. (6.5.3)

Proof. Since $U(z)$ is harmonic in Ω and $\lim_{z \to \zeta} U(z) = f(\zeta)$ for each $\zeta \in \partial\Omega$, then $U \in \mathfrak{V}(f)$ holds; hence, $U(z) \le u(z)$ for all $z \in \Omega$.

On the other hand, pick any $v \in \mathfrak{V}(f)$. For $\varepsilon > 0$ and $\zeta \in \partial\Omega$, there is a neighborhood $N(\zeta)$ of ζ so that $v(z) < f(\zeta) + \frac{\varepsilon}{2}$ and $f(\zeta) < U(z) + \frac{\varepsilon}{2}$ whenever $z \in N(\zeta) \cap \Omega$; consequently, $v(z) < U(z) + \varepsilon$ holds if $z \in N(\zeta) \cap \Omega$.

Fix $\varepsilon > 0$ and $v \in \mathfrak{V}(f)$. Let $E = \{z \in \Omega | v(z) \ge U(z) + \varepsilon\}$. As in Step 1 in the proof of (6.5.1), E is a compact subset of Ω. In case $E \ne \phi$, then $v(z) - U(z)$ assumes a maximum on E and hence, a local maximum in Ω. By the maximum principle (see (1) in (6.4.1)), $v(z) - U(z)$ is a constant $\ge \varepsilon$ in Ω. Hence $v(z) > U(z)$ throughout Ω, a contradiction. Therefore $E = \phi$ should hold.

As a consequence, $v(z) < U(z) + \varepsilon$ for all $z \in \Omega$ and $\varepsilon > 0$; this means that $v(z) \le U(z)$ for all $z \in \Omega$ and for each $v \in \mathfrak{V}(f)$. And of course, it follows that $u(z) \le U(z)$ for all $z \in \Omega$.

Combining together, $U(z) = u(z)$ throughout Ω.   □

Section (2) When does the solution of Dirichlet's problem exist?

Dirichlet's problem may not have a solution.

Example 1. (Ref. [1], p. 251). If Ω is the punctured disk $0 < |z| < 1$ and if $f(\zeta)$ is given by $f(\zeta) = 0$ for $|\zeta| = 1$, $f(0) = 1$, then Dirichlet's problem is not solvable.

Solution. If we had a solution $u(z)$, then, by the maximum principle, $u(z)$ is bounded on Ω; and hence, $z = 0$ is a removable singularity of $u(z)$ (see (6.3.1.7)). Again by the maximum–minimum principle, $u(z) = 0$ on $0 \le |z| \le 1$, contradicting to $u(0) = 1$.

However, this example shows that an isolated boundary point of a domain seems to be a violator to the problem. Really it is. As a preparatory in this direction, we have the following concept of

A barrier at a boundary point of a bounded domain. Let ζ_0 be a boundary point of a bounded domain Ω. Suppose there is a continuous function $w(z)$

on $\bar{\Omega}$ satisfying:

1. $w(z)$ is harmonic in Ω;
2. $\lim_{z \to \zeta_0} w(z) = w(\zeta_0) = 0$; and
3. $\lim_{z \to \zeta} w(z) = w(\zeta) > 0$ for all $\zeta \in \partial\Omega$ such that $\zeta \neq \zeta_0$.

Such a function $w(z)$ is called a *barrier* of the domain Ω at the point ζ_0. Under this circumstance and the assumption that $f(\zeta)$ is continuous at ζ_0, the corresponding harmonic function $u(z)$ defined by Perron's method (6.5.1) satisfies

$$\lim_{z \to \zeta_0} u(z) = f(\zeta_0). \tag{6.5.4}$$

A relaxed definition for a barrier will be given in (6.5.5); also, refer to Exercises A(1) and A(2) for related results.

Proof. This is equivalent to show that $f(\zeta_0) - \varepsilon \leq \underline{\lim}_{z \to \zeta_0} u(z) \leq \overline{\lim}_{z \to \zeta_0} u(z) \leq f(\zeta_0) + \varepsilon$ for all $\varepsilon > 0$.

Fix $\varepsilon > 0$. Choose a neighborhood $N(\zeta_0)$ of ζ_0 so that $|f(\zeta) - f(\zeta_0)| < \varepsilon$ if $\zeta \in N(\zeta_0) \cap \partial\Omega$. Set $m = \min w(z)$ for $z \in \bar{\Omega} - N(\zeta_0) \cap \Omega$. Then $m > 0$ by condition 3 and the maximum–minimum principle; moreover, $w(z) > 0$ in Ω. Let $|f(\zeta)| \leq M$ on $\partial\Omega$.

For the second inequality, consider the harmonic function

$$\varphi(z) = f(\zeta_0) + \varepsilon + \frac{w(z)}{m}(M - f(\zeta_0)), \quad z \in \Omega.$$

Observing that

$$\varphi(\zeta) \geq \begin{cases} f(\zeta_0) + \varepsilon > f(\zeta), & \text{if } \zeta \in N(\zeta_0) \cap \partial\Omega \\ f(\zeta_0) + \varepsilon + M - f(\zeta_0) = M + \varepsilon, & \text{if } \zeta \in \partial\Omega - N(\zeta_0) \cap \partial\Omega \end{cases}$$

$\Rightarrow \varphi(\zeta) > f(\zeta), \quad \zeta \in \partial\Omega.$

$\Rightarrow$ (by $\overline{\lim}_{z \to \zeta} v(z) \leq f(\zeta)$ for $v \in \mathfrak{V}(f))v(z) \leq f(\zeta) < \varphi(\zeta)$ in $N(\zeta) \cap \Omega$,

  where $N(\zeta)$ is a neighborhood of a point $\zeta \in \partial\Omega$, and $v \in \mathfrak{V}(f)$

  is a fixed function.

$\Rightarrow$ (by continuity of φ on $\bar{\Omega}$) $v(z) < \varphi(z)$ in $N(\zeta) \cap \Omega$ for each $\zeta \in \partial\Omega$. $(*_3)$

$\Rightarrow$ (by the maximum principle, see the proof of (6.5.3)) $v(z) < \varphi(z)$,

  $z \in \Omega$, for each $v \in \mathfrak{V}(f)$. $(*_4)$

$\Rightarrow u(z) \le \varphi(z), \quad z \in \Omega.$

$\Rightarrow \overline{\lim}_{z \to \zeta_0} u(z) \le \varphi(\zeta_0) = f(\zeta_0) + \varepsilon.$

For the first inequality, consider the harmonic function

$$\psi(z) = f(\zeta_0) - \varepsilon - \frac{w(z)}{m}(M + f(\zeta_0)), \quad z \in \Omega.$$

Observing that

$$\psi(\zeta) \le \begin{cases} f(\zeta_0) - \varepsilon < f(\zeta), & \text{if } \zeta \in N(\zeta_0) \cap \partial\Omega \\ f(\zeta_0) - \varepsilon - (M + f(\zeta_0)) \\ \quad = -M - \varepsilon < -M \le f(\zeta), & \text{if } \zeta \in \partial\Omega - N(\zeta_0) \cap \partial\Omega \end{cases}$$

$\Rightarrow \psi(\zeta) < f(\zeta), \quad \zeta \in \partial\Omega.$

$\Rightarrow$ (by reasoning as in the last paragraph) $\psi(z) \le v(z), z \in \Omega,$

for each $v \in \mathfrak{V}(f).$

$\Rightarrow \psi(z) \le u(z), \quad z \in \Omega.$

$\Rightarrow \underline{\lim}_{z \to \zeta_0} u(z) \ge \underline{\lim}_{z \to \zeta_0} \psi(z) = \psi(\zeta_0) = f(\zeta_0) - \varepsilon.$

Combining together, this ends the proof.   □

(6.5.4) indicates that the solvability of Dirichlet's problem for a domain Ω reduces to the existence of a barrier at each point of its boundary. And it eventually depends on the geometric or topological character of the boundary.

Two useful situations are studied in the following two examples.

Example 2. Let ζ_0 be a boundary point of a domain Ω. In case ζ_0 is the end point of a line segment all of whose points, except ζ_0, lie in exterior of Ω, then ζ_0 is a barrier of Ω.

Solution. Let ζ_1 be the other end point of the line segment. Then $\overline{\zeta_0\zeta_1}$ can be expressed as $z = (1 - t)\zeta_0 + t\zeta_1, 0 \le t \le 1$. Observe that

$$\frac{z - \zeta_0}{z - \zeta_1} = \frac{-t}{1 - t} \le 0 \quad \text{if } z \in \overline{\zeta_0\zeta_1}.$$

$$\Rightarrow \frac{z - \zeta_0}{z - \zeta_1} \in \mathbf{C} - (\infty, 0] \quad \text{for all } z \in \Omega \text{ since } \overline{\zeta_0\zeta_1} \subseteq \Omega^{\sim}.$$

According to (3.3.1.5), $\sqrt{\frac{z-\zeta_0}{z-\zeta_1}}$ has two single-valued branches on Ω. Fix any one of them and denote it by $f(z)$. Then the image $f(\Omega)$ lies in an open half-plane whose boundary is a line passing the origin 0 and having an inclination α with the positive real axis. The function $e^{-i\alpha}f(z)$ has its range domain in the upper half-plane; and thus $w(z) = \text{Im}(e^{-i\alpha}f(z))$ is a barrier at ζ_0.

A careful examination of the proof for (6.5.4) enables us to realize that the key point still lies on the commonly cited maximum principle, as indicated from $(*_3)$ to $(*_4)$, owned by the local subharmonicity of a function rather than the full power of its global harmonicity. Therefore, the definition in (6.5.4) for a barrier can be less restrictive and be redefined in the following: Let ζ_0 be a boundary point of a bounded domain Ω. Suppose there exist a $\rho > 0$ and a continuous function $w(z)$ in $\Omega_\rho(\zeta_0) = \{|z - \zeta_0| < \rho\} \cap \Omega$, satisfying:

1. $w(z) > 0$ and is subharmonic in $\Omega_\rho(\zeta_0)$, and
2. $\lim_{z \to \zeta_0} w(z) = 0$. $\hspace{3cm}$ (6.5.5)

Then $w(z)$ is sometimes called a *barrier* of Ω at ζ_0. Under this generalized version, we have the following extension of (6.5.4).

Example 3. Let ζ_0 be a boundary point of a bounded domain Ω so that ζ_0 lies in a *continuum* component of the complement $\Omega^{\sim} = \mathbf{C} - \Omega$ (namely, a component that is a compact connected set containing more than one point). Then the corresponding harmonic function $u(z)$ by Perron's method (6.5.1) satisfies

$$\lim_{z \to \zeta_0} u(z) = f(\zeta_0)$$

provided that the boundary function $f(\zeta)$ is continuous at ζ_0. $\hspace{1cm}$ (6.5.6)

Proof. By assumption, we are able to choose $0 < \rho < 1$ so small that $z - \zeta_0 \neq 0$ in the simply-connected domain $\Omega_\rho(\zeta_0)$. Hence (see (3.3.1.8)), $\log(z - \zeta_0)$ has single-valued analytic branches in $\Omega_\rho(\zeta_0)$. Pick up such a branch, still denoted by $\log(z - \zeta_0)$. Set $\log(z - \zeta_0) = p(z) + iq(z)$. Then $p(z) = \log|z - \zeta_0| < 0$ on $\Omega_\rho(\zeta_0)$. Let

$$w(z) = \text{Re}\left\{\frac{-1}{\log(z - \zeta_0)}\right\} = \frac{-p(z)}{p(z)^2 + q(z)^2}, \quad z \in \Omega_\rho(\zeta_0).$$

Then $w(z) > 0$ and is harmonic in $\Omega_\rho(\zeta_0)$, and satisfies $\lim_{z \to \zeta_0} w(z) = 0$. And $w(z)$ is then a barrier at ζ_0, according to (6.5.5).

Fix any $0 < r \le \rho$. Then $w(z) \ge \dfrac{-\log r}{(\log r)^2 + 4\pi^2} > 0$ along $|z - \zeta_0| = r$. Choose a constant $c > 0$ so that $\dfrac{-c \log r}{(\log r)^2 + 4\pi^2} \ge 1$. Define

$$
s(z) = \begin{cases} \min(cw(z), 1), & z \in \Omega_r(\zeta_0) = \{|z - \zeta_0| < r\} \cap \Omega \\ 1, & z \in \Omega - \Omega_r(\zeta_0) \end{cases} \qquad (*_5)
$$

This $s(z)$ is continuous, superharmonic and $0 < s(z) \le 1$ in Ω; $s(z) = 1$ on $|z - \zeta_0| = r$ and $\lim_{z \to \zeta_0} s(z) = 0$ holds. *Note* that $s(z)$ is going to play the role of $w(z)$ in (6.5.4).

Since $f(\zeta)$ is continuous at ζ_0, for $\varepsilon > 0$, choose $0 < r \le \rho$ so that $f(\zeta_0) - \varepsilon \le f(\zeta) \le f(\zeta_0) + \varepsilon$ if $\zeta \in \partial\Omega \cap \{|z - \zeta_0| \le r\}$.

Consider the continuous subharmonic function

$$
v(z) = f(\zeta_0) - \varepsilon - Ms(z), \quad z \in \Omega, \qquad (*_6)
$$

where $M > 0$ is a constant. Since $s(z) > 0$,

$$
\overline{\lim}_{z \to \zeta} v(z) \le f(\zeta_0) - \varepsilon < f(\zeta), \quad \zeta \in \partial\Omega \cap \{|z - \zeta_0| \le r\}.
$$

Since $s(z) = 1$ in $\Omega - \Omega_r(\zeta_0)$, if $M > 0$ is chosen large enough, then

$$
\overline{\lim}_{z \to \zeta} v(z) \le f(\zeta_0) - \varepsilon < f(\zeta), \quad \zeta \in \partial\Omega - \partial\Omega \cap \{|z - \zeta_0| \le r\}.
$$

It follows that $v \in \mathfrak{V}(f)$ (see (6.5.1)), and hence $u(z) \ge v(z)$ holds throughout Ω. On the other hand, $\lim_{z \to \zeta_0} s(z) = 0$ implies that $\lim_{z \to \zeta_0} v(z) = f(\zeta_0) - \varepsilon$; hence, there exist $r_1 > 0$ $(r_1 \le r)$ so that

$$
v(z) > f(\zeta_0) - 2\varepsilon, \quad z \in \Omega_{r_1}(\zeta_0)
$$

$$
\Rightarrow u(z) > f(\zeta_0) - 2\varepsilon, \quad z \in \Omega_{r_1}(\zeta_0). \qquad (*_7)
$$

Correspondingly, consider the continuous superharmonic function

$$
\tilde{v}(z) = f(\zeta_0) + \varepsilon + Ms(z), \quad z \in \Omega, \qquad (*_8)
$$

where $M > 0$ is a constant, sufficiently large. Since $\lim_{z \to \zeta_0} \tilde{v}(z) = f(\zeta_0) + \varepsilon$, there exists $r_2 > 0$ $(r_2 \le r)$ so that

$$
\tilde{v}(z) < f(\zeta_0) + 2\varepsilon, \quad z \in \Omega_{r_2}(\zeta_0). \qquad (*_9)
$$

Since $\tilde{v}(z) \ge v(z)$ for *any* $v \in \mathfrak{V}(f)$, not just the one defined in $(*_6)$ (see the *note* below), it follows that $\tilde{v}(z) \ge u(z)$ throughout Ω. Hence

$$
u(z) < f(\zeta_0) + 2\varepsilon, \quad z \in \Omega_{r_2}(\zeta_0). \qquad (*_{10})
$$

Choosing $r_0 = \min(r_1, r_2)$, then $f(\zeta_0) - 2\varepsilon < u(z) < f(\zeta_0) + 2\varepsilon$ if $z \in \Omega_{r_0}(\zeta_0)$ and hence $\lim_{z \to \zeta_0} u(z) = f(\zeta_0)$ holds.

Note. Let $\tilde{v}(z)$ be defined in $(*_8)$. Then

$$\underline{\lim}_{z \to \zeta} \tilde{v}(z) \geq \begin{cases} f(\zeta_0) + \varepsilon > f(\zeta), & \text{if } \zeta \in \partial\Omega \cap \{|z - \zeta_0| \leq r\} \\ f(\zeta_0) + \varepsilon > f(\zeta), & \text{if } \zeta \in \partial\Omega - \partial\Omega \cap \{|z - \zeta_0| \leq r\} \end{cases},$$

where in the lower inequality, $M > 0$ is chosen large enough. Choose *any* fixed function $v \in \mathfrak{V}(f)$. $\tilde{v}(z) - v(z)$ is superharmonic in Ω and, for each $\zeta \in \partial\Omega$, $\underline{\lim}_{z \to \zeta}(\tilde{v}(z) - v(z)) \geq \underline{\lim}_{z \to \zeta} \tilde{v}(z) - \overline{\lim}_{z \to \zeta} v(z) \geq f(\zeta) - f(\zeta) \geq 0$ holds. This implies that $\tilde{v}(z) - v(z) \geq 0$ in Ω (applying (6.4.1) to the subharmonic function $-(\tilde{v}(z) - v(z)))$; therefore $\tilde{v}(z) \geq v(z)$ in Ω for any $v \in \mathfrak{V}(f)$. $\qquad\qquad\square$

In conclusion, we have

The solvability of Dirichlet's problem. Let $f(\zeta)$ be a continuous real function along the boundary of a bounded domain Ω. For each of the following domains Ω:

1. each boundary point of $\partial\Omega$ is the end point of a segment whose other points are exterior to Ω, or
2. the complement $\mathbf{C} - \Omega$ is such that no component reduces to a point, or
3. Ω is bounded by a finite number of Jordan curves,

the Dirichlet's problem is solvable, namely, there *exists* a *unique* harmonic function $u(z)$ in Ω whose boundary value function is coincident with $f(\zeta)$ along $\partial\Omega$.   (6.5.7)

An interesting application is that the Jordan domain in (6.3.1.12) can now be replaced by any (bounded) simply connected domain and the result remains valid (refer to "About Step 1" in Remark 1 of Sec. 6.1).

Exercises A

(1) In addition to conditions 1 and 2 in (6.5.5), suppose $w(z)$ also satisfies that, on the part of $|z - \zeta_0| = r \leq \rho$, contained in Ω, $w(z) \geq c > 0$, where c is a constant depending only on r. Show that the result in (6.5.4) is still valid.

(2) Show that the result in (6.5.4) remains valid even if (6.5.5) is adopted as the definition of a barrier. This indicates that the additional assumption in Exercise (1) is indeed unnecessary.

6.6 Canonical Mappings and Canonical Domains of Finitely Connected Domains

As claimed in Remark 3 of Sec. 2.4 and justified in Sec. 6.1, each simply connected domain in $\mathbf{C}^*$ is *conformally equivalent* to one and only one of the following three *canonical domains*:

1. $\mathbf{C}^*$;
2. $\mathbf{C}$; and
3. the open unit disk $|z| < 1$.   (6.6.1)

The most prominent one, among all, is that any two simply connected domains, other than $\mathbf{C}^*$ and $\mathbf{C}$, are necessarily conformally equivalent. This is not more true for domains of connectivity $n \geq 2$. Take, for instance, the ring domains $0 < r_1 < |z| < r_2 < \infty$ and $0 < R_1 < |w| < R_2 < \infty$ mentioned in Example 4 of Sec. 2.5.4 and Example 6 of Sec. 3.4.6. They are conformally equivalent if and only if $\frac{r_2}{r_1} = \frac{R_2}{R_1}$.

The above example suggests that we need to find a system of *canonical domains* so that each finitely connected domain is *conformally equivalent* to one and only one of them. The choice of canonical domains is to a certain extent arbitrary, and in what follows we list several types of them with equally simple properties in terms of *extremal problems*.

Recall that the Riemann mapping theorem is proved by solving the extremal problem: $\max_{g \in \mathfrak{F}} g'(z_0)$, where $\mathfrak{F}$ is the set of all univalent functions $g(z)$ mapping the simply connected domain Ω into $|w| \leq 1$ that satisfy $g(z_0) = 0$ and $g'(z_0) > 0$ (see $(*_1)$ and $(*_4)$ in Sec. 6.1.1).

Ω will always denote a domain of finite connectivity $n \geq 2$ in what follows, unless otherwise stated.

Type 1. *The parallel slit domain*: A domain in $\mathbf{C}^*$ with a finite number of line segment slits making an angle θ with the positive real axis. See Fig. 6.43(a).

Fix an arbitrary point $z_0 \in \Omega$. Consider the extremal problem:

$\mathfrak{F}$: the family of all univalent analytic functions g on Ω with a simple pole of residue 1 at z_0, namely,

$$g(z) = \frac{1}{z - z_0} + \widehat{g}(z) \text{ where } \widehat{g}(z) \text{ is analytic at } z_0.$$

In case $z_0 = \infty$, $g(z) = z + \widehat{g}(z)$ where $\hat{g}(z)$ is analytic at ∞.

The extremal function $f : \max_{g \in \mathfrak{F}} \mathrm{Re}\{e^{-2i\theta} \hat{g}'(z_0)\}$.   (6.6.2)

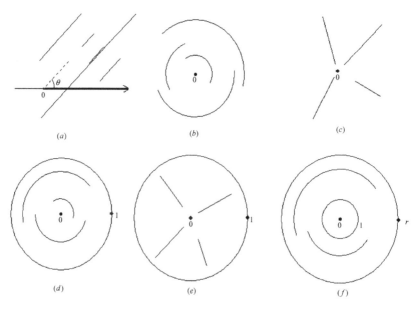

(a)   (b)   (c)

(d)   (e)   (f)

Fig. 6.43

Then $w = f(z)$ is the canonical mapping which maps Ω univalently onto the parallel slit domain. In case $\theta = 0$, we obtain the *horizontal slit domain* by maximizing $\mathrm{Re}\{\hat{g}'(z_0)\}$ for $g \in \mathfrak{F}$; in case $\theta = \frac{\pi}{2}$, the *vertical slit domain* by minimizing $\mathrm{Re}\{\hat{g}'(z_0)\}$, $g \in \mathfrak{F}$.

Type 2. *The concentric circular slit domain*: A domain in $\mathbf{C}^*$ with a finite number of concentric circular slits whose common center is the origin. See Fig. 6.43(b).

Fix two arbitrary but distinct points z_0 and z_1 in Ω. Consider the extremal problem:

$\mathfrak{F}$: The family of univalent analytic functions g on Ω satisfying the conditions:

$g(z_0) = \infty$ and $\mathrm{Re}\,s(g; z_0) = 1$ (namely, a simple pole of residue 1 at z_0);

$g(z_1) = 0$.

The extremal function $f : \max_{g \in \mathfrak{F}} |g'(z_1)|$.   (6.6.3)

Then $w = f(z)$ is the canonical mapping.

Type 3. *The radial slit domain*: A domain in $\mathbf{C}^*$ with a finite number of radial slits pointing at the origin. See Fig. 6.43(c).

The extremal problem is:

$\mathfrak{F}$: the same as in (6.6.3).

The extremal function $f : \min_{g \in \mathfrak{F}} |g'(z_1)|$.   (6.6.4)

Then $w = f(z)$ is the canonical mapping.

Type 4. *The disk with concentric circular slits*: A disk with a finite number of circular slits concentric with the outer circle. See Fig. 6.43(d).

Fix an arbitrary point z_0 in Ω. Consider the extremal problem:

$\mathfrak{F}$: The family of all univalent analytic functions $g(z)$ on Ω satisfying:

$$|g(z)| \leq 1 \quad \text{on } \Omega;$$

$$g(z_0) = 0.$$

The extremal function $f : \max_{g \in \mathfrak{F}} |g'(z_0)|$.   (6.6.5)

Then $w = f(z)$ is the canonical mapping.

Type 5. *The disk with radial slits*: A disk with a finite number of radial slits pointing at the origin. See Fig. 6.43(e).

The extremal problem is:

$\mathfrak{F}$: the same as in (6.6.5).

The extremal function $f : \min_{g \in \mathfrak{F}} |g'(z_0)|$.   (6.6.6)

Then $w = f(z)$ is the canonical mapping.

Type 6. *The annulus with concentric circular slits*: In case $n = 2$, the canonical domain is an annulus, namely, a circular ring $1 < |z| < r$; in case $n \geq 3$, there are some additional circular slits concentric to the origin. See Fig. 6.43(f).

Let $C_1, \ldots, C_n$ be the n components of $\mathbf{C}^* - \Omega$. Let $f_1(z)$ be the map in Type 4 that maps C_1 onto $|w| = 1$ and z_0 into 0; $f_n(z)$ the one in Type 4 that maps C_n onto $|w| = r > 1$ and z_0 into 0. Then

$$f(z) = \frac{f_n(z)}{f_1(z)}   (6.6.7)$$

is the canonical mapping, unique up to an arbitrarily multiplicative constant.

There are three methods to handle finitely connected canonical domains and the associated canonical mappings:

The extremal problem method: See the proof of the Riemann mapping theorem, Exercise B(6) of Sec. 6.6.2 and Exercise B(5) of Sec. 6.6.3.

The Dirichlet problem method: See Secs. 6.6.2, 6.6.3, 6.6.4 and the exercises that follow.

The continuity method: See Ref. [35], p. 234.

References [63], Chap. VII, and [35], Chaps. 5 and 6, to which the readers are referred, have excellent treatment about topics (6.6.2)–(6.6.7) and, even much more.

We will concentrate on the canonical domains of Types 1 and 6, and follow the Dirichlet problem method to accomplish them. The method mainly relies on the existence of certain harmonic functions with particularly simple behavior on the boundary of Ω: the harmonic measure (see Sec. 6.6.1) and Green's function (see Sec. 6.3.1).

Before starting to find the canonical functions and domains, we try to *simplify* the problem and show that Ω, a *domain of connectivity $n > 1$, can be mapped univalently onto a domain bounded by n analytic Jordan closed curves*. Then, the canonical mapping of the latter will induce the one for the original domain Ω.

In case both 0 and ∞ are in Ω, a translation of the plane will kick 0 out of Ω; and the map $z \to \frac{1}{z}$ will transform Ω onto a bounded domain of the same connectivity n.

Henceforth, we may assume that $\infty \notin \Omega$. Let $E_1, \ldots, E_n$ be the n components of the complement $\mathbf{C}^* - \Omega$ so that $\infty \in E_n$, the unbounded component. Moreover, we may also assume that no E_k reduces to a point. Otherwise, a point component would be a removable singularity of any mapping function, according to (4.10.1.1) or (4.10.1.2), and consequently the mapping remains the same if the isolated boundary points are added to the domain Ω.

Under these circumstance, the domain $\Omega_n = \mathbf{C}^* - E_n$, the complement of E_n in $\mathbf{C}^*$, is simply connected. By the Riemann mapping theorem, Ω_n can be mapped univalently onto the unit disk $|z| < 1$ so that

1. E_n is onto the set $1 \le |z| \le \infty$;
2. Ω is onto a domain, still denoted by Ω, in $|z| < 1$; and
3. $E_1, \ldots, E_{n-1}$ are onto $(n-1)$ bounded components of $\mathbf{C}^* - \Omega$ (Ω, the new one), still denoted by $E_1, \ldots, E_{n-1}$, respectively.

The whole new picture about Ω is that it is a bounded domain in $|z| < 1$, whose complement has $n - 1$ bounded components $E_1, \ldots, E_{n-1}$ and the unbounded one E_n is the set $1 \le |z| \le \infty$. The remarkable point is that the boundary of E_n is the unit circle $|z| = 1$, a closed analytic curve. $|z| = 1$, endowed with the counterclockwise direction, is particularly denoted as γ_n and is called the *outer contour* of the (new) domain Ω.

Again, map the simply connected domain $\Omega_1 = \mathbf{C}^* - E_1$ univalently onto $1 < |z| \leq \infty$ so that ∞ is mapped into itself. Keeping the same rotations as above, γ_n turns out to be a positively oriented analytic Jordan closed curve in $|z| > 1$; (the new) E_1 is the set $|z| \leq 1$ with its boundary γ_1, the unit circle $|z| = 1$, endowed with the clockwise direction and called an *inner contour* of Ω. While, the remaining $(n-2)$ bounded components of the complement $\mathbf{C}^* - \Omega$ lie between γ_1 and γ_n. The point is that, now, one more boundary curve γ_1 turns out to be an analytic Jordan closed curve.

Repeat this process until we end up with a domain Ω bounded by

1. an outer contour γ_n, in the positive direction, and
2. $(n-1)$ inner contours $\gamma_1, \ldots, \gamma_{n-1}$, all in the negative directions.

$$(6.6.8)$$

$\gamma_1, \ldots, \gamma_{n-1}$, and γ_n all are *analytic Jordan close curves*. See Fig. 6.44. In what follows, the domain Ω mentioned will always be the one characterized in (6.6.8). It should be mentioned that an inversion with respect to an interior point of E_k will interchange the inner contour γ_k and the outer contour γ_n; consequently, the distinction between the inner and outer contours is coincidental.

According to the basic properties (3.5.2.4) for winding numbers, it is easy to see that, for $1 \leq k \leq n-1$,

$$n(\gamma_k; z) = \begin{cases} -1, & z \in \operatorname{Int} E_k \\ 0, & z \notin \operatorname{Int} E_k \quad \text{and} \quad z \notin \gamma_k \end{cases} ; \qquad (6.6.9)$$

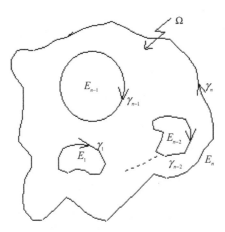

Fig. 6.44

$$n(\gamma_n; z) = \begin{cases} 1, & z \notin \text{Int } E_n \quad \text{and} \quad z \notin \gamma_n \\ 0, & z \in \text{Int } E_n \end{cases}$$

$$\Rightarrow n\left(\sum_{k=1}^{n} \gamma_k; z\right) = \begin{cases} 1, & z \in \Omega \\ 0, & z \notin \Omega \quad \text{and} \quad z \notin \gamma_k \quad \text{for } 1 \le k \le n \end{cases}.$$

$$(6.6.10)$$

It is in this sense that the cycle $\sum_{k=1}^{n} \gamma_k$ is said to *bound* the domain Ω (refer to the paragraph after (4.11.3.2)).

If the orientation of γ_k is changed to the positive one for each k, $1 \le k \le n - 1$, then (6.6.9) indicates that the contours $\gamma_1, \ldots, \gamma_{n-1}$ from a *homology basis* for cycles in Ω (see (4.3.2.5)). Here homology is understood with respect to a larger domain containing $\bar{\Omega}$. That this is possible will be clear in Sec. 6.6.1.

Section 6.6.1 is devoted to the study of the harmonic measures and their periods.

Based on Sec. 6.6.1, Sec. 6.6.2 will realize how the annulus with concentric circular slits can be treated as the canonical domains.

Section 6.6.3 will study the parallel slit domains, using the harmonic measure and Green's function introduced in Sec. 6.3.1.

Section 6.6.4 introduces some other canonical domains. This section is optional.

6.6.1   *Harmonic measures*

Let Ω be a domain of connectivity $n \ge 2$, characterized as in (6.6.8).

For each fixed k, $1 \le k \le n$, according to (6.5.7), there is a unique harmonic function $\omega_k(z)$ in Ω satisfying the condition

$$\omega_k(\gamma_j) = \delta_{jk}, \quad 1 \le j \le n, \tag{6.6.1.1}$$

called the *harmonic measure* of γ_k with respect to the domain Ω. Note that

$$0 < \omega_k(z) < 1 \text{ in } \Omega, \quad \text{and}$$

$$\omega_1(z) + \cdots + \omega_n(z) \equiv 1 \text{ throughout } \Omega. \tag{6.6.1.2}$$

These two relations follow easily by the max–min principle. Since each γ_k, $1 \le k \le n$, is an analytic Jordan curve, according to the symmetry principle (3.4.6.4) or (6.3.2.13), $\omega_k(z)$ can be harmonically continued across each γ_j, $1 \le j \le n$, so that it is still harmonic in a larger domain than the original Ω. This will be tacitly understood in the subsequent discussion. By the way, it is in this sense that the contours $\gamma_1, \ldots, \gamma_{n-1}$, all oriented in the counterclockwise direction form a *homology basis* for cycles in Ω; namely, for each cycle γ in Ω, there are constants $a_1, \ldots, a_{n-1}$ so that γ is homologous to $\sum_{k=1}^{n-1} a_k \gamma_k$ in Ω, in notation $\gamma \sim \sum_{k=1}^{n-1} a_k \gamma_k$. In particular, $\sum_{k=1}^{n-1} a_k \gamma_k \sim 0$ if and only if $a_k = 0$ for $1 \le k \le n-1$.

Consider the conjugate harmonic differential $*d\omega_k$ (see (6.3.1.1)) and its period along γ_j (see (6.3.1.4))

$$\alpha_{kj} = \int_{\gamma_j} *d\omega_k = \int_{\gamma_j} \frac{\partial \omega_k}{\partial n} ds, \quad 1 \le k, j \le n-1. \tag{6.6.1.3}$$

We claim that the period matrix of order $n-1$

$$A = [\alpha_{kj}]_{(n-1) \times (n-1)} \tag{6.6.1.4}$$

is nonsingular (*i.e., invertible*). This is equivalent to say that the homogeneous system of linear equations

$$\bar{\lambda} A = \bar{0}, \quad \text{where } \bar{\lambda} = (\lambda_1, \ldots, \lambda_{n-1}) \in \mathbf{R}^{n-1}$$

has only the trivial solution $\bar{\lambda} = \bar{0}$. And, according to (2) in (6.3.1.3), this amounts to say that *no* linear combination $\sum_{k=1}^{n-1} \lambda_k \omega_k(z)$ with constant coefficients λ_k can have a single-valued conjugate harmonic function unless all the λ_k reduce to zero. This A is even a *symmetric* matrix since $\alpha_{kj} = \alpha_{jk}$, $1 \le j, k \le n-1$, as an easy consequence of (3) in (6.3.1.3).

Proof of (6.6.1.4). Suppose that there are real constants $\lambda_1, \ldots, \lambda_{n-1}$ satisfying $\bar{\lambda} A = \bar{0}$ with $\bar{\lambda} = (\lambda_1, \ldots, \lambda_{n-1})$, namely, $\sum_{k=1}^{n-1} \lambda_k \alpha_{kj} = 0$ for $1 \le j \le n-1$. We try to show that $\lambda_1 = \cdots = \lambda_{n-1} = 0$ is the only possibility.

Consider the harmonic function

$$\omega(z) = \sum_{k=1}^{n-1} \lambda_k \omega_k(z) \tag{$*_1$}$$

on Ω. By assumption, for $1 \le j \le n-1$,

$$\int_{\gamma_j} *d\omega = \sum_{k=1}^{n-1} \lambda_k \alpha_{kj} = 0$$

$\Rightarrow$ (since $\gamma_1, \ldots, \gamma_{n-1}$ is a homology basis for cycles in Ω) for each cycle γ in Ω, $\int_{\gamma} *d\omega = 0$.

According to (2) in (6.3.1.3), $\omega(z)$ has a single-valued harmonic conjugate in Ω; and this, in turn, implies that there exists a single-valued analytic function $f(z)$ in Ω so that $\operatorname{Re} f(z) = \omega(z)$ throughout Ω. As noted before, this $f(z)$ can be analytically continued to the closure domain $\bar{\Omega}$. Under this circumstance, for $z \in \Omega$,

$$\operatorname{Re} f(z) = \omega_j(z) = \lambda_j, \quad \text{if } z \in \gamma_j \quad \text{for } 1 \le j \le n-1;$$

$$\operatorname{Re} f(z) = 0, \quad \text{if } z \in \gamma_n$$

$$\Rightarrow \ f(\gamma_j) \text{ is a } \textit{vertical line segment} \text{ for each } j, 1 \le j \le n.$$

Fix any point w_0, not in $\bigcup_{j=1}^{n} f(\gamma_j)$. Then

$$n(f(\gamma_j); w_0) = \frac{1}{2\pi i} \int_{f(\gamma_j)} \frac{dw}{w - w_0} = \frac{1}{2\pi i} \int_{\gamma_j} \frac{f'(z)dz}{f(z) - w_0}$$

$$= \frac{1}{2\pi i} \int_{f(\gamma_j)} d\log(w - w_0) = 0, \quad 1 \le j \le n,$$

because each line segment $f(\gamma_j)$, considered as a closed curve, does not contain the point w_0 and hence, a single-valued branch of $\arg(w - w_0)$ can be defined on it (see (2.7.1.6)). By applying the argument principle (3.5.3.3) or (3.5.3.4) to $f(z) - w_0$, we have

$$\frac{1}{2\pi i} \int_{\sum_{j=1}^{n} \gamma_j} \frac{f'(z)}{f(z) - w_0} dz = n\left(\sum_{j=1}^{n} f(\gamma_j); w_0\right) = \sum_{j=1}^{n} n(f(\gamma_j); w_0) = 0$$

$$= \text{ the total number of times } f(z) \text{ assumes } w_0 \text{ in } \Omega.$$

This means that $w = f(z)$ never takes values w_0 in Ω for any point $w_0 \notin \bigcup_{j=1}^{n} f(\gamma_j)$. In case $f(z)$ is not a constant, $f(\Omega)$ would be a domain and must contain points outside $\bigcup_{j=1}^{n} f(\gamma_j)$. As a consequence, $f(z)$ is necessarily a constant function; and then so is $\operatorname{Re} f(z) = \omega(z)$. Since $\omega(z) = 0$ along γ_n, hence $\omega(z) \equiv 0$ throughout Ω. When restricting z to γ_j for $1 \le j \le n-1$, it follows that $\lambda_j = 0, 1 \le j \le n-1$.

For another proof, see the *note* near the end of the proof of (6.6.1.9).

□

In passing, let us go back to (6.3.1.14) and consider the special case when $u(z)$ there is Green's function $g(z, z_0)$, where $z_0 \in \Omega$ is a fixed point. Recall that $g(z, z_0)$ is a positive harmonic function in $\Omega - \{z_0\}$, with a logarithmic pole at z_0 and principal part $-\log|z - z_0|$, and has the boundary values 0 along $\partial\Omega = \bigcup_{j=1}^{n} \gamma_j$ (see (6.3.1.11)).

Choose a small disk $|z - z_0| \le \rho$ contained in Ω. Let C be the circle $|z - z_0| = \rho$, oriented in the positive sense. Fix k, $1 \le k \le n$. Since $\sum_{j=1}^{n} \gamma_j - C$ is homologous to 0 in $\Omega - \{z_0\}$ (here $n = 1$ is permitted), then, by (3) in (6.3.1.3),

$$\int_{\sum_{j=1}^{n} \gamma_j - C} \omega_k{}^* dg - g^* d\omega_k = 0$$

$$\Rightarrow \int_{\sum_{j=1}^{n} \gamma_j} \omega_k{}^* dg \quad g^* d\omega_k = \int_C \omega_k{}^* dg - g^* d\omega_k$$

$\Rightarrow$ (by invoking $(*_{12})$–$(*_{13})$ in Sec. 6.3.1 to the last integral or by (1) in (6.3.1.14))

$$\int_{\gamma_k} {}^* dg = -2\pi\omega_k(z_0).$$

And we obtain

The periods of the conjugate differential $^* dg$ of Green's function $g(z, z_0)$. Let Ω be as in (6.6.8) and ω_k, $1 \le k \le n$ ($n = 1$ is permitted), be the harmonic measure of γ_k with respect to Ω. Fix any point $z_0 \in \Omega$. Let $g(z, z_0)$ be Green's function of Ω with respect to the point z_0. Then the periods of $^* dg$ satisfy the relations

$$\omega_k(z_0) = \frac{1}{2\pi} \int_{\gamma_k} {}^* dg(z, z_0), \quad 1 \le k \le n. \tag{6.6.1.5}$$

This is (6.3.1.15). Note that this is also an easy consequence of (2) in (6.3.1.14) by putting $u(z)$ there equal to $\omega_k(z)$ here.

In what follows, we give two applications of the concept of the harmonic measure.

Application 1:
As a corollary to (6.6.1.4), we assert that *the equations $\bar{\lambda}A = (2\pi, 0, \ldots, 0)$ has a unique solution*

$$\bar{\lambda} = (2\pi, 0, \ldots, 0)A^{-1}.$$

Namely, the equations in $\lambda_1, \ldots, \lambda_{n-1}$

$$\sum_{k=1}^{n-1} \lambda_k \alpha_{k1} = 2\pi, \quad \sum_{k=1}^{n-1} \lambda_k \alpha_{kj} = 0, \quad 2 \le j \le n - 1, \tag{6.6.1.6}$$

can be solved uniquely.

Let $\lambda_1, \ldots, \lambda_{n-1}$ be such a unique solution. Consider the *corresponding* harmonic function $\omega(z)$ defined in $(*_1)$. Thus,

$$\int_{\gamma_1} {}^*d\omega = 2\pi;$$

$$\int_{\gamma_j} {}^*d\omega = 0, \quad 2 \le j \le n - 1. \tag{$*_2$}$$

Since γ_n is homologous to $-\sum_{j=1}^{n-1} \gamma_j$ in Ω, it follows that, by homologous form of the Cauchy integral theorem (4.3.3.11) or just referring to (3) in (6.3.1.4),

$$\int_{\gamma_n} {}^*d\omega = -\sum_{j=1}^{n-1} \int_{\gamma_j} {}^*d\omega = -2\pi. \tag{$*_3$}$$

Therefore,

$$\int_{\gamma_j} d\omega + i{}^*d\omega = i \int_{\gamma_j} {}^*d\omega = \begin{cases} 2\pi i, & j = 1 \\ 0, & 2 \le j \le n - 1 \\ -2\pi i, & j = n \end{cases}. \tag{$*_4$}$$

Fix any point $z_0 \in \Omega$ and set

$$F(z) = \omega(z) + i \int_\gamma {}^*d\omega(z) = \int_\gamma (d\omega(z) + i{}^*d\omega(z)) + \omega(z_0), \quad z \in \Omega,$$

$$(6.6.1.7)$$

where γ is any rectifiable curve in Ω, connecting z_0 to $z \in \Omega$. In conclusion, we obtain

A single-valued analytic function on Ω. Let Ω be a domain satisfying conditions in (6.6.8). Let ω_k, $1 \le k \le n - 1$, be the harmonic measures of γ_k with respect to Ω (see (6.6.1.1)), and $\lambda_1, \ldots, \lambda_{n-1}$ be the unique solutions of Eqs. (6.6.1.6). Define the harmonic function $\omega(z) = \sum_{k=1}^{n-1} \lambda_k \omega_k(z)$ and then the function $F(z)$ in (6.6.1.7) throughout Ω. Then

(1) The periods of the conjugate differential ${}^*d\omega$ of $\omega(z)$ are shown in $(*_2)$ and $(*_3)$.
(2) The real part $\operatorname{Re} F(z) = \omega(z)$ has constant value λ_k along γ_k for $1 \le k \le n$, where $\lambda_n = 0$.
(3) $F(z)$ is multiple-valued in Ω, whose values at any fixed point of Ω differ by an integer multiple of $2\pi i$ (see $(*_4)$).

Therefore, the function

$$f(z) = e^{F(z)}$$

is a single-valued analytic function in Ω. $\tag{6.6.1.8}$

This $f(z)$ is a canonical mapping that maps Ω univalently onto the canonical domain of Type 6 (6.6.7). See Sec. 6.6.2 for details.

Application 2:
A slight modification of the method leading to (6.6.1.7) can be adopted to obtain a more general result, better stated as

An existence and uniqueness theorem of a Dirichlet problem for Ω. Let Ω be a domain satisfying conditions in (6.6.8). Let $\varphi(\zeta)$ be a real-valued C^1 function on $\partial\Omega$ so that $\varphi(\zeta)$ and its partial derivatives satisfy the Hölder condition with exponent α (namely, $|\varphi(\zeta_1) - \varphi(\zeta_2)| \leq C|\zeta_1 - \zeta_2|^\alpha$, $C > 0$ a constant and $0 < \alpha < 1$, and so do the partial derivatives of $\varphi(\zeta)$). Set

$$h(\zeta) = \begin{cases} 0, & \zeta \in \gamma_n \\ \lambda_k, & \zeta \in \gamma_k \quad \text{for } 1 \leq k \leq n-1 \end{cases},$$

where $\lambda_1, \ldots, \lambda_{n-1}$ are real constants to be determined by the periods α_{kj}, $1 \leq k, j \leq n-1$ (see (6.6.1.3) and $\varphi(\zeta)$. Fix any point $z_0 \in \gamma_n$ and a real constant β. Then there exists a unique single-valued analytic function $f(z)$ in Ω, C^1 continuous on $\bar\Omega$ and satisfying the boundary conditions:

1. $\operatorname{Re} f(\zeta) = \varphi(\zeta) + h(\zeta)$, $\zeta \in \partial\Omega = \bigcup_{k=1}^n \gamma_n$, and
2. $\operatorname{Im} f(z_0) = \beta$.   (6.6.1.9)

If Ω is bounded by n regular curves $\gamma_1, \ldots, \gamma_n$, we still need to assume that the parametric representation $z(t)$ of each γ_k $(1 \leq k \leq n)$ and its derivatives $z'(t)$ both satisfy the *Hölder condition with exponent* α. Refer to *Kellog's theorem* mentioned right after (6.3.2.9). In our present case, γ_k as an analytic Jordan closed curve, it does satisfy the Hölder condition locally.

Owing to its complicacy, no proof about the continuity of f on $\bar\Omega$ will be given here.

Sketch of proof.
About uniqueness. Suppose $f(z)$ and $g(z)$ are two such solutions. Set $F(z) = f(z) - g(z)$. Then $F(z)$ is analytic in Ω, C^1 continuous on $\bar\Omega$ and satisfies

$$\operatorname{Re} F(\zeta) = \begin{cases} 0, & \zeta \in \gamma_n \\ a_k, & \zeta \in \gamma_k \quad \text{for } 1 \leq k \leq n-1 \end{cases};$$

$$\operatorname{Im} F(z_0) = 0,$$ $(*_5)$

where $a_1, \ldots, a_{n-1}$ are real constants. Let $F(z) = U(z) + iV(z)$. Then

$$\iint_\Omega (U_x^2 + U_y^2)\,dx\,dy = \iint_\Omega [(UU_x)_x + (UU_y)_y]\,dx\,dy$$

$$= \int_{\partial\Omega} UU_x\,dy - UU_y\,dx \quad \text{(by Green's formula (2.9.12))}$$

$$= \int_{\partial\Omega} U\frac{\partial U}{\partial n}\,ds \quad \text{(see ($*_1$) in Sec. 6.3.1)}$$

$$= -\int_{\partial\Omega} U\frac{\partial V}{\partial s}\,ds \quad \text{(see Exercise A(5) of Sec. 3.2.2}$$

and use C^1-continuity of f on $\bar\Omega$)

$$= -\int_{\partial\Omega} U\,dV$$

$$= -\sum_{k=1}^{n-1} a_k \int_{\gamma_k} dV = 0 \quad \text{(each γ_k is a closed curve for}$$

$$0 \le k \le n-1)$$

$$\Rightarrow U_x(z) = U_y(z) = 0 \quad \text{in } \Omega$$

$$\Rightarrow U(z) = \int_{z_0}^z dU = \int_{z_0}^z U_x\,dx + U_y\,dy = 0, \quad z \in \Omega;$$

$$V(z) = \int_{z_0}^z -U_y\,dx + U_x\,dy = 0 \quad \text{(recall that } V(z_0) = 0).$$

Hence, $F(z) = 0$ throughout Ω. And it follows that $f(z) = g(z)$ in Ω.

About existence. By (6.5.7), there exists a unique harmonic function $u(z)$ in Ω satisfying the boundary condition: $u(\zeta) = \varphi(\zeta)$, $\zeta \in \partial\Omega$. Consider the periods

$$\alpha_j = \int_{\gamma_j} {}^*du, \quad 1 \le j \le n-1 \tag{$*_6$}$$

of the conjugate harmonic differential ${}^*du$.

In case $\alpha_j = 0$ for $1 \le j \le n-1$: Since $\gamma_1, \ldots, \gamma_{n-1}$ form a homology basis for Ω, it follows from (2) in (6.3.1.3) that $u(z)$ has a single-valued harmonic conjugate $v(z)$ in Ω, unique up to an additive constant. Set

$$f(z) = u(z) + iv(z). \tag{6.6.1.10}$$

Then $f(z)$ is single-valued, analytic in Ω and $\operatorname{Re} f(\zeta) = u(\zeta) = \varphi(\zeta)$ for $\zeta \in \partial\Omega$. In this case, $\lambda_j = 0$ for $1 \le j \le n-1$.

In case at least one of the α_j's is not equal to zero, consider the nonhomogenous system of linear equations (compare to (6.6.1.6))

$$\sum_{k=1}^{n-1} \lambda_k \alpha_{kj} = -\alpha_j, \quad 1 \le j \le n-1. \tag{6.6.1.11}$$

According to (6.6.1.4) (see also the *note* near the end of this proof), it has a unique solution $(\lambda_1, \dots, \lambda_{n-1}) = (-\alpha_1, \dots, -\alpha_{n-1})A$, where $A = [\alpha_{kj}]$. Construct the harmonic function

$$\omega(z) = u(z) + \sum_{k=1}^{n-1} \lambda_k \omega_k(z), \quad z \in \Omega.$$

Since

$$\int_{\gamma_j} {}^* d\omega = \alpha_j + \sum_{k=1}^{n-1} \lambda_k \alpha_{kj} = 0, \quad 1 \le j \le n-1$$

and $\gamma_1, \dots, \gamma_{n-1}$ form a homology basis for Ω, again by using (6.3.1.3), $\omega(z)$ has a single-valued harmonic conjugate $\tilde{\omega}(z)$ in Ω, unique up to an additive constant. Choose $\tilde{\omega}(z)$ so that $\tilde{\omega}(z_0) = \beta$, the preassigned constant. Then

$$f(z) = \omega(z) + i\tilde{\omega}(z), \quad z \in \Omega \tag{6.6.1.12}$$

is single-valued, analytic in Ω and satisfies the required boundary conditions.

Note. The nonsingularity of $A = [\alpha_{kj}]$ in (6.6.1.4) can be proved alternatively as follows. Suppose on the contrary that there are constants $\lambda_1, \dots, \lambda_{n-1}$, not all equal to zero, so that the function defined by $(*_1)$ has all its periods $\int_{\gamma_k} {}^* d\omega = 0$ for $1 \le k \le n-1$. Let $\tilde{\omega}(z)$ be a single-valued harmonic conjugate of $\omega(z)$ in Ω. Then $f(z) = \omega(z) + i\tilde{\omega}(z)$ is a single-valued analytic function which satisfies the boundary condition that $\mathrm{Re}\, f(\zeta) = \omega(\zeta) = \lambda_k$ for $\zeta \in \gamma_k$ and $1 \le k \le n$, where $\lambda_n = 0$. Define a constant function $g(z)$ by $g(z) = \tilde{\omega}(z_0)$. Then the function $F(z) = f(z) - ig(z)$ satisfies $(*_5)$. By the uniqueness property of the solution, it follows that $F(z) \equiv 0$ in Ω, and hence

$$f(z) = ig(z) \quad \text{in } \Omega.$$

$$\Rightarrow \omega(z) = \sum_{k=1}^{n-1} \lambda_k \omega_k(z) = 0 \quad \text{everywhere in } \Omega.$$

This is not possible and then we prove the nonsingularity of A. $\square$

By adding a complex constant to the solution $f(z)$ in (6.6.1.9),we can rewrite (6.6.1.9) as

A modified version of (6.6.1.9). Let Ω and $\varphi(\zeta)$ be as in (6.6.1.9). Also suppose that $0 \in \Omega$. Then there is a unique analytic function $f(z)$ in Ω, C^1 continuous on $\bar{\Omega}$, which satisfies

$$\operatorname{Re} f(\zeta) = \varphi(\zeta) + c_k, \quad \zeta \in \gamma_k \quad \text{for } 1 \le k \le n;$$

$$f(0) = c,$$

where c_k, $1 \le k \le n$, are suitable real constants and c is a preassigned complex constant.   (6.6.1.13)

$f(z)$ in (6.6.1.9) or (6.6.1.13) can be used to prove the existence of the canonical mapping of a finitely connected domain (include the simply connected one) onto its canonical domain. See Sec. 6.6.3.

Exercises A

(1) Let Ω be as in (6.6.8). Fix a point $z_0 \in \Omega$ and a positive integer $m \ge 1$. Show that there is a meromorphic function $f(z)$ in Ω, continuous on $\bar{\Omega} - \{z_0\}$ and having a pole of order m at z_0.
(2) Try to deduce (6.6.1.8) from (6.6.1.9) as a rather special case.

6.6.2   *Canonical domains: The annuli with concentric circular slits*

Our main result is the following

Canonical mapping and canonical domain of a domain of connectivity $n \ge 2$. Let Ω and $f(z)$ be as in (6.6.1.8). Then $w = f(z)$ maps Ω univalently onto the annulus $1 < |w| < e^{\lambda_1}$ with $n - 2$ disjoint concentric circular slits situated on the circles $|w| = e^{\lambda_k}$ for $2 \le k \le n - 1$. Moreover, this $f(z)$ is unique up to a multiplicative constant of absolute one, namely, if $w = g(z)$ is another such mapping, then there is a constant μ, $|\mu| = 1$, such that $g(z) = \mu f(z)$, $z \in \Omega$.   (6.6.2.1)

Proof. Since the boundary $\partial\Omega$ is composed of analytic Jordan closed curves (see (6.6.8)), $f(z)$ can be assumed to be analytic on $\bar{\Omega}$, as we had claimed in Sec. 6.6.1.

By the definition of $f(z)$ as shown in (6.6.1.7), we have

$$|f(z)| = e^{\operatorname{Re} F(z)} = e^{\omega(z)} = \begin{cases} e^{\lambda_k}, & \text{if } z \in \gamma_k \quad \text{for } 1 \le k \le n-1 \\ e^0 = 1, & \text{if } z \in \gamma_n \end{cases}. \qquad (*_1)$$

In particular, $f(z) \ne 0$ along the boundary $\partial\Omega = \bigcup_{k=1}^n \gamma_k$; and, according to $(*_4)$ in Sec. 6.6.1,

$$\frac{1}{2\pi i}\int_{\gamma_k} \frac{f'(z)}{f(z)}dz = n(f(\gamma_k); 0)$$

$$= \frac{1}{2\pi i}\int_{\gamma_k} F'(z)dz = \frac{1}{2\pi i}\int_{\gamma_k} d\omega + i^* d\omega$$

$$= \begin{cases} 1, & k = 1 \\ 0, & 2 \le k \le n-1 \\ -1, & k = n \end{cases}. \qquad (*_2)$$

Owing to property (4) in (3.5.2.4) for winding numbers, the above relation indicates that

$$n(f(\gamma_1); w) = \begin{cases} 1, & \text{if } |w| < e^{\lambda_1} \\ 0, & \text{if } |w| > e^{\lambda_1} \end{cases} \Rightarrow f(\gamma_1) \text{ fills up the whole circle } |w| = e^{\lambda_1};$$

$$n(f(\gamma_n); w) = \begin{cases} 0, & \text{if } |w| > 1 \\ -1, & \text{if } |w| < 1 \end{cases} \Rightarrow f(\gamma_n) \text{ fills up the whole circle } e^{\lambda_n} = 1,$$

and

$$n(f(\gamma_k); w) = 0 \quad \text{if } |w| \ne e^{\lambda_k} \text{ for } 2 \le k \le n-1. \qquad (*_3)$$

Based on these facts, we try to show that $f(\gamma_1)$ covers $|w| = e^{\lambda_1}$ and $f(\gamma_n)$ covers $|w| = 1$ exactly once; while the remaining $f(\gamma_k)$, $2 \le k \le n-1$, are only proper arcs lying on concentric circles $|w_0| = e^{\lambda_k}$ situated between $|w| = 1$ and $|w| = e^{\lambda_1}$.

Suppose $w = f(z)$ assumes a value w_0 in Ω. *Under the additional assumption that* $|w_0| \ne e^{\lambda_\kappa}$ *for* $1 \le \kappa \le n$, we *claim that*

1. $1 < |w_0| < e^{\lambda_1}$ and, in particular, $\lambda_1 > 0$;
2. w_0 is assumed exactly once in Ω and hence, γ_1 and γ_n are mapped in a one-to-one manner onto $|w| = e^{\lambda_1}$ and $|w| = 1$, respectively, and then
3. $0 < \lambda_k < \lambda_1$ for $2 \le k \le n-1$. $\qquad (*_4)$

To prove these statements, the argument principle (3.5.3.3), and $(*_3)$ together show that

The total number times $f(z)$ assuming w_0 in Ω

$$= \frac{1}{2\pi i} \int_{\partial \Omega} \frac{f'(z)}{f(z) - w_0} dz = \frac{1}{2\pi i} \int_{\sum_{k=1}^{n} \gamma_k} \frac{f'(z)}{f(z) - w_0} dz$$

$$= \sum_{k=1}^{n} n(f(\gamma_k); w_0) = n(f(\gamma_1); w_0) + n(f(\gamma_n); w_0) \geq 1 > 0 \qquad (*_5)$$

$\Rightarrow n(f(\gamma_1); w_0) = 1, \quad n(f(\gamma_n); w_0) = 0$ and the number of times $f(z)$

assuming w_0 in Ω is equal to 1.

$\Rightarrow |w_0| < e^{\lambda_1}$ and $|w_0| > 1$ and $f(z)$ is univalent in Ω.

In case there are two distinct points ζ_1 and ζ_2 on γ_1 such that $f(\zeta_1) = f(\zeta_2) = \eta$ and $|\eta| = e^{\lambda_1}$ hold, by continuity of $f(z)$, there is a point in $|w| < e^{\lambda_1}$, sufficiently close to η, which is assumed in a neighborhood of ζ_1 in Ω and in a neighborhood of ζ_2 in Ω. This contradicts to the univalence of $f(z)$ in Ω. Consequently, $f(\gamma_1)$ covers $|w| = e^{\lambda_1}$ exactly once. Similarly, so does $f(\gamma_n)$ cover $|w| = 1$ exactly once. By the results obtained so far in 1 and 2, $\lambda_k > 0$ is true for each k, $2 \leq k \leq n - 1$. On the other hand, if $\lambda_k > \lambda_1$ for some k ($2 \leq k \leq n - 1$), then by $(*_3)$ $w = f(z)$ cannot assume these w whose absolute values are sufficiently near e^{λ_k} (namely, $e^{\lambda_1} < |w| < e^{\lambda_k}$); and hence, by continuity of $f(z)$, $f(z)$ does not assume values on $|w| = e^{\lambda_k}$ along γ_k, contradicting to (2) in (6.6.1.8). Therefore $\lambda_k \leq \lambda_1$ should hold for $2 \leq k \leq n - 1$. It is not possible that $\lambda_k = \lambda_1$ for some k ($2 \leq k \leq n - 1$), for this would contradict to the univalence of $f(z)$ in Ω, too.

What is left is to handle the cases that $|w_0| = e^k$ for $2 \leq k \leq n-1$, and to show that $f(\gamma_k)$ are disjoint proper arcs lying on $|w| = e^{\lambda_k}$, $2 \leq k \leq n - 1$. To achieve this purpose, it is still more instructive and simpler to apply a *generalized version* of the argument principle. (In what follows, readers are advised to refer to (4.11.3.3), Exercise A(12) of Sec. 4.11.3, and the proof and the content of the argument principle (3.5.3.3), and above all the concept of Cauchy principle value introduced in Sec. 4.7, (4.11.3.4), and (4.11.3.5).)

Let $|w_0| = e^{\lambda_k}$, $1 \leq k \leq n$.

Fix k, $1 \leq k \leq n$. According to (2.7.1.4), a single-valued continuous branch of $\arg f(z)$ can be defined along the contour γ_k. Choose a fixed branch $\arg f(z)$ in the sequel and a fixed value $\arg w_0$ ($0 \leq \arg w_0 < 2\pi$).

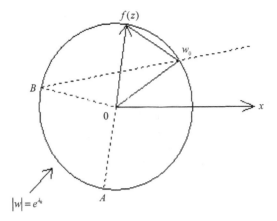

Fig. 6.45

Figure 6.45 shows that a single-valued continuous branch of $\arg\left(f(z)-w_0\right)$ along γ_k can be chosen as

$$\arg(f(z)-w_0) = \frac{\pi - \arg f(z) - \arg w_0}{2} + \arg f(z) - \frac{1}{2}(\pi + \arg f(z) - \arg w_0)$$

whenever $f(z) \neq w_0$. This implies that.

$$d\arg(f(z)-w_0) = \frac{1}{2}d\arg f(z), \quad z \in \gamma_k \quad \text{and} \quad f(z) \neq w_0. \qquad (*_6)$$

An alternative way to obtain $(*_6)$ is to observe that, in Fig. 6.45, the circular angle $\angle f(z)w_0B$ is half that of the central angle $\angle f(z)OB$. The result follows by letting B to approach $f(z)$ along the circle $|w| = e^{\lambda_k}$. By the very definition of the Cauchy principal value of a contour integral (see (4.7.8)), $(*_6)$ shows that

$$\text{P.V.} \int_{\gamma_k} \frac{f'(z)}{f(z)-w_0}dz = \frac{1}{2}\int_{\gamma_k}\frac{f'(z)}{f(z)}dz. \qquad (*_7)$$

Now, *under the assumption* that $|w_0| = e^{\lambda_k}$, $1 \leq k \leq n$, $(*_7)$ implies that $(*_2)$ above are replaced by

$$\text{P.V.}\frac{1}{2\pi i}\int_{\gamma_k}\frac{f'(z)}{f(z)-w_0}dz = \begin{cases} \dfrac{1}{2}, & \text{if } k=1 \\[2mm] 0, & \text{if } 2 \leq k \leq n-1 \text{,} \\[2mm] -\dfrac{1}{2}, & \text{if } k=n \end{cases} \qquad (*_8)$$

which, in turn, implies that $(*_5)$ becomes

$$\text{P.V.} \frac{1}{2\pi i} \int_{\partial\Omega} \frac{f'(z)}{f(z) - w_0} dz = \begin{cases} \dfrac{1}{2}, & \text{if } |w_0| = e^{\lambda_1} \\ -\dfrac{1}{2}, & \text{if } |w_0| = e^0 = 1 \\ n_0, & \text{if } 1 < |w_0| < e^{\lambda_1} \\ & \quad \text{for } 2 \le k \le n-1 \end{cases}, \qquad (*_9)$$

where n_0 denotes the sum of the orders of these zeros of $f(z) - w_0$ in $\bar{\Omega}$ (see n_j in (4.11.3.4)).

As a consequence, by recalling the meaning of (4.11.3.5), each value on the circles $|w| = 1$ and $|w| = e^{\lambda_1}$ is taken exactly only once on the boundary γ_n and γ_1, respectively; this proves that γ_1 and γ_n are mapped by $w = f(z)$ in one-to-one manner onto $|w| = e^{\lambda_1}$ and $|w| = 1$, respectively, and that $0 < \lambda_k < \lambda_1$ for $2 \le k \le n-1$.

On the other hand, in case $1 < |w_0| < e^{\lambda_1}$ for $2 \le k \le n-1$, n_0 is a positive integer. If $|w_0| \ne e^{\lambda_k}$ for $2 \le k \le n-1$, $(*_4)$ and $(*_5)$ had shown that $n_0 = 1$. If $|w_0| = e^{\lambda_k}$ for $2 \le k \le n-1$, then $n_0 = 2$ and hence w_0 is taken either twice on the boundary γ_k or only once on γ_k with the multiplicity 2. It is not possible for $n_0 \ge 3$ because we have shown that $w = f(z)$ is univalent in Ω (see $(*_4)$). As we noted right ahead of $(*_6)$, a branch of $\arg f(z)$ can be defined along γ_k $(2 \le k \le n-1)$, and the values of multiplicity 2 correspond to the (relative) maxima and minima of such a branch. The compactness of γ_k and univalence of $w = f(z)$ together show that there is exactly only one global maximum and only one global minimum along γ_k; moreover, the difference between the maximum and the minimum must be strictly less than 2π. Therefore, the image curve $f(\gamma_k)$ for each k $(2 \le k \le n-1)$ is a proper arc on the circle $|w| = e^{\lambda_k}$. Again the univalence of $w = f(z)$ on Ω shows that the arcs corresponding to different contours are disjoint.

This finishes the proof of the *existence* of such a canonical mapping $w = f(z)$.

As for the *uniqueness*, suppose $w = g(z)$ is a univalent analytic function mapping Ω onto the annulus $1 < |w| < R$ minus concentric circular slits such that $g(\gamma_1)$ is the circle $|w| = R$ and $g(\gamma_n)$ the circle $|w| = 1$. We try to show that

1. $R = e^{\lambda_1}$, and
2. there is a constant μ, $|\mu| = 1$, so that $g = \mu f$. $\qquad (*_{10})$

Remember that such a $g(z)$ can be analytically extended to $\bar{\Omega}$. Since $\beta_k = \log|g(z)|$ is a constant if $z \in \gamma_k$ for each k, $1 \le k \le n$, it follows that

$$\log|g(z)| = \sum_{k=1}^{n-1} \beta_k \omega_k(z), \quad z \in \Omega.$$

Set up linear equations in $\beta_1, \ldots, \beta_{k-1}$ as in (6.6.1.6):

$$n(g(\gamma_1); 0) = 1 = \frac{1}{2\pi}\int_{\gamma_1} {}^*d\log|g(z)| = \frac{1}{2\pi}\sum_{k=1}^{n-1}\beta_k\alpha_{k1};$$

$$n(g(\gamma_j); 0) = 0 = \frac{1}{2\pi}\int_{\gamma_j} {}^*d\log|g(z)| = \frac{1}{2\pi}\sum_{k=1}^{n-1}\beta_k\alpha_{kj}, \quad 2 \le j \le n-1$$

$$\Rightarrow \sum_{k=1}^{n-1}\beta_k\alpha_{k1} = 2\pi, \quad \sum_{k=1}^{n-1}\beta_k\alpha_{kj} = 0, \quad 2 \le j \le n-1.$$

Since the coefficient matrix $A = [\alpha_{kj}]$ is nonsingular (see (6.6.1.4)), it follows that the above equations have unique solutions

$$\beta_k = \lambda_k, \quad 1 \le k \le n$$

as (6.6.1.5) did. Now, as $z \in \gamma_1$, then

$$\log|g(z)| = \log R = \beta_1 = \lambda_1 \Rightarrow R = e^{\lambda_1}.$$

Hence,

$$\log|g(z)| = \log|f(z)|, \quad z \in \Omega$$

$$\Rightarrow \left|\frac{g(z)}{f(z)}\right| = 1, \quad z \in \Omega$$

$$\Rightarrow \left(\text{since } \frac{g(z)}{f(z)} \text{ is analytic in } \Omega \text{ and, by max–min principle}\right)$$

$$\frac{g(z)}{f(z)} = \mu, \text{ a constant.}$$

$$\Rightarrow g(z) = \mu f(z), \quad z \in \Omega \text{ with } |\mu| = 1.$$

This ends the proof. $\qquad\qquad\qquad\qquad\qquad\qquad\qquad\qquad\qquad\qquad\qquad\square$

Exercises B

In the following exercises, Ω will always denote an n-connected domain satisfying conditions (6.6.8), unless otherwise stated.

Let θ and c be two real constants. The plane curve $\operatorname{Im}(e^{-i\theta}\log z) = c$ is a *logarithmic spiral* approaching 0 and having its intersection with any ray emanating from 0 the constant angle θ (called the *inclination angle* of the spiral). In case $\theta = 0$, it is a ray emanating from the origin 0; in case $\theta = \frac{\pi}{2}$, it is a circle with center at 0.

Imitate the context and do the following problems.

(1) Suppose $0 \in \Omega$ and $z_0 \in \Omega$, $z_0 \neq 0$, is a fixed point. Let θ be a real constant. Then there is a *unique univalent meromorphic* function $w = f(z)$ on Ω satisfying the following properties:

1. $f(z_0) = 0$;
2. $f(z)$ has a simple pole at 0 with residue $\operatorname{Res}(f; 0) = 1$, namely, in a neighborhood of 0, $f(z)$ has the Laurent series expansion

$$f(z) = \frac{1}{z} + \sum_{m=0}^{\infty} a_m z_m;$$

3. $w = f(z)$ maps Ω univalently onto $\mathbf{C}^*$ with n logarithmic spiral slits, called a *domain with spiral slits* for simplicity.

Try the following steps:

(a) According to (6.6.1.13), let $F(z)$ be the unique analytic function satisfying the boundary conditions:

$$\operatorname{Re} F(\zeta) = \operatorname{Re}\left\{ie^{-i\theta}\log\frac{z_0\zeta}{z_0 - \zeta}\right\} + c_k, \quad \zeta \in \gamma_k \text{ for } 1 \le k \le n;$$

$$F(0) = 0,$$

where c_k, $1 \le k \le n$, are undetermined real constants and $\log\frac{z_0\zeta}{z_0-\zeta}$, $\zeta \in \gamma_k$ ($1 \le k \le n$) is a branch along γ_k (see (2.7.1.4)).

(b) Set $w = f(z) = \frac{z_0 - z}{z_0 z} e^{-ie^{i\theta}F(z)}$, $z \in \Omega$. Note that $f(z)$ is continuous on $\bar{\Omega} - \{0\}$ and $\operatorname{Re}(ie^{-i\theta}\log f(z)) = c_k$ along γ_k for $1 \le k \le n$. To prove the univalence of $f(z)$ in Ω, fix any point $w \notin \bigcup_{k=1}^{n} f(\gamma_k)$ and observe that $\Delta_{\partial\Omega} \arg(f(z) - w) = 0$. Since $f(z)$ has a simple pole at $z = 0$, there is a unique point $z \in \Omega$ so that $f(z) - w = 0$ due to the argument principle (3.5.3.4). One also needs to show that $w = f(z)$ cannot assume any point on $\bigcup_{k=1}^{n} f(\gamma_k)$ inside Ω. See the context.

(c) Let $g(z)$ be another such mapping. Then $f(z)$ and $g(z)$ are continuous on $\bar{\Omega} - \{0\}$ and

$$\mathrm{Re}(ie^{-i\theta}\log f(z)) = c_k, \quad \mathrm{Re}(ie^{-i\theta}\log g(z)) = c_k', \quad 1 \le k \le n.$$

Then 0 and z_0 are removable singularities of each branch of $\log \frac{f(z)}{g(z)}$ in Ω. Choose such a branch $G(z)$. Along γ_k, $\mathrm{Re}(ie^{-i\theta}G(z)) = c_k - c_k'$ for $1 \le k \le n$. This means that $G(z)$ does not assume values outside $\bigcup_{k=1}^{n} f(\gamma_k)$ (just prove it by the argument principle). Consequently, $G(z)$ is a constant; and since $\frac{f(z)}{g(z)} \to 1$ as $z \to 0$, it follows that $f(z) \equiv g(z)$ in Ω.

(2) (continued from Exercise (1)) Let the canonical mapping $w = f(z)$ in Exercise (1) be denoted by $f_\theta(z)$. Show that

$$\log f_\theta(z) = e^{i\theta}(\log f_0(z) \cdot \cos\theta - i\log f_{\frac{\pi}{2}}(z) \cdot \sin\theta), \quad z \in \Omega.$$

Make sure what are the n-connected domains $f_0(\Omega)$ and $f_{\frac{\pi}{2}}(\Omega)$. Refer to (6.6.3.6) if necessary.

(3) Let Ω be such that its outer contour γ_n is the unit circle $|z| = 1$ and Ω is contained in $|z| < 1$ and $0 \in \Omega$. Then there is a unique univalent analytic function $w = f(z)$ on Ω satisfying the following properties:

1. $w = f(z)$ maps $\gamma_n : |z| = 1$ onto $|w| = 1$, $f(0) = 0$ and $f(1) = 1$, and

2. $w = f(z)$ maps Ω univalently onto $|w| < 1$ with $n - 1$ concentric circular slits with common center at 0, called a *disk with concentric circular slits* (see (6.6.5)).

Try the following steps:

(a) By (6.6.1.13), there is a unique analytic function $F(z)$ in Ω satisfying:

$$\mathrm{Re}\,F(\zeta) = \begin{cases} 0, & \zeta \in \gamma_n \\ -\log|\zeta| + \lambda_k, & \zeta \in \gamma_k \text{ if } 1 \le k \le n-1 \end{cases} ; \quad \mathrm{Im}\,F(1) = 0,$$

where $\lambda_1, \ldots, \lambda_{n-1}$ are undetermined real constants.

(b) Set $w = f(z) = ze^{F(z)}$ and prove its univalence in Ω (see Exercise (1)(b)).

(c) Uniqueness?

(4) Use the symmetry principle and Exercise (1) to prove the existence of the mapping stated in Exercise (3).

(5) Let Ω be an n-connected domain contained in $r < |z| < 1$ such that the outer contour γ_n is the circle $|z| = 1$ and an inner contour, say γ_1, is the circle $|z| = r$. Then there exists a unique univalent analytic function $w = f(z)$ on Ω satisfying the properties:

1. $w = f(z)$ maps γ_1 onto $|w| = R$, where $0 < R < 1$, and γ_n onto $|w| = 1$ with $f(1) = 1$, and

2. $w = f(z)$ maps Ω univalently onto the ring domain $R < |w| < 1$ with $(n-2)$ concentric circular slits with common center at 0, called an *annulus with concentric circular slits* (see (6.6.7)).

Try the following steps:

(a) Let $w = f(z) = ze^{F(z)}$ as in Exercise (3)(b).

(b) Observe that $f(\gamma_1)$ is the circle $|w| = R$. In case $R > 1$, choose $w_0 \notin \bigcup_{k=1}^{n} f(\gamma_k)$ and $1 < |w_0| < R$. Then $\frac{1}{2\pi}\Delta_{\partial\Omega}\arg(f(z) - w_0) = -1$ implies that $f(z) - w_0$ has a simple pole in Ω (see (3.5.3.4) and (4.11.3.5)), which contradicts to the analyticity of $f(z)$ in Ω. If $R = 1$, choose $w_0 \notin \bigcup_{k=1}^{n} f(\gamma_k)$ and $|w_0| < 1$. Then $\frac{1}{2\pi}\Delta_{\partial\Omega}\arg(f(z) - w_0) = 0$ implies that $f(z)$ does not assume such w_0 in Ω which, in turn, implies that $f(z)$ is constant in Ω by the open mapping theorem (see (3.4.4.4)). This result contradicts to the analyticity of $F(z) = \log\frac{f(z)}{z}$ in Ω. In conclusion, $0 < R < 1$ holds.

(c) Univalence? Uniqueness?

(6) Let $\mathfrak{F}$ be the family of univalent meromorphic functions $g(z)$ in $|z| > R$, having the following properties: $g(z_0) = 0$ (z_0 fixed and $R < |z_0| < \infty$), $g(\infty) = \infty$ and, $g(z) = z + a_0 + \frac{a_1}{z} + \cdots$ at ∞. Then the function $w = f(z)$ that minimizes $\mathrm{Re}\{e^{-2i\theta}\log g'(z_0)\}$ maps $|z| > R$ univalently onto a *domain with spiral slits* of inclination angle θ (see Exercise (1)). Here $\log g'(z)$ is the branch determined by the condition that $\log g'(z) \to 0$ as $z \to \infty$.

6.6.3   *Canonical domains: The parallel slit domains*

Still denote by Ω an n-connected domain ($n \geq 1$) satisfying conditions in (6.6.8).

Fix a point $z_0 \in \Omega$. Recall (see (6.3.1.13)) that Green's function $g(z, z_0)$ of the domain Ω with pole z_0 is given by

$$g(z, z_0) = G(z, z_0) - \log|z - z_0|, \quad z \in \Omega,$$

where $G(z, z_0)$ is symmetric and harmonic in z and z_0; and, as a function of z, has the boundary values $\log |\zeta - z_0|$ as $z \in \Omega \to \zeta \in \partial\Omega$. Equivalently, $g(z, z_0)$ is symmetric in z and z_0, $z \neq z_0$; positive and harmonic, as a function of z, in $\Omega - \{z_0\}$, and has the continuous boundary values zero as $z \in \Omega \to \zeta \in \partial\Omega$.

Set $z_0 = x_0 + iy_0$. We *claim* that

$$\frac{\partial}{\partial x_0} G(z, z_0) \tag{$*_1$}$$

is harmonic in Ω and continuous on $\bar{\Omega}$, as a function of z, and has the boundary values $-\mathrm{Re}\frac{1}{\zeta - z_0}$ as $z \in \Omega \to \zeta \in \partial\Omega$. To prove this, choose h real and $|h| > 0$ sufficiently small and consider the harmonic function in z:

$$Q(z, h) = \frac{1}{h}[G(z, z_0 + h) - G(z, z_0)], \quad z \in \Omega. \tag{$*_2$}$$

Since the boundary values $Q(\zeta, h) = \frac{1}{h}[\log|\zeta - z_0 - h| - \log|\zeta - z_0|]$ approach $\frac{\partial}{\partial x_0} \log|\zeta - z_0| = -\mathrm{Re}\frac{1}{\zeta - z_0}$ *uniformly* on $\partial\Omega$ as $h \to 0$ (see Exercise A(1)), it follows by the max–min principle that $Q(z, h) \to \frac{\partial}{\partial x_0} G(z, z_0)$ uniformly on the whole compact set $\bar{\Omega}$ as $h \to 0$. According to (6.3.2.16) (also refer to Exercise A(11) of Sec. 6.3.2), the claim follows.

As a consequence of $(*_1)$, it follows that, with $z_0 \in \Omega$ fixed,

$$u(z, z_0) = \frac{\partial}{\partial x_0} g(z, z_0) = \frac{\partial}{\partial x_0} G(z, z_0) + \mathrm{Re}\frac{1}{z - z_0}, \quad z \in \Omega, \tag{6.6.3.1}$$

is harmonic in $z \neq z_0$, continuous on $\bar{\Omega} - \{z_0\}$, and has the boundary values zero along $\partial\Omega$. Set the periods of $*du$ along γ_j (compare to $(*_2)$ and $(*_6)$ in Sec. 6.6.1)

$$\alpha_j = \int_{\gamma_j} {}^*du, \quad 1 \leq j \leq n - 1$$

and consider the inhomogeneous system of equations (compare to (6.6.1.6) and (6.6.1.11))

$$\sum_{k=1}^{n-1} \lambda_k \alpha_{kj} = -\alpha_j, \quad 1 \leq j \leq n - 1.$$

Let $\lambda_1, \ldots, \lambda_{n-1}$ be the unique solution of this system (see (6.6.1.4)). Define the harmonic function

$$\omega(z) = u(z) + \sum_{k=1}^{n-1} \lambda_k \omega_k(z), \quad z \in \Omega,$$

where $\omega_k(z)$ is the harmonic measure of γ_k with respect to Ω, for $1 \leq k \leq n - 1$ (see (6.6.1.2)). Now $\int_{\gamma_j} {}^*d\omega = 0$ for $1 \leq j \leq n - 1$ imply that

$\omega(z)$ has a single-valued harmonic conjugate $\widetilde{\omega}(z)$ in Ω, unique up to a *constant*. Choosing this constant adequately, we obtain a function (compare to (6.6.1.12))

$$p(z) = \omega(z) + i\widetilde{\omega}(z), \quad z \in \Omega \tag{6.6.3.2}$$

which is single-valued and analytic in Ω, except a simple pole at z_0 with the residue 1. Moreover, its real part $\operatorname{Re} p(z) = \omega(z)$ is constant λ_k on each γ_k for $1 \le k \le n$, where $\lambda_n = 0$; namely, the image curve $f(\gamma_k)$ lies on the vertical line $\operatorname{Re} w = \lambda_k$.

Such a $p(z)$ is unique up to a constant; geometrically, this means that the image domain $p(\Omega)$ is unique except for a parallel translation. To see this, let $\widetilde{p}(z)$ be another such a meromorphic function in Ω. Then the function $f(z) = p(z) - \widetilde{p}(z)$ is analytic throughout Ω and has its real part a constant along each contour γ_k, $1 \le k \le n$. By the symmetry principle (see (3.4.6.3) or (6.1.2.3)), $f(z)$ can be analytically continued to $\bar{\Omega}$; then, by the argument principle (refer to Exercise B(1)(b) of Sec. 6.6.2), $f(z)$ does not assume values not lying on $\bigcup_{k=1}^{n} f(\gamma_k)$ and, by the open mapping theorem, $f(z)$ turns out to be a constant.

As for the univalence of $p(z)$ in Ω: Fix $w_0 \in \mathbf{C}$. $(*_5)$ and $(*_9)$ in Sec. 6.6.2 now become, in the present case,

$$\frac{1}{2\pi i} \int_{\partial\Omega} \frac{p'(z)}{p(z) - w_0} dz = \sum_{k=1}^{n} \frac{1}{2\pi i} \int_{\gamma_k} \frac{p'(z)}{p(z) - w_0} dz$$

$$= \{\text{the number of zeros of } p(z) - w_0 \text{ in } \Omega\}$$

$$- \{\text{the number of poles of } p(z) - w_0 \text{ in } \Omega\}$$

$$= 0.$$

Note that whenever $w_0 \in \bigcup_{k=1}^{n} f(\gamma_k)$, a boundary value, the principle values of the integrals concerned should be adopted as in $(*_7)$–$(*_9)$ of Sec. 6.6.2:

$$\text{P.V. } \frac{1}{2\pi i} \int_{\gamma_k} \frac{p'(z)}{p(z) - w_0} dz = 0 \quad \text{if } p(z) \text{ assumes } w_0 \text{ on } \gamma_k \text{ for } 1 \le k \le n.$$

Since $p(z) - w_0$ has only one simple pole z_0 in Ω, it follows that $p(z)$ assumes w_0 only once in Ω, twice on the boundary, or once on the boundary with the multiplicity 2. Exactly the same reasoning as in Sec. 6.6.2 will finish the rest of the proof.

In conclusion, we summarize as the

Canonical mapping and the vertical slit domain. Let Ω be an n-connected domain as in (6.6.8), where $n \ge 1$. Fix a point $z_0 \in \Omega$. Then there is a

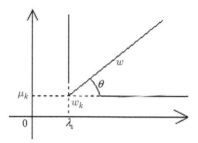

Fig. 6.46

univalent meromorphic function $w = f(z)$ on Ω satisfying the following properties:

1. $p(z)$ has only one simple pole at z_0 with residue equal to 1;
2. $w = p(z)$ maps Ω univalently onto $\mathbf{C}^*$ with n vertical segment slits.

Such a canonical mapping $w = p(z)$ is unique up to a constant. In case it is required that $p(z)$ has the Laurent series expansion

$$\frac{1}{z - z_0} + \sum_{n=1}^{\infty} a_n z^n$$

at z_0, then it is unique.   (6.6.3.3)

Instead of (6.6.3.1), consider the harmonic function

$$v(z, z_0) = \frac{\partial}{\partial y_0} g(z, z_0) = \frac{\partial}{\partial y_0} G(z, z_0) - \operatorname{Im} \frac{1}{z - z_0}, \quad z \in \Omega \qquad (6.6.3.4)$$

in $z \neq z_0$, which is continuous on $\bar{\Omega} - \{z_0\}$ with the boundary values zero along $\partial\Omega$. Similar argument will lead to the

Canonical mapping and the horizontal slit domain. Let Ω and z_0 be as in (6.6.3.3). There is a univalent meromorphic function $w = q(z)$ on Ω satisfying the following properties:

1. $q(z)$ has only one simple pole at z_0 with residue equal to 1;
2. $w = q(z)$ maps Ω univalently onto $\mathbf{C}^*$ with n horizontal segment slits.

Such a canonical mapping $w = q(z)$ is unique up to a constant.   (6.6.3.5)

Let $w = p(z)$ be as in (6.6.3.3) and $w = q(z)$ be as in (6.6.3.5).

Fix k, $1 \le k \le n$. Suppose the vertical segment slit $p(\gamma_k)$ lies on the line $\operatorname{Re} w = \lambda_k$, and the horizontal segment slit $q(\gamma_k)$ on the line $\operatorname{Im} w = \mu_k$. Set $w_k = \lambda_k + i\mu_k$. Fix a real θ, $0 \le \theta < 2\pi$. Then the equation of the line passing w_k and having the inclination θ to the positive real axis is given by (see (2) in (1.4.3.1))

$$\mathrm{Im}\frac{w - w_k}{e^{i\theta}} = 0$$

$$\Rightarrow \mathrm{Im}\, e^{-i\theta} w = \mathrm{Im}\, e^{-i\theta}(\lambda_k + i\mu_k) = \mu_k \cos\theta - \lambda_k \sin\theta,$$

which is the imaginary part of $q(z)\cos\theta - ip(z)\sin\theta$ on the contour γ_k. This observation helps to formulate the following

Canonical mapping and the parallel slit domain. Let Ω and z_0 be as in (6.6.3.3). For any fixed θ $(0 \le \theta < 2\pi)$, the univalent meromorphic function

$$f_\theta(z) = e^{i\theta}(q(z)\cos\theta - ip(z)\sin\theta), \quad z \in \Omega$$

has the following properties:

1. $f_\theta(z)$ has only one simple pole at z_0 with the residue equal to 1;
2. $w = f_\theta(z)$ maps Ω univalently onto $\mathbf{C}^*$ with n parallel segment slits whose inclinations to the positive real axis are equal to the same θ.

Such a canonical mapping is unique up to a constant.   (6.6.3.6)

For detailed account, refer to Ref. [6], pp. 175–185.

Exercises A

(1) In $(*_2)$, let $0 < h < \mathrm{dist}(z_0, \partial\Omega)$. Then $0 < h < |\zeta - z_0|$ for all $\zeta \in \partial\Omega$. Show that $\frac{1}{h}\mathrm{Log}\left(1 - \frac{h}{\zeta - z_0}\right) \to -\frac{1}{\zeta - z_0}$ uniformly on $\partial\Omega$ as $h \to 0$. Then try to use (2) in (1.7.3) to conclude that $Q(\zeta, h) \to -\mathrm{Re}\frac{1}{\zeta - z_0}$ uniformly on $\partial\Omega$ as $h \to 0$.
(2) Prove (6.6.3.5) in detail.
(3) Prove (6.6.3.6) in detail.

Exercises B

In this exercises, Ω always denotes an n-connected domain satisfying conditions (6.6.8), unless otherwise stated.

(1) Let $p(z)$ and $q(z)$ be as in (6.6.3.3) and (6.6.3.5), respectively. Show that $w = p(z) + q(z)$ maps Ω univalently onto a domain bounded by n convex contours γ_k, $1 \le k \le n$. Here, a closed curve is called *convex* if it intersects every straight line at most twice. Try the following steps: Set $f(z) = p(z) + q(z)$, $z \in \Omega$ and $p(z) = p_1(z) + ip_2(z)$, $q(z) = q_1(z) + iq_2(z)$.

(a) That $f(\gamma_k)$ is convex means that, for each θ, the function $\operatorname{Re}(p(z)+$ $q(z))e^{-i\theta} = (p_1+q_1)(z)\cos\theta+(p_2+q_2)(z)\sin\theta$ takes no value more than once on γ_k. But this function differs from $\operatorname{Re}(q(z)\cos\theta - ip(z)\sin\theta) = q_1(z)\cos\theta + p_2(z)\sin\theta$ only by a constant (note that, on γ_k, $p_1(z)\cos\theta + q_2(z)\sin\theta = \lambda_k\cos\theta + \mu_k\sin\theta$. Refer to Exercise B (1)(c) of Sec. 6.6.2). Try to derive the conclusion from the mapping properties of the function in (6.6.3.6).

(b) Use the argument principle to show that $\Delta_{\gamma_k}(f(z);w_0) = 0$ for any $w_0 \in f(\Omega)$ and $1 \le k \le n$. This implies, in particular, that the convex curves $f(\gamma_k)$, $1 \le k \le n$, lie outside of each other.

In Exercises (2)–(5) below, we try to use the extremal method (see (6.6.2)) to give (6.6.3.6) another proof.

(2) Let $\mathfrak{F}$ be the set of all univalent meromorphic functions of the form $z + \frac{a_1}{z} + \cdots$ in $0 < R < |z| < \infty$. Suppose θ is a fixed real number. Then there is a unique function

$$f_0(z) = z + \frac{R^2e^{2i\theta}}{z}$$

that maximizes $\operatorname{Re}(e^{-2i\theta}a_1)$, $f \in \mathfrak{F}$, which is equal to R^2. In this case, $w = f_0(z)$ maps $|z| > R$ univalently onto $\mathbf{C}^*$ with the segment slits connecting $2\operatorname{Re}^{i\theta}$ and $-2\operatorname{Re}^{i\theta}$. Try the following steps:

(a) Set $F(z) = \frac{1}{R}f(Rz) = z + \frac{a_1}{R^2z} + \cdots$, $f \in \mathfrak{F}$. On applying the area theorem (see Exercise B(2) of Sec. 4.8), $|a_1| \le R$ holds with equality if and only if $F(z) = z + \frac{e^{i\alpha}}{z}$ for some $\alpha \in \mathbf{R}$, which maps $|z| > 1$ univalently (see Sec. 2.5.5) onto $\mathbf{C}^* - [-2e^{i\alpha/2}, 2e^{i\alpha/2}]$.

(b) In case $\operatorname{Re}(e^{-2i\theta}a_1) = R^2$, it follows that $|a_1| \ge R^2$. By (a), $|a_1| = R^2 \Leftrightarrow f(z) = z + \frac{R^2e^{i\alpha}}{z}$ and $a_1 = R^2e^{i\alpha}$. Try to show that $\alpha = 2\theta$.

(3) Let $f(z) = z + a_0 + \frac{a_1}{z} + \cdots$ be univalent and meromorphic in $|z| > 1$. Then the complement of $f(1 < |z| < \infty)$ in $\mathbf{C}^*$ is contained in $|w-a_0| \le 2$; and $w = f(z)$ does not assume a point on $|w - a_0| = 2$ if and only if $f(z) = z + a_0 + \frac{e^{2i\alpha}}{z}$ for some real α.
Hint: Try to apply Exercise B(3) of Sec. 4.8 to $F(z) = \frac{1}{f(z^{-1})-w_0}$ if $f(z)$ does not take the value w_0 in $|z| > 1$.

(4) Let $f(z) = z + a_0 + \frac{a_1}{z} + \cdots$ be univalent and meromorphic in $|z| > R$. Show that

$$|f(z) - a_0| \le 2|z|, \quad |z| > R.$$

Hint: Fix any point ζ, $|\zeta| > R$, and apply Exercise (3) to $F(z) =$

$\frac{1}{\zeta}[f(\zeta z) - a_0] = z + \frac{a_1}{\zeta^2 z} + \cdots$ which is meromorphic in $|z| > 1 > \frac{R}{|\zeta|}$.
Note that $F(z)$ does not assume $F(1) = \frac{1}{\zeta}[f(\zeta) - a_0]$ in $|z| > 1$ because,
otherwise, $F(1) = F(z)$ for some z ($|z| > 1$), or equivalently, $f(\zeta z) = f(z)$, would contradict to the univalence.

(5) Let Ω be an n-connected domain in $\mathbf{C}^*$, containing the point ∞, and θ be a fixed real number. Then there is a unique univalent meromorphic function $w = f(z)$ satisfying the following properties:

1. $f(\infty) = \infty$ and $f(z)$ has the Laurent series expansion

$$f(z) = z + \frac{a_1}{z} + \cdots$$

in a neighborhood of ∞;

2. $w = f(z)$ maps Ω univalently onto $\mathbf{C}^*$ with n parallel segment slits, all having the inclination θ with the positive real axis.

In case $n = 1$, the Riemann mapping theorem (6.1.1) and the Joukowski function (see Sec. 2.5.5) together will realize the existence of such a mapping. For $n \geq 2$, try to follow the steps (and better recall how we proved the Riemann mapping theorem in Sec. 6.1.1):

(a) Denote by $\mathfrak{F}$ the family of univalent meromorphic functions in Ω with the Laurents expansions $z + \frac{a_1}{z} + \cdots$ at ∞. Note that the identity function $z \in \mathfrak{F}$. Consider the *extremal problem*

$$M = \max_{g \in \mathfrak{F}} \operatorname{Re}(e^{-2i\theta} a_1).$$

Show that $\mathfrak{F}$ is locally uniformly bounded in $\Omega - \{\infty\}$. To see this: For $R < |z| < K < \infty$, use Exercise (4) to show that $|g(z)| \leq 2R \leq 2K$ for $g \in \mathfrak{F}$; for $\Omega \cap \{|z| \leq R\}$, apply Exercise (3) to $\frac{1}{R}g(Rz)$ to obtain $|g(z)| \leq 2R < 2K$ for $g \in \mathfrak{F}$. By Exercise (2), note that $M \leq R^2$ for some $R > 0$.

(b) Using normality of $\mathfrak{F}$ (see (5.8.1.3) and (5.8.1.5)), there is a sequence $f_n \in \mathfrak{F} \to f$ locally uniformly in $\Omega - \{\infty\}$. In $|z| > R$, $f_n(z) = z + \frac{a_1^{(n)}}{z} + \cdots$, $n \geq 1$. It follows by coefficient integrals (4.9.1.1) or Weierstrass's double series theorem (5.3.1.2) that $f(z) = z + \frac{a_1}{z} + \cdots$. By Hurwitz's theorem (5.3.2.2), $f \in \mathfrak{F}$; moreover, $\operatorname{Re}(e^{-2i\theta} a_1^{(n)}) \to M = \operatorname{Re}(e^{-2i\theta} a_1)$ holds.

(c) Suppose one of the components of $\mathbf{C}^* - f(\Omega)$ is neither a line segment with inclination θ nor a single point. Let it be Γ. Let $\zeta = \zeta(w)$ be a univalent meromorphic function mapping the simply connected domain $\mathbf{C}^* - \Gamma$ onto $\mathbf{C}^* - L$, where L is a line segment whose inclination with the real axis is θ, and $\zeta(w) = w + \frac{b_1}{w} + \cdots$ in a neighborhood of ∞. Try to normalize $\zeta = \zeta(w)$ as follows: Let $w = w(\eta) = \eta + c_0 + \frac{c_1}{\eta} + \cdots$ be a univalent meromorphic function mapping $|\eta| > r$ onto $\mathbf{C}^* - \Gamma$. Then $\zeta = g(w(\eta)) - c_0 = \eta + \frac{b_1 + c_1}{\eta} + \cdots$ is the extremal function in Exercise (2); and hence, $\mathrm{Re}\,(e^{-2i\theta}c_1) < \mathrm{Re}\,(c^{-2i\theta}(b_1 + c_1))$ holds. It follows that $\mathrm{Re}\,(e^{-2i\theta}b_1) > 0$ is true. Finally, $\zeta - y(f(z)) = z + \frac{a_1 + b_1}{z} + \cdots$ is in $\mathfrak{F}$ and then, $\mathrm{Re}\,(e^{-2i\theta}(a_1 + b_1)) > \mathrm{Re}\,(e^{-2i\theta}a_1) = M$, a contradiction. This proves that $w = f(z)$ has the required properties.

(d) *Uniqueness.* Here, we may suppose Ω is as in (6.6.8). Let $w = f_1(z)$ and $w = f_2(z)$ be two such functions. They are continuous (even, analytic) up to $\bar{\Omega} - \{\infty\}$. Suppose

$$\mathrm{Im}\,(e^{-i\theta}f_1(\zeta)) = \alpha_k \quad \text{and} \quad \mathrm{Im}\,(e^{-i\theta}f_2(\zeta)) = \beta_k, \ \zeta \in \gamma_k \text{ for } 1 \le k \le n.$$

Therefore $F(z) = f_1(z) - f_2(z)$ is analytic in $\bar{\Omega}$, $F(\infty) = 0$ and $F(\gamma_k)$ is a line segment for each k $(1 \le k \le n)$. Use the argument principle to show that $w = F(z)$ does not assume, in Ω, values not in $\bigcup_{k=1}^{n} F(\gamma_k)$. Then the open mapping theorem says that $F(z)$ is a constant and $F(z) \equiv 0$ in Ω.

Note. The proofs for (a)–(c) show that the result is still valid for infinitely connected domain except the uniqueness requirement.

*6.6.4   Other canonical domains

Using the so-called *continuity method*, the domain Ω mentioned in (6.6.8) and the associated Fig. 6.44 can be improved to the following

Canonical domain bounded by n circles. Let Ω be an n-connected domain containing the point ∞. Then there exists a unique univalent meromorphic function $w = f(z)$ satisfying the following properties:

1. $f(\infty) = \infty$ and, in a neighborhood of ∞, $f(z) = z + \frac{a_1}{z} + \cdots$;
2. $w = f(z)$ maps Ω univalently onto a domain whose boundary consists of pairwise disjoint circles with one of them the unit circle $|w| = 1$.

(6.6.4.1)

See Fig. 6.47. By performing a bilinear transformation, we even can have the boundary circle $f(\gamma_n)$, the unit circle $|w| = 1$, and the other boundary circles $f(\gamma_k)$ are contained in $|w| < 1$ for $2 \le k \le n$. See Fig. 6.48 and compare with Fig. 6.44. For details, see Ref. [35], p. 237; in this book, one can also find generalizations of (6.6.3.6) and Exercise B(2) of Sec. 6.6.2:

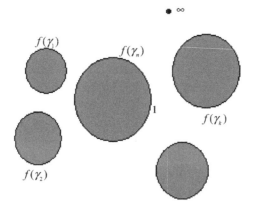

Fig. 6.47

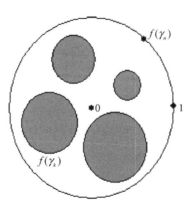

Fig. 6.48

in both cases, the same inclination angle θ is allowed to be chosen as the presumed different angles $\theta_1, \ldots, \theta_n$.

However, based on (6.6.8), we still have another elementary

Canonical domain obtained as a horizontal strip with $(n-1)$ parallel segment slits. Let Ω be a bounded n-connected domain with n boundary curves $\gamma_1, \ldots, \gamma_{n-1}, \gamma_n$, where γ_n is a Jordan closed curve. Then there exists a unique univalent analytic function $w = f(z)$ having the properties:

1. $w = f(z)$ maps three points $z_1, z_2, z_3 \in \gamma_n$, in the counterclockwise direction with respect to Ω, to the boundary points $+\infty$, $\frac{\pi}{2}i$, $-\infty$, respectively;

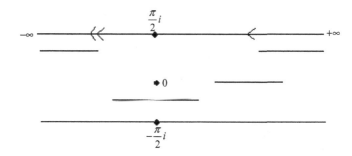

Fig. 6.49

2. $w = f(z)$ maps Ω univalently onto the horizontal strip $-\frac{\pi}{2} < \text{Im } z < \frac{\pi}{2}$
   with $(n-1)$ segment slits parallel to the real axis.   (6.6.4.2)

   Recall that the boundary lines $\text{Im} = \frac{\pi}{2}$ and $\text{Im} = -\frac{\pi}{2}$ both pass through
the point at infinity ∞ (see the words after (1.6.10)); and they together
should be viewed as the Jordan closed boundary curve $f(\gamma_n)$ in $\mathbf{C}^*$ (refer
to Fig. 1.43). See Fig. 6.49.

Sketch of proof. We may just assume that the original domain Ω is the
canonical domain obtained in (6.6.4.1), with γ_n the unit circle $|z| = 1$ and
all other boundary circles γ_k, $1 \le k \le n-1$, contained in $|z| < 1$ (refer to
Fig. 6.48 in which $f(\gamma_k)$ is renamed as γ_k for $1 \le k \le n$). Moreover, we
may assume that $z_1 = 1$, $z_2 = i$, and $z_3 = -1$. As a matter of fact, the
domain Ω in (6.6.8) will work, too.

   Designate by $\text{Log}\frac{1+z}{1-z}$ the single-valued analytic branch of $\log\frac{1+z}{1-z}$, uni-
quely determined by $\log 1 = 0$ in $|z| < 1$ (see Example 4 in Sec. 2.7.3). It
maps $|z| < 1$ univalently onto the strip $-\frac{\pi}{2} < \text{Im } w < \frac{\pi}{2}$ and the points 1,
i, -1 onto $+\infty$, $\frac{\pi}{2}i$, $-\infty$, respectively.

   Consider the Dirichlet problem with the boundary value conditions:

$$\text{Re } F(\zeta) = \begin{cases} 0, & \zeta \in \gamma_n \\ -\text{Im Log}\dfrac{1+\zeta}{1-\zeta} + \lambda_k, & \zeta \in \gamma_k \text{ for } 1 \le k \le n-1 \end{cases} ;$$

$$F(i) = 0,$$

where $\lambda_1, \ldots, \lambda_{n-1}$ are undetermined constants. According to (6.6.1.9) or
(6.6.1.13), there is a unique analytic function $F(z)$ in Ω, continuous on $\bar{\Omega}$,
that solves this problem.

Set

$$f(z) = \text{Log}\frac{1+z}{1-z} + iF(z), \quad z \in \Omega.$$

Then $f(z)$ is analytic in Ω, continuous on $\bar{\Omega}$ in the generalized sense (see (2.2.2)), such that

$$\text{Re } f(i) = -\text{Im } F(i) + \text{Re Log}\frac{1+i}{1-i} = 0;$$

$$\text{Im } f(\zeta) = \text{Re } F(\zeta) + \text{Im Log}\frac{1+\zeta}{1-\zeta}$$

$$= \begin{cases} \dfrac{\pi}{2}, & \zeta \in \gamma_{n1} = \{z | |z| = 1 \text{ and } 0 < \text{Arg } z < \pi\} \\[2mm] -\dfrac{\pi}{2}, & \zeta \in \gamma_{n2} = \{z | |z| = 1 \text{ and } -\pi < \text{Arg } z < 0\} \\[2mm] \lambda_k, & \zeta \in \gamma_k \text{ for } 1 \leq k \leq n-1 \end{cases} \cdot$$

As a nonconstant function, $w = f(z)$ maps Ω onto a domain $f(\Omega)$. And the above boundary value relations indicate that

$f(\gamma_{n1})$: the line $\text{Im } w = \frac{\pi}{2}$ with $f(1) = +\infty$, $f(i) = \frac{\pi}{2}i$, and $f(-1) = -\infty$;

$f(\gamma_{n2})$: the line $\text{Im } w = -\frac{\pi}{2}$, and

$f(\gamma_k)$: a line segment on the horizontal line $\text{Im } w = \lambda_k$ for $1 \leq k \leq n-1$.

To prove the *univalence* of $w = f(z)$ in Ω: Fix any point $w_0 \notin \bigcup_{k=1}^{n} f(\gamma_k)$ and $|\text{Im } w_0| > \frac{\pi}{2}$. Then

$$\Delta_{\partial\Omega} \arg(f(z) - w_0) = 0 \quad \text{(by the argument principle)}$$
$$\Rightarrow f(z) - w_0 \neq 0 \quad \text{throughout } \Omega.$$

Hence $f(\Omega)$ is a domain in $|\text{Im } w| < \frac{\pi}{2}$ and $f(\gamma_k)$ lies in $|\text{Im } w| \leq \frac{\pi}{2}$ for each k $(1 \leq k \leq n)$. In case $w_0 \notin \bigcup_{k=1}^{n} f(\gamma_k)$ and $|\text{Im } w_0| < \frac{\pi}{2}$, then

$$\Delta_{\partial\Omega} \arg(f(z) - w_0) = 2\pi \quad \text{(by the argument principle)}$$

$$\Rightarrow f(z) - w_0 \quad \text{has exactly only one simple zero in } \Omega.$$

This means that $w = f(z)$ is univalent in Ω. Moreover, $f(z)$ cannot assume, in Ω, values lying on $\bigcup_{k=1}^{n} f(\gamma_k)$ owing to this univalence; and the line segments $f(\gamma_k)$, $1 \leq k \leq n-1$, all lie in $|\text{Im}| < \frac{\pi}{2}$ (refer to the established argument in the proof of (6.6.2.1)).

As for *uniqueness*: Let $g(z)$ be another such function. As usual, set $G(z) = f(z) - g(z)$. Then $G(z)$ is analytic in Ω and satisfies the boundary conditions:

$$\text{Im}\, G(\zeta) = \alpha_k \quad \text{(a constant)}, \quad \zeta \in \gamma_k \text{ for } 1 \le k \le n \ (\zeta \ne \pm 1, \text{ if } k = n);$$

$$G(i) = 0$$

$$\Rightarrow G(z) \quad \text{cannot assume values outside} \bigcup_{k=1}^{n} G(\gamma_k)$$

(by the argument principle).

$$\Rightarrow G(z) = 0 \quad \text{in } \Omega \text{ (by the open mapping theorem)}.$$

Therefore, $f(z) \equiv g(z)$ in Ω. □

As a final result of this section (and this chapter, too), we have the following

Mapping of an n-connected domain onto an n-sheeted open strip. Let Ω be as in (6.6.8). Fix three distinct points $a_n, b_n, c_n \in \gamma_n$ (the outer contour) in the counterclockwise direction and fix a point $a_k \in \gamma_k$ for each k, $1 \le k \le n - 1$. Then there exists a unique analytic function $w = f(z)$ in Ω having the following properties:

1. $w = f(z)$ maps Ω univalently onto an *n-sheeted vertical strip* over $0 < \text{Re}\, w < 1$;
2. $f(a_n) = +i\infty$, $f(b_n) = -i\infty$, $f(c_n) = 1$, and $f(a_k) = +i\infty$ for $1 \le k \le n - 1$. \hfill (6.6.4.3)

By an *n-sheeted vertical strip* over $0 < \text{Re}\, w < 1$, we mean a Riemann surface R_w lying entirely over the strip $0 < \text{Re}\, w < 1$ in $\mathbf{C}^*$ and covering each point of it exactly n times (counting multiplicity at a branch point); and, in this case, $w = f(z)$ maps Ω univalently onto R_w. In other words, the mapping $w = f(z)$ realizes the correspondence between Ω and the strip $0 < \text{Re}\, w < 1$ in n-to-one manner; namely, $f(z) - w_0$ has exactly n zeros in Ω for any w_0 in $0 < \text{Re}\, w < 1$ (with a zero of order m counted m times) and has no zero at all in Ω for any w_0 in $\text{Re}\, w < 0$ or $\text{Re}\, w > 1$. Meanwhile the points $1 + i\infty$, $1 - i\infty$, and 1 lie on the boundary line $\text{Re}\, w = 1$ of the ordinary strip $0 < \text{Re}\, w < 1$. See Fig. 6.50.

Sketch of proof (due to Grunsky, 1937). It is harmless to assume that Ω is the canonical domain in Fig. 6.48: the outer contour is the unit circle $|z| = 1$ with $a_n = 1$, $b_n = -1$, and $c_n = -i$, for simplicity; and the inner

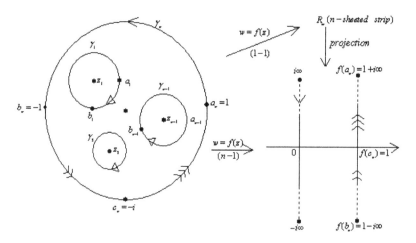

Fig. 6.50

contour γ_k, for each k $(1 \leq k \leq n-1)$, is the whole circle $|z - z_k| = r_k$ with $a_k = z_k + r_k$.

Choose another point $b_k \in \gamma_k$, $b_k \neq a_k$, for $1 \leq k \leq n-1$. Denote by $\widehat{a_k b_k} = \gamma_k'$ the open arc from a_k to b_k along the circle γ_k in the clockwise direction (namely, the positive direction with respect to Ω) and $\widehat{b_k a_k} = \gamma_k''$ the open arc from b_k to a_k in the same direction (see Fig. 6.50), $1 \leq k \leq n-1$. In case $k = n$, the arcs $\widehat{a_n b_n} = \gamma_n'$ and $\widehat{b_n a_n} = \gamma_n''$ are both in the counterclockwise direction. As a consequence, if γ_n is represented by $z = z_n(\theta) = e^{i\theta}$, $0 \leq \theta \leq 2\pi$, and γ_k by $z = z_k(\theta) = z_k + r_k e^{i\theta}$, $0 \leq \theta \leq 2\pi$, then $a_n = z_1(0)$, $b_n = z_1(\pi)$ while $a_k = z_k(2\pi)$ and $b_k = z_k(\theta_k)$ for $1 \leq k \leq n-1$.

According to (6.3.2.8), there is a harmonic function $u_n(z)$ in $|z| < 1$ with the boundary values:

$$u_n(\zeta) = \begin{cases} 0, & \zeta \in \gamma_n' \\ 1, & \zeta \in \gamma_n'' \end{cases}.$$

Indeed, this $u_n(z)$ can be evaluated explicitly by (refer to Exercise A(8) of Sec. 6.3.2)

$$P_{u_n}(z) = u_n(z)$$

$$= \text{Re} \left\{ \frac{1}{2\pi i} \int_{\gamma_n} u_n(\zeta) \frac{\zeta + z}{\zeta - z} \cdot \frac{d\zeta}{\zeta} \right\}$$

$$= \operatorname{Re}\left\{\frac{1}{2\pi i}\int_{\gamma_n''}\left(\frac{2}{\zeta-z}-\frac{1}{\zeta}\right)d\zeta\right\} = \operatorname{Re}\left\{\frac{1}{\pi i}\left[\operatorname{Log}\frac{z-1}{z+1}-\frac{\pi}{2}i\right]\right\}$$

$$= \frac{1}{\pi}\operatorname{Arg}\frac{z-1}{z+1}-\frac{1}{2}, \tag{$*_1$}$$

where $\operatorname{Arg}(-1) = \pi$. Similarly, the harmonic function $u_k(z)$ in $|z - z_k| > r_k$ with the boundary values

$$u_k(\zeta) = \begin{cases} 0, & \zeta \in \gamma_k' \\ 1, & \zeta \in \gamma_k'' \end{cases}$$

is given by

$$P_{u_k}(z) = u_k(z)$$

$$= \operatorname{Re}\left\{\frac{1}{2\pi i}\int_{\gamma_k}u_k(\zeta)\frac{\zeta+z-2z_k}{\zeta-z}\cdot\frac{d\zeta}{\zeta-z_k}\right\} \quad \text{(in clockwise direction)}$$

$$= \operatorname{Re}\left\{\frac{1}{2\pi i}\int_{\gamma_k''}\left(\frac{2}{\zeta-z}-\frac{1}{\zeta-z_k}\right)d\zeta\right\}$$

$$= \operatorname{Re}\left\{\frac{1}{\pi i}\left[\operatorname{Log}\frac{z-z_k-r_k}{z-z_k-r_ke^{i\theta_k}}+\frac{\theta_k i}{2}\right]\right\}$$

$$= \frac{1}{\pi}\operatorname{Arg}\frac{z-z_k-r_k}{z-z_k-r_ke^{i\theta_k}}+\frac{\theta_k}{2\pi}, \quad 1 \le k \le n-1. \tag{$*_2$}$$

Finally, according to (6.5.7), there is a *unique* harmonic function $u_0(z)$ in Ω, satisfying the boundary conditions:

$$-\sum_{\substack{k=1\\k\ne j}}^{n}u_k(\zeta), \quad \zeta \in \gamma_j \quad \text{for } 1 \le j \le n. \tag{$*_3$}$$

Then the function

$$u(z) = \sum_{k=0}^{n}u_k(z), \quad z \in \Omega \tag{$*_4$}$$

is the unique harmonic function in Ω, satisfying the boundary conditions:

$$u(\zeta) = \begin{cases} 0, & \zeta \in \displaystyle\bigcup_{k=1}^{n}\gamma_k' \\[3mm] 1, & \zeta \in \displaystyle\bigcup_{k=1}^{n}\gamma_k'' \end{cases}.$$

As we have experienced so much in the previous cases, what remains is to make sure if it is possible for $u(z)$ to possess a single-valued conjugate harmonic function $v(z)$ in Ω. And this will lead to the existence of the required mapping $w = f(z) = u(z) + iv(z)$. That possibility comes true depends mainly on how suitable points $b_k \in \gamma_k$, $1 \le k \le n-1$, can be chosen. More precisely, we try to choose $n-1$ real numbers $\theta_1, \ldots, \theta_{n-1} \in [0, 2\pi]$ and hence the points $b_k = z(\theta_k) \in \gamma_k$, $1 \le k \le n-1$. Let $u(z) = u(z, \theta_1, \ldots, \theta_{n-1})$ be the associated harmonic function obtained in $(*_4)$ in order to have its periods along γ_k (recall that γ_k is positively oriented with respect to Ω)

$$\omega_k = \omega_k(\theta_1, \ldots, \theta_{n-1}) = \int_{\gamma_k} {}^* du(z) = \int_{\gamma_k} u_x(z)dy - u_y(z)dx, \quad 1 \le k \le n$$

$$(*_5)$$

all equal to zero. Then (see (6.6.8), (6.6.9), and (2) in (6.3.1.3)) $u(z)$ will have a single-valued harmonic conjugate $v(z)$ in Ω, unique up to an additive constant.

Before we start, one needs the following properties (1)–(4) for ω_k:

(1) $\omega_k(\theta_1, \ldots, \theta_{n-1})$ is a continuous function of each of the variables $\theta_1, \ldots, \theta_{n-1} \in [0, 2\pi]$, for $1 \le k \le n$. To see this, fix any point $z \in \Omega$ and consider Green's function $g(\zeta, z)$ of Ω with pole at z. For each k $(1 \le k \le n)$, according to the symmetry principle (6.3.2.13) or (3.4.6.4), $g(\zeta, z)$ can be harmonically extended across the arc γ_k'' (discarding the end points b_k and a_k); hence, let $h_k(\zeta, z)$ be a local single-valued harmonic conjugate of $g(\zeta, z)$ near γ_k''. By using (2) in (6.3.1.14),

$$u(z) = \frac{-1}{2\pi} \sum_{k=1}^{n} \int_{\gamma_k''} u(\zeta)^* dg(\zeta; z) = \frac{-1}{2\pi} \sum_{k=1}^{n} \int_{\gamma_k''} dh_k(\zeta, z)$$

$$= \frac{1}{2\pi} \sum_{k=1}^{n} (h_k(b_k, z) - h_k(a_k, z)).$$

This shows that $u(z) = u(z, \theta_1, \ldots, \theta_{n-1})$, $\frac{\partial u}{\partial x}$ and $\frac{\partial u}{\partial y}$ are continuous in $z \in \Omega$ and $\theta_1, \ldots, \theta_{n-1} \in [0, 2\pi]$; in turn, by $(*_5)$, the claim follows.

(2) $\omega_k(\theta_1, \ldots, \theta_{n-1})$ is a decreasing function of $\theta_j \in [0, 2\pi]$, where $j \ne k$ and $1 \le k, j \le n-1$. To prove this, choose $0 \le \theta_j < \theta_j' \le 2\pi$ and set

$$\Delta_{k,j} = \omega_k(\theta_1, \ldots, \theta_j', \ldots, \theta_{n-1}) - \omega_k(\theta_1, \ldots, \theta_j, \ldots, \theta_{n-1}).$$

Let $v(z)$ be the harmonic function in Ω with the boundary values

$$v(\zeta) = \begin{cases} 0, & \sigma_j \\ 1, & \gamma_j - \sigma_j \end{cases},$$

where $\sigma_j = \widehat{b_j b_j'}$ is the arc along γ_j from $b_j = z(\theta_j)$ to $b_j' = z(\theta_j')$ in the positive direction with respect to Ω. Then $\Delta_{k,j} = \int_{\gamma_j} {}^* dv \ (= \int_{\gamma_j - \sigma_j} {}^* dv)$. Since $v(z) > 0$ in Ω and $\frac{\partial v}{\partial n} < 0$ (see $(*_1)$ in Sec. 6.3.1 and (6.3.1.4); and refer to $(*_6)$ in Sec. 6.1.2), it follows that

$$\Delta_{k,j} = \int_{\gamma_j} {}^* dv = \int_{\gamma_j} \frac{\partial v}{\partial n} |d\zeta| < 0.$$

So is the claim.

(3) $\omega_k(\theta_1, \ldots, \theta_{n-1})$ is an increasing function of $\theta_k \in [0, 2\pi]$, for $1 \le k \le n - 1$. Since $\gamma_1, \ldots, \gamma_{n-1}$ form a homology basis for Ω and γ_n is homologous to $-\sum_{k=1}^{n-1} \gamma_k$ in Ω, it follows that

$$\int_{\gamma_n} {}^* du = -\sum_{k=1}^{n-1} \int_{\gamma_k} {}^* du \quad \text{(see (6.3.1.3))}$$

$$\Rightarrow \sum_{k=1}^{n} \omega_k(\theta_1, \ldots, \theta_{n-1}) = 0. \tag{$*_6$}$$

By property (2), the result follows.

(4) $\omega_k(\theta_1, \ldots, \theta_k, \ldots, \theta_{n-1})$ is negative at $\theta_k = 0$ and positive at $\theta_k = 2\pi$ if the other θ_j ($j \ne k$ and $1 \le j \le n - 1$) is neither 0 nor 2π. To show this, recall that $b_k = z_k(\theta_k)$, $0 \le \theta \le 2\pi$, and observe that, if $\theta_k = 0$ (then $\gamma_k' = \gamma_k$), then $u(\zeta) = 0$ on γ_k but $u(z) > 0$ in Ω; if $\theta_k = 2\pi$ (then $\gamma_k'' = \gamma_k$), $u(\zeta) = 1$ on γ_k but $u(z) < 1$ in Ω. Therefore

$$\frac{\partial u}{\partial n} \begin{cases} < 0, & \text{if } \theta_k = 0 \\ > 0, & \text{if } \theta_k = 2\pi \end{cases}.$$

$$\Rightarrow \omega_k(\theta_1, \ldots, \theta_k, \ldots, \theta_{n-1}) = \int_{\gamma_k} \frac{\partial u}{\partial n} |d\zeta| \begin{cases} < 0, & \text{if } \theta_k = 0 \\ > 0, & \text{if } \theta_k = 2\pi \end{cases}.$$

After these preparatory works, the *claim* concerned $(*_5)$ is: *The equations*

$$\omega_k(\theta_1, \ldots, \theta_{n-1}) = 0, \quad 1 \le k \le n \tag{$*_7$}$$

have a (unique) solution.

To start with, observe that, for a given set of values $\theta_1, \theta_2, \ldots, \theta_{n-1}$, properties (1) and (4) show that there is a unique $\theta_1 = t_1(\theta_2, \ldots, \theta_{n-1}) \in (0, 2\pi)$ so that $\omega_1(t_1, \theta_2, \ldots, \theta_{n-1}) = 0$ holds. From properties (1), (2), and (3), the identity $\omega_1(t_1(\theta_2, \ldots, \theta_{n-1}), \theta_2, \ldots, \theta_{n-1}) \equiv 0$ implicitly defines $t_1(\theta_2, \ldots, \theta_{n-1})$ as a decreasing function in each of θ_k $(2 \le k \le n-1)$ and as a continuous function of $\theta_2, \ldots, \theta_{n-1} \in [0, 2\pi]$.

Consider the functions

$$\omega'_k(\theta_2, \ldots, \theta_{n-1}) = \omega_k(t_1(\theta_2, \ldots, \theta_{n-1}), \theta_2, \ldots, \theta_{n-1}), \quad 2 \le k \le n-1.$$

$$(*_8)$$

They are continuous in $\theta_2, \ldots, \theta_{n-1} \in [0, 2\pi]$. ω'_k is decreasing in θ_j $(j \ne k, \ 2 \le j \le n-1)$, by properties (2); and, by $(*_6)$ that $\sum_{k=1}^{n} \omega_k(\theta_1, \ldots, \theta_{n-1}) = \sum_{k=2}^{n} \omega'_k(\theta_2, \ldots, \theta_{n-1}) = 0$, ω'_k is increasing in θ_k $(2 \le k \le n-1)$; and, by property (4), ω'_k is negative at $\theta_k = 0$ and positive at $\theta_k = 2\pi$. Reasoning similarly to $(*_8)$, there is a unique $\theta_2 = t_2(\theta_3, \ldots, \theta_{n-1}) \in [0, 2\pi]$ so that $\omega'_2(t_2, \theta_3, \ldots, \theta_{n-1}) = 0$ holds, and $t_2(\theta_3, \ldots, \theta_{n-1})$ is decreasing in each of θ_k $(3 \le k \le n-1)$ and is continuous in $\theta_3, \ldots, \theta_{n-1} \in [0, 2\pi]$. Continue this process $(n-1)$ times and we get a unique system of values

$$\theta_1 = t_1(\theta_2, \ldots, \theta_{n-1}), \quad \theta_2 = t_2(\theta_3, \ldots, \theta_{n-1}), \ldots,$$

$$\theta_{n-2} = t_{n-2}(\theta_{n-1}), \quad \theta_{n-1} = t_{n-1},$$

all lie in $(0, 2\pi)$ and satisfy $(*_7)$. This satisfies the claim.

Let us go back to $(*_5)$ after confirming the validity of the claim in $(*_7)$.

The *existence* of such a $f(z) = u(z) + iv(z)$: Choose a single-valued harmonic conjugate $v(z)$ of $u(z)$ in Ω. Set $w = f(z) = u(z) + iv(z)$, $z \in \Omega$.

To see if this $f(z)$ assumes each value in $0 < \text{Re}\, w < 1$ exactly n times, let us start with the investigation of the local behaviors of $f(z)$ at a_k and b_k. Map the simply connected domain $\mathbf{C}^* - \overline{\text{Int}}\, \gamma_k$ univalently onto the upper half-plane $\text{Im}\, \zeta > 0$, say by $\zeta = \varphi(z)$, so that $\varphi(z)$ is analytic up to γ_k and $\varphi(a_k) = 0$ and $\varphi(b_k) = \infty$. This can be done by a bilinear fractional transformation if γ_k is the circle $|z - z_k| = r_k$, or by a univalent function provided by the Riemann mapping theorem (6.1.1) and the extension theorem (6.1.2.3) if γ_k is just an analytic Jordan closed curve in Fig. 6.44. The inverse function $z = \varphi^{-1}(\zeta)$ is analytic at $\zeta = 0$ and can be expanded into the Taylor series $a_k + \alpha_1 \zeta + \cdots$ with $\alpha_1 \ne 0$. According to the harmonic invariance under analytic function (see (3) in (3.4.3.5)), $u(\varphi^{-1}(\zeta))$ is harmonic in a domain of the ζ-plane that contains the real axis as part of its boundary and

$$u(\varphi^{-1}(\zeta)) = \begin{cases} 0, & \text{if } \zeta \text{ lies on the positive real axis } (0,+\infty) \\ 1, & \text{if } \zeta \text{ lies on the negative real axis } (-\infty,0) \end{cases}.$$

Recall that $\frac{1}{\pi}\text{Arg}\,\zeta$ has this same property (see (6.3.2.8) and (6.3.2.11)). In accordance with the symmetry principle (3.4.6.4) or (6.3.2.13), $u(\varphi^{-1}(\zeta)) - \frac{1}{\pi}\text{Arg}\,\zeta$ can be harmonically extended across $\mathbf{R} - \{0\} = (-\infty,0) \cup (0,+\infty)$. In other words, $f(\varphi^{-1}(\zeta)) - \frac{1}{\pi i}\text{Log}\,\zeta$ can be analytically continued across $\mathbf{R} - \{0\}$. In conclusion,

$$f(z) - \frac{1}{\pi i}\text{Log}(z - a_k) \text{ is analytic in a neighborhood of } a_k.$$

$\Rightarrow$ Its imaginary part $v(z) + \dfrac{1}{\pi}\text{Log}|z - a_k|$ is harmonic in a

neighborhood of a_k.

$\Rightarrow \quad \lim_{z \in \Omega \to a_k} v(z) = +\infty$ uniformly in the manner as

$-\dfrac{1}{\pi}\text{Log}|z - a_k|$ does. $(*_9)$

Similar argument leads to the following fact:

$$f(z) + \frac{1}{\pi i}\log(z - b_k) \text{ is analytic in a neighborhood of } b_k.$$

$\Rightarrow \quad \lim_{z \in \Omega \to b_k} v(z) = -\infty$ uniformly in the manner as

$\dfrac{1}{\pi}\log|z - b_k|$ does. $(*_{10})$

$(*_9)$ and $(*_{10})$ are valid for $1 \le k \le n$.

Fix any point w_0 in $0 < \text{Re}\,w < 1$. By the argument principle (3.5.3.4), the number of zeros of $f(z) - w_0$ in Ω

$$= \frac{1}{2\pi i}\int_{f(\partial\Omega)} \frac{dw}{w - w_0} = \frac{1}{2\pi i}\int_{\partial\Omega} \frac{f'(z)dz}{f(z) - w_0} = \sum_{k=1}^{n} \frac{1}{2\pi i}\int_{\gamma_k} \frac{f'(z)dz}{f(z) - w_0}$$

$$= \frac{1}{2\pi}\sum_{k=1}^{n} \Delta_{f(\gamma_k)}\arg[f(z) - w_0] \tag{$*_{11}$}$$

= the number of times the curve $f(\partial\Omega)$ surrounding the point w_0.

The answer to $(*_{11})$ is the integer n. Geometrically speaking, this means that the curve $f(\partial\Omega)$ winds around w_0 exactly n times; more precisely, each $f(\gamma_k)$, $1 \le k \le n$, surrounds w_0 only once. Owing to $(*_9)$ and $(*_{10})$,

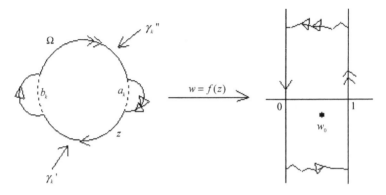

Fig. 6.51

$f(z) \neq w_0$ holds in sufficiently small neighborhoods of a_k and b_k; and, moreover, when z moves around γ_k in the positive direction with respect to Ω, the path is dented to small arcs contained in Ω in neighborhoods of a_k and b_k as shown in Fig. 6.51. The very definition of $w = f(z)$ and its properties (see, in particular, $(*_4)$, $(*_9)$, and $(*_{10})$) show that, as indicated graphically in Fig. 6.51,

$$\Delta_{f(\gamma_k)} \arg[f(z) - w_0] = 2\pi, \quad 1 \le k \le n.$$

This justifies the claim.

It is routine to check that $\Delta_{f(\gamma_k)} \arg[f(z) - w_0] = 0$, $1 \le k \le n$, if $\operatorname{Re} w_0 < 0$ or $\operatorname{Re} w_0 > 1$. Hence $f(z)$ does not assume values outside $0 \le \operatorname{Re} w \le 1$.

The *uniqueness*: Recall that $u(z)$ in $(*_4)$ is already uniquely defined. On the other hand, the harmonic conjugate $v(z)$ of $u(z)$ in Ω is uniquely defined by assigning $v(-i) = 0$. Then $f(a_n) = f(1) = +i\infty$, $f(b_n) = f(-1) = -i\infty$, and $f(c_n) = f(-i) = u(-i) = 1$.

This finishes the proof.   □

By performing the following successive elementary mappings

$$z \to \zeta = \pi i(2z - 1) \to \eta = e^{\zeta} \to \tau = \sqrt{\eta} \to w = \frac{\tau + 1}{\tau - 1},$$

we are able to transform (6.6.4.3) into the

Mapping of an n-connected domain onto an n-sheeted open disk. Let Ω, a_n, b_n, $c_n \in \gamma_n$ and $a_k \in \gamma_k$, $1 \le k \le n - 1$, be as in (6.6.4.3). Then, there exists a unique analytic function $w = f(z)$ in Ω, having the following properties:

1. $w = f(z)$ maps Ω univalently onto an n-sheeted disk over $|w| < 1$;
2. $f(a_n) = 1$, $f(b_n) = -1$, $f(c_n) = -i$, and $f(a_k) = 1$ for $1 \le k \le n - 1$.

$$(6.6.4.4)$$

Ahlfors, in 1947, proved this result via the extremal method to which Ref. [35], p. 468, is referred.

CHAPTER 7

Riemann Surfaces (Abstract)

Introduction

The descriptive Riemann surfaces, as we saw in Sec. 2.7, originated as a mean of dealing with the problem of explicit multiply valued functions. It was the idea of Riemann to replace the domain (in $\mathbf{C}$) of the function by a many-sheeted covering of it, with the number of sheets as many as the number of values the function assumed at a point. The meaning of the term Riemann surface is then used two-fold. One is that, on it, the original multiply valued function turns out to be single-valued and they will eventually lead to the notion of *abstract Riemann surface*, just forgetting the fact that the surface is "spread over" the complex plane. The other is that, when observing the relative positions of the surface so constructed and the classical complex plane, this leads to the topological notion of *covering surface*. Recall Definition 2 (see (2.7.2.8)) for a concise description of abstract Riemann surface and Definition 3 (see (2.7.2.10)) for that of covering surface, and read again Examples 1 and 2 in Sec. 3.3.3 to consolidate these notions.

A general procedure to obtain multiply valued functions is by analytically continuing a given (analytic) function element along different curves (or paths) which, in general, leads to different branches of that function. See Secs. 3.3.3 and 5.2 for details. The many-sheeted covering of the complex plane so constructed that it has as many points lying over any given point in the plane as there are function elements at that point is the Riemann surface, in the classic literature, on which the analytic function becomes single-valued. For a complete description in this direction, read Chap. 3 in Ref. [75].

The purpose of this ending chapter is going to give a formal, rigorous yet concise introduction to abstract Riemann surfaces.

For convenience, we collect the following concepts in general topology: Let X be a topological space (refer to Exercise B(2) of Sec. 1.8).

1. X is *Hausdorff* if any two distinct points in X have disjoint open neighborhoods each (see Exercise B(2) of Sec. 1.8).
2. X is *connected* if X cannot be expressed as the union of two disjoint nonempty open (or closed) subsets (refer to (1.8.10)).
3. X is *path-connected* or *arcwise connected* if for any two distinct points a and b in X, there is a continuous curve (see 6 below), also called *path*, $\gamma : [0, 1] \to X$ so that $\gamma(0) = a$ and $\gamma(1) = b$ (refer to (2) in (1.8.11)).
4. X is *locally path-connected* if each open neighborhood of each point in X contains another open neighborhood which is path-connected.
5. A connected subset (with the induce topology) of $\mathbf{X}$ is called a *component* if it is not contained in any other connected subset of X (see Sec. 1.8). Path-connected component is similarly defined.
   Let X and Y be topological spaces, and $f : X \to Y$ be a function.
6. f is said to be *continuous* if $f^{-1}(O)$ is open in X for each open set O in Y (refer to (2.2.3)).
7. f is a *homeomorphism* between X and Y if f is one-to-one, onto, and both f and f^{-1} are continuous (see Sec. 2.2).

Basic point-set terminologies, such as closed set, interior, compact set, etc., are defined similarly as in Secs. 1.8 and 1.9. We will feel free to use them if needed in the sequel. Refer to Ref. [27] for details.

Sketch of the Content

Section 7.1 rephrases Definitions (2.7.2.7) and (2.7.2.8) for two-dimensional topological manifolds, surfaces, and abstract Riemann surfaces. In addition to these descriptive examples of Riemann surfaces of multiple-valued functions, this section also proves that the infinite cylinder and the torus are Riemann surfaces in abstract sense, and explains how to realize a smooth surface in $\mathbf{R}^3$ as a Riemann surface.

These properties of analytic functions, with strong topological characters, can be easily extended to analytic mappings or meromorphic functions between Riemann surfaces. *Local representations* of analytic mappings and meromorphic functions at *regular* and *branched points* are studied in Sec. 7.2. The concept of *conformal equivalence* between Riemann surfaces is formally introduced. Then Sec. 7.2 tries to use these basic concepts to *unify* so many individual descriptive Riemann surfaces scattered from Sec. 2.7 on up to the end of Chap. 6. And Sec. 7.2 tries to figure out which one of them can be conformally equivalent to the sphere $\mathbf{C}^*$ (elliptic type), the finite or punctured plane (parabolic type) or the open disk (hyperbolic type). The most simplest compact Riemann surfaces (hyperelliptic type) are emphasized, too.

Perron's method (refer to Sec. 6.5) is adopted to construct *harmonic measure* and *Green's function* on a hyperbolic Riemann surface in Sec. 7.3. Besides, the method of regular exhaustion is also sketched. The validity of the *maximum principle* on a nonhyperbolic surface enables us to construct a particular *harmonic function* with a prescribed simple pole. These functions are essential to the proof of the uniformization theorem in Sec. 7.6.

Contents of Secs. 7.4 and 7.5 are somewhat deviated from pure complex analyses and can be studied in the realm of surface (or algebraic) topology. They are supplementary to our main purpose in Sec. 7.6. The concept of the *fundamental group* (Sec. 7.4) is used as an *algebraic system* to characterize a covering surface (rephrasing (2.7.2.9)) and the covering transformations on it (Sec. 7.5). What is important to Sec. 7.6 is the concept of the universal covering surface of a surface, detailed in Sec. 7.5.5.

Finally, Sec. 7.6 highlights the proof of the *uniformization theorem* of Riemann surfaces, formally stated in (7.6.1).

7.1   Riemann Surface: Definition and Examples

We repeat Definitions 1 and 2 in (2.7.2.7) and (2.7.2.8) for easier reference.

Definition. A connected Hausdorff space R together with a collection of *charts* $\{O_\lambda, \varphi_\lambda\}_{\lambda \in \Lambda}$ is called a *Riemann surface* if the following conditions are satisfied:

1. $\{O_\lambda\}_{\lambda \in \Lambda}$ forms an open covering of R, and each φ_λ maps O_λ homeomorphically onto an open set in the complex plane $\mathbf{C}$.
2. If $O_{\lambda_1} \cap O_{\lambda_2} \neq \phi$, the composite map

$$\varphi_{\lambda_2} \circ \varphi_{\lambda_1}^{-1} : \varphi_{\lambda_1}(O_{\lambda_1} \cap O_{\lambda_2}) \to \varphi_{\lambda_2}(O_{\lambda_1} \cap O_{\lambda_2})$$

   is (univalently) analytic.   (7.1.1)

   See Fig. 7.1.

Remarks. (1) R, together with $\{O_\lambda, \varphi_\lambda\}_{\lambda \in \Lambda}$, satisfying only condition 1 is termed as a *topological manifold*. The topology of R is completely determined by the homeomorphisms φ_λ which carry local topology properties in $\mathbf{C}$ to these on R. For instance, R is then locally compact and locally

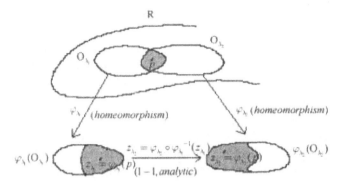

Fig. 7.1

path-connected. The connectedness of R is a separate requirement which, together with local path-connectedness, turns R into a path-connected space.

In short, *locally* a topological manifold is nothing but an open set in the complex plane.

(2) It is the condition 2 that makes a topological manifold a Riemann surface R. In this case, the collection $\{O_\lambda, \varphi_\lambda\}_{\lambda \in \Lambda}$ is said to define a *complex* or *conformal structure* on R.

A point p in O_λ is uniquely determined by the complex number $z_\lambda = \varphi_\lambda(p)$. As a consequence, a chart $\{O_\lambda, \varphi_\lambda\}$ is usually called a (local) *coordinate system* on R, where O_λ is a (local) *coordinate neighborhood* of any point in it and the associated φ_λ is referred to as a *local coordinate (variable)* or *local uniformizing parameter*. While the composite mapping $\varphi_{\lambda_2} \circ \varphi_{\lambda_1}^{-1}$ is called a *local coordinate* or *parameter transformation* on $O_{\lambda_1} \cap O_{\lambda_2}$ if this set is not empty.

Complex structure is tacitly understood to be *maximal* in the following sense. If O is an open set in R and φ is a homeomorphism mapping O onto an open set in $\mathbf{C}$ so that $\varphi \circ \varphi_\lambda^{-1}$ is always analytic whenever $O \cap O_\lambda \neq \phi$, then $\{O, \varphi\}$ is also considered as a chart in the given collection. Therefore, a chart $\{O, \varphi\}$ with a given point p_0 in O can, for convenience, be always so chosen that $\varphi(O)$ is an open disk with center at $\varphi(p_0)$.

Moreover, *the complex structure ensures that the topology of a Riemann surface has a countable base.* For a proof, see Ref. [6], pp. 142–144.

Some simplifications:

We just speak of a Riemann surface R without mentioning its complex structure if it is understood which complex structure is referred to.

The subscripts in $\{O_\lambda, \varphi_\lambda\}$ are dropped in most cases if no confusion might arise.

A point p on the Riemann surface R is quite occasionally identified with its value $z = \varphi(p)$ under the action of a local parameter φ. Any given point in R is contained in many different coordinate neighborhoods (namely, charts) and no one of these should be distinguished from the other. For this reason, only those notions from complex analysis in the complex plane that remain invariant under univalent analytic (simply called *conformal*) mappings can be carried over to Riemann surfaces. Typical ones are the notions of analytic functions (see (4) in (3.1.6)), harmonic functions (see (2) in (3.4.3.5)), residues (see (4.11.1.4)), and subharmonic functions (see (5) in (6.4.3)), etc. Therefore the identification of p and $z(p)$ as we did in so many previous concrete Riemann surfaces causes harmlessly. As a consequence, the set $|z - z_0| < \rho$ can be treated as a disk in $\mathbf{C}$ or as its inverse image on R and one can talk about harmonic functions defined on it.

(3) In condition 2, if $\varphi_{\lambda_2} \circ \varphi_{\lambda_1}^{-1}$ is only required to be $\mathbf{C}^k (1 \leq k \leq \infty)$ differentiable in the real sense or real analytic, the resulted R is called a *two-dimensional $\mathbf{C}^k$-manifold* or a *$\mathbf{C}^\omega$-manifold*, simply called a *$\mathbf{C}^k$-surface* or *$\mathbf{C}^\omega$-surface*, respectively. Every Riemann surface is a *$\mathbf{C}^\omega$-surface*, and, of cause, a *$\mathbf{C}^k$-surface*, for $1 \leq k \leq \infty$. Sometimes, a two-dimensional topological manifold is also termed as a *$\mathbf{C}^0$-surface* or simply a *surface*. $\square$

Examples of Riemann surfaces

The complex plane C. Let id: $\mathbf{C} \to \mathbf{C}$ be the identity map. The complex structure $\{\mathbf{C}, \mathrm{id}\}$ defines $\mathbf{C}$ as a Riemann surface.

Domains in a Riemann surface. A nonempty connected open subset Ω, called a domain as usual, is a Riemann surface with the induced complex structure, namely, the collection of charts $\{(O, \varphi)\}$, where $O \subseteq \Omega$.

The Riemann surfaces of these multiply valued analytic functions mentioned in Sec. 2.7. Read Sec. 3.3.3 for formal discussions and Chap. 3 in Ref. [75] for thorough treatment.

We formulate three other basic examples in the following.

Example 1. *The Riemann sphere S or the extended complex plane $\mathbf{C}^*$* (see Sec. 1.6 and (1.8.13)).

Explanation. We choose charts $\{O_k, z_k\}_{k=1,2}$ as follows:

$$O_1 = S - \{(0,0,1)\} \quad \text{and} \quad z_1 = \frac{x_1 + ix_2}{1 - x_3}, \text{ where } (x_1, x_2, x_3) \in S;$$

$$O_2 = S - \{(0,0,-1)\} \quad \text{and} \quad z_2 = \frac{x_1 - ix_2}{1 + r_3}.$$

Recall that, as the stereographic projection, $z_1 : O_1 \to \mathbf{C}$ is not only homeomorphic but also conformal (see Exercise B(1) of Sec. 1.6). So is $z_2 : O_2 \to \mathbf{C}$. Observe that $z_2 = \frac{1}{z_1}$ on the connected set $O_1 \cap O_2$. Hence these two charts form the complex structure for S being a Riemann surface. S is compact as a space by itself.

Example 2. The *infinite cylinder* $S^1 \times \mathbf{R}^1$ or the *punctured plane* (the plane $\mathbf{C}$ with a single point removed).

Explanation. Fix a nonzero complex number ω. Two complex numbers z_1 and z_2 are defined to be equivalent if $z_1 - z_2 = n\omega$ for some integer n and is denoted as $z_1 \sim z_2$. It is an equivalence relation. Let $\mathbf{C}/\sim$ denote the set of all such equivalence classes $[z]$, $z \in \mathbf{C}$, and $\pi : \mathbf{C} \to \mathbf{C}/\sim$ the natural projection defined by $\pi(z) = [z]$.

As usual, a set O in $\mathbf{C}/\sim$ is open if its inverse image $\pi^{-1}(O)$ is open in $\mathbf{C}$. For intuitive sake, see Fig. 7.2. Then $\mathbf{C}/\sim$ is a Hausdorff space and π is continuous.

As an image, under π, of the strip $o \le \operatorname{Re} z < \operatorname{Re}\omega$ (it is assumed here that $\operatorname{Re}\omega > 0$), $\mathbf{C}/\sim$ is connected but not compact. Obviously π is

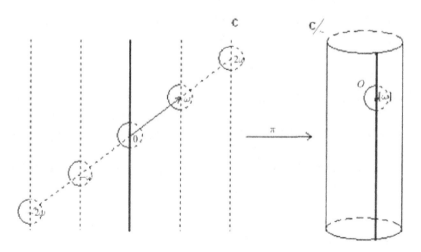

Fig. 7.2

also open, since, for each open set U in $\mathbf{C}$, $\pi^{-1}(\pi(U)) = \cup(n\omega + U)$ is a countable union of open sets $n\omega + U$ in $\mathbf{C}$.

Choose an open set U in $\mathbf{C}$, containing no two equivalent points under $\sim$. Then $\pi\colon U \to \pi(U)$ is a homeomorphism between U and the open set $\pi(U)$. In this case, define $\varphi : \pi(U) \to U$ by the inverse of π restricted to U. Then $\{\pi(U), (\pi|U)^{-1}\}$ forms a chart on $\mathbf{C}/\!\sim$ and all such charts make $\mathbf{C}/\!\sim$ a topological manifold.

To see all such charts forming a complex structure on $\mathbf{C}/\!\sim$, choose two charts $\{\pi(U_1), (\pi|U_1)^{-1}\}$ and $\{\pi(U_2), (\pi|U_2)^{-1}\}$ so that $\pi(U_1)\cap\pi(U_2) \neq \phi$. Set $\varphi_k = (\pi|U_k)^{-1}$, for $k = 1, 2$. For each $z \in \varphi_1(\pi(U_1) \cap \pi(U_2))$, then

$$z \xrightarrow{\varphi_1^{-1}} \varphi_1^{-1}(z) = \pi(z) = [z] \xrightarrow{\varphi_2} \pi^{-1}([z]) = z \xrightarrow{\pi} \varphi_1^{-1}(z) = \pi(z)$$

$$\Rightarrow \pi(\varphi_2 \circ \varphi_1^{-1})(z) = \pi(z)$$

$$\Rightarrow (\varphi_2 \circ \varphi_1^{-1})(z) - z \quad \text{is in the discrete set } \{n\omega | n = 0, \pm 1, \pm 2, \ldots\}.$$

Since $(\varphi_2 \circ \varphi_1^{-1})(z) - z$ is continuous, it follows that $(\varphi_2 \circ \varphi_1^{-1})(z) - z$ is constant on each component of $\varphi_1(\pi(U_1) \cap \pi(U_2))$ and hence $\varphi_2 \circ \varphi_1^{-1}$ is univalently analytic.

In conclusion, $\mathbf{C}/\!\sim$ is a noncompact Riemann surface with the complex structure $\{\pi(U), (\pi|U)^{-1}\}$.

Note that $\pi : \{0 \leq \operatorname{Re} z < \operatorname{Re}\omega\} \to \mathbf{C}/\!\sim$ is a homeomorphism. This means that one can identify each point a along the line $\operatorname{Re} z = 0$ with the point $a + \omega$ on the line $\operatorname{Re} z = \operatorname{Re}\omega$ to constitute the infinite cylinder $\mathbf{C}/\!\sim$. And the half open strip then can be considered as a *topological model* for $\mathbf{C}/\!\sim$. Recall the following successive univalent mappings:

$$\{0 \leq \operatorname{Re} z < \operatorname{Re}\omega\} \xrightarrow{\zeta = \frac{2\pi i z}{\operatorname{Re}\omega}} \{0 \leq \operatorname{Im}\zeta < 2\pi\} \xrightarrow{w = e^\zeta} \mathbf{C} - \{0\}.$$

It follows that the punctured plane $\mathbf{C} - \{0\}$ also acts as a *topological model* for $\mathbf{C}/\!\sim$. As a matter of fact, $\mathbf{C} - \{0\}$ is *conformally* equivalent to $\mathbf{C}/\!\sim$ since π is a conformal mapping between the strip and $\mathbf{C}/\!\sim$ as we will see in Remark 1 of Sec. 7.2.

Now, $\mathbf{C}/\!\sim$ can be realized as the set $\{[x+iy]|0 \leq x < 2\pi, y \in \mathbf{R}\}$. Then define $\Phi\colon \mathbf{C}/\!\sim \to S^1 \times \mathbf{R}$ by

$$\Phi([x + iy]) = (e^{ix}, y).$$

Φ is a homeomorphism. Conventionally, $\mathbf{C}/\!\sim$ is then denoted as

$$S^1 \times \mathbf{R}$$

and is called the *infinite cylinder*, conformally equivalent to the *punctured plane* $\mathbf{C} - \{0\}$.

Example 3. *The torus $S^1 \times S^1$.*

Explanation. The description is similar to that of Example 2 and we sketch as follows.

Let ω_1 and ω_2 be two nonzero complex numbers whose ratio $\frac{\omega_1}{\omega_2}$ is not real, namely, linearly independent over the real $\mathbf{R}$. Consider the lattice

$$\Gamma = \{m_1\omega_1 + m_2\omega_2 | m_1 \text{ and } m_2 \text{ are integers}\}$$

spanned by ω_1 and ω_2. Two complex numbers z_1 and z_2 are called equivalent mod Γ if $z_1 - z_2 \in \Gamma$. This indeed is an equivalence relation and the set of all equivalence classes (z), $z \in \mathbf{C}$, is denoted by $\mathbf{C}/\Gamma$. Let $\pi\colon \mathbf{C} \to \mathbf{C}/\Gamma$ be the natural projection defined by $\pi(z) = (z)$.

Endowed with the quotient topology, $\mathbf{C}/\Gamma$ is a connected Hausdorff space. It is even compact. To see these, recall that a set O in $\mathbf{C}/\Gamma$ is open if its inverse image $\pi^{-1}(O)$ is open in $\mathbf{C}$. See Fig. 7.3.

This makes $\mathbf{C}/\Gamma$ a Hausdorff space and π continuous, and hence $\pi(\mathbf{C}) = \mathbf{C}/\Gamma$ is connected. Since $\mathbf{C}/\Gamma$ is the image, under π, of the compact parallelogram $\{\lambda_1\omega_1 + \lambda_2\omega_2 | 0 \le \lambda_1, \lambda_2 \le 1\}$, it is compact. Moreover π is open. To see this, choose any open set U in $\mathbf{C}$ and we have to show that $\pi(U)$ is open in $\mathbf{C}/\Gamma$, namely, $\pi^{-1}(\pi(U))$ is open in $\mathbf{C}$. But $\pi^{-1}(\pi(U)) = \bigcup_{\omega \in \Gamma}(\omega + U)$ which is the union of open sets $\omega + U$. This is done.

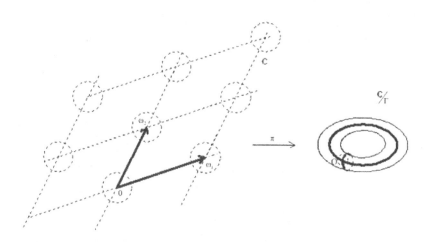

Fig. 7.3

To define the complex structure on $\mathbf{C}/\Gamma$, let U be any open set in $\mathbf{C}$, containing no two equivalent points mod Γ. Then $\pi: U \to O = \pi(U)$ is a homeomorphism between U and the open set $\pi(U)$. In this case, the associated local parameter φ is defined as the inverse of π restricted to U; namely, $\{\pi(U), (\pi|U)^{-1}\}$ is a chart. The collection of these charts makes $\mathbf{C}/\Gamma$ into a topological manifold.

Let $\{\pi(U_1), (\pi|U_1)^{-1}\}$ and $\{\pi(U_2), (\pi|U_2)^{-1}\}$ be two charts so that $\pi(U_1) \cap \pi(U_2) \neq \phi$. Set $\varphi_k = (\pi|U_k)^{-1}, k = 1, 2$, for simplicity. For each $z \in \varphi_1(\pi(U_1) \cap \pi(U_2))$,

$$\pi(\varphi_2 \circ \varphi_1^{-1})(z) = \varphi_1^{-1}(z) = \pi(z)$$
$$\Rightarrow (\varphi_2 \circ \varphi_1^{-1})(z) - z \in \Gamma.$$

Since Γ is discrete, so $(\varphi_2 \circ \varphi_1^{-1})(z) - z$ is constant on each component of $\varphi_1(\pi(U_1) \cap \pi(U_2))$ and hence $\varphi_2 \circ \varphi_1^{-1}$ is univalently analytic.

Hence $\mathbf{C}/\Gamma$ is a compact Riemann surface with the complex structure $\{\pi(U), (\pi|U)^{-1}\}$.

Define a mapping $\Phi: \mathbf{C}/\Gamma \to S^1 \times S^1$, where S^1 is the unit circle $|z| = 1$, by

$$\Phi([z]) = (e^{2\pi i x}, e^{2\pi i y}) \quad \text{if } z = x + iy.$$

Φ is a homeomorphism (refer to Example 3 of Sec. 1.7). Henceforth, $\mathbf{C}/\Gamma$ is conventionally denoted by

$$S^1 \times S^1$$

and is called the *torus*, topologically viewed as a sphere with one handle.

That the sphere S, the infinite cylinder $S^1 \times \mathbf{R}$, and the torus $S^1 \times S^1$ are Riemann surfaces is not accidental but is a rather special case of a remarkable general result to be further explained below. Henceforth readers are assumed to have basic differential geometric knowledge of concrete $\mathbf{C}^k$-surfaces in $\mathbf{R}^3$, where $1 \leq k \leq \infty$ or $k = \omega$.

Let $\sum$ be a $\mathbf{C}^k$-surface in $\mathbf{R}^3$.

Orientation: Let $\{O, z\}$ and $\{U, w\}$, be two charts on $\sum$ with $O \cap U \neq \phi$. Fix a point $p_0 \in O \cap U$. Denote by $f = (f_1, f_2, f_3)$ the inverse $z^{-1}: z(O) \to O$ and $g = (g_1, g_2, g_3)$ the inverse $w^{-1}: w(U) \to U$ of these two local parameters z and w, respectively. Also, set $z = x + iy$ and $w = u + iv$ and let $z(p_0) = w(p_0) = 0$.

The ordered basis $\{df_0(1, 0), df_0(0, 1)\}$ determines an orientation of the tangent space $T_{p_0}(\sum)$ of the surface $\sum$ at p_0, and $\{dg_0(1, 0), dg_0(0, 1)\}$ another orientation of $T_{p_0}(\sum)$. How are these two orientations related to each other? The local parameter transformation $w \circ z^{-1}$ now is $g^{-1} \circ f$, namely, $(u, v) = (g^{-1} \circ f)(x, y)$. Observing that $f = g \circ (g^{-1} \circ f)$, we have

$$df_0 = dg_0 \circ d(g^{-1} \circ f)_0, \quad \text{where } 0 = 0 + i0 = (0, 0) \text{ in notation.}$$

$$\Rightarrow df_0(1, 0) = dg_0 \left(\frac{\partial u}{\partial x}, \frac{\partial v}{\partial x} \right) = \frac{\partial u}{\partial x} dg_0(1, 0) + \frac{\partial v}{\partial x} dg_0(0, 1),$$

$$df_0(0, 1) = dg_0 \left(\frac{\partial u}{\partial y}, \frac{\partial v}{\partial y} \right) = \frac{\partial u}{\partial y} dg_0(1, 0) + \frac{\partial v}{\partial y} dg_0(0, 1) \qquad (7.1.2)$$

$\Rightarrow$ (the vector products in $\mathbf{R}^3$)

$$df_0(1, 0) \wedge df_0(0, 1) = \frac{\partial(u, v)}{\partial(x, y)} dg_0(1, 0) \wedge dg_0(0, 1), \quad \text{where}$$

$$\frac{\partial(u, v)}{\partial(x, y)} = \begin{vmatrix} \dfrac{\partial u}{\partial x} & \dfrac{\partial v}{\partial x} \\[2mm] \dfrac{\partial u}{\partial y} & \dfrac{\partial v}{\partial y} \end{vmatrix} = \det d(g^{-1} \circ f)_0 \quad \text{is the Jacobian of } g^{-1} \circ f \text{ at } 0.$$

$$(7.1.3)$$

This shows that these two orientations of $T_{p_0}(\Sigma)$ is *consistent* if and only if $\det d(g^{-1} \circ f)_0 > 0$ holds, and then the local parameters z and w are said to define the same *orientation* of the tangent space $T_{p_0}(\Sigma)$ of Σ at p_0.

A C^k-surface Σ is said to be *orientable* if Σ admits a collection of charts $\{O_\lambda, \varphi_\lambda\}$ such that, whenever $O_{\lambda_1} \cap O_{\lambda_2} \neq \phi$, the Jacobian determinant of the local parameter transformation $\varphi_{\lambda_1} \circ \varphi_{\lambda_2}^{-1}$ always satisfies

$$\det d(\varphi_{\lambda_1} \circ \varphi_{\lambda_2}^{-1})_{\varphi_{\lambda_2}(p)} > 0, \quad p \in O_{\lambda_1} \cap O_{\lambda_2}. \qquad (7.1.4)$$

Such a collection of charts is called an *orientation* of Σ.

Metric (or the first fundamental form) on Σ. Let $t \rightarrow (x(t), y(t))$ be a differentiable curve passing the point $z(p)$. Then $f(t) = (f_1(x(t), y(t)), f_2(x(t), y(t)), f_3(x(t), y(t)))$ is a differentiable curve in Σ, passing the point p and having the *tangent vector* at p

$$\frac{df}{dt} = df_{z(p)} \left(\frac{dx}{dt}, \frac{dy}{dt} \right) = df_{z(p)}(1, 0) \frac{dx}{dt} + df_{z(p)}(0, 1) \frac{dy}{dt}$$

$\Rightarrow$ (in differential form, see Sec. 2.8)

$$df = df_{z(p)}(1, 0) dx + df_{z(p)}(0, 1) dy, \quad \text{where}$$

$$df_{z(p)}(1, 0) = \left(\frac{\partial f_1}{\partial x}, \frac{\partial f_2}{\partial x}, \frac{\partial f_3}{\partial x} \right)(z(p)), \quad \text{etc.} \qquad (7.1.5)$$

The inner product $<,>$ on $T_p(\sum)$ is the one inherited from that of $\mathbf{R}^3$. The square of the length $|df|$ of the tangent vector $df \in T_p(\sum)$ is then given by

$$\langle df, df \rangle = E dx^2 + 2F dx dy + G dy^2$$

$$= [dx \ dy] \begin{bmatrix} E & F \\ F & G \end{bmatrix} \begin{bmatrix} dx \\ dy \end{bmatrix}, \tag{7.1.6}$$

where $E = \langle df_{z(p)}(1,0), df_{z(p)}(1,0) \rangle$

$$= \sum_{i=1}^{3} \left(\frac{\partial f_i}{\partial x} \right)^2 (z(p)),$$

$$F = \langle df_{z(p)}(1,0), df_{z(p)}(0,1) \rangle \quad \text{and} \quad G = \langle df_{z(p)}(0,1), df_{z(p)}(0,1) \rangle. \tag{7.1.7}$$

Note that the symmetric matrix of order two is *positive definite* since f is a C^k map with $df_{z(p)}$ having rank equal to 2 everywhere. Moreover, *the quadratic form* (7.1.6) *is invariant under change of local parameters*. To see this, let (U, w) be another chart with $U \cap O \neq \phi$. For $p \in U \cap O$, the tangent vector at p in the chart (U, w) is given by

$$dg = dg_{w(p)}(1,0) du + dg_{w(p)}(0,1) dv.$$

We have to compute $\langle dg, dg \rangle$ and show that it indeed is equal to $\langle df, df \rangle$, given by (7.1.6). Observing that (7.1.2) implies

$$\begin{bmatrix} df_{z(p)}(1,0) \\ df_{z(p)}(0,1) \end{bmatrix} = J \begin{bmatrix} dg_{w(p)}(1,0) \\ dg_{w(p)}(0,1) \end{bmatrix}, \quad \text{where } J = \begin{bmatrix} \dfrac{\partial u}{\partial x} & \dfrac{\partial v}{\partial x} \\ \dfrac{\partial u}{\partial y} & \dfrac{\partial v}{\partial y} \end{bmatrix} (z(p))$$

is the Jacobian matrix of $w \circ z^{-1}$ at $z(p)$

and that

$$[dx \ dy] = [du \ dv] J^{-1},$$

it follows that

$$\langle df, df \rangle = [dx \ dy] \begin{bmatrix} df_{z(p)}(1,0) \\ df_{z(p)}(0,1) \end{bmatrix} \begin{bmatrix} df_{z(p)}(1,0) \\ df_{z(p)}(0,1) \end{bmatrix}^{*} \begin{bmatrix} dx \\ dy \end{bmatrix}$$

(where * denotes the transpose operation)

$$= [du \ dv] J^{-1} \cdot J \begin{bmatrix} dg_{w(p)}(1,0) \\ dg_{w(p)}(0,1) \end{bmatrix} \begin{bmatrix} dg_{w(p)}(1,0) \\ dg_{w(p)}(0,1) \end{bmatrix}^{*} J^{*}(J^{*})^{-1} \begin{bmatrix} du \\ dv \end{bmatrix}$$

$$= \langle dg, dg \rangle.$$

The claim follows.

The *metric* on Σ is defined locally by the arc length element $|df|$, traditionally denoted as ds, where

$$ds^2 = E dx^2 + 2F dx dy + G dy^2$$

$$= \lambda^2 |dz + \mu d\bar{z}|^2, \quad \text{in complex notation in which } dz = dx + idy, \quad \text{and}$$

$$\lambda^2 = \frac{1}{4}(E + G + 2\sqrt{EG - F^2}) \quad \text{and} \quad \mu = \frac{E - G + iF}{2\lambda} \quad (\text{with } |\mu|^2 < 1).$$

$$(7.1.8)$$

An abstract surface endowed with such a (Riemannian) metric is called a *two-dimensional Riemannian manifold*.

Conformal mapping and isothermal coordinate system:
The *angle* between two curves passing a point p on Σ is defined to be the one between their tangent vectors in $T_p(\Sigma)$.

$f : z(O) \to O$ is said to be *conformal* if, for each $p \in O$, $df_{z(p)} : T_{z(p)}(z(O)) \to T_p(\Sigma)$ preserves angles between corresponding vectors. In this case,

$$F = \langle df_{\varphi(p)}(1,0), df_{\varphi(p)}(0,1) \rangle$$

$$= |df_{\varphi(p)}(1,0)| \| df_{\varphi(p)}(0,1)| \langle (1,0), (0,1) \rangle = 0, \quad \text{and}$$

$$\frac{\langle df_{\varphi(p)}(1,1), df_{\varphi(p)}(1,0) \rangle}{|df_{\varphi(p)}(1,1)| \| df_{\varphi(p)}(1,0)|}$$

$$= \frac{E}{\sqrt{E + G}\sqrt{E}} = \frac{1}{\sqrt{2}} \quad \text{which implies that } E = G.$$

Moreover, $\mu = 0$ in (7.1.7) and then $ds = \lambda |dz|$. In conclusion, we detain

Local isothermal coordinate on C^k-surface Σ in $\mathbf{R}^3$. Let $\{O, z\}$ be a chart on Σ and $f : z(O) \to O$ be the inverse map of $z : O \to z(O)$. Then

   (1) f is conformal on O (or on $z(O)$).
$\Leftrightarrow$ (2) The classical Gaussian quantities E, F, G in (7.1.7) satisfy: $E = G$ and $F = 0$ on $z(O)$.
$\Leftrightarrow$ (3) $\mu = 0$ in (7.1.7).
$\Leftrightarrow$ (4) The metric $ds = \lambda |dz|$, where λ is a positive C^{k-1} function on $z(O)$.

In this case, $\{O, z\}$ is called an *isothermal coordinate system* on Σ with z the (local) *isothermal parameter*.   (7.1.8)

A similar argument will lead to a slight generalization of (7.1.8) in the following

Conformal mapping between two C^k-surfaces in $\mathbf{R}^3$. Let $\Phi : \sum_1 \to \sum_2$ be a differentiable mapping between two C^k-surfaces $\sum_1$ and $\sum_2$ in $\mathbf{R}^3$ (modeling the definition in (7.2.1)). Then

(1) Φ is *conformal,* namely, for each $p \in \sum$, $d\Phi_p : T_p(\sum_1) \to T_{\Phi(p)}(\sum_2)$ is conformal.

$\Leftrightarrow$ (2) There is a positive C^{k-1} function λ on $\sum_1$ so that, for each $p \in \sum_1$,
$$\langle d\Phi_p(d\gamma), d\Phi_p(d\sigma)\rangle = \lambda(p)\langle d\gamma, d\sigma\rangle, d\gamma, d\sigma \in T_p(\sum_1). \tag{7.1.9}$$

Make certainty what $d\Phi_p$ means and try to prove this statement. Note that, in this case, the ratio $E : F : G$ is uniquely determined.

The realization of Σ as a Riemann surface:
Suppose $\{U, w\}$ is another isothermal coordinate system on $\sum$ so that $U \cap O \neq \emptyset$, where $\{O, z\}$ is as in (7.1.8). Let $ds = \tau|dw|$, where τ is a positive C^{k-1} function on $W(U)$. The composite map $w \circ z^{-1} : z(O \cap U) \to w(O \cap U)$ then satisfies

$$\tau|dw| = \lambda|dz|.$$

By (7.1.9), this shows that $w \circ z^{-1}$ is either conformal or indirectly conformal (see Exercise (6) in Appendix C). If $\sum$ is presumed to be orientable, then all such $w \circ z^{-1}$ may be adjusted to be conformal. According to (2) in (3.2.3.3) or (4) in (3.2.2), $w \circ z^{-1}$ is analytic.

We summarize as part (1) of the following

Realization of an orientable surface in $\mathbf{R}^3$ as a Riemann surface.

(1) Suppose an orientable C^1-surface admits a collection of charts $\{O, z\}$, composed of isothermal coordinate systems. Then the collection defines a natural complex (or conformal) structure to make the surface a Riemann surface.

(2) Every orientable C^2-surface in $\mathbf{R}^3$ admits *such* a collection of charts and hence can be made into a Riemann surface.   (7.1.10)

The key to the proof of (2) lies on the fact that every z-coordinate on $\sum$ can be homeomorphically transformed into an isothermal coordinate. For a proof, refer to Refs. [10], pp. 15–35; [18], p. 217; [53], p. 134, in which the existence theorem for Beltrami equations and the concept of quasiconformal mappings are invoked; Ref. [6], p. 125, under a stronger hypothesis or Ref. [75], pp. 18–23.

Illustrations of (7.1.10):

Fix a point $\vec{p}_0$ and two orthogonal unit vectors $\vec{v}_1$ and $\vec{v}_2$ in $\mathbf{R}^3$. The plane $\sum$ in $\mathbf{R}^3$, passing $\vec{p}_0$ and containing $\vec{v}_1$ and $\vec{v}_2$, has the equation

$$f(x, y) = \vec{p}_0 + x\vec{v}_1 + y\vec{v}_2, \quad x, y \in \mathbf{R}.$$

The metric on $\sum$ is given by $ds^2 = dx^2 + dy^2$ and $z(f(x, y)) = (x, y)$ is the isothermal parameter. Hence $(\sum, f^{-1})$ is an isothermal coordinate system on $\sum$ and $\sum$ is then a Riemann surface.

Let S be the Riemann sphere in Example 1 above. Then f_1, the inverse of the stereographic projection z_1, is given by (see (1.6.7))

$$f_1(x, y)$$
$$= \left(\frac{2x}{x^2 + y^2 + 1}, \frac{2y}{x^2 + y^2 + 1}, \frac{x^2 + y^2 - 1}{x^2 + y^2 + 1} \right), \quad z_1 = (x, y) \in \mathbf{R}^2 \text{ or } \mathbf{C}.$$

In this case, the arc length element on $S - \{(0, 0, 1)\} = O_1$ is

$$ds = \frac{2|dz_1|}{1 + |z_1|^2}$$

(see Exercise B(1) of Sec. 1.6). While f_2, the inverse of z_2, is

$$f_2(x, y) = \left(\frac{2x}{x^2 + y^2 + 1}, \frac{-2y}{x^2 + y^2 + 1}, \frac{x^2 + y^2 - 1}{x^2 + y^2 + 1} \right),$$
$$z_2 = (x, y) \in \mathbf{R}^2 - \{(0, 0)\}$$

and the arc length element on $S - \{(0, 0, -1)\} = O_2$ is given by

$$ds = \frac{2|dz_2|}{1 + |z_2|^2}.$$

Hence, S can be covered by the isothermal coordinate systems (O_1, z_1) and (O_2, z_2) and, according to (7.1.10), can be made into a Riemann surface.

In Example 2, the infinite cylinder $x_1^2 + x_1^2 = 1$, $-\infty < x_3 < \infty$, has the parametric representation

$$f(x, y) = (\cos x, \sin x, y), \quad 0 \le x < 2\pi, -\infty < y < \infty.$$

The associated arc length element is $ds^2 = |dz|^2 = dx^2 + dy^2$. Set O the open set obtained from $S^1 \times \mathbf{R}$ by deleting the line $(1, 0, y)$ $-\infty < y < \infty$. Then (O, z) is an isothermal coordinate system. Allow $-\pi < x < \pi$ and we will get another such system. Therefore $S^1 \times \mathbf{R}$ is a Riemann surface.

The torus $S^1 \times S^1$ in Example 3 can be parametrized by

$$f(x,y) = ((a + r \cos x) \cos y, (a + r \cos x) \sin y, r \sin x),$$

$$0 < x < 2\pi, \quad 0 < y < 2\pi$$

except for a meridian and a parallel, where $0 < r < a$. Direct calculation shows that $ds^2 = r^2 dx^2 + (r \cos x + a)^2 dy^2$. Since $S^1 \times S^1$ is an oriented C^2-surface (as a matter of fact, a C^ω one), the theory says that this local parameter (x, y) can be changed homeomorphically into an isothermal one. Readers are urged to decide how many such parameterized coordinates can cover the whole torus.

Exercises A

(1) Prove (7.1.9) in detail.
(2) Refer to one of the books mentioned after (7.1.10) and read the proof for (2) there.

7.2   Analytic Mappings and Meromorphic Functions on Riemann Surfaces

The concept of local variables (or charts) is the main tool to carry over the theory of functions in the classical plane to Riemann surfaces.

Definition. A *continuous* mapping f from a Riemann surface R_1 into another Riemann surface R_2 is called *an analytic mapping* if, for every pair of charts $\{O, \varphi\}$ on R_1 and $\{U, \psi\}$ on R_2, the function

$$\psi \circ f \circ \varphi^{-1} : \varphi(O \cap f^{-1}(U)) \to \psi(U)$$

is analytic in the usual sense. See Fig. 7.2.

$f : R_1 \to R_2$ is called a *conformal mapping* if f is one-to-one, onto, and both f and f^{-1} are analytic; in this case, R_1 and R_2 are said to be *conformally equivalent*.

In case $R_2 = \mathbf{C}$, the complex plane, an analytic mapping f is particularly called an *analytic function* for distinction.

In case $R_2 = \mathbf{C}^*$, the Riemann sphere, then either f is identically equal to ∞ or $f^{-1}(\infty)$ consists of isolated points and $f : R_1 - f^{-1}(\infty) \to \mathbf{C}$ is an analytic function. In the latter case, $f : R_1 \to \mathbf{C}$ is called a *meromorphic function*. Points in $f^{-1}(\infty)$ are called poles of f.   (7.2.1)

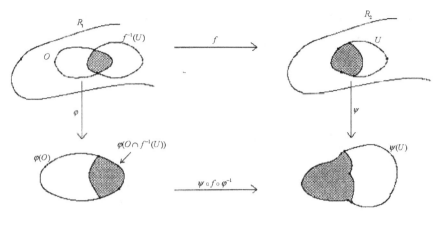

Fig. 7.4

See Fig. 7.4.

Remarks. (1) The continuity assumption of f is used to guarantee that $f^{-1}(U)$ is open and so is $O \cap f^{-1}(U)$. As a matter of fact, the analyticity can also be defined pointwise in the following sense.

f is said to be *analytic at a point* $p \in R_1$ if there is a chart $\{O, \varphi\}$ at p and a chart $\{U, \psi\}$ at $f(p)$ so that $\psi \circ f \circ \varphi^{-1} : \varphi(O \cap f^{-1}(U)) \to \psi(U)$ is analytic in the classical sense. This definition is independent of the charts chosen. Suppose $\{\tilde{O}, \tilde{\varphi}\}$ is another chart at p and $\{\tilde{U}, \tilde{\psi}\}$ is another one at $f(p)$. Then the relation

$$\tilde{\psi} \circ f \circ \tilde{\varphi}^{-1} = (\tilde{\psi} \circ \psi^{-1}) \circ (\psi \circ f \circ \varphi^{-1}) \circ (\varphi \circ \tilde{\varphi}^{-1}) \tag{7.2.2}$$

and the conformality of both $\varphi \circ \tilde{\varphi}^{-1}$ and $\tilde{\psi} \circ \psi^{-1}$ together show that $\tilde{\psi} \circ f \circ \tilde{\varphi}^{-1}$ and $\psi \circ f \circ \varphi^{-1}$ are analytic simultaneously. Then we say f is *analytic on an open set* of R_1 if it is analytic at each of its points. *Conformality* can be similarly extended to functions analytic between open sets of R_1 and R_2.

According to this definition, the local variable φ in each chart $\{O, \varphi\}$ is a (local) *conformal* mapping from O to $\varphi(O) \subseteq \mathbf{C}$. This fact enables us to extend many local concepts in classical complex analysis to abstract Riemann surfaces.

(2) *Local representation of an analytic mapping*

Let $\{O, z\}$ be a chart at a point p where f is analytic so that $z(p) = 0$, and $\{U, w\}$ be a chart at $f(p)$ so that $w(f(p)) = 0$. In terms of these local coordinates, f (as a matter of fact, $w \circ f \circ z^{-1}$) can be expressed as

$$w = f(z) = \sum_{n \geq k} a_n z^n, \quad k \geq 1 \text{ and } a_k \neq 0 \text{ (see (3.4.2.6))}$$

$$= z^k g(z), \quad \text{where } g(z) = \sum_{n=k}^{\infty} a_n z^{n-k} \text{ with } g(0) = a_k \neq 0$$

$$= z^k (h(z))^k, \quad \text{where } h(z) \text{ is a single-valued branch of } \sqrt[k]{g(z)} \text{ in an open}$$

disk neighborhood of 0 on which $g(z) \neq 0$ (see (3.3.1.5) or (4.4.2)).

$$\Rightarrow w = f(z) = \zeta^k, \quad \text{where } \zeta = z h(z) \text{ is another local coordinate at } p \text{ with}$$

$$\zeta(p) = 0. \tag{7.2.3}$$

For details, see (3.5.1.8). Owing to (7.2.2), the number k in (7.2.3) is clearly independent of the local coordinates chosen and is called the *ramification number* or the *multiplicity* of f at p, namely, the number of times f taking on the value $f(p)$ at p. In case $k = 1$, p is called a *regular point* of f; if $k \geq 2$, a *branch point of order* $k - 1$. In either case, the number $b_f(p) = k - 1$ is unifiedly called the *branch number* of f at p.

(3) *Local representation of a meromorphic function at a pole*

In the definition of a meromorphic function, we define $f(p) = \infty$ at a pole p of f. Recall again that $f : R_1 \to \mathbf{C}^*$ is analytic between Riemann surfaces R_1 and $\mathbf{C}^*$, while $f : R_1 - f^{-1}(\infty) \to \mathbf{C}$ is a complex-valued analytic function. This definition is suitable for meromorphic functions just defined on open subsets of R_1.

Choose a chart $\{O, z\}$ at a pole p of f so that $z(p) = 0$. In terms of this local coordinate and the global identity coordinate on $\mathbf{C}$, f can be written as

$$w = f(z) = \sum_{n=-k}^{\infty} a_n z^n, \quad k \geq 1 \text{ and } a_{-k} \neq 0 \text{ (see (4.10.2.1))}$$

$$= \frac{g(z)}{z^k}, \quad \text{where } g(z) = a_{-k} + a_{-k+1} z + \cdots + a_{-1} z^{k-1}$$

$$+ \sum_{n=0}^{\infty} a_n z^{n+k} \text{ with } g(0) = a_{-k} \neq 0.$$

$$= \frac{(h(z))^{-k}}{z^k} \quad \text{(see (3.3.1.5) or (4.4.2))}$$

$\Rightarrow w = f(z) = \dfrac{1}{\zeta^k}$, where $\zeta = zh(z)$ is a another local coordinate with

$$\zeta(p) = 0. \tag{7.2.4}$$

The number k in (7.2.4) is well-defined and is called the *order* of f at the pole p. In case $k \geq 2$, p is also a *branch point of order* $k - 1$ of f (see (3.5.1.11)). The number $b_k(p) = k - 1$, $k \geq 1$, is called the *branch number* of f at p.

(4) *If a Riemann surface* **R** *carries a (single) nonconstant meromorphic function* f *on it, then its complex structure is uniquely determined by such an* f. To see this, fix a point $p_0 \in R$ with the branch number $k - 1$ and consider a local chart $\{O, \varphi\}$ at p_0 with $\varphi(p_0) = 0$. In case $f(p_0) \neq \infty$, then by (7.2.3),

$$f \circ \varphi^{-1}(z) = f(p_0) + z^k \quad \text{in a neighborhood of } 0.$$

$\Rightarrow$ (setting $\varphi^{-1}(z) = p \in O$) $\quad f(p) = f(p_0) + \varphi(p)^k$.

$\Rightarrow \varphi(p) = (f(p) - f(p_0))^{1/k}, \quad p \in O$ or $z = (w - w_0)^{1/k}$ if $z = \varphi(p)$ and

$$w = f(p). \tag{7.2.5}$$

In case $f(p_0) = \infty$, then by (7.2.4),

$$f \circ \varphi^{-1}(z) = \dfrac{1}{z^k} \quad \text{in } s \text{ neighborhood of } \infty.$$

$\Rightarrow$ (setting $\varphi^{-1}(z) = p \in O$) $\quad f(p) = \dfrac{1}{\varphi(p)^k}$

$$\Rightarrow \varphi(p) = \dfrac{1}{f(p)^{1/k}} = f(p)^{-1/k}, \quad p \in O. \tag{7.2.6}$$

The local coordinate neighborhood O of p_0 can be so chosen that its image under φ is the open disk $|z| < \sqrt[k]{r}(r > 0)$ or $\sqrt[k]{r} < |z| \leq \infty$.

As a vivid illustration, try to review Example 1 in Sec. 3.3.3 where, in this case, $R = R_{\sqrt{z}}$ and $f(z) = z^2$ (note that z in $f(z)$ is identified as a point in $R_{\sqrt{z}}$). Note that $w = \sqrt{z}$ realizes a conformal mapping between $R_{\sqrt{z}}$ and the w-plane $\mathbf{C}^*$. In case $R_{\sqrt{z}}$ is identified with w-plane $\mathbf{C}^*$, then $f(z) = z$ turns out to be $z = w^2$. And this is the reason why we call $R_{\sqrt{z}}$ the Riemann surface of $\sqrt{z}$ or $z = w^2$. For more examples, see Sec. 2.7.2 or Section (2) below. $\qquad\qquad\square$

Section (1) Some basic theorems

In what follows, we try to extend some of the famous theorems, especially these with strong topological characters, to analytic mappings between

Riemann surfaces. Remind the readers that the idea of local charts is our main tool in the process as we have already experienced in the Remarks above.

As an extension of (3.4.2.17) and (4.10.1.1), we have

Removable singularity theorem. *Let U be an open subset of a Riemann surface and $p_0 \in U$ a point. Suppose $f : U - \{p_0\} \to \mathbf{C}$ is a bounded analytic function. Then p_0 is a removable singularity of f, namely, there is an analytic function $F : U \to \mathbf{C}$ so that $f(p) = F(p)$ for each $p \in U - \{p_0\}$.*
$$(7.2.7)$$

Choose a chart $\{O, \varphi\}$ at p_0. Then $f \circ \varphi^{-1} : \varphi(O) - \{\varphi(p_0)\} \to \mathbf{C}$ is bounded and analytic. And the results follow easily from (3.4.2.17).

As an extension of (4) in (3.4.2.9), we have the

Interior uniqueness theorem. *Let f and g be analytic mappings from a Riemann surface R_1 to another Riemann surface R_2. Suppose the set $A = \{p \in R_1 | f(p) = g(p)\}$ has a limit point in R_1. Then $f \equiv g$ on R_1.*
$$(7.2.8)$$

Proof. Consider the set $\Omega = \{p \in R_1 | p$ has an open neighborhood on which f and g coincide$\}$. Ω is nonempty. To see this, let p_0 be a limit point of the set A. Choose a chart $\{O, \varphi\}$ at p_0 and a chart $\{U, \psi\}$ at $f(p_0) = g(p_0)$ (by continuity of f and g). Then $\psi \circ f \circ \varphi^{-1}(z) = \psi \circ g \circ \varphi^{-1}(z)$ for a sequence of points $z \in \varphi(O)$ converging to $\varphi(p_0) \in \varphi(O)$. By the interior uniqueness theorem in the plane, $\psi \circ f \circ \varphi^{-1}(z) = \psi \circ g \circ \varphi^{-1}(z)$ throughout $\varphi(O)$; this in turn implies that $f(p) = g(p)$ for all $p \in O$. Hence $p_0 \in \Omega$ and $\Omega \neq \phi$ holds.

Ω is, by definition, open.

Ω is also closed. For, suppose p_0 is a boundary point of Ω. The $f(p_0) = g(p_0)$ holds by continuity. Choose a chart $\{O, \varphi\}$ at p_0 so that O is a *connected* subset of R_1, and a chart $\{U, \psi\}$ at $f(p_0)$ so that $f(O) \subseteq U$ and $g(O) \subseteq U$ both hold. Then $\psi \circ f \circ \varphi^{-1}(z) = \psi \circ g \circ \varphi^{-1}(z)$ holds on $\varphi(O \cap \Omega)$, and hence, by the interior uniqueness theorem, they coincide on $\varphi(O)$, a connect set. Therefore $f(p) = g(p)$ for $p \in O$, and $p_0 \in \Omega$. Then Ω is closed.

Since R_1 is connected, $\Omega = R_1$ holds and the claim follows. $\qquad\square$

Let us go back to observe (7.2.3). Geometrically it means that, for each fixed point p_0 of R_1, there exist neighborhoods O of p_0 and U of $f(p_0)$ so that the set $f^{-1}(q) \cap O$ contains exactly k points for each point $q \in U$. Consequently, we have the following extension of (3.4.4.4).

Open mapping theorem. *Let* f *be a nonconstant analytic mapping from a Riemann surface* R_1 *to a Riemann surface* R_2. *Then* f *is open, namely, the image of every open set under* f *is open.* (7.2.9)

In case f is one-to-one and onto R_2, then $k = 1$ in (7.2.3) for each point $p \in R_1$ and this means $f^{-1} : R_2 \to R_1$ is also analytic (note that the openness of f shows that f^{-1} is continuous on R_2). Therefore we extend (3.5.1.9) to

Conformal mapping theorem. *Let* f *be a univalent analytic mapping from* R_1 *onto* R_2. *Then* f *is conformal.* (7.2.10)

In (7.2.9), suppose $R_2 = \mathbf{C}$, the classical plane, in case there is a point $p_0 \in R_1$ so that $|f(p)| \le |f(p_0)|$ for all $p \in R_1$, then $f(R_1) \subseteq \{|w| \le |f(p_0)|\}$ with the interior point $f(p_0)$ of the open set $f(R_1)$ belonging to the boundary circle $|w| = |f(z_0)|$. This is a contradiction. And we have extended (3.4.4.1) to the

Maximum modulus principle. Let $f : R$ (a Riemann surface) $\to \mathbf{C}$ be a nonconstant analytic function. Then $|f|$ does not attain its (local or global) maximum in R. (7.2.11)

Another implication of (7.2.9) happens if R_1 is compact. In this case, the image $f(R_1)$ is both compact (in particular, closed) and open in R_2. Then either $f(R_1)$ is a constant or $f(R_1) = R_2$ holds. This leads to the following theorem, an extension of (1) in (3.5.1.14).

Analytic mapping on a compact Riemann surface. Let $f : R_1$ (a *compact* Riemann surface) $\to R_2$ (a Riemann surface) be analytic. Then f is either constant or onto the whole surface R_2, namely, $f(R_1) = R_2$ and R_2 is compact, too. (7.2.12)

In particular, if $R_2 = \mathbf{C}$, which is not compact, then we obtain the fact:

Analytic function on a compact surface. Every analytic function on a compact Riemann surface is a constant. (7.2.13)

This also follows from the maximum principle (7.2.11). According to (7.2.7), a bounded entire function in $\mathbf{C}$ has the point at infinity ∞ as a removable singularity, and hence is analytic on the compact surface $\mathbf{C}^*$. By (7.2.13), we have

Liouville's theorem. *Every bounded entire function on* $\mathbf{C}$ *is a constant.* (7.2.14)

A nonconstant polynomial has a pole at ∞. When considering such a polynomial as an analytic mapping on $\mathbf{C}^*$ to itself, (7.2.12) says that its range must cover the whole space $\mathbf{C}^*$; in particular, it must assume the value zero. This observation leads to the

Fundamental theorem of algebra. *Every nonconstant polynomial has at least one zero.* (7.2.15)

We carry (7.2.12) one step further to the following result, extending (2) in (3.5.1.14).

The covering properties of an analytic mapping between compact Riemann surfaces. Let $f : R_1$ (a compact Riemann surface) $\rightarrow R_2$ (compact) be a *nonconstant* analytic mapping. Then there exists a (unique) positive integer m, called the *degree* of f, so that f assumes each point q in R_2 exactly m times (counting multiplicities) in R_1. Namely, in terms of the branch numbers introduced in (7.2.3) and (7.2.4), we have

$$\sum_{p \in f^{-1}(q)} [b_f(p) + 1] = m, \quad q \in R_2. \tag{7.2.16}$$

Proof is exactly the same as that of (3.5.1.14), except that now we have to use the fact that a Riemann surface has a countable basis as we claimed in Sec. 7.1 and hence is separable. Also refer to (7.5.2.5) and its proof, where such an f is called an *m-sheeted covering map* of R_2 by R_1 and m is the *degree* of f. This result explains and unifies so many phenomena appeared in compact Riemann surfaces we constructed before. See Examples 1, 2, and 3 below.

Section (2) Examples for compact Riemann surfaces (review of Sec. 2.7.2)

The illustrative Riemann surfaces we constructed from Chaps. 2–6 for particularly chosen multiple-valued functions provide vivid and fruitful examples of the abstract concepts introduced above. Recall that the Riemann surface $R_{f(z)}$ of a multiple-valued function $w = f(z)$ is so designed in order to turn $f(z)$ out to be single-valued on it, provided that a point p in $R_{f(z)}$ is identified with its value $z = \varphi(p)$ under a local parameter φ (refer to Sec. 3.3.3; in particular (3.3.3.2) and examples there).

Compact Riemann surfaces are usually termed as *closed* Riemann surfaces in the literature; while, the noncompact ones as *open* Riemann surfaces. In the afore-mentioned examples, it is easy to recognize which one is compact by observing that such a surface composes of finitely many sheets or its line complex contains finitely many "o" and "$\times$". We had pointed out after (3.5.6.10) that the Riemann surface of an arbitrary algebraic function is a finitely sheeted covering of the sphere with only a finite number of algebraic branch points and hence is compact. A remarkable result in the theory of compact Riemann surfaces is that *every abstract compact Riemann surface can be realized as the Riemann surface of an algebraic function*. See Ref. [75], p. 289, for a proof.

Example 1. *Simply-connected compact Riemann surfaces: Riemann sphere (elliptic type).*

Explanation. Readers are asked to get familiar with the materials in Sec. 2.7.2 and Exercise A(3) of Sec. 3.5.6.

(a) $z = w^n$ and $w = \sqrt[n]{z}$ ($n \geq 2$) (see Sec. 2.7.2): $z = w^n$ maps the compact Riemann surface $\mathbf{C}^*$ (the w-plane), in n-to-1 manner, onto the compact Riemann surface $\mathbf{C}^*$ (the z-plane). It has n zeros at $w = 0$ as well as a pole of order n at $w = \infty$. Figure 2.49 indicates how the w-plane $\mathbf{C}^*$ covers the z-plane $\mathbf{C}^*$ n-times just like the Riemann surface $R_{\sqrt[n]{z}}$ does. Note that $z = w^n : \mathbf{C}^* \to R_{\sqrt[n]{z}}$ is univalently analytic, or equivalent, $w = \sqrt[n]{z} : R_{\sqrt[n]{z}} \to \mathbf{C}^*$ is a univalent single-valued analytic mapping. According to (7.2.10), $w = \sqrt[n]{z}$ is conformal and hence, $R_{\sqrt[n]{z}}$ and $\mathbf{C}^*$ (the w-plane) are conformally equivalent. Therefore, $R_{\sqrt[n]{z}}$ is realized, in conformality, as the Riemann sphere $\mathbf{C}^*$ or S. This surface is simply connected and compact.

(b) $w = \sqrt{z^2 - 1}$ (see Example 2 in Sec. 2.7.2): This function $w = \sqrt{z^2 - 1}$ sets up a conformal mapping between the Riemann surface R_w and the w-plane $\mathbf{C}^*$, both cover the z-plane $\mathbf{C}^*$ twice. See Fig. 2.60.

(c) $z = \frac{1}{2}(w + \frac{1}{w})$ and *its inverse* $w = z + \sqrt{z^2 - 1}$ (see Figs. 2.61 and 2.62): Same as $w = \sqrt{z^2 - 1}$.

(d) $z = w^3 - 3w$ and *its inverse* (see Example 4 in Sec. 2.7.2): Fig. 2.66 (interchanging z and w) indicates that the w-plane $\mathbf{C}^*$ covers the z-plane $\mathbf{C}^*$ exactly three times just like the Riemann surface R_z of the inverse function $w = w(z)$ does. Now $w = w(z)$ is single-valued on R_z, and, in fact, $w = w(z)$ maps R_z conformally onto the w-plane $\mathbf{C}^*$ and, vice versa, $z = w^3 - 3w$ maps the w-plane $\mathbf{C}^*$ conformally onto R_z. Hence the Riemann sphere $\mathbf{C}^*$ or S plays not only a topological model for R_z but also a conformal model.

(e) $z = \frac{1}{2}(w^2 + \frac{1}{w^2})$ and *its inverse* $w = \sqrt{z + \sqrt{z^2 - 1}}$ (see Exercise B(5)(j) of Sec. 2.7.2): Solve $z' = w - \frac{1}{w^3} = 0$ and we get $w = \pm 1, \pm i$. Since $z''(\pm 1) \neq 0$ and $z''(\pm i) \neq 0$, so $z(\pm 1) = 1$ and $z(\pm i) = -1$ are algebraic

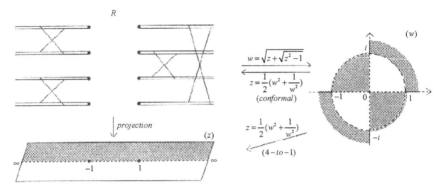

Fig. 7.5

branch points of order 1. Note that $w = 0$ and $w = \infty$ both are poles of order 2. As a consequence, $z = \infty$ is also an algebraic branch point of order 1. The Riemann surface R_z of $w = \sqrt{z + \sqrt{z^2 - 1}}$ is illustrated in Fig. 7.5 and its line complex is in Fig. 2.69. There are two algebraic branch points of order 1 over -1, 1, and ∞ each. In short, R_z is a four-sheeted covering surface of the sphere branched at six points. Each sheet of R_z can be realized topologically as the Riemann sphere with a slit through the north pole and four such slit Riemann spheres can be pasted topologically to form a complete sphere. Hence the Riemann sphere is a topological model of R_z. Meanwhile (7.2.12) and (7.2.10) say that R_z is indeed conformally equivalent to the w-plane $\mathbf{C}^*$. And it follows that the Riemann sphere also plays as a conformal model of R_z. On the other hand, the line complex shown in Fig. 2.69 indicates that the genus g of R_z must satisfy the relation $6 = 2(4 + g - 1)$ (see Exercise A(3) of Sec. 3.5.6 or (7.2.22)) and hence $g = 0$. This confirms again that the Riemann sphere is indeed a topological model of R_z.

(f) $z = \frac{w^4 - w^2 + 1}{w^4 + w^2 + 1}$ and *its inverse* (see Exercise B(6)(b) of Sec. 2.7.2 or Exercise A(3)(b) of Sec. 3.5.6): Note that

$$z = \frac{\zeta - 1}{\zeta + 1} \quad \text{where } \zeta = w^2 + \frac{1}{w^2}.$$

So the Riemann surface of the inverse function is the same as in Fig. 7.5, subject to a conformal mapping. See also Fig. 2.71 and compare to Fig. 2.69.

Example 2. *A compact Riemann surface that admits a nonconstant meromorphic function with precisely two poles (counting multiplicity):*

Hyperelliptic surfaces, topologically equivalent to a sphere with g handles ($g \geq 0$).

Explanation. A compact Riemann surface that admits a nonconstant meromorphic function with precisely one simple pole should be, according to (7.2.12), (7.2.16), and (7.2.10), conformal to $\mathbf{C}^*$. In this case, linear fractional transformations provide such nonconstant meromorphic functions (see Example 6 and Exercise A(6) of Sec. 4.10.2).

Let R be a compact Riemann surface on which there is a nonconstant meromorphic function $z = z(p)$ with exactly two poles. According to (7.2.16), $z = z(p)$ assumes every point in $\mathbf{C}^*$ exactly twice in R, and R is a two-sheeted covering of the sphere with $z = z(p)$ acting as a two-sheeted covering map (projection). In such an R, any branch point, if exists, is of order 1 which is the branch number of $z = z(p)$ at that branch point.

According to Remark 4, $z = z(p)$ uniquely determines the complex structure of R. That is to say, the local parameter $\varphi(p)$ at p_0 is given by

$$
\varphi(p) = \begin{cases}
z(p) - z(p_0), & \text{if } p_0 \text{ is a regular point and } z(p_0) \in \mathbf{C} \\[2mm]
\dfrac{1}{z(p)}, & \text{if } p_0 \text{ is a simple pole of } z \\[2mm]
\sqrt{z(p) - z(p_0)}, & \text{if } p_0 \text{ is a branch point of order 1 and} \\
& \quad z(p_0) \in \mathbf{C} \ (\text{see } (7.2.5)). \\[2mm]
\dfrac{1}{\sqrt{z(p)}}, & \text{if } p_0 \text{ is a branch point of order 1 and } z(p_0) = \infty \\
& (\text{see } (7.2.6))
\end{cases}
$$

$$(7.2.17)$$

Let α be the sum of the orders of the branch points, which is equal to the number of the branch points in this case. Then R is topologically equivalent to a sphere with g handles, where g is given by $\alpha = 2(2 + g - 1) = 2(g + 1)$ (see Exercise A(3) of Sec. 3.5.6 or (7.2.22) below). g is called the *genus* of the surface R.

Case 1. The genus $g = 0$.
Suppose $z = z(p)$ has a pole of order 2 at p_0. Since $\alpha = 2$ in this case, $z = z(p)$ has exactly another branch point of order 1 at some other point, say at q_0, and we may assume, no loss of its generality, that $z(q_0) = 0$. To construct a *concrete model* for R, observe first that $z^{-1}(\zeta)$ consists of precisely two points in R for each $\zeta \in \mathbf{C}^* - \{0, \infty\}$. Take two copies of $\mathbf{C}^*$ and label them sheets I and II. On each sheet, cut along the nonpositive real

axis from $z(p_0) = \infty$ to $z(q_0) = 0$ or joining ∞ to 0 by any Jordan curve (called a *cut*) in $\mathbf{C}^*$. Paste sheets I and II crosswise along the cut $[\infty, 0]$ as we did in Fig. 2.49 for $n = 2$ (see the explanation after it) and we obtain a simply connected two-sheeted compact Riemann surface $R_{\sqrt{z}}$. $R_{\sqrt{z}}$ is the Riemann surface of the two-valued function $w = \sqrt{z}$ when z ranges in the z-plane and w in the w-plane. Recall that $w = \sqrt{z}$ is single-valued on $R_{\sqrt{z}}$; as a matter of fact, it is a univalent conformal mapping between $R_{\sqrt{z}}$, the model of R, and the extended w-plane $\mathbf{C}^*$. When $R_{\sqrt{z}}$ is identified with $\mathbf{C}^*$, then $z = z(p)$ turns out to be the mapping $z = w^2$.

It is possible that $z = z(p)$ has two distinct simple poles at p_1 and p_2 each, where $p_1 \neq p_2$ are points in R. In this case, $z = z(p)$ has two distinct branch points q_1 and q_2, each of order 1. We may assume that $z(q_1) = -1$ and $z(q_2) = 1$. Now, $z^{-1}(\zeta)$ consists of precisely two distinct points for each $\zeta \in \mathbf{C}^* - \{-1, 1\}$; in particular, $z^{-1}(\infty) = \{p_1, p_2\}$. Exactly like the process in the last paragraph, we can construct a *concrete model* for R, namely, $R_{\sqrt{z^2-1}}$ as shown in Fig. 2.60, the Riemann surface of $w = \sqrt{z^2 - 1}$. When $R_{\sqrt{z^2-1}}$ is deemed as R conformally, $w = \sqrt{z^2 - 1}$ is single-valued on $R_{\sqrt{z^2-1}}$ on which $p_1 = \infty$ and $p_2 = \infty$ are the simple poles of the projection $z = z(p)$. ∞ is what we called the node point of $\sqrt{z^2 - 1}$ in Example 2 of Sec. 2.7.2.

Case 2. The genus $g = 1$.

The ramification index $\alpha = 2(2 + 1 - 1) = 4$ and there are four branch points of order 1 each on R. Let p_1, p_2, p_3, and p_4 be the four branch points of $z = z(p)$. Set $z(p_j) = e_j$ for $1 \leq j \leq 4$.

In case p_4 is a pole of order 2, then $e_4 = \infty$ and $e_1, e_2, e_3 \in \mathbf{C}$ are distinct. Since $z^{-1}(\zeta)$ is a two-point set for each $\zeta \in \mathbf{C}^* - \{e_1, e_2, e_3, \infty\}$, take two copies of $\mathbf{C}^*$, taped as sheets 1 and 2, and cut from e_1 to e_2 along a Jordan $\widehat{e_1 e_2}$ curve and from e_3 to ∞ along another disjoint curve $\widehat{e_3 \infty}$ on each sheet. Paste two sheets together crosswise along the cuts $\widehat{e_1 e_2}$ and $\widehat{e_3 \infty}$ to form a Riemann surface $R_{\sqrt{}}$. $R_{\sqrt{}}$ is the Riemann surface of $w = \sqrt{(z - e_1)(z - e_2)(z - e_3)}$, on which it turns out to be single-valued (see Exercise B(3) of Sec. 2.7.2). $R_{\sqrt{}}$ is a *concrete model* for R. See Fig. 7.6. Draw a Jordan closed curve a around the cut $\widehat{e_1 e_2}$ in sheet 1 as indicated in Fig. 7.6, and a Jordan closed curve b, starting from a point on the upper side of the cut $\widehat{e_1 e_2}$ in sheet 1 and going to a point in the upper side of the cut $\widehat{e_3 \infty}$ and then returning to its starting point through sheet 2. Intuitively both a and b cannot be deformed continuously to a point in $R_{\sqrt{}}$, and this fact suggests that $R_{\sqrt{}}$ is not simply connected. More accurately, sheet 1 can

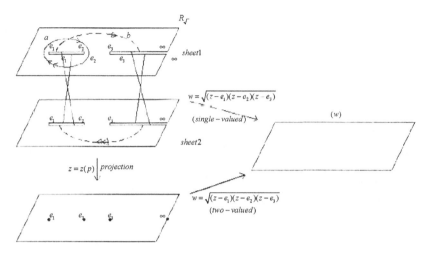

Fig. 7.6

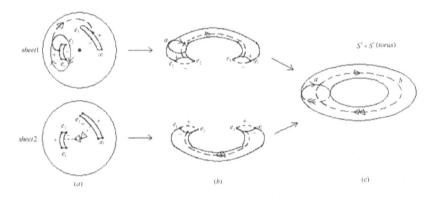

Fig. 7.7

be deformed topologically into the top figure in Fig. 7.7(a) and then into the top figure in Fig. 7.7(b), while sheet 2 into the bottom figures in Fig. 7.7(a) and 7.7(b), correspondingly, and finally, topologically pasting into the torus $S^1 \times S^1$ (refer to Example 3 in Sec. 7.1). This process convinces us that the torus is a topological model of $R_{\sqrt{}}$. See Remark 5 below.

In case $z = z(p)$ has two simple poles, then e_1, e_2, e_3, and e_4 are distinct points in $\mathbf{C}$. Now $z^{-1}(\zeta)$ consists of precisely two distinct points for each $\zeta \in \mathbf{C}^* - \{e_1, e_2, e_3, e_4\}$. The same procedure will lead to a *concrete model* $R_{\sqrt{}}$ of R. This $R_{\sqrt{}}$ is the Riemann surface of $w = \sqrt{\prod_{k=1}^4 (z - e_k)}$ on

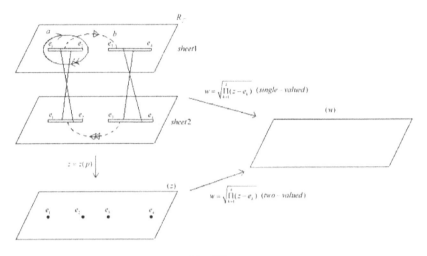

Fig. 7.8

which it becomes single-valued (see Exercise B(3) of Sec. 2.7.2), and can be deformed topologically into the torus $S^1 \times S^1$, too. See Fig. 7.8.

Note. The zeros and poles of $w = w(z)$ on R (or $R_{\sqrt{}}$).

Consider the case $w = \sqrt{\prod_{k=1}^{3}(z - e_k)}$ on $R_{\sqrt{}}$: According to (7.2.17), $R_{\sqrt{}}$ has the local parameter $\zeta = (z - e_k)^{1/2}$ at p_k for $k = 1, 2, 3$. Then $z = \zeta^2 + e_k$ and, then, in a neighborhood of 0 in the ζ-plane,

$$w \circ z^{-1}(\zeta) = \zeta\sqrt{(\zeta^2 + e_k - e_l)(\zeta^2 + e_k - e_j)}, \quad k \neq l \neq j$$

shows that $w = w(z)$ has a simple zero at p_k, $k = 1, 2, 3$. How about the pole? $R_{\sqrt{}}$ has the local parameter $\zeta = z^{-\frac{1}{2}}$ at p_4 (recall that p_4 is a pole of order 2 of $z = z(p)$). Then $z = \zeta^{-2}$ and, in a neighborhood of $\zeta = 0$,

$$w \circ z^{-1}(\zeta) = \zeta^{-3}\sqrt{\prod_{k=1}^{3}(1 - e_k\zeta^2)}$$

shows that $w = w(z)$ has a pole of order 3 at p_4. As a meromorphic function on $R_{\sqrt{}}$, $w = w(z)$ has the same number 3 of zeros and poles, and acts as a three-sheeted covering (or projection) of $R_{\sqrt{}}$ onto the w-plane $\mathbf{C}^*$.

Similar argument concludes that $w = \sqrt{\prod_{k=1}^{4}(z - e_k)}$ has four simple zeros at p_1, p_2, p_3, p_4, and two poles of order 2 at q_1, q_2 which are the simple poles of $z = z(p)$.

Case 3. The genus $g \geq 2$.

The ramification number now is $\alpha = 2(g + 2 - 1) = 2(g + 1)$, and R has exactly α branch points of order 1, say $p_1, p_2, \ldots, p_{2g+1}$, and p_{2g+2}. Set $e_k = z(p_k)$, $1 \leq k \leq 2g + 2$.

In case p_{2g+2} is a pole of order 2 of $z = z(p)$, then $e_1, \ldots, e_{2g+1}$ are distinct points in $\mathbf{C}$. R can be realized as the Riemann surface $R_{\sqrt{}}$ of the multiple-valued function

$$w = \sqrt{\prod_{k=1}^{2g+1} (z - e_k)}.$$

On $R_{\sqrt{}}$, $w = w(z)$ is single-valued and has simple zeros at $p_1, \ldots, p_{2g+1}$ and a pole of order $2g + 1$ at p_{2g+2}. As a meromorphic function $w = w(z)$: $R_{\sqrt{}} \rightarrow \mathbf{C}^*$, it covers the w-plane $\mathbf{C}^*$ totally $2g + 1$ times. $R_{\sqrt{}}$ covers the z-plane $\mathbf{C}^*$ twice via $z = z(p)$, namely, $R_{\sqrt{}}$ is two-sheeted and is topologically equivalent to a sphere with g handles. See Fig. 7.9 for $g = 2$. Note that $R_{\sqrt{}}$ is compact but not simply connected.

If $z = z(p)$ has two simple poles at q_1 and q_2, then $e_k \in \mathbf{C}$ for $1 \leq k \leq 2g + 2$. In this case R is realized as the Riemann surface $R_{\sqrt{}}$ of the function

$$w = \sqrt{\prod_{k=1}^{2g+2} (z - e_k)}.$$

$R_{\sqrt{}}$ is two-sheeted over the z-plane, not simply connected but is compact, and $R_{\sqrt{}}$ is topologically equivalent to a sphere with g handles. The meromorphic function $w = w(z)$: $R_{\sqrt{}} \rightarrow \mathbf{C}^*$ (the w-plane) is single-valued and covers the w-plane totally $2g + 2$ times.

Case 4. $w^{2g+1} = (z - a_1)(z - a_2)$, $a_1 \neq a_2$ and $g \geq 1$.

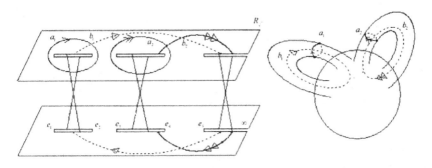

Fig. 7.9

Let us go back to Example 3 of Sec. 2.7.2 and consider $w = \sqrt[3]{z^2 - 1}$. ± 1 and ∞ are algebraic branch points of order 2. The ramification index is $\alpha = 3 \cdot 2 = 6$. The genus g of the Riemann surface $R_{\sqrt[3]{}}$ is given by the relation $6 = 2(g + 3 - 1)$, namely, $g = 1$. Refer to Figs. 2.63 and 2.64. $R_{\sqrt[3]{}}$ is a three-sheeted covering surface of the classical z-plane $\mathbf{C}^*$. The local parameter $\psi(q)$ at a point q_0 on $R_{\sqrt[3]{}}$ is given by (see (7.5.5) and (7.2.6))

$$
\psi(q) = \begin{cases}
z - z(q_0), & \text{if } q_0 \text{ is a regular point} \\
\sqrt[3]{z - 1}, & \text{if } z(q_0) = 1 \\
\sqrt[3]{z + 1}, & \text{if } z(q_0) = -1 \\
\dfrac{1}{\sqrt[3]{z}}, & \text{if } z(q_0) = \infty
\end{cases}
$$

Here we treat $z = z(q)$ as a nonconstant meromorphic function on $R_{\sqrt[3]{}}$ of degree 3. As a consequence, $w = w(z)$ has two simple zeros and a pole of order 2 on $R_{\sqrt[3]{}}$, and hence $w = w(z) : R_{\sqrt[3]{}} \to \mathbf{C}^*$ (the classical w-plane) is a nonconstant meromorphic function of degree 2, covering $\mathbf{C}^*$ twice.

On the other hand, the inverse function of $w = w(z)$ is $z = z(w) = \sqrt{w^3 + 1} = \sqrt{(w - e^{\frac{\pi i}{3}})(w + 1)(w - e^{\frac{5\pi i}{3}})}$. According to our discussion in Case 2, $w = w(p)$ indeed is a meromorphic function on a two-sheeted Riemann surface $R_{\sqrt{}}$ spread over the w-plane, and $z = z(w) : R_{\sqrt{}} \to \mathbf{C}^*$ (the classical z-plane) is meromorphic of degree 3 and hence, covers $\mathbf{C}^*$ three times. See Fig. 7.10. In conclusion, $w = \sqrt[3]{z^2 - 1} : R_{\sqrt[3]{}} \to R_{\sqrt{}}$ is univalently conformal. $R_{\sqrt[3]{}}$ and $R_{\sqrt{}}$ are then conformally equivalent and both

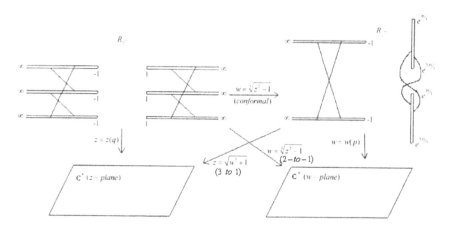

Fig. 7.10

are topologically equivalent to a sphere with a handle, namely, the torus $S^1 \times S^1$. This particular case justifies (d)3 in (7.2.23) below. It is in this sense that we still view $R_{\sqrt[3]{}}$ as a hyperelliptic surface. Note that, by setting an intermediate parameter $t = w^3 = z^2 - 1$, we still can obtain the same conclusion by considering $t = w^3$ and $t = z^2 - 1$ separately and combining the results together.

The general case $w^{2g+1} = (z - a_1)(z - a_2)$ can be treated similarly to obtain the following facts:

1. The Riemann surface $R_{2g+\sqrt{}}$ of $w = {}^{2g+1}\!\!\sqrt{(z - a_1)(z - a_2)}$ is a $(2g + 1)$-sheeted covering surface of genus g of the z-plane. On it, z is a meromorphic function of degree $2g + 1$ while w is a meromorphic function of degree 2.

2. The Riemann surface $R_{\sqrt{}}$ of the inverse function $z = z(w)$ is a two-sheeted covering surface of genus g of the w-plane. $w = w(z)$ is a meromorphic function of degree 2 on $R_{\sqrt{}}$ and z is the one with degree $2g + 1$.

Moreover, $w = {}^{2g+1}\!\!\sqrt{(z - a_1)(z - a_2)}$ sets up a conformal mapping between $R_{2g+\sqrt{}}$ and $R_{\sqrt{}}$. $R_{2g+\sqrt{}}$ is then hyperelliptic.

Remark. (5) Some surface topology are needed for a better understanding of the abstract theory of Riemann surfaces. For instance, a Riemann surface is orientable, triangulable, and has a countable basis. A general reference for these is Ref. [6], or [75], p. 112 and p. 239. Also, see (7.3.8) and the last paragraph of this chapter.

Any triangulation of a compact surface is necessarily finite, say n, in the number of triangles. Such a triangulation can be mapped topologically onto a planar $(n + 2)$-gon with even number of sides, treated as a topological model of the surface. Via the notion of linked edges, such a model can be further simplified into the so-called *normal form* of a compact orientable surface R: a polygon whose symbol is

$$aa^{-1}(\text{a surface of genus zero}), \text{ or}$$

$$a_1 b_1 a_1^{-1} b_1^{-1} \cdots a_g b_g a_g^{-1} b_g^{-1} \quad (\text{a surface of genus } g \geq 1). \quad (7.2.18)$$

Note that the genus $g \geq 0$ is a topological invariant. A surface of genus 0 is both topologically and conformally the Riemann sphere; a surface of genus 1 is topologically a torus and, in general, a surface of genus $g \geq 1$ is topologically a sphere with g *handles* (see Ref. [75], p. 124).

(6) Let R be a triangulable, orientable two-dimensional real manifold, simply called a *surface*. For a triangulation of R, we call the vertices $\{P_1, P_2, P_3, \ldots\}$ *zero-simplices*, the edges *one-simplices*, and the triangles *two-simplices*. The oriented edge from P_1 to P_2 is denoted by $\langle P_1, P_2 \rangle$ and the oriented triangle $\langle P_1, P_2, P_3 \rangle$ is bounded by the oriented edges $\langle P_1, P_2 \rangle$, $\langle P_2, P_3 \rangle$, $\langle P_3, P_1 \rangle$, subject to the conventions:

$$\langle P_2, P_1 \rangle = -\langle P_1, P_2 \rangle; \langle P_3, P_2, P_1 \rangle = -\langle P_1, P_2, P_3 \rangle.$$

An *n-chain* $(n = 0, 1, 2)$ is a finite linear combination of n-simplices with integral coefficients. The set of n-chains forms an abelian group under addition, denoted by C_n. Designate $C_n = \{0\}$ for $n > 2$. The *boundary operator* $\partial : C_n \to C_{n-1}$ $(n \geq 1)$ is defined by:

For $n = 0 : \partial \langle P \rangle = 0$;

For $n = 1 : \partial \langle P_1, P_2 \rangle = P_2 - P_1$;

For $n = 2 : \partial \langle P_1, P_2, P_3 \rangle = \langle P_1, P_2 \rangle + \langle P_2, P_3 \rangle + \langle P_3, P_1 \rangle$,

and then extend ∂ to n chains by linearity. ∂ is a group homomorphism from C_n into C_{n-1}, with kernel $Z_n = \mathrm{Ker} \partial$ a normal subgroup of C_n and its image $B_{n-1} = \partial(C_n)$ a subgroup of C_{n-1}, so that C_n/Z_n is group isomorphic to B_{n-1}. Members in Z_n are called *n-cycles* or *closed n-chains* and members in B_n are called *n-boundaries* or *exact n-chains*. $B_2 = 0$ since $C_n = 0$ for $n > 2$ and B_n is a normal subgroup of Z_n since $\partial^2 = \partial \circ \partial = 0$. What is interested to us is the quotient group

$$H_n(R) = H_n = Z_n/B_n, \quad n = 0, 1, 2 \tag{7.2.19}$$

called the *n-th simplicial homology group* of R with respect to the given triangulation.

Denote by $[z]$ the equivalence class in H_n of $z \in Z_n$. $[z_k]$, $k = 1, 2, \ldots, b_n$, form a *basis* for H_n if $\sum_{k=1}^{b_n} \alpha_k[z_k] = 0$ (α_k, integers) always implies $\alpha_k = 0$ for $1 \leq k \leq b_n$, and any element of H_n can be expressed as an integral linear combination of these $[z_k]$. The number b_n is called the *nth Betti number* of the triangulated surface.

An important fact is that both H_n and b_n $(n = 0, 1, 2)$ are topological invariants. As a matter of fact, it can be shown that

(1) $b_o = 1$ (for any surface) and H_0 is isomorphic to the integers.
(2) $b_2 = 0$ for noncompact surface and $H_2 = 0$;
    $b_2 = 1$ for compact surface and H_2 is isomorphic to integers.

(3) H_1 is isomorphic to the abelianized fundamental group of R.

$$(7.2.20)$$

See Ref. [75], pp. 129–139 for a proof of (3).

In particular, for a compact surface with normal form $a_1 b_1 a_1^{-1} b_1^{-1} \cdots a_g b_g a_g^{-1} b_g^{-1}$ (see (7.2.18)), a_k and b_k ($1 \leq k \leq g$) are simplicial one-cycles, and all of them form a homology basis for the first homology group. For a proof, see Ref. [75], pp. 139–142.

(7) For a compact surface R of genus g, its *Euler–Poincaré characteristic* is given by

$$\chi = 2 - 2g$$

$$= \alpha_0 - \alpha_1 + \alpha_2 \quad (\alpha_k \text{ is the number of } k\text{-simplices in a triangulation}$$

$$\text{of } R, \text{ for } k = 0, 1, 2)$$

$$= b_o - b_1 + b_2 \quad (b_k \text{ is the } k\text{-th Betti number of the}$$

$$\text{triangulated surface}). \tag{7.2.21}$$

This is a topological invariant, too. See Ref. [75], pp. 142–143.

As an application of (7.2.21), we have the following remarkable

Riemann–Hurwitz relation. Let $f : R_1 \to R_2$ be a nonconstant analytic mapping from a compact Riemann surface R_1 of genus g_1 onto a compact Riemann surface R_2 of genus g_2. Let

$$n = \text{the degree of } f \text{ (see (7.2.16)), and}$$

$$B = \text{the ramification index (or the total branch number) of } f$$

$$= \sum_{p \in R_1} b_f(p) \quad (\text{see (7.2.3) and (7.2.4)}).$$

Then

$$g_1 = n(g_2 - 1) + 1 + \frac{B}{2}. \tag{7.2.22}$$

In particular, in case $g_2 = 0$, then $B = 2(g_1 + n - 1)$ is the one mentioned in Exercise A(3) of Sec. 3.5.6.

Sketch of proof. Since R_2 is compact and f is nonconstant, the set $S = \{f(p) | p \in R_1 \text{ and } b_f(p) > 0\}$ is a finite set. Triangulate R_2 so that each point in S is a vertex of the triangulation, say with α_k k-simplices for $k = 0, 1, 2$. Then the mapping f lifts this triangulation to a triangulation of

R_1, with $(n\alpha_0 - B)$ zero simplices, $n\alpha_1$ one-simplices and $n\alpha_2$ two-simplices. On applying (7.2.21) to both R_1 and R_2, we have

$$\alpha_0 - \alpha_1 + \alpha_2 = 2 - 2g_2,$$

$$(n\alpha_0 - B) - n\alpha_1 + n\alpha_2 = 2 - 2g_1$$

$$\Rightarrow\ g_1 = n(g_2 - 1) + 1 + \frac{B}{2}.$$

For another yet analytic proof, see Ref. [32], pp. 75–76, as an immediate consequence of the important Riemann–Roch theorem.

From (7.2.22), it follows easily the following

Corollaries. *Adopt notations in* (7.2.22).

(a) *The ramification index $B \geq 0$ is always even.*
(b) *If $g_1 = 0$, then $g_2 = 0$ and thus $B = 2(n-1)$.*
(c) *If $g_1 = g_2 \geq 1$, then either $n = 1$ and thus $B = 0$ or $g_1 = 1$ and thus $B = 0$.*
(d) *In case $B = 0$, namely, $f:R_1 \to R_2$ is an uniramified covering, then:*

    1. *$g_1 = 0 \Rightarrow n = 1$ and $g_2 = 0$.*
    2. *$g_1 = 1 \Rightarrow g_2 = 1$ and n is arbitrary.*
    3. *$g_1 \geq 2 \Rightarrow g_1 = g_2$ for $n = 1$;*

$$g_1 > g_2 > 1 \text{ for } n > 1 \text{ and } n \text{ dividing } g - 1. \tag{7.2.23}$$

These results do help to clarify many concrete examples we constructed before.

(8) What we talked about hyperelliptic surfaces seems superficial. There are remarkable theoretical results in this field. They involve important topics in the theory of compact Riemann surface, far beyond the scope of this introductory text. Take, for instance, that the g-differentials

$$\frac{dz}{w}, \frac{z\,dz}{w}, \ldots, \frac{z^{g-1}}{w}dz, \quad \text{where } w = \sqrt{\prod_{k=1}^{2g+1}(z - e_k)} \text{ or } \sqrt{\prod_{k=1}^{2g+2}(z - e_k)}$$

$$\tag{7.2.24}$$

form a basis for the abelian differentials of the first kind on $R_{\sqrt{\,}}$. Minded readers are urged to read Ref. [32], pp. 93–103 and the last Chap. VII there or Ref. [75], pp. 292–298.

(9) Using the Riemann–Roch theorem, one can prove that *every compact Riemann surface of genus g is conformally equivalent to a $(g+1)$-sheeted*

covering surface of the sphere (as explained in (7.2.16)). For a proof, see Ref. [75], p. 275.

This result shows that, from the function-theoretic point of view, the introduction of abstract concept in the case of compact surface gives rise to nothing more than the branched coverings of the sphere with a finite number of sheets. In particular, the Riemann sphere is the only compact Riemann surface of genus zero.   □

Example 3. *Other (nonsimply connected) compact Riemann surfaces.*

Explanation. Here we consider five concrete functions.

(a) $w = \sqrt[3]{\frac{z-1}{z^2}} + \sqrt{z-i}$ (see Exercise A(9) of Sec. 2.7.2):

Let $z \neq 0,\ 1, i, \infty$ and set $z - 1 = r_1 e^{i\theta_1}$, $z - i = r_2 e^{i\theta_2}$, $z_3 = r_3 e^{i\theta_3}$ where $0 \leq \theta_k \leq 2\pi$. Then $w = w(z)$ has the following six distinct values:

$$w_k = \sqrt[3]{r_1/r_3^2}\, e^{i(\theta_1 - 2\theta_3 + 2k\pi)/3} + \sqrt{r_2}\, e^{i\theta_2/2}, \quad k = 1, 2, 3;$$

$$w_{3+k} = \sqrt[3]{r_1/r_3^2}\, e^{i(\theta_1 - 2\theta_3 + 2k\pi)/3} + \sqrt{r_2}\, e^{i(\theta_2 + 2\pi)/2}, \quad k = 1, 2, 3.$$

Choose the branch cuts $[1,0]$, $[0,i]$, and $[i, +i\infty]$. Hence $w = w(z)$ has six single-valued branches on $\mathbf{C}^* - [1, 0] - [0, i] - [i, +i\infty]$. The Riemann surface R_z is a six-sheeted covering of the sphere, with double branch points of order 2 over each point of $z = 0$ and 1, a triple branch point of order 1 over i and a branch point of order 5 over ∞. Its Riemann surface R_z can be pictured as in Fig. 7.11 or, more concretely, in Fig. 7.12. According to (7.2.22), the genus of R_z is given by $6(0 - 1) + 1 + \frac{16}{2} = 3$.

The local parameter $\varphi(p)$ at a point $p_0 \in R_z$ is given by

$$\varphi(p) = \begin{cases} z - z(p_0), & \text{if } p_0 \text{ is a regular point} \\ \sqrt[3]{z - 1}, & \text{if } z(p_0) = 1 \\ \sqrt{z - i}, & \text{if } z(p_0) = i \\ \sqrt[3]{z}, & \text{if } z(p_0) = 0 \\ \dfrac{1}{\sqrt[6]{z}}, & \text{if } z(p_0) = \infty \end{cases},$$

where $z = z(p)$ is treated as a meromorphic function on R_z of degree 6. Set $\zeta = \sqrt[3]{z} = \varphi(p)$ in a coordinate neighborhood of the point p_0 where

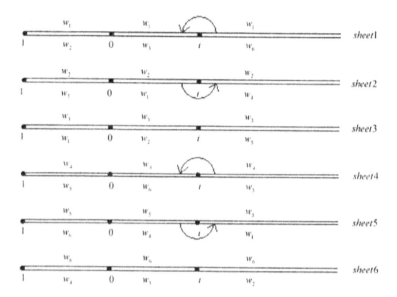

Fig. 7.11

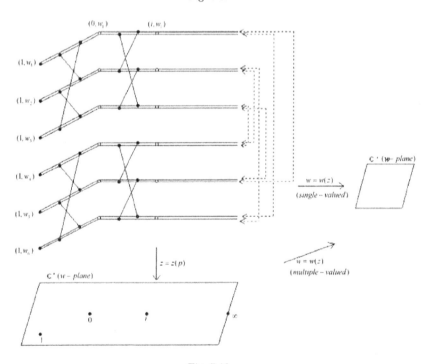

Fig. 7.12

$z(p_0) = 0$. Then $z = \zeta^3$ and

$$w \circ \varphi^{-1}(\zeta) = \frac{1}{\zeta^2} \sqrt[3]{\zeta^3 - 1} + \sqrt{\zeta^3 - i}.$$

This means that $w = w(z)$, as a function on R_z, has a pole of order 2 at the point p_0 where $z(p_0) = 0$. Similarly, set $\zeta = \frac{1}{\sqrt[6]{z}} = \varphi(p)$ in a neighborhood of the point p_0 where $z(p_0) = \infty$ and we have $w \circ \varphi^{-1}(\zeta) = \zeta^2 \sqrt[3]{1 - \zeta^6} + \frac{1}{\zeta^3}\sqrt{1 - i\zeta^6}$. Thus $w = w(z)$ has a pole of order 3 at the point p_0 where $z(p_0) = \infty$. As a whole, $w = w(z)$ has two poles on R_z with order 5 in total and, according to (7.2.16), $w = w(z) : R_z \to \mathbf{C}^*$ (the w-plane) is a single-valued meromorphic function of degree 5. The inverse function $z = z(w) : \mathbf{C}^* \to R_z$ is nevertheless multiple-valued, and it becomes single-valued on the five-sheeted Riemann surface R_w over the w-plane. This R_w is of genus 3, and $w = w(z) : R_z \to R_w$ is univalently conformal.

   (b) $w = \sqrt[3]{z^4 - 1}$ (see Exercise B(5)(c) of Sec. 2.7.2):

   The Riemann surface $R_{\sqrt[3]{}}$ has an algebraic branch point of order 2 over each of the points $1, -1, i, -i$, and ∞. Hence $R_{\sqrt[3]{}}$ is three-sheeted and of genus 3. $z = z(p)$, as a meromorphic function from $R_{\sqrt[3]{}}$ onto the extended z-plane, is of degree 3. $w = w(z)$ is a single-valued meromorphic function on $R_{\sqrt[3]{}}$ of degree 4 (it has four distinct zeros and a single pole of order 4).

   The inverse function $z = \sqrt[4]{w^3 + 1}$ is single-valued on the Riemann surface $R_{\sqrt[4]{}}$ over the w-planes. $R_{\sqrt[4]{}}$ is four-sheeted, has an algebraic branch point of order 3 over each of the points -1, $e^{\pi i/3}$, $e^{5\pi i/3}$, and ∞, and hence is of genus 3, too. $z = z(w)$ is single-valued, meromorphic on $R_{\sqrt[4]{}}$ of degree 3 (it has a pole of order 3).

   Therefore $w = w(z) : R_{\sqrt[3]{}} \to R_{\sqrt[4]{}}$ is univalently conformal. $w = w(z)$ maps the point p_0, where $z(p_0) = \infty$, onto the point q_0, where $w(q_0) = \infty$ (recall that $w = w(q) : R_{\sqrt[4]{}} \to \mathbf{C}^*$ is of degree 4). Take the local parameter $\zeta = \frac{1}{\sqrt[3]{z}}$ at p_0 and the local parameter $\eta = \frac{1}{\sqrt[4]{w}}$ at q_0. Then

$$\eta \circ w \circ \zeta^{-1}(\zeta) = \eta\left(\sqrt[3]{\zeta^{-12} - 1}\right) = \eta\left(\frac{1}{\zeta^4}\sqrt[3]{1 - \zeta^{12}}\right) = \frac{\zeta}{\sqrt[12]{1 - \zeta^{12}}}$$

shows that $w = w(z)$ is indeed univalently analytic at p_0.

   (c) $w = \sqrt{\sqrt[3]{z} - 1}$ (see Exercise B(5)(d) of Sec. 2.7.2):

   $w = w(z)$ can be decomposed as $\tau = \sqrt[3]{z}$ and $w = \sqrt{\tau - 1}$. The Riemann surface R_τ of $\tau = \sqrt[3]{z}$ is a three-sheeted covering surface, pasted crosswise along the branch cuts $[\infty, 0]$ and having a branch point of order 2 over $z = 0$ and ∞ each. Meanwhile, the Riemann surface $R_{w(\tau)}$ of $w = \sqrt{\tau - 1}$ is a two-sheeted covering surface over the τ plane, pasted crosswise along the branch

cut $[1, \infty]$ and having a branch point of order 1 over $\tau = 1$ and ∞ each. Take
two copies of R_τ and paste them crosswise along the cut $[1, \infty]$, and we get
the Riemann surface R_z of $w = w(z)$. R_z is six-sheeted and has a double
branch point of order 2 over $z = 0$, a branch point of order 1 over $z = 1$
and a branch point of order 5 over $z = \infty$. Its genus is $g = 0$ (according to
(7.2.22) with $g = g_1$, $n = 6$, $g_2 = 0$, and $B = 2 \times 2 + 1 \times 1 + 5 \times 1 = 10$)
and R_w is thus conformally and topologically equivalent to the sphere. See
the right figure for R_z in Fig. 7.13.

   An alternative way to construct R_z is to consider the inverse function
$z = (1 + w^2)^3$. Observing that $z' = 6w(1 + w^2)^2 = 0 \Leftrightarrow w = 0, \pm i$, and
$z''(0) \neq 0$, $z''(\pm i) = 0$, and $z'''(\pm i) \neq 0$, we know that $z = z(\pm i) = 0$ is
a double branch point of order 2 and $z = z(0) = 1$ is the one of order 1.
Since $z = (1 + w^2)^3$ has a pole of order 6 at ∞, it follows that $z = \infty$
is an algebraic branch point of order 5. Set $z = x + iy$ and $w = u + iv$,
and they are related by $x = (1 + u^2 - v^2)^3 - 12u^2v^2(1 + u^2 - v^2)$ and
$y = 6uv(1 + u^2 - v^2 - \frac{2}{\sqrt{3}}uv)(1 + u^2 - v^2 + \frac{2}{\sqrt{3}}uv)$. Follow the steps as
indicated in Example 4 of Sec. 2.7.2 to find the fundamental domains for
$z = z(w)$ (see (2.7.4)). See Fig. 7.13 for the Riemann surface R_z, where
in the left figure, each of the shaded portions are mapped univalently onto

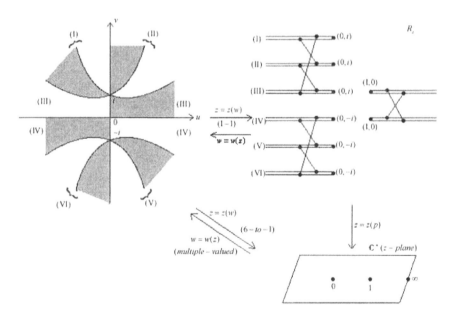

Fig. 7.13

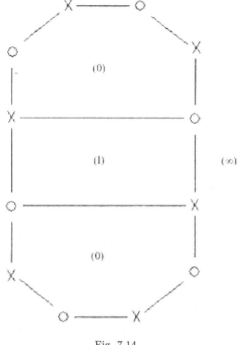

Fig. 7.14

the lower half-plane in the z-plane. Its line complex is in Fig. 7.14. R_z is compact and simply connected ($g = 0$).

The local parameter of R_z at the point p_0, where $z(p_0) = 1$, is given by $\zeta = \varphi(p) = \sqrt{z - 1}$. Then $z = 1 + \zeta^2$ and

$$w \circ \varphi^{-1}(\zeta) = \sqrt{\sqrt[3]{1 + \zeta^2} - 1}$$

$$= \zeta \sqrt{\sum_{n=1}^{\infty} C_n^3 \zeta^{n-1}} \text{ (see Example 2 in Sec. 3.3.2 with } \sqrt[3]{1} = 1)$$

shows that $w = w(z)$ is locally univalent at p_0. On the other hand, the local parameter of R_z at the point p_∞, where $z(p_\infty) = \infty$, is given by $\zeta = \varphi(p) = \frac{1}{\sqrt[6]{z}}$. In this case, $z = \zeta^{-6}$ and $w \circ \varphi^{-1}(\xi) = \sqrt{\sqrt{\zeta^{-2} - 1}} = \frac{1}{\zeta}\sqrt{1 - \zeta^2}$ again indicates that $w = w(z)$ is locally univalently at p_∞, too. As a whole, (7.2.16) guarantees that $w = w(z) : R_z \to \mathbf{C}^*$ (the w-plane) is univalently conformal. Note that $z = z(p) : R_z \to \mathbf{C}^*$ (the z-plane) is meromorphic and is of degree 6.

(d) $w^n = \frac{z^n}{(1+z^n)^2}$ ($n \geq 3$ is odd): Set temporarily $\zeta = z^n$, $\eta = w^n$, and then $\eta = \frac{\zeta}{(1+\zeta)^2}$ holds.

Via the composite maps: $z \to \zeta = z^n \to \eta = \frac{\zeta}{(1+\zeta)^2} \to w = \sqrt[n]{\eta}$, we can construct the n-sheeted Riemann surface R_z, over the z-plane, of

$$w = \frac{z}{\sqrt[n]{(1+z^n)^2}} = \frac{z}{\sqrt[n]{\prod_{k=0}^{n-1}(z-z_k)^2}}, \quad \text{where}$$

$$z_k = e^{\frac{\pi i}{n}} \omega^k \text{ for } 0 \leq k \leq n-1 \quad \text{and}$$

$$\omega = e^{\frac{2\pi i}{n}}.$$

R_z has an algebraic branch point of order $n-1$ over each of the points z_k, $0 \leq k \leq n-1$. Note that ∞ is not a branch point but a node point (see (2.7.3) and Exercise B(4) of Sec. 2.7.2) since n divides $2n$. Adopting the rays $\overline{z_k \infty}$, $0 \leq k \leq n-1$, as branch cuts, R_z is graphed as in Fig. 7.15. R_z is of genus $\frac{1}{2}(n-1)(n-2)$. $z = z(p) : R_z \to \mathbf{C}^*$ (the z-plane) is a meromorphic function of degree n.

What is the degree of $w = w(z)$ as a meromorphic function from R_z onto the w-plane? Fix k ($0 \leq k \leq n-1$), the local parameter of R_z at the point p_k, where $z = z(p_k) = z_k$, is given by $\zeta = \varphi(p) = \sqrt[n]{z - z_k}$. In terms of ζ, $w = w(z)$ becomes

$$w \circ \varphi^{-1}(\zeta) = \frac{\zeta^n + z_k}{\sqrt[n]{\zeta^{2n} \prod_{\substack{l=0 \\ l \neq k}}^{n-1}(\zeta^n + z_k - z_l)^2}}$$

$$= \left(\zeta^{n-2} + \frac{z_k}{\zeta^2}\right)\left\{\prod_{\substack{l=0 \\ l \neq k}}^{n-1}(\zeta^n + z_k - z_l)^2\right\}^{-\frac{1}{2}}.$$

Hence, $w = w(z)$ has a pole of order 2 at p_k for $0 \leq k \leq n-1$; in total, $w = w(z)$ has $2n$ poles on R_z. According to (7.2.16), $w = w(z)$ is of degree $2n$. It follows that $w = w(z)$ is single-valued on R_z and covers the w-plane $2n$ times. This fact allows us to expect a $2n$-sheeted covering surface R_w, over the w-plane, so that it is conformally equivalent to R_z. What might R_w look like?

To construct R_w: Solve $w^n = \frac{z^n}{(1+z^n)^2}$ and we get the $2n$-valued inverse function

$$z = \sqrt[n]{\frac{-(2w^n - 1) + \sqrt{1 - 4w^n}}{2w^n}}$$

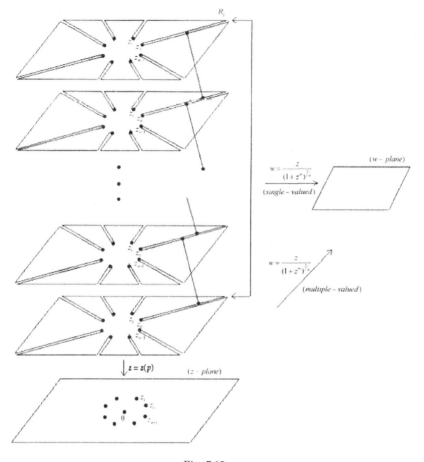

Fig. 7.15

except at $w_k = \frac{1}{\sqrt[n]{4}}e^{2k\pi i/n}$, $0 \le k \le n-1$, and $w = \infty$. This suggests that R_w should be of $2n$-sheeted covering of the w-plane with branch points over w_k ($0 \le k \le n-1$) and ∞. Since $z = z(w_k) = \sqrt[n]{1}$ assumes n distinct values $z_l = e^{2\pi l i/n}$ ($0 \le l \le n-1$) for each k ($0 \le k \le n-1$), R_w has n branch points of order 1 over each w_k and, hence, has n^2 branch points of order 1 in total over these w_k. Also $z(\infty) = \sqrt[n]{-1} = e^{\pi i/n}z_l$ ($0 \le l \le n-1$) shows that R_w has another n branch points of order 1 over $w = \infty$. Therefore R_w has its genus equal to $\frac{1}{2}(n-1)(n-2)$.

Better considering consecutively the maps $z \to \zeta = z^n \to \eta = \frac{\zeta}{(1+\zeta)^2} \to \sqrt[n]{\eta}$, it is easy to see that *the branch* of $w = \frac{z}{\sqrt[n]{(1+z^n)^2}}$, uniquely determined by the positive $\frac{1}{\sqrt[n]{4}}$ at $z = 1$, maps the sector domain $|z| < 1$, $|\text{Arg } z| < \frac{\pi}{n}$, univalently conformally onto the angular domain $|\text{Arg } z| < \frac{\pi}{n}$ with the cut $\left[\frac{1}{\sqrt[n]{4}}, \infty\right]$ along the positive real axis deleted: $w(0) = 0, w(1) = \frac{1}{\sqrt[n]{4}}$, and $w(e^{\pi i/n}) = w(e^{-\pi i/n}) = \infty$. Observing that $w(\frac{1}{z}) = w(z)$ or by using the reflection principle, we obtain that *the branch* maps the angular domain $|\text{Arg } z| < \frac{\pi}{n}$ univalently and conformally onto a two-sheeted angular domain $|\text{Arg } z| < \frac{\pi}{n}$, pasted together crosswise along the cuts $\left[\frac{1}{\sqrt[n]{4}}, \infty\right]$. See Fig. 7.16. As a whole, *the branch* maps the z-plane univalently and conformally onto a two-sheeted covering Riemann surface, pasted together crosswise along the cuts $\left[\frac{1}{\sqrt[n]{4}}e^{2k\pi i/n}, \infty\right]$ for $0 \le k \le n-1$. See Fig. 7.17. We intentionally call this two-sheeted surface *Model* 1 induced by *the* (*positive*) *branch* of $w = w(z)$, determined by $w(v) = \frac{1}{\sqrt[n]{4}}$. Applying similar process to the remaining $(n-1)$ branches determined by $\frac{1}{\sqrt[n]{4}}e^{2k\pi i/n}$, $1 \le k \le n-1$, we will obtain *Model* 2, ..., *Model* n. These n models are completely congruent.

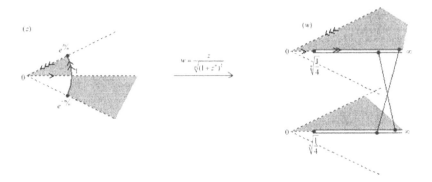

Fig. 7.16

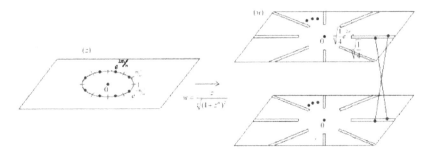

Fig. 7.17

Finally paste the second sheet of Model 1 and the first sheet of Model 2 crosswise along the cuts $\left[\frac{1}{\sqrt[n]{4}}e^{2k\pi i/n}, \infty\right]$ for $o \le k \le n-1, \ldots$, and the second sheet of Model n to the first sheet of Model 1. The resulted surface is R_w. See Fig. 7.18.

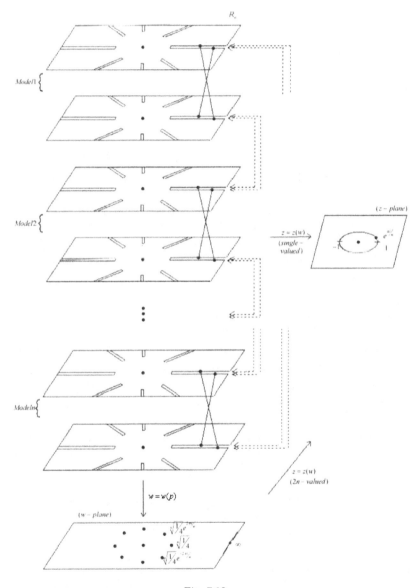

Fig. 7.18

As a conclusion, the function $w = w(z)$ defined by $w^n = \frac{z^n}{(1+z^n)^2}$ or its inverse function $z = z(w)$ provides a univalent conformal mapping between R_z and R_w.

(e) $w^n = \frac{z^n}{(1+z^n)^2}$ ($n \geq 2$ is even):

R_z is an n-sheeted covering Riemann surface of the z-plane, has a double algebraic branch point of order $\frac{n}{2} - 1$ over each of the points $z_k = e^{\pi i/n}\omega^k$ ($0 \leq k \leq n - 1$) where $\omega = e^{2\pi i/n}$ (refer to Exercise B(4)(a) of Sec. 2.7.2). And $w = w(z)$ has n distinct branches

$$w_k = w_0 e^{k\pi i/n}(0 \leq k \leq n - 1), \quad \text{where}$$

$$w_0(z) = \frac{z}{\sqrt[n/2]{1 + z^n}} \quad \text{with} \quad \sqrt[n/2]{1} = 1.$$

According to (7.2.22), R_z has its genus equal to $g = n(0 - 1) + 1 + \frac{1}{2} \cdot 2n \cdot (\frac{n}{2} - 1) = \frac{n^2}{2} - 2n + 1$ for $n \geq 4$. What happens to $n = 2$ (something to do with the Joukowski function).

Readers are asked to take pains to construct R_z and R_w, the Riemann surface of the inverse function $z = z(w)$. See Exercise A(2).

Section (3) Open Riemann surface conformally equivalent to the finite plane C or the puncture plane C − {0} (review of Secs. 2.7.3 and 2.7.4)

The most basic example in this direction is provided by the Riemann surface $R_{\log z}$ of $\log_z$, as indicated by Fig. 2.73 and in Example 2 of Sec. 3.3.3. Remind that $R_{\log z}$ is an infinite-sheeted smooth covering of the classical z-plane $\mathbf{C}^*$; has two logarithmic branch points over $z = 0$ and $z = \infty$, and is conformally equivalent to the finite w-plane $\mathbf{C}$. Note that no point in $R_{\log z}$ lies over $z = 0$ and $z = \infty$.

$w = \log(z + \sqrt{z^2 - 1}) = \cosh^{-1} z$ (see Example 3 of Sec. 2.7.3): Its Riemann surface R_w (see Fig. 2.79) has infinitely many algebraic branch points of order 1 over the points $z = \pm 1$ and two logarithmic branch points over $z = \infty$. No points in R_w lies over ∞. The local parameter is given by

$$\varphi(p) = \begin{cases} z - z(p_0), & \text{if } p_0 \text{ is a regular point} \\ \sqrt{z - 1}, & \text{if } z(p_0) = 1 \\ \sqrt{z + 1}, & \text{if } z(p_0) = -1 \end{cases},$$

where $z = z(p) : R_z \to \mathbf{C}^*$ is the projection. Let $D : |z - 1| < r$ ($0 < r < 1$) be a Euclidean disk with center at 1. What is then $z^{-1}(D)$ in R_z?

Referring to Fig. 3.14, $z^{-1}(D)$ is a countable disjoint union of "open disks" in R_z with center p_0, where $z(p_0) = 1$, and radius r. Fix such a point p_0 and a corresponding "open disk" $N_r(p_0)$. Consider the local parameter $\zeta = \sqrt{z-1}$ on $N_r(p_0)$ and the local parameter $\eta = z-1$ on D. Then, in terms of ζ and η, the projection $z = z(p)$ can be expressed as $\eta = \zeta^2$. This fact justifies that p_0 is a branch point of order 1 of the surface R_z. In fact R_z is an infinitely-sheeted branched covering surface of the classical z-plane. On the other hand, at a point p_0 where $z(p_0) = 1$, we have, by setting $\zeta = \varphi(p) = \sqrt{z-1}$

$$w \circ \varphi^{-1}(\zeta) = \log(1 + \zeta^2 + \zeta\sqrt{2+\zeta^2})$$

in a sufficiently small neighborhood of $\zeta = 0$. Fix any branch of $w \circ \varphi^{-1}(\zeta)$ in such a neighborhood and we have $(w \circ \varphi^{-1})'(0) = \log\sqrt{2} \neq 0$. This indicates that $w \circ \varphi^{-1}(\zeta)$ is univalently conformal there. As a whole, $w = w(z) : R_z \to \mathbf{C}$ (the finite w-plane) is univalently conformal (refer to the construction of Fig. 2.79).

Check (2.7.1) or review Sec. 2.7.4, and we confirm that the Riemann surfaces of $\cos^{-1} z$, $\sin^{-1} z$, and $\sinh^{-1} z$ are conformally equivalent to the finite complex plane.

$w = \log \frac{z-a}{z-b}, a \neq b$ (see Example 4 in Sec. 2.7.3): Since $\zeta = \frac{z-a}{z-b}$ is univalently conformal between the extended z-plane and the ζ-plane, and maps the line segment $\overline{ab}$ onto ray $[0, \infty]$, its Riemann surface (see Fig. 2.80 for $a = 1$ and $b = -1$) is conformally equivalent to that of $\log z$, which is known to be conformally equivalent to the finite complex plane.

Recall (2.7.1) or Sec. 2.7.4, and we again confirm that the Riemann surfaces of $\tan^{-1} z$, $\cot^{-1} z$, $\tanh^{-1} z$, and $\coth^{-1} z$ are conformally equivalent to the finite complex plane.

The Riemann surface R_w of the inverse function $z = z(w)$ of $w = z + e^z$ (see Example 2 of Sec. 6.2.3): The Riemann surface R_w (see Fig. 6.30) has an algebraic branch point of order 1 over each of the points $-1 + (2n+1)\pi i$, $n = 0, \pm 1, \ldots$, in the w-plane and a logarithmic branch point over the point ∞. It is an infinitely-sheeted branched covering surface of the w-plane and $w = w(p) : R_w \to \mathbf{C}$ is the projection map.

The inverse function $z = z(w)$ itself provides a univalent conformal mapping between R_w and the classical finite z-plane as can be seen in the construction of R_w. We check its local univalence only at a branch point, say p_0, where $w(p_0) = w_0 = -1 + \pi i$. Set $z_0 = z(p_0) = \pi i$. According to (3.5.6.6), $z = z(w)$ has the local Lagrange series representation

$$z = z_0 + \sum_{n=1}^{\infty} \frac{1}{n!} \frac{d^{n-1}}{dz^{n-1}} \left\{ \frac{z - \pi i}{\sqrt{z + e^z - (-1 + \pi i)}} \right\}^n \Bigg|_{z=\pi i} (w - w_0)^{\frac{n}{2}}$$

$$= z_0 - \sqrt{2}i(w - w_0)^{\frac{1}{2}} - \frac{1}{3}(w - w_0) + \cdots, \quad |w - w_0| < 2\pi.$$

The local parameter of R_w at p_0 is given by $\zeta = \sqrt{w - w_0}$. Choose the local parameter of $\mathbf{C}$ (the z-plane) at z_0 by $\eta = z - z_0$. Then, in terms of ζ and η, we have

$$\eta \circ z(w) \circ \zeta^{-1}(\zeta) = -\sqrt{2}i\zeta - \frac{1}{3}\zeta^2 + \cdots$$

which indicates that $z = z(w)$ is indeed univalent in a neighborhood of z_0.

The Riemann surface R_w of the inverse function $z = z(w)$ of $w = ze^{-z}$ (see Example 4 of Sec. 3.5.1): R_w (see Fig. 3.57) is conformally equivalent to the finite complex plane, too.

The Riemann surface R_z of $w = \sqrt{e^z}$ where $-\pi < \operatorname{Im} z < \pi$ (see Exercise B(2)(g) of Sec. 2.7.3): Note that $w = \sqrt{e^z}$ is the composite of the maps $z \to \zeta = e^z \to w = \sqrt{\zeta}$. The double-valuedness of $w = \sqrt{e^z}$ originates from that of $w = \sqrt{\zeta}$. It is well-known that $\sqrt{\zeta}$ has an algebraic branch point of order 1 over each of $\zeta = 0$ and ∞. Since $\zeta = e^z \neq 0, \infty$ on $\mathbf{C}$, it follows that $w = \sqrt{e^z}$ possesses two single-valued branches in any (open disk) neighborhood of each point in $\mathbf{C}$. This means that $w = \sqrt{e^z}$ does not have any branch point and, hence, any branch cut.

What domains in the z-plane can be mapped univalently, under a branch of $w = \sqrt{e^z}$, onto the domain $\mathbf{C}^* - \{0, \infty\} = \mathbf{C} - \{0\}$? We just restrict ourself to the strip $-\pi < \operatorname{Im} z < \pi$ because it is mapped univalently, under $\zeta = e^z$, onto the slit domain $\mathbf{C}^* - [\infty, 0]$. And recall that the Riemann surface $R_{\sqrt{\zeta}}$ is the two-sheeted surface composed of two copies of $\mathbf{C}^* - [\infty, 0]$, pasted crosswise along the cut $[\infty, 0]$.

Now take two copies of the strip $-\pi < \operatorname{Im} z < \pi$. Identify the upper side $\operatorname{Im} z = \pi$ and the lower side $\operatorname{Im} z = -\pi$ of the first sheet with the lower side $\operatorname{Im} z = -\pi$ and the upper side $\operatorname{Im} z = \pi$ of the second sheet, respectively. Let R_z denote this resulted surface. Then $\zeta = e^z$ maps R_z univalently onto $R_{\sqrt{\zeta}} - \{0, \infty\}$ which, in turn, is mapped univalently onto $\mathbf{C} - \{0\}$ via the map $w = \sqrt{\zeta}$. In conclusion, the Riemann surface R_z is conformally equivalent to $\mathbf{C} - \{0\}$ under the map $w = \sqrt{e^z}$. See Fig. 7.19. Note that, as a matter of fact, R_z is topologically equivalent to the infinite cylinder $S^1 \times R$ (see Example 2 in Sec. 7.1) and this, in turn, is conformally equivalent to $\mathbf{C} - \{0\}$.

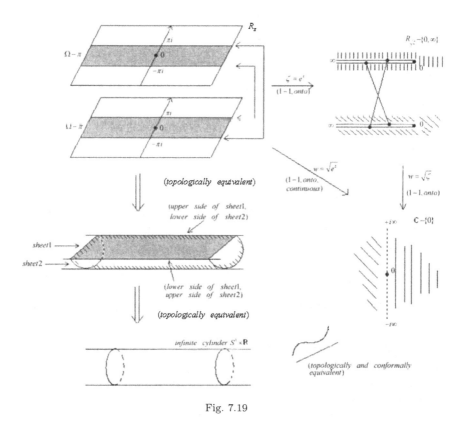

Fig. 7.19

The Riemann surface R_z of the extended function $w = f(z)$ which maps the upper half-plane $\mathrm{Im}\, z > 0$ univalently onto a rectangle (see Note 2 after Example 3 of Sec. 3.5.5) : R_z (see Figs. 3.80 and 3.81) is infinitely-sheeted and has infinitely many algebraic branch points of order 2 over each of the points $\pm 1, \pm \frac{1}{k}$. R_z is conformally equivalent to **C** under the map $w = f(z)$.

Recall that the inverse function $z = f^{-1}(w)$ is a doubly periodic meromorphic function, the simplest nonconstant elliptic function.

For the present, we take a break for a

Remark 9 (The meromorphic function on the torus C/Γ). We adopt notations used and concepts developed in Example 3 of Sec. 7.1. Any doubly periodic function f with periods ω_1 and ω_2 induces a function $F : \mathbf{C}/\Gamma \to \mathbf{C}^*$ such that $f = F \circ \pi$, where π is the natural projection. See the diagram. Let $\{\pi(U), (\pi|U)^{-1}\}$ be any chart on $\mathbf{C}/\Gamma$. Then

$$F \circ (\pi|U)(z) = (F \circ \pi)(z) = f(z), \quad z \in U$$

shows that F is meromorphic on $\mathbf{C}/\Gamma$. Conversely, a meromorphic function $F : \mathbf{C}/\Gamma \to \mathbf{C}^*$ gives rise to a meromorphic function $f = F \circ \pi : \mathbf{C} \to \mathbf{C}^*$, doubly periodic with periods ω_1 and ω_2. In conclusion, we have

The meromorphic functions on the torus $\mathbf{C}/\Gamma$. These are precisely those meromorphic functions on $\mathbf{C}$ that are doubly periodic with periods ω_1 and ω_2, where $\Gamma = \{m_1\omega_1 + m_2\omega_2 | m_1$ and m_2 are integers$\}$. Hence, for any doubly periodic meromorphic function $f : \mathbf{C} \to \mathbf{C}^*$, either f is a constant (in $\mathbf{C}^*$) or f has its range equal to $\mathbf{C}^*$. (7.2.25)

Section (4) Open Riemann surfaces conformally equivalent to the open unit disk (hyperbolic type)

Recall the elliptic modular function $z = \psi(\zeta) : \{|\zeta| < 1\} \to \mathbf{C}^*(0, 1, \infty) = \mathbf{C} - \{0, 1\}$ in (5.8.3.5). Its inverse function $\zeta = \psi^{-1}(z)$ is infinitely valued. The Riemann surface R_z of $\zeta = \psi^{-1}(z)$ is an infinitely sheeted smooth covering surface of the classical z-plane, pasted properly together along the branch cuts $(0, 1)$, $(1, \infty)$, and $(\infty, 0)$ and having branch points of infinite order over each of the points $0, 1$, and ∞. Note that no point in R_z is projected to $0, 1$, and ∞. Moreover, $\zeta = \psi^{-1}(z) : R_z \to \{|\zeta| < 1\}$ is univalently conformal. Hence R_z is conformally equivalent to the open unit disk. See Figs. 5.31–5.33.

Other open Riemann surfaces of this kind are provided by the Schwarzian curvilinear triangles $\Delta w_1 w_2 w_3$ with interior angles $\alpha_1\pi, \alpha_2\pi$, and $\alpha_3\pi$, respectively, where $\frac{1}{\alpha_1}, \frac{1}{\alpha_2}$, and $\frac{1}{\alpha_3}$ are positive integers or ∞ so that $\alpha_1 + \alpha_2 + \alpha_3 < 1$ holds. For a sketch, see Section (3) in Sec. 6.2.2; for detail, see Ref. [63], pp. 308–316.

Exercises A

(1) Let p_0 be a fixed point in a compact Riemann surface R. Suppose $f : R - \{p_0\} \to \mathbf{C}$ is an analytic function. Show that f is either a constant or a nonconstant function whose image comes arbitrarily close to every point in $\mathbf{C}$.

(2) Complete the discussion in Example 3(e).
(3) Do as many exercises as possible in Exercises B of Sec. 2.7.2–2.7.4, and
Exercises A(3)–(5) of Sec. 3.5.6 if you missed them previously. In doing
so, please enrich the problems required in Exercise A(3) of Sec. 3.5.6
to the following: For the multiple-valued function $w = f(z)$ or/and
the inverse multiple-valued function $z = f^{-1}(w)$ of the nonunivalent
function $w = f(z)$ (such as $w = e^{z^n}, n \geq 1$):

1. Construct the Riemann surface R_z for $w = f(z)$ or R_w for $z = f^{-1}(w)$. Prove that they are indeed abstract Riemann surfaces
   according to (7.1.1). How many branch points (algebraic, logarith-
   mic or transcendental)? Where? Orders? Draw the line complexes if
   possible.
2. Determine if R_z or R_w is a smooth or a branched covering surface
   (of how many sheets?) over the z-plane or w-plane, respectively.
3. $w = f(z)$ turns out to be single-valued on R_z and $z = f^{-1}(w)$ single-
   valued on R_w. Try to decide if they are univalent. If not, try to find
   out another Riemann surface which is conformally equivalent to it
   (see (7.2.23) and 5 below for closed surfaces).
4. Decide if R_z or R_w is closed or open (parabolic or hyperbolic).
5. In case R_z or R_w is closed. Try to determine its number of sheets,
   the sum of the orders of its branched points and then, its genus
   according to (7.2.22). Is it hyperelliptic? If yes, say something more
   about it as in Example 2.
6. In case R_z or R_w is open and is conformally equivalent to either the
   finite complex plane or the open unit disk, try to determine if it does
   have the following properties:

   1. The existence of Green's function (see Sec. 7.3.2).
   2. The existence of the harmonic measure (see Sec. 7.3.1).
   3. The validity of the maximum principle (see Sec. 7.3.1).

Alternatively, just test if **C** or the disk $|z| < 1$ enjoys these properties.

7.3 Harmonic Functions and the Maximum Principle on Riemann Surfaces

As we have experienced in Chap. 6, the crucial step to map a simply
connected Riemann surface univalently onto the sphere, the finite complex plane or the open unit disk lies barely on how to construct a *global*
meromorphic function on it to realize the mapping process (see Sec. 7.4).
As far as the techniques concerned with Riemann surfaces, all we have for

the moment are *local* ones, such as local parameters and local coordinates. The problem thus reduces to how to use these local techniques to carry over to the Riemann surfaces what we obtained in the classical complex plane and, eventually, leads to the global functions sought.

Even though the harmonicity and subharmonicity are local properties in character and are conformally invariant, the later has less restrictive operational properties (see (6.4.3)) and hence, is easier to be constructed for particular purposes. Consequently we choose subharmonic functions as our starting point.

A real-valued continuous function v on a Riemann surface R is *subharmonic* if, for each local coordinate system $\{O, \varphi\}$, the composite $v \circ \varphi^{-1}(z)$ is subharmonic on the open set $\varphi(O)$ (see (6.4.1)). For some cases, functions that tend to $-\infty$ at isolated points are included in this category. It is easy to see that subharmonic functions on R also enjoy properties listed in (6.4.1) and (6.4.3). In particular, v is said to be superharmonic on R if $-v$ is subharmonic on R.

A *Perron family* $\mathfrak{V}$ on R is a nonempty family of subharmonic functions on R, satisfying the following conditions:

1. $\max (v_1, v_2) \in \mathfrak{V}$ whenever $v_1, v_2 \in \mathfrak{V}$.
2. For every Jordan disk D on R and every $v \in \mathfrak{V}$, there is a $v_D \in \mathfrak{V}$ such that $v_D|D$ is harmonic and $v_D \geq v$ on R.   (7.3.1)

By a *Jordan disk* D on R, we mean an open set in R whose closure $\bar{D}$ lies in a single local coordinate neighborhood O, where $\{O, \varphi\}$ is a coordinate system, such that $\varphi(\bar{D})$ is a closed disk. v_D in condition 2 is usually chosen to be the subharmonic function

$$v_D(z) = \begin{cases} v(z), & z \in R - D \\ P_v(z), & z \in D \end{cases},$$ (7.3.2)

where a point p in D is identified with $z = \varphi(p)$, and $P_v(z)$ is Poisson's integral of $v(z)$ along ∂D (see (4) in (6.4.3)). Note that such a v_D always exists on R.

Then (6.5.1) can be extended to the

Perron's existence theorem (principle). *If $\mathfrak{V}$ is a Perron family on a Riemann surface R, then the function u defined by*

$$u(p) = \sup_{v \in \mathfrak{V}} v(p), \quad p \in R$$

is either identically $+\infty$ or harmonic on R.   (7.3.3)

Proof. According to (6.5.1), more precisely steps 2 and 3 in its proof, the theorem is valid for any Jordan disk on R. The alternative of either $+\infty$ or harmonicity comes from Harnack's principle (6.3.2.15).

Moreover, the connectivity of R will assert that $u(z)$ is harmonic on all Jordan disks once it is harmonic on a Jordan disk. To see this, suppose $u(z)$ is harmonic on the disk D_0 and D is another disk. Then, there is a chain of Jordan disks $D_0 = D_1, D_2, \ldots, D_n = D$ such that $D_i \cap D_{i+1} \neq \phi$ for $1 \leq i \leq n-1$. Since $u(z)$ is known to be harmonic on D_1 and $u(z)$ is either $+\infty$ or harmonic on D_2, the fact that $u(z)$ is harmonic on $D_1 \cap D_2$ shows that $u(z)$ should be harmonic throughout D_2, too. The inductive process will justify our claim.   □

By a *barrier* w at a boundary point $p \in \partial\Omega$ of a domain Ω in R, we mean that there is a relatively compact neighborhood N of p and a continuous function $w : \overline{\Omega \cap N} \to \mathbf{R}$ such that w is superharmonic on $\Omega \cap N$, $w(p) = 0$ and $w(q) > 0$ for all $q \in \overline{\Omega \cap N}$ with $q \neq p$ (compare to (6.5.5)).

Let Ω be a relatively compact domain in a Riemann surface R. *Dirichlet's problem* reads as: For a given continuous real-valued function f on $\partial\Omega$, try to construct a continuous function u on $\bar{\Omega}$ such that

1. $u|\Omega$ is harmonic, and
2. $u|\partial\Omega = f$ holds.   (7.3.4)

If such a u does exists, we say that Dirichlet's problem on Ω is *solvable*.

As an extension of (6.5.4), (6.5.6), and (6.5.7), we have

The solvability of Dirichlet's problem. Let Ω be a relatively compact domain in a Riemann surface. Then the following are equivalent:

(1) Every point of $\partial\Omega$ has a barrier at it (such a boundary point is called a *regular point* for Dirichlet's problem)
(2) Dirichlet's problem is always solvable on Ω for arbitrary continuous function f on $\partial\Omega$.

Moreover, if u is such a solution, then

$$\min_{p \in \partial\Omega} f(p) \leq u(q) \leq \max_{p \in \partial\Omega} f(p), \quad \text{for all } q \in \bar{\Omega}$$

and such a solution u is unique.   (7.3.5)

The existence of a barrier at a boundary point is a local property. Hence, as a consequence of (6.5.7), for each of the following relatively compact domains Ω in R:

1. each boundary point of $\partial\Omega$ can be reached by an analytic arc (namely, the image of a straight line under univalent analytic mapping) with no points in common with $\bar{\Omega}$, or
2. the complement $R\text{–}\Omega$ is such that no component reduces to a point, or
3. Ω is bounded by finitely many Jordan (analytic) curves,   (7.3.6)

Dirichlet's problem is solvable.

Sketch of proof for (7.3.5). $(2) \Rightarrow (1)$: Fix any point $p_0 \in \partial\Omega$. Construct a continuous function f on $\partial\Omega$ with the properties that $f(p_0) = 0$ and $f(p) > 0$ for all $p \in \partial\Omega$ other than p_0. Then the solution of Dirichlet's problem for this f is evidently a barrier at p_0.

$(1) \Rightarrow (2)$: The proof is a slight modification of the proof of (6.5.4), or more accurately, the proof of Example 3 of Sec. 6.5.

Let $\mathfrak{V}(f)$ be the set of continuous functions v on $\bar{\Omega}$, subharmonic in Ω, $m = \min_{p\in\partial\Omega}f(p) \le v(q) \le \max_{p\in\partial\Omega}f(p) = M$ for all $q \in \Omega$, and $v(p) \le f(p)$ for all $p \in \partial\Omega$. $\mathfrak{V}(f) \ne \phi$ since the constant function m is in it. $\mathfrak{V}(f)$ obviously is a Perron family on Ω. By (7.3.3), the function

$$u(q) = \sup_{v\in\mathfrak{V}(f)}v(q), \quad q \in R$$

is harmonic in Ω and $m \le u(q) \le M$ holds for all $q \in \bar{\Omega}$.

Fix any point $p \in \partial\Omega$. As usual we have to show that, for $q \in \Omega$

$$\overline{\lim}_{q\to p}u(q) \le f(p) \le \underline{\lim}_{q\to p}u(q)$$

and hence $\lim_{q\to p}u(q) = f(p)$ holds.

To prove the second inequality: We may suppose that $m < f(p)$. Choose $\varepsilon > 0$ so that $f(p) - m > \varepsilon$. By the continuity of f at p, there is a neighborhood N of p so that $f(p) - \varepsilon < f(q)$ for all $q \in N \cap \partial\Omega$. Let w be a barrier at p. For any smaller neighborhood N_1 of p with $\overline{N_1} \subseteq N$, set $m_1 = \min w(q)$ on the set $\overline{(N - N_1)} \cap \bar{\Omega}$. Note that $m_1 > 0$. Set

$$s(q) = \begin{cases} \min\left\{\dfrac{w(q)}{m_1}, 1\right\}, & q \in N \cap \Omega \\ 1, & q \in \bar{\Omega} - N \end{cases}$$

(compare to $s(z)$ in Example 3 of Sec. 6.5). Then, try to show that

$$v(q) = f(p) - \varepsilon - (f(p) - m - \varepsilon)s(q), \quad q \in \bar{\Omega}$$

is in $\mathfrak{V}(f)$. This would imply that $v(q) \le u(q)$, $q \in \bar{\Omega}$ and hence $\underline{\lim}_{q\to p}u(q) \ge v(p) = f(p) - \varepsilon$.

To prove the first inequality: Consider instead the function

$$\tilde{v}(q) = f(p) + \varepsilon + (M - f(p) - \varepsilon)s(q), \quad q \in \bar{\Omega}.$$

Try to show that, for any $v \in \mathfrak{V}(f)$, the subharmonic function $v(q) - (M - f(p) - \varepsilon)s(q) \leq f(p) + \varepsilon$ holds on Ω. Hence $v(q) \leq \tilde{v}(q), q \in \bar{\Omega}$ and thus $u(q) \leq \tilde{v}(q), q \in \bar{\Omega}$, holds. It follows that $\overline{\lim}_{q \to p} u(q) = \tilde{v}(p) = f(p) + \varepsilon$.

$\square$

Remark (The relaxation of the relative compactness of the domain Ω in Dirichlet's problem). (7.3.5) (and (7.3.6)) and its proof are still valid if Ω is just a domain, not necessarily relatively compact and having every point of its boundary $\partial\Omega$ a regular point. In this case, the boundary function f should be presumed to be *bounded* and *continuous* on $\partial\Omega$, and the solution u now satisfies

$$\inf_{p \in \partial\Omega} f(p) \leq u(q) \leq \sup_{p \in \partial\Omega} f(p), \quad \text{for all } q \in \bar{\Omega}. \tag{7.3.7}$$

Note that *nothing can be said now about the uniqueness of the solution*, owing to the fact that general maximum principle for harmonic function on a general domain in a Riemann surface is not available for the moment.

An alternative way to justify the above claim is as follows. Let R be an open Riemann surface. R can be compactified by adding a single point "at infinity", also referred to as the *ideal boundary* and denoted as ∞, whose neighborhoods are sets with compact complement as we did to $\mathbf{C}$ to obtain $\mathbf{C}^*$ (see the Remark in Sec. 1.9). Set $R^* = R \cup \{\infty\}$, the *one-point compactification* of R. Let Ω be a domain in R, each of its boundary points being regular. The boundary $\partial^*\Omega$ of Ω in R^* is $\partial\Omega \cup \{\infty\}$ which is compact. Choose f any (bounded) continuous function in $\partial\Omega$ and let $f(\infty)$ be an arbitrary finite number. Then Perron's method yields a harmonic function u on Ω, satisfying $\lim_{q \to p} u(q) = f(p)$ for each $p \in \partial\Omega$.

This observation leads to the fact that, *for any Jordan disk D in a Riemann surface R, there exists a nonconstant harmonic function on R–$\bar{D}$.*

This fact, in turn, can be used to prove the following important results.

Countable basis. Every Riemann surface has a countable basis. (7.3.8)

Regular exhaustion. On every Riemann surface R, there exists a sequence of regular domains Ω_n such that

1. $\overline{\Omega_n} \subseteq \Omega_{n+1}, n \geq 1$, and
2. $R = \bigcup_{n=1}^{\infty} \Omega_n$.

Moreover, each compact connected set in R is contained in a regular domain.

(7.3.9)

A domain Ω in R is called *regular* if $\bar{\Omega}$ is compact and the boundary $\partial\Omega$ consists of finitely many analytic Jordan closed curves. For the proofs of (7.3.8) and (7.3.9), see Ref. [6], pp. 142–146.

As an application of (7.3.5), we just *prove* the last statement in (7.3.9). Let K be a compact connected set in an open Riemann surface R. Finitely many Jordan disks will cover K. Let D be the union of these disks, which we may assume that they intersect only in pairs and nontangentially (why this can be done!). D is a relatively compact domain. Let Δ be a small open disk in the domain $D - K$. Set $\Omega_0 = D - \bar{\Delta}$, a domain with all its boundary points regular. Let u be the solution of Dirichlet's problem on Ω_0 with boundary values 1 on $\partial\Delta$ and 0 on ∂D. Note that u is not a constant and hence, the point p where $du(p) = 0$ is fulfilled only at isolated points. By choosing $\varepsilon > 0$ properly, where $\varepsilon < \min_{p \in K} u(p)$, let Ω be the component containing K of the set $\{p \in \Omega_0 | u(p) > \varepsilon > 0\}$ so that $du(p) \neq 0$ along the level curve $u(p) = \varepsilon$. This Ω is then a required regular domain containing K.

$\square$

Three subsections are divided.

Section 7.3.1 studies the existence of harmonic measure.

Section 7.3.2 considers Green's function and develops its basic properties.

Section 7.3.3 uses Green's function, the harmonic measure, the maximum principle, and the existence of a negative nonconstant superharmonic function to classify the Riemann surfaces into their mutually exclusive classes: elliptic, parabolic, and hyperbolic.

7.3.1   *Harmonic measure*

Recall the role played by the harmonic measure (see Sec. 6.6.1) of a planar domain bounded by finitely many analytic Jordan closed curve in the study of the canonical domains (see Sec. 6.6). Here we try to extend this concept to the ideal boundary (see Remark in Sec. 7.3) of an open Riemann surface.

Let R be an open Riemann surface and let K be a compact subset of R whose complement $R - K$ is connected. A real-valued continuous function ω_K on $\overline{R - K}$ is called the *harmonic measure of K* if

1. ω_K is harmonic in $R - K$.
2. $0 < \omega_K < 1$ on $R - K$.
3. $\omega_K = 1$ on $\partial(R - K)$.

4. ω_K is the smallest among all continuous functions having properties 1–3. Namely, if $\tilde{\omega}$ is another such function, then $\tilde{\omega} \geq \omega_K$ throughout $R - K$.

$$(7.3.1.1)$$

Condition 4 is assumed to guarantee the uniqueness of ω_K if it exists (the readers are urged to compare this to the one defined in Sec. 6.6.1 for planar domain!); it can be replaced by the property $\varliminf_{p \to \infty} \omega_K(p) = 0$ (see (7.3.2.15)).

Before we start, let us observe that there is an open Riemann surface which carries no harmonic measures for its compact subsets. Take, for instance, the finite complex plane $\mathbf{C}$ with $\boldsymbol{K}$ any fixed closed disk $|z| \leq r$. Suppose these would exist the harmonic measure ω_K of K. Since $0 < \omega_K(z) < 1$ on $r < |z| < \infty$, it follows that ∞ is a removable singularity of ω_K (see (6.3.1.7) and Exercise A(25) of Sec. 3.4.3). After defining $\omega_K(\infty) = 0$, then $\omega_K(z)$ turns out to be harmonic on $r < |z| \leq \infty$ with constant value 1 along the boundary $|z| = r$ and hence, is a constant on $r < |z| \leq \infty$, a contradiction (see (3.4.3.7) and (2) in (3.4.3.5)).

Section (1) How to construct the harmonic measure?

Let R and K be as above. Let $\mathfrak{V}_K$ be the family of functions v satisfying the following properties:

1. v is defined and subharmonic in $R - K$.

2. $v \leq 1$ on $R - K$.

3. $\overline{\lim}_{p \to \infty} v(p) \leq 0$; namely, for any $\varepsilon > 0$, there is a compact set K_ε in R such that $v(p) < \varepsilon$ whenever $p \in R - K_\varepsilon$. $(7.3.1.2)$

$\mathfrak{V}_K$ is nonempty since 0 belongs to it.

It is easy to check that $\mathfrak{V}_K$ is a Perron family. Condition 2 and the maximum principle for subharmonic functions shows that members of this $\mathfrak{V}_K$ are uniformly bounded from above. Hence,

$$\omega_K(p) = \sup_{v \in \mathfrak{V}_K} v(p), \quad p \in R - K \qquad (7.3.1.3)$$

always exists as a harmonic function and $0 \leq \omega_K \leq 1$ on $R - K$. However, it could be either $\omega_K \equiv 0$ or $\omega_K \equiv 1$.

In case the interior Int K of K is not empty, then $\omega_K > 0$ holds on $R - K$. To see this, fix a point $p_0 \in$ Int K. Choose a local parameter $z(p)$ at p_0 with range $|z| < 1$ so that, for some constant $0 < r < 1$, the disks $|z - z(p_0)| < r$, and $|z - z(p_0)| < 2r$ are contained in $|z| < 1$ but the smaller one is contained in the set $z(K)$ while the larger one is not. Define a function v on R by

$$
v(p) = \begin{cases} \dfrac{1}{\log 2} \log \dfrac{2r}{|z(p) - z(p_0)|}, & \text{if } z(p) \text{ is defined and} \\[2ex] & r < |z(p) - z(p_0)| < 2r \text{ holds} \\[2ex] 0, & \text{otherwise} \end{cases} \cdot
$$

Consider the restriction $v|_K$ which is not identically zero in $R - K$. Since $v|_K$ is in $\mathfrak{V}_K$, it follows that $\omega_K \geq v|_K$ holds, and hence, $\omega_K > 0$ on $R - K$.

As a consequence of the last paragraph, then either $0 < \omega_K < 1$ or $\omega_K \equiv 1$ holds throughout $R - K$ if Int $K \neq \emptyset$. In case $\omega_K \equiv 1$ does happen, we say that the harmonic measure of K does *not* exist.

Now suppose $0 < \omega_K < 1$ does hold on $R - K$. To check condition 4 in (7.3.1.1), let $\tilde{\omega}$ be any function satisfying properties 1–3 in (7.3.1.1). For any function $v \in \mathfrak{V}_K$, $v - \tilde{\omega}$ is defined and subharmonic in $R - K$, and $v(p) - \tilde{\omega} \leq 1 - 1 = 0$ on ∂K. By the maximum principle for subharmonic functions, then $v(p) \leq \tilde{\omega}(p)$ on $R - K$ and hence, $\omega(p) \leq \tilde{\omega}(p)$ holds on $R - K$.

In conclusion, we have

The existence of the harmonic measure of a compact set K (with nonempty interior) in an open Riemann surface R. Consider the Perron family $\mathfrak{V}_K$ in (7.3.1.2) and define ω_K as in (7.3.1.3). In case $0 < \omega_K < 1$ holds on $R - K$, then ω_K is the harmonic measure of K. (7.3.1.4)

Note that, $\omega_K|\partial K = 1$ does hold since ω_K is not a constant on the domain $R - K$. Later in (7.3.2.11), we will see that the alternative $0 < \omega_K < 1$ or $\omega_K \equiv 1$ is independent of the compact sets K chosen, but dependent only on the open surface R itself. It is in this sense that the existence of the harmonic measure is a property of the ideal boundary and ω_K in (7.3.1.4) can be termed as *the harmonic measure of the ideal boundary ∞ of R with respect to $R_n - K$.* Also, we will show $\varliminf_{p \to \infty} \omega_K(p) = 0$ in (7.3.2.15).

Remark (An alternate method to construct the harmonic measure). We take for granted the concept of regular exhaustion of an open Riemann surface as stated in (7.3.9).

Let K be a regular domain in R and Ω_n, $n \geq 1$, be a sequence of regular domains of R such that $\bar{K} \subseteq \Omega_1$. For each fixed $n \geq 1$, according to

(7.3.5) and (7.3.6), there exists a unique continuous function ω_n on $\overline{\Omega_n} - K$ so that

1. ω_n is harmonic in $\Omega_n - \bar{K}$;
2. $0 < \omega_n < 1$ in $\Omega_n - \bar{K}$; and
3. $\omega_n|\partial K = 1$ and $\omega_n|\partial \Omega_n = 0$ $\hspace{2cm}$ (7.3.1.5)

ω_n is called *the harmonic measure of R with respect to $\Omega_n - \bar{K}$*, or simply *the harmonic measure of $\bar{K}$*. By the maximum principle, we have $\omega_n < \omega_{n+1}$ in $\Omega_n - \bar{K}$ for each $n \geq 1$; and hence, by Harnack's principle (see (6.3.2.15))

$$\omega(p) = \lim_{n \to \infty} \omega_n(p), \hspace{2cm} (7.3.1.6)$$

locally uniformly in $R - \bar{K}$. ω is harmonic and $0 \leq \omega \leq 1$ in $R - \bar{K}$. In case $\omega \equiv 1$, we say that the harmonic measure does not exist; in case $0 < \omega < 1$, the only other alternative, ω is called *the harmonic measure of the ideal boundary ∞ of R with respect to $R_n - \bar{K}$* or simply *the harmonic measure of $\bar{K}$*. Note that $\omega|\partial K = 1$ holds.

What should be mentioned is that $\omega \equiv 0$ or $0 < \omega < 1$ is independent of the regular exhaustion chosen. Let R_n, $n \geq 1$, be another regular exhaustion of R and $\tilde{\omega}_n$ the corresponding harmonic measure of R with respect to $R_n - \bar{K}$. Set $\tilde{\omega} = \lim_{n \to \infty} \tilde{\omega}_n$ (locally uniformly in $R - \bar{K}$). There is a sequence n_i, $i \geq 1$, of positive integers so that

$$\overline{\Omega_i} \subseteq R_{n_i}, \quad i \geq 1$$
$$\Rightarrow \omega_i < \tilde{\omega}_{n_i} \text{ in } \Omega_i - \bar{K} \text{ for } i \geq 1$$
$$\Rightarrow \omega(p) \leq \tilde{\omega}(p) \text{ in } R - \bar{K}.$$

Similarly, $\tilde{\omega} \leq \omega$ holds in $R - \bar{K}$. Hence $\bar{\omega} \equiv \omega$ in $R - \bar{K}$.

As a matter of fact, the ω defined in (7.3.1.6) is exactly the one $\omega_{\bar{K}}$ defined in (7.3.1.3). For reference and clarity, we have

The equivalence of the definitions (7.3.1.3) and (7.3.1.6). Let K be a regular domain in an open Riemann surface R and Ω_n, $n \geq 1$, be a regular exhaustion of R such that $\bar{K} \subseteq \Omega_1$. Let $\omega_{\bar{K}}$ be the harmonic measure defined in (7.3.1.3) and ω the one defined in (7.3.1.6).

(1) Let u_n be the unique harmonic function in $\Omega_n - \bar{K}(n \geq 1)$ so that $u_n|\partial K = 0$ and $u_n|\partial \Omega_n = \omega_{\bar{K}}$ (the existence of such a u_n is guaranteed by (7.3.5) and (7.3.6)). Then

$$\omega_n(p) = \omega_{\bar{K}}(p) - u_n(p), \quad p \in \Omega_n - \bar{K}$$

is the harmonic measure of $\Omega_n - \bar{K}$ as defined in (7.3.1.5).

(2) Hence, $\omega_{\bar{K}}(p) = \omega(p) = \lim_{n \to \infty} \omega_n(p)$ holds in $R - \bar{K}$.   (7.3.1.7)

Only (2) is needed to be *proved*. By condition 4 in (7.3.1.1), $\omega_{\bar{K}}(p) \le \omega(p)$ holds in $R - \bar{K}$. Since $u_n(p) > 0$ on $\Omega_n - \bar{K}$, it follows that $\omega_n(p) < \omega_{\bar{K}}(p)$ on $\Omega_n - \bar{K}$ and hence, by local uniform consequence, $\omega(p) \le \omega_{\bar{K}}(p)$ in $R - \bar{K}$. This proves the claim.   □

Section (2) What kind of open surfaces does carry the harmonic measure?

As we already observed above, the finite complex plane C does not carry the harmonic measure (of any closed disk). Now let us consider the following *problem*: Suppose $u(z)$ is bounded and harmonic in $1 < |z| < \infty$, and is continuous on $1 \le |z| < \infty$ with boundary value along $|z| = 1$ identically equal to 1. Is it true that $u(z) \equiv 1$ on $1 < |z| < \infty$? Yes, it is, according to Exercise A(25) of Sec. 3.4.3 and (6.3.1.7). But this can also be justified by using the concept of harmonic measure. Let ω_n be the harmonic measure of $1 < |z| < n$ with $\omega_n = 1$ on $|z| = 1$ and $\omega_n = 0$ on $|z| = n$. If there is a point z_0 in $|z| > 1$ so that $u(z_0) > 1 + \varepsilon$ for some $\varepsilon > 0$. The set $O = \{z \,|\, u(z) > 1 + \frac{\varepsilon}{2}\}$ is open and $u(z) = 1 + \frac{\varepsilon}{2}$ along ∂O. For sufficiently large n, by the maximum principle, we have $u(z) \le M(1 - \omega_n) + 1 + \frac{\varepsilon}{2}$ on the set $O \cap \{|z| \le n\}$, where $u(z) \le M$ (a constant) on $1 < |z| < \infty$. Since $\lim_{n \to \infty} \omega_n \equiv 1$ holds, therefore $u(z) \le 1 + \frac{\varepsilon}{2}$ on O which is a contradiction. Hence $u(z) \le 1$ on $1 < |z| < \infty$. Similarly, by a considering point z_0 where $u(z_0) < 1 - \varepsilon$ for some $0 < \varepsilon < 1$, we will obtain that $u(z) \ge 1$ on $1 < |z| < \infty$. Consequently $u(z) \equiv 1$ on $1 < |z| < \infty$.

What we talked about in last paragraph can be universally extended to

The maximum principle on an open Riemann surface without carrying the harmonic measure. Let R be such a surface and K be a nonempty compact set in it. Let $u(z)$ be harmonic and *bounded above* on $R - K$ such that $\overline{\lim}_{p \to K} u(p) \le M < +\infty$. Then

$$u(p) \le M \quad \text{on } R - K.$$

In particular, if $-\infty < m \le \underline{\lim}_{p \to K} u(p) \le \overline{\lim}_{p \to K} u(p) \le M < +\infty$, then $m \le u(p) \le M$ on $R - K$; and if $u \equiv$ constant on K, then $u \equiv$ constant on $R - K$.   (7.3.1.8)

Later in (7.3.2.11), we will show that the validity of this maximum principle depends only on R but not on K.

A proof can be given by using the method of regular exhaustion and definition (7.3.1.6) as we did in the above example. However we present another proof, based on definition (7.3.1.3).

Another proof. Choose any nonempty compact set K and another compact set K' with nonempty interior.

In case $K' \subseteq K$: Let u be harmonic in $R - K$ and bound above by 1 there so that $\overline{\lim}_{p \to K} u(p) \leq 0$ holds. Take any $v \in \mathfrak{V}_{K'}$ (see (7.3.1.2)). Since

$$\overline{\lim}_{p \to \infty} (u(p) + v(p)) \leq 1 + 0 = 1, \quad \text{and}$$

$$\overline{\lim}_{p \to K} (u(p) + v(p)) \leq 0 + 1 = 1$$

$\Rightarrow$ (by the maximum principle for subharmonic functions) $u(p) + v(p) \leq 1$

on $R - K$.

If the harmonic measure ω_K in (7.3.1.3) does not exists, namely, $\omega_K \equiv 1$ in $R - K$, then v can be chosen so that, for any fixed p in $R - K$, $v(p)$ is arbitrarily closed to 1. Under this circumstance $u(p) \leq 0$ holds and the maximum principle follows.

In case K' is not contained in K: Choose a compact set K'' so that $K \cap K' \subseteq \text{Int}\, K''$, the interior of K''. Let u be as before. Then $u(p) \leq \max_{\partial K''} u$ on $R - K''$ since the maximum principle is known to be valid there by the last paragraph. We *claim* that $u(p) \leq 0$ along $\partial K''$. For if $\max_{\partial K''} u > 0$ were true, by the maximum principle for harmonic functions, $u(p) \leq \max_{\partial K''} u$ also holds on $K'' - K$. This means that u would attain its maximum in $R - K$ on some point along $\partial K''$, which is an interior point. This is a contradiction if u is not a constant. Consequently, by the assumption that $\overline{\lim}_{p \to K} u(p) \leq 0$ and again by the maximum principle for harmonic functions, $u(p) \leq 0$ in $K'' - K$. In conclusion we obtain that $u(p) \leq 0$ in $R - K$. This proves the claimed maximum principle. $\square$

Before we draw up our main result (7.3.1.9), let us consider the special case that R is the open unit disk $|z| < 1$. For any compact set K in $|z| < 1$, with $\text{Int}\, K \neq \phi$ and having a connected complement in $|z| < 1$, (6.5.7) shows that there does exist a harmonic function $\omega(z)$ on $R - K$, $0 < \omega(z) < 1$, $\omega|\partial K = 1$ and $\omega = 0$ along $|z| = 1$. This $\omega(z)$ is the harmonic measure of K.

On the other hand, maximum principle stated in (7.3.1.8) is not more true in $|z| < 1$. Note that $u(z) = \mathrm{Re}\, z$ is harmonic and bounded in $0 < |z| < 1$. Even though $\lim_{z \to 0} u(z) = 0$ holds, $u(z) \leq 0$ does not hold in $0 < |z| < 1$.

We have the following universal

Criteria for an open Riemann surface carrying the harmonic measure. Let R be an open Riemann surface. Then,

   (1) Harmonic measure exists on R.

$\Leftrightarrow$ (2) (Ahlfors, 1952) There exists a nonconstant continuous positive superharmonic function on R.

$\Leftrightarrow$ (3) The maximum principle (as stated in (7.3.1.8)) is not valid in R.

$$(7.3.1.9)$$

(7.3.1.8) proved (3) $\Rightarrow$ (1). The proof of (1) $\Rightarrow$ (3) is postponed to (7.3.2.10).

Proof. (1) $\Rightarrow$ (2): Let D be a Jordan disk in R. Let $\omega(= \omega_D)$ be the harmonic measure of R with respect to $\bar{D}$: ω is harmonic in $R - \bar{D}$, $0 < \omega < 1$, and $\omega = 1$ along ∂D. Define

$$v_0(p) = \begin{cases} 1, & p \in \bar{D} \\ \omega(p), & p \in R - \bar{D} \end{cases}.$$

Then v_0 is a required superharmonic function on R.

(2) $\Rightarrow$ (1): Let v_0 be a nonconstant continuous positive superharmonic function on R. Let D be a Jordan disk. Consider the family $\mathfrak{U}_D$ of functions v satisfying the properties:

1. v is continuous, positive, and superharmonic in $R - D$.
2. $v \geq 1$ along ∂D.

$\mathfrak{U}_D \neq \phi$ because the constant function 1 is in it. According to (6.5.2), the function

$$\omega(p) = \inf_{v \in \mathfrak{U}_D} v(p), \quad p \in R - \bar{D}$$

is harmonic in $R - \bar{D}$, $0 \leq \omega \leq 1$, and $\omega = 1$ along ∂D (by (7.3.6)). In particular, $\omega \not\equiv 0$.

Set $m = \min_{p \in \partial D} v_0(p)$. Then $\frac{v_0}{m} \in \mathfrak{U}_D$. Hence

$$\frac{v_0}{m} \geq \omega \quad \text{in } R - D$$

⇒ *In case* $\omega \equiv 1$, *then* $v_0 \geq m$ in $R - D$.

⇒ (since v_0 is superharmonic in D) $v_0 \geq m$ in D.

The continuity of v_0 shows that v_0 would attain its minimum at some point along ∂D, an interior point of R. Hence v_0 turns out to be a constant, a contradiction to our assumption. Therefore $\omega \not\equiv 1$ on $R - \bar{D}$. What remains is that $0 < \omega < 1$ and $\omega|\partial D = 1$. By (7.3.1.8), ω is the harmonic measure of $\bar{D}$. □

Exercises A

(1) Prove (7.3.1.8) by using the definition (7.3.1.6).
(2) Suppose R is an open Riemann surface, carrying no harmonic measure. Show that there does not exist a nonconstant bounded harmonic function on R.
   Note. Refer to Exercises B(3) and B(4) of Sec. 3.4.3.
(3) Suppose R is an open Riemann surface, carrying no harmonic measure. Show that there does not exist a nonconstant positive harmonic function on R.
   Note. Refer to Exercise B(6) of Sec. 3.4.3 and Example 3(1) in Sec. 6.3.2.

Exercises B

(1) Suppose R is an open Riemann surface, carrying the harmonic measure and $\tilde{R}$ is its smooth covering surface. Show that there exists the harmonic measure on $\tilde{R}$.

7.3.2 *Green's function*

Readers are asked to review Section (2) in Sec. 6.3.1 for the concept (and its basic properties) of Green's function for a classical domain in C.

Let R be a Riemann surface and $p_0 \in R$, a fixed point. A real-valued function g on $R - \{p_0\}$ is called *a Green's function for R with pole at p_0* if the following conditions are satisfied:

1. g is harmonic in $R - \{p_0\}$.

2. $g > 0$ in $R - \{p_0\}$.

3. In terms of a local parameter $z(p)$ at p_0 with $z(p_0) = 0$, $g(p) + \log |z(p)|$ has a harmonic extension to a neighborhood of p_0 (and hence, g has a *logarithmic pole* at p_0).

4. g is the smallest among all the functions having properties 1, 2, and 3. Namely, if $\tilde{g}$ is another such function, then $\tilde{g} \geq g$ throughout $R - \{p_0\}$.

$$(7.3.2.1)$$

This definition makes sense on a Riemann surface because *condition* 3 is independent of the choice of local parameters vanishing at p_0. To see this, suppose $\tilde{z}(p)$ is another such local parameter. Then $z \circ \tilde{z}^{-1}(\tilde{z}) = \tilde{z}\varphi(\tilde{z})$, where $\varphi(\tilde{z})$ is analytic at 0 and $\varphi(0) \neq 0$. Set $\tilde{z}^{-1}(\tilde{z}) = p$. Then $\tilde{z}(p) = \tilde{z}$ and $z(p) = \tilde{z}(p)\varphi(\tilde{z}(p))$ holds in a neighborhood of p_0. Since $\log \left| \frac{z(p)}{\tilde{z}(p)} \right| = \log |\varphi(\tilde{z})|$ is bounded in the neighborhood, it follows from (6.3.1.7) that $\log \left| \frac{z(p)}{\tilde{z}(p)} \right|$ has a removable singularity at p_0. Hence $g(p) + \log |\tilde{z}(p)|$ and $g(p) + \log |\tilde{z}(p)|$ are simultaneously harmonic at p_0. Moreover, *condition* 4 is used to guarantee the uniqueness of Green's function if it exists; it is adopted to replace the requirement that $\lim_{z \to \zeta} g(z, z_0) = 0$ for any $\zeta \in \partial\Omega$ in the definition (6.3.1.11) and can be replaced by the property $\underline{\lim}_{p \to \infty}(p, p_0) = 0$ in (7.3.2.12) below. See Exercise A(5).

Section (1) How to construct a Green's function for R with pole at p_0?

Let $z = z(p)$ be a local parameter at p_0 with $z(p_0) = 0$. Let $\mathfrak{V}_{p_0}$ be the family of functions v satisfying the following properties:

1. v is subharmonic in $R - \{p_0\}$.

2. $v \equiv 0$ outside a compact set.

3. $v(p) + \log |z(p)|$ is subharmonic in a neighborhood of p_0. $(7.3.2.2)$

Same argument for condition 3 in (7.3.2.1) shows that condition 3 in (7.3.2.2) is still independent of the local parameters chosen.

$\mathfrak{V}_{p_0}$ is nonempty. To see this, suppose that the disk $|z| \leq r_0$ is contained in the range of $z(p)$. Define

$$v_0(p) = \begin{cases} \log \dfrac{r_0}{|z(p)|} & , \text{if } |z(p)| \leq r_0 \\ 0, & \text{otherwise} \end{cases} \qquad (7.3.2.3)$$

Then $v_0 \in \mathfrak{V}_{p_0}$ holds. It is easily seen that $\mathfrak{V}_{p_0}$ is indeed a Perron family. By (7.3.3),

$$g(p, p_0) = \sup_{v \in \mathfrak{V}_{p_0}} v(p), \quad p \in R - \{p_0\} \qquad (7.3.2.4)$$

is either identically $+\infty$ or harmonic on $R - \{p_0\}$.

In case $g(p, p_0) \equiv +\infty$ on R, we say that there exists no Green's function on R. This case does happen for some Riemann surfaces; for example, the Riemann sphere $\mathbf{C}^*$ and the finite complex plane $\mathbf{C} = \mathbf{C}^* - \{\infty\}$ are the cases. See *Section* (2) below.

In what follows, we always consider the case that $g(p, p_0)$ is harmonic on $R - \{p_0\}$. Since $0 \in \mathfrak{V}_{p_0}$, so $g(p, p_0) > 0$ holds throughout $R - \{p_0\}$. Moreover, $g(p, p_0) \geq v_0(p)$ indicates that $\lim_{p \to p_0} g(p, p_0) = +\infty$ and, in particular, $g(p, p_0)$ is not a constant.

That $g(p, p_0)$ satisfies condition 3 in (7.3.2.1) will be postponed to Section (2).

Let $\tilde{g}(p, p_0)$ be another function satisfying conditions 1–3 in (7.3.2.1). For any function $v \in \mathfrak{V}_{p_0}$, $\tilde{g} - v$ is superharmonic. Then $\tilde{g} - v \geq 0$ on $R - \mathrm{supp}\, v$, where $\mathrm{supp}\, v = \{p \in R | v(p) \geq 0\}$, the compact support of v. By the minimum principle for superharmonic functions, $\tilde{g} - v \geq 0$ also holds on $\mathrm{supp}\, v$. Hence $\tilde{g} - v \geq 0$ on $R - \{p_0\}$ which implies that $\tilde{g} \geq g$ on $R - \{p_0\}$; namely, either $\tilde{g} > g$ or $\tilde{g} \equiv g$ on $R - \{p_0\}$.

In conclusion, we have

The existence of Green's function $g(p, p_0)$ for a Riemann surface R with a logarithmic pole at p_0. In case $g(p, p_0)$ defined by (7.3.2.4) is harmonic on $R - \{p_0\}$, it is the (unique) Green's function of R with logarithmic pole at p_0.   (7.3.2.5)

In (7.3.2.11), we will show that the existence of $g(p, p_0)$ is independent of p_0 but dependent only on the surface R itself. See *Section* (3) for some basic properties of Green's function.

Remark (An alternate way to construct Green's function). Let Ω be a regular domain in a Riemann surface and $p_0 \in \Omega$ a fixed point. Let $\mathfrak{V}_{p_0}$ be the Perron family in (7.3.2.2) with Ω replacing R. Then

$$g_\Omega(p, p_0) = \sup_{v \in \mathfrak{V}_{p_0}} v(p), \quad p \in \Omega - \{p_0\} \qquad (7.3.2.6)$$

is either harmonic or identically $+\infty$ on $\Omega - \{p_0\}$. Note that $g_\Omega(p, p_0) > 0$ since $o \in \mathfrak{V}_{p_0}$ and $\lim_{p \to p_0} g_\Omega(p, p_0) = +\infty$ since $v_0 \in \mathfrak{V}_{p_0}$ where v_0 is defined in (7.3.2.3).

To prove that $g_\Omega(p, p_0)$ exists as a harmonic function on $\Omega - \{p_0\}$, it suffices to show that members of $\mathfrak{V}_{p_0}$ are uniformly bounded above on a compact subset of Ω. Choose a local coordinate system $\{U, z\}$ at p_0 so that $\bar{U} \subseteq \Omega$ and $z(U)$ is the disk $|z| < 1$ with $z(p_0) = 0$. Let $0 < r_1 < r_2 < 1$ and set $K_1 = z^{-1}(|z| \leq r_1)$ and $K_2 = z^{-1}(|z| \leq r_2)$. Note that $K_1 \subseteq K_2 \subseteq U$.

According to (7.3.1.5), the *harmonic measures* ω_{K_1} and ω_{K_2} of K_1 and K_2 exist, respectively, and $\omega_{K_1} < \omega_{K_2}$ on $\Omega - \bar{U}$. Consider $v \in \mathfrak{V}_{p_0}$ and observe that $v^+ = \max\{v, o\}$ is also in $\mathfrak{V}_{p_0}$. Note that

$$v^+(p) \le (\max_{K_1} v^+)\omega_{K_1}(p), \quad p \in \partial\Omega \quad \text{and} \quad p \in \partial K_1$$

$\Rightarrow$ (by the maximum principle for subharmonic functions)

$$v^+(p) \le (\max_{K_1} v^+)\omega_{K_1}(p), \quad p \in \Omega - K_1; \text{ in particular,}$$

$$\max_{\partial K_2} v^+ \le (\max_{\partial K_1} v^+)(\max_{\partial K_2} \omega_{K_1}). \tag{$*_1$}$$

On the other hand, $v^+(p) + \log|z(p)|$ is subharmonic in K_2 and continuous in $\overline{K_2}$. Therefore, its maximum is taken along ∂K_2, and we have

$$\max_{\partial K_1}(v^+(p) + \log|z(p)|) = \max_{\partial K_1} v^+ + \log r_1 \le \max_{\partial K_2}(v^+(p) + \log|z(p)|)$$

$$= \max_{\partial K_2} v^+ + \log r_2 \tag{$*_2$}$$

$\Rightarrow$ (by using $(*_1)$) $\max_{\partial K_1} v^+ \le (1 - \max_{\partial K_2} \omega_{K_1})^{-1} \log \dfrac{r_2}{r_1}$ $\qquad(*_3)$

$\Rightarrow$ (since $v^+ = 0$ outside a compact set and by the maximum principle

for subharmonic functions)

$$\max_{\Omega - K_1} v^+ \le (1 - \max_{\partial K_2} \omega_{K_1})^{-1} \log \frac{r_2}{r_1}, \quad v \in \mathfrak{V}_{p_0}. \tag{$*_3'$}$$

Observe that $v \le v^+$. Hence members of $\mathfrak{V}_{p_0}$ are uniformly bounded above on a compact subset of Ω and then $g_\Omega(p, p_0)$ exists as a finite number for each $p \in \Omega - \{p_0\}$.

To show that $g_\Omega(p, p_0) + \log|z(p)|$ *has a removable singularity at* p_0, where $z(p)$ is still a local parameter with $z(p_0) = 0$: By $(*_2)$, for $0 < r_1 < r_2 < 1$ and $v \in \mathfrak{V}_{p_0}$

$$\max_{K_1}(v(z(p)) + \log|z(p)|)$$

$$= \max_{\partial K_1} v(p) + \log r_1$$

$$\le \frac{1}{1 - \max_{\partial K_2} \omega_{K_1}} \log \frac{r_2}{r_1} + \log r_1 \quad \text{(by using $(*_3)$)}$$

$$\le \frac{- \max_{\partial K_2} \omega_{K_1}}{1 - \max_{\partial K_2} \omega_{K_1}} \log r_1 \quad \text{(replacing r_2 by 1)}$$

$$\Rightarrow \log r_1 \le g_\Omega(p, p_0) + \log |z(p)| \le M \quad \text{(a constant) on } |z(p)| < r_1.$$
$$(*_4)$$

The first inequality in $(*_4)$ comes from the fact that the function defined by

$$v(p) = \begin{cases} -\log |z(p)| + \log r_1, & |z(p)| < r_1 \\ 0, & \text{elsewhere} \end{cases}$$

belongs to $\mathfrak{V}_{p_0}$. According to (6.3.1.7), p_0 is a point of a removable singularity of $g(p, p_0) + \log |z(p)|$.

Finally we want *to show that* $\lim_{p \to \partial\Omega} g_\Omega(p, p_0) = 0$ *holds.* Fix any boundary point $\zeta_0 \in \partial\Omega$. Since $\partial\Omega$ consists of finitely many analytic Jordan closed curves, there exists a barrier $\alpha(p)$ at ζ_0 (see Sec. 7.3), namely, there is a continuous function $\alpha(p)$ on the set $\overline{\Omega \cap N}$ (where N is a neighborhood of ζ_0), superharmonic on $\Omega \cap N$, $\alpha(\zeta_0) = 0$ and $\alpha(p) > 0$ for $p \in \overline{\Omega \cap N}$ and $p \ne \zeta_0$. Set

$$m(r) = \text{the minimum of } \alpha(p) \text{ along the set } \Omega \cap \{|z(p)| = r\}, \quad 0 < r < 1,$$

where $\{N, z\}$ is supposed to be a local coordinate system at ζ_0 with $z(N)$ being the open disk $|z| < 1$ and $z(\zeta_0) = 0$. Note that $m(r) > 0$. Let $g(p, p_0) \le M$ (a constant) in $\bar{N} \cap \Omega$ (so $p_0 \notin \bar{N}$). For any fixed $v \in \mathfrak{V}_{p_0}$, then the superharmonic function

$$\frac{M}{m(r)} \alpha(p) - v(p) \begin{cases} \ge M - v(p) \ge 0, & \text{along } \Omega \cap \{|z(p)| = r\} \\ = \frac{M}{m(r)} \alpha(p) \ge 0, & \text{along } \partial\Omega \cap \{|z(p)| < r\} \\ & \text{(for } v \text{ has a compact support in } \Omega) \end{cases}$$

$$\Rightarrow \text{(by the maximum principle) } v(p) \le \frac{m(r)}{M} \alpha(p), \quad p \in \Omega \cap \{|z(p)| < r\}$$

and $v \in \mathfrak{V}_{p_0}$

$$\Rightarrow g_\Omega(p, p_0) \le \frac{m(r)}{M} \alpha(p), \quad p \in \Omega \cap \{|z(p)| < r\}$$

$$\Rightarrow \lim_{p \to \zeta_0} g_\Omega(p, p_0) = 0. \tag{$*_5$}$$

In conclusion, we have (compare to (7.3.2.1) and (6.3.1.11))

The existence and uniqueness of Green's function $g_\Omega(p, p_0)$ *on a regular domain* Ω *in a Riemann surface.* Let Ω be a regular domain in a Riemann surface and $p_0 \in \Omega$. Then there exists a unique function $g_\Omega(p, p_0)$ on

$\Omega - \{p_0\}$, satisfying the properties:

1. $g_\Omega(p, p_0)$ is harmonic in $\Omega - \{p_0\}$.
2. $g_\Omega(p, p_0) > 0$ in $\Omega - \{p_0\}$.
3. In terms of a local parameter $z(p)$ at p_0 with $z(p_0) = 0$, $g_\Omega(p, p_0) + \log|z(p)|$ has a harmonic extension to a neighborhood of p_0 (and hence, p_0 is a *logarithmic pole* of $g(p, p_0)$).
4. For each $q \in \partial\Omega$, $\lim_{p \to q} g_\Omega(p, p_0) = 0$. \hfill (7.3.2.7)

Property 4 characterizes the uniqueness of such a function having properties 1–3. See Exercise A(2) for another proof; see Exercise A(3) for an extension.

Let Ω_n, $n \geq 1$, be a sequence of regular exhaustion for a Riemann surface R (see (7.3.9)). Fix a point $p_0 \in \Omega_1$. Let $g_n(p, p_0)$ be Green's function on Ω_n, $n \geq 1$, with pole at p_0 as stated in (7.3.2.7). Then, by the maximum principle

$$g_n(p, p_0) < g_{n+1}(p, p_0) \text{ on } \Omega_n, \quad n \geq 1$$

$\Rightarrow$ (by Harnack's principle) $\lim_{n \to \infty} g_n(p, p_0) \equiv +\infty$ on $R - \{p_0\}$, or

$$\lim_{n \to \infty} g_n(p, p_0) = g(p, p_0), \quad \text{harmonic on } R - \{p_0\}, \hfill (7.3.2.8)$$

where the convergence is carried out locally uniformly in $R - \{p_0\}$. In case the former happens, we say that there exists no Green's function on R; and in the latter case, $g(p, p_0)$ is called *Green's function of R*, with pole at p_0.

However, that $g(p, p_0) \equiv +\infty$ or $g(p, p_0)$ is harmonic is *independent of the choice of p_0 and the regular exhaustion*.

First, let R_n, $n \geq 1$, be another regular exhaustion of R and $\tilde{g}_n(p, p_0)$ the associated Green's function of R_n with pole at p_0 (here, without loss of its generality, we may assume that $p_0 \in R_1$, too). Set $\tilde{g}(p, p_0) = \lim_{n \to \infty} \tilde{g}_n(p, p_0)$, locally uniformly in $R - \{p_0\}$. Choose a sequence of positive integers n_i, $i \geq 1$, so that

$$\overline{\Omega}_i \subseteq R_{n_i}, \quad i \geq 1$$

$\Rightarrow$ (by the maximum principle) $g_i(p, p_0) < g_{n_i}(p, p_0)$ on Ω_i, $i \geq 1$

$\Rightarrow g(p, p_0) \leq \tilde{g}(p, p_0)$, $R - \{p_0\}$.

Similarly, $\tilde{g}(p, p_0) \leq g(p, p_0)$. Hence $\tilde{g}(p, p_0) = g(p, p_0)$ holds on $R - \{p_0\}$.

Second, let p_1 and p_2 be two distinct points in R so that $p_0 \neq p_1$ and $p_0 \neq p_2$ hold. By Harnack's inequality applying to $g_n(p, p_0)$ (for all sufficiently large n), we can find two constants $m > 0$ and $M > 0$ (see Example 3(2) of Sec. 6.3.2) so that

$$mg_n(p_1, p_0) \le g_n(p_2, p_0) \le Mg_n(p_1, p_0)$$

$$\Rightarrow \text{(see (2) in (6.3.1.13) or (7.3.2.13) below) } mg_n(p_0, p_1)$$

$$\le g_n(p_0, p_2) \le Mg_n(p_0, p_1) \qquad (*_6)$$

$$\Rightarrow m \lim_{n \to \infty} g_n(p_0, p_1) \le \lim_{n \to \infty} g_n(p_0, p_2) \le M \lim_{n \to \infty} g_n(p_0, p_1).$$

Therefore, $\lim_n g_n(p_0, p_1) = \infty$ or $< \infty$, according to $\lim_n g_n(p_0, p_2) = \infty$ or $< \infty$, respectively.

Indeed, we do have (compare to (7.3.1.7))

The equivalence of definitions (7.3.2.4) *and* (7.3.2.8). Let p_0 be a fixed point in a Riemann surface R and Ω_n, $n \ge 1$, be a regular exhaustion of R so that $p_0 \in \Omega_1$. Let $g(p, p_0)$ be the harmonic function defined in (7.3.2.4).

(1) Let u_n be the unique harmonic function in Ω_n, continuous on $\overline{\Omega}_n$, with the boundary value $u_n | \partial \Omega_n = g(., p_0)$ (this fact is assured by (7.3.5) and (7.3.6)). Then

$$g_n(p, p_0) = g(p, p_0) - u_n(p), \quad p \in \Omega_n - \{p_0\}$$

is the (unique) Green's function of Ω_n with pole at p_0 as stated in (7.3.2.7).

(2) (In case $g(p, p_0) < \infty$) Then $g(p, p_0) = \lim_{n \to \infty} g_n(p, p_0)$ in the sense (7.3.2.8), for each p in $R - \{p_0\}$.   (7.3.2.9)

Proof. To prove (2), set $\tilde{g}(p, p_0) = \lim_{n \to \infty} g_n(p, p_0)$ for the moment. Since $\tilde{g}(p, p_0)$ has properties 1–3 in (7.3.2.1) (why property 3 does hold for $\tilde{g}(p, p_0)$?), we have already shown that $\tilde{g}(p, p_0) \ge g(p, p_0)$ in $R - \{p_0\}$. Define $g_n(p, p_0) \equiv 0$ on $R - \Omega_n$, $n \ge 1$. Then $g_n(\cdot, p_0) \in \mathfrak{V}_{p_0}$ (defined in (7.3.2.2)) and $g_n(p, p_0) \le g(p, p_0)$ on $R - \{p_0\}$. It follows that $\tilde{g}(p, p_0) \le g(p, p_0)$ holds on $R - \{p_0\}$. This finishes the proof. $\qquad \square$

See Exercise A(4) for another definition of $g(p, p_0)$.

Section (2) What kind of surface does Green's function possess?

Green's function cannot exist on a closed Riemann surface R. Otherwise $g(p, p_0)$ would obtain its global minimum at a point other than p_0. This cannot happen since $g(p, p_0)$ is not constant and is harmonic everywhere in $R - \{p_0\}$. An *alternate* way to justify this claim is as follows. Suppose $g(p, p_0)$ does exist. Let D be a Jordan disk, containing z_0, so that $z(\bar{D})$ is the closed disk $|z| \leq 1$ under a local parameter $z(p)$ with $z(p_0) = 0$. According to (2) in (6.3.1.14) or by a direct computation (noting that $g(p, p_0) + \log|z(p)|$ is harmonic in $\bar{D}$),

$$\int_{\partial D} \frac{\partial g}{\partial n} ds = -2\pi.$$

On the other hand, $g(p, p_0)$ is harmonic in $R - D$. Hence (see $(*_3)$ in Sec. 6.3.1)

$$\int_{\partial D} \frac{\partial g}{\partial n} ds = 0.$$

These contradictory facts justify our claim.

Green's function cannot exist on the open Riemann surface $R - \{q\}$, where R is a closed Riemann surface and q is a fixed point in it. Suppose on the contrary that there does exist Green's function $g(p, p_0)$ in $R - \{q\}$, where $p_0 \neq q$. Since $g(p, p_0)$ is bounded in a neighborhood of q, q is a removable singularity of $g(p, p_0)$. Hence, $g(p, p_0)$ is Green's function of R with pole at p_0, contradicting to the fact obtained in the last paragraph.

In particular, both C^* and C do not carry Green's function.

How about the open unit disk? It is, as we have observed in $(*_{10})$ of Sec. 6.3.1. As suggested by $(*_1)$–$(*_4)$, (6.3.1.10), and (7.3.1.9) can be extended to the following universal

Criteria for an open Riemann surface carrying Green's function. Let R be an open Riemann surface. Then

(1) Green's function in (7.3.2.4) exists on R.
⇔ (2) Harmonic measure in (7.3.1.3) exists on R.
⇔ (3) The maximum principle in (7.3.1.8) is not valid in R.   (7.3.2.10)

Recall the following notations: $g(p, p_0)$ for Green's function with pole at p_0 and $\omega_K(p)$ for the harmonic measure of a compact set K with nonempty interior. The scheme to prove (7.3.2.10) is the following implications:

(3) ⇒ (2) for all compact sets K and K' with Int $K' \neq \emptyset$;

(2) ⇒ (1) if $p_0 \in$ Int K; and

(1) ⇒ (3) if $p_0 \in K$.   (7.3.2.11)

Proof. (3) ⇒ (2) is the one in proving (7.3.1.8). This proof also indicates that once the maximum principle is not valid for *a single compact set*

K, then the harmonic measure $\omega_{K'}$ exists for *any* compact set K' with Int $K' \neq \emptyset$.

(2) $\Rightarrow$ (1) is the one given from ($*_1$) to ($*_4$) if the regular domain Ω there is now replaced by the open Riemann surface R. Since any point p_0 in R can always be an interior point of a compact set, this proof indicates that the existence of the harmonic measure ω_K for any compact K with Int $K \neq \emptyset$ guarantees the existence of Green's function for *any* point $p_0 \in R$. What remains is to prove (1) $\Rightarrow$ (3). Once the existence of Green's function $g(p, p_0)$ for any $p_0 \in R$ is known, this proof indicates that the maximum principle is not valid for *any* compact set K because it is always possible to arrange the pole p_0 as a point belonging to K.

The proof of (1) $\Rightarrow$ (3): the negative $-g(p, p_0)$ of Green's function has a maximum M on K. On the other hand, $-g(p, p_0) < 0$ holds on $R - K$. If the maximum principle were valid on $R - K$, then $-g(p, p_0) \leq M$ on $R - K$. Hence $-g(p, p_0) \leq M$ throughout the whole R and $-g(p, p_0)$ attains its global maximum M at an interior point of the open Riemann surface $R - \{p_0\}$. This contradicts to classical maximum principle since $-g(p, p_0)$ is not a constant. □

Section (3) Some further properties

(2) in (6.3.1.11) and (6.3.1.13) can be extended to

Two basic properties of Green's function. Let R be an open Riemann surface carrying Green's function $g(p, p_0)$. Then

(1) $\varliminf_{p \to \infty} g(p, p_0) = 0$, where ∞ is the ideal boundary of R.

(2) $g(p, p_0)$ is symmetric as a function of p and p_0 in R, namely,

$$g(p, p_0) = g(p_0, p), p, p_0 \in R \quad \text{and} \quad p \neq p_0. \qquad (7.3.2.12)$$

Proof. (1) Suppose on the opposite that $g(p, p_0) \geq \alpha > 0$ on R. Choose any $v \in \mathfrak{V}_{p_0}$, defined in (7.3.2.2) and (7.3.2.4). Then $g(p, p_0) - \alpha - v(p)$ is superharmonic not only in $R - \{p_0\}$ but also in a neighborhood of p_0. This is because $g(p, p_0) - \alpha - v(p) = g(p, p_0) + \log |z(p)| - \alpha - [v(p) + \log |z(p)|]$ holds where $z(p)$ is a local parameter at p_0 with $z(p_0) = 0$. Let supp$v$ be the compact support of v. Then

$g(p, p_0) - \alpha - v(p) = g(p, p_0) - \alpha \geq 0$, if p lies on the boundary of a relatively compact domain containing supp$v$.

$\Rightarrow$ (by the maximum principle for superharmonic functions) $g(p, p_0) - \alpha - v(p) \geq 0$ on supp$v$.

Since $g(p, p_0) - \alpha - v(p) = g(p, p_0) - \alpha \geq 0$ on $R - \mathrm{supp}\, v$, it follows that

$$v(p) \leq g(p, p_0) - \alpha, \quad v \in \mathfrak{V}_{p_0} \quad \text{and} \quad p \in R - \{p_0\}$$

$$\Rightarrow g(p, p_0) \leq g(p, p_0) - \alpha, \quad p \in R - \{p_0\}$$

$$\Rightarrow \alpha \leq 0, \quad \text{a contradiction.}$$

Hence $\underline{\lim}_{p \to \infty} g(p, p_0) = 0$ holds.

If we adopt (7.3.2.8) as our definition of $g(p, p_0)$, an easier proof is as follows. Still suppose $g(p, p_0) \geq \alpha > 0$ holds on R. Then $g(p, p_0) - \alpha - g_n(p, p_0)$ is harmonic on Ω_n. Since $g_n(p, p_0) = 0$ along $\partial \Omega_n$, it follows by the maximum principle that $g(p, p_0) - \alpha - g_n(p, p_0) > 0$ on Ω_n. Hence $g_n(p, p_0) < g(p, p_0) - \alpha$ on Ω_n for $n \geq 1$, and this in turn implies that $g(p, p_0) \leq g(p, p_0) - \alpha$, a contradiction.

(2) In the proof of $(*_6)$, we have used the fact:

The symmetric property of Green's function $g_\Omega(p, p_0)$ *defined in* (7.3.2.7).

$$g_\Omega(p, p_0) = g_\Omega(p_0, p), \quad p, p_0 \in \Omega \quad \text{and} \quad p \neq p_0. \tag{7.3.2.13}$$

Its proof is essentially the same as that given for (2) in (6.3.1.13). The only change needed is to use *Green's formula* (refer to Exercise B(1) of Sec. 2.9)

$$\iint_\Omega (u_1 \Delta u_2 - u_2 \Delta u_1) = \int_{\partial \Omega} (u_1 * du_2 - u_2 * du_1) \tag{7.3.2.14}$$

to replace (3) in (6.3.1.3), where u_1 and u_2 are C^1-functions in a neighborhood of the compact set $\bar{\Omega}$. Since $\bar{\Omega}$ can be covered by finitely many triangles on a Riemann surface, by using Stokes's theorem (see (2.9.12)') directly or by invoking Exercise B(1) of Sec. 2.9, (7.3.2.14) follows easily. Try it, and if a difficulty might arise, refer to Ref. [75], p. 167, or [32], p. 28.

Take (7.3.2.13) for granted, and then $g(p, p_0) = g(p_0, p)$ follows by noting the definition (7.3.2.8) and its equivalence to that given in (7.3.2.4) (see (7.3.2.9)). □

Based on (7.3.2.12), we have

A property of the harmonic measure. Let R be an open Riemann surface carrying the harmonic measure. Let K be a compact set in R, with $\mathrm{Int}\, K \neq \emptyset$. Then

$$\underline{\lim}_{p \to \infty} \omega_K(p) = 0,$$

where ∞ is the ideal boundary of R. $\tag{7.3.2.15}$

As a *corollary*, $\omega_K(p) \equiv 1$ happens only if R does not carry Green's function, or correspondingly, the harmonic measure.

Proof. We adopt the definition (7.3.1.6) and the related notations with $K = \bar{\Omega}_1$.

Set $M_n = \max_{p \in \partial \Omega_1} g_n(p, p_0)$ for $n \geq 1$. Then $M_n > 0$, $n \geq 1$. Let ω_n, $n \geq 2$, be the harmonic measure of $\bar{\Omega}_n$, namely, ω_n is harmonic in $\Omega_n - \bar{\Omega}_1$ for $n \geq 2$, $0 < \omega_n < 1$, $\omega_n|\partial\Omega_1 = 1$ and $\omega_n|\partial\Omega_n = 0$ (see (7.3.1.5)). By the maximum principle for harmonic functions,

$$g_n(p, p_0) \leq \omega_n(p) M_n, \quad p \in \Omega_n - \overline{\Omega_1} \quad \text{for } n \geq 2$$

$$\Rightarrow \max_{p \in \partial \Omega_2} g_n(p, p_0) \leq M_n \max_{p \in \partial \Omega_2} \omega_n(p), \quad n \geq 3. \qquad (*_7)$$

On the other hand, since $g_n(p, p_0) - g_2(p, p_0)$, $n \geq 3$, is harmonic in Ω_2, again by the maximum principle,

$$\max_{p \in \partial \Omega_1} [g_n(p, p_0) - g_2(p, p_0)] \leq \max_{p \in \partial \Omega_2} [g_n(p, p_0) - g_2(p, p_0)], \quad n \geq 3$$

$$\Rightarrow M_n - \alpha \leq \max_{p \in \partial \Omega_2} g_n(p, p_0) \text{ for some constant } \alpha > 0 \quad \text{and} \quad n \geq 3$$

$$\Rightarrow (\text{by}(*_7)) \; M_n - \alpha \leq M_n \max_{p \in \partial \Omega_2} \omega_n(p), \quad n \geq 3$$

$$\Rightarrow 0 \leq 1 - \max_{p \in \partial \Omega_2} \omega(p) \leq 1 - \max_{p \in \partial \Omega_2} \omega_n(p) \leq \frac{\alpha}{M_n} \quad \text{for } n \geq 3.$$

In case R does not carry Green's function, then $\lim_{n \to \infty} g_n(p, p_0) = \infty$ implies that $\lim_{n \to \infty} M_n = \infty$, and hence $\max_{p \in \partial \Omega_2} \omega(p) = 1$. Since $\max_{p \in \partial \Omega_2} \omega(p) < 1$ if $\omega(p) \not\equiv 1$, it follows that $\omega(p) \equiv 1$ should hold. This proves the corollary.

Set $m = \min_{p \in \partial \Omega_1} g(p, p_0)$ if $g(p, p_0)$ exists. By the maximum principle,

$$0 \leq \omega_n(p) \leq \frac{g(p, p_0)}{m} \quad \text{on } \Omega_n - \bar{\Omega}_1 \text{ for } n \geq 2$$

$$\Rightarrow 0 \leq \omega(p) \leq \frac{g(p, p_0)}{m} \quad \text{on } R - \bar{\Omega}_1$$

$$\Rightarrow (\text{by}(7.3.2.12)) \; \underline{\lim}_{p \to \infty} \omega(p) = 0.$$

This finishes the proof. $\qquad \qquad \square$

Exercises A

(1) Prove in detail Green's formula (7.3.2.14) and then prove (7.3.2.13) carefully.

(2) Adopt the following steps to give (7.3.2.7) another proof.

(a) Take a Jordan disk D containing p_0 so that $D \subseteq \Omega$ and $z(D)$ is the unit disk $|z| < 1$, where $z(p)$ is a local parameter with $z(p_0) = 0$. Let $1 > r_0 > r_1 > \cdots > r_n > 0$ so that $\lim_{n \to \infty} r_n = 0$ and set $D_n = \{p \in D \, | \, |z(p)| < r_n\}$ for $n \geq 0$. By (7.3.5), let $u_n(p)$ be the harmonic function in $\Omega - \overline{D_n}$ such that $u_n|\partial\Omega = 0$ and $u_n|\partial D_n = \log\frac{1}{r_n}$ for $n \geq 0$. Applying Green's formula (7.3.2.14) to u_n, u_0 on $\Omega - \bar{D}_0$, we have (see (6.3.1.4) for $*du = -\frac{\partial u}{\partial n}|dz| = -r\frac{\partial u}{\partial r}d\theta = -\frac{\partial u}{\partial r}ds$ along $|z| = r$, where $\frac{\partial u}{\partial n}$ is the *inward* normal derivative of u)

$$\int_{\partial D_0} u_n \frac{\partial u_0}{\partial r} ds = \log\frac{1}{r_0}\int_{\partial D_0} \frac{\partial u_n}{\partial r} ds; \qquad (*_8)$$

to u_n, $\log\frac{1}{r}$ on $D_0 - \overline{D_n}$, we have

$$\int_{\partial D_0}\left(u_n\frac{\partial}{\partial r}\log\frac{1}{r} - \log\frac{1}{r_0}\frac{\partial u_n}{\partial r}\right) ds$$

$$= \int_{\partial D_n}\left(u_n\frac{\partial}{\partial r}\log\frac{1}{r} - \log\frac{1}{r_n}\frac{\partial u_n}{\partial r}\right) ds;$$

$$\Rightarrow \left(\text{noting that } \int_{\partial D_n}\frac{\partial u_n}{\partial r}ds = \int_{\partial D_0}\frac{\partial u_n}{\partial r}ds\right)$$

$$\int_{\partial D_0}\left(-\frac{\partial u_n}{\partial r}\right) ds \leq \frac{2\pi\log\frac{1}{r_1}}{\log\frac{r_0}{r_1}}, \quad n \geq 1. \qquad (*_9)$$

Combining $(*_8)$ and $(*_9)$, we obtain

$$\int_{\partial D_0} u_n\left(-\frac{\partial u_0}{\partial r}\right) ds \leq 2\pi M,$$

$$\text{where } M = \frac{(\log r_0)(\log r_1)}{\log\frac{r_0}{r_1}}, \text{ for } n \geq 1. \qquad (*_{10})$$

Try to show that $-\frac{\partial u_0}{\partial r} \geq \alpha$ (a constant) > 0 along ∂D_0 (refer to $(*_2)$ in Sec. 7.3.3). Hence, by $(*_{10})$,

$$\min_{\partial D_0} u_n \leq \frac{M}{\alpha r_0} \text{ for } n \geq 1. \qquad (*_{11})$$

(b) Choose $0 < \rho < r_0$ and set $D_\rho = \{p \in D | |z(p)| \le \rho|\}$. Then $u_n(p) > 0$ on $\Omega - D_\rho$ and, by Example 3(2) of Sec. 6.3.2 and $(*_{11})$, there is a constant $K(\rho)$, depending only on ρ, so that

$$0 < u_n(p) \le K(\rho) \quad \text{on } \Omega - \bar{D}_\rho \text{ for all sufficiently large } n. \qquad (*_{12})$$

According to (6.3.2.16), by passing to a subsequence if necessary, we obtain that $u_n(p)$ converges to a harmonic function $g(p, p_0)$ locally uniformly on $\Omega - \{p_0\}$.

(c) To prove $g(p, p_0) = 0$ along $\partial\Omega$: By $(*_{12})$,

$$0 < u_n(p) \le K \quad \text{(a constant) on } \Omega - \bar{D}_0 \text{ for all } n \ge 0. \qquad (*_{13})$$

In particular $K \ge \log\frac{1}{r_0}$. By the maximum principle for harmonic function,

$$0 < u_n(p) \le K\frac{u_0(p)}{\log\frac{1}{r_0}} \quad \text{on } \Omega - \bar{D}_0 \text{ for } n \ge 0.$$

$$\Rightarrow 0 < g(p, p_0) \le K\frac{u_0(p)}{\log\frac{1}{r_0}} \quad \text{on } \Omega - \bar{D}_0.$$

The result follows.

(d) To prove that $g(p, p_0) + \log|z(p)|$ is harmonic at p_0: Since $u_n(p) + \log|z(p)| = 0$ along ∂D_n, it follows that, on $D_0 - \bar{D}_n$,

$$|u_n(p) + \log|z(p)|| \le \max_{\partial D_0}|u_n(p) + \log|z(p)|| \le K - \log\frac{1}{r_0} \le K$$

$$\Rightarrow |g(p, p_0) + \log|z(p)|| \le K \text{ in } D_0 - \{p_0\}.$$

(3) Show that (7.3.2.7) is still valid even if Ω is a relatively compact domain in a Riemann surface, having its boundary $\partial\Omega$ composed of finitely many disjoint Jordan curves.

(4) Let Ω be a regular domain in an open Riemann surface R carrying Green's function $g(p, p_0)$ so that $p_0 \in \Omega$. Extend Green's function $g_\Omega(p, p_0)$ (see (7.3.2.7)) to be identically zero in $R - \Omega$. The resulted function, still denoted as $g_\Omega(p, p_0)$, is subharmonic in R.

(a) Let $\mathfrak{V}$ be the family of all such subharmonic functions $g_\Omega(p, p_0)$. Show that $\mathfrak{V}$ is a Perron family.

(b) Show that $g(p, p_0) = \sup_{\mathfrak{V}} g_\Omega(p, p_0)$.

(5) Show that Green's function $g(p, p_0)$, if exists, is the unique harmonic function having the properties 1, 2, 3 in (7.3.2.1) and the additional property $\underline{\lim}_{p\to\infty} g(p, p_0) = 0$ (see (7.3.2.12) and compare to property 4 in (7.3.2.1)).

(6) Let Ω be a simply connected domain in a closed Riemann surface, whose boundary $\partial\Omega$ is a Jordan closed curve (even a continuum). Show that Ω can be mapped univalently conformally onto the open unit disk. *Note.* This is an extension of (6.3.1.12) to domains in a closed Riemann surface.

Exercises B

(1) Let R be a closed Riemann surface of genus zero. Then R can be mapped univalently conformally onto the extended complex plane $\mathbf{C}^*$. Observe the following steps.

(a) Fix a point $p_0 \in R$. Choose $1 \geq r_1 > r_2 > \cdots > r_n > \cdots$ and $r_n \to 0$ so that $D_n = \{p \mid |z(p)| < r_n\}, n \geq 1$, are Jordan disks, where $z(p)$ is a local parameter with $z(p_0) \neq 0$. Let $g_n(p, p_0)$ be Green's function on $R - \bar{D}_n$ and $h_n(p, p_0)$ be its harmonic conjugate. Set

$$\varphi_n(p) = e^{-(g_n(p,p_0)+ih_n(p,p_0))}, \quad \varphi_n(p_0) = 0 (n \geq 1)$$

(here, p is identified with $z(p) = z$). Why is $R - D_n$ simply connected (see (7.2.18))? By Exercise A(6), φ_n maps $R - \bar{D}_n$ conformally onto $|w| < 1$. Here (see Remark 2 in Sec. 6.1.1)

$$f_n(p) = \frac{\varphi_n(p)}{\varphi_n'(p_0)}, \quad f_n(p_0) = 0, \quad f_n'(p_0) = 1 \quad (n \geq 1)$$

maps $R - D_n$ conformally onto the disk $|w| < R_n$, where $R_n = \frac{1}{|\varphi_n'(p_0)|}$.

(b) Since $g_n(p, p_0) < g_{n+1}(p, p_0)$ on $R - \bar{D}_n$ for $n \geq 1$, it follows that $R_n \leq R_{n+1}$. Hence $\lim_{n\to\infty} R_0 = R$ exists where $0 < R_0 \leq \infty$. According to (5.8.2.3), $\{f_n\}$ is a normal family. We may assume that

$$\lim_{n\to\infty} f_n(z) = f(z), \quad f(p_0) = 0, \quad f'(p_0) = 1$$

locally uniformly in $R - \{p_0\}$. Hence $w = f(p)$ maps $R - \{p_0\}$ conformally onto $|w| < R_0$.

(c) In case $R_0 < \infty$, say reason why there would exist Green's function on $R - \{p_0\}$, which is a contradiction (see the second paragraph in Section (2)).

(d) Hence $R_0 = \infty$ holds. This completes the proof. *Note.* This result unifies so many individual cases in Example 1 of Sec. 7.2.

7.3.3   A classification of Riemann surfaces

A closed (=compact) Riemann surface is called *elliptic*.

As for open (noncompact) Riemann surface, we combine (7.3.1.9) and (7.3.2.10) into the following remarkable

Properties on an open Riemann surface R. The following are equivalent:

(1) There exists a nonconstant continuous positive superharmonic function on R.
(2) Harmonic measure (see (7.3.1.1)) exists on R.
(3) Green's function (see (7.3.2.1))) exists on R.
(4) The maximum principle (see (7.3.1.8)) is not valid in R.   (7.3.3.1)

An open Riemann surface with one, and hence all the properties listed in (7.3.3.1), is called *hyperbolic*. While, an open Riemann surface carrying no any, and hence all these properties is called *parabolic*.

Classification theory of Riemann surfaces stands on its own right, and interested readers might refer to Ref. [72]. Our main and restricted purpose in this direction is to show that any (simply connected) Riemann surface is conformally equivalent to exactly one of the above three types (see Sec. 7.6). As we claimed at the beginning of Sec. 7.3, we need to construct a global meromorphic function on the surface to realize the mapping. In hyperbolic case, the existence of Green's function is what we need because we can construct local harmonic conjugates of Green's function and then paste them into a global single-valued meromorphic function. The lack of such a global harmonic or subharmonic function on a nonhyperbolic surface causes our special attention to construct one based on a less direct method. And this is what we are going to do in the sequel.

It is worthy to mention, before we start, the influence of the existence or nonexistence of the maximum principle (7.3.1.8) on

The uniqueness or nonuniqueness of the solution of the Dirichlet problem on an open Riemann surface R.

(1) On a parabolic Riemann surface, the Dirichlet problem has at most one bounded solution (see (7.3.7) and (7.3.1.8)).
(2) On a hyperbolic Riemann surface, the Dirichlet problem does not always have a unique solution even if it is solvable. In particular, if K is a regular domain in R and u is any solution to a Dirichlet problem for $R - \bar{K}$, then so is $(1 - \omega_{\bar{K}}) + u$, where $\omega_{\bar{K}}$ is the harmonic measure of $\bar{K}$ (see (7.3.1.1) and (7.3.1.7)).   (7.3.3.2).

The main result essential to our proof of the uniformization theorem in Sec. 7.6 is the following

Existence of a harmonic function with a singularity on a nonhyperbolic Riemann surface. Let R be an elliptic or a parabolic Riemann surface and $p_0 \in R$ be a fixed point. Choose a local parameter $z(p)$ at p_0 so that $z(p_0) = 0$ and the range of $z(p)$ contains the disk $|z| \le 1$. For $0 < \rho \le 1$, set $D_\rho = \{p||z(p)| < \rho\}$.

(1) There exists a unique bounded harmonic function u_ρ on $R - \bar{D}_\rho$, with the boundary values $\operatorname{Re}\frac{1}{z(p)}$.
(2) $u = \lim_{\rho \to 0} u_\rho$ exists as a harmonic function on $R - \{p_0\}$, satisfying the properties:

   (a) u is bounded outside of every D_ρ, $0 < \rho < 1$.
   (b) $\lim_{p \to p_0} [u(p) - \operatorname{Re}\frac{1}{z(p)}] = 0$; in particular, p_0 is a removable singularity of $u - \operatorname{Re}\frac{1}{z}$.

Such a u is uniquely determined by properties (a) and (b).   (7.3.3.3)

Farkas and Kra [32] present a slight generalization of this result on p. 172, where $\frac{1}{z(p)}$ is replaced by any function analytic on $R - \{p_0\}$, and its proof lasts up to p. 178. We will adopt part of this proof as our proof of (7.3.3.3) below.

Proof. (1) Consider the family $\mathfrak{V}_\rho$ of bounded subharmonic functions v on $R - \bar{D}_\rho$ that satisfy $v(p) \le \operatorname{Re}\frac{1}{z(p)}$ along the boundary ∂D_ρ of D_ρ. $\mathfrak{V}_\rho$ is nonempty because $\max\{0, \operatorname{Re}\frac{1}{z(p)}\}$ belongs to it. For each $v \in \mathfrak{V}_\rho$,

$$v(p) \le \operatorname{Re}\frac{1}{z(p)} \le \max_{p \in \partial D_\rho} \operatorname{Re}\frac{1}{z(p)} < \infty \text{ along } \partial D_\rho$$

$\Rightarrow$ (by the maximum principle (7.3.1.8)) v is uniformly bounded above in $R - \bar{D}_\rho$.

Once $\mathfrak{V}_\rho$ being a Perron family, according to (7.3.5) and (7.3.6)

$$u_\rho = \sup_{v \in \mathfrak{V}_\rho} v$$

exists as a bounded harmonic function on $R - \bar{D}_\rho$ with the boundary values $\operatorname{Re}\frac{1}{z(p)}$ along ∂D_ρ. Also such a u_ρ is unique as a consequence of the maximum principle.

Note. One can refer to Remark in Sec. 7.3 for the existence of u_ρ. In case R is an elliptic surface, only the classical maximum principle is needed instead of that stated in (7.3.1.8).

(2) Suppose these are two harmonic functions u_1 and u_2 satisfying properties (a) and (b).

Then $u_1 - u_2$ has a removable singularity at p_0 and hence, is bounded everywhere in R. Again the maximum principle implies that $u_1 - u_2$ is a constant which is zero because $[u_1 - u_2](p_0) = 0$ holds by (b). Consequently $u_1 = u_2$ on R.   □

Note. One might use Exercise A(2) of Sec. 7.3.1 to give another proof.

In order to prove the *existence* of such a harmonic u, we digress from the subject for a moment to the following

Lemma. Let u_ρ be the harmonic function on $R - \bar{D}_\rho$ as stated in (7.3.3.3). Then

$$\int_{\partial D_\rho} *du_\rho = r \int_0^{2\pi} \frac{\partial}{\partial r} u_\rho(re^{i\theta})d\theta = 0,$$

where the second integral expression is independent of r, $\rho \leq r \leq 1$.

(7.3.3.4)

Fix an r, $\rho \leq r \leq 1$. The closed disk $\bar{D}_r = \{p \,|\, |z(p)| \leq r\}$ is contained in the coordinate neighborhood where the local parameter $z(p)$ is defined. Identify $u_\rho(p)$ with $u_\rho(z(p)), p \in \bar{D}_r$. Then, according to (6.3.1.4),

$$*du_\rho = \frac{\partial u_\rho}{\partial n}|dz| = r\frac{\partial u_\rho}{\partial r}d\theta \text{ along } |z| = r, \qquad (7.3.3.5)$$

where $\frac{\partial u_\rho}{\partial n}$ denotes the directional derivative of u_ρ in the outward normal to the circle $|z| = r$, called the *normal derivative* of u_ρ.

Proof of the Lemma. Choose a regular domain D so that $\bar{D}_\rho \subseteq D$ holds (see (7.3.9)). Let ω be the harmonic measure of ∂D_ρ with respect to $D - \bar{D}_\rho$, namely, ω is harmonic in $D - \bar{D}_\rho$, $0 < \omega < 1$, $\omega|\partial D_\rho = 1$ and $\omega|\partial D = 0$ (see (7.3.1.5)). Applying Green's formula (7.3.2.14) to u_ρ and ω on $D - \bar{D}_\rho$, we have

$$\iint_{D-\bar{D}_\rho} (\omega \Delta u_\rho - u_\rho \Delta \omega) = \int_{\partial D - \partial D_\rho} (\omega *du_\rho - u_\rho *d\omega) = 0$$

$$\Rightarrow \int_{\partial D_\rho} \left(\omega\frac{\partial u_\rho}{\partial n} - u_\rho\frac{\partial \omega}{\partial n} \right) ds = \int_{\partial D} \left(\omega\frac{\partial u_\rho}{\partial n} - u_\rho\frac{\partial \omega}{\partial n} \right) ds, \qquad (*_1)$$

where the second line integral and the normal derivatives within can be expressed and explained in terms of local variables as we did in (7.3.3.5).

We *claim* that $\frac{\partial \omega}{\partial n}$ has constant sign on ∂D_ρ as well as on ∂D (refer to $(*_{11})$ in Sec. 7.3.2).

We digress again to justify this claim as follows. Subject to a suitable bilinear transformation, this problem is equivalent to the following statement: $\omega(z)$ is harmonic in $|z| < r$ such that $\omega > 0$ for $y > 0$ and $\omega = 0$ for $y = 0$, where $z = x + yi$, then

$$\left.\frac{\partial \omega}{\partial y}\right|_{z=0} > 0 \qquad (*_2)$$

holds. Obviously $\left.\frac{\partial \omega}{\partial y}\right|_{z=0} \geq 0$ holds. Let $\omega(re^{i\theta}) = \sum_{n=1}^{\infty} a_n r^n \sin n\theta$ be its Fourier series expansion (see (4) in (6.3.1.7)). Then $a_1 = \left.\frac{\partial \omega}{\partial y}\right|_{z=0} \geq 0$. In case $a_1 = 0$ and $k \geq 2$ is the least positive integer such that $a_k \neq 0$. Note that $\sin k\theta > 0$ for $0 < \theta < \frac{\pi}{k}$, while $\sin k\theta < 0$ for $\frac{\pi}{k} < \theta < \frac{2\pi}{k}$. If r_0 is chosen sufficiently small, it is always possible to have θ_0, $0 < \theta_0 \leq \pi$, so that $\omega(r_0 e^{i\theta_0}) < 0$ holds, a contradiction. Hence $a_1 > 0$ is true.

Returning to our subject, observe that $|u_\rho| \leq \frac{1}{\rho}$ on $R - \overline{D}_\rho$. By the maximum principle and noting that $\omega|\partial D_\rho = 1$ and $\omega|\partial D = 0$, it follows from $(*_1)$ that

$$\left| \int_{\partial D_\rho} \frac{\partial u_\rho}{\partial n} ds \right| \leq \left| \int_{\partial D_\rho} u_\rho \frac{\partial \omega}{\partial n} ds \right| + \left| \int_{\partial D} u_\rho \frac{\partial \omega}{\partial n} ds \right|$$

$$\leq \frac{1}{\rho} \left| \int_{\partial D_\rho} \frac{\partial \omega}{\partial n} ds \right|$$

$$+ \frac{1}{\rho} \left| \int_{\partial D} \frac{\partial \omega}{\partial n} ds \right| \quad \text{(by the claim in the last paragraph)}$$

$$= \frac{2}{\rho} \left| \int_{\partial D_\rho} \frac{\partial \omega}{\partial n} ds \right|$$

$$\left(\text{Why is } \int_{\partial D_\rho - \partial D} \frac{\partial \omega}{\partial n} ds = 0? \text{ See (2) in (6.3.1.3)} \right).$$

$$(*_3)$$

Since $\omega|\partial D_\rho = 1$, by the symmetry principle (3.4.6.4) or (6.3.2.13), the harmonicity of ω can be extended, across ∂D_ρ, into D_ρ. Now expand D (see (7.3.9)). Since R is nonhyperbolic, $\overline{D}_\rho$ does not have harmonic measure.

This means that, using ω_D to represent ω above temporarily,

$$\lim_{D \to R} \omega_D = 1$$

locally uniformly in R (see the corollary after (7.3.2.15)). In particular, $\omega = \omega_D \to 1$ uniformly in a neighborhood of ∂D_ρ and hence, $\frac{\partial \omega}{\partial n} \to 0$ uniformly on ∂D_ρ (see Exercise A(11) of Sec. 6.3.2).

We conclude from (*$_3$) that

$$\int_{\partial D_\rho} \frac{\partial u_\rho}{\partial n} ds = 0. \tag{*$_4$}$$

The above argument is applicable to any $\bar{D}_r$, where $\rho \le r \le 1$, and then $\int_{\partial D_r} \frac{\partial u_\rho}{\partial n} ds = 0$ holds, too. Alternatively, observe that $\int_{\partial D_\rho - \partial D_r} \frac{\partial \omega}{\partial n} ds = 0$ and this will indicate that (*$_4$) holds for ∂D_r, $\rho \le r \le 1$. □

The continuation of proof of (2) in (7.3.3.3)

Step 1. Fix $0 < \rho < 1$. We try to show that there is a constant $C(r)$, dependent only on r ($\rho < r < 1$) but not on ρ, so that

$$|u_\rho| \le C(r) \quad \text{on } R - \bar{D}_r. \tag{*$_5$}$$

Since $|u_\rho|$ is subharmonic in $R - \bar{D}_\rho$, it suffices to show that $|u_\rho| \le C(r)$ on the boundary ∂D_r.

Owing to (7.3.3.4), we are able to define an analytic function on $\rho < |z| \le 1$ by

$$f_\rho(z) = u_\rho(z_0) + \int_{z_0}^z (du_\rho + i^* du_\rho) = u_\rho(z) + i \int_{z_0}^z {}^* du_\rho, \tag{*$_6$}$$

where z_0 is a fixed point in $\rho < |z| \le 1$. As usual, a point $p \in \bar{D}_1 - \bar{D}_\rho$ is identified with $z = z(p)$, $\rho < |z| \le 1$. The definition of f_ρ imitates that defined in (6.6.1.7).

Both f_ρ and $\frac{1}{z}$ are analytic on $\rho < |z| \le 1$. $f_\rho - \frac{1}{z}$ has the Laurent series expansion

$$f_\rho(z) - \frac{1}{z} = \sum_{n=-\infty}^{\infty} a_n z^n, \quad \rho < |z| \le 1.$$

Set $z = te^{i\theta}$ ($0 \le \theta \le 2\pi$), $a_n = \alpha_n(\rho) - i\beta_n(\rho)$, $n = \pm 1, \pm 2, \ldots$ and $\text{Re}\, a_0 = \frac{\alpha_0(\rho)}{2}$, and then take the real parts of both sides (refer to (*$_6$) in

Sec. 6.3.1). Hence, on $\rho < |z| \le 1$,

$$u_\rho(z) - \mathrm{Re}\frac{1}{z} = \frac{\alpha_0(\rho)}{2} + \sum_{n=1}^{\infty} \left(\alpha_n(\rho)t^n + \alpha_{-n}(\rho)t^{-n}\right)\cos n\theta$$

$$+ \sum_{n=1}^{\infty} \left(\beta_n(\rho)t^n - \beta_{-n}(\rho)t^{-n}\right)\sin n\theta \qquad (*_7)$$

with its coefficients given by

$$\alpha_0(\rho) = \frac{1}{\pi}\int_0^{2\pi}\left[u_\rho(te^{i\theta}) - \mathrm{Re}\frac{1}{te^{i\theta}}\right]d\theta;$$

$$\alpha_n(\rho)t^n + \alpha_{-n}(\rho)t^{-n} = \frac{1}{\pi}\int_0^{2\pi}\left[u_\rho(te^{i\theta}) - \mathrm{Re}\frac{1}{te^{i\theta}}\right]\cos n\theta d\theta, \quad n \ge 1;$$

$$\beta_n(\rho)t^n + \beta_{-n}(\rho)t^{-n} = \frac{1}{\pi}\int_0^{2\pi}\left[u_\rho(te^{i\theta}) - \mathrm{Re}\frac{1}{te^{i\theta}}\right]\sin n\theta d\theta, \quad n \ge 1,$$

$$(*_8)$$

where $\rho < t \le 1$. Note that $u_\rho(z) - \mathrm{Re}\frac{1}{z} = 0$ along $|z| = \rho$. It follows by the symmetry principle (6.3.2.13) that $u_\rho(z) - \mathrm{Re}\frac{1}{z}$ can be harmonically extended beyond the circle $|z| = \rho$. As a consequence, setting $t = \rho$ in $(*_8)$, we obtain

$$\alpha_0(\rho) = 0, \quad \alpha_n(\rho)\rho^n + \alpha_{-n}(\rho)\rho^{-n} = 0 \quad \text{and}$$

$$\beta_n(\rho)\rho^n + \beta_{-n}(\rho)\rho^{-n} = 0 \text{ for } n \ge 1. \qquad (*_9)$$

On setting $t = 1$ in $(*_8)$, we obtain

$$|\alpha_n(\rho) + \alpha_{-n}(\rho)| \le 2(M(\rho) + 1),$$

$$|\beta_n(\rho) + \beta_{-n}(\rho)| \le 2(M(\rho) + 1), \quad \text{where } M(\rho) = \max_{|z|=1}|u_\rho(z)|. \qquad (*_{10})$$

Combining $(*_9)$ and $(*_{10})$, we have

$$\left.\begin{array}{l}|\alpha_n(\rho)|(1 - \rho^{2n})\\ |\beta_n(\rho)|(1 - \rho^{2n})\end{array}\right\} \le 2(M(\rho) + 1)$$

$$\Rightarrow |\alpha_n(\rho)|, \ |\beta_n(\rho)| \le 4(M(\rho) + 1) \quad \text{if } n \ge 0 \text{ and } 0 < \rho < \frac{1}{2}, \text{ and similarly,}$$

$$|\alpha_{-n}(\rho)|, |\beta_{-n}(\rho)| \le 4(M(\rho) + 1)\rho^{2n} \quad \text{if } 0 < \rho < \frac{1}{2}. \qquad (*_{11})$$

On substituting these results in $(*_7)$, we obtain

$$\max_{|z|=r} |u_\rho(z)| \leq \max_{|z|=r} \left| \text{Re} \frac{1}{z} \right| + 8 \sum_{n=1}^{\infty} (M(\rho) + 1) \left(r^n + \frac{\rho^{2n}}{r^n} \right)$$

$$\leq \frac{1}{r} + 16(M(\rho) + 1) \cdot \frac{r}{1-r} \quad \left(\text{by observing } \frac{\rho^2}{r} < r \right).$$

Since $|u_\rho|$ is subharmonic in $R - \bar{D}_\rho$, by the maximal principle,

$$M(\rho) \leq \max_{|z|=r} |u_\rho(z)|$$

$$\Rightarrow M(\rho) \leq \frac{1}{r} + 16(M(\rho)+1)\frac{r}{1-r}, \quad \text{and hence } M(\rho) \leq \frac{1}{1-16s}$$

$$\times \left(\frac{1}{r} + 16s \right), \quad \text{if } r \text{ is so small that } s = \frac{r}{1-r} < \frac{1}{16} \left(\text{say } r < \frac{1}{17} \right).$$

$$\Rightarrow \max_{|z|=r} |u_\rho(z)| \leq \frac{1}{r} + 16 \left[\frac{1}{1-16s} \left(\frac{1}{r} + 16s \right) + 1 \right] s = C(r).$$

This *proves* $(*_5)$ in case $0 < \rho < r < 1$, and r and ρ are sufficiently small. By the way, from $(*_{11})$, we have

$$|\alpha_n(\rho)|, \quad |\beta_n(\rho)| \leq 4(C(r) + 1) \text{ for } n \geq 1, \text{ and}$$

$$|\alpha_{-n}(\rho)|, \quad |\beta_{-n}(\rho)| \leq 4(C(r) + 1)\rho^{2n} \text{ for } n \geq 1. \qquad (*_{12})$$

Step 2. In what follows, let $\rho > 0$ be small enough so that there exists r satisfying $\rho < r < \frac{1}{17}$. Now, the family of harmonic functions u_ρ, indexed by $\rho < r$, is uniformly bounded along ∂D_r by $(*_5)$ and hence, by the maximum principle, in $R - \bar{D}_r$. According to (6.3.2.16), a sequence from this family converges *uniformly* on $R - \bar{D}_r$ to a harmonic function (as $\rho \to 0$).

Step 3. A Cantor diagonal process argument.

Choose $1 > r_1 > r_2 > \cdots > r_n > \cdots$ so that $\lim_{n\to\infty} r_n = 0$. Choose a sequence $r_1 > \rho_{11} > \rho_{12} > \cdots > \rho_{1n} > \cdots$ with $\lim_{n\to\infty} \rho_{1n} = 0$. Set, according to Step 2,

$$\lim_{n\to\infty} u_{\rho_{1n}} = u \quad \text{harmonic on } R - \bar{D}_{r_1}.$$

By inductive assumption, there is a subsequence $\cdots < \rho_{k-1,n} < \cdots < \rho_{k-1,2} < \rho_{k-1,1} < r_{k-1}$ so that $\lim_{n\to\infty} u_{\rho_{k-1,n}} = u$ holds on $R - \bar{D}_{r_{k-1}}$ for $k \geq 2$. Note that this new harmonic function u (in $R - \bar{D}_{r_{k-1}}$) coincides with the previous u on $R - \bar{D}_{r_{k-2}}$. Then there is subsequence $\cdots < \rho_{kn} < \cdots < r_k$

of the preceding subsequence $\rho_{k-1,n}$, $n \geq 1$, so that $\lim_{n\to\infty} u_{\rho_{kn}} = u$ holds on $R - \bar{D}_{r_k}$.

Choose the diagonal element $u_{\rho_{kk}}$ for each $k \geq 1$, from the sequence $u_{\rho_{kn}}$, $n \geq 1$. Then

$$\lim_{k\to\infty} u_{\rho_{kk}} = u, \quad \text{uniformly on } R - D_r \text{ for all } r > 0. \qquad (*13)$$

Hence u is harmonic on $R - \{p_0\}$. From the set of inequalities in $(*12)$, after passing to a subsequence if necessary, we may assume that $\lim_{\rho\to 0} \alpha_n(\rho) = \alpha_n$ and $\lim_{\rho\to 0}\beta_n(\rho) = \beta_n$ exist for each $n \geq 1$, and $\lim_{\rho\to 0}\alpha_{-n}(\rho) = \lim_{\rho\to 0}\beta_{-n}(\rho) = 0$ for each $n \geq 1$. Putting these results together with $(*13)$ into $(*7)$, we obtain

$$\left(u - \operatorname{Re}\frac{1}{z}\right)(te^{i\theta}) = \sum_{n=1}^{\infty} (\alpha_n \cos n\theta + \beta_n \sin n\theta)t^n \quad \text{for } 0 < t \leq 1$$

$$\Rightarrow \lim_{p\to p_0} \left[u(p) - \operatorname{Re}\frac{1}{z(p)}\right] = 0.$$

According to $(6.3.1.7)$, p_0 is a removable singularity of $u - \operatorname{Re}\frac{1}{z}$.

The boundedness of u outside of every D_ρ, $0 < \rho < 1$, follows easily from $(*5)$ and $(*13)$.

Note that the uniqueness of u shows that the process of taking subsequences in the above discussion was completely unnecessary. $\qquad \square$

To the end of this section, we have an important fact.

The existence of meromorphic function on a Riemann surface. Every Riemann surface carries nonconstant meromorphic functions. $\qquad (7.3.3.6)$

This result can be used to introduce a C^∞-Riemannian metric, consistent with the conformal structure, on any Riemann surface; to metrize and triangulate any Riemann surface, and to guarantee that every Riemann surface has a countable basis for its surface topology. For all of these, read Ref. [32], pp. 179, 191, and 204. Note that Mittag–Leffler's theorem (see $(5.4.1.1)$ and Exercises B(2) and B(4) there) is a rather special case of $(7.3.3.6)$.

Before we start to prove $(7.3.3.6)$, let us make the following *observation.* Let $g(p, p_0)$ be Green's function with pole at p_0 on a (hyperbolic) Riemann surface. If $z = z(p)$ is a local variable at p_0 with $z(p_0) = 0$, then $g(p, p_0) + \log|z(p)| = \varphi(p)$ is harmonic in a coordinate neighborhood O of p_0. Identify p with $z = z(p)$ and we have $g(z, 0) = \varphi(z) - \log|z|$ in some open disk

$|z| < r$, where $z = x + iy$. Then

$$\frac{\partial g}{\partial x} - i\frac{\partial g}{\partial y} = \frac{\partial \varphi}{\partial x} - i\frac{\partial \varphi}{\partial y} - \frac{1}{z} \quad \text{in } 0 < |z| < r \qquad (*14)$$

is meromorphic with a pole of order ≥ 1 at $z = 0$ (see (1) in (3.4.3.4)). In case $z = z(p)$ is a local variable at a point $q \neq p_0$, then $\frac{\partial g}{\partial x} - i\frac{\partial g}{\partial y}$ is always analytic in a neighborhood of q. As a nonconstant function, the set of critical points (those $p \in R$ with $dg(p, p_0) = 0$) is a discrete set.

Proof. Let $p_1, p_2 \in R$ with $p_1 \neq p_2$. Let $z = x + iy$ be an arbitrary local parameter on R.

In case R is hyperbolic: Let $g_k(p, p_k)$ be Green's function with pole at p_k for $k = 1, 2$. Then

$$f(p) = \frac{\frac{\partial g_1}{\partial x}(p, p_1) - i\frac{\partial g_1}{\partial y}(p, p_1)}{\frac{\partial g_2}{\partial x}(p, p_2) - i\frac{\partial g_2}{\partial y}(p, p_2)} \qquad (*15)$$

is a required meromorphic function on R, with a pole of order ≥ 1 at p_1 and a zero of order ≥ 1 at p_2.

In case R is nonhyperbolic: Let $h_k(p, p_k)$ be the harmonic function in $R - \{p_k\}$ whose existence is confirmed in (7.3.3.3), for $k = 1, 2$. In this case, if $z = z(p)$ is a local parameter at p_k, then $h_k(p, p_k) = \mathrm{Re}\frac{1}{z(p)} + u_k(p)$ where $u_k(p)$ is harmonic in a coordinate neighborhood of p_k; and

$$\frac{\partial h_k}{\partial x} - i\frac{\partial h_k}{\partial y} = -\frac{1}{z^2} + \frac{\partial u_k}{\partial x} - i\frac{\partial u_k}{\partial y} \quad \text{in } 0 < |z| < r \quad \text{(for some } r > 0\text{)},$$

$$(*16)$$

which is meromorphic with a pole of order ≥ 2 at $z = 0$ (namely, p_k). Define $f(p)$ as in $(*15)$ with $g_k(p, p_k)$ replaced by $h_k(p, p_k)$, for $k = 1, 2$. This $f(p)$ is a required one, with a pole of order ≥ 2 at p_1 and a zero of order ≥ 2 at p_2.

7.4 The Fundamental Group

The essence of this section is to use an algebraic system (here, a group) to characterize a surface.

Let M be s surface (namely, a two-dimensional topological manifold, see Remarks 1 and 3 in Sec. 7.1). The concept of homotopy between planar curves introduced in Sec. 4.2.2 will be extended to surface curves. Only sketch, barely needed in Sec. 7.5, is presented.

Two curves $\gamma_1 : [0, 1] \rightarrow M$ and $\gamma_2 : [0, 1] \rightarrow M$ with common end points, $\gamma_1(0) = \gamma_2(0)$ and $\gamma_1(1) = \gamma_2(1)$, are said to be *homotopic*, denoted as

$$\gamma_1 \simeq \gamma_2, \tag{7.4.1}$$

if there is a continuous mapping $H : [0,1] \times [0,1] \to M$ satisfying $H(t,0) = \gamma_1(t)$ and $H(t,1) = \gamma_2(t)$ for $0 \le t \le 1$, $H(0,s) = \gamma_1(0)$ and $H(1,s) = \gamma_2(1)$ for $0 \le s \le 1$ (refer to Fig. 4.8). Such an H is called a *homotopy* between γ_1 and γ_2 or simply a *deformation* of γ_1 into γ_2 in M.

Remark (About the parametric interval). The restriction of the parametric interval to $[0,1]$ is just for convenience. Via the linear change $\tau = a + (b-a)t$, $[0,1]$ is mapped onto $[a,b]$ and vice versa. If $\gamma_1 \simeq \gamma_2$, so is $\gamma_1 \simeq \tilde{\gamma}_2$ where $\tilde{\gamma}_2(t) = \gamma_2\left(\frac{t-a}{b-a}\right)$ with $a \le t \le b$.

On the other hand, let $\tau : [0,1] \to [0,1]$ be a change of parameter, namely an increasing homeomorphism. Then $H(t,s) = \gamma((1-s)t + s\tau(t))$ provides a deformation of γ into $\gamma \circ \tau$. $\qquad\qquad\square$

Homotopy is an equivalence relation between curves with common end points. The homotopy $H(t,s) = \gamma(t)$ deforms γ into itself. If $H(t,s)$ deforms γ_1 into γ_2, then $H(t,1-s)$ deforms γ_2 into γ_1. If $H_1(t,s)$ deforms γ_1 into γ_2 and $H_2(t,s)$ deforms γ_2 into γ_3, then

$$H(t,s) = \begin{cases} H_1(t,2s), & 0 \le s \le \dfrac{1}{2} \\[2mm] H_2(t,2s-1), & \dfrac{1}{2} \le s \le 1 \end{cases}$$

deforms γ_1 into γ_3. The equivalence classes are called *homotopy classes*. The homotopy class of γ is defined as

$$[\gamma]. \tag{7.4.2}$$

The product $[\gamma_1][\gamma_2] = [\gamma_1\gamma_2]$ where $\gamma_2(0) = \gamma_1(1)$ is required. The product $\gamma_1\gamma_2$ of the curves γ_1 followed γ_2 is defined by

$$(\gamma_1\gamma_2)(t) = \begin{cases} \gamma_1(2t), & 0 \le t \le \dfrac{1}{2} \\[2mm] \gamma_2(2t-1), & \dfrac{1}{2} \le t \le 1 \end{cases}.$$

In case $\gamma_1 \simeq \gamma_1'$ by $H_1(t,s)$ and $\gamma_2 \simeq \gamma_2'$ by $H_2(t,s)$, then

$$H(t,s) = \begin{cases} H_1(2t,s), & 0 \le t \le \dfrac{1}{2} \\[2mm] H_2(2t-1,s), & \dfrac{1}{2} \le t \le 1 \end{cases}$$

deforms $\gamma_1\gamma_2$ into $\gamma_1'\gamma_2'$. Thus the definition of the *product* $[\gamma_1][\gamma_2]$ as $[\gamma_1\gamma_2]$ is well-defined.

Note that the product $\gamma_1\gamma_2$ is not always defined except γ_2 beginning where γ_1 ending. However, if $(\gamma_1\gamma_2)\gamma_3$ is defined, then $(\gamma_1\gamma_2)\gamma_3 \simeq \gamma_1(\gamma_2\gamma_3)$ holds. In order to make multiplication possible for all pairs of curves with common end points, we will *consider only curves that begin and end at a point $p_0 \in R$ in the sequel.*

The homotopy class of the constant map $\gamma(t) = p_0$, $0 \leq t \leq 1$, is specifically denoted as

$$1 \quad (\text{or } 1_{p_0} \text{ for distinction}). \tag{7.4.3}$$

It acts as a multiplication unit, namely, $1[\gamma] = [\gamma]1 = [\gamma]$ for all such curves γ. The inverse curve γ^{-1} of γ is given by $\gamma^{-1}(t) = \gamma(1-t)$, $0 \leq t \leq 1$. Consequently, every homotopy class $[\gamma]$ has an inverse $[\gamma]^{-1} = [\gamma^{-1}]$, for $[\gamma][\gamma^{-1}] = 1_{p_0}$.

In conclusion, we have

The fundamental group $\pi_1(M, p_0)$ or $\pi_1(M)$. Let M be a surface.

(1) Fix a point $p_0 \in M$. The set of the homotopy classes

$$\pi_1(M, p_0) = \{[\gamma] | \gamma \text{ is a closed curve from } p_0 \text{ (in M)}\}$$

is a group under the product operation of curves. It is called *the fundamental group* of M with the *base point p_0.*

(2) Let $p_1 \in M$ be another point and let σ be a curve from p_0 to p_1 (this is permissible since M is arcwise connected). Then $[\gamma_1] \in \pi_1(M, p_1) \to [\sigma\gamma_1\sigma^{-1}] = [\sigma][\gamma_1][\sigma]^{-1} \in \pi_1(M, p_0)$ sets up a group isomorphism. Since $\pi_1(M, p_1)$ is isomorphic to $\pi_1(M, p_0)$ for each $p_1 \in M$, as an *abstract* group, the fundamental group $\pi_1(M, p_0)$ is denoted by

$$\pi_1(\text{M}),$$

called the *fundamental group* of M. (7.4.4)

The concept of the fundamental group is suitable for any topological space, owing to the fact that the definitions concerned are simple and natural. Yet it is not abelian in general, it is difficult to compute in most cases (see Exercises B) and hence is different from the simplicial homology group. When we are only interested in the algebraic structure, $\pi_1(M)$ is usually adopted as the fundamental group; but when the mapping problem is concerned, the base point p_0 in $\pi_1(M, p_0)$ cannot be ignored.

(2) in (7.4.4) suggests the following

Topological invariance of the fundamental group. If a surface M is homeomorphic to a surface M', then the fundamental group $\pi_1(M)$ is isomorphic to $\pi_1(M')$. (7.4.5)

Proof. Let $f : M \to M'$ be the homeomorphism of M onto M'. A closed curve γ from p_0 in M is mapped into a closed curve $f \circ \gamma$ from $f(p_0)$ in M'. This sets up a mapping $\Phi : \pi_1(M, p_0) \to \pi_1(M', f(p_0))$ defined by

$$\Phi([\gamma]) = [f \circ \gamma].$$

In case $H(t, s)$ deforms γ_1 into γ_2, the $f \circ H(t, s)$ deforms $f(\gamma_1)$ into $f(\gamma_2)$. Hence Φ is well-defined. It is easy to check that $\Phi([\gamma_1][\gamma_2]) = \Phi([\gamma_1 \gamma_2]) = [f \circ \gamma_1 \gamma_2] = [f \circ \gamma_1][f \circ \gamma_2] = \Phi([\gamma_1])\Phi([\gamma_2])$; in particular, $\Phi([\gamma_1][\gamma_2]^{-1}) = \Phi([\gamma_1])\Phi([\gamma_2])^{-1}$ and $\Phi[1_{p_0}] = 1_{f(p_0)}$. Thus Φ is a homomorphism of $\pi_1(M, p_0)$ into $\pi_1(M', f(p_0))$. Under f^{-1}, a closed curve γ' from $f(p_0)$ in M' is mapped into a closed curve $f^{-1} \circ \gamma'$ from p_0 in M. And this indicates that Φ should be a one-to-one and onto mapping, and hence is an isomorphism. $\qquad\square$

A surface M is called *simply connected* if its fundamental group $\pi_1(M)$ reduces to the unit element. Recall (2.4.14) for the definition of a plane domain to be simply connected. Thus every closed curve in a simply connected surface can be deformed continuously to a point (refer to Example 3 in Sec. 4.2.2). (7.4.5) tells us:

Simple connectivity is a topological invariant. $\hfill$ (7.4.6)

Exercises A

(1) Prove (2) in (7.4.4) in detail.

Exercises B
Review Sec. 4.2.2 and extend various types of homotopy to an abstract topological space. Refer to Refs. [57, 74] if necessary. The fundamental group $\pi_1(X, p_0)$ or $\pi_1(X)$ can also be defined on a topological space X, and it is topologically invariant, too. Usually there are four ways to compute a fundamental group:

1. *Homotopic invariant of the fundamental group.* Let $f : X \to Y$ be a *homotopic equivalence* (namely, $f : X \to Y$ and $g : Y \to X$ are continuous so that $f \circ g \simeq id_Y$ and $g \circ f \simeq id_X$ hold). If $p_0 \in X$, then f induces an isomorphism $f_* : [\gamma] \to f_*([\gamma]) = [f \circ \gamma]$ between $\pi_1(X, p_0)$ and $\pi_1(Y, f(p_0))$.

2. *Direct expression of the fundamental group of a product space.* Let $p_0 \in X$ and $q_0 \in Y$. Then $\pi_1(X \times Y, (p_0, q_0))$ is isomorphic to $\pi_1(X, p_0) \oplus \pi_1(Y, q_0)$.

3. *Van Kampen theorem.* A special case reads as follows: Let X_1 and X_2 are nonempty open subsets of X so that $X = X_1 \cup X_2$ and $X_1 \cap X_2$ is path-connected. Fix $x_0 \in X_1 \cup X_2$. Suppose the *inclusion maps* $i_{X_1} : X_1 \to X$ (defined by $i_{X_1}(p) = p$ for $p \in X_1$) and $i_{X_2} : X_2 \to X$ induce the trivial homomorphisms $(i_{X_1})_* : \pi_1(X_1, p_0) \to \pi_1(X, p_0)$ (namely, $(i_{X_1})_*([\gamma]) = 1_{p_0}$, the multiplication unit, for any $[\gamma]$) and $(i_{X_2})_* . \pi_1(X_2, p_0) \to \pi_1(X, p_0)$. Then

$$\pi_1(X, p_0) = \{1_{p_0}\}.$$

4. *Via the concept of covering spaces* (to be explained in Sec. 7.5 for covering surfaces).

Try to apply the above methods to prove the following statements.

(1) Two homeomorphic topological spaces are homotopic equivalent but not vice versa. For instance, let $X = S^1$ (the unit circle) and $Y = S^1 \cup [1, 2]$. Then X and Y are homotopic equivalent but not homeomorphic.

(2) The fundamental group of a *contractible space* (namely, a space X on which the identity map $i_X : X \to X$ is homotopic to a constant map) reduces to the unit element.
Note. Hence, a contractible space is simply connected but not vice versa. For instance,

$$\pi_1(S^n) = \{1\} \quad \text{for } n \geq 2, \text{ where } S^n = \{\vec{x} \in \mathbf{R}^n \| \vec{x} = 1\}$$

is the unit sphere.

(3) $\pi_1(S^1) \cong$ (isomorphism) $\mathbf{Z}$, the set of integers.
Note. This fact can be used to prove the following remarkable results:

1. *The fundamental theorem of algebra.*
2. S^1 is not a *contractible kernel* of the closed disk $D^2 : |z| \leq 1$ (namely, it is not possible to find a continuous map $r : D^2 \to S^1$ so that $r \circ i = id_{S^1} : S^1 \to S^1$ is the identity map, where $i : S^1 \to D^2$ is the inclusion map).
3. *Brouwer's fixed point theorem* in $\mathbf{R}^2$ (namely, any continuous map $f : D^2 \to D^2$ has a fixed point).

(4) $\pi_1(\mathbf{C} - \{0\}) = \pi_1(\{1 \leq |z| \leq 2\}) \cong \mathbf{Z}$.
(5) $\pi_1(S^1 \times S^1) \cong \mathbf{Z} \oplus \mathbf{Z}$, where $S^1 \times S^1$ is the torus.
(6) $\pi_1(S^1 \times \mathbf{R}) \cong \mathbf{Z}$, where $S^1 \times \mathbf{R} = \{(x, y) | x \in S^1, y \in \mathbf{R}\}$ is the cylinder.
(7) $\pi_1(S^n) = \{1\}$ for $n \geq 2$.

(8) $\pi_1(P^2) \cong \mathbf{Z}_2$; the group of two elements, where P^2 is the projective plane.

(9) $\pi_1(X)$ is not abelian where X is the figure-eight space.

(10) $\pi_1(D^2 \times S^1) \cong \mathbf{Z}$, where $D^2 \times S^1$ is the solid cylinder.

(11) $\pi_1(M) \cong \mathbf{Z}$, where M is the Mobius band.

7.5   Covering Spaces (or Surfaces) and Covering Transformations

We experienced many concrete examples such as the Riemann surfaces of multiple-valued analytic functions as the covering surfaces of the extended complex plane $\mathbf{C}^*$. A definition of the covering surface for our previous descriptive purpose is the one presented in (2.7.2.9). A slightly general definition in order to provide a wider class of examples will be adopted in the following discussion.

The study of the theory of covering spaces has two-fold purposes: One is to use the properties of covering spaces to further study the fundamental groups; the other is to use the fundamental groups to classify the covering spaces and to study their invariants. Space does not permit us to go into details in both directions, and we will sketch mainly what is needed in the proof of the uniformization theorem in Sec. 7.6. Interested readers should consult books on algebraic topology, such as Refs. [57, 74], etc.

Five subsections are divided.

Section 7.5.1 gives the definitions needed, along with some examples.

Section 7.5.2 studies the basic properties of covering spaces (or surfaces), including the monodromy theorem in its homotopic form.

Section 7.5.3 devotes to the relations between the fundamental group and the covering spaces of a space. These are concerned with the classification of covering spaces and their existence.

Section 7.5.4 studies the group of covering transformations of a covering *surface* and its relation to the subgroups of the fundamental group that determines the surface.

The study of the universal covering surface is in Sec. 7.5.5. This concept is most wanted in our proof of the uniformization theorem in Sec. 7.6.

7.5.1   *Definitions and examples*

Let X and Y be topological spaces and $F : Y \to X$ be a continuous map. F is said to be a *local homeomorphism* if every point in Y has a neighborhood V such that the restriction of F to V is a homeomorphism. In this case, we call

the pair (Y, F) a *covering space* of X;
F the *projection* (map) of Y into X;
X the *base space*.   (7.5.1.1)

The point $F(y)$ is also called the *projection* of the point $y \in Y$ and y is said to *lie over* $F(y)$. For $x \in X$, the set $F^{-1}(x)$ is called the *fiber* of F *over* x.

A covering space (Y, F) of X is called *complete* if every point in X has an open neighborhood U so that $F^{-1}(U)$ can be expressed as a union $\bigcup_\lambda V_\lambda$ of disjoint open sets V_λ and all the mappings $F|_{V_\lambda} : V_\lambda \to \mathbf{C}$ are homeomorphic, namely, such a U is *evenly covered* by (Y, F). Such a U is called a *fundamental neighborhood* of each point in it.

Remark 1 (The base space X can be realized as the quotient topology of its complete covering space (Y, F) under the projection F). Observe that the fiber $F^{-1}(x)$ is discrete in Y for each $x \in X$. F is a continuous, open (local homeomorphic) map from Y onto X.

Two points y_1 and y_2 are defined to be equivalent if $F(y_1) = F(y_2)$ holds. The equivalence class of y is denoted by $[y]$. Let $Y/\sim = \{[y] | y \in Y\}$ and $\pi : Y \to Y/\sim$ be the natural map given by $\pi(y) = [y]$. A set O in $Y/\sim$ is open if $\pi^{-1}(O)$ is open in Y and this definition makes $Y/\sim$ a topological space, called the *quotient topology (space)* of Y *under* F. $Y/\sim$ enjoys the following properties:

1. $\pi : Y \to Y/\sim$ is continuous.
2. The quotient topology $Y/\sim$ is the finest topology making π to be continuous.
3. The map $f : Y/\sim \to Z$ (a topological space) is continuous if and only if $f \circ \pi : Y \to Z$ is continuous.
4. The map $F^* : Y/\sim \to X$ defined by $F^*([y]) = F(y)$ is a one-to-one and onto continuous map (since F is onto and continuous), and hence is a homeomorphism (since F is open).

In short, X can be realized as $Y/\sim$ topologically. Details are left as Exercise A(1).   □

Remark 2 (The Cartesian product of two projections is still a projection). Let (Y_k, F_k) be a covering space of X_k for $k = 1, 2$. Define $F_1 \times F_2 : Y_1 \times Y_2 \to X_1 \times X_2$ by

$$(F_1 \times F_2)(y_1 \times y_2) = (F_1(y), F_2(y)).$$

Then $(Y_1 \times Y_2, F_1 \times F_2)$ is a covering space of $X_1 \times X_2$. $F_1 \times F_2$ is called the *Cartesian product* of F_1 and F_2. Proof is left as Exercise A(2). □

What is really interested to us is the following more restricted case: Both Y and X are two-dimensional topological manifolds, simply called *surfaces*; when this is the case, Y and X are usually denoted as M^* and M, respectively, and (M^*, F) is called a *smooth covering surface* of M. Moreover, $F : M^* \to M$ is not necessarily always assumed to be locally one-to-one; in this case, (M^*, F) may be considered as a *branched* or *ramified covering surface* especially when both M^* and M are Riemann surfaces. This is precisely the definition (2.7.2.9) we gave for a covering surface.

In what follows, we raise some basic examples to consolidate these concepts introduced.

Example 1. Covering spaces of S^1, the unit circle.

Explanation. Define the exponential map $F : \mathbf{R} \to S^1$ by $F(t) = e^{2\pi it}$. Then $(\mathbf{R}, F)$ is a complete covering space of S^1. As a matter of fact, for any fixed point $z \in S^1$, any proper open circular arc U is a fundamental neighborhood of z. Take, for instance $z = 1$ and U the open right half circle, then $F^{-1}(U) = \bigcup_{n \in Z} (n - \frac{1}{4}, n + \frac{1}{4})$ and the restriction of F to each $(n - \frac{1}{4}, n + \frac{1}{4})$ is a homeomorphism. See Fig. 7.20.

Fix any positive integer n and define $F_n : S^1 \to S^1$ by $F_n(z) = z^n$. Then (S^1, F_n) is a complete covering space of S^1. To see this, fix a point $z \in S^1$ and choose an open arc U, containing z and having its central angle θ less than 2π. Then $F_n^{-1}(U)$ composes of n disjoint open circular arcs, each

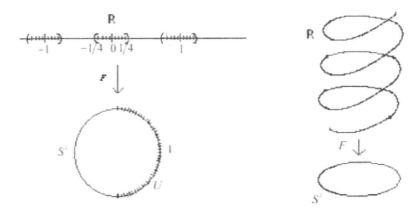

Fig. 7.20

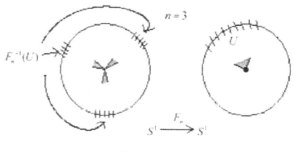

Fig. 7.21

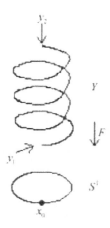

Fig. 7.22

having its central angle $\frac{\theta}{n}$ and containing exactly only one n-th root of z. F_n maps each such arc homeomorphically onto U. See Fig. 7.21.

On the other hand, let Y be the finite open circular helix (regardless of its endpoints) over S^1 as shown in Fig. 7.22 and let $F : Y \to S^1$ be the usual orthogonal projection. Then (Y, F) is a covering space of S^1 but it is not complete, because the point x_0 does not have a fundamental neighborhood. In case the end points y_1 and y_2 (lying over x_0) of the helix is identified as a single point, then the resulted (Y, F) turns out to be a complete covering space of S^1. Say the reason why!

Example 2. Covering surfaces of an annulus $r_1 < |w| < r_2$ $(0 < r_1 < r_2)$.

Explanation. Let M^* be the strip $\log r_1 < \operatorname{Re} z < \log r_2$. Denote the annulus $r_1 < |w| < r_2$ by M. Define $F : M^* \to M$ by $F(z) = e^z$. F is a local homeomorphism. For each point w_0 in M, fix an angle θ_0

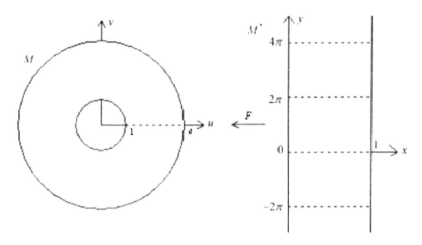

Fig. 7.23

$(0 \le \theta_0 < 2\pi)$ so that $\operatorname{Arg} w_0 \ne \theta_0$ and set $U = \{w \in M | \operatorname{Arg} w \ne \theta_0\}$. Then w_0 is in the open set U and $F^{-1}(U) = \bigcup_{n=-\infty}^{\infty} V_n$, where $V_n = \{z \in M^* | \theta_0 + 2n\pi < \operatorname{Im} z < \theta_0 + 2(n+1)\pi\}$ for $n = 0, \pm 1, \pm 2, \ldots$. Then $F|_{V_n} : V_n \to U$ is homeomorphic for each n. This shows that (M^*, F) is a complete smooth covering surface of M. See Fig. 7.23. Topologically M^* might wind around to form an ascending right circular helicoid over the annulus M and the projection F would be the ordinary orthogonal projection. See Fig. 7.24. Choose $z^* = e^{x+iy}$ as the local parameter at each point $(x, y) \in M^*$ and $w = re^{i\theta}$ as the local parameter at each point $(r, \theta) \in M$. Then, in terms of these parameters, F is nothing but $w = z^*$, and hence (M^*, F) is indeed a smooth covering surface of M according to our classical definition (2.7.2.9).

If we consider only n rectangles $\log r_1 < x < \log r_2$, $0 < y \le 2\pi$ and $2\pi k \le y < 2(k+1)\pi$ for $k = 1, \ldots, n-2, n-1$, to form the surface M_n^* for $n \ge 1$. Then (M_n^*, F) is a smooth covering surface of M but it is not complete. This is because points along the ray $r_1 < u < r_2$, $v = 0$ ($w = u + iv$) do not have fundamental neighborhoods. If we instead consider the n rectangles $\log r_1 < x < \log r_2$, $2\pi k \le y < 2(k+1)\pi$ for $k = 0, 1, \ldots, n-2, n-1$, with the terminal end $y = 2n\pi$ identified with the initial end $y = 0$. Then the resulted $(\widetilde{M_n^*}, F)$ is still a complete smooth covering surface of M. Note that $\tilde{M}_n^*$ is n-sheeted, namely, it covers each point of M equally and evenly n times, while M^* constructed above covers M countably infinitely many times.

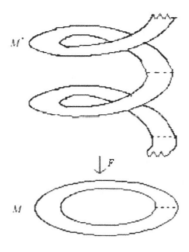

M^*

F

M

Fig. 7.24

Example 3. Covering surfaces of the punctured plane $\mathbf{C} - \{0\}$ and the punctured unit disk $0 < |z| < 1$.

Explanation. In Example 2, let $r_2 = 1$ and $r_1 \to 0$. Then M^* is the left open half-plane $H : \operatorname{Re} z < 0$. Then $(H, \exp)$ is a complete smooth covering surface of the punctured disk $0 < |z| < 1$. Readers are asked to give a direct proof. On the other hand, letting $r_1 \to 0$ and $r_2 \to \infty$, we obtain that $(\mathbf{C}, \exp)$ is a complete smooth covering surface of $\mathbf{C} - \{0\}$.

From (2.6.1.3), we know that $\exp : \mathbf{C} \to \mathbf{C} - \{0\}$ is onto and locally homeomorphic. $(\mathbf{C}, \exp)$ is indeed a complete smooth covering surface of the punctured plane $\mathbf{C} - \{0\}$. To see this, for any point $z \in \mathbf{C} - \{0\}$, the open disk $U_z = \{\zeta \in \mathbf{C} | |\zeta - z| < |z|\}$ acts as a fundamental neighborhood of z. This is because, for each component V of $\exp^{-1}(U_z)$, choose a single-valued branch $f(z)$ of $\log z$ in U_z (see (2.7.1.1) and (2.7.1.2)) and then $\exp \circ f = 1_{U_z}$ and $f \circ \exp = 1_V$ hold.

An alternate way to realize this is as follows: Set $z = x + iy = (x, y)$. Then $\exp z = e^x(\cos y + i \sin y) = re^{iy}$. Therefore, topologically $\mathbf{C} \cong \mathbf{R} \times \mathbf{R}$ and $\mathbf{C} - \{0\} \cong R_+ \times S^1$, where R_+ denotes the set of positive real numbers; and, at the same time, the exponential map $\exp$ can be considered as the Cartesian product $p \times q$, where $p(x) = e^x$ and $q(y) = (\cos y, \sin y)$. According to Remark 2, $(\mathbf{R} \times \mathbf{R}, p \times q)$ is a complete covering surface of $\mathbf{C} - \{0\}$.

Example 4. A covering surface of the torus $S^1 \times S^1$.

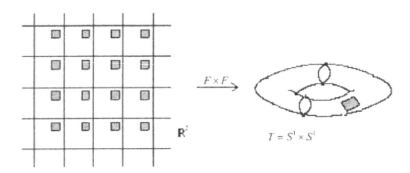

Fig. 7.25

Explanation. Let $F(t) = e^{2\pi it}, t \in \mathbf{R}$, be as in Example 1. Consider the Cartesian product $F \times F : \mathbf{R}^2 \to S^1 \times S^1$ defined as $(F \times F)(t_1, t_2) = (e^{2\pi it_1}, e^{2\pi it_2})$. Then $(\mathbf{R}^2, F \times F)$ is a complete smooth covering surface of the torus $S^1 \times S^1$. See Fig. 7.25.

As a matter of fact, the natural (or canonical) projection $\pi : \mathbf{C} \to \mathbf{C}/\Gamma$ as defined in Example 3 of Sec. 7.1 is known to be a local homeomorphism. The argument there for the openness of π shows that $(\mathbf{C}, \pi)$ is a complete smooth covering surface of the torus $\mathbf{C}/\Gamma$.

Example 5. A covering surface of the projective plane P^2.

Explanation. The projective plane P^2 is defined as the quotient space of S^2, the two-dimensional sphere in $\mathbf{R}^3$, by identifying a pair of two antipodal points as a single point. If $x \in S^2$, the point $[x] = \{x, -x\} \in P^2$. P^2 is a surface (why?). Define $F : S^2 \to P^2$ by $F(x) = [x]$. Then (S^2, F) is a complete smooth covering surface of P^2. This can be done as follows. Fix a point $[x] \in P^2$. Choose an open neighborhood V of x such that V is connected and does not contain any pair of antipodal points. The continuity and openness of F indicate that $F(V)$ is an open connected neighborhood of $[x]$. Then $F^{-1}(F(V)) = V \cup V_-$, where $V_- = \{y \mid -y \in V\}$. Note that the restriction of F on either V or V_- is one-to-one, and hence is homeomorphic to $F(V)$. Hence $F(V)$ is a fundamental neighborhood of $[x]$.

Example 6. Branched (or ramified) covering surfaces.

Explanation. Suppose $w = f(z)$ is a nonconstant analytic function. Suppose $f'(z_0) = \cdots = f^{(k-1)}(z_0) = 0$ for $k \geq 2$ and $f^{(k)}(z_0) \neq 0$. Then, according to (3.5.1.8), the point $w_0 = f(z_0)$ is an algebraic branch point of order $k - 1$ of the multiple-valued function $z = f^{-1}(w)$. There would be a branched point of order $k - 1$ over w_0 on the Riemann surface of $f^{-1}(w)$. Such a Riemann surface is a branched covering surface of the z-plane. Most Riemann surfaces of multiple-valued analytic functions we raised from Sec. 2.7 all the way down to Sec. 7.2 are of this kind. While the Riemann surface of $\log z$ (see Fig. 2.73) is an exception; it is a complete smooth covering surface of $\mathbf{C}-\{0\}$ (refer to Example 2 in Sec. 3.3.3).

Let $f : R_1 \to R_2$ be a nonconstant analytic mapping from a Riemann surface R_1 to another Riemann surface R_2. Then (R_1, f) is a covering surface of R_2, probably branched. See (7.2.1)–(7.2.6) and (7.2.16) for detailed account.

Exercises A

(1) Prove statements in Remark 1.

(2) Prove the statement in Remark 2.

(3) Let D be the open unit disk $|z| < 1$ and $F_n : D \to \mathbf{C}$ be defined by $F_n(z) = z^n$, where $n \geq 2$ is an integer. Show that (D, F_n) is a branched covering surface of $\mathbf{C}$ but it is not complete. Also show that $(\mathbf{C} - \{0\}, F_n)$ is a complete smooth covering surface of $\mathbf{C} - \{0\}$.

(4) Let $M = \mathbf{C} - \{\frac{\pi}{2} + n\pi | n = 0, \pm 1, \ldots\}$ and $F(z) = \sin z$. Show that $(\mathbf{C} - \{\pm 1\}, F)$ is a complete smooth covering surface of M. Is $(\mathbf{C}, F)$ a branched covering surface of $\mathbf{C}$?

(5) Show that $(\mathbf{C}, \tan)$ is a smooth covering surface of $\mathbf{C}^*$ and is a complete covering surface of $\mathbf{C}^* - \{\pm i\}$. Set $\Omega = \mathbf{C} - \{iy | y \in \mathbf{R}$ and $|y| \geq 1\}$. Show that, for each integer n, there is a unique analytic function $(\tan^{-1} z)_n : \Omega \to \mathbf{C}$ satisfying $\tan \circ (\tan^{-1} z)_n = z, z \in \Omega$, and $(\tan^{-1} 0)_n = n\pi$ (refer to (2.7.4.7)).

(6) Suppose $F : R_1 \to R_2$ is a nonconstant analytic mapping from a Riemann surface R_1 to a Riemann surface R_2.

  (a) Show that F is *open*, namely, F maps open sets onto open sets.

  (b) Show that F is *discrete*, namely, the fiber $F^{-1}(q)$ of every point $q \in R_2$ is a discrete set of R_1 (i.e., every point x in $F^{-1}(q)$ has a neighborhood V such that $V \cap F^{-1}(q) = \{x\}$).

  Note. Any analytic (or meromorphic) function $f : R_1 \to \mathbf{C}$ (or $\mathbf{C}^*$) may be considered as a *multiple-valued* analytic (or meromorphic) function g on R_2. The reasoning is that, for $q \in R_2$ and all possible $p \in F^{-1}(q)$, which may be a singleton or empty, then $f(p)$ are the different values $g(q)$ of this multiple-valued function g at the point q so that $g \circ F = f$ holds. See the left diagram below.

$$R_1 \xrightarrow{\ F\ } R_2 \qquad\qquad \mathbf{C} \xrightarrow{\ \exp\ } \mathbf{C}\text{-}\{0\}$$

$$f \downarrow \ \ \swarrow_g \qquad\qquad\qquad \text{id} \searrow \ \ \swarrow_{\log}$$

$$\mathbf{C} \ (\text{or } \mathbf{C}^*) \qquad\qquad\qquad \mathbf{C}$$

(c) Take, for instance, $R_1 = \mathbf{C}$, $R_2 = \mathbf{C} - \{0\}$, and $F = \exp : \mathbf{C} \to \mathbf{C} - \{0\}$. Show that the identity map id: $\mathbf{C} \to \mathbf{C}$ gives rise to the multiple-valued logarithm $\log z$ on $\mathbf{C} - \{0\}$. See the right diagram above.

Note. F is said to have a *branch point* or *ramification point* at p if there is no neighborhood V of p so that the restriction $F|_V$ is univalent. Algebraic branch points defined in (7.2.3) and (7.2.4) are just special cases of this kind.

(d) Show that $F : R_1 \to R_2$ is an *unbranched* or *smooth* map, namely, a map without branch points, if and only if F is a local homeomorphism.

**7.5.2   *Basic properties of covering spaces (or surfaces):
The lifting of mappings and the monodromy theorem***

Section (1) The lifting of the complex structure

Our main result is the following

Induced complex structure on a covering surface of a Riemann surface. Let R^* be a surface and R be a Riemann surface so that (R^*, F) is a covering surface of R (see definition (2.7.2.9)). Then there exists a unique complex structure on R^* so that R^* is a Riemann surface and $F : R^* \to R$ is analytic. In case (R^*, F) is a smooth covering surface of R, F is then locally conformal.   (7.5.2.1)

This result is still valid if R^* is just a connected Hausdoff space and (R^*, F) is a covering space of R. See Exercise A(1).

Proof. Consider the collection of these charts $\{U^*, \varphi^*\}$ on R^* satisfying the properties:

1. φ^* is heomorphic on U^*, and
2. the composite map $\varphi \circ F \circ \varphi^{*-1}$ is analytic for each local parameter φ on R whenever it is defined.

This kind of charts does exist according to the definition of (R^*, F) being a covering surface of R. All such U^* form a covering of R^*.

Suppose $\{U_1^*, \varphi_1^*\}$ and $\{U_2^*, \varphi_2^*\}$ are two such charts so that $U_1^* \cap U_2^* \neq \phi$. We assume that the local parameter φ on R has its coordinate neighborhood containing $F(U_1^* \cap U_2^*)$. In case F is homeomorphic on $U_1^* \cap U_2^*$, then

$$\varphi_2^* \circ \varphi_1^{*^{-1}} = \left(\varphi \circ F \cup \varphi_2^{*^{-1}}\right)^{-1} \circ \varphi \circ F \cap F^{-1} \circ \varphi^{-1} \circ \left(\varphi \cup F \cup \varphi_1^{*^{-1}}\right)$$

$$= \left(\varphi \circ F \circ \varphi_2^{*^{-1}}\right)^{-1} \circ \left(\varphi \circ F \circ \varphi_1^{*^{-1}}\right) \tag{$*_1$}$$

is analytic on $\varphi_1^*(U_1^* \cap U_2^*)$. In case F is not homeomorphic on $U_1^* \cap U_2^*$, $(*_1)$ still holds on a neighborhood of any point in $\varphi_1^*(U_1^* \cap U_2^*)$ which is not the image of any branch point of F. Consequently $\varphi_2^* \circ \varphi_1^{*^{-1}}$ is analytic in $\varphi_1^*(U_1^* \cap U_2^*)$ except for isolated singularities. These singularities are removable because $\varphi_2^* \circ \varphi_1^{*^{-1}}$ is known to be continuous on $\varphi_1^*(U_1^* \cap U_2^*)$. This proves the analyticity of $\varphi_2^* \circ \varphi_1^{*^{-1}}$ on $\varphi_1^*(U_1^* \cap U_2^*)$. And the charts $\{U^*, \varphi^*\}$ constitute a complex structure Φ on R^*.

Suppose Ψ is another such complex structure on R^* so that $F : R^* \to R$ is analytic. The identity map $id : R^*$ (with structure Ψ) $\to R^*$ (with structure Ψ) is locally conformal except at isolated branched points of F. The branched points of F and their orders are identical no matter with respect to Φ or Ψ, owing to the fact that (R^*, F) is assumed to be a covering surface of R. Hence id is a global conformal map from R^* onto itself. $\square$

Section (2) The lift of mapping from the base space to the covering space

Let X, Y, and Z be topological spaces and $F : Y \to X$ and $f : Z \to X$ be continuous maps. If there is a continuous map $g : Z \to Y$ satisfying that $F \circ g = f$, then g is called a *lift* of f *with respect to* F and *lies over* f. See the diagram.

$$Y$$
$$\begin{array}{ccc} & g \nearrow & \downarrow F \\ Z & \xrightarrow{\quad f \quad} & X \end{array}$$

In case (Y, F) is a covering space of X, we are particularly interested in the lifting of a curve $\gamma : [0, 1] \to X$ to the *lift curve* $\gamma^* : [0, 1] \to Y$ such that $F \circ \gamma^*(t) = \gamma(t)$ for all $t \in [0, 1]$; and the lifting of a homotopy $H : [0, 1] \times [0, 1] \to X$ to the *lift homotopy* $H^* : [0, 1] \times [0, 1] \to Y$ such that $F \circ H^*(t, s) = H(t, s), t, s \in [0, 1]$.

The lifting of a curve to the covering space is not always possible. Let X be the anaulus $1 < |w| < e$ and Y the rectangle $0 < \operatorname{Re} z < 1, 0 < \operatorname{Im} z < 3\pi$. Then $(Y, \exp)$ is a smooth covering surface of X but it is not complete. Consider the curve $\gamma(t) = 2e^{3\pi it/2}$, $0 \le t \le 1$. Pick the point $(\log 2, 2\pi)$ which lies over 2. Suppose there is a curve $\gamma^*(t) = x(t) + iy(t)$ in Y with the initial point $(\log 2, 2\pi)$ so that $\exp \circ \gamma^*(t) = \gamma(t)$, then it is necessary that $x(t) = \log 2$ and $y(t) = 2\pi + \frac{3\pi}{2}t$. But the curve $\gamma^*(t) = \log 2 + i(2\pi + \frac{3\pi}{2}t)$ has $[0, \frac{2}{3})$ as its defining interval, not the whole $[0, 1]$. Therefore, there is no lift γ^* from $(\log 2, 2\pi)$ on Y so that $\exp \circ \gamma^*(t) = \gamma(t)$ for all $t \in [0, 1]$.

A covering space (Y, F) of X is said to have the *curve lifting property* if every curve $\gamma : [0, 1] \to X$ can be lifted to a curve γ^* from any initial point q_0 over p_0, the initial point of γ. In short, γ^* is said to *continue* from q_0 over the curve γ.

First we have

The uniqueness and existence of the lift curve. Let (M^*, F) be a smooth covering surface of the surface M.

(1) If two curves $\gamma_1^*, \gamma_2^* : [0, 1] \to M^*$ lie over the same curve $\gamma : [0, 1] \to M$, then they either coincide or are disjoint.

(2) A complete smooth covering surface has *the curve lifting property*; the continuation over any curve is always possible and unique (from the same initial point).   (7.5.2.2)

See Exercises A(3) and (4) for general settings. For a converse of (2), see Exercise B(2); for generalizations, see (7.5.2.7) and Exercise B(3).

Proof. (1) This can be easily seen by observing that, for any point p^* lying over a fixed point $\gamma(t)$ of the curve, there is an open neighborhood V of p^* which is mapped *homeomorphically* onto an open neighborhood U of $\gamma(t)$ under F. Hence the portion $U \cap \gamma$ of the curve γ can only be uniquely continued from p^* in V. This fact indicates that the set $I = \{t \in [0, 1] | \gamma_1(t) = \gamma_2(t)\}$ is relatively open as well as closed. Consequently either $I = [0, 1]$ or $I = \phi$ can happen.

(2) Let $I = \{t \in [a, b] | \gamma|_{[a,t]}$ can be uniquely lifted to $\gamma^*|_{[a,t]}$ with the initial point $p^*\}$, where p^* is any point lying over $\gamma(a)$. $I \ne \phi$ because $a \in I$. Suppose $t \in I$. Choose an evenly covered neighborhood U of $\gamma(t)$ and a $\delta > 0$ such that $\gamma[t, t + \delta] \subseteq U$. The point $\gamma^*(t)$ is in a component V of $F^{-1}(U)$. Since $F : V \to U$ is homeomorphic, $\gamma[t, t + \delta]$ can only be uniquely continued from $\gamma^*(t)$ to obtain the lift curve $\gamma^*[t, t + \delta]$. This

argument shows that I is relatively open and closed in $[0, 1]$. Hence $I = [0, 1]$ holds.   □

A lift curve γ^* of a closed curve γ is not necessarily closed. For instance, $(\mathbf{C} - \{0\}, F)$, where $F(z) = z^2$, is a complete smooth covering surface of $\mathbf{C} - \{0\}$. The closed curve $\gamma(t) = c^{2\pi i t}, 0 \leq t < 1$, is lifted to the $\gamma^*(t) = e^{\pi i t}, 0 \leq t \leq 1$, which is not closed.

However, we have

The homotopy lifting property of a complete smooth covering surface. Suppose (M^*, F) is a complete smooth covering surface of a surface M.

(1) Let $\gamma^* : [0, 1] \to M^*$ be a lift curve of the curve $\gamma : [0, 1] \to M$. Suppose $H : [0, 1] \times [0, 1] \to M$ is a homotopy of γ, namely, $H(t, 0) = \gamma(t)$ for $t \in [0, 1]$. Then there exists a unique lift homotopy $H^* : [0, 1] \times [0, 1] \to M^*$ such that $H^*(t, 0) = \gamma^*(t)$ for $t \in [0, 1]$.

(2) (*The monodromy theorem*) Let $\gamma_1, \gamma_2 : [0, 1] \to M$ be two curves with the same initial point $\gamma_1(0) = \gamma_2(0) = p_0$. Let p_0^* be any point in $F^{-1}(p_0)$ and $\gamma_1^*, \gamma_2^* : [0, 1] \to M^*$ be the lift curves from p_0^* of γ_1, γ_2, respectively. Then,

γ_1 and γ_2 are homotopic with fixed end points in M (see Sec. 4.2.2).
$\Leftrightarrow \gamma_1^*$ and γ_2^* are homotopic with fixed end points in M^*.

In particular, $\gamma_1^*(0) = \gamma_2^*(0) = p_0^*$ and $\gamma_1^*(1) = \gamma_2^*(1)$ hold.   (7.5.2.3)

For a generalization to topological space, see Exercise A(4).

Proof. For simplicity, set $I = [0, 1]$.

(1) In case $H(I \times I)$ is contained in an evenly covered neighborhood U, pick the component V of $F^{-1}(U)$ that contains p_0^*. Then $H^* = (F|_V)^{-1}(H)$ is the required lift homotopy.

In the general case, choose any fixed point $t \in I$. Then $\{H^{-1}(U) | U$ is a fundamental neighborhood in $M\}$ forms an open covering of the compact set $\{t\} \times I$. It is thus possible to choose a $\delta > 0$ to form the open interval $I_t = (t - \delta, t + \delta)$ and $n + 1$ points $0 = s_0 < s_1 < \cdots < s_{n-1} < s_n = 1$ (depending on t) so that $H(I_t \times [s_{k-1}, s_k])$ is contained in a fundamental neighborhood in M for each k, $1 \leq k \leq n$. Applying the method and the result in the last paragraph to each $I_t \times [s_{k-1}, s_k]$, $1 \leq k \leq n$, there is a homotopy $H_t^* : I_t \times I \to M^*$ satisfying $F \circ H_t^* = H|_{I_t \times I}$ and $H_t^*(t', 0) = \gamma^*(t')$ for $t' \in I_t$.

We claim that for any $t_1, t_2 \in I, H_{t_1}^* = H_{t_2}^*$ on $(I_{t_1} \cap I_{t_2}) \times I$. Suppose $t' \in I_{t_1} \cap I_{t_2}$. Define curves $\sigma : I \to M$ and $\sigma^*, \sigma^{**} : I \to M^*$ by

$$\sigma(s) = H(t', s) \quad \text{and} \quad \sigma^*(s) = H_{t_1}^*(t', s), \quad \sigma^{**}(s) = H_{t_2}^*(t', s), s \in I,$$

respectively. Both σ^* and σ^{**} are lift curves from $\gamma^*(t') = \sigma^*(0) = \sigma^{**}(0)$ of γ. By the uniqueness of the lift curve, it follows that $\sigma^* = \sigma^{**}$ and hence $H_{t_1}^*(t', s) = H_{t_2}^*(t', s)$, for all $t' \in I_{t_1} \cap I_{t_2}$ and all $s \in I$.

Finally, define $H^* : I \times I \to M^*$ by $H^*(t', s) = H_t^*(t', s)$ if $t' \in I_t$ for some $t \in I$ and for all $s \in I$. This H^* is well-defined and satisfies the property that $F \circ H^* = H$ and $H^*(t', 0) = \gamma^*(t')$ for all $t' \in I$. Uniqueness of such an H^* is a consequence of (1) in (7.5.2.2).

(2) Only the necessity is needed to be proved.

Let $H : I \times I \to M$ be the homotopy with fixed end points between γ_1 and γ_2, namely, $H(t, 0) = \gamma_1(t), H(t, 1) = \gamma_2(t)$ for all $t \in I$ and $H(0, s) = \gamma_1(0) = \gamma_2(0), H(1, s) = \gamma_1(1) = \gamma_2(1)$ for all $s \in I$. By (2) in (7.5.2.2) and part (1), there exists a (unique) lift homotopy $H^* : I \times I \to M^*$ of H such that $H^*(t, 0) = \gamma_1^*(t)$ for $t \in I$.

To show that H^* is a homotopy with fixed end points, set $\rho_k : I \to M^*$ by $\rho_k(s) = H^*(k, s)$ for $k = 0, 1$. Then $F \circ \rho_k(s) = F \circ H^*(k, s) = H(k, s) = \gamma_1(k) = \gamma_2(k)$ for $k = 0, 1$ and for all $s \in I$. This means that $F \circ \rho_k$ is a constant closed curve from $\gamma_1(k)$ for $k = 0, 1$. By the uniqueness of the lift curve, both ρ_k are constant closed curves from $\gamma_1^*(k)$ for $k = 0, 1$. Hence $H^*(k, s) = \gamma_1^*(k)$ for $k = 0, 1$ and all $s \in I$. And H^* is then a homotopy with fixed end points.

Finally, let $\sigma : I \to M^*$ be given by $\sigma(t) = H^*(t, 1)$. Then $F \circ \sigma(t) = F \circ H^*(t, 1) = H(t, 1) = \gamma_2(t)$. Since $\sigma(0) = H^*(0, 1) = \gamma_1^*(0) = \gamma_2^*(0)$, again by the uniqueness of the lift curve, $\sigma = \gamma_2^*$ holds. Therefore γ_2^* is homotopic to γ_1^* under H^*. $\qquad\square$

As an easy application of (7.5.2.3), we have

The induced homomorphism of the projection. Let (M^*, F) be a complete smooth covering surface of a surface M. Fix $p_0 \in M$ and $p_0^* \in F^{-1}(p_0)$. Then $F_*: \pi_1(M^*, p_0^*) \to \pi_1(M, p_0)$ defined by

$$\mathbf{F}_*([\gamma^*]) = [F \circ \gamma^*]$$

is a one-to-one homomorphism between the fundamental groups.

$$(7.5.2.4)$$

See Exercise B(3) for another result concerning fundamental groups.

Section (3) The number of sheets of the covering

The number of times that a nonconstant meromorphic function $f\colon \mathbf{C}^* \to \mathbf{C}^*$ assumes every point in $\mathbf{C}^*$ is a fixed constant (see (3.6.1.14)). This result as well as its proof can be extended to nonconstant analytic mapping between compact Riemann surfaces (see (7.2.16)). Both of them are special cases of the following fact.

The number of sheets of the covering surfaces. Suppose (M^*, F) is a covering surface of a surface M, having the curve lifting property. Then only one of the following two alternatives can hold.

(1) If there is a point $p_0 \in M$ so that the cardinal number of $F^{-1}(p_0)$ is a finite number n (counting multiplicity at a branched point), then every point of M is covered the same number of times n.

(2) If there is a point $p_0 \in M$ so that the cardinal number c of $F^{-1}(p_0)$ is not finite, then every point of M is covered the same number of times c.

The natural number n or the cardinal number c is called *the number of sheets* of the covering surface (M^*, F), and (M^*, F) is called an *n-sheeted* or *c-sheeted* (or *infinitely-sheeted*) covering surface of M.   (7.5.2.5)

See Exercises A(7) and (8) for some related results.

Proof. The former part of the proof for (3.6.1.14) is still valid for our present case, but the latter part cannot be adopted here because we do not know whether the surface M^* has a countable base for its topology. It is here the curve lifting property of a complete covering surface that plays its role. Recall that $b_F(p^*)$ denotes the branch number of F at $p^* \in M^*$ (see (7.2.3) and (7.2.4)). For each integer $n \geq 1$, set as in (3.6.14) or (7.2.16)

$$A_n = \{p \in M \mid \sum_{p* \in F^{-1}(p)} [b_F(p^*) + 1] \geq n\}.$$

Obviously A_n is open by (3.5.1.8) or (7.2.3) and (7.2.4).

To see that $\tilde{A}_n = M - A_n$ is open: Fix a point $p_0 \in \tilde{A}_n$. Then $F^{-1}(p_0) = \{p_1^*, \ldots, p_m^*\}$, where $1 \leq m \leq n-1$. Choose an open connected neighborhood U_j^* of p_j^*, pairwise disjoint for $1 \leq j \leq m$, so that $U_j = F(U_j^*)$ is an open neighborhood of p_0 and each point in U_j is covered exactly $b_F(p_j^*)+1$ times by points in U_j^* for each $1 \leq j \leq m$. Note that $\sum_{j=1}^m [b_F(p_j^*) + 1] \leq n - 1$. Choose a local coordinate disk D with range $|w| < 1$ and center at p_0 so that

$$D \subseteq U = \bigcap_{j=1}^m U_j$$

holds. We *claim* that $F^{-1}(q)$ contains at most m points for each point $q \in D$, and hence $D \subseteq \tilde{A}_n$ and $\tilde{A}_n$ is open. To see this, observe firstly that $F^{-1}(q)$ contains at least m points by counting the multiplicities, if necessary. There are no other points than these m points that lie over q. For if $q^* \in F^{-1}(q)$, then construct a curve $\gamma : [0,1] \to M$ with $\gamma(0) = q$ and $\gamma(1) = p_0$ and lift γ to a curve γ^* in M^* with $\gamma^*(0) = q^*$ and $\gamma^*(1)$ is one of the points p_j^*. This is possible since M^* has the curve lifting property and q^* lies in some U_j^*. Hence γ^* lies entirely in some U_j^* and is one of the m points already accounted for.

Since M is connected, either $A_n = M$ of $A_n = \phi$ holds. If there is an integer n so that $A_n \neq \phi$ and $A_{n+1} = \phi$, then $A_n = M$ and the proof of (1) is finished. If no such n exists, namely, $A_n = M$ for all $n \geq 1$, then each point of M is covered infinitely many times by points of M^*.

In case (M^*, F) is restricted to be smooth and complete, an alternate yet easier proof can be given as follow. Take two distinct points p and q in M. We need to show that there is a one-to-one mapping from the fiber $F^{-1}(p)$ to the fiber $F^{-1}(q)$. Let $\gamma : [0,1] \to M$ be a curve with $\gamma(0) = p$ and $\gamma(1) = q$, To each $p^* \in F^{-1}(p)$, lift γ to a curve $\gamma_{p^*}^* : [0,1] \to M^*$ with initial point p^* (see (2) in (7.5.2.2)). Define $\Phi : F^{-1}(p) \to F^{-1}(q)$ by setting $\Phi(p^*) = \gamma_{p^*}^*(1)$. By the uniqueness of the lift curve (see (1) in (7.5.2.2)), it follows that Φ is one-to-one; moreover, $F \circ \gamma_{p^*}^*(1) = \gamma(1) = q$ shows that $\Phi(F^{-1}(p)) \subseteq F^{-1}(q)$ holds. Similarly, the inverse curve $\gamma^{-1} : [0,1] \to M$ defines another one-to-one mapping $\Psi : F^{-1}(q) \to F^{-1}(p)$. Since $\Psi \circ \Phi = 1_{F^{-1}(p)}$ and $\Phi \circ \Psi = 1_{F^{-1}(q)}$ hold, therefore Φ is onto, too. This shows that $F^{-1}(p)$ and $F^{-1}(q)$ have the same cardinal number. $\square$

When applying (7.5.2.5) to Riemann surfaces, we have a (see also Example 2 in Sec. 7.5.5)

Finitely-sheeted covering Riemann surface of a Riemann surface. Let R^* and R be Riemann surfaces, and $F : R^* \to R$ be a *proper* nonconstant analytic mapping. Then there exists a positive integer n such that F assumes each point in R exactly n times (counting multiplicities). Hence (R^*, F) is an n-sheeted covering surface of R; it is complete and smooth except branched points.   (7.5.2.6).

A continuous mapping $\mathbf{F}: Y$ (locally compact space) $\to X$ (locally compact space) is called *proper* if the preimage of every compact set in X is compact in Y. A *locally compact space* is a Hausdorff space in which every point has a compact neighborhood. Some related results are in Exercises A(9) and (10). (7.5.2.6) unifiedly explains many common phenomena appeared in our previous descriptive construction of Riemann surfaces of multiple-valued functions. See examples in Section (4).

Proof. Let B be *the set of branch points* (of order ≥ 1) of F. According to (7.2.3) and (7.2.4), B is closed and discrete in R^*; the latter means that every point p^* in B has a neighborhood V such that $V \cap B - \{p^*\}$ holds. Again by (7.2.3) and (7.2.4), *the set $F(B)$ of critical values* is also discrete. And $F(B)$ is closed, too. This is because F is proper and, in a locally compact space, a subset is closed precisely if its intersection with every compact set is compact. Note that, for each $p \in R$, $F^{-1}(p)$ is a *finite* set since F is proper and nonconstant.

Set $Y = R^* - B$ and $X = R - F(B)$. Then $F|_Y : Y \to X$ is a proper local homeomorphism and hence, $(Y, F|_Y)$ is a complete smooth covering surface of R. According to (7.5.2.5), (R^*, F) is a finitely-sheeted covering surface of R. □

Section (4) Application and examples

As a combined application of this section, we have

The lift mapping of a mapping into a complete smooth covering surface. Suppose (M^*, F) is a complete smooth covering surface of M and Z is a simply connected, locally path-connected space. For any continuous mapping $f : Z \to M$, each point $z_0 \in Z$, and any $p_0^* \in F^{-1}(f(z_0))$, there exists a unique lift map $f^* : Z \to M^*$ of f such that $f^*(z_0) = p_0^*$ holds. (7.5.2.7)

The proof given below uses only the local homeomorphism and curve lifting properties of F. Hence the result remains valid even if X and Y are Hausdorff spaces and the map $F : Y \to X$ is a local homeomorphism having the curve lifting property. See Exercises B(1) and B(3). Moreover, (7.5.2.7) is also suitable for n-sheeted covering surface (R^*, F) of a Riemann surface as stated in (7.5.2.6).

Proof. *Definition of f^*:* Pick any $z \in Z$ and construct a curve $\sigma : [0,1] \to Z$ with $\sigma(0) = z_0$ and $\sigma(1) = z$. Then $\gamma = f \circ \sigma : [0,1] \to M$ is a curve connecting $f(z_0)$ to $f(z)$. γ can be uniquely lifted to a curve $\gamma^* : [0,1] \to M$, with $\gamma^*(0) = p_0^*$. Define $f^* : Z \to M^*$ by

$$f^*(z) = \gamma^*(1).$$

f^* is *well-defined* in the sense of being independent of the choice of such an curve σ. For, suppose $\tilde{\sigma}$ is homotopic to σ (recall that Z is simply connected), and then, by the continuity of, $f \circ \tilde{\sigma}$ is also homotopic to $f \circ \sigma$.

By the monodromy theorem (7.5.2.3), the lift curve $\tilde{\gamma}^*$ of $f \circ \tilde{\sigma}$ is homotopic to γ^*, and hence $\tilde{\gamma}^*(1) = \gamma^*(1)$ holds.

The continuity of f^*: Let $z \in Z$ and set $p^* = f^*(z)$. Let V be an open neighborhood of p^*, and without loss of its generality, V may be assumed to satisfy that $F : V \to F(V) = U$ is homeomorphic with U an open neighborhood of $F(p^*) = f(z)$ in M.

Since f is continuous and Z is locally path-connected, there exists a path-connected neighborhood O of z so that $f(O) \subseteq U$.

It remains to show that $f^*(O) \subseteq V$, and hence f is continuous at z. To see this, choose any point $z' \in O$ and let $\rho : [1, 2] \to O$ be a curve with $\rho(1) = z$ and $\rho(2) = z'$. Let σ, $\gamma = f \circ \sigma$ and γ^* be as before. Then the curve $f \circ \rho$ lies entirely in U and its lift curve $(f \circ \rho)^* = (F|_U)^{-1}(f \circ \rho)$ has its initial point at p^* and lies in V. The product curve $\gamma^*(f \circ \rho)^*$ is thus a lift curve of $\gamma(f \circ \rho)$, with the initial point at p_0^*. Hence $f^*(z') = \gamma^*(f \circ \rho)^*(2) = (\gamma(f \circ \rho))^*(2) \in V$. $\qquad\square$

To the end, we present some examples.

Example 1. Describe the branched covering surface of $\mathbf{C}^*$ determined by the mapping $f(z) = 1 + z^2 + z^4$ (see Example 3 of Sec. 3.5.1).

Explanation. $w = f(z) : \mathbf{C}^* \to \mathbf{C}^*$ is a proper nonconstant analytic mapping between Riemann surfaces $\mathbf{C}^*$. Since $f^{-1}(3) = \{1, -1, \sqrt{2}i, -\sqrt{2}i\}$, according to (7.5.2.6), $(\mathbf{C}^*, f)$ is a (complete) branched four-sheeted covering surface of $\mathbf{C}^*$. This fact is better illustrated as Figs. 3.54 and 3.55. According to (7.2.3) and (7.2.4), $w = f(z)$ has an algebraic branch point of order 1 at $z = 0$, $\frac{i}{\sqrt{2}}$, $-\frac{i}{\sqrt{2}}$ each, and an algebraic point of order 3 at ∞.

Traditionally, we construct the Riemann surface R_w, pictured in Fig. 3.55, of the inverse function $z = f^{-1}(w) = \sqrt{\frac{1}{2}(-1 + \sqrt{4w - 3})}$ on which it turns out to be single-valued. As a matter of fact, $z = f^{-1}(w)$: $R_w \to \mathbf{C}^*$ is one-to-one, onto, and hence is conformal. This is equivalent to say that the *lift function* $f^*(z) = 1 + z^2 + z^4$: $\mathbf{C}^* \to R_w$ of $w = f(z)$ is conformal, while the original $w = f(z)$ is four-to-one except at branched points $0, \pm\frac{i}{\sqrt{2}}$, and ∞. Recall that R_w is a complete four-sheeted covering surface of $\mathbf{C}^*$, with an algebraic branch point of order 1 over the point $f(0) = 1$, two algebraic branch points of order 1 over the point $f(\pm\frac{i}{\sqrt{2}}) = \frac{3}{4}$, and an algebraic branch point of order 3 over $f(\infty) = \infty$. As usual, $\mathbf{C}^*$ and R_w deem to be identical.

Consider the curve $\gamma : |w - 3| = 2$ in the w-plane. Note that $f^{-1}(1) = \{0, 0, i, -i\}$. There are two lift curves from $p_0^* = 0$ (a double point) over γ, and there is a single lift curve from i and $-i$ each over γ because both i and $-i$ are regular points of $f(z)$. See Fig. 3.54. Where are the lift curves from $F^{-1}(1)$ in R_w, where $F : R_w \to \mathbf{C}^*$ is the orthogonal projection? Try to trace out the lift curves γ^* of a closed curve γ passing $\frac{3}{4}$ and 1.

Example 2. Describe the covering surface of $\mathbf{C}$ determined by $f(z) = ze^{-z}$ (see Example 1 of Scc. 3.5.1).

Explanation. $w = f(z) = ze^{-z} : \mathbf{C} - \{1\} \to \mathbf{C} - \{e^{-1}\}$ is a local homeomorphism and hence, $(\mathbf{C}-\{1\}, f)$ is a smooth covering surface of $\mathbf{C}-\{e^{-1}\}$. While, $(\mathbf{C}, f)$ is a branched covering surface of $\mathbf{C}$, with $z = 1$ as an algebraic branch point of order 1 (see Figs. 3.57, 3.59, and 3.82). Observe that $f^{-1}(0) = \{0\}$ and $f^{-1}(w)$ is countably infinite for each $w \neq 0$. According to (2) in (7.5.2.5), $(\mathbf{C}, f)$ is not complete as a covering surface of $\mathbf{C}$.

Let R_w be the Riemann surface of the inverse function $z = f^{-1}(w)$ of $w = f(z)$. See Fig. 3.57. Let $p : R_w \to \mathbf{C}$ (the w-plane) be the projection. Then the fiber $p^{-1}(0)$ is a singleton. R_w has a logarithmic branch point over each of $w = 0$ and, ∞ and an algebraic branch point of order one over $w = e^{-1}$. As usual, these two logarithmic branch points are not considered to lie on the surface R_w itself. Hence, as the Riemann surface $R_{\log z}$ of $\log z$ (see Figs. 2.73 and 2.74), R_w is simply connected. The lift curves $p^{-1}(\gamma)$ of the circle $\gamma : |w| = e^{-1}$ consists of two disjoint parts; one is a "usual" circle with its upper half lying on the upper half-plane of Ω_0^* and its lower half in the lower half-plane of Ω_0^* (see Fig. 3.82), the other is an endless screw circle as shown in Fig. 2.73. What are the components of $p^{-1}(\{|w| < e^{-1}\})$? Is each of them homeomorphic to $|w| < e^{-1}$ under p?

The lift map $f^* : \mathbf{C}$ (the z-plane) $\to R_w$ of the function $w = f(z)$ is the function f itself (when a point on R_w is identified with its image point under p). Then f^* maps $\mathbf{C}$ conformally onto R_w.

Note. This example shows that, in (7.5.2.7), even (M^*, F) is a noncomplete branched covering surface of M, some continuous mapping $f : Z \to M$ might still have a lift map to M^*.

Example 3. Describe the covering surface of $\mathbf{C}^*$ determined by $f(z) = z^3 - 3z$ (see Example 4 of Sec. 2.7.2 and Example 2 of Sec. 3.5.6).

Explanation. $(\mathbf{C}^*, f)$ is a (complete) branched three-sheeted covering surface of $\mathbf{C}^*$ (the w-plane). 1 and -1 are algebraic branch points of order 1, while ∞ is an algebraic branch point of order 2.

On the other hand, the Riemann surface R_w of the inverse function $z = f^{-1}(w)$ of $w = f(z)$ is a three-sheeted covering surface of the w-plane

$\mathbf{C}^*$ (see Figs. 2.65 and 2.66). It has an algebraic branch point of order 1 over each of $w = -2$ and 2, and an algebraic branch point of order 2 over ∞. The lift map $f^* : \mathbf{C}^* \to R_w$ of $w = f(z)$ is f itself, but now f^* turns out to be a conformal mapping.

Readers are urged to look back Figs. 3.86–3.88, and then try to interpret various lift curves both on $\mathbf{C}^*$ (the z-plane) and on R_w of a curve in the w-plane.

Example 4. In (7.5.2.7), let $M^* = \mathbf{C}$, $F = \exp$ (the exponential map) and $M = \mathbf{C} - \{0\}$. Suppose $Z = R$ is a simply connected Riemann surface. What is the lift map $f^* : R \to M^*$?

Explanation. Such a lift map f^* always exists since R is simply connected and locally path-connected (refer to Exercise B(3)); moreover, f^* is analytic (see Exercise A(2)). Since $\exp \circ f^* = f$, it follows that $f^* = \log f$, a logarithmic function of f, and any two of them differs by an additive constant $2n\pi i, n = 0, \pm 1, \pm 2, \ldots$. Try to review (3.3.1.8).

Exercises A

(1) Let (R^*, f) be a covering space of a Riemann surface R, where R^* is a connected Hausdorff space. Show that there exists a unique complex structure on R^* to make R^* a Riemann surface and $F : R^* \to R$ a local conformal map.

(2) Suppose R_1, R_2, and R_3 are Riemann surfaces, $F : R_1 \to R_2$ is an unbranched analytic mapping and $f : R_3 \to R_2$ is analytic. Show that every lift $g : R_3 \to R_1$ of f is analytic.

(3) Let (Y, F) be a covering space of X.

   (a) Suppose X and Y are Hausdorff spaces. If $\gamma_1^*, \gamma_2^* : [0, 1] \to Y$ are two lift curves of a curve $\gamma : [0, 1] \to X$ and $\gamma_1^*(t_0) = \gamma_2^*(t_0)$ for some $t_0 \in [0, 1]$, then $\gamma_1^* = \gamma_2^*$.

   (b) Suppose (Y, F) is a complete covering space of X. Then (Y, F) has the curve lifting property.
   Note. If both X and Y are Hausdorff, by (a), any such a lift curve is uniquely determined.

(4) Let X be a path-connected space, and Y a connected and locally path-connected space. Suppose (Y, F) is a complete covering space of X.

   (a) Let Z be a connected space and $f : Z \to X$ be a continuous map. If $g_1, g_2 : Z \to Y$ are two lifts of f with respect to F and $g_1(z_0) = g_2(z_0)$ for some $z_0 \in Z$, then $g_1 = g_2$.

   (b) (Y, F) has the (unique) curve lifting property.

(c) Suppose Z is a locally connected space, $z_0 \in Z$, and a continuous map $f : Z \to X$ with $f(z_0) = x_0$ has a lift $f^* : Z \to Y$ with $f^*(z_0) \in F^{-1}(x_0)$. Let $H : Z \times I \to X$ be a homotopy of f, namely, $H(z, 0) = f(z)$ for all $z \in Z$. Then there is a unique lift homotopy $H^* : Z \times I \to Y$ so that $H^*(z, 0) = f^*(z)$, $z \in Z$.

(d) (The *monodromy theorem*) Show that (2) in (7.5 2.3) is still valid in this case, with M and (M^*, F) replaced by X and (Y, F), respectively.

(5) Prove (7.5.2.4) in detail.

(6) Let (M^*, F) be a complete smooth covering surface of M. Define $\Phi : \pi_1(M, p_0) \to F^{-1}(p_0)$ by $\Phi([\gamma]) = \gamma^*(1)$. Show that Φ is onto, and Φ is one-to-one if $\pi_1(M^*, p_0^*) = \{1\}$, where $p_0^* \in F^{-1}(p_0)$. Try to use this result to prove Exercise B(8) of Sec. 7.4: $\pi_1(P^2) \cong \mathbf{Z}_2$.

(7) Let (Y, F) be a complete covering space of X as defined in Exercise (4). Show the fiber $F^{-1}(x)$ has the same cardinal for all $x \in X$.

(8) Let X and Y be Hausdorff spaces with X path-connected. Suppose (Y, F) is a complete covering space of X. Show that $F^{-1}(x)$ has the same cardinal for all $x \in X$. In particular, F is an onto mapping if Y is nonempty.

(9) Suppose $F : Y$ (locally compact space) $\to X$ (locally compact space) is proper and *discrete*, the latter means that $F^{-1}(x)$ is a discrete set for each $x \in X$. Show that:

   (a) $F^{-1}(x)$ is a finite set for each $x \in X$.

   (b) Suppose $x \in X$ and V is a neighborhood of $F^{-1}(x)$. Then there exists a neighborhood U of x such that $F^{-1}(U) \subseteq V$ holds.

(10) Let X and Y be locally compact spaces and $F : Y \to X$ be a proper local homeomorphism. Show that (Y, F) is a complete covering space of X.

(11) Try to imitate Example 4 to define single-valued branches of $\cos^{-1} z$ and $\tan^{-1} z$.

Exercises B

(1) Let X and Y be Hausdorff spaces and $F : Y \to X$ be a local homeomorphism having the curve lifting property (in particular, if (Y, F) is a complete covering space of X). Suppose Z is a simply connected and locally path-connected space, and $f : Z \to X$ is a continuous map. Show that, for each $z_0 \in Z$ and $y_0 \in F^{-1}(f(z_0))$, there exists a unique lift map $g : Z \to Y$ of f such that $g(z_0) = y_0$ holds.

(2) Suppose M is a surface and Y is a Hausdorff space. If $p : Y \to M$ is a local homeomorphism having the curve lifting property, then (Y, p) is a complete covering space of M. In case Y is also connected, then (Y, p) turns out to be a compete covering *surface* of M (see Exercise A(1)).

(3) Let (Y, F) be a complete covering space of X, where X and Y are defined as in Exercise A(4), or X and Y are Hausdorff spaces. Let Z be a connected and path-connected space, and $f : Z \to X$ be a continuous map. Let $z_0 \in Z$ and $y_0 \in F^{-1}(f(z_0))$. Show that there exists a lift map $g : Z \to X$ of f with $g(z_0) = y_0$ if and only if

$$f_* \pi_1(Z, z_0) \subseteq F_* \pi_1(Y, y_0).$$

See the diagram. In case the lift g does exist, it is unique. In particular, if Z is simply-connected and path-connected, then every continuous map $f : Z \to X$ can always be lifted to Y.

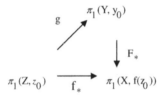

7.5.3 *Characteristics and classifications of covering surfaces: The existence theorem*

Section (1) Characteristics

Let us start with (7.5.2.4). $F_* \pi_1(M^*, p_0^*)$ is a *subgroup* of the fundamental group $\pi_1(M, p_0)$. Members of the subgroup $F_* \pi_1(M^*, p_0^*)$ compose of these homotopy classes $[\gamma]$ so that the lift curve γ^* from p_0^* and over such a curve γ is a closed curve and hence $[\gamma^*] \in \pi_1(M^*, p_0^*)$ with $F \circ \gamma^* = \gamma$ and $F_*[\gamma^*] = [\gamma]$ (henceforth, ( ) in $F_*([\gamma^*])$ is sometimes omitted for simplicity).

The subgroup $F_* \pi_1(M^*, p_0^*)$ depends on the choice of $p_0^* \in F^{-1}(p_0)$. Choose another point $p_1^* \in F^{-1}(p_0)$. To figure out *possible connection* between $F_* \pi_1(M^*, p_0^*)$ and $F_* \pi_1(M^*, p_1^*)$, let σ be a curve from p_0^* to p_1^*. Then σ induces the isomorphism $\sigma_\# : \pi_1(M^*, p_0^*) \to \pi_1(M^*, p_1^*)$ (see (2) in (7.4.4)), defined by $\sigma_\#[\gamma^*] = [\sigma^{-1}\gamma^*\sigma]$, for $[\gamma^*] \in \pi_1(M^*, p_0^*)$. Set $\rho = F(\sigma)$, a closed curve passing p_0. $\rho_\#$ is similarly defined. See the diagram.

$$\pi_1(M^*, p_0^*) \xrightarrow{\;\sigma_\#\;} \pi_1(M^*, p_1^*)$$
$$F_* \downarrow \qquad\qquad F_* \downarrow$$
$$\pi_1(M, p_0) \xrightarrow{\;\rho_\#\;} \pi_1(M, p_0)$$

For each $[\gamma^*] \in \pi_1(M^*, p_0^*)$,

$$F_*\sigma_\#([\gamma^*]) = F_*([\upsilon^{-1}\gamma^*\sigma]) = [(F\sigma^{-1})(F\gamma^*)(F\sigma)] = [\rho^{-1}(F\gamma^*)\rho)]$$
$$= \rho_\#([F\gamma^*]) = \rho_\# F_*([\gamma^*]).$$

Namely, $F_*\sigma_\# = \rho_\# F_*$ holds and the diagram commutes. Since $\sigma_\#$ and $\rho_\#$ are isomorphisms, it follows that

$$F_*\pi_1(M^*, p_1^*) = F_*\sigma_\#\pi_1(M^*, p_0^*)$$
$$= \rho_\# F_*\pi_1(M^*, p_0^*) = [\rho]^{-1} F_*\pi_1(M^*, p_0^*)[\rho]. \qquad (*_1)$$

Hence $F_*\pi_1(M^*, p_1^*)$ is conjugate to $F_*\pi_1(M^*, p_0^*)$ as subgroups of $\pi_1(M, p_0)$.

On the other hand, suppose $H = [\rho]^{-1} F_*\pi_1(M^*, p_0^*)[\rho]$ is a conjugate subgroup of $F_*\pi_1(M^*, p_0^*)$, where ρ is a closed curve from p_0. Let σ be the unique lift curve from p_0^* of ρ. Let $p_1^* = \sigma(1)$. From the last paragraph, it is easily seen that

$$F_*\pi_1(M^*, p_1^*) = [F\sigma]^{-1} F_*\pi_1(M^*, p_0^*)[F\sigma] = [\rho]^{-1} F_*\pi_1(M^*, p_0^*)[\rho] = H$$
$$(*_2)$$

This means that H is one of the groups $F_*\pi_1(M^*, p_0^*), p_0^* \in F^{-1}(p_0)$.

Two subgroups G_1 and G_2 of a group G is said to be *conjugate* if there is an element $g \in G$ so that $G_2 = g^{-1}G_1 g$ holds. Conjugacy is an equivalence relation among subgroups, and the resulted equivalent classes are called *the classes of conjugate subgroup*.

In conclusion, we obtain

The characteristic of a covering surface. Let (M^*, F) be a complete smooth covering surface of a surface M. Then, for each point $P_0 \in M$,

$$\{F_*\pi_1(M^*, p_0^*)|p_0^* \in F^{-1}(p_0)\}$$

is a *class of conjugate subgroups* of the fundamental group $\pi_1(M, p_0)$, called the *characteristic* (class) of the covering surface (M^*, F). $\qquad (7.5.3.1)$

In particular, if $\pi_1(M)$ is abelian, then the characteristic of each complete smooth covering surface of M consists of exactly only one subgroup.

If the complete covering space (Y, F) of X is defined as in Exercise A(4) of Sec. 7.5.2, then (7.5.3.1) is still valid in this case. Readers should try to figure this out by themselves.

We raise some examples.

Example 1. It is known that $\pi_1(S^1) \cong \mathbf{Z}$ (see Exercise B(3) of Sec. 7.4). Try to determine the characteristics of the covering spaces $(\mathbf{R}, \exp)$ and (S^1, F_n) of S^1, where $F_n(z) = z^n$ for integer $n \geq 1$ (see Example 1 in Sec. 7.5.1).

Explanation. Since $\pi_1(\mathbf{R}) = \{1\}$, it follows that the characteristic of $(\mathbf{R}, \exp)$ is the set $\{0\}$ of conjugate subgroups of $\mathbf{Z}$, determined by the identity element 0.

We *claim* that the characteristic of (S^1, F_n) is the set of conjugate subgroups of $\mathbf{Z}$, determined by $n\mathbf{Z}$, and simply denoted as $[n\mathbf{Z}]$. To see this, we introduce the *degree* of a closed curve γ from 1 in S. Let γ^* be the unique lift curve from 0 in the covering space $(\mathbf{R}, \exp)$. Then, geometrically, $\gamma^*(1)$ is the number of times the interval $[0,1]$ winding around the circle S^1 under exp. The degree of γ is defined and denoted by

$$\deg \gamma = \gamma^*(1).$$

According to (7.5.2.2) and (7.5.2.3), $\deg \gamma$ is well-defined and is dependent only on the homotopy class $[\gamma]$. Since $\exp \circ \gamma^*(1) = \gamma(1) = 1$, it follows that $\deg \gamma$ is an integer. This enables us to define

$$\deg_\# : \pi_1(S^1, 1) \to \mathbf{Z} \text{ by } \deg_\#([\gamma]) = \deg \gamma.$$

It can be shown that $\deg_\#$ is an *isomorphism* (why?). And this proved again that $\pi_1(S^1, 1) \cong \mathbf{Z}$. Define $\rho_n : \mathbf{Z} \to \mathbf{Z}$ by $\rho_n = \deg_\# \circ F_{n_*} \circ \deg_\#^{-1}$. See the diagram. For any

$$
\begin{array}{ccc}
 & F_{n_*} & \\
\pi_1(S^1, 1) & \longrightarrow & \pi_1(S^1, 1) \\
\deg_\# \downarrow & \rho_n & \deg_\# \downarrow \\
\mathbf{Z} & \longrightarrow & \mathbf{Z}
\end{array}
$$

integer m, let $\gamma(t) = e^{2m\pi i t}$, $0 \leq t \leq 1$. Then $[\gamma] \in \pi_1(S^1, 1)$ and $F_n(\gamma)(t) = e^{2mn\pi i t}$. Since $\deg_\#([\gamma]) = m$, it follows that

$$\rho_n(m) = \deg_\# \circ F_{n_*} \circ \deg_\#^{-1}(m) = \deg_\# \circ F_{n_*}([\gamma]) = \deg_\#([F_n(\gamma)]) = mn.$$

This justifies that $\rho_n(\mathbf{Z}) = n\mathbf{Z}$, and hence the claim is true.

Example 2. Try to describe the covering surfaces of $\mathbf{C}-\{0\}$. It is known that $\pi_1(\mathbf{C} - \{0\}) = \mathbf{Z}$ (see Exercise B(4) of Sec. 7.4). Also, refer to Example 1 of Sec. 7.5.5.

Explanation. The classes of conjugate subgroups of $\mathbf{Z}$ are $[0], [\mathbf{Z}], [2\mathbf{Z}], \ldots,$ $[n\mathbf{Z}], \ldots$.

$[0]$: In this case, the complete smooth covering surface (R^*, F) is necessarily simply connected because the induced homomorphism F_* is one-to-one. Both $(\mathbf{C}, \text{cxp})$ and $(R_{\log z}, \log z)$ are the candidates for the covering surfaces (see Example 3 of Sec. 7.5.1, and Fig. 2.73 and Example 2 of Sec. 3.3.3). Note that the exponential map exp sets up a homeomorphism (actually, a conformal mapping) between $\mathbf{C}$ and Riemann surface $R_{\log z}$ of the logarithmic function $\log z$. Then $\mathbf{C}$ and $R_{\log z}$ are equivalent or called *isomorphic* and are regarded essentially the same (see Section (2) below). They are termed *the universal covering surface* of $\mathbf{C} - \{0\}$ in Sec. 7.5.5.

$[\mathbf{Z}]$: The complete smooth covering surface of $\mathbf{C} - \{0\}$, in this case, is itself $(\mathbf{C} - \{0\}, \text{id})$ or its isomorphic (conformally equivalent) surface $S^1 \times \mathbf{R}$, the cylinder. See Example 2 in Sec. 7.1 and Exercise B(6) of Sec. 7.4.

$[n\mathbf{Z}]$, where $n \geq 2$ is a fixed integer: Based on the construction of the n-sheeted covering surface (M_n^*, F) of $M = \{r_1 < |z| < r_2\}$ in Example 2 of Sec. 7.5.1, similar argument as in the above Example 1 shows that the Riemann surface $R_{\sqrt[n]{z}}$ of $\sqrt[n]{z}$, together with the orthogonal projection F, constitutes a branched covering surface of $\mathbf{C}$ (see Fig. 2.49). Hence, $(R_{\sqrt[n]{z}} - \{0\}, F)$ is a complete smooth covering surface of $\mathbf{C}-\{0\}$ whose fundamental group is $n\mathbf{Z}$. Since $R_{\sqrt[n]{z}}-\{0\}$ is conformally equivalent to $\mathbf{C}-\{0\}$ under the mapping $\sqrt[n]{z}$, $(\mathbf{C} - \{0\}, z^n)$ can also be considered as a required candidate.

$[n\mathbf{Z}]$, $n \geq 0$, range all possible classes of conjugate subgroups of $\mathbf{Z}$. Note that $m\mathbf{Z}$ is a subgroup of $n\mathbf{Z}$ if n divides m. What is the possible relation between $(\mathbf{C} - \{0\}, z^m)$ and $(\mathbf{C} - \{0\}, z^n)$? Geometric intuition tells us that $(\mathbf{C} - \{0\}, z^m)$ seems to be "larger" than or to cover $(\mathbf{C} - \{0\}, z^n)$. Strictly speaking, $(\mathbf{C} - \{0\}, z^m)$ is said to be *stronger than* $(\mathbf{C} - \{0\}, z^n)$ because there is a local homeomorphism $F_{mn}(z) = z^{\frac{m}{n}} : (\mathbf{C} - \{0\}, z^m) \rightarrow (\mathbf{C}-\{0\}, z^n)$ so that $F_m(z) = F_n \circ F_{mn}(z)$ and $(\mathbf{C}-\{0\}, F_{mn})$ is a complete smooth covering surface of $(\mathbf{C} - \{0\}, z^n)$. See Fig. 7.26. This concept will be formally introduced in Section (2).

Example 3. Try to describe the covering surfaces of P^2, the projective plane.

Explanation. $\pi_1(P^2) \cong \mathbf{Z}_2$ (see Exercise B(8) of Sec. 7.4). $\mathbf{Z}_2$ has two classes of conjugate subgroups: $[0]$ and $[\mathbf{Z}_2]$. The corresponding complete smooth covering surfaces under isomorphism are (S^2, F), where $F(x) = [x] = \{x, -x\}$, and (P^2, id).

Section (2) Equivalence (or classification) of covering surfaces

The concept of a covering surface stronger than another covering surface introduced in the last paragraph inside Example 2 can be adopted to compare covering surfaces of a given surface. *All covering surfaces are understood to be complete and smooth, henceforth.*

Let (M_1^*, F_1) and (M_2^*, F_2) be two covering surfaces of a given surface M. (M_2^*, F_2) is said to be *stronger than* (M_1^*, F_1) if there is a local homeomorphism $F_{21}: M_2^* \to M_1^*$ satisfying the properties that (M_2^*, F_{21}) is a covering surface of M_1^* and $F_2 = F_1 \circ F_{21}$ holds. Such an F_{21} is called a *homomorphism* from M_2^* onto M_1^*. Suppose, vice versa, (M_1^*, F_1) is said to be stronger than (M_2^*, F_2), namely, there is a local homeomorphism $F_{12}: M_1^* \to M_2^*$ so that (M_1^*, F_{12}) is a covering surface of M_2^* and $F_1 = F_2 \circ F_{12}$ holds. Then $F_2 = F_2 \circ F_{12} \circ F_{21}$ implies that $F_{12} \circ F_{21} = id_{M_2^*}$. Similarly, $F_{21} \circ F_{12} = id_{M_1^*}$ holds. Under this circumstance, $F_{12} = F_{21}^{-1}$ and F_{21} is a homeomorphism from M_2^* onto M_1^*. In this case, (M_1^*, F_1) and (M_2^*, F_2) are said to be *equivalent* or *isomorphic*, and the homeomorphism F_{12} or F_{21} is called an *isomorphism*.

The relation "strong than" is transitive and defines a partial ordering. And isomorphism is indeed an equivalent relation. Covering surfaces in the same equivalence class are classified to be in the same class or are considered to be identical.

It is possible that $M_1^* \neq M_2^*$ but (M_1^*, F_1) and (M_2^*, F_2) might be equivalent (see Fig. 7.26). Even through $M_1^* = M_2^*$, but (M_1^*, F_1) and (M_2^*, F_2) might be not equivalent owing to different styles of covering presented by F_1^* and F_2^*. See Fig. 7.27, where $\mathbf{C}/\Gamma$ is the torus discussed in Example 3 of Sec. 7.1.

Our main result is in the following

Invariance of the characteristics under equivalence (isomorphism). Let (M_1^*, F_1) and (M_2^*, F_2) be complete smooth covering surfaces of a surface M. Then (M_1^*, F_1) and (M_2^*, F_2) are equivalent if and only if their characteristics are identical.   (7.5.3.2)

The latter means that, for any fixed point $P_0 \in M$, there are $p_1^* \in F_1^{-1}(p_0)$ and $p_2^* \in F_2^{-1}(p_0)$ so that $F_{1*}\pi_1(M_1^*, p_1^*) = F_{2*}\pi_1(M_2^*, p_2^*)$ holds

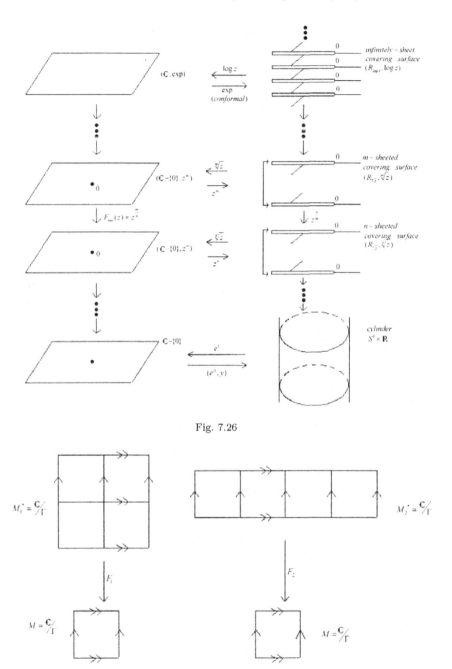

Fig. 7.26

Fig. 7.27

(see (7.5.2.4) and (7.5.3.1)). In case $\pi_1(M, p_0)$ is abelian, then for *arbitrary* $p_1^* \in F_1^{-1}(p_0)$ and $p_2^* \in F_2^{-1}(p_0)$, these two subgroups are identical.

Proof. ($\Rightarrow$) Let $f : M_1 \to M_2$ be the isomorphism. For any $p_1^* \in F_1^{-1}(p_0)$, set $p_2^* = f(p_1^*)$ which lies in $F_2^{-1}(p_0)$. The relation $F_1 = F_2 \circ f$ shows that $F_{1*} = F_{2*} \circ f_*$ holds. This means the following diagram is commutative. Hence $F_{1*}\pi_1(M_1^*, p_1^*) = F_{2*}f_*\pi_1(M_1^*, p_1^*) = F_{2*}\pi_1(M_2^*, p_2^*)$, a common subgroup belonging to the two classes of conjugate subgroups. Consequently they should be identical.

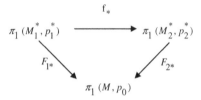

($\Leftarrow$)$\{F_{1*}\pi_1(M_1^*, p^*)|p^* \in F_1^{-1}(p_0)\} = \{F_{2*}\pi_1(M_2^*, p^*)|p^* \in F_2^{-1}(p_0)\}$ means that there exist $p_1^* \in F_1^{-1}(p_0)$ and $p_2^* \in F_2^{-1}(p_0)$ so that $F_{1*}\pi_1(M_1^*, p_1^*) = F_{2*}\pi_2(M_2^*, p_2^*)$ holds. By the map lifting property (refer to (7.5.2.7) and see Exercise B(3) of Sec. 7.5.2), there are lift maps $f : M_1^* \to M_2^*$ of F_1 and $g : M_2^* \to M_1^*$ of F_2 such that $F_1 - F_2 \circ f$ and $f(p_1^*) = p_2^*$, and $F_2 = F_1 \circ g$ and $g(p_2^*) = p_1^*$, hold. Therefore $F_1 = F_1 \circ g \circ f$, and since $g \circ f(p_1^*) = g(p_2^*) = p_1^*$, by the uniqueness of the lift map, $g \circ f = id_{M_{1*}}$. Similarly $f \circ g = id_{M_{2*}}$. It follows that f is a homeomorphism and $F_1 = F_2 \circ f$ holds. Therefore f is an isomorphism. $\quad\square$

Section (3) The existence (or construction) of covering surfaces

The converse of (7.5.3.1) is true. As a matter of fact, we have

The existence theorem of covering surfaces. Suppose M is a surface and $\pi_1(M, p_0)$ is its fundamental group with base point $p_0 \in M$. The statement that we made in (7.5.3.1) sets up a one-to-one correspondence between the classes of equivalent (complete smooth) covering surfaces of M and the classes of conjugate subgroups of $\pi_1(M, p_0)$. Moreover, if the class $[G]$ of conjugate subgroups and the covering surface (M^*, F) correspond to each other, then the fundamental group $\pi_1(M^*)$ is isomorphic to G or any element in $[G]$.   (7.5.3.3)

See Exercise B(1) for a general setting. *In case M is a Riemann surface, then all its covering surfaces turn out to be Riemann surfaces* owing to (7.5.2.1).

(7.5.3.1) and (7.5.3.2) together proved *the necessity*.

Proof of the sufficiency. Let G be a subgroup of $\pi_1(M, p_0)$. We try to construct a corresponding complete smooth covering surface (M^*, F) of M so that $\{F_*\pi_1(M^*, p^*)|p^* \in F^{-1}(p_0)\} = [G]$, the class of conjugate subgroups of $\pi_1(M, p_0)$ that contains the subgroup G.

Definition of (M^*, F). Two curves γ_1 and γ_2 on M from p_0 are said to be equivalent if

1. γ_1 and γ_2 have the same end point, i.e., $\gamma_1(1) = \gamma_2(1)$, and

2. $[\gamma_1\gamma_2^{-1}] \in G.$ (*1)

It is indeed an equivalent relation. The equivalence class determined by γ is denoted as $[\gamma]$. Condition 2 actually means $[\gamma_1\gamma_2^{-1}] = [\gamma_1][\gamma_2]^{-1} \in G$. Set

$$M^* = \{\text{The equivalence classes } [\gamma] \text{ of curves } \gamma \text{ satisfying conditions 1 and}$$

$$2 \text{ in } (*1)\}$$ (*2)

and define $F : M^* \to M$ by

$$F([\gamma]) = \text{the common terminal point of all } \gamma \in [\gamma].$$ (*3)

Topological structure of M^*: Let $\{O_\lambda, \varphi_\lambda\}_{\lambda \in \Lambda}$ be a collection of charts for the surface topology of M. Each O_λ might be assumed to be a topological disk, i.e., $\varphi_\lambda(O_\lambda)$ is a planar disk (see Remarks 2 and 3 in Sec. 7.1). Take any O_λ and fix any point $p \in O_\lambda$. For any $p^* = [\gamma_0] \in F^{-1}(p)$, let

$$O_\lambda^*(p^*) = \{[\gamma_0\gamma]|\gamma \text{ is a curve in } O_\lambda \text{ connecting } p = \gamma_0 \ (1) \text{ to some}$$

$$\text{point in } O_\lambda.\}.$$ (*4)

Recall that γ_0 is a curve in M connecting p_0 to $p = \gamma_0(1)$. Hence $\gamma_0\gamma$ is the product of γ_0 followed by γ and is well-defined. The point $[\gamma_0\gamma]$ is independent of the choice of γ in O_λ, from p to the same terminal point. For, if σ is another such a curve, then $\gamma\sigma^{-1}$ is homotopic to the constant curve p since O_λ is a topological disk (simply connected), and so is $\gamma_0\gamma$ homotopic to $\gamma_0\sigma$. Therefore $\gamma_0\gamma(\gamma_0\sigma)^{-1} = \gamma_0\gamma\sigma^{-1}\gamma_0^{-1}$ is homotopic to the constant curve p_0 and hence is in G. By (*1), $[\gamma_0\gamma] = [\gamma_0\sigma]$ holds.

The collection of sets $\{O_\lambda^*(p^*)|\lambda \in \Lambda \text{ (index set) and } p \in O_\lambda\}$ acts as the base for the topology on M^*. In particular, $O_\lambda^*(p^*)$ is an open neighborhood of each point in it.

M^* is Hausdorff. Let p_1^* and p_2^* be two distinct points in M^*.

In case $F(p_1^*) = p_1$ is different from $F(p_2^*) = p_2$: Since M is Hausdorff, there exists O_{λ_1} and O_{λ_2} so that $p_1 \in O_{\lambda_1}$, $p_2 \in O_{\lambda_2}$, and $O_{\lambda_1} \cap O_{\lambda_2} = \phi$. Then $O_{\lambda_1}^*(p_1^*) \cap O_{\lambda_2}^*(p_2^*) = \phi$.

The case that $F(p_1^*) = F(p_2^*)$: Suppose $p_1^* \in O_{\lambda_1}^*(q_1^*)$, $p_2^* \in O_{\lambda_2}^*(q_2^*)$. Then $p_1^* = [\gamma_1 \sigma_1]$ and $p_2^* = [\gamma_2 \sigma_2]$ (see $(^*4)$); moreover, $\gamma_1 \sigma_1 (\gamma_2 \sigma_2)^{-1}$ is a closed curve from p_0, but its homotopy class is not in G because $p_1^* \neq p_2^*$. Suppose $O_{\lambda_1}^*(q_1^*) \cap O_{\lambda_2}^*(q_2^*) \neq \phi$ and p^* belongs to this common part. Then p^* can also be written as $[\gamma_1 \tau_1] = [\gamma_2 \tau_2]$, and $[\gamma_1 \tau_1 (\gamma_2 \tau_2)^{-1}]$ is in G. If σ_1 and τ_1 have their terminal points in the same component of $O_{\lambda_1} \cap O_{\lambda_2}$, then $\sigma_1 \sigma_2^{-1}$ and $\tau_1 \tau_2^{-1}$ are homotopic which, in turn, implies that $\gamma_1 \sigma_1 (\gamma_2 \sigma_2)^{-1}$ and $\gamma_1 \tau_1 (\gamma_2 \tau_2)^{-1}$ are homotopic. This would show that $[\gamma_1 \sigma_1 (\gamma_2 \sigma_2)^{-1}]$ is in G, a contradiction. Hence $O_{\lambda_1}^*(q_1^*) \cap O_{\lambda_2}^*(q_2^*) = \phi$.

M^* *is path-connected* (and hence is connected). Let p_1^* and p_2^* be two points in M^*. Set $p_1^* = [\gamma_1]$ and $p_2^* = [\gamma_2]$ (see $(^*2)$). Set $\gamma = \gamma_1^{-1} \gamma_2 :$ $[0,1] \to M$, a curve joining $p_1 = F(p_1^*)$ to $p_2 = F(p_2^*)$, where

$$
\gamma(t) = \begin{cases} \gamma_1(1 - 2t), & 0 \leq t \leq \dfrac{1}{2} \\[2mm] \gamma_2(2t - 1), & \dfrac{1}{2} \leq t \leq 1 \end{cases}.
$$

Define, for each s, $0 \leq s \leq 1$, a curve $\sigma_s(t)$ by

$$
\sigma_s(t) = \begin{cases} \gamma_1(t(1 - 2s)), & 0 \leq s \leq \dfrac{1}{2}, \ \ 0 \leq t \leq 1 \\[2mm] \gamma_2(t(2s - 1)), & \dfrac{1}{2} \leq s \leq 1, \ \ 0 \leq t \leq 1 \end{cases}.
$$

Each σ_s $(0 \leq s \leq 1)$ is a curve connecting p_0 to $\gamma(s)$. Therefore $\Gamma : [0,1] \to M^*$ defined by $\Gamma(s) = [\sigma_s]$ is a curve connecting $\Gamma(0) = [\sigma_0] = [\gamma_1] = p_1^*$ to $\Gamma(1) = [\sigma_1] = [\gamma_2] = p_2^*$.

Surface structure of M^*. The charts $\{O_\lambda^*(p^*), \varphi_\lambda \circ F\}$, where $p \in O_\lambda$ and $\lambda \in \Lambda$, define the surface structure on M^*. Observe that, for each $p \in O_\lambda$ and $p^* \in F^{-1}(p)$ (see $(^*3)$), the restriction map $F|O_\lambda^*(p^*) : O_\lambda^*(p^*) \to F(O_\lambda^*(p^*)) = O_\lambda$ is one-to-one. Hence $\varphi_\lambda \circ F : O_\lambda^*(p^*) \to (\varphi_\lambda \circ F)(O_\lambda^*(p^*))$ is identical with $\varphi_\lambda : O_\lambda \to \varphi_\lambda(O_\lambda)$ which is homeomorphic. In case $O_{\lambda_1}^*(p_1^*) \cap O_{\lambda_2}^*(p_2^*) \neq \phi$, then $(\varphi_{\lambda_2} \circ F) \circ (\varphi_{\lambda_1} \circ F)^{-1} = \varphi_{\lambda_2} \circ \varphi_{\lambda_1}^{-1} : \varphi_{\lambda_1}(O_{\varphi_{\lambda_1}} \cap O_{\varphi_{\lambda_2}}) \to \varphi_{\lambda_2}(O_{\lambda_1} \cap O_{\lambda_2})$ is a C^k diffeomorphism $(0 \leq k \leq \omega)$.

M^* *is a complete smooth covering surface of M.* By the very definition of the surface structure of M^*, $F : M^* \to M$ is a local homeomorphism. In the proof of M^* being Hausdorff, we acquired that any two $O_\lambda^*(p^*)$ with the same projection O_λ are either disjoint or identical. Hence, these distinct $O_\lambda^*(p^*)$ are just components of $F^{-1}(O_\lambda)$, and each of them is homeomorphic to O_λ under F. Thus (M^*, F) is complete.

(M^*, f) *is the covering surface that corresponds to the given subgroup G or one of its conjugate subgroups.* This means that $F_*\pi_1(M^*, p_0^*) = G$ or its conjugate group. For simplicity, choose $p_0^* = [1]$, the equivalence class of the constant curve from p_0. Let $\gamma : [0, 1] \to M$ be a closed curve from p_0. Set the restriction $\gamma_t = \gamma|[0, t]$ for $0 \le t \le 1$. Define $\gamma^* : [0, 1] \to M^*$ by $\gamma^*(t) = [\gamma_t]$. Then γ^* is a closed curve from p_0^*, namely, $\gamma^*(1) = [\gamma_1] - [1]$ holds, if and only if $[\gamma] \subset G$. This proves the claim. The choice of p_0^* as other points in $F^{-1}(p_0)$ will result that (M^*, F) corresponds to a conjugate subgroup of G.

$\pi_1(M^*)$ *is isomorphic to G or one of its conjugate subgroups.* By the last paragraph, F_* is onto, and hence, by (7.5.2.4), $F_* : \pi_1(M^*, p_0^*) \to G$ is an isomorphism. □

One of two extreme cases is that the given subgroup G is the whole group $\pi_1(M, p_0)$, the other is that G equals to the trivial group $\{1\}$. The latter is fully discussed in Sec. 7.5.5. In the former case, only condition 1 in (*1) is needed to be considered. And two curves from p_0 are equivalent as soon as they have the same terminal point. The projection $F : M^* \to M$ defined in (*3) is a homeomorphism and can be considered as the identity mapping in the sense that all such M^* are equivalent among themselves. This observation leads to

The weakest covering surface of a surface M. Let id: $M \to M$ be the identity map. Then (M, id) is the weakest complete smooth covering surface of M.

(7.5.3.4)

Exercises A

(1) Suppose (M_1^*, F_1) and (M_2^*, F_2) are two complete smooth covering surfaces of a surface M, and $F_{21} : M_2^* \to M_1^*$ is a continuous map such that $F_2 = F_1 \circ F_{21}$ holds. Show that (M_2^*, F_{21}) is a complete smooth covering surface of M_1^*.

Note. This result is still valid for complete covering spaces defined in Exercise A(4) of Sec. 7.5.2.

(2) Let (Y, F) be a complete covering space of X as defined in Exercise A(4) of Sec. 7.5.2. Suppose $x_0 \in X$ and $y_0 \in F^{-1}(x_0)$. In case $F_*\pi_1(Y, y_0)$ is always the identity element of $\pi_1(X, x_0)$, namely, $\pi_1(Y, y_0) = \{1\}$ and Y is simply connected, show that there is a neighborhood U of x_0 so that every closed curve from x_0 in U is homotopic to the constant curve at x_0 in X.

(3) Try to use the torus $S^1 \times S^1$ to justify (7.5.3.3). One might refer to Examples 1 and 2 for imitation.

Exercises B

A topological space X is called *semi-locally simply connected* if, for each $x \in X$, there always exists a neighborhood U of x so that the induced homomorphism $i_* : \pi_1(U, x) \to \pi_1(X, x)$ of the inclusion map $i : U \to X$ is trivial, i.e., $i_*([\gamma]) = 1$ for each $[\gamma] \in \pi_1(U, x)$. Refer to Exercise A(2).

(1) Suppose X is a connected, locally path-connected, and semi-locally simply connected topological space. Then, for each $x_0 \in X$ and every subgroup G of $\pi_1(X, x_0)$, there exist a path-connected space Y and a continuous map $F : Y \to X$ satisfying the properties:

1. (Y, F) is a complete covering space of X.
2. For each $y_0 \in F^{-1}(x_0)$, then

$$F_* \pi_1(Y, y_0) = G \quad \text{or a conjugate subgroup of } G.$$

Namely, the characteristic of (Y, y_0) is the class $[G]$ of conjugate subgroups determined by G.

7.5.4 *Covering transformations*

In this section, all covering surfaces are supposed to be complete and smooth.

Let (M^*, F) be a (complete smooth) covering surface of a surface M. A homeomorphism $\Phi : M^* \to M^*$ is called a *covering transformation* of M^* over M if $F \circ \Phi = F$ holds, that is to say, if Φ is fiber-reserving, namely $F(\Phi(p^*)) = F(p^*)$ for all $p^* \in M^*$. Using the composite operation of mappings, the set of all covering transformations of M^* over M, denoted as

$$G(M^*, F), \qquad (7.5.4.1)$$

forms a group. The covering surface (M^*, F) is called *normal* (or *regular* or *Galois*) if there is a unique (see (7.5.4.2) below) Φ in $G(M^*, F)$ which carries any point in $F^{-1}(p_0)$ into any other prescribed point in $F^{-1}(p_0)$, for each point $p_0 \in M$. In this case, the group $G(M^*, F)$ is said to be *transitive*. On a normal covering surface, points with the same projection cannot be distinguished from each other.

Section (1) Basic properties and examples

First, we have

Basic properties of covering transformations. Let (M^*, F) be a complete smooth covering surface of a surface M.

(1) Each Φ in $G(M^*, F)$ is uniquely determined by a pair of points p_1^* and p_2^* such that $\Phi(p_1^*) = p_2^*$.
(2) Each Φ in $G(M^*, F)$, other than the identity, has no fixed points.
(3) If V is an open set in M^* so that $F|_V : V \to F(V)$ is a homeomorphism, then for each Φ in $G(M^*, F)$, other than the identity, $V \cap \Phi(V) = \emptyset$ always holds. (7.5.4.2)

Owing to property (3), the group of covering transformations $G(M^*, F)$ is said to be *properly discontinuous* on M^*.

Proof. (1) This follows easily from (7.5.2.7) by observing that Φ is the unique lift map of $F : M^* \to M$ with given p_1^*, $p_2^* \in f^{-1}(p_0)$ for some point p_0 in M.

A direct proof using the curve lifting property is as follows. Pick any point p^* in M^*. Construct a curve γ_1^* from p_1^* to p^* in M^*. Set $\gamma_1 = F(\gamma_1^*)$. Since Φ carries p_1^* into p_2^*, Φ also carries γ_1^* into a curve γ_2^* from p_2^* which also lies over γ_1. By the uniqueness of the lift curve, the terminal point of γ_2^*, which is the image $\Phi(p^*)$, is thus uniquely defined by p^*.

(2) Suppose Φ is not the identity and $\Phi(p_0^*) = p_0^*$ for some $p_0^* \in M^*$. p_0^* has an open neighborhood V so that $F|_V : V \to F(V)$ is a homeomorphism. Choose another neighborhood $O \subseteq V$ of p_0^* so that $\Phi(O) \subseteq V$. For any $p^* \in O$, $F(\Phi(p^*)) = F(p^*) \in F(V)$. It is known that $\Phi(p^*)$ and $p^*$are in V, and it follows that $\Phi(p^*) = p^*$. This proves that the set $\{p^* \in M^* | \Phi(p^*) = p^*\}$ is open. That this set is closed is obvious. The connectedness of M^* says that the set is M^* itself and hence Φ is the identity, a contradiction.

Note that (1) is an easy consequence of (2).

(3) Suppose that there is a point $p^* \in V \cap \Phi(V)$, the $p^* = \Phi(q^*)$ for some $q^* \in V$. Now $F(\Phi(q^*)) = F(q^*) = F(p^*)$ shows that $p^* = q^*$ since F is known to be one-to-one on V. Therefore $p^* = \Phi(p^*)$ is a fixed point. By (2), Φ is the identity. □

The concept of covering transformation can also be similarly defined on a covering space (Y, f) of a space X.

In particular, we have

The conformality of covering transformation. Let (R^*, F) be a complete smooth covering surface of a Riemann surface R. Then R^* carries a natural complex structure to make it a Riemann surface (see (7.5.2.1)); and, in this case, each covering transformation $\Phi : R^* \to R^*$ is a conformal mapping. (7.5.4.3)

Recall that, in this case, the projection $F : R^* \rightarrow R$ is locally conformal, too.

Proof. For any $p \in R$, choose a chart (U, φ) at p so that $F^{-1}(U) = \cup \, U_\lambda^*$ where U_λ^*'s are disjoint components and each $F|U_\lambda^* : U_\lambda^* \rightarrow U$ is a homeomorphism. Suppose $p^* \in F^{-1}(p)$ and $p^* \in U_1^*$. According to (3) in (7.5.4.1), $U_1^* \cap \Phi(U_1^*) = \emptyset$ holds. Since $\Phi(U_1^*)$ is connected, it is contained in some U_λ^*, say U_2^*. Then $(U_1^*, \Phi \circ F)$ and $(\Phi(U_1^*), \Phi \circ F)$ are local charts at p^* and $\Phi(p^*)$, respectively. The composite map

$$(\Phi \circ F) \circ \Phi \circ (\Phi \circ F)^{-1} = \Phi \circ F \circ \Phi \circ F^{-1} \circ \Phi^{-1} = \Phi \circ \mathrm{id}_U \circ \Phi^{-1} = \mathrm{id}_U$$

shows that Φ is indeed conformal on U_1^*; in particular, at p^*. $\square$

We present some basic examples.

Example 1. Describe the covering transformations of $(\mathbf{R}, \exp)$ over S^1 (as a continuation of Example 1 of Sec. 7.5.1 and Example 1 of Sec. 7.5.3).

Explanation.
$(\mathbf{R}, \exp)$: it is known that $\pi_1(\mathbf{R}) = \{1\}$. Suppose $\Phi \in G(\mathbf{R}, \exp)$, a covering transformation of $\mathbf{R}$ over S^1. According to $\exp \circ \Phi = \exp$, it follows that $\exp 2\pi i \Phi(t) = \exp 2\pi i t$, $t \in \mathbf{R}$ and hence, $\Phi(t) - t$ is an integer for each $t \in \mathbf{R}$. Therefore the continuity of $\Phi(t) - t$ and the connectedness of $\mathbf{R}$ show that there is an integer n so that $\Phi(t) = t + n, t \in \mathbf{R}$ (refer to Remark 2 in Sec. 2.2). And vice versa, for each $n \in \mathbf{Z}$, $\Phi(t) = t + n$ defines a covering transformation of $(\mathbf{R}, \exp)$ over S^1. These observations set up a group isomorphism between $G(\mathbf{R}, \exp)$ and $\mathbf{Z}$. In conclusion

$$G(\mathbf{R}, exp) \cong Z \cong \pi_1(S^1).$$

(S^1, F_n), where $F_n(z) = z^n$: it is known that (S^1, F_n) is a complete smooth n-sheeted covering space of S^1, with its characteristic equal to $[nZ]$. Given $\Phi \in (S^1, F_n)$. Then $F_n \circ \Phi = F_n$ shows that $(\frac{\Phi(z)}{z})^n = 1$ for all $z \in S^1$. As in the last paragraph, the continuity of $\frac{\Phi(z)}{z}$ and the connectedness of S^1 shows that there is an n-th root w_n of unity so that $\Phi(z) = w_n z$ for all $z \in S^1$. Vice versa, any such an n-th root w_n determines a unique covering transformation $z \rightarrow w_n z$ of S^1 over S^1. This sets up the following group isomorphism:

$$G(S^1, F_n) \cong \{1, w, w^2, \ldots, w^{n-1}\}, \quad \text{the multiplication group generated by}$$

$$w = e^{2\pi i / n}$$

$$\cong \mathbf{Z}_n = \{0, 1, 2, \ldots, n-1\}, \quad \text{the group of residue classes}$$

$$\text{modulo } n\mathbf{Z}$$

$$\cong \mathbf{Z}/n\mathbf{Z}, \quad \text{the quotient group of } \mathbf{Z} \text{ over its subgroup } n\mathbf{Z}.$$

Example 2. Describe the covering transformations of $(\mathbf{C}, \exp)$ over $\mathbf{C} - \{0\}$ (as a continuation of Example 3 of Sec. 7.5.1 and Example 2 of Sec. 7.5.3).

Explanation. Refer to Fig. 7.26.
$(\mathbf{C}, \exp)$: Exactly the same argument as in the first paragraph of Example 1 shows that

$$G(\mathbf{C}, \exp) \cong \mathbf{Z} \cong \pi_1(\mathbf{C} - \{0\})$$

holds. Note that $\pi_1(\mathbf{C}) = \{0\}$ and $\mathbf{Z} \cong \mathbf{Z}/\{0\}$.
$(\mathbf{C} - \{0\}, F_n)$, where $F_n(z) = z^n$: $(\mathbf{C} - \{0\}, F_n)$ or $(R_{\sqrt[n]{z}} - \{0\}, \sqrt[n]{z})$ is a complete smooth n-sheeted covering surface of $\mathbf{C} - \{0\}$. Analogous to Example 1, we still obtain the following group isomorphism:

$$G(\mathbf{C} - \{0\}, F_n) \cong \mathbf{Z}/n\mathbf{Z}.$$

Recall that the characteristic of $(\mathbf{C} - \{0\}, F_n)$ is given by $[n\mathbf{Z}]$. Moreover $\mathbf{C} - \{0\}$ is conformal equivalent to the cylinder $\mathbf{R} \times S^1$ (see Example 2 in Sec. 7.1).

Example 3. Describe the covering transformations of $(\mathbf{R}^2, F \times F)$ of the torus $S^1 \times S^1$, where $F = \exp : \mathbf{R} \to S^1$ (as a continuation of Example 4 of Sec. 7.5.1 and Exercise A(3) of Sec. 7.5.3).

Explanation. Recall that $\pi_1(S^1 \times S^1) \cong \mathbf{Z} \oplus \mathbf{Z}$ (see Exercise B(5) of Sec. 7.4).
$(\mathbf{R}^2, F \times F)$: Essentially analogous to Example 1, one can show that

$$G(\mathbf{R}^2, F \times F) \cong \mathbf{Z} \oplus \mathbf{Z}/\{0\} \cong \mathbf{Z} \oplus \mathbf{Z},$$

where $\pi_1(\mathbf{R}^2) = \{0\}$. An alternate way is as follows. Recall that $(\mathbf{C}, \pi)$ is a complete smooth covering surface of $\mathbf{C}/\Gamma \simeq S^1 \times S^1$, where $\pi : \mathbf{C} \to \mathbf{C}/\Gamma$ is the natural projection (see Example 3 in Sec. 7.1). For a given $\Phi \in G(\mathbf{C}, \pi)$, the relation $\pi \circ \Phi = \pi$ shows that $[\Phi(z)] = [z]$. This means that $\Phi(z) - z$ is in the lattice set $\Gamma = \{m_1\omega_1 + m_2\omega_2 | m_1, m_2 \in \mathbf{Z}\}$. Since Γ is a discrete set, the continuity of $\Phi(z) - z$ shows that there exist unique integers m_1 and m_2 so that $\Phi(z) = m_1\omega_1 + m_2\omega_2 + z$ for all z in $\mathbf{C}$. And any mapping of this kind is obviously a covering transformation of $\mathbf{C}$ over $\mathbf{C}/\Gamma$. This justifies the claim.

$(S^1 \times S^1, F_m \times F_n)$, where $F_k(z) = z^k$ for positive integers $k = m, n$: A covering transformation $\Phi : S^1 \times S^1 \to S^1 \times S^1$ satisfies $(F_m \times F_n) \circ \Phi = F_m \times F_n$, which in turn implies that $(\frac{\Phi_1(z_1, z_2)}{z_1})^m = 1$ and $(\frac{\Phi_2(z_1, z_2)}{z_2})^n = 1$, where $\Phi = (\Phi_1, \Phi_2)$. Therefore $\Phi_1(z_1, z_2) = \omega_m' z_1$ and $\Phi_2(z_1, z_2) = \omega_n'' z_2$ for $z_1, z_2 \in S^1$, where ω_m' and ω_n'' are m-th and n-th roots of unity, respectively. This shows that

$$G(S^1 \times S^1, F_m \times F_n) \cong \mathbf{Z} \oplus \mathbf{Z}/m\mathbf{Z} \oplus n\mathbf{Z} \cong \mathbf{Z}_m \oplus \mathbf{Z}_n.$$

Note that $F_* \pi_1(S^1 \times S^1) = m\mathbf{Z} \oplus n\mathbf{Z}$ (see Exercise B of Sec. 7.4). An alternate way is to consider the parallelogram $[0, m\omega_1] \times [0, n\omega_2] = \{\alpha_1 \omega_1 + \alpha_2 \omega_2 | 0 \le \alpha_1 \le m$ and $0 \le \alpha_2 \le n\}$ on which the projection is the usual canonical mapping π (see Fig. 7.27). Try to figure out the details.

Example 4. Describe the covering transformations of (S^2, F) over the projective plane P^2, where $F : S^2 \to P^2$ is given by $F(x) = [x]$ (as a continuation of Example 5 in Sec. 7.5.1).

Explanation. Recall that $\pi_1(P^2) = \mathbf{Z}_2$ (see Exercise B(8) of Sec. 7.4). A given covering transformation $\Phi \in G(S^2, F)$ satisfies $F \circ \Phi = F$, namely, $[\Phi(x)] = [x]$ for all $x \in S^2$. The connectedness of S^2 and the continuity of Φ show that either $\Phi(x) = x$ or $\Phi(x) = -x$ for all $x \in S^1$. Hence, such a Φ is the identity or the antipodal projection. This shows that

$$G(S^2, F) \cong \mathbf{Z}_2 \cong \mathbf{Z}_2/\{0\},$$

where $\pi_1(S^2) = \{0\}$ because S^2 is simply connected.

Example 5. Describe the covering transformations of (R^*, f) over R, where $R^* = \mathbf{C} - \{\pm 1, \pm 2\}$, $R = \mathbf{C} - \{\pm 2\}$, and $f(z) = z^3 - 3z$: $R^* \to R$ (as a continuation of Example 3 in Sec. 2.7.2 and Example 2 in Sec. 3.5.6).

Explanation. It is known already that $(\mathbf{C}^*, f)$ is a branched three-sheeted covering surface of $\mathbf{C}^*$. Note that f has algebraic branch points at $z = \pm 1$ of order 1 (see (7.2.3)). The projection $f(\pm 1) = \mp 2$ have their fibers $f^{-1}(2) = \{-1, 2\}$ and $f^{-1}(-2) = \{1, -2\}$. Consequently, (R^*, f) is a complete smooth three-sheeted covering surface of R.

  Let $\Phi \in G(R^*, f)$ be a covering transformation. According to (7.5.4.2), $\Phi : R^* \to R^*$ is univalently conformal. Points $\pm 1, \pm 2$, and ∞ in $\mathbf{C}^*$ are isolated singularities of Φ, and are either removable singularities or a simple pole. The extended $\Phi : \mathbf{C}^* \to \mathbf{C}^*$ is thus a meromorphic function with only one simple pole, and hence, $\Phi(z) = \frac{az+b}{cz+d}$ is a bilinear transformation (see

Example 6 in Sec. 4.10.2 and Exercise A(6) there). Now $f \circ \Phi = f$ means that

$$\left(\frac{az+b}{cz+d}\right)^3 - 3\left(\frac{az+b}{cz+d}\right) = z^3 - 3z, \quad z \in R^*$$

$\Rightarrow$ (by considering $z \to \infty$ in both sides) $c = 0$

$\Rightarrow$ (without loss of its generality, assuming $d - 1$) $(az+b)^3 - 3(az+b) = z^3 - 3z$, $z \in R^*$

$\Rightarrow$ (by equating coefficients of equal power terms of z in both sides)

$$a^3 = 1, \quad 3a^2b = 0, \quad 3ab^2 - 3a = -3, \quad \text{and} \quad b^3 - 3b = 0.$$

$\Rightarrow$ $a = 1$ and $b = 0$ are the only possibilities.

Hence, $\Phi(z) = z$ and the identity map: $R^* \to R^*$ is the only covering transformation. And, of course, the covering surface (R^*, f) is *not* regular.

Example 6. Describe the covering transformations of (R^*, f) over R, where $R^* = \mathbf{C}^* - \{0, \pm 1, \pm\sqrt{2}\}$, $R = \mathbf{C} - \{0, 1\}$ and $f(z) = (z^2 - 1)^2$ (compare to Exercise A(12) of Sec. 2.7.2).

Explanation. Solve $f'(z) = 2z(z^2 - 1) = 0$ and we get $z = 0, \pm 1$. Since $f''(0) \neq 0$ and $f''(\pm 1) \neq 0$ hold, it follows that $f(z)$ has an algebraic branch point of order 1 at each of the points $0, \pm 1$, and an algebraic branch point of order 3 at ∞ (see (7.2.3) and (7.2.4)). $(\mathbf{C}^*, f)$ is then a branched four-sheeted covering surface of $\mathbf{C}^*$. Observe that $f(0) = 1$, $f(\pm 1) = 0$ and the fibers $f^{-1}(1) = \{0, \pm\sqrt{2}\}$, $f^{-1}(0) = \pm 1$. Therefore (R^*, f) is a complete smooth four-sheeted covering surface of R.

Take a covering transformation $\Phi : R^* \to R^*$. By reasoning exactly the same way as in Example 5, $\Phi(z) = \frac{az+b}{cz+d}$ is a bilinear transformation. The relation $f \circ \Phi = f$ means that

$$\left[\left(\frac{az+b}{cz+d}\right)^2 - 1\right]^2 = (z^2 - 1)^2, \quad z \in R^*$$

$\Rightarrow$ (setting $z \to \infty$) $\left(\frac{a}{c}\right)^2 - 1 = \infty$ means that $c = 0$.

$\Rightarrow$ (setting $d = 1$)$[(az+b)^2 - 1]^2 = (z^2 - 1)^2$, $z \in R^*$

$\Rightarrow$ (equating coefficients of equal power terms of z)

$$a^4 = 1, \quad a^3b = 0, \quad 4a^2b^2 + 2a^2(b^2 - 1) = 2, \quad \text{and} \quad (b^2 - 1)^2 = 1.$$

$\Rightarrow$ $a = \pm 1$ and $b = 0$ are the only possibilities.

This shows that the group $G(R^*, f) = \{id, \Phi\}$, where $\Phi(z) = -z$. (R^*, f) is *not* a regular covering surface. For instance, $f^{-1}(-1) = \{\pm\sqrt[4]{2}e^{\pi i/8}, \pm\sqrt[4]{2}e^{-\pi i/8}\}$ and there is no covering transformation carrying $\sqrt[4]{2}e^{\pi i/8}$ into $\sqrt[4]{2}e^{-\pi i/8}$.

Section (2). Covering transformation group and the characteristic of a covering surface

One main result is the following

Relation between the covering transformation group and the characteristic of a covering surface. Let (M^*, F) be a complete smooth covering surface of a surface M.

(1) The group $G(M^*, F)$ of covering transformations of (M^*, F) over M is isomorphic to $N/F_*\pi_1(M^*, p_0^*)$ where N is the normalizer of $F_*\pi_1(M^*, p_0^*)$ in $\pi_1(M, p_0)$, $p_0^* \in F^{-1}(p_0)$, and $\{F_*\pi_1(M^*, p_0^*)|p_0^* \in F^{-1}(p_0)\}$ is the characteristic of the covering surface (M^*, F) (see (7.5.3.1)).

(2) (M^*, F) is regular if and only if $F_*\pi_1(M^*, p_0^*)$ is a normal subgroup of $\pi_1(M, p_0)$. In this case, the normalizer $N = \pi_1(M, p_0)$.

(3) In case (M^*, F) is regular, then

$$G(M^*, F) \cong \pi_1(M, p_0)/F_*\pi_1(M^*, p_0^*)$$

and the *sheet number* of (M^*, F) as a covering surface of M (see (7.5.2.5)) is the order of the group $G(M^*, F)$ which, in turn, is equal to the *index* of $F_*\pi_1(M^*, p_0^*)$ in $\pi_1(M, p_0)$.   (7.5.4.4)

Proof. (1) For simplicity, set $D = F_*\pi_1(M^*, p_0^*)$ temporarily. Also recall that the *normalizer* $N = \{[\gamma] \in \pi_1(M, p_0)|[\gamma]D = D[\gamma]\}$. We try to set up a group isomorphism between the quotient group N/D and the group $G(M^*, F)$ of covering transformations.

Choose a closed curve γ from p_0 so that $[\gamma] \in N$. Let σ be a curve connecting p_0 to a point $p \in M$ and lift σ to the curve σ^* connecting p_0^* to a fixed point $p^* \in F^{-1}(p)$. Define a mapping $\Phi_\gamma : M^* \to M^*$ by

$$\Phi_\gamma(p^*) = \text{the terminal point of the lift curve } (\gamma\sigma)^*. \qquad (*_1)$$

Φ_γ is *well-defined*, namely, $\Phi_\gamma(p^*)$ is independent of the choice of the curve σ^*. Suppose another curve σ_1 connecting p_0 to p gives rise to another such a curve σ_1^*. Then $\sigma^*\sigma_1^{*-1}$ is a closed curve from p_0^* and hence its projection $\sigma\sigma_1^{-1}$ is a closed curve from p_0. This means that the homotopic class $[\sigma\sigma_1^{-1}] \in D$ holds. By the assumption that $[\gamma] \in N$ and the monodromy theorem (7.5.2.3), $(\gamma\sigma)^*$ and $(\gamma\sigma_1)^*$ thus have the same terminal point. Moreover Φ_γ *is a covering transformation.* Φ_γ is one-to-one because of the uniqueness of the lifting curve, and Φ_γ is onto because M is path-connected. The continuity of both Φ_γ and Φ_γ^{-1} follows from the complete covering property of (M^*, F).

By the very definition $(*_1)$, $\Psi_{\gamma\gamma'} = \Phi_\gamma \circ \Phi_{\gamma'}$ follows for any two closed curves γ, γ' from p_0 and satisfying $[\gamma]$, $[\gamma'] \in N$. Moreover Φ_γ is the identity map on M^* if and only if, in $(*_1)$, $(\gamma\sigma)^*$ is a closed curve from p^* for any $p \in M$. The latter means that $[\gamma] \in D$ because the terminal point of $(\gamma\sigma)^*$ is known to be independent of the choice of the curve σ^* and hence the constant curve from p_0^* is a choice. In conclusion, the mapping

$$[\gamma] \in N \to \Phi_\gamma \in G(M^*, F) \tag{$*_2$}$$

is a group homomorphism with its kernel identical with D. Therefore, the quotient group N/D is isomorphic to a subgroup of $G(M^*, F)$.

It remains to show that N/D is isomorphic *onto* the whole group $G(M^*, F)$. Let $\Phi \in G(M^*, F)$. Then both p_0^* and $\Phi(p_0^*)$ lies over the same point p_0. A curve γ^* from p_0^* to $\Phi(p_0^*)$ is projected into a closed curve γ from p_0. Construct the covering transformation Φ_γ as in $(*_1)$. Since $\Phi_\gamma(p_0^*) = \Phi(p_0^*)$, it follows that $\Phi_\gamma \circ \Phi^{-1}$ has the fixed point p_0^*. Hence $\Phi = \Phi_\gamma$ by (2) in (7.5.4.2). This proves the claim.

(2) Still set $D = F_*\pi_1(M^*, p_0^*)$, where $p_0^* \in F^{-1}(p_0)$ is a fixed point.

For each closed curve γ from p_0, let γ^* be any lift curve from a point $p^* \in F^{-1}(p_0)$. Then γ^* is a closed curve from p^*, namely, $[\gamma^*] \in \pi_1(M^*, p^*)$, if and only if $F_*[\gamma^*] = [F \circ \gamma^*] = [\gamma]$ is in $F_*\pi_1(M^*, p^*)$. In case D is normal, then all the conjugate subgroups $F_*\pi_1(M^*, p^*)$ of D are identical with D, and the characteristic is D itself. Under this circumstance, it follows that *either* $[\gamma] \in D$, in which case all the lift curves γ^* are closed curves for all $p^* \in F^{-1}(p_0)$ *or* $[\gamma]$ is not in D, in which case all such lift curves γ^* are not closed. In case D is *not* normal, there are p_1^*, $p_2^* \in F^{-1}(p_0)$ such that $F_*\pi_1(M^*, p_1^*) \neq F_*\pi_1(M^*, p_2^*)$ holds. Then there is a $[\gamma]$ in $F_*\pi_1(M^*, p_1^*)$ but not in $F_*\pi_1(M^*, p_2^*)$. Then the lift curve γ_1^* of γ from p_1^* is closed but the lift curve γ_2^* of γ from p_2^* is not closed. We conclude that

D is normal

$\Leftrightarrow$ for each closed curve γ from p_0, all the lift curves γ^* from any $p^* \in F^{-1}(p_0)$ are simultaneously closed or simultaneously not closed. $\quad(*_3)$

Suppose now that D is not normal. Pick any point $p_0 \in M$ and each closed curve γ from p_0. There is a lift curve γ_1^* from $p_1^* \in F^{-1}(p_0)$ which is closed, and there is another lift curve γ_2^* from $p_2^* \in F^{-1}(p_0)$ which is not closed. Suppose that there were a covering transformation Φ mapping p_1^* into p_2^*. The uniqueness of such a Φ (see (1) in (7.5.4.1)) shows that Φ must carry the closed curve γ_1^* into a nonclosed curve γ_2^*. Because Φ is a homeomorphism, this is a contradiction. Hence $G(M^*, F)$ is not regular.

On the other hand, if D is normal, we want to show that there is a covering transformation Φ carrying each point $p_1^* \in F^{-1}(p_0)$ into any pre-scribed point $p_2^* \in F^{-1}(p_0)$. In this case, the normalizer N of D is the whole group $\pi_1(M, p_0)$. Connect p_1^* to p_2^* by a curve γ^* whose projection is a closed curve γ from p_0. Construct the covering transformation Φ_γ as in $(*_1)$. Then, as shown before, Φ_γ is the required one.

(3) Let us go back to Exercise A(6) of Sec. 7.5.2. Fix any point $p_0 \in M$. Define a mapping $\psi : \pi_1(M, p_0) \to F^{-1}(p_0)$ by

$$\psi([\gamma]) = \gamma^*(1),$$

where γ^* is a lift curve of γ from a fixed point p_0^* in $F^{-1}(p_0)$. ψ is *well-defined.* For, if σ is homotopic to γ, then by the monodromy theorem, σ^* is homomorphic to γ^* with the same terminal points.

ψ is *onto.* Choose any point $p^* \in F^{-1}(p_0)$. Let γ^* be a curve connecting p_0^* to p^*, and $\gamma = F \circ \gamma^*$ be the projected curve on M. γ is a closed curve from p_0 and hence $[\gamma] \in \pi_1(M, p_0)$. This shows that $\psi([\gamma]) = \gamma^*(1) = p^*$.

Suppose $\psi([\gamma]) = \psi([\sigma])$ holds for some closed curves γ, σ from p_0. $\gamma^*(1) = \sigma^*(1)$ shows that $\gamma^* \sigma^{*^{-1}}$ is a closed curve from p_0* and hence $[\gamma^* \sigma^{*^{-1}}] \in \pi_1(M^*, p_0^*)$. Therefore the projection $F_*([\gamma^* \sigma^{*^{-1}}]) = [\gamma][\sigma]^{-1} \in F_* \pi_1(M^*, p_0^*)$, or equivalently, $[\gamma]$ is in the coset of $F_* \pi_1(M^*, p_0^*)$ that con-tains $[\sigma]$. In particular, $\psi([\gamma]) = p_0^*$ for all $[\gamma] \in F_* \pi_1(M^*, p_0^*)$. Reverse argument indicates that ψ maps each $[\gamma]$ into a fixed point $\sigma^*(1)$, where $[\gamma]$ ranges over all the elements in the coset of $F_* \pi_1(M^*, p_0^*)$ determined by $[\sigma]$.

In conclusion, we claim that ψ sets up a one to one and onto corre-spondence between $\pi_1(M, p_0)/F_* \pi_1(M^*, p_0^*)$ and the fiber $F^{-1}(p_0)$. And the result follows by recalling (7.5.2.5).

□

Exercises A

(1) Let $R^* = \mathbf{C} - \{0, \pm 1\}$, $R = \mathbf{C} - \{\pm 1\}$, and $f(\gamma) = \frac{1}{2}(z + \frac{1}{z})$. Show that (R^*, f) is a complete smooth covering surface of R. Compute the group

$G(R^*, f)$ of the covering transformations of (R^*, f) over R. Is (R^*, f) regular?

(2) (Continuation of Example 1 in Sec. 7.5.2) Suppose $R^* = \mathbf{C} - \{0, \pm i, \pm \frac{i}{\sqrt{2}}\}$, $R = \mathbf{C} - \{1, \frac{3}{4}\}$, and $f(\bar{z}) = 1 + z^2 + z^4$. Do problems as Exercise (1).

(3) (See Exercise B(6)(a) of Sec. 2.7.2) Set $R^* = \mathbf{C}^* - \{+1\}$, $R - \mathbf{C}^*$ $\{3, \frac{1}{3}\}$, and $f(z) - \frac{z^2 - z + 1}{z^2 + z + 1}$. Do problems as Exercise (1).

(4) (See Exercise B(6)(b) of Sec. 2.7.2) Find Riemann surfaces R^* and R so that (R^*, f) is a complete smooth covering of R, where $f(z) = \frac{z^4 - z + 1}{z^4 + z + 1}$. Is (R^*, f) regular?

(5) (See Sec. 2.7.4 and Exercise A(4) of Sec. 7.5.1) Let $R^* = \mathbf{C} - \{\frac{\pi}{2} + n\pi | n = 0, \pm 1, \ldots\}$, $R = \mathbf{C} - \{\pm 1\}$, and $f(z) = \sin z$. Show that the covering transformation group consists precisely of the following transformations:

1. $\Phi_n(z) = z + 2n\pi$, $\quad n = 0, \pm 1, \pm 2, \ldots$, and

2. $\Phi_n(z) = -z + (2n + 1)\pi$, $\quad n = 0, \pm 1, \pm 2, \ldots$.

Is (R^*, f) regular?

(6) Let R_z be the Riemann surface of $\log \frac{z-1}{z+1}$ (see Example 4 of Sec. 2.7.3 and Fig. 2.80) and $F : R_z \to \mathbf{C}^* - \{\pm 1\}$ be the orthogonal projection. Show that (R_z, F) is a complete smooth covering surface of $\mathbf{C}^* - \{\pm 1\}$. Compute the covering transformation group $G(R_z, F)$.

(7) (See Sec. 2.7.4 and Exercise A(5) of Sec. 7.5.1) Compute the covering transformation group $G(\mathbf{C}, \tan)$ of $(\mathbf{C}, \tan)$ over $\mathbf{C}^* - \{\pm 1\}$.

(8) Show that every complete smooth two-sheeted covering surface (M^*, F) of a surface is always regular.

Exercises B

(1) Suppose (Y, F) is a complete smooth covering space of a space X as defined in Exercise A(4) of Sec. 7.5.2. (Y, F) is called *regular* if $F_* \pi_1(Y, y)$ is a normal subgroup of $\pi_1(X, x)$, where $y \in F^{-1}(x)$. In this case, show that (7.5.4.3) is still valid. Namely, the group of covering transformations

$$G(Y, F) \cong \pi_1(X, x) / F_* \pi_1(Y, y)$$

and the number of sheets of (Y, F) is equal to the index of $F_* \pi_1(Y, y)$ in $\pi_1(X, x)$.

7.5.5   *The universal covering surface of a surface*

Suppose (M^*, F) is a complete smooth covering surface of a surface M.

The covering surface $(\tilde{M}, \tilde{F})$ of M that corresponds to the trivial subgroup $\{1\}$ of $\pi_1(M)$ is called the *universal covering surface* of M. Its existence is acquired by using (7.5.3.3) and, at the same time, its fundamental group $\pi_1(\tilde{M}) = \{1\}$. Hence the universal covering surface is simply connected. According to (7.5.3.2), it is unique up to homeomorphism.

For reference, we list

Equivalent conditions for a covering surface to be universal. Let $(\tilde{M}, \tilde{F})$ be a complete smooth covering surface of a surface M. then

- (1) $(\tilde{M}, \tilde{F})$ is the universal covering surface of M (unique up to homeomorphism).
- ⇔ (2) $\tilde{M}$ is simply connected, namely, $\pi_1(\tilde{M}) = \{1\}$.
- ⇔ (3) A curve $\tilde{\gamma}$ on $\tilde{M}$ is closed if and only if its projection $\tilde{F} \circ \tilde{\gamma}$ is homotopic to a constant path on M.
- ⇔ (4) (*The strongest* or *the universal covering property*) For any complete smooth covering surface (M^*, F) of M, there exists a continuous map $f : \tilde{M} \to M^*$ such that $(\tilde{M}, f)$ is a complete smooth covering surface of M^* and $\tilde{F} = F \circ f$ holds. In short, $\tilde{F}$ can be lifted to a local homeomorphism $f : \tilde{M} \to M^*$ with respect to F.

Moreover, for each $p_0 \in M$,

1. the group $G(\tilde{M}, \tilde{F})$ of covering transformations of $(\tilde{M}, \tilde{F})$ over M is isomorphic to the fundamental group $\pi_1(M, p_0)$ of the surface M based at p_0, and
2. the number of the sheets of $(\tilde{M}, \tilde{F})$ is equal to the order of $\pi_1(M, p_0)$.   (7.5.5.1)

As a matter of fact, condition 4 can be strengthened to the *fact*: For each point $p_0 \in M$, any point $\tilde{p} \in \tilde{F}^{-1}(p_0)$ and any point $p^* \in F^{-1}(p_0)$, there is a *unique* continuous mapping $f : \tilde{M} \to M^*$ such that $(\tilde{M}, f)$ is a complete smooth covering surface of M^*, $\tilde{F} = F \circ f$, and $f(\tilde{p}) = p^*$. The property of being $\tilde{F} = F \circ f$ is also termed as the *fiber-preserving property* of f.

Proof. (1) ⇔ (2) ⇔ (3) are obvious. Note that, in (3), if $\tilde{\gamma}$ is closed on $\tilde{M}$ and is characterized by the fact that the projected curve $\tilde{F}_*(\tilde{\gamma}) = [\tilde{F} \circ \tilde{\gamma}] = 1$, then $[\tilde{\gamma}] = 1$ should hold because $\tilde{F}_*$ is a one-to-one homomorphism (see

(7.5.2.4)). Therefore $\pi_1(\tilde{M}) = \{1\}$ holds. On the other hand, $\pi_1(\tilde{M}) = \{1\}$ indicates that every closed curve $\tilde{\gamma}$ on M is homotopic to the constant and them, so is its projected curve $\tilde{F} \circ \tilde{\gamma}$, and vice versa, by the monodromy theorem.

(1) $\Rightarrow$ (4): The existence and uniqueness of f follows easily from (7.5.2.7) (see also Exercise B(1) there). It remains to show that $f : \tilde{M} \to M^*$ is local homeomorphism so that $(\tilde{M}, f)$ is a complete smooth covering surface of M^*. But this is Exercise A(1) of Sec. 5.7.3. The detail is left as Exercise A(1).

(4) $\Rightarrow$ (1): According to (7.5.3.3), construct a complete smooth covering surface (M^*, F) of M, corresponding to the unite element $\{1\}$ of $\pi_1(M)$. By assumption $\tilde{F}$ can be uniquely lifted to a local homeomorphism $g : \tilde{M} \to M^*$ with respect to F so that $g(\tilde{p}) = p^*$, where $\tilde{p} \in \tilde{F}^{-1}(p_0)$ and $p^* \in F^{-1}(p_0)$ for a point $p_0 \in M$. But it is known that this particular (M^*, F) enjoys conditions (1)–(3); specially, (1) $\Rightarrow$ (4) is valid for this space. Therefore F can also be uniquely lifted to a local homeomorphism $f : M^* \to \tilde{M}$ with respect to $\tilde{F}$ so that $f(p^*) = \tilde{p}$ holds. The composite maps

$$f \circ g : \tilde{M} \to \tilde{M} \quad \text{and} \quad g \circ f : M^* \to M^*$$

are local homeomorphisms satisfying $f \circ g(\tilde{p}) = \tilde{p}$ and $g \circ f(p^*) = p^*$, respectively. Observe that $\tilde{F} \circ (f \circ g) = (\tilde{F} \circ f) \circ g = F \circ g = \tilde{F}$ which means that $f \circ g$ is a lift map of $\tilde{F}$. By the uniqueness of the lifting map, it follows that $f \circ g = id_{\tilde{M}}$, the identity map on $\tilde{M}$. Similarly $g \circ f = id_{M^*}$. In conclusion, $f : M^* \to \tilde{M}$ is a fibre-preserving homeomorphism, and hence (M^*, F) and $(\tilde{M}, \tilde{F})$ are equivalent and are considered identical. This proves the universality of $(\tilde{M}, \tilde{F})$. $\qquad\square$

Remark (An alternate way to compute the fundamental group). Suppose that the universal covering surface of a surface, along with its group of covering transformations, is known. Then (7.5.5.1) says that the fundamental group of the base surface is, up to isomorphism, identical with the group of covering transformations.

See also (7.5.5.4) below. $\qquad\square$

From Examples 1–3 in Sec. 7.5.4, we acknowledged the following

Basic examples.

(1) $(\mathbf{R}, \exp)$ is the universal covering space of S^1 with $\pi_1(S^1) \cong \mathbf{Z}$.

(2) $(\mathbf{C}, \exp)$ is the universal covering surface of the punctured plane $\mathbf{C} - \{0\}$ or the cylinder $S^1 \times \mathbf{R}$, with $\pi_1(\mathbf{C} - \{0\}) \cong \mathbf{Z}$.

(3) $(\mathbf{R}^2, F \times F)$, where $F = \exp$; or $(\mathbf{C}, \pi)$, where π is the canonical projection, is the universal covering surface of the torus $S^1 \times S^1$, with $\pi_1(S^1 \times S^1) = \mathbf{Z} \oplus \mathbf{Z}$.

(4) {the left half-plane $\operatorname{Re} z < 0, \exp$} is the universal covering surface of the punctured disk $0 < |z| < 1$, with its fundamental group given by $\mathbf{Z}$.

(5) {the strip $\log r_1 < |\operatorname{Re} z| < \log r_2, \exp$} is the universal covering surface of the annulus $r_1 < |z| < r_2$, with its fundamental group given by $\mathbf{Z}$.   (7.5.5.2)

For (4) and (5), refer to Examples 2 and 3 in Sec. 7.5.1.

As an easy application of (7.5.5.1), we have the

Characteristic properties of a simply connected surface. Let M be the surface. Then

   (1) M is simply connected.

$\Leftrightarrow$ (2) The universal covering surface of M is M itself, up to homeomorphism.

$\Leftrightarrow$ (3) The only complete smooth covering surface of M is M itself.

$\leftrightarrow$ (4) Every complete smooth covering surface of M has only one sheet.   (7.5.5.3)

Proof is left as Exercise A(2).

As a second application, we have

The covering transformation group of a complete smooth covering surface (M^*, F) *of a surface* $\mathbf{M}$. Let M, $(\tilde{M}, \tilde{F})$ and (M^*, F) be as in condition (4) in (7.5.5.1) and $f : \tilde{M} \to M^*$ be a local homeomorphism which lifts $\tilde{F}$ to M^* so that $(\tilde{M}, f)$ is a complete smooth (or the universal) covering surface of M^*. Then the group $G(\tilde{M}, f)$ of covering transformations of $(\tilde{M}, f)$ over M^* satisfies:

1. $G(\tilde{M}, f)$ is a subgroup of $G(\tilde{M}, \tilde{F})$.

2. $G(\tilde{M}, f) \cong \pi_1(M^*)$ and the number of sheets of (M^*, F) is equal to the order of $\pi_1(M)/F_*\pi_1(M^*) = G(\tilde{M}, \tilde{F})/G(\tilde{M}, f)$.

3. Two points $\tilde{p}_1, \tilde{p}_2 \in \tilde{M}$ satisfy $f(\tilde{p}_1) = f(\tilde{p}_2)$ in M^* if and only if there is a $\Phi \in G(\tilde{M}, f)$ so that $\Phi(\tilde{p}_1) = \tilde{p}_2$ holds. In short, $(\tilde{M}, f)$ is regular.   (7.5.5.4)

Proof. Suppose $\Phi \in G(\tilde{M}, f)$. Then $f \circ \Phi = f$ holds. Since $\tilde{F} = F \circ f$, it follows that $F \circ f = F \circ f \circ \Phi = \tilde{F} \circ \Phi = \tilde{F}$. Hence $\Phi \in G(\tilde{M}, \tilde{F})$ and $G(\tilde{M}, f)$ is thus a subgroup of $G(\tilde{M}, \tilde{F})$.

Since $\tilde{M}$ is simply connected, by (7.5.5.3), $(\tilde{M}, f)$ is the universal covering surface of M^* and then, by (7.5.5.1), $G(\tilde{M}, f) \cong \pi_1(M^*)$ holds. $(\tilde{M}, f)$ is the covering surface corresponding to the unit element of $\pi_1(M^*)$, and then, according to (2) in (7.5.4.3) it is regular. Therefore, for any two points $\tilde{p}_1, \tilde{p}_2 \in \tilde{M}$ with $f(\tilde{p}_1) = f(\tilde{p}_2)$ in M^*, there is a (unique) $\Phi \in G(\tilde{M}, f)$ so that $\Phi(\tilde{p}_1) = \tilde{p}_2$ holds. $\square$

Using (7.5.5.4), we can recapture Example 2 of Sec. 7.5.4 and we reset it as

Example 1. Let R be a Riemann surface and $f : R \to \mathbf{C} - \{0\}$ be an unbranched analytic mapping so that (R, f) is a complete smooth covering surface of $\mathbf{C} - \{0\}$. Then,

(1) if R is infinitely-sheeted, (R, f) is equivalent to $(\mathbf{C}, \exp)$ or $(R_{\log z}, \log z)$;
(2) if R is n-sheeted, (R, f) is equivalent to $(\mathbf{C} - \{0\}, z^n)$ or $(R_{\sqrt[n]{z}} - \{0\}, \sqrt[n]{z})$.

In other words, $(\mathbf{C}, \exp)$ is the universal covering surface of $\mathbf{C} - \{0\}$ and all other covering surfaces are of the forms $(\mathbf{C} - \{0\}, z^n)$ for integers $n \geq 0$.

Proof. The group $G(\mathbf{C}, \exp) = \{z + 2k\pi i | k \in \mathbf{Z}\}$ is isomorphic to $\mathbf{Z}$. While the group $G(\mathbf{C} - \{0\}, z^n) = \{e^{2k\pi i/n} | k = 0, 1, 2, \ldots, n - 1\}$ is isomorphic to $Z/n\mathbf{Z} = \mathbf{Z}_n = \{0, 1, 2, \ldots, n - 1\}$ (see Example 2 in Sec. 7.5.4 and (7.5.4.3)). According to (7.5.5.4), there is a lift map $g : \mathbf{C} \to R$ of $\exp$ so that $\exp = f \circ g$. *Note* that $G(\mathbf{C}, g)$ is a subgroup of $G(\mathbf{C}, \exp)$ but $G(\mathbf{C} - \{0\}, z^n)$ is *not* so.

(1) In case $G(\mathbf{C}, g) = \{1\}$, then g is one-to-one and hence is a conformal mapping between $\mathbf{C}$ and R. This means that (R, f) is equivalent to $(\mathbf{C}, \exp)$.
(2) In case $G(\mathbf{C}, g)$ is a subgroup of $G(\mathbf{C}, \exp)$, other than the unity, there is a positive integer $n \geq 1$ so that $G(\mathbf{C}, g) = \{z + 2kn\pi i | k \in \mathbf{Z}\} \cong$

$n\mathbf{Z}$. Since the quotient group $G(\mathbf{C},\exp)/G(\mathbf{C},g)$ is of order n, R is an n-sheeted covering surface of $\mathbf{C}-\{0\}$. Define a map $f_n:\mathbf{C}\to\mathbf{C}-\{0\}$ by

$$f_n(z) = \exp\frac{z}{n}.$$

Observe that $f_n(z_1) = f_n(z_2)$ if and only if there is a Φ in $G(\mathbf{C},g)$ so that $\Phi(z_1) = z_2$ holds.

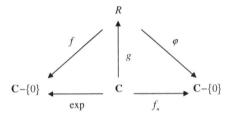

Recall that $g(z_1) = g(z_2)$ if and only if $\Phi(z_1) = z_2$ for some $\Phi \in G(\mathbf{C},g)$. Therefore we can define a map $\varphi: R \to \mathbf{C} - \{0\}$ by $\varphi(g(z)) = f_n(z)$. This φ is one-to-one and onto, and $f_n = \varphi \circ g$ holds on $\mathbf{C}$. Since both f_n and g are locally analytic, φ is then analytic and hence is conformal. Note that $\exp \circ f_n^{-1}(z) = z^n$. Consequently, (R,f) and $(\mathbf{C} - \{0\}, z^n)$ are equivalent. The sheet number of R or $\mathbf{C} - \{0\}$ is equal to the order of $G(\mathbf{C},\exp)/G(\mathbf{C},g) \cong \mathbf{Z}/n\mathbf{Z}$, which is n. □

As a continuation of (7.5.2.6), we restate it as

Example 2. Let $f:R$ (Rieman surface) $\to \mathbf{C}$ be a proper nonconstant analytic mapping so that $(R - f^{-1}(0), f)$ is a complete smooth covering surface of the deleted open plane $\mathbf{C} - \{0\}$. Show that there is a positive integer n so that covering surfaces (R, f) and $(\mathbf{C}, z^n)$ of $\mathbf{C}$ are equivalent.

Proof. Note that $f:R - f^{-1}(0) \to \mathbf{C} - \{0\}$ is proper, too.

By (2) in Example 1, there is a positive integer n and a conformal mapping $\varphi: R - f^{-1}(0) \to \mathbf{C} - \{0\}$ which sets up the equivalence of $(R - f^{-1}(0), f)$ and $(\mathbf{C} - \{0\}, z^n)$ over $\mathbf{C} - \{0\}$.

$$
\begin{array}{ccc}
R-f^{-1}(0) & \xrightarrow{\ \varphi\ } & \mathbf{C}-\{0\} \\
 & f\searrow \quad \swarrow z^n & \\
 & \mathbf{C}-\{0\} &
\end{array}
$$

Suppose $f^{-1}(0) = \{p_1,\ldots,p_k\}$ for $k \geq 2$. There exist an open disk $D_r = \{|z| < r\}$ for some $r > 0$, and disjoint open neighborhoods O_j of p_j, $1 \leq j \leq k$, so that $f^{-1}(D_r) \subseteq \bigcup_{j=1}^{k} O_j$. This is possible because $(R - f^{-1}(0), f)$ is complete and smooth. Set $\tilde{D}_r = D_r - \{0\} = \{0 < |z| < r\}$. Since $f^{-1}(\tilde{D}_r)$ is homeomorphic to the inverse image of $\tilde{D}_r$ under φ which is a connected set, it is connected, too. Hence $f^{-1}(D_r)$ is also connected because each point p_j is a limit point of $f^{-1}(\tilde{D}_r)$. This contradicts to $f^{-1}(D_r) \cap O_j \neq \phi$, $1 \leq j \leq k$, and the pairwise disjointedness of $O'_j s$. Consequently $f^{-1}(0) = \{p\}$ is a singleten.

The point p is a removable similarity of φ and we are able to define $\varphi(p) = 0$ so that φ turns out to conformal on R.   □

Exercises A

(1) Prove the implication $(1) \Rightarrow (4)$ in detail in $(7.5.5.1)$.
(2) Prove $(7.5.5.3)$ in detail.
(3) Use $(7.5.5.4)$ and imitate Example 1 to redo Example 3 of Sec. 7.5.4.
(4) Let $(\tilde{M}, \tilde{F})$ be the universal covering surface of a surface M. Suppose G is a subgroup of the covering transformation group $G(\tilde{M}, \tilde{F})$. Two points $\tilde{p}_1$ and $\tilde{p}_2$ are said to be equivalent if there is a $\Phi \in G$ so that $\Phi(\tilde{p}_1) = \tilde{p}_2$. Let $\tilde{M}/G$ be the associated quotient space and $F : \tilde{M}/G \to M$ be the induced map, namely, $F([\tilde{p}]) = \tilde{F}(\tilde{p})$ where $[\tilde{p}]$ is the equivalence class containing $\tilde{p}$. Show that

1. $(\tilde{M}/G, F)$ is a complete smooth covering surface of M;
2. $(\tilde{M}/G, F)$ is regular if and only if G is a normal subgroup of $G(\tilde{M}, \tilde{F})$, and
3. in this case, $G(\tilde{M}/G, F) \cong G(\tilde{M}, \tilde{F})/G$.

(5) Adopt notation in Example 3 of Sec. 7.1. Let Γ and Γ' be two lattices. Suppose $F : \mathbf{C}/\Gamma \to \mathbf{C}/\Gamma'$ is a nonconstant analytic mapping from $\mathbf{C}/\Gamma$ onto $\mathbf{C}/\Gamma'$. Let $\pi : \mathbf{C} \to \mathbf{C}/\Gamma$ and $\pi' : \mathbf{C} \to \mathbf{C}/\Gamma'$ be the natural projections.

$$
\begin{array}{ccc}
\mathbf{C} & \xrightarrow{\;f\;} & \mathbf{C} \\
\pi \downarrow & & \downarrow \pi' \\
\mathbf{C}/\Gamma & \xrightarrow{\;F\;} & \mathbf{C}/\Gamma'
\end{array}
$$

(a) Show that there is a unique constant $\alpha \in \mathbf{C} - \{0\}$ such that the map $f(z) = \alpha z : \mathbf{C} \to \mathbf{C}$ makes the diagram commutative. In this case, $\alpha \Gamma \subseteq \Gamma'$ holds.
(b) Show that $(\mathbf{C}/\Gamma, F)$ is a complete smooth covering surface of $\mathbf{C}/\Gamma'$, and $G(\mathbf{C}/\Gamma, F) \cong \Gamma'/\alpha\Gamma$.

Exercises B

(1) Let (Y, F) be a complete smooth covering space of a space X as defined in Exercise A(4) of Sec. 7.5.2. Try to extend results in (7.5.5.1), (7.5.5.3) and (7.5.5.4), to (Y, F).

7.6 The Uniformization Theorem of Riemann Surfaces

We end this book in highlighting the proof of this famous theorem, due to Koebe in 1901.

The uniformization theorem of Riemann surface. *A Riemann surface*

1. *with an elliptic universal covering surface is conformally equivalent to the Riemann sphere S;*
2. *with a parabolic universal covering surface is conformally equivalent to the finite complex plane* **C**, *the punctured plane (infinite cylinder)* **C** − {0}, *or a torus, and*
3. *with a hyperbolic universal covering surface is conformally equivalent to a quotient space D/Γ, where D is the open unit disk $|z| < 1$, and Γ is a uniquely defined discontinuous group of fixed-point free non-Euclidean motions of the hyperbolic plane D.* (7.6.1)

It does for Riemann surface exactly what the Riemann mapping theorem (6.1.1) does for planar domains. But it must be admitted, as the Riemann mapping theorem has shown, that the reduction to these special cases does not always simplify matters.

The simplified and manageable classical proof of this theorem relies on following aspects:

1. Perron's method for solving Dirichlet problem, ending up the main results (7.3.3.1) and (7.3.3.3).
2. Repeated applications of the maximum principle.
3. The concept of the universal covering surface (see Sec. 7.5.5), in order to reduce the study of arbitrary Riemann surfaces to that of the simply connected ones.
4. The existence of a single-valued branch out of a multiple-valued function defined on a simply connected surface (see (7.6.3) below).
5. A special argument due to Heins [42]. (7.6.2)

Most recently in 2005, Chen *et al.* [17] gave a purely differential geometric one-page proof of this theorem by using Ricci flow.

For the proof of the Riemann mapping theorem, one needs to make sure the *existence* of a univalent analytic function on the simply connected domain whose image is the open unit disk, even though such a function cannot be explicitly written out in most cases. Exactly the same situation happens in our coming proof of the uniformization theorem. The most concrete functions we can handle on arbitrarily simply connected Riemann surface are Green's functions for hyperbolic surfaces (see (7.3.3.1)) and the rather special harmonic functions with single singularity for nonhyperbolic surfaces (see (7.3.3.3)). A *global* harmonic function such as these can only have *local* harmonic conjugates. And we have to formulate a *global* analytic mapping out of the so many resulted *local* analytic mappings. The possibility to do so lies mainly on the fact that the Riemann surfaces being considered are simply connected. This is what we are going to do in (7.6.3).

Before we formally do so, let us go back to review the situation we encountered in Sec. 2.7.1. $\arg z$ is a multiple-valued function on $\mathbf{C} - \{0\}$. To each point $\zeta \in \mathbf{C} - \{0\}$, let O_ζ be the open disk $|z - \zeta| < |\zeta|$. On each O_ζ, there are countably infinitely many single-valued continuous branches φ_ζ of $\arg z$ (see (2.7.1.1)). Any two of them differ by an integer multiple of 2π, and any one of them is uniquely determined by its single value assumed at a point in O_ζ. These O_ζ and the associated $\varphi_\zeta's$ enjoy the following properties:

1. On a nonempty component $O_{\zeta_1 \zeta_2}$ of $O_{\zeta_1} \cap O_{\zeta_2}$, either $\varphi_{\zeta_1}(z) = \varphi_{\zeta_2}(z)$ for all $z \in O_{\zeta_1 \zeta_2}$ or $\varphi_{\zeta_1}(z) \neq \varphi_{\zeta_2}(z)$ or all $z \in O_{\zeta_1 \zeta_2}$.
2. If $O_{\zeta_1 \zeta_2}$ is a nonempty component of $O_{\zeta_1} \cap O_{\zeta_2}$ and φ_{ζ_1} is a given branch associated with O_{ζ_1}, then there always exists a φ_{ζ_2} associated with O_{ζ_2} so that $\varphi_{\zeta_1}(z) = \varphi_{\zeta_2}(z)$ for all $z \in O_{\zeta_1 \zeta_2}$.

Then we are able to continuously extend, starting from any φ_ζ in each O_ζ, along a Jordan curve γ to obtain a uniquely determined single-valued continuous branch of $\arg z$ along γ (see (2.7.1.4)). Even though we are unable to get a single-valued branch on the whole $\mathbf{C} - \{0\}$, but this process eventually leads us to get one on any *simply connected subdomain* of $\mathbf{C} - \{0\}$, and any two of them differ by an integer multiple of 2π and a specified one can be obtained by preassigning a value of $\arg z$ at a particular point (see (2.7.1.6)).

Think about how we can obtain single-valued analytic branches of $\sqrt[n]{z}$ and $\log z$ on simply connected subdomains of $\mathbf{C} - \{0\}$.

This process repeats when we study the analytic continuation of analytic function elements (germs) in Sec. 5.2.1 which results in the monodromy theorem in Sec. 5.2.2.

All these can be condensed into the following fairly general theorem.

The existence of single-valued branches of a multiple-valued function on a simply connected surface. Suppose M is a simply connected surface. Let $\{O_\lambda\}_{\lambda \in \Lambda}$ be a covering of M by open connected sets. For each O_λ, let there be given a family Φ_λ of functions defined on O_λ with the following conditions:

1. If $\varphi_{\lambda_1} \in \Phi_{\lambda_1}, \varphi_{\lambda_2} \in \Phi_{\lambda_2}$ and if $O_{\lambda_1 \lambda_2}$ is a nonempty component of $O_{\lambda_1} \cap O_{\lambda_2}$, then either $\varphi_{\lambda_1}(p) = \varphi_{\lambda_2}(p)$ for all $p \in O_{\lambda_1 \lambda_2}$ or $\varphi_{\lambda_1}(p) \neq \varphi_{\lambda_2}(p)$ for all $p \in O_{\lambda_1 \lambda_2}$.
2. If $\varphi_{\lambda_1} \in \Phi_{\lambda_1}$ and $O_{\lambda_1 \lambda_2}$ is a nonempty component of $O_{\lambda_1} \cap O_{\lambda_2}$, then there exists a $\varphi_{\lambda_2} \in \Phi_{\lambda_2}$ such that $\varphi_{\lambda_1} = \varphi_{\lambda_2}$ on $O_{\lambda_1 \lambda_2}$.

Then there exists a (single-valued) function φ on M whose restriction to each O_λ belongs to Φ_λ. And such a φ is uniquely determined by its restriction to a single O_λ.   (7.6.3).

Conditions 1 and 2 grant us to adopt the method of analytic continuation we experienced in Sec. 5.2 to prove this theorem. Refer to Ref. [32], pp. 180–181, for a formal discussion. Yet the significant implications of these two conditions lie on the fact that they can be used to *construct a complete smooth covering surface* $\tilde{M}$ of M, where none was originally present. Then *the simply connected assumption* of M assures that $\tilde{M}$, as the universal covering surface of M, should be conformally equivalent to M itself; in particular, it is *one-sheeted* (see (7.5.5.3)). This in turn guarantees the existence of a global single-valued function out of these local $\varphi_\lambda's$.

Proof. For each $\lambda \in \Lambda$ (the index set), consider all possible pairs (p, φ) where $\varphi \in \Phi_\lambda$ and $p \in O_\lambda$. Two pairs (p, φ) and (q, ψ) are said to be equivalent if $p = q$ and $\varphi(p) = \psi(q)$. This is indeed an equivalent relation, and the equivalence class of (p, φ) is denoted as $[p, \varphi]$. Let

$$M^* = \{[p, \varphi] | p \in O_\lambda \quad \text{and} \quad \varphi \in \Phi_\lambda \text{ for all } \lambda \in \Lambda\}, \quad \text{and}$$

$$F : M^* \to M \text{ be defined by } F([p, \phi]) = p. \qquad (*_1)$$

Note that F is not necessarily one-to-one.

Hausdorff topological structure of M^*. For a given $\lambda \in \Lambda$ and a fixed function $\varphi \in \Phi_\lambda$, set

$$O_\lambda^*(\varphi) = \{[p, \varphi] | p \in O_\lambda\}. \qquad (*_2)$$

This set $O_\lambda^*(\varphi)$ might be imagined as the set $\varphi(O_\lambda)$. Moreover, by virtue of condition 1, the sets $O_\lambda^*(\varphi)$ that correspond to different φ in Φ_λ are either identical or disjoint. The map F sets up a one-to-one correspondence between $O_\lambda^*(\varphi)$ and O_λ, and the $O_\lambda^*(\varphi)'s$ are then considered to be open sets in M^*. This induces a topology on M^*. To see that M^* is Hausdorff, consider two points (p, φ) and (q, ψ) in it. Then

$$[p, \varphi] \neq \lfloor q, \psi \rfloor \Leftrightarrow \text{(a) } p \neq q, \quad \text{or}$$

$$\text{(b) } p = q \text{ but } \varphi(p) \neq \psi(q).$$

In case $p \in O_{\lambda_1}$ and $q \in O_{\lambda_2}$, we may arrange them so that $O_{\lambda_1} \cap O_{\lambda_2} = \phi$; and then $O_{\lambda_1}^*(\varphi) \cap O_{\lambda_2}^*(\psi) = \phi$ holds. If $p \neq q$ and $p, q \in O_\lambda$, we may choose two disjoint open connected sets U_1 and U_2, contained in O_λ and $p \in U_1, q \in U_2$; and hence $U_1^*(\varphi) \cap U_2^*(\psi) = \phi$. If (b) happens, we already observed that $O_{\lambda_1}^*(\varphi) \cap O_{\lambda_2}^*(\psi) = \phi$.

M^* is not necessarily connected. Let $\tilde{M}$ be a *component* of M^*. Then $\tilde{M}$ is a connected Hausdorff space. The map F would induce a surface topology on $\tilde{M}$ (refer to the argument in the proof of (7.5.2.1)). Hence, $\tilde{M}$ becomes a *smooth covering surface* of M because F is locally one-to-one and hence is a local homeomorphism.

$(\tilde{M}, F)$ is a *complete smooth covering surface* of M, and hence, is the *universal covering surface* of M since the latter is simply connected (see (7.5.5.3)). As a matter of fact, each O_λ is a fundamental domain. To see this, set $p^* = [p, \varphi] \in F^{-1}(O_\lambda)$. Then $\varphi \in \Phi_{\lambda'}$ and $p \in O_\lambda \cap O_{\lambda'}$, for some $\lambda' \in \Lambda$. According to Condition 2, there exists a $\psi \in \Phi_\lambda$ so that $\varphi(p) = \psi(p)$. This means that $p^* \in O_\lambda^*(\psi)$. Hence $F^{-1}(O_\lambda) \subseteq \bigcup_{\varphi \in \Phi_\lambda} O_\lambda^*(\varphi)$ holds. On the other hand, each $O_\lambda^*(\varphi)$ is contained in $F^{-1}(O_\lambda)$ because F maps each such a set onto O_λ. This proves that

$$F^{-1}(O_\lambda) = \bigcup_{\varphi \in \Phi_\lambda} O_\lambda^*(\varphi). \tag{$*_3$}$$

Our construction shows that each $O_\lambda^*(\varphi)$ is open and connected, and, by Condition 1, each $O_\lambda^*(\varphi)$ is indeed a component of $F^{-1}(O_\lambda)$ and is homeomorphic to O_λ. These $O_\lambda^*(\varphi)$ that are contained in $\tilde{M}$ constitute the components of $F^{-1}(O_\lambda) \cap \tilde{M}$.

That $(\tilde{M}, F)$ is the universal covering surface of M shows that $F^{-1}(O_\lambda) \cap \tilde{M} = O_\lambda^*(\varphi_\lambda)$ for a single $\varphi_\lambda \in \Phi_\lambda$ because $\tilde{M}$ is one-sheeted. In case $O_\lambda \cap O_{\lambda'} = O \neq \phi$ and $\varphi_\lambda(p) \neq \varphi_{\lambda'}(p)$ for all $p \in O$, then $F^{-1}(O_\lambda) \cap \tilde{M} = O_\lambda^*(\varphi_\lambda) \cup O_{\lambda'}^*(\varphi_{\lambda'})$ and $O_\lambda^*(\varphi_\lambda) \cap O_{\lambda'}^*(\varphi') = \phi$ holds which contradicts to the one-sheetedness of $\tilde{M}$. Hence $\varphi_\lambda = \varphi_{\lambda'}$ on O. This fact enables us to define a global single-valued function φ on M by

$$\varphi(p) = \varphi_\lambda(p) \quad \text{if} \quad p \in O_\lambda. \tag{$*_4$}$$

The choice of $\tilde{M}$ as the component of M^* that contains $O^*_{\lambda_0}(\varphi_{\lambda_0})$ is then to make φ coincide with a given φ_{λ_0} on O_{λ_0} and is thus completely determined by such a φ_{λ_0}. $\qquad\qquad\square$

As an application, we have a

Corollary. *Suppose M is a simply connected surface.*

(1) *Suppose a map $f\colon M \to \mathbf{C} - \{0\}$ is continuous. Then $\log f$ has a single-valued continuous branch on M, satisfying $e^{\log f} = f$.*
(2) *Suppose $u\colon M \to \mathbf{R}$ is harmonic. Then u has a single-valued harmonic conjugate on M.* $\qquad\qquad (7.6.4)$

Proofs are left to readers. For (1), compare to Example 4 in Sec. 7.5.2.

Two subsection is divided. Section 7.6.1 is devoted to the proof of the uniformization theorem for simply connected Riemann surfaces, while Sec. 7.6.2 is for arbitrary Riemann surfaces.

7.6.1   *Simply connected Riemann surfaces*

Our aim is to prove

The uniformization theorem for simply connected Riemann surfaces. *Every simply connected Riemann surface is conformally equivalent to one and only one of the following three types:*

1. *The Riemann sphere $\mathbf{C}^*$;*
2. *The finite complex plane $\mathbf{C}$; and*
3. *The open unit disk $|z| < 1$.*

The corresponding surfaces are elliptic, parabolic, and hyperbolic, respectively. $\qquad\qquad (7.6.1.1)$

These three types are mutually exclusive to each other. Elliptic type is characterized by the compactness of the surface. Since the existence of Green's functions is a conformally equivalent property (see (1) in (6.3.1.13)), the classification of open Riemann surfaces into parabolic and hyperbolic types is also a conformally equivalent property. And a Riemann surface can be conformally equivalent to a disk only if it is hyperbolic, and to the finite complex plane only if it is parabolic. $\mathbf{C}^*$ is not topologically equivalent to both $\mathbf{C}$ and $|z| < 1$. Even though $\mathbf{C}$ and $|z| < 1$ are topologically equivalent, but they are conformally disjoint by Liouville's theorem.

Proof. The proof will be divided into hyperbolic and nonhyperbolic cases.

The hyperbolic case.
Suppose R is a hyperbolic simply connected Riemann surface. $p_0 \in R$ is a fixed point.

Let $g(p, p_0)$ be Green's function on R with pole at p_0. Each point $p \neq p_0$ has a neighborhood O_p, not containing the point p_0 and conformally equivalent a disk. Let h_p be a harmonic conjugate of $g(p_1, p_2)$ in O_p. Such an h_p does exist and is unique up to an arbitrary additive constant (see the paragraph beneath (3.4.3.1)). The function $f_p = e^{-(g+ih_p)}$ is then analytic in O_p and is uniquely determined up to a constant factor of absolute value 1. Set Φ_p be the set of all such functions f_p on O_p.

Recall that $g(p, p_0) + \log|z(p)|$ is harmonic in a neighborhood O_{p_0} of p_0 on which the local parameter $z = z(p)$ is chosen so that $z(p_0) = 0$. Let h_{p_0} be the harmonic conjugate of $g(p, p_0) + \log|z(p)|$ in O_{p_0}, still unique up to an additive constant. Let

$$f_{p_0}(p) = \begin{cases} e^{-[g(p,p_0)+ih_{p_0}(p)]}, & p \in O_{p_0} - \{p_0\} \\ 0, & p = p_0 \end{cases} \qquad (*_1)$$

It is unique up to a constant factor of absolute value 1. Let Φ_{p_0} be the set of all such analytic functions on O_{p_0}.

O_p, where $p \in R$, forms an open covering of R and, associated with each O_p, a family Φ_p of analytic function has been defined. In order to apply (7.6.3), we need to check if conditions 1 and 2 there are satisfied by O_p and Φ_p for $p \in R$:

1. Suppose $O_p \cap O_q \neq \emptyset$. For each $f_p \in \Phi_p$ and each $f_q \in \Phi_q$, the quotient $\frac{f_p}{f_q}$ has constant absolute value 1. Hence, on each component of $O_p \cap O_q$, $f_p = cf_q$ with $|c| = 1$, where either $c = 1$ or $c \neq 1$. It follows that either f_p and f_q are everywhere equal or everywhere not equal on each component of $O_p \cap O_q$.
2. In case a certain f_p is defined in a given component of $O_p \cap O_q$, the constant in f_q can be adjusted so that $f_p = f_q$ there.

According to (7.6.3), there exists a global analytic function $f(p, p_0)$ on R satisfying the following properties:

(a) $\log|f(p, p_0)| = -g(p, p_0)$, $p \in R$ and $p \neq p_0$.
(b) $f(p, p_0) = 0$ if and only of $p = p_0$.
(c) $|f(p, p_0)| < 1$ on R. $\qquad\qquad (*_2)$

One might also refer to Ref. [32], p. 181, for another treatment. It remains to show that $f(p, p_0)$ is univalent on R. If this is done, then the image of R under $f(p, p_0)$ is a simply connected domain contained in $|z| < 1$. By the Riemann mapping theorem, such a domain is conformally equivalent to the open unit disk, and the composite map will then finish the proof. As a matter of fact, the standard proof of the Riemann mapping theorem (see step 3 within its proof in Sec. 6.1.1) would show that $f(p, p_0)$ is indeed *onto* the whole disk $|z| < 1$. And we will acknowledge that *the Riemann mapping theorem is just a special case of* (*7.6.1.1*).

The univalence of $f(p, p_0)$:
For $p_1 \neq p_0$, it is sufficient to show that $f(p, p_0) = f(p_1, p_0)$ only for $p = p_1$.
   Note that $|f(p, p_0)| < 1$. Composite with the disk-self bilinear transformation $\frac{z - f(p_1, p_0)}{1 - \overline{f(p_1, p_0)} z}$, and set

$$F(p) = \frac{f(p, p_0) - f(p_1, p_0)}{1 - \overline{f(p_1, p_0)} f(p, p_0)}, \quad p \in R. \tag{*3}$$

F is analytic on R, $|F(p)| < 1$ and $F(p_1) = 0$. Recall Perron's family $\mathfrak{V}_{p_1}$ in (7.3.2.2) with p_0 replaced by p_1. Choose a local parameter $z_1 = z_1(p)$ at p_1 so that $z_1(p_1) = 0$. Since $F(p_1) = 0$, it follows that $\frac{F(p)}{z_1(p)}$ is analytic at p_1. For each $v \in \mathfrak{V}_{p_1}$, the condition that $v(p) + \log |z_1(p)|$ is subharmonic in a neighborhood of p_1 (and hence $\overline{\lim}_{p \to p_1}[v(p) + \log |z_1(p)|] < \infty$) then guarantees that, for $\varepsilon > 0$,

$$\overline{\lim}_{p \to p_1}[v(p) + (1 + \varepsilon)\log |F(p)|]$$

$$= \overline{\lim}_{p \to p_1}\left[v(p) + \log |z_1(p)| + \log \left|\frac{F(p)}{z_1(p)}\right| + \varepsilon \log |F(p)|\right] = -\infty$$

$\Rightarrow$ (by the maximum principle) $v(p) + (1 + \varepsilon) \log |F(p)| \leq 0$ on $R - \{p_1\}$, for $\varepsilon > 0$

$\Rightarrow$ (letting $\varepsilon \to 0$) $v(p) + \log |F(p)| \leq 0$ on $R - \{p_1\}$ for all $v \in \mathfrak{V}_{p_1}$

$\Rightarrow g(p, p_1) + \log |F(p)| \leq 0$ or $|F(p)| \leq |f(p, p_1)|$ on $R - \{p_1\}$ $\quad$ (*4)

$\Rightarrow$ (setting $p = p_1$) $|f(p_1, p_0)| \leq |f(p_0, p_1)|$, $p_0 \neq p_1$

$\Rightarrow$ (interchanging the role of p_0 and p_1) $|f(p_1, p_0)| = |f(p_0, p_1)|$, $p_0, p_1 \in R$ and $p_1 \neq p_0$. $\quad$ (*5)

Note that the equality in (*5) also follows from the symmetry of Green's function (see (7.3.2.12)).

$(*_5)$ shows that the equality in $(*_4)$ is obtained at $p = p_0$. Again by the maximum principle, the harmonic function $g(p, p_1) + \log |F(p)|$ is identically zero on $R - \{p_1\}$. Hence $|F(p)| = |f(p, p_1)|$ throughout $R - \{p_1\}$ and hence $F(p) = cf(p, p_1)$ with constant c ($|c| = 1$). This shows that $F(p) = 0$ if and only if $p = p_1$, and this in turn implies that $f(p, p_0) = f(p_1, p_0)$ holds only if $p = p_1$.

This ends the proof for the hyperbolic case.

The nonhyperbolic case.

Now suppose R is either a compact or a parabolic Riemann surface. What should be emphasized in advance is that the maximum principle (7.3.1.8) holds in these two kinds of surfaces; as a matter of fact, the classical maximum principle (7.2.11) is valid in compact surfaces.

Fix a point $p_0 \in R$. This time we need to use (7.3.3.3): The existence of a global harmonic function u on $R - \{p_0\}$ satisfying the properties that it is bounded outside every neighborhood of p_0 and that $u(p) - \text{Re}\frac{1}{z(p)}$ is harmonic in a neighborhood of p_0 and vanishes at p_0 for a local parameter $z = z(p)$ at p_0 with $z(p_0) = 0$.

Existence of f:

Observe that every point on R, including p_0, has a neighborhood on which u has a harmonic conjugate v which is uniquely determined up to a additive constant. By invoking (7.6.3) and using the same process leading to $(*_2)$, we obtain a global analytic function $f(p)$ on $R - \{p_0\}$ satisfying the following properties:

(a) $f(p) - \frac{1}{z(p)}$ is analytic in a neighborhood of p_0 and vanishes at p_0, and

(b) $\text{Re}\, f(p) = u(p)$ for $p \in R - \{p_0\}$, and hence $\text{Re}\left\{f(p) - \frac{1}{z(p)}\right\} = u(p) - \text{Re}\frac{1}{z(p)}$ in a neighborhood of p_0, \hfill $(*_6)$

where $z = z(p)$ is a local parameter at p_0 with $z(p_0) = 0$. *In short*, the global meromorphic function $f = u + iv$ on R has a simple pole at p_0 with the Laurent series expansion

$$f(p) = \frac{1}{z} + a_1 z + \cdots \qquad (*_7)$$

in terms of $z = z(p)$ with $z(p_0) = 0$. Properties (a) and (b) in $(*_6)$ determine f uniquely.

Correspondingly, if we replace $z = z(p)$ by the local parameter $-iz$, we obtain uniquely another meromorphic function $\tilde{f} = \tilde{u} + i\tilde{v}$ on R, with a single simple pole at p_0 and the expansion

$$\widetilde{f}(p) = \frac{1}{-iz} + a_1'(-iz) + \cdots$$

$$= \frac{i}{z} + b_1 z + \cdots \tag{*8}$$

in terms of the original parameter $z = z(p)$. Of course, the real part $\operatorname{Re}\widetilde{f}(p) = \widetilde{u}(p)$ is harmonic in $R - \{p_0\}$, bounded outside every neighborhood of p_0, $\widetilde{u}(p) - \operatorname{Re}\frac{1}{-iz(p)}$ is harmonic at p_0 and vanishes there.

We want to show that not only $\operatorname{Re} f$, but also f *itself is bounded outside every neighborhood of* p_0. It suffices to prove that $\widetilde{f} = if$. For, if this is the case, then $\operatorname{Im} f = -\operatorname{Re}\widetilde{f}$ implies that $\operatorname{Im} f$ is also bounded outside neighborhoods of p_0.

The proof of $\widetilde{f} = if$:
Let $D_\rho = \{p \mid |z(p)| < \rho\}$ be as in (7.3.3.3), where $0 < \rho \leq 1$ so that D_1 is contained in the range of $z = z(p)$. Assume that $|\operatorname{Re} f| \leq M$ and $|\operatorname{Re}\widetilde{f}| \leq M$ outside D_ρ. Without loss of its generality, we may also assume that both f and $\widetilde{f}$ are one-to-one on D_1 (see (3.5.1.11)). Choose a point p_1 in D_ρ, $z(p_1) \neq 0$ so that $|\operatorname{Re} f(p_1)| > 2M$ and $|\operatorname{Re}\widetilde{f}(p_1)| > 2M$ hold. Then $f(p) \neq f(p_1)$ outside D_ρ. Moreover $f(p) - f(p_1)$ has p_1 as its only zero, this zero being simple. The same applies to $\widetilde{f}$. Therefore

$$F(p) = \frac{1}{f(p) - f(p_1)} \quad \text{and} \quad \widetilde{F}(p) = \frac{1}{\widetilde{f}(p) - \widetilde{f}(p_1)} \tag{*9}$$

satisfy simultaneously the following properties:

1. both are analytic on $R - \{p_1\}$ and have a simple pole at p_1.
2. In terms of the local parameter $z = z(p)$,

$$F(p) = \frac{a}{z - z_1} + a_0 + a_1(z - z_1) + \cdots,$$

$$\widetilde{F}(p) = \frac{\widetilde{a}}{z - z_1} + \widetilde{a_0} + \widetilde{a_1}(z - z_1) + \cdots, \quad \text{where } z_1 = z(p_1). \tag{*10}$$

Note. Instead, by choosing p_1 sufficiently close to p_0 with $\arg z(p_1) = \frac{\pi}{4}$, it is possible to choose p_1 in D_ρ so that $\operatorname{Re} f(p_1) > M$ and $\operatorname{Re}\widetilde{f}(p_1) > M$. Now $\operatorname{Re}[f(p) - f(p_1)] < 0$ along ∂D_ρ; by the argument principle that $\frac{1}{2\pi i}\int_{\partial D_\rho} \frac{f'(z(p))}{f(z(p)) - f(z(p_1))} dz = 0 = N - P$ with $P = 1$, it follows that $N = 1$ and hence $f(p) - f(p_1)$ has only one zero at p_1, a simple one. Here, we do not need to presume that f and $\widetilde{f}$ are one-to-one on D_1.

Outside of D_ρ, $|f(p) - f(p_1)| \geq |\operatorname{Re} f(p) - \operatorname{Re} f(p_1)| \geq |\operatorname{Re} f(p_1)| - |\operatorname{Re} f(p)| \geq 2M - M = M$. Hence $|F(p)| \leq \frac{1}{M}$ outside D_ρ. Similarly $|\widetilde{F}(p)| \leq \frac{1}{M}$ outside D_ρ. Therefore the linear combination $\tilde{a}F - a\widetilde{F}$ is bounded outside D_ρ. Since it has a removable singularity at p_1, $\tilde{a}F - a\widetilde{F}$ is bounded and analytic on the whole surface R. Since R is nonhyperbolic, it follows (for compact surfaces, by the maximum principle; for parabolic surfaces, see Exercise A(?) of 7.3.1 or consider the one-point compactification of R) that

$$\tilde{a}F - a\widetilde{F} = b, \text{ a constant}$$

$\Rightarrow \tilde{f}(p) = \frac{\alpha f(p) + \beta}{\gamma f(p) + \delta}, \quad \alpha\delta - \beta\gamma \neq 0$ $\hspace{2cm}$ $(*_{11})$

$\Rightarrow$ (since $\tilde{f}(p_1) = f(p_1) = \infty$) $\gamma = 0$

$\Rightarrow$ (owing to the Laurent expansions $(*_7)$ and $(*_8)$) $\tilde{f}(p) = if(p)$, $p \in R - \{p_0\}$.

This proves the claim.

The univalence of f:
We have established the boundedness of f outside every neighborhood of p_0. Suppose $|f(p)| < M$ outside of D_ρ. Choose a point p_1 in D_ρ, $p_1 \neq p_0$ so that $|f(p_1)| > 2M$. Then the function F, as defined in $(*_9)$, is bounded outside of D_ρ and has the properties stated in $(*_{10})$.

Fix a local parameter $z = z(p)$ at p_0 and denote our original function $f(p)$ by $f(p, p_0)$ for the mere purpose of emphasizing its dependence on p_0 (and hence on the chosen $z(p)$). For $p_1 \in D_\rho$, let $f(p, p_1)$ be the corresponding function, still using the same $z(p)$. Observe that $f(p, p_1)$ and $F(p)$ have the same single simple pole at p_1. The argument leading to $(*_{11})$ is available and we still get

$$F(p) = \alpha f(p, p_1) + \beta \quad \text{for constants } \alpha \text{ and } \beta$$
$$\Rightarrow f(p, p_1) = L \circ f(p, p_0),$$
$$\text{where } L : \mathbf{C}^* \to \mathbf{C}^* \text{ is a bilinear transformation.} \hspace{1cm} (*_{12})$$

This result is still valid even if p_0 and p_1 are not in the same Jordan disk D_ρ, for we can always find intermediate points $\zeta_0, \zeta_1, \ldots, \zeta_{n-1}$ to form a sequence $p_0 = \zeta_0, \zeta_1, \ldots, \zeta_{n-1}, \zeta_n = p_1$ of points so that each pair ζ_k, ζ_{k+1} of consecutive points lies in a single Jordan disk.

Suppose now that $(*_{12})$ is true for any two fixed distinct points p_1, p_0 in R. Then

$$\Rightarrow f(p, p_0) = f(p_1, p_0)$$
$$\Rightarrow f(p, p_1) = L \circ f(p, p_0) = L \circ f(p_1, p_0) = f(p_1, p_1) = \infty$$
$$\Rightarrow \text{(since } p_1 \text{ is the only pole of } f(p, p_1)) \; p = p_1.$$

This proves that f is indeed one-to-one.

The fulfillment of the proof:
In case R is closed, then $f(R)$ is a compact as well as an open subset of $\mathbf{C}^*$. Since $f(R)$ is connected, therefore $f(R) = \mathbf{C}^*$ should hold. Hence $f : R \to \mathbf{C}^*$ realizes as a conformal mapping between R and $\mathbf{C}^*$.

In case R is parabolic, $f(R)$ is then an open simply connected subset of $\mathbf{C}^*$; namely, $f(R)$ is a simply connected domain in $\mathbf{C}^*$. If $\mathbf{C}^* - f(R)$ contains at least two distinct points, then $f(R)$ is conformally equivalent to the open disk by the Riemann mapping theorem, and this would show that R is hyperbolic. The only possibility is that $\mathbf{C}^* - f(R)$ is a singleton $\{z_0\}$. Then $z \to \frac{1}{z - z_0}$ will map $f(R)$ conformally onto $\mathbf{C}$. $\qquad\qquad\square$

7.6.2   *Arbitrary Riemann surfaces*

Let R be an arbitrary Riemann surface.

Let $(\tilde{R}, \tilde{F})$ be its universal covering surface. It is unique up to conformal equivalence (see (7.5.3.2)). The projection $\tilde{F} : \tilde{R} \to R$ is then an analytic mapping (see (7.5.2.1)). Moreover, each covering transformation $\Phi : \tilde{R} \to \tilde{R}$ of $\tilde{R}$ over R is a conformal homeomorphism (see (7.5.4.3)).

We can apply (7.6.1.1) to conclude that $\tilde{R}$ is conformally equivalent to either the Riemann sphere $\mathbf{C}^*$, the finite complex plane $\mathbf{C}$, or the open unit disk $|z| < 1$. Since the relevant properties of Riemann surfaces are invariant under conformal mappings, we just assume that $\tilde{R}$ is one of these three surfaces and that points on $\tilde{R}$ can be regarded as the complex numbers z, possibly including $z = \infty$. Under this circumstance, the projection $\tilde{F}(z)$ can be regarded as an analytic function on $\tilde{R}$ with values in R.

No matter for $\mathbf{C}^*, \mathbf{C}$ or $|z| < 1$, all conformal self-mappings are given by bilinear transformations $\Phi(z) = \frac{az+b}{cb+d}, ad - bc \neq 0$ (for $\mathbf{C}^*$ and $\mathbf{C}$, see Exercise A(6) of Sec. 4.10.2; for $|z| < 1$, see Example 1 of Sec. 3.4.5). Any such transformation has at least one and at most two distinct fixed points in $\mathbf{C}^*$ except the identical mapping $w = z$ (see (2.5.4.13)). But, if Φ *acts* *as a covering transformation* over R, it cannot have any fixed point except the identity (see (2) in (7.5.4.2)). More precisely, we have:

1. If $\tilde{R} = \mathbf{C}^*$, Φ can only be the identity.
2. If $\tilde{R} = \mathbf{C}$, the only fixed point of Φ must be ∞, and this implies that $\Phi(z) = z + b$, a translation.
3. If $\tilde{R} = \{|z| < 1\}$, the fixed points of Φ must lie on the unit circle $|z| = 1$. In this case, Φ is either a non-Euclidean parallelism (hyperbolic type)

$$\frac{\Phi(z) - z_1}{\Phi(z) - z_2} = k \frac{z - z_1}{z - z_2}, \quad z_1 \neq z_2, \quad |z_1| = |z_2| = 1 \quad \text{and} \quad k > 0,$$

or a non-Euclidean limit rotation (parabolic type)

$$\frac{1}{\Phi(z) - z_0} = \frac{1}{z - z_0} + \lambda, \quad |z_0| = 1, \quad \lambda \in \mathbf{C}$$

(see (5.3.4.5) or (B.41) in Appendix B). (7.6.2.1)

Recall that the group $G(\tilde{R}, \tilde{F})$ of covering transformations of $\tilde{R}$ over R is isomorphic to the fundamental group $\pi_1(R)$ of R (see (7.5.4.3)).

Three types are considered separately.

The elliptic type $\tilde{\mathbf{R}} = \mathbf{C}^*$. Since the group $G(\tilde{R}, \tilde{F}) = \{1\}$, it follows that $\pi_1(R) = \{1\}$ and hence R is simply connected. By the uniformization theorem (7.6.1.1), R is conformally equivalent to the Riemann sphere $\mathbf{C}^*$ and can be identified with its universal covering surface $\tilde{R}$.

Before we proceed to the remaining types, we digress a little from the subject under discussion to (refer to Remark 1 in Sec. 7.5.1 for general setting).

An equivalent model of a surface. Let $(\tilde{M}, \tilde{F})$ be the universal covering surface of a surface M. Then $(\tilde{M}, \tilde{F})$ is regular (see (2) in (7.5.4.4)). Two points $\tilde{p}_1, \tilde{p}_2 \in \tilde{M}$ are said to be equivalent if there is a $\Phi \in G(\tilde{M}, \tilde{F})$, the group of covering transformations of $\tilde{M}$ over M, so that $\Phi(\tilde{p}_1) = \tilde{p}_2$. Set $G = G(\tilde{M}, \tilde{F})$, for simplicity, and

$[\tilde{p}]$: the equivalence class determined by $\tilde{p}$;

$\tilde{M}/G = \{[\tilde{p}] | \tilde{p} \in \tilde{M}\}$, and a mapping

$F : \tilde{M}/G \to M$ defined by $F([\tilde{p}]) = \tilde{F}(\tilde{p})$.

Then

1. $(\tilde{M}/G, F)$ is a complete smooth covering surface of M.
2. F is a homeomorphism between $\tilde{M}/G$ and M.

3. The covering transformation group $G(\tilde{M}/G, F)$ is isomorphic to $G/G = \{1\}$.

In short, $(\tilde{M}/G, F)$ is equivalent to (M, id) (see Sec. 7.5.3), and $\tilde{M}/G$ is then considered as a model of M.   (7.6.2.2)

The above is a rather special case of Exercise A(4) of Sec. 7.5.5 if G there is chosen to be the whole group $G(\tilde{M}, \tilde{F})$. In particular, when taken (7.5.2.1) into consideration, we have

A conformally equivalent model of a Riemann surface. Let $(\tilde{R}, \tilde{F})$ be the universal covering surface of a Riemann surface R. Then the quotient space, defined as in (7.6.2.2),

$$\tilde{R}/G(\tilde{R}, \tilde{F})$$

is conformally equivalent to the Riemann surface R.   (7.6.2.3)

A connected subset of $\tilde{R}$ which contains exactly only one point from each equivalence class with respect to $G(\tilde{R}, \tilde{F})$ is called a *fundamental domain*. It is essentially a "sheet" of $\tilde{R}$.

Let us go back to our original subject.

The parabolic type $\tilde{R} = \mathbf{C}$.
The group $G(\mathbf{C}, \tilde{F})$ can consist only of translations $z + b$ of the plane. The important thing to recall is that, as a group of covering transformations, $G(\mathbf{C}, \tilde{F})$ is *properly discontinuous* on $\mathbf{C}$ (see (3) in (7.5.4.2)).

Set $G = G(\mathbf{C}, \tilde{F})$, for simplicity. There are only three possibilities for G:

Case 1. $G = \{id\}$, consisting only of the identity map $w = z$. According to (7.6.2.2), $\mathbf{C}/\{id\} = \mathbf{C}$ is conformally equivalent to R.

Case 2. G consists of the translations generated by $T(z) = z + b$ for some $b \neq 0$, namely,

$$G = \{T_n(z) = z + nb | n = 0, \pm 1, \pm 2, \ldots\}.$$

The fundamental domain, in this case, is an infinite strip of width $|b|$, which contains one of its boundary straight lines. Then $\mathbf{C}/G$ is conformally equivalent, which can be realized by $z \to e^{2\pi i z/b}$, to the punctured finite complex plane or the infinite cylinder $S^1 \times R^1$. See Example 2 of Sec. 7.1 for details.

Case 3. G consists of the translations generated by $z \to z + \omega_1$ and $z \to z + \omega_2$ where ω_1 and ω_2 are nonzero complex numbers with $\text{Im}\frac{\omega_1}{\omega_2} \neq 0$.

Then

$$G = \{z + m_1\omega_1 + m_2\omega_2 \mid m_1, m_2 = 0, \pm1, \pm2, \ldots\}.$$

The corresponding fundamental domain is the parallelogram shown in Fig. 7.3, with vertices $0, \omega_1, \omega_2, \omega_1 + \omega_2$, and including the two half-open edges $[0, \omega_1)$ and $[0, \omega_2)$. Then $\mathbf{C}/G$ is conformally equivalent to the torus $S^1 \times S^1$. For details, see Example 3 in Sec. 7.1.

Remark (These are the only possibilities because G should be properly discontinuous). Set $\sum = \{m_1\omega_1 + m_2\omega_2 \mid m_1, m_2 = 0, \pm1, \pm2, \ldots\}$, the set of all possible equivalent points under G. $\sum$ does not have limit points in $\mathbf{C}$. Suppose there is a real number α so that $\omega_2 = \alpha\omega_1$. Let $\omega = n_1\omega_1 + n_2\omega_2$ be the point on the line passing $0, \omega_1$, and ω_2 which is closest to 0. $\omega \in \sum$ and $\omega \neq 0$, otherwise 0 would be a limit point of $\sum$. Then there is an integer k so that $|k\omega - \omega_1| < |\omega|$ if $|\omega| < |\omega_1|$. This means that $|(kn_1 - 1)\omega_1 + kn_2\omega_2| < |\omega|$ which is contradictory to the choice of ω unless $k\omega - \omega_1 = 0$. Hence $\omega_1 = k\omega$ should hold. Similarly, $\omega_2 = l\omega$ for some integer l. It follows that G is indeed generated by a single translation $z \to z + \omega$. And this is Case 2.

Is it possible for G to be generated by more than two independent translations while being kept properly discontinuous? The answer is definitely not so. For details, see Ref. [48], pp. 58–63.

The hyperbolic case $\tilde{R} = \{|z| < 1\}$.
In this case, G is a properly discontinuous group of fixed-point free bilinear self-mappings of the disk onto itself (see 3 in (7.6.2.1)). Then (7.6.2.3) says that R is conformally equivalent to D/G, where D is the open unit disk.

This ends the proof of the uniformization theorem (7.6.1).   □

For more subtle discussions about the uniformization theorem and its consequences see Ref. [75], Chap. 9; [32], pp. 188–240.

However, some immediate *applications* of (7.6.1.1) and then (7.6.1) are as follows:

(1) Every Riemann surface satisfies the second countability for its surface topology.
(2) Every Riemann surface is triangulizable in such a way that each edge of the triangles is an analytic arc.
(3) The theory of analytic functions on a hyperbolic Riemann surface turns out to be the theory of automorphic functions under the group of covering transformations.

(4) The hyperbolic metric (see 2 in $(*_{12})$ of Sec. 5.8.4) of the disk car-
    ries over to Poincaré's metric on a hyperbolic Riemann surface with
    constant curvature -1. In particular, every simply connected planer
    domain whose boundary contains at least two distinct points carries a
    Poincaré's metric (see $(*_{13})$ in Sec. 5.8.4).

Appendix

Appendix B: Parabolic, Elliptic, and Hyperbolic Geometries

In Exercises A(9) and B of Sec. 2.5.4, a linear transformation $w = w(z)$ is called a *parabolic type* if it has only one fixed point; an *elliptic type* if it has two distinct fixed points and $|k| = 1$, a *hyperbolic type* if $k > 0$ and a *loxodromic type* if otherwise. In the sequel, we will present a series of problems which eventually lead to the construction of the *parabolic, elliptic, and hyperbolic geometries*. General references in this direction are Refs. [14], Vol. 1, Chaps. 2 and 3; [46], Chaps. 1 and 2; and [56], Vol. 2, Secs. 5.11 and 5.12 in which purely linear algebraic method is adopted.

Let

$$w = T(z) = \frac{az + b}{cz + d}, \quad ad - bc = 1;$$

$$K = \frac{a + d - \sqrt{(a+d)^2 - 4}}{a + d + \sqrt{(a+d)^2 - 4}}$$

$$= \text{the ratio of the two eigenvalues of the coefficient}$$

$$\text{matrix } \begin{bmatrix} a & b \\ c & d \end{bmatrix}. \tag{B.1}$$

K is called the *multiplier* of T. Note that $K + \frac{1}{K} = (a+d)^2 - 2$.

Exercise (1) Show that T is

1. Parabolic $\Leftrightarrow a + d = \pm 2$.
   $\qquad\qquad \Leftrightarrow K = 1$.
   In this case, the
   general form: $\frac{1}{w - z_0} = \frac{1}{z - z_0} \pm c$ (plus sign if $a + d = 2$; minus sign if $a + d = -2$);
   canonical form: $w = z \pm c$ (a translation).

631

2. Elliptic $\Leftrightarrow a + d$ is real and $|a + d| < 2$.
   $\qquad \Leftrightarrow |K| = 1$ and $K \neq 1$.

   In this case, the

   general form: $\frac{w - z_1}{w - z_2} = K \frac{z - z_1}{z - z_2}$ $(z_1 \neq z_2)$;

   canonical form: $w = Kz$ (a rotation).

3. Hyperbolic $\Leftrightarrow a + d$ is real and $|a + d| > 2$.
   $\qquad \Leftrightarrow K > 0$ and $K \neq 1$.

   In this case, the

   general form: $\frac{w - z_1}{w - z_2} = K \frac{z - z_1}{z - z_2}$ $(z_1 \neq z_2)$;

   canonical form: $w = Kz$ (a stretching).

4. Loxodromic $\Leftrightarrow a + d$ is a nonreal complex number.
   $\qquad\qquad \Leftrightarrow K = |K|e^{i\theta}$, where $|K| > 0$ and $|K| \neq 1$, $\theta \neq 2n\pi$
   $\qquad\qquad$ (n integers).

   In this case, the

   general form: $\frac{w - z_1}{w - z_2} = K \frac{z - z_1}{z - z_2}$ $(z_1 \neq z_2)$;

   canonical form: $w = Kz$ (the composite of a stretching and a rotation).

Moreover, in case T has two distinct fixed points z_1 and z_2, show that

$$K = (T(z), z; z_1, z_2) = \frac{(T(z) - z_1)(z - z_2)}{(T(z) - z_2)(z - z_1)}, \quad z \in \mathbf{C}^* \quad \text{and} \quad z \neq z_1, z_2.$$

In particular, $K = (T(\infty), \infty; z_1, z_2) = \frac{a - cz_1}{b - cz_2}$.

Exercise (2) Define $T^n = T \circ T^{n-1}$ inductively for integer n. T^n is called the n-th iterate of T itself.

(a) In case T is not the identity transformation that satisfies $T^n(z) = z$ for some integer n, then T is necessarily elliptic.

(b) In case T is either hyperbolic or loxodromic, then for any $z \in \mathbf{C}^*$,

$$\lim_{n \to \infty} T^n(z) = \begin{cases} z_1, & \text{if} \quad 0 < |K| < 1 \\ z_2, & \text{if} \quad |K| > 1 \end{cases}.$$

One of the limit points z_1 and z_2 is called *attractive*, while the other *repellent*.

(c) In case T is parabolic, for any $z \in \mathbf{C}$,

$$\lim_{n \to \infty} T^n(z) = z_0.$$

Note. The behavior of the circular nets formed by Steiner's circles (see (1.4.4.2) and Fig. 1.28) has interesting properties under various types of linear transformations. Read Ref. [1], pp. 84–88 for a sketch or *Ford's Automorphic Functions (Chelsea, N.Y., 1929)* for details.

In what follows, seven sections are divided.

Section (1) Bundles of Circles

The totality of circles on the Riemann sphere S whose planes pass through same fixed point P_0 (in $\mathbf{R}^3$ or the point at infinity) is called a *bundle of circles* on the sphere S determined by P_0. It is said to be *parabolic, elliptic,* or *hyperbolic* according to whether the points P_0 lies on the sphere, inside or outside the sphere, respectively. The bundle of circles on S determined by $P_0 = (\zeta_0, \eta_0, \xi_0)$ is given by

$$\begin{cases} ax_1 + bx_2 + cx_3 - (a\zeta_0 + b\eta_0 + c\xi_0) = 0 \\ x_1^2 + x_2^2 + x_3^2 = 1 \end{cases}, \qquad (B.2)$$

where a, b, and c are *real parameters*, not equal to zero simultaneously. The stereographic projection (1.6.7) maps (B.2) into a bundle of circles on $\mathbf{C}^*$:

$$c(|z|^2 - 1) + (a - bi)z + (a + bi)\bar{z} - (a\zeta_0 + b\eta_0 + c\xi_0)(|z|^2 + 1) = 0, \quad \text{or}$$

$$(c - a\zeta_0 - b\eta_0 - c\xi_0)|z|^2 + (a - bi)z + (a + bi)\bar{z}$$
$$+ (-c - a\zeta_0 - b\eta_0 - c\xi_0) = 0. \qquad (B.3)$$

It represents a real circle if and only if $c(1 - \xi_0) - a\zeta_0 - b\eta_0 \neq 0$ and $a^2 + b^2 + c^2 > (a\zeta_0 + b\eta_0 + c\xi_0)^2$ (see (1.4.3.6)).

Exercise (3) *Analytic expression of bundles of circles*

(a) *Parabolic bundle*: Set $\zeta_0 = \eta_0 = 0$ and $\xi_0 = 1$. Then (B.3) turns out to be

$$(a - bi)z + (a + bi)\bar{z} - 2c = 0 \quad \text{or} \quad ax + by + c = 0 \ (z = x + iy). \qquad (B.4)$$

This is the totality of straight lines in $\mathbf{C}$; the point at infinity ∞ together with the lines $x = 0$ and $y = 0$ generate the entire bundle.

(b) *Elliptic bundle*: Set $\zeta_0 = \eta_0 = \xi_0 = 0$. Then (B.3) becomes

$$c(|z|^2 - 1) + (a - bi)z + (a + bi)\bar{z} = 0 \quad \text{or}$$
$$c(x^2 + y^2) + 2ax + 2by = c. \qquad (B.5)$$

Let $c = 1$ and $a = b = 0$, and we get the fixed circle $|z| = 1$. All other circles or lines in the bundle always pass through the end points of some diameters of $|z| = 1$; moreover, there is no point circle. The circle $|z| = 1$ and the lines $x = 0$ and $y = 0$ generate the bundle.

(c) *Hyperbolic bundle*: Set $\zeta_0 = \eta_0 = 0$ and choose $\xi_0 > 1$. Then (B.3) becomes

$$c(|z|^2 - 1) + (a - bi)z + (a + bi)\bar{z} - c\xi_0(|z|^2 + 1) = 0, \quad \text{or}$$

$$c[(1 - \xi_0)(x^2 + y^2) - (1 + \xi_0)] + 2ax + 2by = 0. \tag{B.6}$$

Set $a = b = 0$ and $c = 1$, and we get the imaginary circle $|z|^2 = \frac{1+\xi_0}{1-\xi_0}$ which, together with $x = 0$ and $y = 0$, generate the bundle. For a given circle $\alpha|z|^2 + \beta\bar{z} + \bar{\beta}z + \gamma = 0$ to be orthogonal to each member of the bundle, it is necessary and sufficient that

$$\alpha(-c - c\xi_0) + \gamma(c - c\xi_0) = 2(a\beta_1 + b\beta_2), \quad \text{where } \beta = \beta_1 + i\beta_2,$$

holds for all real parameters a, b, and c. In particular, choose $a = b = 0$ and $c \neq 0$, then $\frac{\gamma}{\alpha} = -\frac{\xi_0+1}{\xi_0-1}$. Hence the circle (choosing $\beta = 0$)

$$|z|^2 = \frac{\xi_0 + 1}{\xi_0 - 1} \tag{B.7}$$

is *orthogonal* to the bundle, namely, the bundle is composed of those circles or lines that are orthogonal to the circle (B.7).

The other way to describe the bundle is as follows. In (B.3), replacing a, b, c by $a\xi_0, b\xi_0, -c$, respectively, and then dividing by ξ_0, we get

$$d(|z|^2 + 1) + (a - bi)z + (a + bi)\bar{z} - \frac{c}{\xi_0}(|z|^2 - 1) = 0.$$

$$\Rightarrow \text{ (Letting } \xi_0 \to \infty) \ d(|z|^2 + 1) + (a - bi)z + (a + bi)\bar{z} = 0 \tag{B.8}$$

where a, b, d are real parameters, not all equal to zero. The imaginary circle $|z|^2 = -1$ and the lines $x = 0$ and $y = 0$ generate the bundle. The bundle consists precisely of these circles or lines that are orthogonal to the unit circle $|z| = 1$, obtaining by letting $\xi_0 \to \infty$ in (B.7).

In conclusion, we may redefine bundle of circles on $\mathbf{C}^*$ as follows: The totality of circles (lines included) that pass a fixed point is called a *parabolic bundle* of circles (see Fig. B.1); the totality of circles that pass the end points of diameters of a fixed circle (under a fixed linear transformation) is called an *elliptic bundle* (see Fig. B.2), while the totality of circles that are orthogonal to a fixed circle is called a *hyperbolic bundle* (see Fig. B.3).

*Section (2) Bundles of circles (linearly algebraic viewpoint)

Hermitian matrices of order 2 can be used to study bundles (or pencils) of circles in the plane which, in turn, provides a vivid yet concise geometric background for the study of the former.

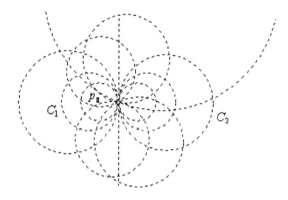

Fig. B.1

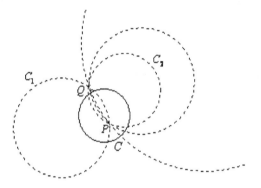

Fig. B.2

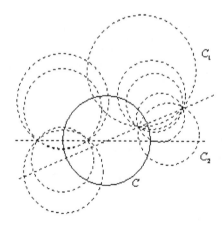

Fig. B.3

To each circle $a|z|^2 + b\bar{z} + \bar{b}z + c = 0$, a, $c \in \mathbf{R}$ and $b \in \mathbf{C}$, there corresponds a Hermitian matrix

$$H = \begin{bmatrix} a & b \\ \bar{b} & c \end{bmatrix}. \tag{B.9}$$

If H_1 is another Hermitian matrix so that $H = \lambda H_1$ holds for some real $\lambda \neq 0$, then we say that H_1 and H *represent* the same circle. It is easy to see that the determinant det $H = ac - |b|^2 > 0$, $= 0$, < 0 represents an imaginary, a point or a real circle, respectively (see (1.4.3.6)). Let

$$F = \begin{bmatrix} 0 & 1 \\ -1 & 0 \end{bmatrix}. \tag{B.10}$$

Then $FHF\bar{H}$ has its *trace*

$$\mathrm{tr}(FHF\bar{H}) = 2(|b|^2 - ac) = -2\det H.$$

Hence $\mathrm{tr}(FHF\bar{H}) < 0$, $= 0$, $> 0 \Leftrightarrow H$ represents an imaginary circle, a point, a real circle, respectively.

A bilinear transformation

$$w = \frac{\alpha z + \beta}{\gamma z + \delta}, \quad P = \begin{bmatrix} \alpha & \beta \\ \gamma & \delta \end{bmatrix} \quad \text{and det } P = \alpha\delta - \beta\gamma \neq 0,$$

maps a circle $(a|w^2| + b\bar{w} + \bar{b}w + c = 0)$ represented by H into a circle represented by

$$\bar{P}^*HP = \begin{bmatrix} |\alpha|^2 a + \bar{\alpha}\gamma b + \alpha\bar{\gamma}\bar{b} + |\gamma|^2 c & \bar{\alpha}\beta a + \bar{\alpha}\delta b + \beta\bar{\gamma}\bar{b} + \delta\bar{\gamma}c \\ \alpha\bar{\beta}a + \alpha\bar{\delta}b + \bar{\beta}\gamma b + \bar{\delta}\gamma c & |\beta|^2 a + \bar{\beta}\delta b + \beta\bar{\delta}\bar{b} + |\delta|^2 c \end{bmatrix}.$$

Owing to the fact that det $\bar{P}^*HP = |\det P|^2\det H$, it follows that a bilinear transformation preserves an imaginary circle, a point circle, or a real circle. Consequently a circle can always be transformed into one of the following three canonical types:

1. Imaginary circle: $\begin{bmatrix} 1 & 0 \\ 0 & 1 \end{bmatrix}$;

2. Point circle: $\begin{bmatrix} 1 & 0 \\ 0 & 0 \end{bmatrix}$; and   (B.11)

3. Real circle: $\begin{bmatrix} 1 & 0 \\ 0 & -1 \end{bmatrix}$.

Two circles

$$H = \begin{bmatrix} a & b \\ \bar{b} & c \end{bmatrix} \quad \text{and} \quad H_1 = \begin{bmatrix} a_1 & b_1 \\ \bar{b}_1 & c_1 \end{bmatrix}$$

are *orthogonal* to each other if and only if

$$ac_1 + a_1 c = 2 \left(\operatorname{Re} b \cdot \operatorname{Re} b_1 + \operatorname{Im} b \cdot \operatorname{Im} b_1 \right) = b \bar{b}_1 + \bar{b} b_1$$

$$\Leftrightarrow \operatorname{tr}(FHF H_1) = 0. \tag{B.12}$$

By invoking the following relations:

$$PFP^* = \begin{bmatrix} \alpha & \beta \\ \gamma & \delta \end{bmatrix} \begin{bmatrix} 0 & 1 \\ -1 & 0 \end{bmatrix} \begin{bmatrix} \alpha & \gamma \\ \beta & \delta \end{bmatrix} = (\det P) F;$$

$$\operatorname{tr}(F(\bar{P}^* H P) F(P^* \overline{H_1} \bar{P})) = |\det P|^2 \operatorname{tr}(FHF \overline{H_1}),$$

it is easy to check that a bilinear transformation preserves the orthogonality of two circles.

Exercise (4) The set

$$\{ H \,|\, \operatorname{tr}(FHF \bar{H}_0) = 0 \}$$

of these circles H that are orthogonal to a given circle $H_0 = \begin{bmatrix} a_0 & b_0 \\ \bar{b}_0 & c_0 \end{bmatrix}$ is called a *bundle of circles* (generated by H_0). In case H_0 is a point circle, it is understood that a circle orthogonal to H_0 means a circle passing the point represented by H_0. By (B.12), $H = \begin{bmatrix} a & b \\ \bar{b} & c \end{bmatrix}$ is orthogonal to H_0 if and only if $a_0 c + a c_0 = b_0 \bar{b} + \bar{b}_0 b$ holds. If $a \neq 0$, then this condition can be rewritten as $c_0 + a_0 \lambda = b_0 \bar{\mu} + \bar{b}_0 \mu$, where $\lambda = \frac{c}{a} \in \mathbf{R}$ and $\mu = \frac{b}{a} \in \mathbf{C}$ are two parameters. Similar explanation holds if $c \neq 0$.

(a) *Parabolic bundle.* Choose H_0 as the point circle in (B.11), and we obtain, by (B.12), the parabolic bundle

$$\begin{bmatrix} a & b \\ \bar{b} & 0 \end{bmatrix} = a \begin{bmatrix} 1 & 0 \\ 0 & 0 \end{bmatrix} + (\operatorname{Re} b) \begin{bmatrix} 0 & 1 \\ 1 & 0 \end{bmatrix}$$

$$+ (\operatorname{Im} b) \begin{bmatrix} 0 & i \\ -i & 0 \end{bmatrix}, a, \operatorname{Re} b, \operatorname{Im} b \in \mathbf{R}. \tag{B.13}$$

Note. The circles $a|z|^2 + b\bar{z} + \bar{b}z = 0$ in the bundle pass the point $(0,0)$ represented by the point circle H_0 (see Fig. B.4). Also the bundle is generated by the point circle $|z| = 0$ and the lines $\operatorname{Im} z = 0$ and $\operatorname{Re} z = 0$.

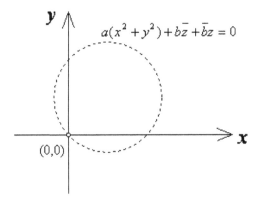

Fig. B.4

(b) *Elliptic bundle.* Choose H_0 as the real circle in (B.11), and we obtain the elliptic bundle

$$\begin{bmatrix} a & b \\ \bar{b} & -a \end{bmatrix} = a \begin{bmatrix} 1 & 0 \\ 0 & -1 \end{bmatrix} + (\operatorname{Re} b) \begin{bmatrix} 0 & 1 \\ 1 & 0 \end{bmatrix}$$

$$+ (\operatorname{Im} b) \begin{bmatrix} 0 & i \\ -i & 0 \end{bmatrix}, a, \operatorname{Re} b, \operatorname{Im} b \in \mathbf{R}. \qquad (\mathrm{B}.14)$$

Note. This bundle is generated by the unit circle $|z| = 1$ and the lines $\operatorname{Im} z = 0$ and $\operatorname{Re} z = 0$. The circle $a|z|^2 + b\bar{z} + \bar{b}z - a = 0$ in the bundle always intersects the circle $|z| = 1$ at the two endpoints of the diameter of the latter which lies on the line $b\bar{z} + \bar{b}z = 0$ (see Fig. B.5).

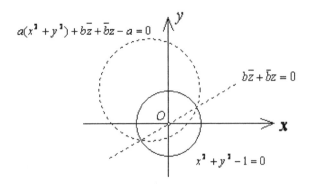

Fig. B.5

(c) *Hyperbolic bundle.* Choose H_0 as the imaginary circle in (B.11), and we get the hyperbolic bundle

$$\begin{bmatrix} a & b \\ \bar{b} & a \end{bmatrix} = a \begin{bmatrix} 1 & 0 \\ 0 & 1 \end{bmatrix} + (\operatorname{Re} b) \begin{bmatrix} 0 & 1 \\ 1 & 0 \end{bmatrix}$$

$$+ (\operatorname{Im} b) \begin{bmatrix} 0 & i \\ -i & 0 \end{bmatrix}, a, \operatorname{Re} b, \operatorname{Im} b \in \mathbf{R} \qquad (B.15)$$

Note. The bundle is generated by the imaginary circle $|z|^2 + 1 = 0$ and the lines $\operatorname{Im} z = 0$ and $\operatorname{Re} z = 0$. The circles $a|z|^2 + b\bar{z} + \bar{b}z + a = 0$ in the bundle always intersect orthogonally the unit circle $|z| = 1$ (see Fig. B.6).

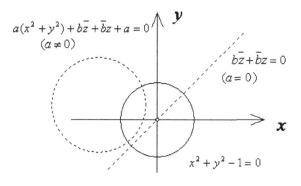

Fig. B.6

Given a *pair* of circles H_1 and H_2, represented by

$$H_j = \begin{bmatrix} a_j & b_j \\ \bar{b}_j & c_j \end{bmatrix}, \quad j = 1, 2.$$

The totality of circles $\lambda_1 H_1 + \lambda_2 H_2$, $\lambda_1, \lambda_2 \in \mathbf{R}$, is said to form a *pencil of circles* generated by the pair H_1 and H_2. If a circle $H = \begin{bmatrix} a & b \\ \bar{b} & c \end{bmatrix}$ intersects orthogonally with H_1 and H_2, then it intersects orthogonally with each member of the pencil $\lambda_1 H_1 + \lambda_2 H_2$; in this case, we have (see (B.12))

$$\operatorname{tr}(FHF\overline{H_1}) = 0 \Rightarrow aa_1 + cc_1 = b\bar{b}_1 + \bar{b}b_1;$$
$$\operatorname{tr}(FHF\overline{H_2}) = 0 \Rightarrow aa_2 + cc_2 = b\bar{b}_2 + \bar{b}b_2.$$

This indicates that the elements a, b, and c of H should satisfy the above two linear relations. Let

$$\Delta = \begin{vmatrix} a_1 & c_1 \\ a_2 & c_2 \end{vmatrix} = a_1 c_2 - a_2 c_1.$$

In case $\Delta \neq \mathbf{0}$, then H can be written in terms of b and H_1, H_2 as follows

$$H = \begin{bmatrix} \dfrac{b}{\Delta}\begin{vmatrix} \overline{b_1} & c_1 \\ \overline{b_2} & c_2 \end{vmatrix} + \dfrac{\overline{b}}{\Delta}\begin{vmatrix} b_1 & c_1 \\ b_2 & c_2 \end{vmatrix} & b \\[3mm] \overline{b} & \dfrac{b}{\Delta}\begin{vmatrix} a_1 & \overline{b_1} \\ a_2 & \overline{b_2} \end{vmatrix} + \dfrac{\overline{b}}{\Delta}\begin{vmatrix} a_1 & b_1 \\ a_2 & b_2 \end{vmatrix} \end{bmatrix}$$

$$= (\operatorname{Re} b) H^{(1)} + (\operatorname{Im} b) H^{(2)}, \operatorname{Re} b, \operatorname{Im} b \in \mathbf{R} \tag{B.16}$$

where $H^{(1)}$ and $H^{(2)}$ are two fixed circles. *In case* $\Delta = \mathbf{0}$, a similar expression can also be obtained (try it out). As a consequence, the totality of these circles H that are orthogonal to both H_1 and H_2 forms a *pencil of circles, conjugate* to (the pencil of circles generated by) the pair of circles H_1 and H_2: it is called *parabolic* if H_1 and H_2 are tangent to each other (see Fig. B.7); is *elliptic* if H_1 and H_2 does not intersect each other (see Fig. B.8), and is *hyperbolic* if H_1 and H_2 intersects at two distinct points (see Fig. B.9).

Note that each of the right figures in Figs. B.7–9 is the canonical form and is obtained via an obvious bilinear transformation (see Exercise (1) above) from the left figure. Refer to Fig. B.10.

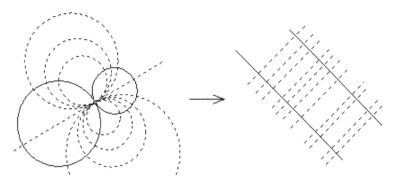

Fig. B.7

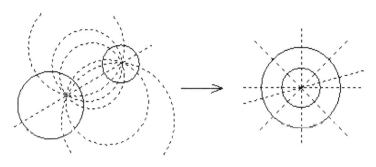

Fig. B.8

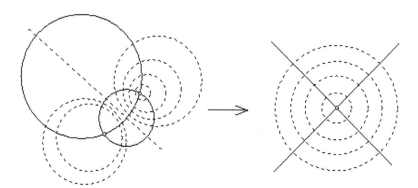

Fig. B.9

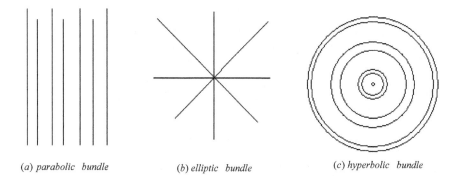

(*a*) *parabolic bundle*   (*b*) *elliptic bundle*   (*c*) *hyperbolic bundle*

Fig. B.10

Exercise (5) To find the number of the point circles in the pencil $\{\lambda_1 H_1 + \lambda_2 H_2 | \lambda_1, \lambda_2 \in \mathbf{R}\}$ is equivalent to solves

$$\operatorname{tr}\left(F(\lambda_1 H_1 + \lambda_2 H_2)F(\lambda_1 \bar{H}_1 + \lambda_2 \bar{H}_2\right) = -2\det(\lambda_1 H_1 + \lambda_2 H_2)$$
$$= -2[(\lambda_1 a_1 + \lambda_2 a_2)(\lambda_1 c_1 + \lambda_2 c_2) - |\lambda_1 b_1 + \lambda_2 b_2|^2] = 0. \qquad (*_1)$$

This results in the following quadratic form in λ_1 and λ_2

$$Q(\lambda_1, \lambda_2) = (a_1 c_1 - |b_1|^2)\lambda_1^2 + (a_1 c_2 + a_2 c_1 - \bar{b}_1 b_2 - b_1 \bar{b}_2)\lambda_1 \lambda_2$$
$$+ (a_2 c_2 - |b_2|^2)\lambda_2^2 = \lambda A \lambda^*, \quad \text{where } \lambda = (\lambda_1 \lambda_2) \in \mathbf{R}^2 \quad \text{and}$$

$$A = \begin{bmatrix} a_1 c_1 - |b_1|^2 & \frac{1}{2}(a_1 c_2 + a_2 c_1 - \bar{b}_1 b_2 - b_1 \bar{b}_2) \\ \frac{1}{2}(a_1 c_2 + a_2 c_1 - \bar{b}_1 b_2 - b_1 \bar{b}_2) & a_2 c_2 - |b_2|^2 \end{bmatrix}.$$

$$\text{(B.17)}$$

Then (recall what $\det H > 0$, $=0$, <0 mean in (B.9)), we have:

$\det A = 0 \Leftrightarrow (*_1)$ has a real (coincident) solution. $\Leftrightarrow$ a real point circle;
$\det A > 0 \Leftrightarrow (*_1)$ does not have real solution. $\Leftrightarrow$ an imaginary point circle;
$\det A < 0 \Leftrightarrow (*_1)$ does have two distinct real solutions. $\Leftrightarrow$ two real point circles,

and the corresponding pencils are parabolic, elliptic, and hyperbolic, respectively.

(a) Try to use bilinear transformations to show that

 1. *Parabolic pencil* (see Fig. B.10(a)):

$$\lambda_1 \begin{bmatrix} 0 & 0 \\ 0 & 1 \end{bmatrix} + \lambda_2 \begin{bmatrix} 0 & 1 \\ 1 & 0 \end{bmatrix}, \quad \lambda_1, \lambda_2 \in \mathbf{R}. \qquad \text{(B.18)}$$

 2. *Elliptic pencil* (see Fig. B.10(b)):

$$\lambda_1 \begin{bmatrix} 0 & 1 \\ 1 & 0 \end{bmatrix} + \lambda_2 \begin{bmatrix} 0 & i \\ -i & 0 \end{bmatrix}, \quad \lambda_1, \lambda_2 \in \mathbf{R}. \qquad \text{(B.19)}$$

 3. *Hyperbolic pencil* (see Fig. B.10(c))

$$\lambda_1 \begin{bmatrix} 1 & 0 \\ 0 & 0 \end{bmatrix} + \lambda_2 \begin{bmatrix} 0 & 0 \\ 0 & 1 \end{bmatrix}, \quad \lambda_1, \lambda_2 \in \mathbf{R}. \qquad \text{(B.20)}$$

(b) Two pencil of circles are said to be *conjugate* to each other if they are orthogonal, namely, each member of one pencil is always orthogonal to every member of the other. Show that 1 in (a) is conjugate to the pencil

$$\lambda_1 \begin{bmatrix} 0 & 0 \\ 0 & 1 \end{bmatrix} + \lambda_2 \begin{bmatrix} 0 & i \\ -i & 0 \end{bmatrix}, \quad \lambda_1, \lambda_2 \in \mathbf{R};$$

while 2 and 3 are conjugate to each other.

Exercise (6) Similar procedure as in Exercise (5) can be used to find the number of point circles in a bundle of circles $\lambda_1 H_1 + \lambda_2 H_2 + \lambda_3 H_3$, $\lambda_1, \lambda_2, \lambda_3 \in \mathbf{R}$ (see Exercise (4)). That is to solve

$$tr(F(\lambda_1 H_1 + \lambda_2 H_2 + \lambda_3 H_3)F(\lambda_1 \overline{H_1} + \lambda_2 \overline{H_2} + \lambda_3 \overline{H_3}))$$

$$= -2 \det(\lambda_1 H_1 + \lambda_2 H_2 + \lambda_3 H_3) = 0.$$

And hence, we introduce the quadratic form in $\lambda_1, \lambda_2, \lambda_3$

$$Q(\lambda_1, \lambda_2, \lambda_3) = \det(\lambda_1 H_1 + \lambda_2 H_2 + \lambda_3 H_3)$$

$$= \lambda A \lambda^*, \quad \text{where} \quad \lambda = (\lambda_1, \lambda_2, \lambda_3) \in \mathbf{R}^3 \quad \text{and}$$

$$A = \begin{bmatrix} a_1 c_1 - |b_1|^2 & \begin{aligned} &\tfrac{1}{2}(a_1 c_2 + a_2 c_1 \\ &-b_1\overline{b_2} - \overline{b_1}b_2) \end{aligned} & \begin{aligned} &\tfrac{1}{2}(a_1 c_3 + a_3 c_1 \\ &-b_1\overline{b_3} - \overline{b_1}b_3) \end{aligned} \\ \begin{aligned} &\tfrac{1}{2}(a_1 c_2 + a_2 c_1 \\ &-b_1\overline{b_2} - \overline{b_1}b_2) \end{aligned} & a_2 c_2 - |b_2|^2 & \begin{aligned} &\tfrac{1}{2}(a_2 c_3 + a_3 c_2 \\ &-b_2\overline{b_3} - \overline{b_2}b_3) \end{aligned} \\ \begin{aligned} &\tfrac{1}{2}(a_1 c_3 + a_3 c_1 \\ &-b_1\overline{b_3} - \overline{b_1}b_3) \end{aligned} & \begin{aligned} &\tfrac{1}{2}(a_2 c_3 + a_3 c_2 \\ &-b_2\overline{b_3} - \overline{b_2}b_3) \end{aligned} & a_3 c_3 - |b_3|^2 \end{bmatrix}.$$

$$\text{(B.21)}$$

Then

$\det A = 0 \Leftrightarrow$ only one real point circle;

$\det A > 0 \Leftrightarrow$ no point circle;

$\det A < 0 \Leftrightarrow$ infinitely many real point circles.

The associated bundles are parabolic, elliptic and hyperbolic, respectively.

(a) Via bilinear transformations, show that

1. *Parabolic bundle* (see Figs. B.1 and B.4):

$$\lambda_1 \begin{bmatrix} 1 & 0 \\ 0 & 0 \end{bmatrix} + \lambda_2 \begin{bmatrix} 0 & 1 \\ 1 & 0 \end{bmatrix} + \lambda_3 \begin{bmatrix} 0 & i \\ -i & 0 \end{bmatrix} = \begin{bmatrix} \lambda_1 & \lambda_2 + i\lambda_3 \\ \lambda_2 - i\lambda_3 & 0 \end{bmatrix},$$

$$\lambda_1, \lambda_2, \lambda_3 \in \mathbf{R}. \qquad \text{(B.22)}$$

In this case,

$$Q(\lambda_1, \lambda_2, \lambda_3) = (\lambda_1\lambda_2\lambda_3) \begin{bmatrix} 0 & 0 & 0 \\ 0 & -1 & 0 \\ 0 & 0 & -1 \end{bmatrix} \begin{bmatrix} \lambda_1 \\ \lambda_2 \\ \lambda_3 \end{bmatrix} = -\lambda_2^2 - \lambda_3^2;$$

and $-\lambda_2^2 - \lambda_3^2 = 0 \Leftrightarrow \lambda_2 = \lambda_3 = 0$. The only real point circle is $|z|^2 = 0$.

2. *Elliptic bundle* (see Figs. B.2 and B.5):

$$\lambda_1 \begin{bmatrix} 1 & 0 \\ 0 & -1 \end{bmatrix} + \lambda_2 \begin{bmatrix} 0 & 1 \\ 1 & 0 \end{bmatrix} + \lambda_3 \begin{bmatrix} 0 & i \\ -i & 0 \end{bmatrix} = \begin{bmatrix} \lambda_1 & \lambda_2 + i\lambda_3 \\ \lambda_2 - i\lambda_3 & -\lambda_1 \end{bmatrix},$$

$$\lambda_1, \lambda_2, \lambda_3 \in \mathbf{R}. \tag{B.23}$$

In this case,

$$Q(\lambda_1, \lambda_2, \lambda_3) = (\lambda_1\lambda_2\lambda_3) \begin{bmatrix} 1 & 0 & 0 \\ 0 & 1 & 0 \\ 0 & 0 & 1 \end{bmatrix} \begin{bmatrix} \lambda_1 \\ \lambda_2 \\ \lambda_3 \end{bmatrix} = \lambda_1^2 + \lambda_2^2 + \lambda_3^2;$$

and $Q(\lambda_1, \lambda_2, \lambda_3) = 0 \Leftrightarrow \lambda_1 = \lambda_2 = \lambda_3 = 0$. There is no real point circle.

3. *Hyperbolic bundle* (see Figs. B.3 and B.6):

$$\lambda_1 \begin{bmatrix} 1 & 0 \\ 0 & 1 \end{bmatrix} + \lambda_2 \begin{bmatrix} 0 & 1 \\ 1 & 0 \end{bmatrix} + \lambda_3 \begin{bmatrix} 0 & i \\ -i & 0 \end{bmatrix} = \begin{bmatrix} \lambda_1 & \lambda_2 + i\lambda_3 \\ \lambda_2 - i\lambda_3 & \lambda_1 \end{bmatrix},$$

$$\lambda_1, \lambda_2, \lambda_3 \in \mathbf{R}. \tag{B.24}$$

In this case,

$$Q(\lambda_1, \lambda_2, \lambda_3) = (\lambda_1\lambda_2\lambda_3) \begin{bmatrix} -1 & 0 & 0 \\ 0 & 1 & 0 \\ 0 & 0 & 1 \end{bmatrix} \begin{bmatrix} \lambda_1 \\ \lambda_2 \\ \lambda_3 \end{bmatrix} = -\lambda_1^2 + \lambda_2^2 + \lambda_3^2;$$

and $Q(\lambda_1, \lambda_2, \lambda_3) = 0 \Leftrightarrow \lambda_1^2 = \lambda_2^2 + \lambda_3^2$. There are infinitely many real point circles $(x - \frac{\lambda_2}{\sqrt{\lambda_2^2+\lambda_3^2}})^2 + (y - \frac{\lambda_3}{\sqrt{\lambda_2^2+\lambda_3^2}})^2 = 0, \lambda_2, \lambda_3 \in \mathbf{R}$ (at least one of λ_2 and λ_3 is not equal to zero).

(b) "Passing a point" means "orthogonal to a point circle". Under this convention, these circles in a bundle that pass a point, other than the common point of the parabolic bundle, form a pencil of circles. Show that

1. Parabolic bundle contains parabolic and elliptic pencils.
2. Elliptic bundle contains only elliptic pencils.

3. Hyperbolic bundle contains parabolic, elliptic, and hyperbolic pencils.

Section (3) Rigid motions

Recall (2.5.4.12): The *symmetric motion* (or inversion or reflection) of a point z with respect to a circle $\alpha|z|^2 + \bar{\beta}z + \beta\bar{z} + \gamma = 0$ is given by

$$z \to z^* = \frac{-\beta\bar{z} - \gamma}{\alpha\bar{z} + \bar{\beta}}, \tag{B.25}$$

which is the composite map of $z \to \zeta = \bar{z}$ and $\zeta \to z^* = \frac{-\beta\zeta - \gamma}{\alpha\bar{\zeta} + \bar{\beta}}$. And two such consecutive motions result in a linear transformation.

Our main scheme is the following

Parabolic, elliptic, hyperbolic geometries and their respective groups of rigid motions. The totality of these linear transformations obtained by performing two consecutive symmetric motions with respect to members of a fixed bundle of circles forms a *subgroup* of the group of all linear transformations (see (1) in (2.5.4.13)). As a matter of fact, if the bundle is

1. *Parabolic*, the corresponding subgroup is the *Euclidean (motion) group*

$$\{az + b \,|\, a, b \in \mathbf{C} \quad \text{and} \quad |a| = 1\}.$$

2. *Elliptic*, the subgroup is the *unitary group*

$$G_e = \left\{ e^{i\theta} \frac{z - z_0}{1 + \bar{z}_0 z} \,\middle|\, \theta \in \mathbf{R} \quad \text{and} \quad z_0 \in \mathbf{C}^* \right\} \quad \text{or}$$

$$\left\{ \frac{az - b}{\bar{b}z + \bar{a}} \,\middle|\, a, b \in \mathbf{C} \quad \text{and} \quad |a|^2 + |b|^2 \neq 0 \right\}.$$

3. *Hyperbolic*, the subgroup is the *noneuclidean (motion) group* or the *Lobachevski group*

$$G_h = \left\{ e^{i\theta} \frac{z - z_0}{1 - \bar{z}_0 z} \,\middle|\, \theta \in \mathbf{R} \quad \text{and} \quad |z_0| < 1 \right\} \quad \text{or}$$

$$\left\{ \frac{az + \bar{b}}{bz + \bar{a}} \,\middle|\, a, b \in \mathbf{C} \quad \text{and} \quad |a|^2 - |b|^2 > 0 \right\}.$$

The *invariants* under the action of the subgroup on members of the associated bundle of circles form a *geometry*, and in this geometry, the subgroup is termed as the *group of rigid motions*. Consequently 1, 2, and 3 result in respectively the *parabolic* (or *Euclidean*) *geometry*, the *elliptic* (or *spherical*) *geometry*, and the *hyperbolic* (or *Lobachevski* or *noneuclidean*) *geometry*. (B.26)

When performing two consecutive symmetric motions with respect to two circles (or lines) in a bundle, it can be shown that one of the circles can be chosen arbitrarily but still remains in the same bundle. For the proofs of this and the former part of (B.26), refer to Ref. [14], pp. 41–42 and p. 49.

Proof of the latter part of (B.26). We choose the two consecutive symmetric motions as follows:

1. first, the symmetric motion $z \to \bar{z}$ with respect to the real axis;
2. second, the symmetric motion $z \to z^*$ as indicated in (B.25) with respect to an arbitrary circle (line) in the same bundle on $\mathbf{C}^*$.   (B.27)

To write out explicitly the motion $z \to z^*$ in 2, we adopt the equation shown in (B.3):

$$a(z + \bar{z}) + ib(\bar{z} - z) + c(|z|^2 - 1) + d(|z|^2 + 1) = 0,$$

$$d = -a\zeta_0 - b\eta_0 - c\xi_0$$

$$\Rightarrow \text{ (under the symmetric motion } z \to z^*)$$

$$a(z^* + \bar{z}) + ib(\bar{z} - z^*) + c(z^*\bar{z} - 1) + d(z^*\bar{z} + 1) = 0 \qquad (*_2)$$

where a, b, c, and d are real parameters, not all equal to zero. Therefore the *resulted linear transformation* under (B.27) is then (using $\bar{\bar{z}} = z$ to replace $\bar{z}$, w to replace z^* in $(*_2)$)

$$a(w + z) + ib(z - w) + c(wz - 1) + d(wz + 1) = 0$$

$$\Rightarrow w = \frac{-(a + ib)z - (-c + d)}{(c + d)z + (a - ib)}, \quad a, b, c, d \in \mathbf{R} \text{ but not all}$$

equal to zero. $(*_3)$

$$\Rightarrow \text{ (under the additional assumption that a fixed point } z_0 \neq 0$$
$$\text{is mapped into zero)}$$

$$w = \frac{\bar{z_0}}{z_0} \frac{(c - d)(z_0 - z)}{(c - d) + (c + d)\bar{z_0}z}, \quad \text{where } z_0 = |z_0|e^{i\theta_0}. \qquad (B.28)$$

Fix θ_0 and let $z_0 \to 0$ while $c - d \neq 0$ (namely, the original circle in (B.3) does not pass the origin), and we obtain $w = -e^{-2i\theta_0}z$. This means that the rigid motion keeping the origin o fixed should be of the form

$$w = e^{i\varphi}z, \quad \varphi \in \mathbf{R}. \qquad (B.29)$$

In conclusion, the most general linear transformation obtained under (B.27) is the composite of the following two mappings:

1. a rotation keeping o fixed, such as (B.29);
2. a linear transformation mapping a point $z_0 \neq 0$ into o, such as (B.28) by dropping

$$\frac{\overline{z_0}}{z_0} = e^{-2i\theta_0}.$$ (B.30)

In short, (B.28) itself is such a mapping.   □

Exercise (7) To finish the proof of (B.26), do the following:

1. *Parabolic bundle*: Choosing $d = -c$ in (B.3) (see (B.4)), then (B.28) becomes

$$w = -\frac{\overline{z_0}}{z_0}(z - z_0) = e^{i\theta}(z - z_0), \quad \text{where} \quad e^{i\theta} = -\frac{\overline{z_0}}{z_0}.$$

2. *Elliptic bundle*: Choosing $d = 0$ in (B.3) (see (B.5)), then (B.28) becomes

$$w = -\frac{\overline{z_0}}{z_0}\frac{z - z_0}{1 + \overline{z_0}z} = e^{i\theta}\frac{z - z_0}{1 + \overline{z_0}z}, \quad \text{where} \quad e^{i\theta} = -\frac{\overline{z_0}}{z_0}$$

$$= \frac{az - b}{\overline{b}z + \overline{a}}, \quad \text{where} \quad e^{i\theta} = \frac{a}{\overline{a}} \text{ and } z_0 = \frac{b}{a}.$$

3. *Hyperbolic bundle*: Choosing $c = 0$ in (B.3) (see (B.6)), then (B.28) becomes

$$w = -\frac{\overline{z_0}}{z_0}\frac{z - z_0}{1 - \overline{z_0}z} = e^{i\theta}\frac{z - z_0}{1 - \overline{z_0}z}, \quad \text{where} \quad e^{i\theta} = -\frac{\overline{z_0}}{z_0}$$

$$= \frac{az + b}{\overline{b}z + \overline{a}}, \quad \text{where} \quad e^{i\theta} = \frac{a}{\overline{a}} \text{ and } z_0 = -\frac{b}{a}.$$

Why $|z_0| < 1$ and $|a|^2 - |b|^2 > 0$ hold? This is because $-(a + ib)z_0 = d$ and $c = 0$ in (*3). Since $d^2 < a^2 + b^2$ in (*3), it follows that $|z_0| < 1$ and hence $|z_0| = \frac{|b|}{|a|} < 1$ in the new defining relation $z_0 = -\frac{b}{a}$.

Section (4) Parabolic geometry

An Euclidean motion $w = e^{i\theta}z + b$ preserves distance between two points z_1 and z_2, namely, $|w_1 - w_2| = |z_1 - z_2|$, where $w_k = e^{i\theta}z_k + b$ for $k = 1, 2$. Conversely, if $|w_1 - w_2| = |z_1 - z_2|$ holds for points z_1 and z_2, w_1 and w_2, then the Euclidean motion $w = e^{i\theta}(z - z_1) + w_1$, $\theta = \text{Arg}(w_2 - w_1)/(z_2 - z_1)$, maps z_1 and z_2 into w_1 and w_2, respectively. This property of distance-preserving

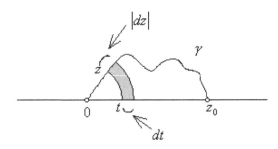

Fig. B.11

also shares by the symmetric motion $z \to \bar{z}$. Consequently, the Euclidean motion group and the symmetric motion $z \to \bar{z}$ together generate a group, called the *generalized Euclidean (motion) group*.

As a matter of fact, the Euclidean distance $|z_1 - z_2|$ between two points z_1 and z_2 is the *shortest* in length among the lengths of all rectifiable curves connecting z_1 and z_2. Owing to its invariance under generalized Euclidean motion (namely, under translations and rotations), we may suppose that $z_1 = 0$ and $z_2 = z_0 > 0$. Let γ be a differentiable curve connecting 0 to z_0 (see Fig. B.11). Take a point z on γ and set $|z| = t$. The arc length element $|dz| = \sqrt{dx^2 + dy^2}$ at z satisfies

$$|dz| \ge dt,$$

because $dt = d|z| = d(\sqrt{x^2 + y^2}) = \dfrac{xdx + ydy}{\sqrt{x^2 + y^2}}$ and then

$(\sqrt{x^2 + y^2})^2 |dz|^2 \ge (xdx + ydy)^2$ holds. Hence

$$\text{the arc length of } \gamma = \int_\gamma |dz| \ge \int_0^{z_0} dt = z_0 - 0$$

$$= \text{the length of the segment connecting 0 to } z_0.$$

$$(B.31)$$

This proves the claim.

We summarize as the following

Parabolic (Euclidean) geometry.

Space: the finite complex plane $\mathbf{C}$.

Group: the generalized Euclidean group G_p generated by $\{e^{i\theta}z + b | \theta \in \mathbf{R} \text{ and } b \in \mathbf{C}\}$ and the symmetric motion $z \to \bar{z}$.

(A) points:

1. Points in $\mathbf{C}$ are *transitive* (namely, for any two points $z_1, z_2 \in \mathbf{C}$, there is a $T \in G_p$ so that $T(z_1) = z_2$ holds).
2. The *distance* $|z_1 - z_2|$ between two points z_1 and z_2 is the only *invariant* of z_1 and z_2; moreover, $|z_1 - z_2|$ is the shortest length among lengths of curves connecting z_1 and z_2.

(B) Lines:

1. There is only one straight line, called *geodesic*, passing two distinct points.
2. The totality of lines is *transitive* (namely, for any two lines L_1, L_2 in $\mathbf{C}$, there is a $T \in G_p$ so that $T(L_1) = L_2$; or, circles in a parabolic bundle are always transformed to circles in the bundle under G_p).
3. The *angle* between two lines is the only *invariant* of these two lines: In particular,

   a. $w = e^{i\theta} + b$ preserves the magnitude and sense of an angle.

   b. $w = \bar{z}$ preserves the magnitude but reverses the sense of an angle.

4. (Parallelism axiom) There exists only one line passing a given point that does not intersect a given line. (B.32)

In short, length and angle constitute the essential elements in the content of the Euclidean geometry.

Section (5) Elliptic geometry

Let us make certainty what geometric mapping properties a member

$$w = \frac{az - b}{\bar{b}z + \bar{a}} = e^{i\varphi}\frac{z - z_0}{1 + \overline{z_0}z} \quad \left(e^{i\varphi} = \frac{a}{\bar{a}}, z_0 = \frac{b}{a}\right) \tag{B.33}$$

of the unitary group G_e (see (B.26)) might have. In case $b = 0$, $w = e^{i\varphi}z$ represents a rotation with fixed points 0 and ∞. If $b \neq 0$, $w = w(z)$ has two distinct fixed points $\alpha_0 = \frac{-(\bar{a}-a)+\sqrt{(\bar{a}-a)^2-4|b|^2}}{2b}$ and $-\frac{1}{\overline{\alpha_0}}$; and, in this case, $w = w(z)$ can be rewritten as

$$\frac{w - \alpha_0}{1 + \overline{\alpha_0}w} = e^{i\theta}\frac{z - \alpha_0}{1 + \overline{\alpha_0}z}, \quad \text{where} \quad e^{i\varphi} = e^{-i\theta}\frac{|z_0|^2 + e^{i\theta}}{|z_0|^2 + e^{-i\theta}}. \tag{B.33}'$$

Note that the preimages P and P' on the Riemann sphere of α_0 and $-\overline{\alpha_0}^{-1}$ (0 and ∞ are included) under the stereographic projection are antipodal points of the sphere. Consequently (B.33)$'$ or (B.33) represents geometrically the rotation of the sphere with the diameter PP' as axis and through the angle θ (see Exercise B(2) of Sec. 1.6 and Fig. 1.36). For fixed α_0 and varying $\theta \in \mathbf{R}$ in (B.33)$'$, the set of all such mappings constitute a group of rotations of the sphere with fixed axis PP'. This observation leads to

Exercise (8) *The realization of the unitary group G_e as the rotation group of the Riemann sphere.*

The elliptic bundle $\gamma(|z|^2 - 1) + (\alpha - i\beta)z + (\alpha + i\beta)\bar{z} = 0$ on $\mathbf{C}^*$ (see (B.5)), where α, β, γ are real parameters not all equal to zero, turns out to be the elliptic bundle on the Riemann sphere (see (B.2)):

$$x_1^2 + x_2^2 + x_3^2 = 1,$$

$$\alpha x_1 + \beta x_2 + \gamma x_3 = 0 \text{ (a plane passing } (0,0,0))$$

under the stereographic projection. Then

(a) A symmetric motion (B.25) on $\mathbf{C}^*$ with respect to a member of the elliptic bundle on $\mathbf{C}^*$ is equivalent to a *symmetric motion* on the sphere with respect to a member of the elliptic bundle on the sphere.

(b) The bilinear transformation (B.33) resulted from two such consecutive motions with respect to two circles on $\mathbf{C}^*$ is equivalent to a rotation (B.33)$'$ of the sphere with the two antipodal points of its axis corresponding to the two points of intersection of the two circles.

By a *symmetric motion* (or orthogonal reflection) of a point (x_1, x_2, x_3) with respect to a plane $\alpha x_1 + \beta x_2 + \gamma x_3 = 0$, we mean the mapping

$$(x_1, x_2, x_3) \rightarrow (x_1, x_2, x_3) \cdot \frac{2}{\alpha^2 + \beta^2 + \gamma^2}$$

$$\begin{bmatrix} \dfrac{-\alpha^2 + \beta^2 + \gamma^2}{2} & -\alpha\beta & -\alpha\gamma \\ -\alpha\beta & \dfrac{\alpha^2 - \beta^2 + \gamma^2}{2} & -\beta\gamma \\ -\alpha\gamma & -\beta\gamma & \dfrac{\alpha^2 + \beta^2 - \gamma^2}{2} \end{bmatrix}.$$

For details, refer to Ref. [56], Vol. 1, pp. 624–626.

What are possible invariants under the unitary group G_e?

Suppose $w = w(z)$ is a mapping in G_e that maps a given point z_1 into another given point w_1. The transformations $\zeta = \frac{z - z_1}{1 + \bar{z}_1 z}$ and $\eta = \frac{w - w_1}{1 + \bar{w}_1 w}$

are in G_e, too, and map z_1 into 0 in the z-plane and w_1 into 0 in the η-plane, respectively. The composite map $\eta = w(z^{-1}(\zeta)) = \eta(\zeta)$ is in G_e and satisfies $\eta(0) = 0$. Hence $\eta = e^{i\psi}z$ for some real ψ; and then the original $w = w(z)$ can be expressed as

$$\frac{w - w_1}{1 + \overline{w_1}w} = e^{i\psi}\frac{z - z_1}{1 + \overline{z_1}z}, \quad z \in \mathbf{C}^*, \quad \psi \in \mathbf{R}$$

$$\Rightarrow \text{(set } z = z_2 \text{ and } w_2 = w(z_2)) \left|\frac{w_2 - w_1}{1 + \overline{w_1}w_2}\right| = \left|\frac{z_2 - z_1}{1 + \overline{z_1}z_2}\right|.$$

This means that the quantity

$$\left|\frac{z_1 - z_2}{1 + \overline{z_2}z_1}\right|, \quad z_1, z_2 \in \mathbf{C}^* \left(\text{it is } \frac{1}{|z_1|} \text{ in case } z_2 = \infty\right) \tag{B.34}$$

is *invariant* under the unitary group G_e. Denote this as $d(z_1, z_2)$ temporarily. Observe that $d(0, 1) + d(1, 2) = 1 + \frac{1}{3} = \frac{4}{3} < 2 = d(0, 2)$, so $d(z_1, z_2)$ does not satisfy the triangle inequality in general and cannot be adopted as the distance between z_1 and z_2 (as against the Euclidean distance $|z_1 - z_2|$ in (B.32)).

Suppose there is a function $f : \mathbf{R} \to \mathbf{R}$ so that

$$d_e(z_1, z_2) = f\left(\frac{|z_1 - z_2|}{|1 + \overline{z_2}z_1|}\right), \quad z_1, z_2 \in \mathbf{C}^*$$

satisfied the requirement that, for all $\alpha \geq 0$, $\beta \geq 0$,

$$d_e(0, \alpha + \beta) = d_e(0, \alpha) + d_e(\alpha, \alpha + \beta), \text{ namely,}$$

$$f(\alpha + \beta) = f(\alpha) + f\left(\frac{\beta}{1 + (\alpha + \beta)\alpha}\right).$$

Set $\alpha = \beta = 0$ and we get $f(0) = 0$. Let $t = \frac{\beta}{1 + (\alpha + \beta)\alpha}$. The above equation can be rewritten as

$$\frac{f(\alpha + \beta) - f(\alpha)}{\beta} = \frac{f(t)}{t} \cdot \frac{1 - t\alpha}{1 + \alpha^2}$$

$\Rightarrow$ (under the additional assumption that f is differentiable, and letting

$$t \to 0 \text{ (so does } \beta \to 0)) f'(\alpha) = \frac{1}{1 + \alpha^2}f'(0), \alpha \geq 0$$

$\Rightarrow f(\alpha) = f'(0)\tan^{-1}\alpha$

where $f'(0)$ could be any real constant. Since the sphere has its diameter equal to 2, let us choose $f'(0) = 2$. In conclusion, the quantity

$$d_e(z_1, z_2) = \begin{cases} 2\tan^{-1}\dfrac{|z_1 - z_2|}{|1 + \overline{z_2}z_1|}, & z_1, z_2 \in \mathbf{C} \\[2ex] 2\tan^{-1}\dfrac{1}{|z_1|}, & z_1 \in \mathbf{C} \quad \text{and} \quad z_2 = \infty \end{cases} \tag{B.35}$$

is also an *invariant* under G_e. Let Q_1 and Q_2 be the preimages of z_1 and z_2 on the Riemann sphere under the stereographic projection. Then

$d_e(z_1, z_2) =$ the arc length of the smaller circular arc $\overparen{Q_1Q_2}$ lying on

the *great circle* along which the plane passing Q_1, Q_2, and

the center intersects the sphere

$=$ the substanted angle of the spherical chord $\overline{Q_1Q_2}$

with respect to the center   (B.35)$'$

and is called the *spherical surface distance* of z_1 and z_2 (see Exercise B(3) of Sec. 1.6 and Fig. 1.37).

Note that

$$0 \le d_e(z_1, z_2) \le \pi, \quad \text{and}$$

$$d_e(z_1, z_2) = \pi$$

$\Leftrightarrow$ $z_1\overline{z_2} = -1$, namely, the corresponding points Q_1 and Q_2 of z_1 and z_2 are antipodal points on the sphere.

In general, we have

$$d_e(z_1, z_3) \le d_e(z_1, z_2) + d_e(z_2, z_3), \quad z_1, z_2, z_3 \in \mathbf{C}^*, \text{ and the}$$
equality holds

$\Leftrightarrow$ the preimages Q_1, Q_2, Q_3 of z_1, z_2, z_3 (in this ordering) on the

sphere lie on a great circle.   (B.36)

Owing to be invariant under G_e, the sufficiency is a byproduct of the argument leading to (B.35). As for the necessity, see the next paragraph.

Adopt notation in (B-35)$'$ and we try to show that

$d_e(z_1, z_2) =$ the shortest length among the lengths of all rectifiable curves

lying on the Riemann sphere and connecting Q_1 to Q_2. (B.35)$''$

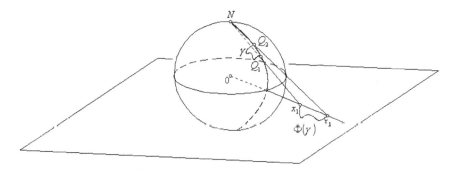

Fig. B.12

Proof. By invoking a sphere rotation, it is harmless to assume that Q_1 and Q_2 lie on the great circle which passes the north pole and is stereographically projected onto the real axis so that z_1 and z_2 are nonnegative reals x_1 and x_2 with $0 \le x_1 \le x_2$. See Fig. B.12. Let γ be such a rectifiable (or continuously differentiable) curve on the sphere. Then its image curve $\Phi(\gamma)$ under the stereographic projection Φ is the one connecting x_1 to x_2. Recall that the spherical arc length element dS and the Euclidean arc length element ds in the plane are related as $dS = \frac{2ds}{1+|z|^2}$ (see Exercise B(1) of Sec. 1.6). Prove this by taking differentials of both sides of (B.33) and compare to (B.42)$'$). Therefore,

$$\text{the length of } \gamma = \int_\gamma dS$$

$$= \int_{\Phi(\gamma)} \frac{2ds}{1+|z|^2} \, (ds = |dz|)$$

$$\ge \int_{x_1}^{x_2} \frac{2dt}{1+t^2} \text{ (by reasoning as in (B.31) and}$$

referring to Fig. B.11)

$$= 2(\tan^{-1} x_2 - \tan^{-1} x_1) = 2\tan^{-1} \frac{x_2 - x_1}{1 + x_1 x_2}$$

$$= 2\tan^{-1} \frac{|z_1 - z_2|}{|1 + \bar{z}_2 z_1|} = d_e(z_1, z_2). \tag{B.38}$$

Note that equality in (B.37) holds $\Leftrightarrow \Phi(\gamma)$ is a line segment. $\Leftrightarrow \gamma$ is a circular arc on a great circle.

In conclusion, we have the

Elliptic (spherical) geometry.

Space: the extended complex plane $\mathbf{C}^*$ or the Riemann sphere S.

Group: the unitary group G_e or the group generated by the rotation group on the sphere (see (B.26) and Exercise (8)) and the symmetric motion with respect to a plane passing the center of the sphere.

Invariant arc length element (see also (B.34)):

$$dS = \frac{2ds}{1+|z|^2}, \quad z \in \mathbf{C} \quad \text{and } ds \text{ is the Euclidean arc length}$$

element in $\mathbf{C}$. $\square$

(A) Points:

1. Points on $\mathbf{C}^*$ (or S) are *transitive*.
2. The *spherical surface distance* $d_e(z_1, z_2)$ (see (B.35) and (B.35)$'$) is the only *invariant* of z_1 and z_2; moreover, in case $z_1 \bar{z}_2 \neq -1$, $d_e(z_1, z_2)$ is the shortest length among lengths of curves connecting z_1 and z_2 (see (B.35)$''$ and (B.38)); in case $z_1 \bar{z}_2 = -1$, $d_e(z_1, z_2) = \pi$ (see (B.36)).
3. $(\mathbf{C}^*, d_e(,))$ is a *complete metric space*:

   $$d_e(z_1, z_2) \geq 0, \quad \text{and } = 0 \Leftrightarrow z_1 = z_2;$$

   $$d_e(z_1, z_2) = d_e(z_2, z_1);$$

   $$d_e(z_1, z_3) \leq d_e(z_1, z_2) + d_e(z_2, z_3), \text{ and equality holds} \Leftrightarrow z_1, z_2, z_3$$

   lie on a great circle (see (B.37)).

(B) Lines (great circles or geodesics):

1. There is only one great circle, also called *geodesic*, passing two distinct points (other than a pair of antipodal points).
2. The totality of lines is *transitive* (namely, circles in an elliptic bundle are always transformed into circles in the bundle under G_e).
3. The *angle* between two lines is the only *invariant* of these two lines.
4. Any two lines cannot be parallel to each other (they are either coincident or intersecting at two distinct points which are antipodal).   (B.39)

Section (6) Hyperbolic geometry

The procedure is similar to that of Section (5), and if such cases do happen, only sketches will be given.

Fix a member of the noneuclidean group G_h (see (B.26))

$$w = \frac{az + b}{\bar{b}z + \bar{a}} = e^{i\varphi}\frac{z - z_0}{1 - \bar{z}_0 z} \quad \left(|a|^2 - |b|^2 > 0 \quad \text{and} \quad e^{i\varphi} = \frac{a}{\bar{a}}, z_0 = -\frac{b}{a}\right).$$

$$\text{(B.40)}$$

$w = w(z)$ maps $|z| < 1$ onto $|w| < 1$. In case $b = 0, w = e^{i\varphi}z$ has two fixed points 0 and ∞. If $b \neq 0$, $w = w(z)$ has fixed points $z_1 = \frac{\sqrt{|b|^2 - (\operatorname{Im} a)^2} + i\operatorname{Im} a}{b}$ and $z_2 = \frac{-\sqrt{|b|^2 - (\operatorname{Im} a)^2} + i\operatorname{Im} a}{b}$. Note that $z_1 z_2 = -\frac{b}{\bar{b}}$. If $(\operatorname{Im} a)^2 > |b|^2$, then $z_2 = \frac{1}{\bar{z}_1}$ and $|z_1| \neq 1$ holds; therefore, either $|z_1| < 1$ or $|z_1| > 1$ is true. If $(\operatorname{Im} a)^2 \leq |b|^2$, then $|z_1| = |z_2| = 1$ but it is possible that $z_1 \neq z_2$ or $z_1 = z_2$. These observations lead to the

Classifications of the noneuclidean motions. Elements in G_h are classified into three types:

1. *Non-Euclidean limit rotation* (parabolic type with only one fixed point on $|z| = 1$):

   Canonical form: $w = z + \lambda$ (λ is real and ∞ is the fixed point).

   General form: $\frac{1}{w - z_0} = \frac{1}{z - z_0} + \lambda$.

   Invariant lines: circles in $|z| < 1$ which are tangent to $|z| = 1$ at z_0 (see Fig. B.13(a)).

2. *Non-Euclidean rotation* (elliptic type with one fixed point in $|z| < 1$):

   Canonical form: $w = e^{i\theta}z$ (θ is real, and 0 and ∞ are fixed points.).

   General form: $\frac{w - z_0}{1 - \bar{z}_0 w} = e^{i\theta}\frac{z - z_0}{1 - \bar{z}_0 z}$ ($|z_0| < 1$).

   Invariant lines: non-Euclidean circles $\left|\frac{z - z_0}{1 - \bar{z}_0 z}\right| = \lambda$ with center at z_0 (see Fig. B.13(b)).

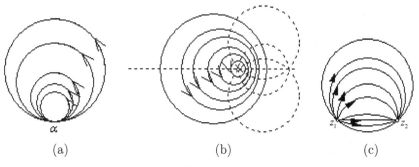

α

(a)   (b)   (c)

Fig. B.13

3. *Non-Euclidean parallelism* (hyperbolic type with two distinct fixed points on $|z| = 1$):

Canonical form: $w = kz$ ($k > 0$, and 0 and ∞ are fixed points.).

General form: $\frac{w - z_1}{w - z_2} = k \frac{z - z_1}{z - z_2}$ ($|z_1| = |z_2| = 1$, $z_1 \neq z_2$).

Invariant lines: hypercircles $\arg \frac{z - z_1}{z - z_2} = \theta$ with z_1 and z_2 as limit points, including the (geodesic) line connecting z_1 and z_2 as the axis (see Fig. B.13(c)). $\hspace{2em}$ (B.41)

By considering the mapping properties of canonical forms or just referring to the properties of Steiner's circles in (1.4.4.2), it is easy to figure out these circles (arcs) that are invariant under the action of the group G_h. See Fig. B.13.

The mapping (B.40) that maps z_1 into w_1 (and hence maps $\frac{1}{\bar{z}_1}$ into $\frac{1}{\bar{w}_1}$) can be rewritten as

$$\frac{w - w_1}{1 - \bar{w}_1 w} = e^{i\psi} \frac{z - z_1}{1 - \bar{z}_1 z}, \quad \psi \in \mathbf{R}, \quad z \in \mathbf{C}^*.$$

This implies that the quality

$$\left| \frac{z_1 - z_2}{1 - \bar{z}_2 z_1} \right|, \quad |z_1| < 1, \quad |z_2| < 1 \hspace{2em} (B.42)$$

is *invariant* under the non-Euclidean group G_h. As a matter of fact, for any $0 < \rho < 1$, there are infinitely many pairs of distinct points z_1 and z_2 in $|z| < 1$ so that $\left| \frac{z_1 - z_2}{1 - \bar{z}_2 z_1} \right| = \rho$ holds.

Suppose there is a differentiable function $f : [0, \infty) \to [0, \infty)$ so that

$$d_h(z_1, z_2) = f\left(\frac{|z_1 - z_2|}{|1 - \bar{z}_2 z_1|} \right), \quad |z_1| < 1, \quad |z_2| < 1$$

satisfies the property: for all $\alpha \geq 0$, $\beta \geq 0$, and $\alpha + \beta < 1$,

$$d_h(0, \alpha + \beta) = d_h(0, \alpha) + d_h(\alpha, \alpha + \beta)$$

always holds. Note that $f(0) = 0$. Set $t = \frac{\beta}{1 - (\alpha + \beta)\alpha}$. Then

$$\frac{f(\alpha + \beta) - f(\alpha)}{\beta} = \frac{f(t)}{t} \cdot \frac{1 + \alpha t}{1 - \alpha^2}$$

$$\Rightarrow \text{(letting } \beta \to 0) \quad f'(\alpha) = f'(0)\frac{1}{1 - \alpha^2}$$

$$\Rightarrow \text{(by choosing } f'(0) = 2) \quad f(\alpha) = \log \frac{1 + \alpha}{1 - \alpha}.$$

Therefore, the quantity

$$d_h(z_1, z_2) = \log \frac{|1 - \bar{z}_2 z_1| + |z_1 - z_2|}{|1 - \bar{z}_2 z_1| - |z_1 - z_2|}, \quad |z_1| < 1, \quad |z_2| < 1 \hspace{2em} (B.43)$$

is also *invariant* under G_h. It enjoys the remarkable property:

$$d_h(z_1, z_3) \le d_h(z_1, z_2) + d_h(z_2, z_3), \quad |z_1| < 1, \quad |z_2| < 1, |z_3| < 1,$$

and the equality holds

$$\Leftrightarrow z_1, z_2, z_3, \quad \text{in this ordering, lie on a circle arc in } |z| < 1 \text{ which is}$$

orthogonal to the circle $|z| = 1$ when extended. $\hspace{2cm}$ (B.44)

Proof. To prove the inequality, we may suppose $z_1 = z$, $z_2 = 0$, and $z_3 = r$ $(0 < r < 1)$. Then the inequality can be simplified to $-|z| \le \operatorname{Re} z$ which is always true (try this out !), with equality if and only if z is a nonpositive real number. Of course, the sufficient of the equality is also contained in the process of the derivation of (B.43). See also (B.44)′ for another proof.

$\hspace{14cm}$ □

On the other hand, the *invariance* of (B.42) can be infinitesimally expressed as

$$\frac{|dw|}{1 - |w|^2} = \frac{|dz|}{1 - |z|^2}, \quad w = w(z) \in G_h. \tag{B.42′}$$

Proof. A rigorous proof is as follow: Take the differential of (B.40), and we obtain

$$dw = e^{i\phi} \frac{1 - |z_0|^2}{(1 - \bar{z}_0 z)^2} dz$$

$$\Rightarrow |dw| = \frac{1 - |z_0|^2}{|1 - \bar{z}_0 z|^2} |dz|.$$

Observe that $1 - |w|^2 = \frac{(1-|z_0|^2)(1-|z|^2)}{|1-\bar{z}_0 z|^2}$, and the result (B.42)′ follows.

$\hspace{14cm}$ □

(B.42)′ gives the *non-Euclidean arc length element*

$$dS = \frac{2ds}{1 - |z|^2}, \quad |z| < 1, \quad ds = |dz| \tag{B.45}$$

which is *invariant* under G_h. Now, choose any two distinct points z_1 and z_2 in $|z| < 1$, and a differentiable curve in $|z| < 1$ which connects them. There is a mapping $T \in G_h$ so that $T(z_1) = 0$ and $T(z_2) = \rho$ $(0 < \rho < 1)$. Under this circumstance, $T(\gamma)$ is a differential curve connecting 0 to ρ. See Fig. B.14. Hence (compare to (B.38))

the length of $\gamma = \displaystyle\int_\gamma dS$

$$= \int_{T(\gamma)} \frac{2|dz|}{1 - |z|^2}$$

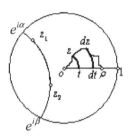

Fig. B.14

$$\geq \int_0^\rho \frac{2dt}{1-t^2} \quad \text{(refer to Fig. B.14)}$$

$$= \log \frac{1+\rho}{1-\rho} (= 2\tanh^{-1}\rho(z_1,z_2)), \quad \text{where } \rho(z_1,z_2) = \left|\frac{z_1-z_2}{1-\bar{z}_2 z_1}\right|$$

$$= \log \frac{|1-\bar{z}_2 z_1| + |z_1-z_2|}{|1-\bar{z}_2 z_1| - |z_1-z_2|} (= d_h(z_1,z_2))$$

$$= \log(z_1, z_2; e^{i\beta}, e^{i\alpha}) \quad \text{(see Fig. B.14)}. \tag{B.46}$$

Moreover, the equality holds in (B.46) if and only if

$$|dz| = dt \quad \text{holds along } T(\gamma).$$

$\Leftrightarrow T(\gamma)$ is a line segment connecting $T(z_1) = 0$ to $T(z_2) = \rho$.

$\Leftrightarrow \gamma = \widehat{z_1 z_2}$ is a circular arc in $|z| < 1$, connecting z_1 to z_2, which lies

on a circle orthogonal to the unit circle $|z| = 1$. \hfill (B.44)'

This reproves (B.44).

As a counterpart of (B.32) and (B.39), we summarize the above as the

Hyperbolic (non-Euclidean or Lobachevski) geometry.

Space: the open unit disk $|z| < 1$.
Group: the noneuclidean group G_h (see (B.26)); more precisely, the group
generated by G_h and the symmetric motion $z \to \bar{z}$.

Invariant arc length element:

$$dS = \frac{2ds}{1-|z|^2}, \quad |z| < 1 \quad \text{and} \quad ds = |dz|.$$

(A) Points:

1. Points on $|z| < 1$ are transitive (moreover, a point with a designated
direction enjoys this property, too).

2. The *non-Euclidean distance* $d_h(z_1, z_2)$ (see (B.43) and (B.46)) is the only *invariant* between z_1 and z_2; moreover, $d_h(z_1, z_2)$ is the shortest length among lengths of curves in $|z| < 1$ which connect z_1 to z_2 (see (B.46)).
3. $|z| < 1$ endowed with $d_h(,)$ is a *complete metric space* (see (B.44) and (B.44)').

(B) Lines (*circular arcs* in $|z| < 1$ whose extended arcs are orthogonal to $|z| = 1$; *geodesics*):

1. There is only one circular arc, called a *geodesic segment*, passing two distinct points, which lies on a circle orthogonal to $|z| = 1$. The whole circular arc of the circle that lies inside $|z| < 1$ is called a *geodesic*, treated as a line.
2. The totality of lines is *transitive*.
3. The *angle* between two intersecting lines is the only *invariant* of these two lines.
4. There are infinitely many lines, passing a given point, which does not intersect a given line (namely, are *parallel* to the given line). See Fig. B.15. (B.47)

Since $|z| < 1$ and the upper half-plane arc conformally equivalent, (B.47) can be transformed to $\operatorname{Im} z > 0$. More precisely, we put it in the following exercise.

Exercise (9) *Poincaré's non-Euclidean geometry.*

Space: the upper half-plane $\operatorname{Im} z > 0$.

Group: the group generated by $\{\frac{az+b}{cz+d} | a, b, c, d \in \mathbf{R} \text{ and } ad - bc > 0\}$ and the symmetric motion $z \to -\bar{z}$ with respect to the imaginary axis.

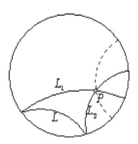

Fig. B.15

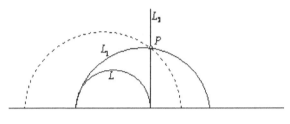

Fig. B.16

Invariant arc length element:

$$dS = \frac{2|dz|}{\text{Im } z}, \quad \text{Im } z > 0.$$

Distance between z_1 and z_2:

$$d_h(z_1, z_2) = \log \frac{|\bar{z}_1 - z_2| + |z_1 - z_2|}{|\bar{z}_1 - z_2| - |z_1 - z_2|}, \quad \text{Im } z_1 > 0, \quad \text{Im } z_2 > 0. \quad (B.48)$$

Try to write out the corresponding details as in (B.47) and prove all of them. See Fig. B.16.

Section (7) The sum of interior angles and the area of a triangle

By a *triangle* $\Delta z_1 z_2 z_3$ with vertices z_1, z_2, and z_3, we mean the figure formed by connecting the vertices by consecutive line segments (geodesics) $z_1 z_2, z_2 z_3$, and $z_3 z_1$.

What we care is

The area and the sum of the interior angles of a triangle $\Delta z_1 z_2 z_3$. Let the interior angles at z_1, z_2, z_3 be $\alpha_1, \alpha_2, \alpha_3$, respectively.

(A) Parabolic geometry: $\alpha_1 + \alpha_2 + \alpha_3 = \pi$.
(B) Elliptic geometry:
   Area: $\alpha_1 + \alpha_2 + \alpha_3 - \pi$;
   Sum of interior angles: $\alpha_1 + \alpha_2 + \alpha_3 > \pi$.
(C) Hyperbolic geometry:
   Area: $\pi - (\alpha_1 + \alpha_2 + \alpha_3)$;
   Sum of interior angles: $\alpha_1 + \alpha_2 + \alpha_3 < \pi$. $\quad$ (B.49)

See Remarks after the proof.

Proof. (B) Treat z_1, z_2, and z_3 as the points A, B, and C on the Riemann sphere whose surface area is equal to 4π. Consider the spherical triangle

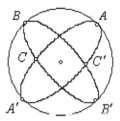

Fig. B.17

$\triangle ABC$ and the antipodal points A', B', C' of its vertices. See Fig. B.17. Then $\triangle ABC(\text{area}) = \triangle A'B'C'(\text{area})$. Now,

$$\triangle ABC + \triangle A'BC + \triangle AB'C + \triangle A'B'C' = \text{the area of half sphere} = 2\pi,$$

where

$\triangle ABC + \triangle A'BC = \text{the area of the lune } ABA'C = 2\angle A;$

$\triangle AB'C = \text{lune } BCB'A - \triangle ABC = 2\angle B - \triangle ABC,$

$\triangle A'B'C = \text{lune } CA'C'A - \triangle A'B'C' = 2\angle C - \triangle ABC.$

Hence

$$2\angle A + (2\angle B - \triangle ABC) + (2\angle C - \triangle ABC) = 2\pi$$

$$\Rightarrow \triangle ABC(\text{area}) = \angle A + \angle B + \angle C - \pi.$$

(C) The non-Euclidean area element is given by

$$dA = \frac{4dxdy}{(1 - |z|^2)^2}, \quad z = x + iy \text{ with } |z| < 1. \tag{B.50}$$

It is *invariant* under G_h, too. We may suppose $z_3 = 0$. Let $r = r(\theta)$ denote the polar equation of the circular arc $\widehat{z_1 z_2}$ which lies on a circle with center at $x_0 > 0$ and radius R. See Fig. B.18. For a point P on $\widehat{z_1 z_2}$, let $re^{i\theta}$ be its

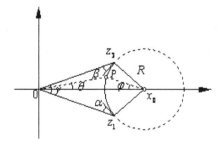

Fig. B.18

polar coordinate with respect to $z_3 = 0$, and $Re^{i\varphi}$ be the one with respect to x_0. Then

$$\tan\theta = \frac{R\sin\varphi}{x_0 - R\cos\varphi}$$

$$\Rightarrow \sec^2\theta d\theta = \frac{R(x_0\cos\varphi - R)}{(x_0 - R\cos\varphi)^2}d\varphi$$

$$\Rightarrow (\text{since } \sec^2\theta = 1 + \tan^2\theta = (x_0 - R\cos\varphi)^{-2}(x_0^2 + R^2 - 2x_0 R\cos\varphi))$$

$$d\theta = \frac{R(x_0\cos\varphi - R)}{x_0^2 + R^2 - 2x_0 R\cos\varphi}d\varphi$$

$$\Rightarrow (\text{since the circle containing } \widehat{z_1 z_2} \text{ is orthogonal to } |z| = 1,$$

$$1 + R^2 = x_0^2 \text{ holds; also } r^2 = x_0^2 + R^2 - 2x_0 R\cos\varphi, \text{ and hence}$$

$$1 - r^2 = 2R(x_0\cos\varphi - R_0))$$

$$d\varphi = \frac{2r^2}{1 - r^2}d\theta.$$

In the quadrilateral $oz_1 x_0 z_2$,

$$\gamma + \left(\alpha + \frac{1}{2}\pi\right) + \angle z_1 x_0 z_2 + \left(\beta + \frac{1}{2}\pi\right) = 2\pi$$

$$\Rightarrow \angle z_1 x_0 z_2 = \pi - (\alpha + \beta + \gamma).$$

By invoking (B.50), we have

the area of $\Delta z_1 z_2 z_3$

$$= \int_{\Delta z_1 z_2 z_3} \frac{4r dr d\theta}{(1 - r^2)^2} = 2\int_{\theta_1}^{\theta_2} \frac{r(\theta)^2}{1 - r(\theta)^2}d\theta = \int_{\varphi_1}^{\varphi_2} d\varphi$$

$$= \angle z_1 x_0 z_2 = \pi - (\alpha + \beta + \gamma).$$

It is interesting to note that this area can also be expressed as 2Arg $(1 - z_1\bar{z}_2)(1 - z_2\bar{z}_3)(1 - z_3\bar{z}_1)$. But why?

$\square$

Finally, two Remarks are provided.

Remark 1 (Gauss curvature). The *Gauss curvature* of a Riemannian metric $ds^2 = \rho^2(dx^2 + dy^2)$ is given by $K(\rho) = -\rho^{-2}\Delta\log\rho$, where $\Delta = \frac{\partial^2}{\partial x^2} + \frac{\partial^2}{\partial y^2}$. It is easy to check that in

1. Parabolic geometry: $K \equiv 0$;

2. Elliptic geometry: $K \equiv 1$; and

3. Hyperbolic geometry: $K \equiv -1$.

(See $(*_{12})$ in Sec. 5.8.4.) By quite elementary method or as an easy consequence of the *Gauss–Bonnet formula*, it can be shown that, on a surface with constant curvature K,

$$K \cdot \text{the area of } \Delta z_1 z_2 z_3 = \alpha_1 + \alpha_2 + \alpha_3 - \pi.$$

Then (B.49) follows obviously.

Remark 2 (Why hyperbolic geometry is useful in complex analysis?). The reason lies on the following

Pick's Lemma. *Analytic functions decrease the non-Euclidean*

1. *distance between two points;*
2. *length of a curve, and*
3. *area of a point set*

in a plane domain (say, in an open disk or hyperbolic Riemann surface).

See (3.4.5.2) and Section (2) of Sec. 5.8.4. It can be used to obtain some remarkable results in complex dynamical system (see Section (3) in Sec. 5.8.5).

Appendix C. Quasiconformal Mappings

Section (1) Some preliminaries

We adopt concepts and notations (such as f_z, $f_{\bar{z}}$, etc.) introduced in Sec. 2.8.

Suppose $w = f(z)$, where $z = x + iy$ and $w = u + iv$, is differentiable in the real sense at a point z_0. The differential (see (2.8.15))

$$dw = f_z dz + f_{\bar{z}} d\bar{z} \tag{C.1}$$

defines infinitesimally an affine transformation from dz to dw. It maps an infinitesimal circle $|dz| = r$ onto a quadratic curve $E(r)$:

$$(v_x^2 + v_y^2)du^2 - 2(u_x v_x + u_y v_y)dudv + (u_x^2 + u_y^2)dv^2 = r^2 J_f^2, \tag{C.2}$$

where $u_x, \ldots, v_y$ are evaluated at z_0, and the Jacobian determinant

$$J_f(z_0) = |f_z(z_0)|^2 - |f_{\bar{z}}(z_0)|^2 = u_x(z_0)v_y(z_0) - u_y(z_0)v_x(z_0). \tag{C.3}$$

Note that $J_f(z_0)^2$ is the discriminant of the coefficients of (C.2).

Let us digress for a moment to the case that (C.2) represents an infinitesimal circle, namely, the conditions $v_x^2 + v_y^2 = u_x^2 + u_y^2 (\neq 0)$ and $u_x v_x + u_y v_y = 0$ hold. Then we recapture (3.2.2), worthy being stated as

Exercise (1) *Geometric characterization of a function differentiable in the complex sense at a point.*

Let $w = f(z)$ be differentiable in the real sense at a point z_0. Then

1. f is differential in the complex sense at z_0 (and $f'(z_0) \neq 0$).
$\Leftrightarrow$ 2. $dw = f_z dz + f_{\bar{z}} d\bar{z}$ maps infinitesimal circles $|dz| = r$ onto infinitesimal circles $|dw| = |f_z(z_0)|r$, namely, $f_{\bar{z}}(z_0) = 0$ and $f'(z_0) = f_z(z_0)$.
$\Leftrightarrow$ 3. The infinitesimal linear operator, where $z = x + iy$ and $w = u + iv$,

$$du = u_x dx + u_y dy$$

$$dv = v_x dx + v_y dy$$

is a conformal mapping on $\mathbf{C}$, namely, there is a $r > 0$ and an orthogonal matrix P so that

$$\begin{pmatrix} u_x & u_y \\ v_x & v_y \end{pmatrix} (z_0) = rP. \tag{C.4}$$

For detailed account, refer to Ref. [56], Vol. 2, Section (2) of Sec. 4.4.

In case $J_f(z_0) \neq 0$, then the quadratic curve (C.2) represents an ellipse and (C.1) is a nonsingular affine transformation. If the additional condition that $J_f(z_0) > 0$ is imposed, then the transformation is a sense-preserving one. See Fig. C.1

An ellipse is characterized completely by the *ratio k* of its major semiaxis to its minor semiaxis, and, if $k > 1$, the *angle* between its major axis and the positive real axis. Henceforth the condition $J_f(z_0) > 0$ is always assumed. And we try to compute the related major and minor semiaxes, and their directions for (C.2).

By (C.1), we have (recall that $|f_z(z_0)| > |f_{\bar{z}}(z_0)|$)

$$(|f_z| - |f_{\bar{z}}|)|dz| \leq |dw| \leq (|f_z| + |f_{\bar{z}}|)|dz|,$$

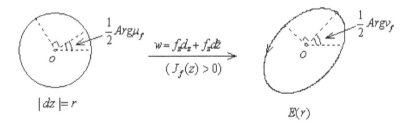

Fig. C.1

and

$$|dw| = (|f_z| + |f_{\bar{z}}|)|dz| \quad \text{(the major semiaxis)}$$

$$\Leftrightarrow \text{(see (1.4.2.2)')} \quad \frac{f_{\bar{z}}d\bar{z}}{f_z dz} \geq 0. \tag{$*_1$}$$

Define the *first complex dilatation* of f at z_0 by

$$\mu_f(z_0) = \frac{f_{\bar{z}}(z_0)}{f_z(z_0)}. \tag{C.5}$$

Then, the *direction* (or angle) of the major axis of $E(r)$ happens if the argument of dz is chosen as

$$\operatorname{Arg} dz = \frac{1}{2}\operatorname{Arg}\mu_f(z_0). \tag{$*_2$}$$

Consequently,

$$dw = f_z dz\left(1 + \frac{f_{\bar{z}}d\bar{z}}{f_z dz}\right)$$

$\Rightarrow$ (by $(*_1)$) the *direction of the major axis* of $E(r)$:

$$\operatorname{Arg} dw = \operatorname{Arg} f_z(z_0) + \operatorname{Arg} dz = \operatorname{Arg} f_z(z_0) + \frac{1}{2}\operatorname{Arg}\mu_f(z_0). \tag{$*_3$}$$

In order to simplify the above expression, we introduce the *second complex dilatation* of f at z_0 by (refer to Exercise B(1)(f) of Sec. 2.8)

$$\upsilon_f(z_0) = \frac{f_{\bar{z}}(z_0)}{\overline{f_z(z_0)}} = \frac{f_{\bar{z}}(z_0)}{\overline{f_{\bar{z}}(z_0)}} = \frac{f_z(z_0)^2}{|f_z(z_0)|^2}\mu_f(z_0). \tag{C.6}$$

Therefore, $(*_3)$ becomes

$$\operatorname{Arg} w = \frac{1}{2}\operatorname{Arg}\upsilon_f(z_0). \tag{$*_3$}'$$

Similar argument leads to

$$|dw| = (|f_z| - |f_{\bar{z}}|)|dz| \quad \text{(the minor semiaxis)}$$

$$\Leftrightarrow \operatorname{Arg} dz = \frac{\pi}{2} + \frac{1}{2}\operatorname{Arg}\mu_f(z_0). \tag{$*_4$}$$

In this case, by the sense-preserving property of the transformation (C.1), it follows that the *direction of the minor axis* of $E(r)$ is given by

$$\operatorname{Arg} dw = \frac{\pi}{2} + \frac{1}{2}\operatorname{Arg}\upsilon_f(z_0). \tag{$*_5$}$$

In conclusion, we summarize as

The infinitesimal mapping properties of $dw = f_z dz + f_{\bar{z}} d\bar{z}$. *Suppose* $w = f(z)$ *is differentiable in the real sense at a point* z_0 *and the Jacobian determinant* $J_f(z_0) = |f_z(z_0)|^2 - |f_{\bar{z}}(z_0)|^2 > 0$ *holds. Then* $dw = f_z(z_0)dz + f_{\bar{z}}(z_0)d\bar{z}$ *is a sense-preserving nonsingular affine transformation, mapping an infinitesimal circle* $|dz| = r$ *one-to-one and onto an infinitesimal ellipse* $E(r)$ *(see (C.2)).*

(a) $E(r)$ has its

$$\text{major semiaxis} = (|f_z(z_0)| + |f_{\bar{z}}(z_0)|)|dz| \quad \text{(in length)};$$

$$\text{minor semiaxis} = (|f_z(z_0)| - |f_{\bar{z}}(z_0)|)|dz|.$$

Define the *dilatation* of f at z_0 as

$$D_f(z_0) = \frac{\text{major semiaxis}}{\text{minor semiaxis}} = \frac{|f_z(z_0)| + |f_{\bar{z}}(z_0)|}{|f_z(z_0)| - |f_{\bar{z}}(z_0)|} \quad (\geq 1).$$

Notice that $D_f \geq 1$ always holds, and $D_f = 1 \Leftrightarrow f_{\bar{z}}(z_0) = 0$, namely, f is differentiable in the complex sense at z_0. Let

$$d_f(z_0) = \frac{|f_{\bar{z}}(z_0)|}{|f_z(z_0)|} \quad (< 1).$$

Then

$$D_f = \frac{1 + d_f}{1 - d_f} \quad \text{and} \quad d_f = \frac{D_f - 1}{D_f + 1}.$$

(b) $E(r)$ has its

direction of the major axis

$$= \frac{1}{2} \text{Arg}\, v_f \quad \text{(the angle inclined to the positive real axis)};$$

$$\text{direction of the minor axis} = \frac{\pi}{2} + \frac{1}{2}\text{Arg}\, v_f,$$

where

the first complex dilatation of f at z_0:

$$\mu_f(z_0) = \frac{f_{\bar{z}}(z_0)}{f_z(z_0)} \quad (\text{and } |\mu_f| = d_f);$$

the second complex dilatation of f at z_0:

$$v_f(z_0) = \frac{f_{\bar{z}}(z_0)}{\overline{f_z(z_0)}} = \frac{f_z(z_0)^2}{|f_z(z_0)|^2}\mu_f(z_0).$$

See Fig. C.1. $(C.6)$

Try to do the following

Exercise (2) Let $\mathbf{R}^2$ be endowed with the natural inner product $\langle\,,\rangle$. Recall that a linear operator $T : \mathbf{R}^2 \to \mathbf{R}^2$ can always be decomposed as $T = U \circ N$, called the *polar decomposition*, where $U : \mathbf{R}^2 \to \mathbf{R}^2$ is an orthogonal operator (i.e., $\langle U\vec{x}, U\vec{y}\rangle = \langle\vec{x}, \vec{y}\rangle$ for all $\vec{x}, \vec{y} \in \mathbf{R}^2$) and $N : \mathbf{R}^2 \to \mathbf{R}^2$ is a positive semidefinite operator (i.e., $\langle N\vec{x}, \vec{y}\rangle = \langle\vec{x}, N\vec{y}\rangle$ for all $\vec{x}, \vec{y} \in \mathbf{R}^2$ and $\langle N\vec{x}, \vec{x}\rangle \geq 0$ for all $\vec{x}$). For details, refer to Ref. [55], Vol. 2, Section (5) of Sec. 4.5. For $r > 0$ and a fixed $\vec{x}_0 \in \mathbf{R}^2$, $\vec{x} \to \vec{x}_0 + rU\vec{x}$ is usually called a *similarity transformation*. Suppose $w = f(z)$ is differentiable in the real sense at z_0.

(a) Suppose $J_f(z_0) \neq 0$. Try to find out similarity transformations so that $z_0 = 0$, $f(z_0) = 0$, and (2.8.1) can be expressed as

$$f(z) = Dx + iy + o(z), \quad D \geq 1.$$

Then, in this case, $f_x(0) = D$ and $f_y(0) = i$. What are $f_{z_0}(0)$, $f_{\bar{z}_0}(0)$, $J_f(0)$, $D_f(0)$, $d_f(0)$, $\mu_f(0)$, and $v_f(0)$.

(b) In case $J_f(z_0) = 0$ and $\max_\theta |df_{z_0}(e^{i\theta})| > 0$, try to find similarity transformations so that $z_0 = 0$, $f(z_0) = 0$, and $f(z) = x + o(y)$. In this case, $f_x(0) = 1$ and $f_y(0) = 0$.

(c) Try to find similarity transformations so that (2.8.2) can be expressed as

$$f(z) = |f_z(0)|z + |f_{\bar{z}}(0)|\bar{z} + o(z).$$

Then, $f_z(0) = |f_z(0)|$ and $f_{\bar{z}}(0) = |f_{\bar{z}}(0)|$.

Section (2) Quasiconformal mappings in the plane

H. Grötzsch: Über einige Extremalprobleme des konformen Abbildung, Ber. Verh. Sächs. Akad. Wiss., Leipzig, 80, 367–376, 497–502 (1928) initiated the concept of quasiconformal mapping, even though it might be traced back to the Beltrami equations (1868, see Exercise (6) below)). A square Q is homeomorphic to a rectangle R which is not a square, but Q and R definitely cannot be conformally equivalent with vertices corresponding to each other (see example in Sec. 6.1.2). The *problem* raised by Grötzsch is: Try to find such a mapping, being as near conformality as possible, that maps Q one-to-one and onto R. How to measure the degree of nearness? It is this idea that starts the concept of the nowaday's quasiconformal mappings. Grötzsch *solved* this *problem* as follows.

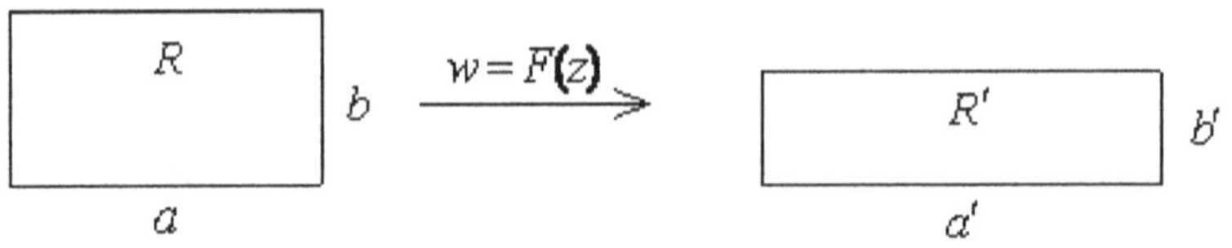

Fig. C.2

Exercise (3) Suppose $w = f(z)$ is a C^1-mapping that maps the rectangle R one-to-one and onto the rectangle R', vertices to vertices, a side to a' side and b side to b' side. See Fig. C.2. We may assume that $\frac{a}{b} \le \frac{a'}{b'}$, otherwise just interchange the ordering of a and b sides. Hence

$$a' \le \int_0^a |df(x + iy)| \le \int_0^a (|f_z| + |f_{\bar{z}}|)dx,$$

$$\Rightarrow a'b \le \int_0^a \int_0^b (|f_z| + |f_{\bar{z}}|)dxdy$$

$$\Rightarrow (a'b)^2 \le \int_0^a \int_0^b \frac{|f_z| + |f_{\bar{z}}|}{|f_z| - |f_{\bar{z}}|}dxdy \cdot \int_0^a \int_0^b (|f_z|^2 - |f_{\bar{z}}|^2)dxdy$$

(by Hölder inequality)

$$= a'b' \int_0^a \int_0^b D_f(z)dxdy$$

$$\Rightarrow \frac{a'}{b'} : \frac{a}{b} \le \frac{1}{ab}\iint_R D_f(z)dxdy \le \sup_{z \in R} D_f(z). \qquad (*_6)$$

To require that $w = f(z)$ should be nearest to a conformal mapping means that $\sup_{z \in R} D_f(z)$ should be as small as possible. And this minimum value $\frac{a'}{b'} : \frac{a}{b}$ is obtained by the affine mapping (regardless of a translation)

$$w = \frac{1}{2}\left(\frac{a'}{a} + \frac{b'}{b}\right)z + \frac{1}{2}\left(\frac{a'}{a} - \frac{b'}{b}\right)\bar{z}.$$

Let

$$M(R) = \frac{a}{b} \quad \text{and} \quad M(R') = \frac{a'}{b'}, \qquad (*_7)$$

called the *moduli* of R and R', respectively. Suppose $K \ge 1$ is a constant so that

$$\sup_{z \in R} D_f(z) \le K; \qquad (*_8)$$

in this case, call $w = f(z)$ a *K-quasiconformal mapping* (see (C.7) below). From $(*_6)$, $(*_7)$, and $(*_8)$, we conclude that *there is a K-quasiconformal* *mapping f from R onto R', if and only if,*

$$\frac{1}{K} \leq \frac{M(R')}{M(R)} \leq K \quad \text{or} \quad \frac{1}{K}M(R) \leq M(R') \leq KM(R). \qquad (*_9)$$

This is the backgroup of *the geometric definition* of a K-quasiconformal mapping on R.

Moreover, this $w - f(z)$ is conformal if and only if

$$\frac{a}{b} = \frac{a'}{b'} \quad \text{or} \quad K = 1 \quad \text{(see Exercise A(6) of Sec. 3.4.6).} \qquad (*_{10})$$

Refer to Ref. [54], p. 16 or [3], pp. 6–8.

Grötzsch came to the following

Definition of a C^1-quasiconformal mapping. Suppose $w = f(z)$ is a C^1-diffeomorphism mapping a planar domain Ω onto a plane domain Ω' and is sense-preserving (i.e., $J_z(f) > 0$ on Ω). If there is a constant $K \geq 1$ so that

$$D_f(z) \leq K \quad \text{on } \Omega \left(\text{or } d_f(z) \leq \frac{K-1}{K+1} \text{ on } \Omega \right)$$

then f is called a regular K-*quasiconformal mapping* on Ω. $\qquad$ (C.7)

It is called a *regular quasiconformal mapping* if no specified K is designated.

Do the following exercises.

Exercise (4) An affine transformation $w = az + b\bar{z} + c$, $|a| > |b|$, is a (regular) quasiconformal mapping. Find μ_f, υ_f, D_f, and d_f.

Exercise (5) $w = f(z) = \frac{z}{1-|z|^2}$ maps $|z| < 1$ C^1-diffeomorphically onto **C**. Show that

$$f_z(z) = \frac{1}{(1-|z|^2)^2}, \quad f_{\bar{z}}(z) = \frac{z^2}{(1-|z|^2)^2}, \quad \mu_f(z) = z^2, \quad \text{and}$$

$$D_f(z) = \frac{1+|z|^2}{1-|z|^2}.$$

Hence $w = f(z)$ is not quasiconformal on $|z| < 1$ but it is on $|z| < r$ where $0 < r < 1$.

Exercise (6) B. Beltrami: *Della variabili complessa sopra una superficie qualunque*, Ann. Mat. 2a. Ser.1 (1867–1868), 329–366, extended the Cauchy–Riemann equations to surfaces. Let a surface S in $\mathbf{R}^3$ be parametrized by $X = f(x,y)$, $Y = g(x,y)$, $Z = h(x,y)$, where f, g, and h

are twice continuously differentiable. Then the arc length element on S is given by

$$ds^2 = E dx^2 + 2F dx dy + G dy^2,$$

where

$$E = f_x^2 + g_x^2 + h_x^2, \quad F = f_x f_y + g_x g_y + h_x h_y,$$

and

$$G = f_y^2 + g_y^2 + h_y^2.$$

(a) Suppose u and v are functions of x and y satisfying

$$du^2 + dv^2 = \lambda(E dx^2 + 2F dx dy + G dy^2),$$

where $\lambda = \lambda(x, y)$ is a real-valued function of x and y. Set

$$H = (EG - F^2)^{1/2}. \tag{$*11$}$$

Beltrami obtained

$$du + i dv = \mu \left(\sqrt{E} dx + \frac{F + iH}{\sqrt{E}} dy \right), \quad \text{where } \mu \bar{\mu} = \lambda$$

$$\Rightarrow u_x + i v_x = \mu \sqrt{E}$$

$$u_y + i v_y = \mu \frac{F + iH}{\sqrt{E}}$$

$$\Rightarrow \text{(by eliminating } \mu) \ E(u_y + i v_y) = (F + iH)(u_x + i v_x)$$

$$\Rightarrow v_x = \frac{1}{H}(F u_x - E u_y),$$

$$v_y = \frac{1}{H}(G u_x - F u_y). \tag{C.8}$$

When treated (x, y) as a coordinate system on S (a concept due to differentiable manifold), then $u + iv$ turns out to be a function on S. Under this circumstance, (C.8) is called the *generalized Cauchy–Riemann equations* or *Beltrami equations*. By elimination v from (C.8), we obtain the so-called *generalized Laplace equation*

$$\frac{\partial}{\partial x}\left[\frac{1}{H}(F u_y - G u_x)\right] + \frac{\partial}{\partial y}\left[\frac{1}{H}(F u_x - E u_y)\right] = 0, \tag{C.9}$$

accompanying a similar equation with u replaced by v.

(b) Suppose $U + iV$ is another such a function as $u + iv$ on S. Then

$$dU^2 + dV^2 = \mu' \left(\sqrt{E} dx + \frac{E + iH}{\sqrt{E}} dy \right)$$

$$\Rightarrow dU^2 + dV^2 = \frac{\mu'}{\mu} (du^2 + dv^2).$$

Consequently (see (7.1.9) or refer to (1) in (3.2.3.3)), $U + iV$ or $U - iV$ is an analytic function of $u + iv$.

(c) *Beltrami equations* (C.8) can be rewritten as

$$f_{\bar{z}} = \eta f_z \tag{C.8$'$}$$

where $u + iv = f(z) = f(x + iy)$ (not to be confused with the f appeared before in the parametric representation of S), and η is a complex parameter. To see this, set $\eta = \alpha + i\beta$ temporarily. Then

$$u_x - v_y + i(v_x + u_y) = (\alpha + i\beta)[(u_x + v_y) + i(v_x - u_y)]$$

$$\Rightarrow v_x = \frac{1}{1 - |\eta|^2} [2\beta u_x - (|\eta|^2 + 2\alpha + 1)u_y],$$

$$v_y = \frac{1}{1 - |\eta|^2} [(|\eta|^2 - 2\alpha + 1)u_x - 2\beta u_y].$$

Suppose $|\eta| < 1$, and set

$$E = |\eta|^2 + 2\alpha + 1, \quad F = 2\beta, \quad G = |\eta|^2 - 2\alpha + 1,$$

and hence $H = 1 - |\eta|^2$. The formula (C.8) follows.

(d) Let $w = f(z)$ as in (c). From (C.8)$'$, it follows that

$$|dw|^2 = |f_z dz + f_{\bar{z}} d\bar{z}|^2 = |f_z|^2 |dz + \eta d\bar{z}|^2. \tag{$*_{12}$}$$

Introduce, in the z-plane, the *Riemannian metric*

$$ds^2 = |dz + \eta d\bar{z}|^2 = (|\eta|^2 + 2\alpha + 1)dx^2 + 4\beta dx dy + (|\eta|^2 - 2\alpha + 1)dy^2$$

$$= E dx^2 + 2F dx dy + G dy^2. \tag{C.10}$$

Then ($*_{12}$) can be rewritten as

$$|dw|^2 = |f_z|^2 ds^2. \tag{C.11}$$

This indicates that $w = f(z)$ in (C.8)$'$ is *conformal according to this new Riemannian metric in the z-plane*. Its dilatation $D_f(z)$ (see (C.6)) is 1.

(e) Choose $\eta = \mu_f$ in (C.8)$'$, where μ_f is the first complex dilatation of f. Then, according to (C.7), a C^1-diffeomorphism $w = f(z) : \Omega \to \Omega'$ which is sense-preserving (namely, $J_f(z) > 0$) and satisfies the conditions:

$$f_{\bar{z}} = \mu_f f_z,$$

where

$$|\mu_f| \leq \frac{K-1}{K+1} \quad (K \geq 1), \qquad (C.12)$$

is a *K-quasiconformal mapping*.

In conclusion, *a K-quasiconformal mapping is conformal with respect to the Riemannian metric* (C.10) *induced by its first complex dilatation.*

For details, refer to Ref. [54], p. 184.

Section (3) Quasiconformal mappings (continued)

Grötzsch's concept about quasiconformal mappings reappeared in *M. A. Lavrentieff : Sur une classe de représentations continues, Rec. Math. 48, 407–423 (1935)*, a work pertaining to partial differential equations. L. A. Ahlfors's work about *the theory of covering surfaces* around 1936 mentioned again this concept. Henceforth it had been received wider attention. *O. Teichmüller: Eine Anwendung quasikonformer Abbildungen auf das Typenproblem. Otsch. Math. 2 (1937)* applied quasiconformal mappings to the type problems of Riemann surfaces and obtained some substantial results which laid the foundation of the so-called **theory of Teichmüller spaces** (see Ref. [3, 53, 55]).

Regular quasiconformal mappings proved to be insufficient in solving some extremal problems, and hence its defining conditions were gradually relaxed as more mathematicians participate in this field. For instance, the mappings are only required to be sense preserving homeomorphisms, even the one-to-one condition is dropped (the *quasiregular mappings* introduced by Yu. G. Reshetnyak in late 1960s), and various geometric and analytic definitions are adopted. See Ref. [54].

Suppose $f : \Omega$ (a domain in $\mathbf{R}^n$, $n \geq 2$) $\to \Omega'$ (a domain in $\mathbf{R}^n$) is a C^1 diffeomorphism so that, for all $\vec{x} \in \Omega$ and all vectors $\vec{h} \in \mathbf{R}^n$, the total differential $df_{\vec{x}}$ of f at $\vec{x}$ satisfies

$$|df_{\vec{x}}(\vec{h})| = \|df_{\vec{x}}\| |\vec{h}|, \qquad (C.13)$$

where $\|df_{\vec{x}}\| = \sup_{|\vec{h}| \leq 1} |df_{\vec{x}}(\vec{h})|$. Such a f is called *a conformal mapping from Ω onto Ω'*.

In case $n \geq 3$ and f is a C^3 conformal mapping, Liouville proved in 1850 that f is nothing but the restriction to Ω of a möbius transformation from $\mathbf{R}^n$ onto itself. By a *möbius transformation* on $\mathbf{R}^n$, we mean that it is a composite of a finite number of the following elementary transformations:

1. Translation: $T(\vec{x}) = \vec{x} + \vec{a}$.
2. Stretch: $T(\vec{x}) = r\vec{x}$, $r > 0$.
3. Orthogonal transformation: $T : \mathbf{R}^n \to \mathbf{R}^n$ is a linear operator such that

$$|T(\vec{x})| = |\vec{x}|, \quad \vec{x} \in R^n.$$

4. Inversion with respect to the sphere $|\vec{x} - \vec{a}| = r$:

$$T(\vec{x}) = \vec{a} + \frac{r^2(\vec{x} - \vec{a})}{|\vec{x} - \vec{a}|^2}. \tag{C.14}$$

Hartman proved in 1959 *Liouville's theorem* under the assumption that f is a C^1 function; R. Nevanlinna gave a simpler proof in 1960 in case f is C^4. The proofs of Liouville's theorem, under much weaker assumptions, last long, even up to 1990s.

Quasiconformal mappings in the plane appeared to be merely a natural extension of conformal mappings. What the last paragraph showed is that the extension of planar conformal mappings to $\mathbf{R}^n$ ($n \geq 3$) reduces to these elementary functions listed in (C.14); as a contrast, the extension to $\mathbf{R}^n$ ($n \geq 3$) of the planar quasiconformal mappings is not just meaningful but also remarkable both in methods and in theory itself. For an introduction, read Ref. [79].

That the theory of quasiconformal mappings turns out to be an active branch of mathematical work is not only due to the seven claimed reasons that appeared in Ref. [3], pp. 1–2, but also due to its theoretical development and applications, such as in elementary particles, superstring, Riemann surfaces and Teichmüller spaces, dynamical system in ordinary differential equations, partial differential equations of elliptic type, complex dynamics, and even in engineering, etc.

References

[1] Ahlfors LV (1979). *Complex Analysis*, 3rd Ed. McGraw-Hill.

[2] Ahlfors LV (1973). *Conformal Invariants, Topics in Geometric Function Theory.* McGraw-Hill.

[3] Ahlfors LV (1966). *Lectures on Quasiconformal Mappings.* Van Nostrand.

[4] Ahlfors LV (1981). *Möbius Transformations in Several Dimensions.* School of Mathematics, University of Minnesota.

[5] Ahlfors LV (1938). An extension of Schwarz's Lemma. *Transaction of the American Mathematical Society*, 43, 359–364.

[6] Ahlfors LV and Sario L (1960). *Riemann Surfaces.* Princepton University Press.

[7] Antimirov MYa, Kolyshkin AA and Vaillancourt R (1998). *Complex Variables.* Academic Press.

[8] Beardon AF (1980). *Complex Analysis, The Angument Principle in Analysis and Topology.* John-Wiley & Sons.

[9] Berenstein CA and Gay R (1991). *Complex Variables, An Introduction.* Springer-Verlag.

[10] Bers L (1957–1958). *Riemann Surfaces.* New York University.

[11] Bochner S and Martin WT (1948). *Several Complex Variables.* Princeton University Press.

[12] Burckel RB (1979). *An Introduction to Classical Complex Analysis.* Birkhäuser Verlag.

[13] Campbell R (1966). *Les Intégrales Eulériennes et Leurs Applications.* Dunod.

[14] Carathéodory C (1954). *Theory of Functions of a Complex Variable,* Vols. I, II. Chelsea Pub. Co.

[15] Carleson L and Gamelin TW (1993). *Complex Dynamics.* Springer-Verlag.

[16] Cartan H (1963). *Elementary Theory of Analytic Functions of One or Several Complex Variables.* Adiwes International Series.

[17] Chen X, Lu P and Tian G (2005). *A Note on Uniformization of Riemann Surfaces by Ricci Flow,* arXiv: Math.DG/0505163 V2 27 May 2005.

[18] Chern SS. Collected Papers.

[19] Collingwood EF and Lohwater AJ (1966). *The Theory of Cluster Sets.* Cambridge University Press.

[20] Copson ET (1955). *An Introduction to the Theory of Functions of a Complex Variable.* Oxford University Press.

[21] Branges de L (1985). A proof of the bieberbach conjecture, *Acta Mathematica*, 154, 137–152.

[22] de Bruijn NG (1958). *Asymptotic Methods in Analysis*. Interscience, Wiley.

[23] Dieudonné J (1960). *Foundations of Modern Analysis*. Academic Press.

[24] Dineen S (1989). *The Schwarz Lemma*. Oxford: Clarendon Press.

[25] Dinghas A (1961). *Vorlesungen über Funktionentheorie*. Springer-Verlag.

[26] Dixion JD (1971). A brief proof of Cauchy's integral theorem, *Procedings of the American Mathematical Society*, 29, 625–626.

[27] Dugundji J (1966). *Topology*. Allyn and Bacon.

[28] Duren PL (1983). *Univalent Functions*. Springer-Verlag.

[29] Evgrafov MA (1962). *Asymptotic Estimates and Entire Functions*. Gordon and Breach.

[30] Evgrafov MA (1966). *Analytic Functions*. Sanders Math. Books.

[31] Evgrafov MA, Sidorov IV, Fedoryuk MV, Shabunin MI and Bezanov KA (1969). *Collection of Problems on the Theory of Analytic Functions*. Moscow: Nauka (in Russian).

[32] Farkas HM and Kra I (1980). *Riemann Surfaces*. Springer-Verlag.

[33] Forster O (1981). *Lectures on Reimann Surfaces*. Springer-Verlag.

[34] Fuchs BA and Shabat BV (1964). *Functions of a Complex Variables and Some of their Application*, Vol. II. Addition-Wesley.

[35] Goluzin GM (1969). *Geometric Theory of a Complex Variable*. AMS.

[36] Gonzáles MO (1992). *Classical Complex Analysis*, Vols. 1, 2. Marcel Dekker.

[37] Grauert H and Reckziegel H (1956). Hermiteschen metriken and normale familien holomorpher abbildungen, *Math. Zeltsch*, 89, 108–125.

[38] Gunning RC (1966). *Lectures on Riemann Surfaces*. Princepton University Press.

[39] Guning RC and Rossi H (1965). *Analytic Functions of Several Complex Variables*. Prentice-Hall, Inc.

[40] Hayman WK (1964). *Meromorphic Functions*. Oxford University Press.

[41] Hayman WK and Kennedy PB (1976). *Subharmonic Functions*. Academic Press.

[42] Heins M (1949). The conformal mapping of simply connected Riemann surfaces, *Annals of Mathematics*, 50, 686–690.

[43] Henrici P (1974). *Applied and Computational Complex Analysis*, Vol. I. John Wiley & Sons; Vol. II (1977), Vol. III (1987).

[44] Hille E (1962). *Analytic Function Theory*, Vol. I. Ginn and Co; Vol. II (1962).

[45] Hörmander L (1973). *An Introduction to Complex Analysis in Several Variables*. North-Holland.

[46] 華羅庚 (Hua LK) (1985). 複變分析導引. 凡異出版社.

[47] Kline M (1972). *Mathematical Thought: From Ancient to Modern Times*. Oxford University Press.

[48] Knopp K (1947). *Theory of Functions*, Vol. II. Dover.

[49] Kobayashi Z (1970). *Hyperbolic Manifolds and Holomorphic Mappings*. Marcel Dekkar.

[50] Kober H (1957). *Dictionary of Conformal Representations*. Dover Pub. Inc.

[51] Krantz SG (1990). *Complex Analysis: The Geometric Viewpoint*. MAA.

[52] Krzyź JG (1971). *Problems in Complex Variable Theory*. Elsevier Pub. Inc.

[53] Lehto O (1987). *Univalent Functions and Teichmüller Spaces*. Springer-Verlag.

[54] Lehto O and Virtanen KI (1970). *Quasiconformal Mappings in the Plane*. Springer-Verlag.

[55] 李忠 (Li Z) (1988). 擬共形映射及其在黎曼曲面論中的應用, 科學出版社.

[56] Lin IH (2005). *Geometric Linear Algebra*, Vol. 1. World Scientific; Vol. 2 (2008).

[57] Massey WS (1991). *A First Course in Algebraic Topology*. Springer-Verlag.

[58] Markushevich AI (1969). *Theory of Functions of a Complex Variable*, Vol. I. Prentice-Hall; Vol. II (1965), Vol. III (1968,1970); Chelsea, 1977.

[59] Minda D and Schober G (1983). Another elementary approach to the theorems of Landau, Montel, Picard and Schottky, *Complex Variables*, 2, 157–164.

[60] Mitrinović DS and Kečkić JD (1984). *The Cauchy Method of Residues, Theory and Applications*, Vol. I. D. Reidel Publ. Co.; Vol. II(1993) Kluwer Academic Press.

[61] Narasimhan R (1985). *Complex Analysis in One Variable*. Birkhäuser.

[62] Needham T (1997). *Visual Complex Analysis*. Clarendon Press.

[63] Nehari Z (1952). *Conformal Mapping*. McGraw-Hill Book Co.

[64] Nevanlinna R (1970). *Analytic Functions*. Springer-Verlag.

[65] Newman MHA (1951). *Elements of the Topology of Plane Sets of Points*. Cambridge University Press.

[66] Palka BP (1991). *An Introduction to Complex Function Theory*. Springer-Verlag.

[67] Pommerenke C (1992). *Boundary Behaviour of Conformal Maps*. Springer-Verlag.

[68] Ratcliffe IG (1994). *Foundations of Hyperbolic Manifolds*. Springer-Verlag.

[69] Robinson RM (1939). A generalization of Picard's and related theorems, *Duke Mathematical Journal*, 5, 118–132.

[70] Rudin W (1966). *Real and Complex Analysis*. McGraw-Hill.

[71] Sansone G and Gerretsen J (1960). *Lectures on the Theory of Functions of a Complex Variable*. Noordhoff.

[72] Sario L and Nakai M (1970). *Classification Theory of Riemann Surfaces*. Springer-Verlag.

[73] Sario L and Noshiro K (1966). *Value Distribution Theory*. D. Van Nostramd Co. Inc.

[74] Spanier EH (1966). *Algebraic Topology*. McGraw-Hill.

[75] Springer G (1957). *Introduction to Riemann Surfaces*. Addison-Wesley.

[76] Titchmarsh EC (1939). *The Theory of Functions*, 2nd Ed. Oxford University Press.

[77] Titchmarsh EC (1937). *The Fourier Transform*. Oxford University Press.

[78] Tsuji M (1959). *Potential Theory in Modern Function Theory.* Tokyo: Maruzen Co. LTD.

[79] Väisälä J (1971). *Lectures on n-Dimensional Quasiconformal mappings,* Lectures Notes in Mathematics 229. Springer-Verlag.

[80] Volkovyskii LV, Lunts GL and Aramanovich IG (1965). *A Collection of Problems on Complex Analysis.* Pergamon Press.

[81] Weyl H (1923). *Die Idee der Riemannschen Flächen,* 2nd Ed. Leipzig: Teubner; New York: Chelsea, 1951.

[82] 楊樂 (Yang L) (1982). 值分布論及其新研究, 科學出版社.

[83] Zalcman L (1975). Heuristic principle in complex function theory, *The American Mathematical Monthly,* 82, 813–817.

[84] 庄圻泰 (Zhuang QT) (1982). 亞純函數的奇異方向, 科學出版社.

Index of Notations

i	imaginary unit: $i = \sqrt{-1}$; $i^2 = -1$
$z = x + iy$ or $x + yi$	complex number where x and y are real numbers
$0 = 0 + i0$	zero complex number
$\operatorname{Re} z$	the real part of z
$\operatorname{Im} z$	the imaginary part of z
iy	pure imaginary number where y is real
$\mathbf{N}$	the natural number system
$\mathbf{Z}$	the integer number system
$\mathbf{Q}$	the rational number system
$\mathbf{R}$	the real number system (field)
$\mathbf{R}^2$	the Euclidean plane
$\mathbf{C}$	the complex number system (field)
$|z|$	the modulus (or absolute value) of z: $\sqrt{x^2 + y^2}$
$\arg z \ (z \neq 0)$	the argument of z: a countably infinite-valued function of z which has single-valued continuous branches in $\mathbf{C} - (\infty, 0]$: $\operatorname{Arg} z + 2n\pi, n = 0, \pm 1, \pm 2, \ldots$
$\operatorname{Arg} z$	the principal argument of z: usually designate its range as $-\pi < \operatorname{Arg} z \leq \pi$; or, $\alpha \leq \operatorname{Arg} z < 2\pi + \alpha$ for any preassigned α
$z = re^{i\theta}$	trigonometric or polar form of z : $r = |z|$, $\theta = \arg z$
$\bar{z}$	conjugate complex number of z
$\overline{ab}$	a line segment connecting a and b
$\Delta a_1 a_2 a_3$	a triangle with vertices at a_1, a_2, and a_3
$\begin{array}{c}|z - z_0| = r \text{ or}\\ z = z_0 + re^{i\theta}\end{array}$	a Euclidean circle with center at z_0 and radius $r > 0$, where $0 \leq \theta \leq 2\pi$
z^*	the symmetric (or reflection) point of z w.r.t a circle $|z - z_0| = r : z_0 + \frac{r^2}{\bar{z} - \bar{z_0}}$; or, w.r.t a line $z = a + bt : a + b^2(\bar{z} - \bar{a})$

$\sqrt[n]{z}$ or $z^{\frac{1}{n}}$ $(n \geq 2)$	n-th root of z; n-th root function of z
∞	the point at infinity or the infinite point
$\mathbf{C}^*$	the extended complex plane: $\mathbf{C} \cup \{\infty\}$
$\{z_n\}_{n=1}^{\infty}$ or $z_n, n \geq 1$ or z_n	a sequence with general term z_n
$\lim_{n \to \infty} z_n = z_0$ or $\lim z_n = z_0$ or $z_n \to z_0$ (as $n \to \infty$)	the sequence z_n converging to the limit z_0 as $n \to \infty$
$\overline{\lim}_{n \to \infty} a_n$ or $\overline{\lim}\, a_n$	the upper limit of the *real* sequence a_n: $\inf_{n \geq 1} \sup_{k \geq n} a_k$
$\underline{\lim}_{n \to \infty} a_n$ or $\underline{\lim}\, a_n$	the lower limit of the *real* sequence a_n: $\sup_{n \geq 1} \inf_{k \geq n} a_k$
$\sum_{n=1}^{\infty} z_n$ or $\sum z_n$	the series with general term z_n
$D_\varepsilon(a)$	the open disk with center a and radius $\varepsilon : |z - a| < \varepsilon$; an open ε-neighborhood of a
A'	derived set of a set A: the set of all the limit points of A
$\bar{A}$	closure of $A : A \cup A'$
$A^\sim$	complement of A (in a larger set): $\mathbf{C} - A$
Int A or A°	the interior of a set A
Bdry A or ∂A	the boundary of A
Ext A or $\bar{A}^\sim$	the exterior of A
$f : A \to \mathbf{C}$ or $w = f(z) : A \to \mathbf{C}$ or $f(z)$ or f	complex-valued function of a complex variable $z \in A \subseteq \mathbf{C}$
Re $f(z)$ or $u(z)$	real part of $f(z) = u(z) + iv(z)$
Im $f(z)$ or $v(z)$	imaginary part of $f(z) = u(z) + iv(z)$
$\lim_{z \in A \to z_0} f(z) = w_0$ or $\lim_{z \to z_0} f(z) = w_0$ or $f(z) \to w_0$ as $z \to z_0$	the limit w_0 of $f(z)$ as z approaches z_0, where z_0 is usually a limit point of the set A
$|f|$	the absolute value function of $f : |f|(z) = |f(z)|$
$\bar{f}$	the conjugate function of $f : \bar{f}(z) = \overline{f(z)}$
$\{f_n\}_{n=1}^{\infty}$ or $f_n, n \geq 1$ or f_n; $\{f_n(z)\}_{n=1}^{\infty}$ or $f_n(z), n \geq 1$	a sequence of functions with general term f_n
$\lim_{n \to \infty} f_n(z) = f(z)$ on A or $f_n \to f$ on A	the sequence $f_n(z)$ converging to the limit function $f(z)$ on a set A as $n \to \infty$

$\sum_{n=1}^{\infty} f_n(z) = f(z)$ on A or $\sum f_n(z) = f(z)$ on A	the series $\sum_{n=1}^{\infty} f_n(z)$ converging to the sum function $f(z)$ on a set A
γ	a (continuous) curve in $\mathbf{C}$, usually with a parametric representation $z(t) = x(t) + iy(t) : [a, b] \to \mathbf{C}$
$\overrightarrow{ab}$	directed line segment from a to b
γ^*	a (continuous) curve in $\mathbf{C}^*$
$\text{Int}\,\gamma$	the interior of a Jordan (or simply) closed curve γ
$\text{Ext}\,\gamma$	the exterior of a Jordan (or simply) closed curve γ
$z^n\,(n \geq 2)$	the n-th power function of z
$\frac{az+b}{cz+d}\,(ad - bc \neq 0)$	linear fractional (or bilinear or Möbius) transformation
$(z_1, z_2; z_3, z_4)$	cross ratio of z_1, z_2, z_3, z_4, in this ordering: $\frac{(z_1-z_3)(z_2-z_4)}{(z_1-z_4)(z_2-z_3)}$
$\frac{1}{2}(z + \frac{1}{z})$	Joukowski function
e^z or $\exp z$	the exponential function with base e : $\lim_{n\to\infty}$ $(1 + \frac{z}{n})^n = \sum_{n=0}^{\infty} \frac{1}{n!} z^n$
$\cos z$	complex cosine function: $\frac{1}{2}(e^{iz} + e^{-iz})$
$\sin z$	complex sine function: $\frac{1}{2i}(e^{iz} - e^{-iz})$
$\tan z, \cot z, \sec z, \csc z$ $\cosh z, \sinh z, \ldots, \operatorname{csc} hz$	
$\Delta_\gamma \arg z$	variation of $\arg z$ along a curve γ
$n(\gamma; z_0)$	the winding number of a closed curve γ around a point z_0 not on γ: $\frac{1}{2\pi}\Delta_\gamma \arg(z - z_0) = \frac{1}{2\pi i}$ $\int_\gamma \frac{1}{z-z_0}dz$ (in case γ is rectifiable)
$\log z\,(z \neq 0)$	the logarithmic function of z, inverse to e^z and hence, multiple-valued: $\log|z| + i\arg z$
$[\infty, 0]$	the nonpositive real exis in $\mathbf{C}^*$; $\{\infty\} \cup \{z \in \mathbf{C}| -\infty < \operatorname{Re} z \leq 0$ and $\operatorname{Im} z = 0\}$
$(\infty, 0] = (-\infty, 0]$	the nonpositive real exis in $\mathbf{C}$
$[0, \infty]$	the nonnegative real exis in $\mathbf{C}^*$
$[0, \infty) = [0, +\infty)$	the nonnegative real exis in $\mathbf{C}$
$[-i, i]$	the segment $\{iy| -1 \leq y \leq 1\}$
$[-i\infty, -i] = [i\infty, -i]$	$\{\infty\} \cup \{iy| -\infty \leq y \leq -1\}$
$[i, +i\infty] = [i, i\infty]$	$\{\infty\} \cup \{iy|1 \leq y \leq +\infty\}$

$\mathbf{C}^* - \gamma$ or $\mathbf{C} - \gamma$	the complement of a curve γ in $\mathbf{C}^*$ or $\mathbf{C}$
$\mathrm{Log}\, z$	principal branch of $\log z : w = \log|z| + i\,\mathrm{Arg}\, z :$ $\mathbf{C} - (\infty, 0] \to \{-\pi < \mathrm{Re}\, w < \pi\}$ or $\mathbf{C} - [o, \infty)$ $\to \{0 < \mathrm{Re}\, w < 2\pi\}$, one-to-one and onto
$\cos^{-1} z$	arc cosine function of z, inverse to $\cos z$ and multiple-valued
$\mathrm{Arc\, cos}\, z$	principal branch: $w = -i\,\mathrm{Log}(z + \sqrt{z^2 - 1}) :$ $\mathbf{C} - (\infty, -1] - [1, \infty) \to \{-\pi < \mathrm{Re}\, w < \pi\}$, 1-to-2, onto positive branch: $w = -i\,\mathrm{Log}(z + \sqrt{z^2 - 1}) :$ $\mathbf{C} - (\infty, -1] - [1, \infty) \to \{0 < \mathrm{Re}\, w < \pi\}$, 1-to-1, onto
$\sin^{-1} z$	arc sine function of z, inverse to $\sin z$ and multiple-valued
$\mathrm{Arc\, sin}\, z$	principal branch: $w = \frac{\pi}{2} - i\,\mathrm{Log}(z + \sqrt{z^2 - 1}) :$ $\mathbf{C} - (\infty, -1] - [1, \infty) \to \{-\frac{\pi}{2} < \mathrm{Re}\, w < \frac{3}{2}\pi\}$, 1-to-2, onto positive branch: $\mathbf{C} - (\infty, -1] - [1, \infty) \to \{\frac{\pi}{2} < \mathrm{Re}\, w < \frac{3\pi}{2}\}$, 1-to-1, onto
$\tan^{-1} z$	arc tangent function of z, inverse to $\tan z$ and multiple-valued
$\mathrm{Arc\, tan}\, z$	principal branch: $w = \frac{-i}{2}\mathrm{Log}\frac{1+iz}{1-iz} : \mathbf{C}^* - \{\pm i\}$ $\to \{0 \le \mathrm{Re}\, w < \pi\}$, 1-to-1, onto
$df_{(x_0,y_0)}$ or $df_{z_0}(z_0 = x_0 + iy_0)$	total differential (in the real sense) of f at $z_0 = x_0 + iy_0$
$[df_{(x_0,y_0)}]$	Jacobian matrix of $df_{(x_0,y_0)}$
$\frac{\partial f}{\partial x}$ or f_x	partial derivative of $f = u + iv$ w.r.t. $x : u_x + iv_x$
$\frac{\partial f}{\partial y}$ or f_y	$u_y + iv_y$
$\frac{\partial}{\partial z}$ or ∂_z or ∂	complex differential operator w.r.t. $z : \frac{1}{2}(\frac{\partial}{\partial x} - i\frac{\partial}{\partial y})$; $\frac{\partial f}{\partial z} = \partial_z f = \partial f = f_z = \frac{1}{2}(f_x - if_y)$
$\frac{\partial}{\partial \bar{z}}$ or $\partial_{\bar{z}}$ or $\bar{\partial}$	complex differential operator w.r.t. $\bar{z} : \frac{1}{2}(\frac{\partial}{\partial x} + i\frac{\partial}{\partial y})$; $\frac{\partial f}{\partial \bar{z}} = \partial_{\bar{z}} f = \bar{\partial} f = f_{\bar{z}} = \frac{1}{2}(f_x + if_y)$
$D_\theta f(z)$	the directional derivative of f at z in the direction $e^{i\theta} : e^{-i\theta}df_z(e^{i\theta}) = f_z(z) + f_{\bar{z}}(z)e^{-2i\theta}$
$C^k(O)$	the vector space of all k-times continuously differentiable functions (in the real sense) on an open set O
$dxdy$	absolute area element

$dx \wedge dy$	oriented area element
$dz \wedge d\bar{z}$	fundamental complex differential form of order 2: $-2i dx \wedge dy = -d\bar{z} \wedge dz$
df	a complex differential form of order 1: $f_x dx + f_y dy = f_z dz + f_{\bar{z}} d\bar{z} = D_\theta f(z) dz$
$d(f dz + g d\bar{z})$	$= (\frac{\partial g}{\partial z} - \frac{\partial f}{\partial \bar{z}}) dz \wedge d\bar{z}$
Δ	the Laplacian operator: $\frac{\partial^2}{\partial x^2} + \frac{\partial^2}{\partial y^2} = r \frac{\partial}{\partial r}(r \frac{\partial}{\partial r}) + \frac{\partial^2}{\partial \theta^2}$; $\Delta u = \frac{\partial^2 u}{\partial x^2} + \frac{\partial^2 u}{\partial y^2} = u_{xx} + u_{yy}$, etc.
$\int_\gamma f(z) dz$	complex integral of f along a curve γ w.r.t. z
$\int_\gamma f(z) d\bar{z}$	complex integral of f along a curve γ w.r.t. $\bar{z}$
$\int_\gamma f(z) |dz|$	complex integral of f along a curve γ w.r.t. arc length $|dz|$
C_n^λ	generalized binomial coefficient: 1, if $n = 0$; $\frac{\lambda(\lambda-1)\cdots(\lambda-n+1)}{n!}$, if $n \geq 1$.
$\mathcal{S}$	the family of the univalent analytic functions $f(z)$ in $|z| < 1$ with the property that $f(0) = 0$ and $f'(0) = 1$
$\gamma \sim 0 \pmod{\Omega}$	a cycle γ homologous to zero in a domain Ω
$\gamma_1 \sim \gamma_2 \pmod{\Omega}$	$\gamma_1 - \gamma_2 \sim 0$ in Ω
$P.V. \int_\gamma \frac{\varphi(\zeta)}{\zeta - z} d\zeta$	Cauchy principal value of the integral $\int_\gamma \frac{\varphi(\zeta)}{\zeta - z} d\zeta$
$\sum_{n=-\infty}^\infty a_n(z - z_0)^n$ or $\sum_{-\infty}^\infty a_n(z - z_0)^n$	the Laurent series at z_0
Res $(f; z_0)$; Res $(f(z); z_0)$	residue of an analytic function $f(z)$ at an isolated singularity $z_0 \in \mathbf{C}^*$
$\widehat{\varphi}(z)$	the Fourier transform of $\varphi(t) : \int_{-\infty}^\infty \varphi(t) e^{-izt} dt$
$(L\varphi)(z)$	the Laplace transform of $\varphi(t) : \int_0^\infty \varphi(t) e^{-zt} dt$
$\Gamma(z)$	Gamma function of z: $\int_0^\infty e^{-t} t^{z-1} dt$, etc. 183, 186
(f, Ω)	analytic function element determined by an analytic function $f(z)$ on a domain Ω 50
(f, D)	(analytic function) germ, where f is defined by a power series on a disk D 51
$B(z, w)$	Beta function of z and w: $\frac{\Gamma(z+w)}{\Gamma(z)\Gamma(w)}$ 202
$\mathbf{C}^*(0, 1, \infty)$ or $\mathbf{C}(0, 1)$	the punctured sphere: $\mathbf{C}^* - \{0, 1, \infty\}$ 270
$g(z, z_0)$	Green's function (of a domain w.r.t a point z_0 or with a logarithmic pole at z_0) 399

The following notations are restricted to Chapter 7 only:

Index